ESSENTIAL ALGEBRA

seventh edition

C. L. JOHNSTON
ALDEN T. WILLIS
JEANNE LAZARIS

PWS PUBLISHING COMPANY • Boston

I(T)P • **An International Thomson Publishing Company**

New York • London • Bonn • Boston • Detroit • Madrid • Melbourne • Mexico City • Paris
Singapore • Tokyo • Toronto • Washington • Albany NY • Belmont CA • Cincinnati OH

*This book is dedicated to our students,
who inspired us to do our best
to produce a book worthy of their time.*

 PWS PUBLISHING COMPANY
20 Park Plaza, Boston, MA 02116-4324

Copyright © 1995 by PWS Publishing Company,
   a division of International Thomson Publishing Inc.
Copyright © 1991, 1988, 1985, 1982, 1978, 1975 by Wadsworth, Inc.

I(T)P™
International Thomson Publishing
The trademark ITP is used under license.
TI-81 is a registered trademark of Texas Instruments, Inc.

Library of Congress Cataloging-in-Publication Data
Johnston, C. L. (Carol Lee).
    Essential algebra / C. L. Johnston, Alden T. Willis, Jeanne
Lazaris. -- 7th ed.
       p.   cm.
    Includes index.
    ISBN 0-534-94494-9
    1. Algebra.   I. Willis, Alden T.   II. Lazaris, Jeanne.
III. Title.
    QA152.2.J63   1994   94-11530
    512.9--dc20   CIP

Sponsoring Editor *Susan McCulley Gay*
Editorial Assistant *Judith A. Mustacchia*
Developmental Editor *Maureen Brooks/Elizabeth Rogerson*
Production Coordinator *Elise S. Kaiser*
Marketing Manager *Marianne C. P. Rutter*
Manufacturing Coordinator *Marcia A. Locke*
Production *Lifland et al., Bookmakers*
Interior/Cover Designer *Elise S. Kaiser*
Interior Illustrator *Scientific Illustrators*
Cover Photo *Al Satterwhite/© The Image Bank*
Compositor *Beacon Graphics Corporation*
Cover Printer *New England Book Components, Inc.*
Text Printer/Binder *R. R. Donnelley & Sons Company/Willard*

Printed and bound in the United States of America.
95 96 97 98 99—10 9 8 7 6 5 4 3 2 1

CONTENTS

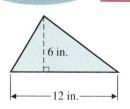
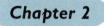

Chapter 3 — Solving Equations and Inequalities   127

Chapter 4 — Applications   177

5 gal

2 gal

Chapter 8 — Rational Expressions   377

Chapter 9 — Radicals   459

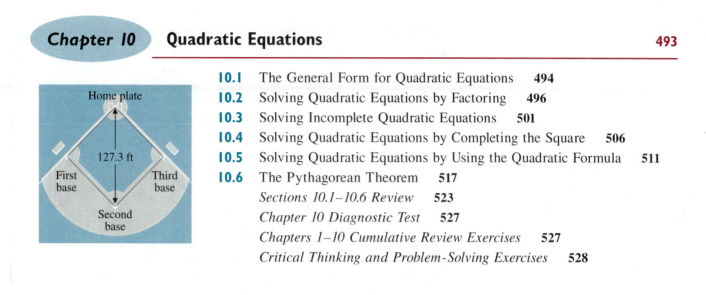
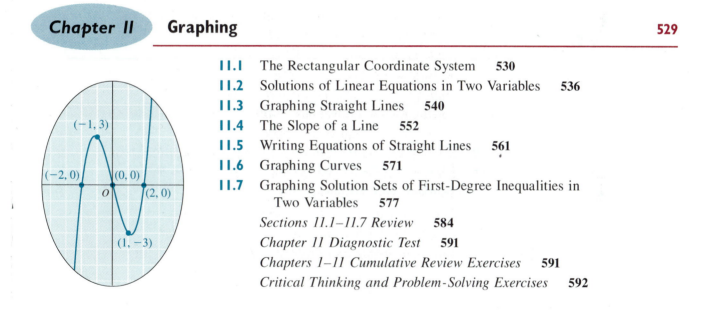

PREFACE

ESSENTIAL ALGEBRA, Seventh Edition, can be used in a beginning algebra course in any community college or four-year college, either with a lecture format, in a learning laboratory setting, or for self-study. The main goal of this book is to prepare students for intermediate algebra and for any other course that requires knowledge of elementary algebra.

Changes in the Seventh Edition

This edition incorporates changes that resulted from many helpful comments from users of the first six editions, as well as from the authors' own classroom experience in teaching from the book. The major changes in the Seventh Edition include the following:

1. Writing Problems have been added throughout the book. We believe that writing about mathematics helps students develop a deeper understanding of the subject. Because communication is one of the essential goals of mathematics, we think that students should be encouraged to express mathematical concepts in writing.

2. Critical Thinking and Problem-Solving Exercises have been added at the end of each chapter. These exercises give the student an opportunity to try types of problems not often seen in elementary mathematics books. Techniques used for solving these problems include, but are not limited to, the following: using logic, looking for patterns or relationships, making organized lists, and guessing-and-checking. Some open-ended critical-thinking questions are asked in these exercises also. Many of the problems are suitable for small-group activities.

3. The objectives for every section are enumerated in an Objectives Checklist found in the printed Test Bank.

4. More material on calculator use has been added whenever appropriate.

5. In Chapter 1 (Real Numbers), factorization of integers is now discussed before finding roots of integers is covered; therefore, the student can find square roots of large numbers by using prime factorization. Simplifying square roots of natural numbers is now discussed in Chapter 1 also.

6. More explanation is provided about how to decide whether a number is rational, irrational, etc., and more instructions are given on graphing numbers on the real number line.

7. The beginning of Chapter 2 (Simplifying and Evaluating Algebraic Expressions) has been reorganized.

8. In solving equations, we now show horizontal addition as the primary method of using the addition property of equality, although vertical addition is still illustrated. (This change was at the request of many users and reviewers of the book.)

Major topics are divided into small manageable sections, as part of the book's one-step, one-concept-at-a-time approach

Boxes enclose important concepts and algorithms, for easy identification and reference

"Notes" to students provide additional information or point out problem-solving hints

---

3.4B Solving Equations That Contain Grouping Symbols

When grouping symbols appear in an equation, first we remove them, and then we solve the resulting equation by the methods discussed in the previous sections.

We suggest the following procedure for solving a first-degree equation (an equation in which the exponent on the variable is understood to be 1) in one variable:

Solving a first-degree equation in one variable

1. If there are any grouping symbols, remove them.
2. At each step, combine like terms (if there are any) on each side of the equation.
3. If the variable appears on both sides of the equation, rewrite the equation so that all terms containing the variable appear on only one side of the equation; do this by adding the additive inverse of one term that contains the variable to both sides of the equation or by subtracting that term from both sides. (If we add the *additive inverse* of the term with the smaller coefficient to both sides of the equation, the coefficient of the sum of the *x*-terms will be positive.)
4. Remove all the numbers that appear on the same side of the equation as the variable, as follows:
   First, remove any numbers being added to (or subtracted from) the term containing the variable by using the addition (or subtraction) property of equality.
   Next, complete the solution by multiplying both sides of the equation by the reciprocal of the coefficient of the variable,
   or complete the solution by using the two-step method (clearing fractions and then dividing both sides of the equation by the coefficient of the variable).
5. Check the apparent solution in the *original* equation.

Note In step 4 in the box above, we suggest using the addition property first and *then* the multiplication or division property. It is possible to use the multiplication (or even the division) property first, but we believe that the order we recommend in step 4 normally makes solving an equation relatively easy.

EXAMPLE 4 Solve and check the equation $10x - 2(3 + 4x) = 7 - (x - 2)$, and graph its solution on the number line.

SOLUTION

$$10x - 2(3 + 4x) = 7 - (x - 2)$$

$10x - 6 - 8x = 7 - x + 2$	Removing the grouping symbols
$2x - 6 = -x + 9$	Combining like terms; the smaller coefficient is -1
$x + 2x - 6 = x + (-x) + 9$	Adding x to both sides to get the x-terms all on one side
$3x - 6 = 9$	We must now remove -6 from the left side
$3x - 6 + 6 = 9 + 6$	Adding $+6$ to both sides
$3x = 15$	The coefficient of x is 3
$\dfrac{3x}{3} = \dfrac{15}{3}$	Dividing both sides by 3
$x = 5$	The apparent solution

Addition shown vertically

$$2x - 6 = -x + 9$$
$$\underline{+x \qquad +x}$$
$$3x - 6 = 9$$
$$\underline{+6 \quad +6}$$
$$3x = 15$$

In special screened boxes labeled "A Word of Caution," students are warned against common errors

The importance of checking solutions is reiterated throughout the book

Two or more approaches to solving a problem are often presented side by side

---

A Word of Caution Notice the difference between the equations (a) $2x = 10$ and (b) $2 + x = 10$. In (a), 2 and *x* are *factors*; that is, 2 is multiplied by *x*. We isolate *x* by dividing both sides of the equation by 2 (or by multiplying both sides of the equation by $\frac{1}{2}$). In (b), 2 and *x* are *terms*; that is, 2 is added to *x*. We isolate *x* by adding -2 to both sides of the equation (or by subtracting 2 from both sides of the equation).

EXAMPLE 2 Solve and check the equation $9x = -27$, and graph the solution on the number line.

SOLUTION

Using the division property	*Using the multiplication property*
We isolate *x* by dividing both sides by 9, the coefficient of *x*	We isolate *x* by multiplying both sides by $\frac{1}{9}$, the reciprocal of 9
$9x = -27$	$9x = -27$
$\dfrac{9x}{9} = \dfrac{-27}{9}$	$\left(\dfrac{1}{9}\right)(9x) = \left(\dfrac{1}{9}\right)(-27)$
$x = -3$	$x = -3$

The apparent solution

Check $9x = -27$
$9(-3) \overset{?}{=} -27$ Substituting -3 for x
$-27 = -27$ A true statement

Therefore, the solution is -3.

Graph

$$\begin{array}{ccccccccc} \bullet \\ \hline -6 & -5 & -4 & -3 & -2 & -1 & 0 & 1 & 2 \end{array}$$

EXAMPLE 3 Solve and check the equation $-12x = 8$, and graph the solution on the number line.

SOLUTION

Using the division property	*Using the multiplication property*
We isolate *x* by dividing both sides by -12, the coefficient of *x*	We isolate *x* by multiplying both sides by $-\frac{1}{12}$, the reciprocal of -12
$-12x = 8$	$-12x = 8$
$\dfrac{-12x}{-12} = \dfrac{8}{-12}$	$\left(-\dfrac{1}{12}\right)(-12x) = \left(-\dfrac{1}{12}\right)(8)$
$x = -\dfrac{2}{3}$	$x = -\dfrac{2}{3}$

The apparent solution

(You can verify that $\frac{8}{-12}$ and $\left(-\frac{1}{12}\right)(8)$ both reduce to $-\frac{2}{3}$.)

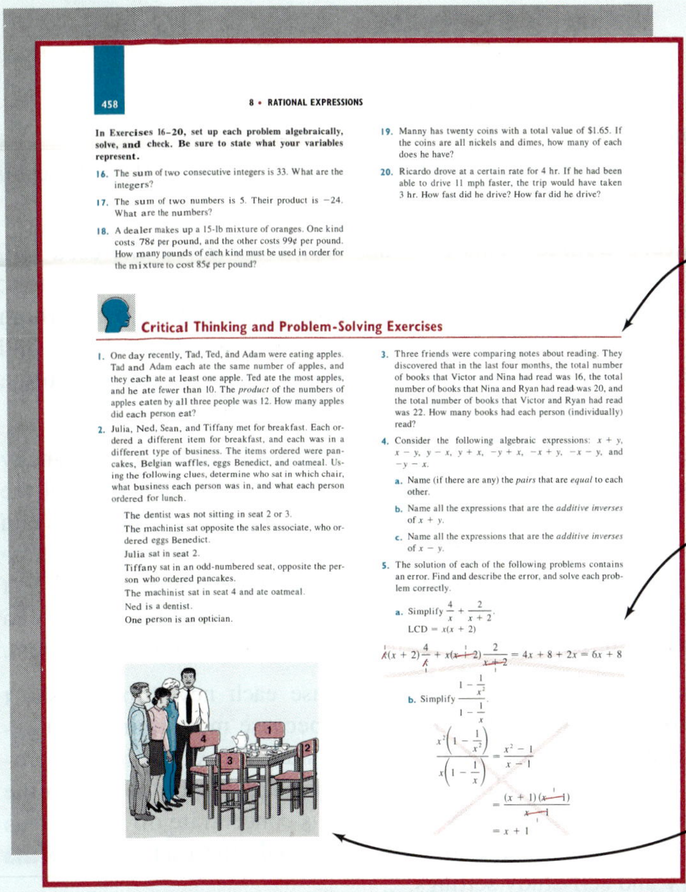

The following shows a sample textbook page (page 458):

458   8 • RATIONAL EXPRESSIONS

In Exercises 16–20, set up each problem algebraically, solve, and check. Be sure to state what your variables represent.

16. The sum of two consecutive integers is 33. What are the integers?

17. The sum of two numbers is 5. Their product is −24. What are the numbers?

18. A dealer makes up a 15-lb mixture of oranges. One kind costs 78¢ per pound, and the other costs 99¢ per pound. How many pounds of each kind must be used in order for the mixture to cost 85¢ per pound?

19. Manny has twenty coins with a total value of $1.65. If the coins are all nickels and dimes, how many of each does he have?

20. Ricardo drove at a certain rate for 4 hr. If he had been able to drive 11 mph faster, the trip would have taken 3 hr. How fast did he drive? How far did he drive?

Critical Thinking and Problem-Solving Exercises

1. One day recently, Tad, Ted, and Adam were eating apples. Tad and Adam each ate the same number of apples, and they each ate at least one apple. Ted ate the most apples, and he ate fewer than 10. The *product* of the numbers of apples eaten by all three people was 12. How many apples did each person eat?

2. Julia, Ned, Sean, and Tiffany met for breakfast. Each ordered a different item for breakfast, and each was in a different type of business. The items ordered were pancakes, Belgian waffles, eggs Benedict, and oatmeal. Using the following clues, determine who sat in which chair, what business each person was in, and what each person ordered for lunch.

  The dentist was not sitting in seat 2 or 3.
  The machinist sat opposite the sales associate, who ordered eggs Benedict.
  Julia sat in seat 2.
  Tiffany sat in an odd-numbered seat, opposite the person who ordered pancakes.
  The machinist sat in seat 4 and ate oatmeal.
  Ned is a dentist.
  One person is an optician.

3. Three friends were comparing notes about reading. They discovered that in the last four months, the total number of books that Victor and Nina had read was 16, the total number of books that Nina and Ryan had read was 20, and the total number of books that Victor and Ryan had read was 22. How many books had each person (individually) read?

4. Consider the following algebraic expressions: $x + y$, $x - y$, $y - x$, $y + x$, $-y + x$, $-x + y$, $-x - y$, and $-y - x$.

  a. Name (if there are any) the *pairs* that are *equal* to each other.
  b. Name all the expressions that are the *additive inverses* of $x + y$.
  c. Name all the expressions that are the *additive inverses* of $x - y$.

5. The solution of each of the following problems contains an error. Find and describe the error, and solve each problem correctly.

  a. Simplify $\dfrac{4}{x} + \dfrac{2}{x + 2}$.
  LCD = $x(x + 2)$

$$x(x+2)\frac{4}{x} + x(x+2)\frac{2}{x+2} = 4x + 8 + 2x = 6x + 8$$

  b. Simplify $\dfrac{1 - \dfrac{1}{x^2}}{1 - \dfrac{1}{x}}$.

$$\frac{x^2\left(1 - \dfrac{1}{x^2}\right)}{x\left(1 - \dfrac{1}{x}\right)} = \frac{x^2 - 1}{x - 1}$$

$$= \frac{(x + 1)(x - 1)}{x - 1}$$

$$= x + 1$$

"Critical Thinking and Problem-Solving Exercises" have been added at the end of every chapter

Large, screened cross-out marks indicate incorrect procedures and anticipate common stumbling blocks for students

All new artwork enhances students' ability to visualize problems and examples

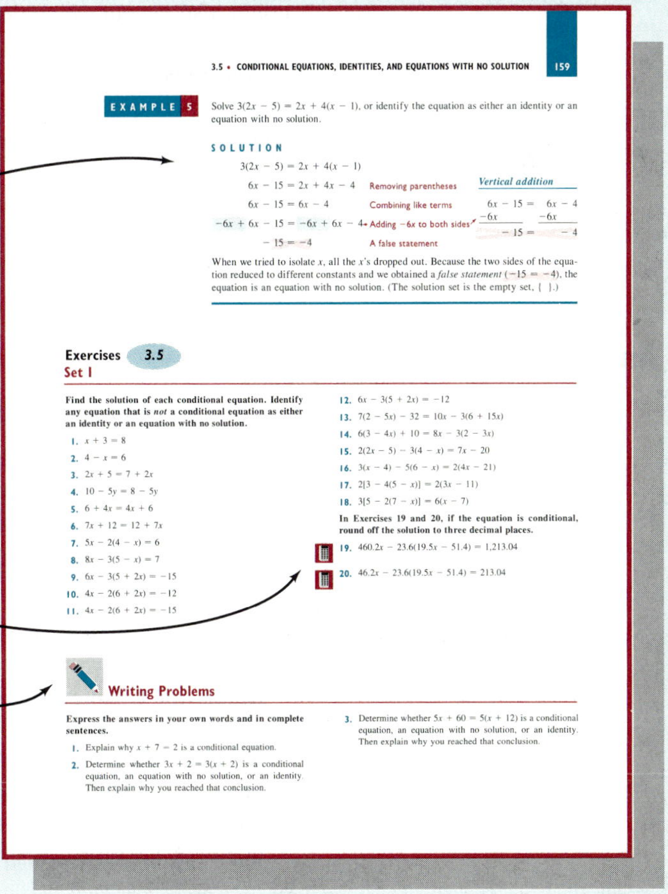

The following shows a sample textbook page (page 159):

3.5 • CONDITIONAL EQUATIONS, IDENTITIES, AND EQUATIONS WITH NO SOLUTION   **159**

EXAMPLE 5   Solve $3(2x - 5) = 2x + 4(x - 1)$, or identify the equation as either an identity or an equation with no solution.

SOLUTION

$$3(2x - 5) = 2x + 4(x - 1)$$
$$6x - 15 = 2x + 4x - 4 \qquad \text{Removing parentheses}$$
$$6x - 15 = 6x - 4 \qquad \text{Combining like terms}$$

Vertical addition
$$6x - 15 = 6x - 4$$
$$\underline{-6x \qquad\quad -6x}$$
$$-15 = -4$$

$$-6x + 6x - 15 = -6x + 6x - 4 \qquad \text{Adding } -6x \text{ to both sides}$$
$$-15 = -4 \qquad \text{A false statement}$$

When we tried to isolate x, all the x's dropped out. Because the two sides of the equation reduced to different constants and we obtained a *false statement* ($-15 = -4$), the equation is an equation with no solution. (The solution set is the empty set, { }.)

Exercises 3.5
Set I

Find the solution of each conditional equation. Identify any equation that is *not* a conditional equation as either an identity or an equation with no solution.

1. $x + 3 = 8$
2. $4 - x = 6$
3. $2x + 5 = 7 + 2x$
4. $10 - 5y = 8 - 5y$
5. $6 + 4x = 4x + 6$
6. $7x + 12 = 12 + 7x$
7. $5x - 2(4 - x) = 6$
8. $8x - 3(5 - x) = 7$
9. $6x - 3(5 + 2x) = -15$
10. $4x - 2(6 + 2x) = -12$
11. $4x - 2(6 + 2x) = -15$
12. $6x - 3(5 + 2x) = -12$
13. $7(2 - 5x) - 32 = 10x - 3(6 + 15x)$
14. $6(3 - 4x) + 10 = 8x - 3(2 - 3x)$
15. $2(2x - 5) - 3(4 - x) = 7x - 20$
16. $3(x - 4) - 5(6 - x) = 2(4x - 21)$
17. $2[3 - 4(5 - x)] = 2(3x - 11)$
18. $3[5 - 2(7 - x)] = 6(x - 7)$

In Exercises 19 and 20, if the equation is conditional, round off the solution to three decimal places.

19. $460.2x - 23.6(19.5x - 51.4) = 1,213.04$
20. $46.2x - 23.6(19.5x - 51.4) = 213.04$

Writing Problems

Express the answers in your own words and in complete sentences.

1. Explain why $x + 7 - 2$ is a conditional equation.
2. Determine whether $3x + 2 = 3(x + 2)$ is a conditional equation, an equation with no solution, or an identity. Then explain why you reached that conclusion.
3. Determine whether $5x + 60 = 5(x + 12)$ is a conditional equation, an equation with no solution, or an identity. Then explain why you reached that conclusion.

Fully worked out solutions are given, with careful one-step-at-a-time explanation provided and problem-solving points highlighted

Calculator icons designate examples and exercises appropriate for solving with a scientific or graphics calculator

"Writing Problems" have been added throughout the text and are denoted by a pencil icon

FOREWORD

To the Student

Mathematics is not a spectator sport! Mathematics is learned by *doing,* not by watching. (The fact that you "understand everything the teacher does on the board" does not mean that you have learned the material!) No textbook can teach mathematics to you, and no instructor can teach mathematics to you; the best that both can do is to help *you* learn.

If you have not been successful in mathematics courses previously and if you really want to succeed now, we strongly recommend that you obtain and study *Mastering Mathematics: How to Be a Great Math Student*, Second Edition, by Richard Manning Smith (Belmont, CA: Wadsworth Publishing Company, 1994). This little book contains *many* fine suggestions for studying, a few of which are listed below:

- Attend all classes. Missing even one class can put you behind in the course by at least two classes. To quote Mr. Smith on this topic:

  > My experience provides irrefutable evidence that you are much more likely to be successful if you attend *all* classes, always arriving on time or a little early. If you are even a few minutes late, or if you miss a class entirely, you risk feeling lost in class when you finally do get there. Not only have you wasted the class time you missed, but you may spend most or all of the first class after your absence trying to catch up. (p. 46)

- Learn as much as you can during class time.
- Sit in the front of the classroom.
- Take complete class notes.
- Ask questions about class notes.
- Solve homework problems on time (*before* the next class).
- Ask questions that deal with course material.
- Organize your notebook.
- Read other textbooks.
- Don't make excuses.

Mastering Mathematics: How to Be a Great Math Student also gives specific suggestions about getting ready for a math course even before the course begins, preparing for tests (including the final exam), coping with a bad teacher, improving your attitude toward math, using class time effectively, avoiding "mental blocks," and so forth.

Two Final Notes (1) If you see a word in this book with which you are not familiar, *look it up in the dictionary.* Learning to use a dictionary effectively is a major part of your education. (2) **It is impossible to overemphasize the value of doing homework!**

It is important for you to realize that, to a very great extent, you can take charge of and have control over your success in mathematics. We believe that studying from *Essential Algebra,* Seventh Edition, can help you be successful.

Real Numbers

CHAPTER 1

Elementary algebra is sometimes considered "generalized arithmetic." It does deal with the six fundamental operations of arithmetic (addition, subtraction, multiplication, division, raising to powers, and extracting roots), but it includes operations on negative as well as positive numbers. In algebra, we often use letters to represent numbers, and we learn to solve algebraic equations. Our major goal is to be able to use algebra to solve the mathematical problems that arise in real-world applications, in higher mathematics, and in science courses. Before we begin the study of algebra, we review a few basic definitions relating to sets and numbers.

1.1   Basic Definitions

Sets   A **set** is a collection of objects or things.

EXAMPLE 1   Examples of sets:

a. A 48-piece set of dishes
b. A collection consisting of an apple, a pillow, and a cat
c. The letters a, b, c, d, e, f, and g

The Elements of a Set   The objects that make up a set are called its **elements**. A set may contain just a few elements, many elements, or no elements at all.

Roster Notation for Representing a Set   A *roster* is a list of the members of a group. To use **roster notation** for representing a set, we list the elements of the set (putting commas between the elements) and we enclose the list within *braces*, { }, *never* parentheses, ( ), or brackets, [ ]. For example, roster notation for the set consisting of the numbers 2, 8, and 12 is $\{2, 8, 12\}$; (4, 5) and [7, 8, 9, 10] do *not* denote sets. The order in which the elements of a set are listed is not important.

EXAMPLE 2   Examples of elements of sets and roster notation:

a. Set $\{1, 2, 3\}$ contains elements 1, 2, and 3.
b. Set $\{a, d, f, h, k\}$ contains elements a, d, f, h, and k.
c. Set {Ben, Kay, Frank, Albert} contains elements Ben, Kay, Frank, and Albert.
d. Set { } contains *no* elements. It is called the **empty set.**

The Meaning of the Equal Sign   The equal sign ($=$) in a statement means that the expression on the left side of the equal sign *has the same value or values* as the expression on the right side of the equal sign. We say that two *sets* are **equal** if they contain exactly the same elements.

Naming a Set   A set is usually named by a capital letter, such as A, N, W, and so on. The expression "$A = \{1, 5, 7\}$" can be read "A is the set whose elements are 1, 5, and 7."

Natural Numbers   The numbers

$$1, 2, 3, 4, 5, 6, 7, 8, 9, 10, 11, 12, \text{ and so on}$$

are called the **natural numbers** (or **counting numbers**).* The largest natural number can never be found, because no matter how far we count there are always larger natural numbers. Since it is impossible to write all the natural numbers, it is customary to represent the set of natural numbers as follows:

$$\{1, 2, 3, 4, \ldots\}$$

Read "and so on"

The three dots to the right of the number 4 indicate that the remaining numbers are to be found by counting in the same way we have begun—namely, by adding 1 to each number to find the next number. We call the set of natural numbers N; that is,

$$N = \{1, 2, 3, 4, \ldots\}$$

The Number Line   Numbers can be represented by numbered points equally spaced along a straight line, as in Figure 1. Such a line is called a **number line**.

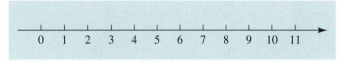

Figure 1   The Number Line

We put an arrowhead at the right, showing the direction in which numbers get larger; some authors put arrowheads at *both* ends of the number line. Natural numbers can be graphed on the number line, and later we will discuss other kinds of numbers—such as fractions, decimals, negative numbers, and irrational numbers—that can be graphed on the number line.

The Graph of a Number   A number is **graphed** by placing a dot on the number line above that number. You will have an opportunity to graph numbers in later sections.

Whole Numbers   When 0 is included with the natural numbers, we have the set of **whole numbers**, which we call W. Figure 2 shows the graphs of the first twelve whole numbers.

$$W = \{0, 1, 2, 3, \ldots\}$$

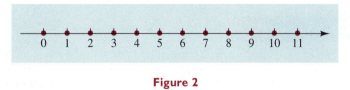

Figure 2

Digits   One important set of numbers is the set of **digits**. This set contains the numbers 0, 1, 2, 3, 4, 5, 6, 7, 8, and 9. These symbols make up our entire number system; *any* number can be written by using some combination of digits.

Numbers are often referred to as *one-digit* numbers, *two-digit* numbers, *three-digit* numbers, and so on. When we wish to refer to the *first*, *second*, or *third* digit of a number, we count from left to right.

---

*In this book, we will not distinguish between a *number* and a *numeral*. Technically, however, a *number* is an abstract idea, or a thought; it is something in our minds. A *numeral* is a symbol we write on paper to represent the number that we have in mind. Some examples of symbols used as numerals are 4, four, ||||, and iv; these particular symbols all represent the following number of objects: ● ● ● ● .

EXAMPLE 3

Examples of the counting of digits:

a. 7 is a one-digit number.
b. 35 is a two-digit number; the first digit is 3, and the second digit is 5.
c. 4,267 is a four-digit number; the first digit is 4, the second digit is 2, the third digit is 6, and the fourth digit is 7.

Fractions   A **fraction** is an indicated division; the fraction $\dfrac{a}{b}$ is equivalent to the division $a \div b$. (Although $\dfrac{a}{b}$ is the preferred form, fractions in a text are often denoted a/b.) In this section, we consider only those fractions in which the **numerator** a is a whole number and the **denominator** b is a natural number. *The denominator of a fraction can never equal zero.* We sometimes call a and b the **terms** of the fraction.

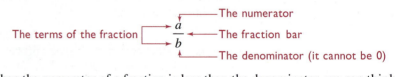

When the numerator of a fraction is less than the denominator, we can think of the fraction as being part of a whole. In this case, the denominator tells us how many equal parts the whole has been divided into, and the numerator tells us how many of those equal parts are being considered.

EXAMPLE 4

An example of the meaning of a fraction:

┌───── The 3 indicates the number of fourths being considered
$\dfrac{3}{4}$
└───── The whole is divided into four equal parts, called fourths

The fraction $\frac{3}{4}$ is equivalent to the division problem $3 \div 4$. The *numerator* is 3, and the *denominator* is 4.

Mixed Numbers   A **mixed number** is made up of both a whole number part and a fraction part. There is an understood plus sign between the whole number part and the fraction part.

EXAMPLE 5

Examples of mixed numbers:

a. $2\frac{1}{2}$ means $2 + \frac{1}{2}$.　　　　　　**b.** $3\frac{5}{8}$ means $3 + \frac{5}{8}$.
c. $5\frac{1}{4}$ means $5 + \frac{1}{4}$.　　　　　　**d.** $12\frac{3}{16}$ means $12 + \frac{3}{16}$.

Decimal Fractions   A **decimal fraction** is a fraction whose denominator is 10, or 100, or 1,000, and so on.

EXAMPLE 6

Examples of decimal fractions:

a. $\frac{4}{10} = 0.4$　　　$\frac{4}{10}$ is in common fraction form, and 0.4 is in decimal form; both are read "four tenths"

b. $\frac{5}{100} = 0.05$　　　Read "five hundredths"

c. $\frac{6}{1,000} = 0.006$　　　Read "six thousandths"

d. $\frac{23}{10} = 2.3$ ←───── Read "two and three tenths"
└───── Read "twenty-three tenths"

Decimal Places   To find the number of **decimal places** in a number, we count the number of digits to the right of the decimal point.

EXAMPLE 7   Examples of the number of decimal places in a number:

a. 75. 14 has two decimal places.   There are two digits to the right of the decimal point

b. 1. 086 has three decimal places.
c. 25 has no decimal places.   There are no digits to the right of the understood decimal point

d. 2. 5000 has four decimal places.

We assume that you remember from arithmetic how to perform the arithmetic operations on fractions, mixed numbers, and decimals, although we will briefly review these topics in later sections.

Real Numbers   Natural numbers, whole numbers, fractions, decimals, and mixed numbers are all elements of a set we call the set of **real numbers**. As we shall see later, there are other kinds of real numbers, also. It is shown in higher level mathematics courses that any number that can be represented by a point on the number line is a real number. Later in this book, we will discuss the real number system in more detail and explain how to graph real numbers on the number line.

"Greater Than" and "Less Than" Symbols   The symbol $>$ is read "is greater than," and the symbol $<$ is read "is less than." These *inequality symbols* are among the symbols that we can use between numbers that are *not* equal to each other. Numbers get larger as we move to the right on the number line and smaller as we move to the left.

EXAMPLE 8 $5 > 2$ is read "5 is greater than 2."

On the number line,
5 is *to the right of* 2; therefore, $5 > 2$

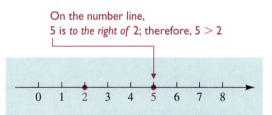

EXAMPLE 9 $2 < 5$ is read "2 is less than 5."

On the number line,
2 is *to the left of* 5; therefore, $2 < 5$

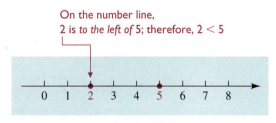

Notice that $5 > 2$ and $2 < 5$ give the same information, even though they are read differently. An easy way to remember the meaning of the symbol is to notice that the wide part of the symbol is next to the larger number. Some people like to think of the symbols $>$ and $<$ as arrowheads that point toward the smaller number.

EXAMPLE 10   Examples of reading inequalities:

a. $7 > 6$ is read "7 is greater than 6."
b. $4 > 1$ is read "4 is greater than 1."
c. $5 < 10$ is read "5 is less than 10."

Another inequality symbol is $\neq$. A slash line drawn through a symbol puts a *not* in the meaning of the symbol.

EXAMPLE 11   Examples of the use of the slash line:

a. $=$ is read "is equal to."
   $\neq$ is read "is *not* equal to."
b. $<$ is read "is less than."
   $\not<$ is read "is *not* less than."
c. $>$ is read "is greater than."
   $\not>$ is read "is *not* greater than."
d. $4 \neq 5$ is read "4 is *not* equal to 5."
e. $3 \not< 2$ is read "3 is *not* less than 2."
f. $5 \not> 6$ is read "5 is *not* greater than 6."

Exercises 1.1
Set I

1. What is the second digit of the number 159?

2. What is the fourth digit of the number 1,975?

3. What is the smallest natural number?

4. What is the smallest digit?

5. What is the largest one-digit natural number?

6. What is the largest two-digit whole number?

7. What is the smallest two-digit natural number?

8. What is the smallest three-digit whole number?

9. Which symbol, $<$ or $>$, should be used to make the statement 10 _?_ 0 true?

10. Which symbol, $<$ or $>$, should be used to make the statement 3 _?_ 15 true?

11. Which symbol, $<$ or $>$, should be used to make the statement 8 _?_ 7 true?

12. Which symbol, $<$ or $>$, should be used to make the statement 0 _?_ 1 true?

13. Is 2.3 a real number?

14. Is $6\frac{7}{8}$ a real number?

15. Is 1.8 a real number?

16. Is $\frac{9}{5}$ a real number?

17. Is 1.8 a natural number?

18. Is $5\frac{3}{4}$ a natural number?

19. Is 15 a digit?

20. Is 15 a natural number?

21. Is $3\frac{1}{2}$ an element of the set of real numbers?

22. Is $1\frac{2}{3}$ an element of the set of natural numbers?

23. How many decimal places are there in the number 7.010?

24. How many decimal places are there in the number 41.0005?

Writing Problems

Express the answers in your own words and in complete sentences.

1. Explain why the number 31 is not a digit.

2. Explain why the number $\frac{3}{8}$ is not a digit.

Exercises 1.1
Set II

1. What is the third digit of the number 3,187?

2. Is 12 a natural number?

3. What is the largest natural number?

4. What is the smallest whole number?

5. What is the largest digit?

6. What is the smallest three-digit natural number?

7. Is 58.4 a real number?

8. What is the smallest one-digit natural number?

9. Which symbol, $<$ or $>$, should be used to make the statement 18 ? 5 true?

10. Which symbol, $<$ or $>$, should be used to make the statement 8 ? 17 true?

11. Which symbol, $<$ or $>$, should be used to make the statement 11 ? 6 true?

12. Is 0 a natural number?

13. Is $5\frac{1}{2}$ a real number?

14. Is $5\frac{1}{2}$ a natural number?

15. Is 0 a real number?

16. Is 58.4 a natural number?

17. Is 3,628 a natural number?

18. Is $\frac{2}{3}$ a real number?

19. Is 12 a digit?

20. Is $\frac{2}{3}$ a digit?

21. Is $8\frac{1}{5}$ an element of the set of real numbers?

22. Is $3\frac{5}{9}$ an element of the set of natural numbers?

23. How many decimal places are there in the number 50.602?

24. How many decimal places are there in the number 23.0?

1.2 Negative Numbers and Rational Numbers

We showed earlier that whole numbers can be represented by points equally spaced along the number line.* We now extend the number line to the left of zero and continue with the equally spaced points.

Numbers used to name the points to the left of zero on the number line are called **negative numbers**. Thus, the point two units to the left of zero is called -2 (read "negative two"), and the point one-half unit to the left of zero is called $-\frac{1}{2}$ (read "negative one-half"). Numbers used to name the points to the right of zero on the number line are called **positive numbers** (see Figure 3). Zero itself is neither positive nor negative. The positive and negative numbers are referred to as **signed numbers**.

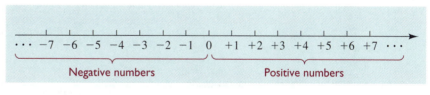

Figure 3

Integers   The set of **integers** includes the set of whole numbers as well as the numbers $-1, -2, -3$, and so on; it can be represented in the following way:

$$\{\ldots, -3, -2, -1, 0, +1, +2, +3, \ldots\}$$

When reading or writing positive numbers, we usually omit the word *positive* and the plus sign. Therefore, when there is no sign in front of a number, it is understood to be positive.

---

*The number line is often called the *real number line*.

EXAMPLE 1

Examples of reading positive and negative integers:

a. −1 is read "negative one."
b. −575 is read "negative five hundred seventy-five."
c. 25 is usually read "twenty-five." It can also be read "positive twenty-five."

EXAMPLE 2

On an unusually cold day in Minnesota, the temperature was −40° F. This means that the temperature was 40° F below 0° F.

In Example 3, m is the abbreviation for meters, a basic unit for measuring length in the metric system.

EXAMPLE 3

Examples of the altitudes of some unusual places on earth:

a. Mt. Everest (Asia)   8,848 m
This means that the peak of Mt. Everest is 8,848 m
(or about 29,029 ft) *above* sea level.
b. Mt. Whitney (California)   4,418 m
c. The lowest point in Death Valley (California)   −86 m
This means that the lowest point in Death Valley is 86 m
(or about 282 ft) *below* sea level.
d. The Dead Sea (Jordan)   −400 m
e. The Mariana Trench (in the Pacific Ocean)   −10,924 m

Using Inequality Symbols with Integers   Any number to the right of a given number on the number line is greater than the given number, and any number to the left of a given number on the number line is less than the given number. These facts are true for negative numbers as well as for positive numbers.

EXAMPLE 4

−3 > −5 is read "negative three is greater than negative five."

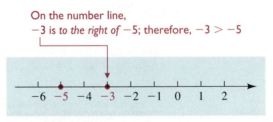

On the number line,
−3 is *to the right of* −5; therefore, −3 > −5

EXAMPLE 5

−5 < −1 is read "negative 5 is less than negative 1."

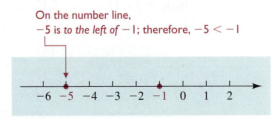

On the number line,
−5 is *to the left of* −1; therefore, −5 < −1

EXAMPLE 6   Examples of verifying inequalities by noting whether, on the number line, the first number of each pair is to the right or left of the second number:

$$-20\ -18\ -16\ -14\ -12\ -10\ -8\ \ -6\ \ -4\ \ -2\ \ 0\ \ \ 2\ \ \ 4\ \ \ 6\ \ \ 8\ \ \ 10$$

a. $6 > 4$ 6 is to the right of 4
b. $0 > -1$ 0 is to the right of -1
c. $-5 < 3$ -5 is to the left of 3
d. $-2 > -5$ -2 is to the right of -5
e. $-20 < -10$ -20 is to the left of -10

Rational Numbers   A **rational number** is any number that can be expressed in the form a/b, where a and b are integers and $b \neq 0$. All integers are rational numbers, and all rational numbers are real numbers. When a rational number is expressed in decimal form, the decimal either terminates (see Example 8) or repeats (see Example 9). (Recall from arithmetic that to convert a fraction to decimal form, we divide the numerator *by* the denominator.)

EXAMPLE 7   Examples of rational numbers:

a. $\frac{3}{4}$, because 3 and 4 are integers and $4 \neq 0$.
b. -5, because $-5 = -\frac{5}{1}$.   Note that $5 = \frac{5}{1}$; therefore, $-5 = -\frac{5}{1}$
c. $1\frac{8}{9}$, because $1\frac{8}{9} = \frac{17}{9}$.
d. -3.7, because $-3.7 = -\frac{37}{10}$.   Note that $3.7 = \frac{37}{10}$; therefore, $-3.7 = -\frac{37}{10}$
e. 0, because $0 = \frac{0}{1}$, or $\frac{0}{5}$, or $\frac{0}{23}$, and so on.

EXAMPLE 8   Examples of rational numbers whose decimals terminate:

a. $\frac{3}{4} = 0.75$ The decimal terminates
b. $\frac{1}{16} = 0.0625$ The decimal terminates

When the decimal form of a rational number is a nonterminating, repeating decimal, it is customary to place a bar over the digit or group of digits that repeats. The bar indicates that that group of digits repeats forever. Thus, $8.1\overline{738} = 8.173817381738\ldots$, and $8.1\overline{73} = 8.1737373\ldots$.

EXAMPLE 9   Examples of rational numbers whose decimals repeat:

a. $\frac{1}{3} = 0.333333\ldots = 0.\overline{3}$ The bar over the 3 indicates that the 3 repeats forever
b. $\frac{4}{33} = 0.121212\ldots = 0.\overline{12}$ The pair of digits "12" repeats forever

Negative Numbers and the Real Number Line   We mentioned earlier that natural numbers, whole numbers, fractions, decimals, and mixed numbers are all real numbers. Negative integers and negative rational numbers are also real numbers, and, as we have seen, negative numbers lie to the left of zero on the real number line. Later in this book, we show how to graph negative integers and other negative numbers on the real number line.

There is no smallest integer, because no matter how far we count to the left of zero, there are always smaller integers. (There is no smallest negative number of *any* kind, in fact.)

Other sets of numbers that are often referred to in applied problems are the following:

Consecutive Numbers   Integers that follow one another in sequence (without interruption) are called **consecutive numbers**. Thus, 5, 6, 7, and 8 are consecutive numbers, because 6 is the next integer after 5, 7 is the next integer after 6, and 8 is the next integer after 7.

Even Integers   Integers that are *exactly* divisible by 2 are called **even integers**. Therefore, 8, −6, and 0 are even integers.

Odd Integers   Integers that are *not* exactly divisible by 2 are called **odd integers**. Therefore, 1, −7, and 5 are odd integers.

Exercises 1.2
Set I

1. Write −75 in words.

2. Write −49 in words.

3. Use digits to write negative fifty-four.

4. Use digits to write negative fourteen.

5. Which is greater, −2 or −4?

6. Which is greater, 0 or −10?

7. Which is greater, −5 or −10?

8. Which is less, −1 or −15?

9. What is the largest negative integer?

10. Is 18 a rational number?

11. Is −3 a rational number?

12. Is 18 a real number?

13. Is −3 a real number?

14. Is $1.\overline{35}$ a rational number?

15. Is 0.74 a rational number?

16. What is the smallest negative integer?

17. Write, in consecutive order, the natural numbers that are < 5.

18. Write, in consecutive order, the digits that are > 6.

19. Write, in consecutive order, the even digits that are < 6.

20. Write, in consecutive order, the odd natural numbers that are < 9.

In Exercises 21–26, determine which symbol, < or >, should be used to make each statement true.

21. 0 _?_ −3   22. −2 _?_ −6   23. −5 _?_ 2

24. −7 _?_ −4   25. −2 _?_ −10   26. −8 _?_ −3

27. A scuba diver descends to a depth of sixty-two feet. Represent this number by an integer.

28. The temperature in Fairbanks, Alaska, was forty-five degrees Fahrenheit below zero. Represent this number by an integer.

Writing Problems

Express the answers in your own words and in complete sentences.

1. Explain why −8 is less than −3.

2. Explain why 5 is a rational number.

Exercises 1.2
Set II

1. Write −17 in words.

2. Write, in consecutive order, the negative integers that are > −5.

3. Use digits to write negative two hundred four.

4. Write, in consecutive order, the even natural numbers that are < 12.

5. Which is greater, −6 or −3?

6. Which is less, −5 or 0?

7. Which is greater, −8 or 5?

8. Which is less, −10 or 2?

9. What is the largest real number?

10. What is the smallest positive integer?

11. Is $\frac{2}{7}$ a rational number?

12. Is $-\frac{2}{15}$ a rational number?

13. Is $\frac{2}{7}$ a real number?

14. Is $0.\overline{62}$ a rational number?

15. Is 2.35 a real number?

16. Is $-\frac{2}{15}$ a real number?

17. Write, in consecutive order, the odd negative integers that are > −7.

18. Write, in consecutive order, the odd digits that are < 8.

19. Write, in consecutive order, the even digits that are > 6.

20. Write, in consecutive order, the even natural numbers that are < 15.

In Exercises 21–26, determine which symbol, < or >, should be used to make each statement true.

21. 0 _?_ −5   **22.** −7 _?_ −3   **23.** 3 _?_ −5

24. −8 _?_ −2   **25.** −6 _?_ −3   **26.** −16 _?_ 0

27. The temperature in Fairbanks, Alaska, was six degrees Fahrenheit below zero. Represent this number by an integer.

28. Nitrogen becomes a liquid at (about) 195 degrees Celsius below zero. Represent this number by an integer.

1.3   Absolute Value; Adding Integers and Other Signed Numbers

Absolute Value

The **absolute value** of a number is the distance between that number and 0 on the number line *with no regard to direction* (see Figure 4). The symbol for the *absolute value* of a real number x is $|x|$.

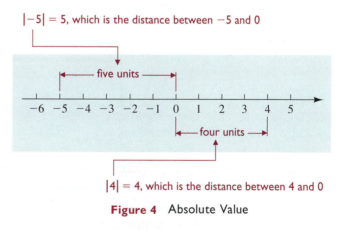

$|-5| = 5$, which is the distance between −5 and 0

five units

four units

$|4| = 4$, which is the distance between 4 and 0

Figure 4   Absolute Value

EXAMPLE 1   Examples of finding the absolute value of a number:

a. $|9| = 9$ A positive number

b. $|0| = 0$ Zero

c. $|-4| = 4$ A positive number

Note that the absolute value of a number can never be negative

A signed number has two distinct parts: its absolute value and its sign.

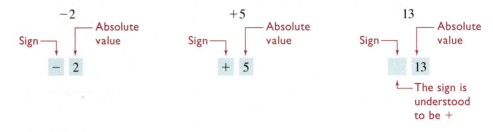

To find the absolute value of a signed number, we can simply drop the sign of the number.

Adding Integers and Other Signed Numbers

Representing Integers by Arrows   We can represent an integer by an arrow whose length represents the absolute value of the number. The arrow must point toward the right if the number is positive (see Example 2) and toward the left if the number is negative (see Example 3). The arrow that represents an integer need not start at zero (see the arrow that represents −7 in Example 4).

EXAMPLE 2   Represent 4 by an arrow.

SOLUTION

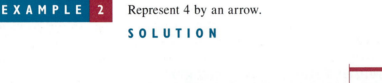

This arrow represents a movement of four units to the *right*.

EXAMPLE 3   Represent −5 by an arrow.

SOLUTION

This arrow represents a movement of five units to the *left*.

Terminology   The answer to an addition problem is called the **sum**. The numbers that are added together to give a sum are called **addends**, or **terms**.

Adding Integers on the Number Line   Before we state the rules for adding integers and other signed numbers, we will give a few examples of adding integers on the number line. To add two numbers on the number line, we locate *one* of the numbers and *start* the arrow for the second number at that point. The arrow that represents the *answer* to the addition problem must go *from* zero *to* the end of the arrow that represents the second number. The answer will have two parts: the *absolute value* and the *sign* (see Examples 4–7).

EXAMPLE 4

Add 5 and −7 on the number line; that is, find 5 + (−7).

SOLUTION We first locate 5 on the number line, considering −7 to be the second number. The arrow that represents −7 must be 7 units long and must point toward the left; we draw this arrow *starting* at 5.

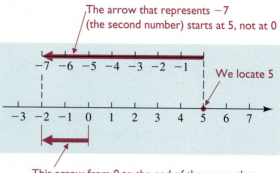

The arrow that represents −7
(the second number) starts at 5, not at 0

We locate 5

This arrow from 0 to the end of the arrow that
represents −7 represents the final answer

Because the arrow that represents the final answer is 2 units long, the absolute value of the answer is 2. Because the arrow that represents the final answer points to the *left*, the answer is *negative*. Therefore, 5 + (−7) = −2.

Note When the problem from Example 4 is written as 5 + (−7), we consider the parentheses around the −7 necessary, as we don't like to see two operation symbols next to each other with no symbol between them. That is, we feel the problem should not be written as 5 + −7 or as ⁺5 + ⁻7. (Some authors and some instructors do not object to these forms.)

You might verify that if we had first located −7 and had then drawn an arrow representing 5, the final answer would still have been −2.

EXAMPLE 5

Add −3 and −4 on the number line; that is, find −3 + (−4).

SOLUTION We first locate −3 on the number line, considering −4 to be the second number. The arrow that represents −4 must be 4 units long and must point toward the left; we draw this arrow *starting* at −3.

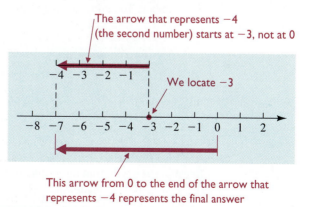

The arrow that represents −4
(the second number) starts at −3, not at 0

We locate −3

This arrow from 0 to the end of the arrow that
represents −4 represents the final answer

Because the arrow that represents the final answer is 7 units long, the absolute value of the answer is 7. Because the arrow that represents the final answer points to the *left*, the answer is *negative*. Therefore, −3 + (−4) = −7.

You can verify that if we had first located −4 and had then drawn an arrow representing −3, the final answer would still have been −7.

EXAMPLE 6   Add −5 and +8 on the number line; that is, find −5 + (+8).

SOLUTION   We first locate −5 on the number line, considering +8 to be the second number. The arrow that represents +8 must be 8 units long and must point toward the right; we draw this arrow *starting* at −5.

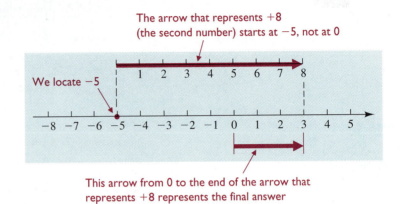

The arrow that represents +8
(the second number) starts at −5, not at 0

We locate −5

This arrow from 0 to the end of the arrow that
represents +8 represents the final answer

Because the arrow that represents the final answer is 3 units long, the absolute value of the answer is 3. Because the arrow that represents the final answer points to the *right*, the answer is *positive*. Therefore, −5 + (+8) = 3.

To see that the result of Example 6 makes sense, consider making a deposit of $8 when your bank balance *had* been −$5 (that is, when you had been overdrawn by $5). Your new balance would be $3.

You can verify for Example 6 that if we had first located 8 and had then drawn an arrow representing −5, the final answer would still have been 3.

EXAMPLE 7   Add −3 and 0 on the number line; that is, find −3 + 0.

SOLUTION   We first locate −3 on the number line, considering 0 to be the second number. To add 0, we then move zero units from that point.

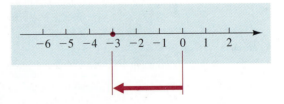

Because the arrow that represents the final answer is 3 units long and points to the left, the answer is −3. Therefore, −3 + 0 = −3.

You might verify that we can use the method discussed in Examples 4–7 to add two positive integers on the number line; for example, you can verify by using arrows and the number line that (+3) + (+2) = 5.

The Additive Identity (Addition Involving Zero)   Because adding zero to a number gives us a number *identical* to the one we started with (see Example 7), we call zero the **additive identity**, or the **identity element for addition**.

The additive identity property	The identity element for addition is 0 . That is, if a represents any real number, then $$a + 0 = a \qquad \text{and} \qquad 0 + a = a$$

The Rules for Adding Integers or Other Signed Numbers   It is easy to understand how to add integers on the number line, but it can be a very slow process. For example, trying to add 8,317,005 and $-17,461,037$ on the number line would be impractical. The following rules give an easier, faster method for adding integers and other signed numbers.

Adding two integers or other signed numbers	**1.** When the numbers have the same sign	**a.** Add the absolute values *and* **b.** Attach (to the left of the result from step a) the sign of the answer; that sign is *positive* when both numbers are positive* and *negative* when both numbers are negative.
	2. When the numbers have different signs	**a.** *Subtract* the smaller absolute value from the larger absolute value *and* **b.** Attach (to the left of the result from step a) the sign of the answer; *the answer has the sign of the number with the larger absolute value.**

*When the sign is positive, it is often dropped.

Notice that *adding* can involve finding a *difference.*
    Examples 8–10 show the addition of integers using the rules given above.

EXAMPLE 8   Find $(-29) + (-35)$.

S O L U T I O N   Because the numbers have the same sign, we use Rule 1 and begin by adding the absolute values.
    Using Rule 1a, we add the absolute values:

$$|-29| + |-35| = 29 + 35 = 64$$

According to Rule 1b, the answer has the sign of both numbers, which is $-$:

$$-64$$

The sign of both numbers is attached to the left of the result from step a

Therefore, $(-29) + (-35) = -64$.

EXAMPLE 9   Find $(-9) + (+23)$.

S O L U T I O N   Because the numbers have different signs, we use Rule 2 and begin by deciding which number has the larger absolute value. $|-9| = 9$ and $|+23| = 23$; because $23 > 9$, the number with the larger absolute value is $+23$.

Using Rule 2a, we subtract the smaller absolute value from the larger absolute value:

$$|+23| - |-9| = 23 - 9 = 14$$

According to Rule 2b, the answer has the sign of the number with the larger absolute value, which is +:

$$+14$$

↑———— The sign of +23 is attached to the left of the result from step a

Therefore, $(-9) + (+23) = +14$, or 14.

In Example 8, the parentheses around the first number were not necessary: $(-29) + (-35) = -29 + (-35)$. In Example 9, the parentheses and one plus sign could have been omitted: $(-9) + (+23) = -9 + 23$.

EXAMPLE 10   Find $-24 + 17$.

SOLUTION   Because the numbers have different signs, we use Rule 2 and begin by deciding which number has the larger absolute value. $|-24| = 24$ and $|17| = 17$; because $24 > 17$, the number with the larger absolute value is -24.

Using Rule 2a, we subtract the smaller absolute value from the larger absolute value:

$$|-24| - |17| = 24 - 17 = 7$$

According to Rule 2b, the answer has the sign of the number with the larger absolute value, which is $-$:

$$-7$$

↑———— The sign of -24 is attached to the left of the result from step a

Therefore, $-24 + 17 = -7$.

As you first use the rules for adding integers, we recommend that you do each problem by using the rules *and* check your results by using arrows and the number line. We also recommend that you practice adding integers until you can write the correct answers quickly. In Examples 11 and 12, we add signed numbers that are not integers.

The Fundamental Property of Fractions   The **fundamental property of fractions** is $\dfrac{a}{b} = \dfrac{ac}{bc}$, where a, b, and c are real numbers, but $b \neq 0$ and $c \neq 0$. This important property is used in *reducing* fractions to lowest terms, where we use it in the form $\dfrac{ac}{bc} = \dfrac{a}{b}$. For example, $\dfrac{15}{18} = \dfrac{5 \cdot 3}{6 \cdot 3} = \dfrac{5}{6}$.

The fundamental property of fractions is also used in *building* fractions so that we can add or subtract them. For example, if we want to add $\frac{1}{2}$ and $\frac{1}{3}$, we must first build both fractions so that they have the same denominator (6), as follows:

$$\frac{1}{2} = \frac{1 \cdot 3}{2 \cdot 3} = \frac{3}{6} \quad \text{and} \quad \frac{1}{3} = \frac{1 \cdot 2}{3 \cdot 2} = \frac{2}{6}$$

Adding Fractions   We will discuss adding algebraic fractions in great detail in a later chapter. For now, we assume that you remember how to find the least common denominator (LCD) of two or more arithmetic fractions. The rule for adding fractions is

$$\frac{a}{c} + \frac{b}{c} = \frac{a + b}{c}$$

EXAMPLE 11

Find $-\frac{5}{2} + \left(-\frac{13}{3}\right)$.

SOLUTION The least common denominator (LCD) is 6. We first find the absolute values:

$$\left|-\frac{5}{2}\right| = \frac{5}{2} \quad \text{and} \quad \left|-\frac{13}{3}\right| = \frac{13}{3}$$

Because the given numbers have the same sign, we use Rule 1a and *add* the absolute values:

$$
\begin{array}{rl}
\frac{5}{2} = & \frac{15}{6} \quad \text{Building } \frac{5}{2}: \ \frac{5}{2} = \frac{5 \cdot 3}{2 \cdot 3} = \frac{15}{6} \\
+\frac{13}{3} = & +\frac{26}{6} \quad \text{Building } \frac{13}{3}: \ \frac{13}{3} = \frac{13 \cdot 2}{3 \cdot 2} = \frac{26}{6} \\
\hline
& \frac{41}{6} \quad \text{Adding the numerators}
\end{array}
$$

According to Rule 1b, the answer has the sign of both numbers, which is $-$:

$$-\frac{41}{6}$$

The sign of both numbers is attached to the left of the result from step a

Therefore, $-\frac{5}{2} + \left(-\frac{13}{3}\right) = -\frac{41}{6}$, or $-6\frac{5}{6}$.

EXAMPLE 12

Find $-2\frac{1}{4} + 8\frac{1}{8}$.

SOLUTION The LCD = 8. We first find the absolute values:

$$\left|-2\frac{1}{4}\right| = 2\frac{1}{4} \quad \text{and} \quad \left|8\frac{1}{8}\right| = 8\frac{1}{8}$$

Because the two numbers have different signs, we use Rule 2. The number with the larger absolute value is $8\frac{1}{8}$. To subtract the smaller absolute value $\left(2\frac{1}{4}\right)$ from the larger, we set up the problem vertically. We build $\frac{1}{4}$ to $\frac{2}{8}$, and we see that the fractional part of the number we're subtracting *from* $\left(\frac{1}{8}\right)$ is *less than* the fractional part of the number we're subtracting $\left(\frac{2}{8}\right)$. Therefore, we will have to borrow, or rewrite $8\frac{1}{8}$ as $7\frac{9}{8}$. (We could, instead, do the subtraction by changing both mixed numbers to improper fractions.)

$$
\begin{array}{rcll}
8\frac{1}{8} = & 8\frac{1}{8} = & 7\frac{9}{8} & \text{Rewriting } 8\frac{1}{8} \text{ as } 7\frac{9}{8} \text{ because } \frac{1}{8} \text{ is less than } \frac{2}{8} \\
-2\frac{1}{4} = & -2\frac{2}{8} = & -2\frac{2}{8} & \text{Building } \frac{1}{4}: \ \frac{1}{4} = \frac{1 \cdot 2}{4 \cdot 2} = \frac{2}{8} \\
\hline
& & 5\frac{7}{8} & \text{Subtracting 2 from 7 and } \frac{2}{8} \text{ from } \frac{9}{8}
\end{array}
$$

According to Rule 2b, the answer has the sign of the number with the larger absolute value, which is $+$:

$$+5\frac{7}{8}$$

The answer is positive, because the sign of $8\frac{1}{8}$ is positive

Therefore, $-2\frac{1}{4} + 8\frac{1}{8} = +5\frac{7}{8}$, or $5\frac{7}{8}$.

EXAMPLE 13

At 6 A.M. the temperature in Alamosa, Colorado, was $-15°$ F. By 10 A.M., the temperature had risen $9°$ F. What was the temperature at 10 A.M.?

SOLUTION We must *add* the rise in temperature $(9°)$ to the original temperature $(-15°)$, and $-15 + (9) = -6$. Therefore, the new temperature is $-6°$ F.

The Additive Inverse of a Number When the sum of two numbers equals 0 (the additive identity), we say that the numbers are the **additive inverses** (or **opposites** or **negatives**) of each other. To find the additive inverse of a signed number, we can simply change the sign of the number. (The additive inverse of 0 is 0.)

The additive inverse property	If a represents any real number, then the additive inverse of a is $-a$. $$a + (-a) = 0 \quad \text{and} \quad -a + a = 0$$

EXAMPLE 14 Examples of naming the additive inverse of a *positive* number:

a. The additive inverse of 5 is -5.
b. The additive inverse of $\frac{8}{3}$ is $-\frac{8}{3}$.

EXAMPLE 15 Examples of naming the additive inverse of a *negative* number:

a. The additive inverse of -10 is 10.
b. The additive inverse of $-\frac{7}{5}$ is $\frac{7}{5}$.

EXAMPLE 16 Remove the absolute value symbols from $-|-12|$.

SOLUTION We must first rewrite $|-12|$ as 12.

$$-\,|-12| = -\,(12) = -12$$

EXAMPLE 17 Add $|-8| + (-3)$.

SOLUTION Before we can perform the addition, we must remove the absolute value symbols, rewriting $|-8|$ as 8.

$$|-8| + (-3) = 8 + (-3) = 5$$

Note There are no subtraction problems in this section, even though we must sometimes subtract absolute values in order to find a *sum*.

Exercises 1.3
Set I

In Exercises 1–4, remove the absolute value symbols.

1. $|-73|$ **2.** $|-55|$ **3.** $-|-48|$ **4.** $-|-26|$

In Exercises 5–52, find the sums.

5. $4 + 5$

6. $6 + 2$

7. $-3 + (-4)$

8. $-7 + (-1)$

9. $-6 + 5$

10. $-8 + 3$

11. $7 + (-3)$

12. $9 + (-4)$

13. $-8 + (-4)$

14. $-5 + (-6)$

15. $3 + (-9)$

16. $4 + (-8)$

17. $-5 + 0$

18. $0 + (-17)$

19. $-7 + 9$

20. $-5 + 8$

21. $-2 + (-11)$

22. $-3 + (-6)$

23. $5 + (-15)$

24. $4 + (-12)$

25. $-8 + 9$

26. $-7 + 13$

27. $-4 + 4$

28. $-9 + 9$

29. $-27 + (-13)$

30. $-42 + (-12)$

31. $-80 + 121$

32. $-69 + 134$

33. $105 + (-73)$

34. $218 + (-113)$

35. $-\frac{3}{2} + \left(-\frac{17}{5}\right)$

36. $-\frac{5}{2} + \left(-\frac{21}{4}\right)$

37. $-\frac{29}{6} + \frac{4}{3}$

38. $-\frac{27}{4} + \frac{17}{8}$

39. $5\frac{1}{4} + \left(-2\frac{1}{3}\right)$

40. $-3\frac{4}{5} + 8\frac{1}{2}$

41. $2\frac{5}{8} + (-8)$

42. $5\frac{3}{5} + (-9)$

43. $6.075 + (-3.146)$

44. $-4.745 + 93.118$

45. $-5.2 + 2.345$

46. $5.325 + (-6.1)$

47. $|-6| + (-2)$

48. $|-8| + (-5)$

49. $-8 + |-17|$

50. $-6 + |-27|$

51. $-9 + |23|$

52. $-1 + |32|$

53. At 6 A.M. the temperature in Hibbing, Minnesota, was $-35°$ F. If the temperature had risen $53°$ F by 2 P.M., what was the temperature at that time?

54. At midnight in Billings, Montana, the temperature was $-50°$ F. By noon the temperature had risen $67°$ F. What was the temperature at noon?

55. Find the additive inverse of -6.

56. Find the additive inverse of $\frac{2}{3}$.

Writing Problems

Express the answers in your own words and in complete sentences.

1. Explain how to add -8 and 5 without using the number line.

2. Explain how to add -9 and -105 without using the number line.

3. Make up an applied problem that involves *adding* two integers, where one number is positive and the other is negative.

Exercises *1.3*
Set II

In Exercises 1–4, remove the absolute value symbols.

1. $|-81|$ **2.** $|-462|$ **3.** $-|-17|$ **4.** $-|-84|$

In Exercises 5–52, find the sums.

5. $8 + 7$

6. $8 + (-5)$

7. $-9 + (-3)$

8. $-19 + (-3)$

9. $-5 + 9$

10. $-28 + 28$

11. $-4 + 7$

12. $0 + (-4,162)$

13. $-2 + (-5)$

14. $18 + (-6)$

15. $6 + (-10)$

16. $145 + (-145)$

17. $-16 + 0$

18. $-\frac{2}{3} + \left(-\frac{1}{5}\right)$

19. $-8 + 5$

20. $-146 + (-362)$

21. $-8 + (-9)$

22. $-1.724 + 3.6$

23. $26 + (-35)$

24. $-4,728 + (-35)$

25. $-29 + 32$

26. $-67 + 28$

27. $-6 + 6$

28. $147 + (-362)$

29. $-38 + (-17)$

30. $-49 + (-38)$

31. $-132 + 261$

32. $-536 + (-2)$

33. $872 + (-461)$

34. $121 + (-517)$

35. $-\frac{4}{3} + \left(-\frac{28}{5}\right)$

36. $-\frac{25}{4} + \left(-\frac{17}{2}\right)$

37. $\frac{11}{3} + \left(-\frac{20}{9}\right)$

38. $3\frac{1}{3} + \left(-5\frac{1}{6}\right)$

39. $-8\frac{1}{6} + 3\frac{1}{3}$

40. $-2\frac{5}{6} + 7\frac{1}{2}$

41. $2\frac{3}{7} + (-5)$

42. $7 + \left(-1\frac{5}{9}\right)$

43. $-18.0164 + 2.281$

44. $-9.6 + (-57.356)$

45. $8.3 + (-3.523)$

46. $-6.127 + 9.3$

47. $(-17) + |-4|$

48. $|-5| + |-2|$

49. $-18 + |-25|$

50. $|-12| + |-17|$

51. $-3 + |-3|$

52. $|-18| + |-25|$

53. In Fairbanks, Alaska, the temperature at 2 A.M. was $-35°$ F. By noon the temperature had risen $27°$ F. What was the temperature at noon?

54. In Duchesne, Utah, the temperature at midnight was $-18°$ F. By 10 A.M. the temperature had risen $12°$ F. What was the temperature at 10 A.M.?

55. Find the additive inverse of $-\frac{1}{5}$.

56. Name the additive inverse of 0.

1.4 Subtracting Integers and Other Signed Numbers

Subtraction is the *inverse* operation of addition; that is, subtraction "undoes" addition.

Terminology   The answer to a subtraction problem is called the **difference**. The number that is being subtracted is called the **subtrahend**, and the number that is being subtracted *from* is called the **minuend**.

The definition of subtraction

$$a - b = a + (-b)$$

In words: To subtract b from a, *add* the additive inverse of b to a.

This definition leads to the following rule for subtracting integers or other signed numbers.

Subtracting one integer or other signed number from another

1. Change the subtraction symbol to an addition symbol, *and* change the sign of the number being subtracted.

2. *Add* the resulting signed numbers, using the rules for adding that were given earlier.

EXAMPLE 1   Find $-2 - (-5)$.

SOLUTION

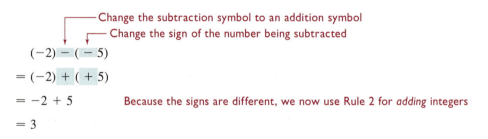

Change the subtraction symbol to an addition symbol
Change the sign of the number being subtracted

$(-2) - (-5)$

$= (-2) + (+5)$

$= -2 + 5$ Because the signs are different, we now use Rule 2 for *adding* integers

$= 3$

In the problem in Example 1, there is just *one* subtraction symbol, as shown below:

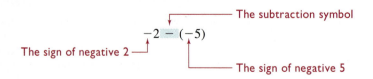

The subtraction symbol

$-2 - (-5)$

The sign of negative 2

The sign of negative 5

We are subtracting -5 from -2.

EXAMPLE 2   Find $13 - (-14)$.

SOLUTION

Change the subtraction symbol to an addition symbol
Change the sign of the number being subtracted

$13 - (-14)$

$= 13 + (+14)$ Because the signs are the same, we now use Rule 1 for *adding* integers

$= 27$

In the problem in Example 2, there is just *one* subtraction symbol, as shown below:

The subtraction symbol

$13 - (-14)$

The sign of negative 14

It is understood that 13 is a positive number and that we are subtracting -14 from (positive) 13.

In Example 3, we use the fact that "subtract b from a" means $a - b$.

EXAMPLE 3   Subtract 9 from 6.

SOLUTION   "Subtract 9 from 6" means $6 - 9$.

Change the subtraction symbol to an addition symbol
Change the sign of the number being subtracted

$6 - (+9)$

$= 6 + (-9)$ Because the signs are different, we now use Rule 2 for *adding* integers

$= -3$

EXAMPLE 4   What is -3 reduced by 6?

SOLUTION   "-3 reduced by 6" means to subtract 6 (the second number) from -3 (the first number).

Change the subtraction symbol to an addition symbol
Change the sign of the number being subtracted

$-3 - (+6)$

$= -3 + (-6)$ Because the signs are the same, we now use Rule 1 for *adding* integers

$= -9$

Subtracting Fractions   The rule for subtracting fractions is $\dfrac{a}{c} - \dfrac{b}{c} = \dfrac{a - b}{c}$, where $c \neq 0$.

EXAMPLE 5

Find $-\frac{7}{2}$ diminished by $+\frac{9}{4}$.

SOLUTION   We must subtract $+\frac{9}{4}$ from $-\frac{7}{2}$.

$$-\frac{7}{2} - \left(+\frac{9}{4}\right)$$

$$= -\frac{14}{4} + \left(-\frac{9}{4}\right)$$

$$= -\frac{23}{4}, \quad \text{or} \quad -5\frac{3}{4}$$

If the *addition* is set up vertically, we have

$$\begin{aligned} -\frac{7}{2} &= -\frac{14}{4} \\ + \left(-\frac{9}{4}\right) &= +\left(-\frac{9}{4}\right) \\ \hline -\frac{23}{4}, &\ \text{or} \ -5\frac{3}{4} \end{aligned}$$

EXAMPLE 6

Find $(-4.56) - (-7.48)$.

SOLUTION   On the right, we show the problem set up vertically.

$$(-4.56) - (-7.48)$$
$$= (-4.56) + (+7.48)$$
$$= 2.92$$

$$\begin{aligned} 7.48 \\ -4.56 \\ \hline +2.92 \end{aligned}$$

EXAMPLE 7

At 4 A.M. the temperature in Missoula, Montana, was $-26°$ F. At noon, the temperature was $-4°$ F. What was the rise in temperature?

SOLUTION   We must find the *difference* in temperatures; that is, we must subtract the *lower* temperature ($-26°$) from the *higher* temperature ($-4°$). The problem then becomes

$$-4 - (-26)$$
$$= -4 + (+26)$$
$$= 22$$

Therefore, the rise in temperature was $22°$ F.

Subtraction Involving Zero   The rules for subtraction involving zero are derived from the rules for addition, since the subtraction $a - b$ has been defined as $a + (-b)$.

Subtraction involving zero

If a represents any real number, then

1. $a - 0 = a$ **2.** $0 - a = 0 + (-a) = -a$

Note   A problem such as $2 - 5$ can be interpreted as a subtraction problem *or* as an addition problem. If it is interpreted as a subtraction problem, it *must* be interpreted as $2 - (+5)$ and then *changed* to the addition problem $2 + (-5)$. If $2 - 5$ is interpreted as an *addition* problem, it is interpreted immediately as $2 + (-5)$. In both cases, the result is -3.

Similarly, the problem $-3 - 7$ can be interpreted as the subtraction problem $-3 - (+7)$ and then changed to $-3 + (-7)$, *or* $-3 - 7$ can be interpreted right away as the addition problem $-3 + (-7)$. In both cases, the result is -10.

Exercises 1.4

Set 1

In Exercises 1–38, find the differences.

1. $6 - (+14)$

2. $8 - (+15)$

3. $-3 - (-2)$

4. $-4 - (-3)$

5. $-6 - (+2)$

6. $-8 - (+5)$

7. $9 - (-5)$

8. $7 - (-3)$

9. $2 - (-7)$

10. $3 - (-5)$

11. $-5 - (-9)$

12. $-2 - (-8)$

13. $-4 - 3$

14. $-7 - 8$

15. $6 - 11$

16. $5 - 9$

17. $-9 - (-4)$

18. $-6 - (-3)$

19. $4 - (-7)$

20. $8 - (-9)$

21. $-15 - 11$

22. $-24 - 16$

23. $0 - 7$

24. $0 - 15$

25. $0 - (-9)$

26. $0 - (-25)$

27. $16 - 0$

28. $10 - 0$

29. $156 - (-97)$

30. $284 - (-89)$

31. $-26.3 - (-3.84)$

32. $-52.64 - (-82.1)$

33. $2.009 - 7$

34. $7.89 - 16$

35. $-\frac{13}{2} - \left(-\frac{11}{3}\right)$

36. $-\frac{25}{3} - \left(-\frac{17}{6}\right)$

37. $6\frac{1}{3} - 8\frac{1}{4}$

38. $12\frac{1}{5} - 9\frac{1}{2}$

39. Subtract (-2) from $(+5)$.

40. Subtract (-10) from (-15).

41. Subtract $\left(2\frac{1}{2}\right)$ from $\left(-5\frac{1}{4}\right)$.

42. Subtract $\left(3\frac{1}{6}\right)$ from $\left(-7\frac{1}{3}\right)$.

43. What is -8 diminished by 5?

44. What is -3 diminished by 6?

45. What is 7 reduced by 12?

46. What is 2 reduced by 12?

47. What is 0 diminished by 4?

48. What is 0 diminished by 8?

49. Mr. Reyes has a balance of $473.29 in his checking account. Find his new balance after he writes a check for $238.43.

50. Ms. Johnson made a $45 deposit on a quadraphonic home music system costing $623.89. What is the balance due?

51. At 5 A.M. the temperature at Mammoth Mountain, California, was $-7°$ F. At noon the temperature was $42°$ F. What was the rise in temperature?

52. At 4 A.M. the temperature in Massena, New York, was $-5.6°$ F. At 1 P.M. the temperature was $37.5°$ F. What was the rise in temperature?

53. A scuba diver descends to a depth of 141 ft below sea level. Her buddy dives 68 ft deeper. What is her buddy's depth at the deepest point of her dive?

54. When Amir checked his pocket altimeter at the seashore on Friday afternoon, it read -150 ft. Saturday morning it read 9,650 ft when he checked it on the peak of a nearby mountain. Allowing for the obvious error in his altimeter reading, what is the correct height of that peak?

55. Mt. Everest (the highest known point on earth) has an altitude of 29,029 ft. The Mariana Trench in the Pacific Ocean (the lowest known point on earth) has an altitude of $-35,840$ ft. Find the difference in altitude of these two places.

56. An airplane is flying 75 ft above the level of the Dead Sea (elevation $-1,312$ ft). How high must it climb to clear a 2,573-ft peak by 200 ft?

Writing Problems

Express the answers in your own words and in complete sentences.

1. Explain how to subtract 18 from 3.

2. Explain how to subtract -6 from 5.

3. Make up an applied problem that involves *subtracting* one integer from another, where the *answer* will be negative.

Exercises 1.4
Set II

In Exercises 1–38, find the differences.

1. $9 - (+17)$ **2.** $5 - (+23)$

3. $-5 - (+8)$ **4.** $7 - (-3)$

5. $-6 - (-8)$ **6.** $-4 - (+6)$

7. $2 - 7$ **8.** $9 - (-7)$

9. $-14 - 10$ **10.** $184 - (-286)$

11. $-473 - 389$ **12.** $-784 - (-528)$

13. $-17 - (-12)$ **14.** $0 - (-12)$

15. $18 - 367$ **16.** $83 - (-5)$

17. $-24 - (-15)$ **18.** $-63 - (-63)$

19. $24 - (-15)$ **20.** $17 - 23$

21. $-28 - 17$ **22.** $1 - 26$

23. $0 - 862$ **24.** $361 - 0$

25. $0 - (-816)$ **26.** $19 - (-26)$

27. $83 - 0$ **28.** $28 - 362$

29. $352 - (-89)$ **30.** $-352 - (-89)$

31. $3.54 - (-28.6)$ **32.** $-48.4 - (-3.75)$

33. $9.794 - 15$ **34.** $5.63 - 25$

35. $-\frac{13}{3} - \left(-\frac{7}{4}\right)$ **36.** $-\frac{46}{5} - \frac{8}{3}$

37. $5\frac{1}{4} - 8\frac{1}{8}$ **38.** $11\frac{1}{6} - 8\frac{1}{2}$

39. Subtract (-120) from (-285).

40. Subtract (-4) from 8.

41. Subtract $\left(3\frac{1}{5}\right)$ from $\left(-5\frac{1}{10}\right)$.

42. Subtract $\left(4\frac{1}{4}\right)$ from $\left(-9\frac{1}{2}\right)$.

43. What is -5 diminished by 6?

44. What is -2 diminished by 8?

45. What is 4 reduced by 10?

46. What is 3 reduced by 8?

47. What is 0 diminished by 6?

48. What is 6 diminished by 0?

49. Juan has a balance of $281.42 in his checking account. Find his new balance after he writes a check for $209.57.

50. Sue has a balance of $563.24 in her checking account. Find her new balance after she writes a check for $347.87.

51. At 2 A.M. the temperature in Burlington, Vermont, was $-3°$ F. At 11 A.M. the temperature was $9°$ F. What was the rise in temperature?

52. At 5 A.M. Don's temperature was $101.8°$ F. By noon his temperature had risen to $103.2°$ F. What was the increase in his temperature?

53. At midnight the temperature in Fairbanks, Alaska, was $-5°$ F. At 10 A.M. the temperature was $24°$ F. What was the rise in temperature?

54. A dune buggy starting from the floor of Death Valley $(-282$ ft$)$ was driven to the top of a nearby mountain that has an elevation of 5,782 ft. What was the change in the dune buggy's altitude?

55. A jeep starting from the shore of the Dead Sea $(-1,312$ ft$)$ was driven to the top of a nearby hill that has an elevation of 723 ft. What was the change in the jeep's altitude?

56. At 2 P.M., Perla's temperature was $103.4°$ F. By 6 P.M., her temperature had dropped to $99.9°$ F. What was the change in her temperature?

1.5   Multiplying Integers and Other Signed Numbers

Terminology   The answer to a multiplication problem is called the **product**. The numbers that are multiplied together to give a product are called the **factors** of that product.

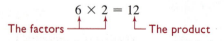

$$6 \times 2 = 12$$

The factors ⌐⌐⌐   └ The product

The numbers 6 and 2 are *factors* of 12; 12 is the *product* of 6 and 2. Similarly, 3 and 4 are factors of 12, because $3 \times 4 = 12$.

The Symbols Used for Multiplication   Multiplication may be shown in a number of different ways:

1. $3 \times 2 = 6$ The "cross" can be used to indicate multiplication.

2. $3 \cdot 2 = 6$ The multiplication dot, ·, can be used to indicate multiplication; it is written a little higher than a decimal point.

3. $3(2) = 6$ When there is no operation symbol between a number (or letter) and parentheses, the operation is understood to be multiplication.

4. $(3)(2) = 6$ The operation is understood to be multiplication.

5. ab When two expressions (other than two *numbers*) are written next to each other in this way, it is understood that they are to be multiplied.

6. $3a$ In this example, it is understood that the value of a is to be *multiplied* by 3. Thus, if a is 7, $3a = 3(7) = 21$.

Multiplication Involving a Positive Integer   Multiplication by a positive integer is a short method for doing repeated addition of the same number. In Example 1, we show multiplication of two positive numbers as repeated addition. Each problem is interpreted as repeated addition in two different ways.

EXAMPLE 1   Examples of multiplication as repeated addition:

a. $3 \times 5 =$ the sum of three 5s $= 5 + 5 + 5 = 15$
or $3 \times 5 =$ the sum of five 3s $= 3 + 3 + 3 + 3 + 3 = 15$
b. $6 \times 2 =$ the sum of six 2s $= 2 + 2 + 2 + 2 + 2 + 2 = 12$
or $6 \times 2 =$ the sum of two 6s $= 6 + 6 = 12$
c. $3 \times 1 =$ the sum of three 1s $= 1 + 1 + 1 = 3$
or $3 \times 1 =$ one 3 $= 3$

In Example 2, we again show multiplication as repeated addition, but this time one of the numbers is positive and one is negative. We have only one way of interpreting each of these problems.

EXAMPLE 2   Examples of multiplication as repeated addition:

a. $3 \times (-2) =$ the sum of three negative 2s $= (-2) + (-2) + (-2) = -6$
b. $(-6) \times 4 =$ the sum of four negative 6s $= (-6) + (-6) + (-6) + (-6) = -24$
c. $1 \times (-8) =$ one negative 8 $= -8$

Example 2 shows that the product of two numbers that have different signs should be negative, since we would always be adding several negative numbers if we interpreted the multiplication as repeated addition.

The Multiplicative Identity   Examples 1c and 2c illustrate the fact that multiplying any real number by 1 gives us back the *identical* number we started with. Because this is true, we call 1 the **multiplicative identity**.

The multiplicative identity property	The identity element for multiplication is 1 . That is, if a represents any real number, then $$a \cdot 1 = a \qquad \text{and} \qquad 1 \cdot a = a$$

Multiplication Involving Zero   Example 3 again shows multiplication as repeated addition, this time where one of the factors is zero.

EXAMPLE 3   Examples of multiplying by zero:

a. $3 \cdot 0 = 0$ $3 \cdot 0 = 0 + 0 + 0 = 0$
b. $4 \cdot 0 = 0$ $4 \cdot 0 = 0 + 0 + 0 + 0 = 0$
c. $0 \cdot 3 = 0$ $0 \cdot 3 = 0 + 0 + 0 = 0$
d. $0 \cdot 4 = 0$ $0 \cdot 4 = 0 + 0 + 0 + 0 = 0$

We see that multiplying a number by zero gives a product that is zero.

The multiplication property of zero

> If a represents any real number, then
> $$a \cdot 0 = 0 \quad \text{and} \quad 0 \cdot a = 0$$

The Product of Two Negative Numbers   To decide what the sign of the product of two negative numbers should be, let's examine the following pattern of products:

These numbers decrease regularly by 1 ⟶ | These numbers increase regularly by 2

$$4\,(-2) = -8$$
$$3\,(-2) = -6 \quad \text{Notice that } -8 + 2 = -6$$
$$2\,(-2) = -4 \quad \text{Notice that } -6 + 2 = -4$$
$$1\,(-2) = -2 \quad \text{Notice that } -4 + 2 = -2$$
$$0\,(-2) = 0 \quad \text{Notice that } -2 + 2 = 0$$
$$-1\,(-2) = \qquad$$
$$-2\,(-2) = \qquad \text{What numbers must go in these blanks if the pattern is to be continued?}$$
$$-3\,(-2) = \qquad$$

If the pattern is to be continued, we must have

$$-1(-2) = \quad 2 \qquad \text{Because } 0 + 2 = 2$$
$$-2(-2) = \quad 4 \qquad \text{Because } 2 + 2 = 4$$
$$-3(-2) = \quad 6 \qquad \text{Because } 4 + 2 = 6$$

Therefore, we want $-1(-2)$ to equal 2, $-2(-2)$ to equal 4, $-3(-2)$ to equal 6, and so on; that is, we would like the *product of two negative numbers* to be *positive*.

Multiplying Integers and Other Signed Numbers   The rule for multiplying two integers or other signed numbers is as follows:

Multiplying two integers or other signed numbers

> Multiply the absolute values *and* attach the correct sign to the left of the product; that sign is
>
> **a.** *positive* when the factors have the same sign;
>
> **b.** *negative* when the factors have different signs.

EXAMPLE 4

Multiply $-7(4)$.

SOLUTION

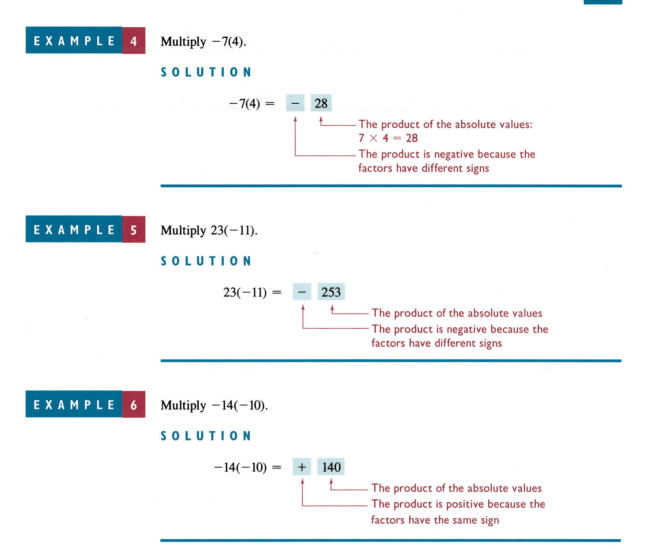

$$-7(4) = \boxed{-}\;\boxed{28}$$

The product of the absolute values:
$7 \times 4 = 28$

The product is negative because the factors have different signs

EXAMPLE 5

Multiply $23(-11)$.

SOLUTION

$$23(-11) = \boxed{-}\;\boxed{253}$$

The product of the absolute values

The product is negative because the factors have different signs

EXAMPLE 6

Multiply $-14(-10)$.

SOLUTION

$$-14(-10) = \boxed{+}\;\boxed{140}$$

The product of the absolute values

The product is positive because the factors have the same sign

Multiplying Fractions and Mixed Numbers Recall from arithmetic that when we *multiply* two fractions, the numerator of the product is the product of the two numerators and the denominator of the product is the product of the two denominators. That is,

$$\frac{a}{b} \cdot \frac{c}{d} = \frac{a \cdot c}{b \cdot d}.$$

We will discuss multiplying *algebraic* fractions in more detail in a later chapter.

Also recall from arithmetic that before we can multiply two *mixed numbers* together, we must change the numbers to improper fractions (see Example 8).

EXAMPLE 7

Multiply $\left(-\frac{1}{2}\right)\left(-\frac{3}{5}\right)$.

SOLUTION We will multiply the numerators and multiply the denominators.

$$\left(-\frac{1}{2}\right)\left(-\frac{3}{5}\right) = +\left(\frac{1}{2}\right)\left(\frac{3}{5}\right) = \frac{1 \cdot 3}{2 \cdot 5} = \frac{3}{10}$$

The product of the absolute values

The product is positive because the factors have the same sign

EXAMPLE 8   Multiply $\left(4\frac{1}{2}\right)\left(-1\frac{1}{3}\right)$.

SOLUTION   We must change both mixed numbers to improper fractions.

$$\left(4\frac{1}{2}\right)\left(-1\frac{1}{3}\right) = \left(\frac{9}{2}\right)\left(-\frac{4}{3}\right) = -\left(\frac{9}{2}\right)\left(\frac{4}{3}\right) = -\frac{9\cdot 4}{2\cdot 3} = -\frac{36}{6} = -6$$

— The product of the absolute values
— The product is negative because the factors have different signs

---

EXAMPLE 9   Multiply $-2.7(-4.6)$.

SOLUTION

— The product is positive because the factors have the same sign
— The product of the absolute values

$$-2.7(-4.6) = +(2.7 \times 4.6) = 12.42$$

The Reciprocal of a Fraction   To find the **reciprocal of a fraction**, we interchange the numerator and denominator; the reciprocal of a positive number is always positive, and the reciprocal of a negative number is always negative. Zero has no reciprocal. In Example 10, we verify that the product of a nonzero number and its reciprocal is 1.

EXAMPLE 10   Name the reciprocal of each of the following numbers, and verify that the product of the number and its reciprocal is 1.

SOLUTION

a. $\frac{5}{6}$ The reciprocal of $\frac{5}{6}$ is $\frac{6}{5}$; observe that $\frac{5}{6}\cdot\frac{6}{5} = \frac{5\cdot 6}{6\cdot 5} = \frac{30}{30} = 1$.

b. -8 The reciprocal of -8 is $-\frac{1}{8}$; observe that $-8\left(-\frac{1}{8}\right) = +\frac{8}{1}\cdot\frac{1}{8} = \frac{8\cdot 1}{1\cdot 8} = \frac{8}{8} = 1$.

c. $-\frac{1}{2}$ The reciprocal of $-\frac{1}{2}$ is -2; observe that $-\frac{1}{2}(-2) = +\frac{1}{2}\cdot\frac{2}{1} = \frac{1\cdot 2}{2\cdot 1} = \frac{2}{2} = 1$.

d. $-\frac{5}{7}$ The reciprocal of $-\frac{5}{7}$ is $-\frac{7}{5}$; observe that $-\frac{5}{7}\left(-\frac{7}{5}\right) = +\frac{5}{7}\cdot\frac{7}{5} = \frac{5\cdot 7}{7\cdot 5} = \frac{35}{35} = 1$.

The Multiplicative Inverse of a Number   When the product of two numbers equals the multiplicative identity (that is, when the product is 1), we say that the numbers are the **multiplicative inverses** of each other. Therefore, the reciprocal of a number can also be called its *multiplicative inverse*.

The multiplicative inverse property

If a represents any real number except 0, then the multiplicative inverse (or reciprocal) of a is $1/a$. (There is no multiplicative inverse for 0.)

$$a\left(\frac{1}{a}\right) = 1 \qquad \text{and} \qquad \frac{1}{a}(a) = 1$$

Exercises  1.5
Set I

In Exercises 1–38, find the products.

1. $3(-2)$ **2.** $4(-6)$ **3.** $-5(2)$

4. $-7(5)$ **5.** $-8(-2)$ **6.** $-6(-7)$

7. $8(-4)$ **8.** $9(-5)$ **9.** $-7(9)$

10. $-6(8)$ **11.** $-10(-10)$ **12.** $-9(-9)$

13. $8(-7)$ **14.** $12(-6)$ **15.** $-26(10)$

16. $-11(12)$ **17.** $-20(-10)$ **18.** $-30(-20)$

19. $75(-15)$ **20.** $86(-13)$ **21.** $-30(+5)$

22. $-50(+6)$ **23.** $-7(-20)$ **24.** $-9(-40)$

25. $\left(-5\frac{1}{2}\right)(0)$ **26.** $\left(-7\frac{2}{7}\right)(0)$ **27.** $\left(\frac{27}{4}\right)\left(-\frac{33}{4}\right)$

28. $\left(\frac{83}{9}\right)\left(-\frac{22}{3}\right)$ **29.** $\left(-3\frac{3}{5}\right)\left(-8\frac{1}{5}\right)$ **30.** $\left(-2\frac{5}{7}\right)\left(-7\frac{1}{5}\right)$

31. $-3.5(-1.4)$ **32.** $-4.7(-1.6)$ **33.** $2.74(-100)$

34. $3.04(-100)$ **35.** $\left(2\frac{1}{3}\right)\left(-3\frac{1}{2}\right)$ **36.** $\left(-5\frac{1}{4}\right)\left(-2\frac{3}{5}\right)$

37. $0\left(-\frac{2}{3}\right)$ **38.** $0\left(-\frac{7}{8}\right)$

39. Find the *multiplicative* inverse of -12.

40. Find the *multiplicative* inverse of $5/7$.

41. Find the *additive* inverse of -12.

42. Find the *additive* inverse of $5/7$.

43. Find the reciprocal of -3.

44. Find the reciprocal of -2.

45. Are 5 and $-\frac{1}{5}$ the multiplicative inverses of each other? Why or why not?

46. Are $\frac{2}{3}$ and $-\frac{3}{2}$ the multiplicative inverses of each other? Why or why not?

Writing Problems

Express the answers in your own words and in complete sentences.

1. Explain how to multiply -3 by 40.

2. Explain how to find the multiplicative inverse (or reciprocal) of a number.

3. Explain why $-\frac{2}{3}$ and $-\frac{3}{2}$ are the multiplicative inverses of each other.

Exercises 1.5
Set II

In Exercises 1–38, find the products.

1. $5(-4)$ **2.** $-6(3)$ **3.** $-7(-3)$

4. $-4(8)$ **5.** $-9(-6)$ **6.** $8(+5)$

7. $-15(10)$ **8.** $15(-10)$ **9.** $-300(-10)$

10. $15(0)$ **11.** $-15(-15)$ **12.** $0(-28)$

13. $9(-12)$ **14.** $-23(-28)$ **15.** $-35(100)$

16. $-100(42)$ **17.** $-56(-100)$ **18.** $83(0)$

19. $27(-81)$ **20.** $-33(42)$ **21.** $-25(+6)$

22. $0(-3,679)$ **23.** $-18(-47)$ **24.** $-73(61)$

25. $\left(-8\frac{1}{3}\right)(0)$ **26.** $\left(-8\frac{1}{3}\right)\left(-2\frac{1}{5}\right)$ **27.** $\left(\frac{36}{7}\right)\left(-\frac{19}{5}\right)$

28. $\left(-\frac{17}{8}\right)\left(\frac{26}{9}\right)$ **29.** $\left(-12\frac{1}{2}\right)\left(-3\frac{1}{4}\right)$ **30.** $\left(-18\frac{1}{2}\right)(0)$

31. $-5.67(10)$ **32.** $-15.1(32)$ **33.** $-0.0032(-100)$

34. $6.07(0)$ **35.** $\left(-2\frac{1}{5}\right)\left(3\frac{3}{4}\right)$ **36.** $\left(8\frac{1}{3}\right)\left(-2\frac{3}{5}\right)$

37. $\left(-5\frac{3}{7}\right)(0)$ **38.** $0\left(-9\frac{1}{2}\right)$

39. Find the *multiplicative* inverse of $-6/11$.

40. Find the *multiplicative* inverse of -1.

41. Find the *additive* inverse of $-6/11$.

42. What is the multiplicative identity?

43. Find the reciprocal of $-\frac{3}{7}$.

44. Find the reciprocal of 5.

45. Are $\frac{3}{5}$ and $-\frac{5}{3}$ the multiplicative inverses of each other? Why or why not?

46. Show that $-\frac{7}{8}$ and $-\frac{8}{7}$ are the multiplicative inverses of each other.

1.6 Dividing Integers and Other Signed Numbers

Division is the *inverse* operation of multiplication; that is, division "undoes" multiplication. For example, $12 \div 4 = 3$ *because* $3 \times 4 = 12$. The division problem $a \div b$ is equivalent to the multiplication problem $a \cdot \dfrac{1}{b}$.

Terminology   The answer to a division problem is called the **quotient**. The number we are dividing *by* is called the **divisor**. The number we are dividing *into* is called the **dividend**. If the divisor does not divide *exactly* into the dividend, the part that is left over is called the **remainder**.

Division may be shown in several ways:

$$12 \div 4 = 12/4 = \frac{12}{4} = 4\overline{)12} \qquad \text{The } \textit{quotient} \text{ in this case is 3}$$

All of these statements are equivalent to the multiplication problem $12 \cdot \frac{1}{4}$. That is, we can solve the division problem $a \div b$ by multiplying the dividend by the multiplicative inverse (the reciprocal) of the divisor.

We check the answer to a division problem as follows:

(Divisor × quotient) + remainder = dividend

If the remainder is zero, then the *divisor* and the *quotient* are both *factors* of the *dividend*, and the problem can be checked as follows:

Divisor × quotient = dividend

EXAMPLE I   Identify the quotient, divisor, dividend, and remainder for $23 \div 6$, and check the solution.

SOLUTION

$$
\begin{array}{r}
\text{Quotient} \\
3 \\
\text{Divisor} \rightarrow 6\overline{)23} \leftarrow \text{Dividend} \\
\underline{18} \\
5 \leftarrow \text{Remainder}
\end{array}
$$

✓ **Check**   (Divisor × quotient) + remainder = $(6 \times 3) + 5 = 18 + 5 = 23$, the dividend.

EXAMPLE 2   Identify the quotient, divisor, and dividend for $12 \div 4$, and check the solution.

SOLUTION

$$
\begin{array}{r}
\text{Quotient} \\
3 \\
\text{Divisor} \rightarrow 4\overline{)12} \leftarrow \text{Dividend}
\end{array}
$$

The remainder is zero. (Because the remainder is zero, 4 and 3 are both *factors* of 12.)

✓ **Check**   Divisor × quotient = $4 \times 3 = 12$, the dividend.

Dividing Integers and Other Signed Numbers

Because of the inverse relation between division and multiplication, the rules for finding the sign of a quotient are the same as the ones used for finding the sign of a product.

Dividing one integer or other signed number by another	Divide the absolute value of the dividend by the absolute value of the divisor *and* attach the correct sign to the left of the quotient; that sign is **a.** *positive* when the numbers have the same sign; **b.** *negative* when the numbers have different signs.

Division Involving Zero   We next consider division in which zero is the dividend or the divisor or both.

Division of zero by a number other than zero is possible, and the quotient is always 0. That is, $0 \div 2 = \frac{0}{2} = 0$, or $2\overline{)0}^{\,0}$, because $2 \times 0 = 0$.

Division of a nonzero number by zero is impossible. Let's try to divide some number (not 0) by 0. For example, let's try $4 \div 0$, or $\frac{4}{0}$. Suppose the quotient is some unknown number we call q. Then $4 \div 0 = q$ means that we must find a number q such that $0 \times q = 4$. But $0 \times q = 4$ is impossible, since 0 times any number is 0. Therefore, no number q exists, and $4 \div 0$ has no answer (or is *undefined*). (Similarly, there is no answer for $1 \div 0$, or $\frac{1}{0}$; it is for this reason that zero has no reciprocal.)

Division of zero by zero cannot be determined. What about $0 \div 0$, or $\frac{0}{0}$? Consider the following examples:

$$0\overline{)0}^{\,1} \quad \text{because } 0 \times 1 = 0 \text{ is true.}$$

$$0\overline{)0}^{\,27} \quad \text{because } 0 \times 27 = 0 \text{ is true.}$$

$$0\overline{)0}^{\,111} \quad \text{because } 0 \times 111 = 0 \text{ is true.}$$

In other words, $0 \div 0 = 1$, 27, and 111. In fact, $0 \div 0$ can be *any* number. Therefore, we don't know what answer to write for $0 \div 0$, and we say that $0 \div 0$ *cannot be determined*.

For these reasons, we say that division by 0 is *undefined*. The important thing to remember about division involving zero is that you cannot divide *by* zero.

Division involving zero can be summarized as follows:

Division involving zero	If a represents any real number except 0, then **1.** $\dfrac{0}{a} = 0 \div a = 0$ **2.** $\dfrac{a}{0} = a \div 0$ is not possible. **3.** $\frac{0}{0} = 0 \div 0$ cannot be determined. **Division by zero is not defined.**

EXAMPLE 3   Find the quotient for $-30 \div 5$.

SOLUTION

The answer is negative because the numbers have different signs

The quotient of the absolute values: $30 \div 5 = 6$

$$-30 \div 5 = -6$$

EXAMPLE 4   Perform the division: $-64 \div (-8)$.

SOLUTION

The answer is positive because the numbers have the same sign

The quotient of the absolute values: $64 \div 8 = 8$

$$-64 \div (-8) = +8$$

EXAMPLE 5   Perform the division: $\dfrac{-42}{8}$.

SOLUTION

The answer is negative because the numbers have different signs

$$\frac{-42}{8} = -\frac{42}{8} = -\frac{21 \cdot 2}{4 \cdot 2} = -\frac{21}{4}, \text{ or } -5\frac{1}{4}, \text{ or } -5.25$$

EXAMPLE 6   Perform the division $0 \div (-3)$, or write "not defined."

SOLUTION   0 divided by any nonzero number is 0.

$$0 \div (-3) = 0$$

EXAMPLE 7   Perform the division $-7 \div 0$, or write "not defined."

SOLUTION   We cannot divide by zero; $-7 \div 0$ is not defined.

Dividing Fractions   Recall from arithmetic that to *divide* one fraction by another, we multiply the dividend by the reciprocal of the divisor; that is,

$$\frac{a}{b} \div \frac{c}{d} = \frac{a}{b} \cdot \frac{d}{c} = \frac{ad}{bc}$$

We will discuss dividing *algebraic* fractions in more detail in a later chapter.

EXAMPLE 8   Perform the division: $\frac{2}{7} \div \left(-\frac{5}{14}\right)$.

SOLUTION   We must multiply the dividend $\left(\frac{2}{7}\right)$ by the reciprocal of the divisor.

$$\frac{2}{7} \div \left(-\frac{5}{14}\right) = \left(\frac{2}{7}\right)\left(-\frac{14}{5}\right) = -\left(\frac{2}{7} \cdot \frac{14}{5}\right) = -\frac{28}{35} = -\frac{4 \cdot 7}{5 \cdot 7} = -\frac{4}{5}$$

The answer is negative because the numbers have different signs

Using a Scientific Calculator

Scientific calculators generally use either **algebraic logic** or **Reverse Polish Notation (RPN)**. (The type of logic used determines the order in which the keys must be pressed.) Below we give some general guidelines *only* for using a scientific calculator that uses algebraic logic. Consult the instruction manual to find out how *your* particular calculator works (how to turn the calculator on, how to clear the memory, how the "inverse," or "shift," or "second function" key works, and so on).

To perform an operation that involves just two numbers (an operation such as addition, subtraction, multiplication, or division) on a scientific calculator that uses algebraic logic, we press the keys in the order in which we would read (or write) the problem. For example, to add 2 and 3, we press $\boxed{2}$ $\boxed{+}$ $\boxed{3}$ $\boxed{=}$. The display is $\boxed{5}$.

Negative Numbers and the Calculator   Scientific calculators generally have a $\boxed{+/-}$ (or $\boxed{\text{CHS}}$) key that is used to change the sign of a number, and it is pressed *after* the number is entered. For example, to enter -8, we press $\boxed{8}$ $\boxed{+/-}$. The display is $\boxed{-8}$ or $\boxed{-8.0000}$.

EXAMPLE 9   Find $3.25 \div (-2.5)$.

SOLUTION   We can determine the *sign* of the answer first and just use the calculator to divide the absolute values, or we can enter the problem as follows:

$$\boxed{3}\ \boxed{.}\ \boxed{2}\ \boxed{5}\ \boxed{\div}\ \boxed{2}\ \boxed{.}\ \boxed{5}\ \boxed{+/-}\ \boxed{=}$$

The display is $\boxed{-1.3}$. Therefore, $3.25 \div (-2.5) = -1.3$.

Exercises 1.6
Set I

Find each of the following quotients, or write "not defined."

1. $-10 \div (-5)$

2. $-12 \div (-4)$

3. $-8 \div 2$

4. $-6 \div 3$

5. $\dfrac{+10}{-2}$

6. $\dfrac{+8}{-4}$

7. $\dfrac{-6}{-3}$

8. $\dfrac{-10}{-2}$

9. $-40 \div 8$

10. $-60 \div 10$

11. $16 \div (-4)$

12. $25 \div (-5)$

13. $-15 \div (-5)$

14. $-27 \div (-9)$

15. $\dfrac{12}{-4}$

16. $\dfrac{24}{-6}$

17. $\dfrac{-18}{-2}$

18. $\dfrac{-49}{-7}$

19. $\dfrac{-150}{10}$

20. $\dfrac{-250}{100}$

21. $36 \div (-12)$

22. $56 \div (-8)$

23. $-45 \div 15$

24. $-39 \div 13$

25. $\dfrac{4}{0}$

26. $\dfrac{8}{0}$

27. $0 \div 0$

28. $\dfrac{0}{0}$

29. $0 \div 7$

30. $0 \div (-12)$

31. $\dfrac{-1}{0}$

32. $\dfrac{-15}{0}$

33. $\dfrac{-15}{6}$

34. $\dfrac{-27}{12}$

35. $\dfrac{7.5}{-0.5}$

36. $\dfrac{1.25}{-0.25}$

37. $\dfrac{-6.3}{-0.9}$

38. $\dfrac{-4.8}{-0.6}$

39. $\dfrac{-367}{100}$

40. $\dfrac{-4,860}{1,000}$

41. $-\frac{3}{8} \div \frac{2}{5}$

42. $-\frac{2}{3} \div \left(-\frac{1}{6}\right)$

43. $23.1616 \div (-4.136)$

44. $-17.0312 \div (-2.792)$

Writing Problems

Express the answers in your own words and in complete sentences.

1. Explain how to divide −45 by 15.

2. Explain how to divide $-\frac{3}{8}$ by $-\frac{2}{5}$.

3. Explain why $-8 \div 0$ is not defined.

Exercises  1.6
Set II

Find each of the following quotients, or write "not defined."

1. $-8 \div (-2)$

2. $-10 \div 5$

3. $12 \div (-6)$

4. $\dfrac{14}{-7}$

5. $\dfrac{-8}{2}$

6. $\dfrac{-50}{-10}$

7. $36 \div (-12)$

8. $-18 \div (-9)$

9. $-16 \div (4)$

10. $\dfrac{-15.6}{10}$

11. $\dfrac{13.8}{-10}$

12. $\dfrac{-6.3}{-0.7}$

13. $-8.4 \div (0.6)$

14. $-9.6 \div (-0.8)$

15. $18.5 \div (-3.7)$

16. $-14 \div 0$

17. $\dfrac{-18}{6}$

18. $-\left(\dfrac{0}{0}\right)$

19. $\dfrac{21}{-14}$

20. $\dfrac{-17}{-34}$

21. $81 \div (-15)$

22. $\dfrac{-51}{17}$

23. $-8 \div 16$

24. $-35 \div (-7)$

25. $\dfrac{13}{0}$

26. $0 \div (-10)$

27. $-8 \div 0$

28. $\dfrac{-12}{-2}$

29. $0 \div 9$

30. $\dfrac{72}{-18}$

31. $\dfrac{-11}{22}$

32. $\dfrac{16}{-24}$

33. $\dfrac{31}{-62}$

34. $-35 \div 0$

35. $\dfrac{-1.2}{0.24}$

36. $\dfrac{37}{-10}$

37. $\dfrac{-42}{-36}$

38. $\dfrac{-8.1}{2.7}$

39. $\dfrac{64}{-12}$

40. $\dfrac{-315}{100}$

41. $\frac{15}{16} \div \left(-\frac{3}{5}\right)$

42. $-4 \div \frac{1}{6}$

43. $-15.327 \div 23.58$

44. $5.63992 \div (-0.0572)$

Sections 1.1–1.6   REVIEW

Sets
1.1
A **set** is a collection of objects or things.

Equal Sign
1.1
The **equal sign** (=) in a statement means that the expression on the left side of the equal sign *has the same value or values* as the expression on the right side of the equal sign.

Natural Numbers
1.1
$\{1, 2, 3, \ldots\}$

Whole Numbers
1.1
$\{0, 1, 2, \ldots\}$

Digits
1.1
$\{0, 1, 2, 3, 4, 5, 6, 7, 8, 9\}$

Fractions 1.1	A **fraction** is a number that can be expressed in the form a/b, where a is a whole number and b is a natural number.
Mixed Numbers 1.1	A **mixed number** is made up of both a whole number part and a fraction part. There is an understood plus sign between the two parts.
Decimal Fractions 1.1	A **decimal fraction** is a fraction whose denominator is 10 or 100 or 1,000, and so on.
"Greater Than" and "Less Than" Symbols 1.1	The symbol $>$ is read "is greater than," and the symbol $<$ is read "is less than." Numbers get larger as we move to the right on the number line and get smaller as we move to the left.
Positive Numbers 1.2	All real numbers that are greater than 0 are **positive numbers**.
Negative Numbers 1.2	All real numbers that are less than 0 are **negative numbers**.
Integers 1.2	$\{\ldots, -3, -2, -1, 0, 1, 2, 3, \ldots\}$
Real Numbers 1.1 and 1.2	All the numbers that can be represented by points on the number line are called **real numbers**.
Rational Numbers 1.2	**Rational numbers** are numbers that can be expressed in the form a/b, where a and b are integers and $b \neq 0$. (All natural numbers, whole numbers, fractions, decimals, and mixed numbers are rational numbers.) All rational numbers are *real* numbers. The *decimal form* of any rational number is always either a *terminating* or a *repeating* decimal.
Absolute Value 1.3	The **absolute value** of a number is the distance between that number and 0 on the number line with no regard to direction. It can never be negative. The absolute value of a real number x is written $\lvert x \rvert$.

Addition of Two Signed Numbers
1.3
	1. When the numbers have the same sign,
	    a. add the absolute values, *and*
	    b. attach (to the left of the result from step a) the sign of the answer; that sign is *positive* when both numbers are positive and *negative* when both numbers are negative.
	2. When the numbers have different signs,
	    a. *subtract* the smaller absolute value from the larger absolute value, *and*
	    b. attach (to the left of the result from step a) the sign of the answer; *the answer has the sign of the number with the larger absolute value.*

Fundamental Property of Fractions 1.3	$\dfrac{a}{b} = \dfrac{ac}{bc}$, where a, b, and c are real numbers, but $b \neq 0$ and $c \neq 0$.
Addition of Fractions 1.3	$\dfrac{a}{c} + \dfrac{b}{c} = \dfrac{a+b}{c}$, where $c \neq 0$.
Additive Identity 1.3	The **additive identity** is 0. $$a + 0 = a \qquad \text{and} \qquad 0 + a = a$$
Additive Inverse 1.3	The **additive inverse** (or **opposite** or **negative**) of a is $-a$. $$a + (-a) = 0 \qquad \text{and} \qquad (-a) + a = 0$$
Subtraction of One Signed Number from Another 1.4	1. Change the subtraction symbol to an addition symbol, *and* change the sign of the number being subtracted. 2. *Add* the resulting signed numbers. $$a - b = a + (-b)$$ Subtraction is the inverse operation of addition.

Subtraction of Fractions
1.4
$\dfrac{a}{c} - \dfrac{b}{c} = \dfrac{a-b}{c}$, where $c \neq 0$.

Subtraction Involving 0
1.4

$$a - 0 = a$$

$$0 - a = -a$$

Multiplication of Two Signed Numbers
1.5
Multiply the absolute values *and* attach the correct sign to the left of the product of the absolute values. That sign is *positive* when the numbers have the same sign and *negative* when the numbers have different signs.

Multiplicative Identity
1.5
The **multiplicative identity** is 1.

$$a \cdot 1 = a \quad \text{and} \quad 1 \cdot a = a$$

Multiplication Property of 0
1.5

$$a \cdot 0 = 0 \quad \text{and} \quad 0 \cdot a = 0$$

Multiplicative Inverse or Reciprocal
1.5
If $a \neq 0$, the **multiplicative inverse**, or **reciprocal**, of a is $1/a$.

$$a\left(\dfrac{1}{a}\right) = 1 \quad \text{and} \quad \dfrac{1}{a}(a) = 1$$

Division of One Signed Number by Another
1.6
Divide the absolute value of the dividend by the absolute value of the divisor *and* attach the correct sign to the left of the quotient of the absolute values. That sign is *positive* when the numbers have the same sign and *negative* when the numbers have different signs.
Division is the inverse operation of multiplication.

Division Involving Zero
1.6
If a represents any real number except 0:

$$\dfrac{0}{a} = 0$$

$$\dfrac{a}{0} \quad \text{is not possible}$$

$$\dfrac{0}{0} \quad \text{cannot be determined}$$

Division *by* 0 is not defined.

Sections 1.1–1.6   REVIEW EXERCISES   Set I

1. Write all the digits that are > 7.

2. Write the smallest two-digit natural number.

3. Write the smallest one-digit integer.

4. Write the largest one-digit integer that is < 0.

5. Which symbol, $<$ or $>$, should be used to make each statement true?
   a. $-3 \ \underline{?} \ 8$
   b. $5 \ \underline{?} \ -2$

6. Which symbol, $<$ or $>$, should be used to make each statement true?
   a. $7 \ \underline{?} \ -8$
   b. $-3 \ \underline{?} \ -9$

7. What is the multiplicative identity?

8. What is the additive inverse of 4?

In Exercises 9–40, perform the indicated operations, or write "not defined."

9. $-2 + (+3)$
10. $-5 + (+4)$
11. $-6 \div (-2)$

12. $-8 \div (-4)$
13. $-5 - (-3)$
14. $-7 - (-2)$

15. $(+5) - |-2|$
16. $(+8) - |-3|$
17. $-3(-4)$

18. $-5(-4)$
19. $-7 - (3)$
20. $-3 - (1.37)$

21. $4 + (-12)$
22. $6 + (-8)$
23. $8(-15)$

24. $5(-18)$
25. $(24) \div (-3)$
26. $-\frac{7}{4} \div \frac{3}{4}$

27. $9 - (-4)$
28. $4 - (-7)$
29. $\left(-2\frac{2}{3}\right)\left(2\frac{1}{2}\right)$

30. $\frac{29}{5} + \left(-\frac{3}{2}\right)$
31. $-6 \div (2)$
32. $-12 \div (4)$

33. $-10 + (-2)$
34. $-5(3.21)$
35. $\dfrac{-25}{-5}$

36. $\dfrac{-16}{-2}$
37. $0 \div (-4)$
38. $(0)(-5)$

39. Subtract (-6) from (-10).

40. Subtract (-8) from (-5).

ANSWERS

Name _____

1. Write all the digits that are > 5.

2. Write all the even whole numbers that are < 12.

3. What is the additive identity?

4. Write all the whole numbers that are < 4.

5. What is the smallest digit?

6. What is the multiplicative inverse of −1/6?

7. Which symbol, < or >, should be used to make each statement true?
   a. −5 _?_ −1   **b.** −2 _?_ 0

8. Which symbol, < or >, should be used to make each statement true?
   a. −3 _?_ −8   **b.** 5 _?_ −2

In Exercises 9–40, perform the indicated operations, or write "not defined."

9. −5 + (+2)   10. −7 − (2)   11. 8 ÷ (−4)   12. −5(+2)

13. −8 ÷ (−2)   14. $-\frac{5}{3} + \left(-\frac{2}{5}\right)$ 15. $|0| - |-5|$ 16. 5 − (−12)

17. −8(−2)   18. −8 − 2   19. −9 − (3)

1. _____

2. _____

3. _____

4. _____

5. _____

6. _____

7a. _____

b. _____

8a. _____

b. _____

9. _____

10. _____

11. _____

12. _____

13. _____

14. _____

15. _____

16. _____

17. _____

18. _____

19. _____

20. $-9(-3)$ **21.** $9 + (-3)$ **22.** $16 \div (-12)$ **20.** _____

                                                                          21. _____

                                                                          22. _____

23. $41(-1)$ **24.** $\frac{8}{9} \div \left(-\frac{1}{2}\right)$ **25.** $\frac{1}{0}$ **23.** _____

                                                                          24. _____

                                                                          25. _____

26. $\dfrac{3}{-12}$ **27.** $24 \div 0$ **28.** $-72 \div (-8)$ **26.** _____

                                                                          27. _____

                                                                          28. _____

29. $\frac{0}{2}$ **30.** $-15 + (-23)$ **31.** $-|-4|$ **29.** _____

                                                                          30. _____

                                                                          31. _____

32. $(-437)(0)$ **33.** $0 \div (-15)$ **34.** $0 - 5$ **32.** _____

                                                                          33. _____

                                                                          34. _____

35. $0(-5)$ **36.** $(-20) - 2.5$ **37.** $(-20)(-2.5)$ **35.** _____

                                                                          36. _____

                                                                          37. _____

38. $\left(-1\frac{7}{8}\right)\left(-3\frac{1}{5}\right)$ **39.** Subtract -12 from 7.   **40.** Subtract 3 from -15.   **38.** _____

                                                                          39. _____

                                                                          40. _____

38

The purpose of this test is to see how well you understand addition, subtraction, multiplication, and division of two integers or other signed numbers. If you will be tested on Sections 1.1–1.6, we recommend that you work this diagnostic test *before* your instructor tests you on this material. Allow yourself about 50 minutes to do this test.

Complete solutions for all the problems on this test, together with section references, are given in the answer section at the end of the book. We suggest that you study the sections referred to for the problems you do incorrectly.

In Problems 1–10, write "true" if the statement is always true; otherwise, write "false."

1. The smallest natural number is 0.

2. 0 is a rational number.

3. $|-6| = -6$

4. $-4 > -3$

5. The additive identity element is 1.

6. $6 > -10$

7. 18 is a digit.

8. 3.762 is an integer.

9. $-\frac{1}{2}$ is a rational number.

10. The reciprocal of -4 is 4.

11. Which symbol, $<$ or $>$, should be used to make the statement "$-5 \ ? \ -1$" true?

12. Which symbol, $<$ or $>$, should be used to make the statement "$2 \ ? \ -8$" true?

13. Numbers that are multiplied together to give a product are called _____ .

14. Express $-|-56|$ without absolute value symbols.

15. Subtract 17 from 3.

16. Subtract -5 from -12.

17. What is 6 diminished by -2?

18. What is the additive inverse of $\frac{5}{8}$?

19. What is the additive inverse of -6?

20. What is the multiplicative inverse of -6?

In Problems 21–50, perform the indicated operation, or write "not defined."

21. $8 + (-26)$

22. $-13 + (-5)$

23. $-21 + (-5)$

24. $-\frac{2}{5} + \frac{3}{10}$

25. $6.16 + (-8.3)$

26. $-\frac{7}{2} + \frac{17}{8}$

27. $-8 - (-3)$

28. $6 - (-12)$

29. $-4 - 1$

30. $8 - 37$

31. $-5\frac{2}{3} - \left(-2\frac{8}{9}\right)$

32. $3\frac{1}{3} - 8\frac{1}{6}$

33. $-2.325 - (-6.3)$

34. $0 - 12$

35. $0(-12)$

36. $-4(-1)$

37. $12(-6)$

38. $-16(0)$

39. $-\frac{5}{6}\left(-\frac{3}{10}\right)$

40. $\left(-\frac{4}{3}\right)\left(\frac{7}{4}\right)$

41. $(6.32)(-0.1)$

42. $8 \div (-2)$

43. $\frac{17}{0}$

44. $\frac{-24}{10}$

45. $-36 \div (-12)$

46. $\frac{18}{-24}$

47. $0 \div (-2)$

48. $\frac{0}{0}$

49. $-54 \div 9$

50. $\frac{-4.9}{-7}$

1.7   The Commutative, Associative, and Distributive Properties of Real Numbers

We have already discussed several important properties of the real number system. Namely, we've mentioned that 0 is the additive identity, that 1 is the multiplicative identity, that the additive inverse of a is $-a$ (where a represents any real number), and that if a is any real number except 0, the multiplicative inverse of a is $1/a$. The *commutative*, *associative*, and *distributive properties* are also important properties of the set of real numbers. *All* of these properties are **axioms** (statements that must be accepted as true without proof).

The Commutative Properties

Addition Is Commutative   Reversing the order of two numbers in an addition problem does not change the sum (see Example 1). This important property is called the **commutative property of addition.**

The commutative property of addition

> If a and b represent any real numbers, then
> $$a + b = b + a$$

Note When we go from home to work and then from work to home (when we go from a to b and then from b to a), we say we "commute." Notice the similarity between the words *commute* and *commutative*.

A Word of Caution Notice that there is no "n" in the word *commutative*. A common error is to spell the word incorrectly as *communative* or *communitive*.

EXAMPLE 1 **a.** Verify that the commutative property holds for the sum of 2 and 3.

SOLUTION $2 + 3 = 5$, and $3 + 2 = 5$. Since $2 + 3$ and $3 + 2$ both equal 5, they must equal each other. Therefore, $2 + 3 = 3 + 2$.

b. Verify that the commutative property holds for the sum of -6 and 2.

SOLUTION $-6 + 2 = -4$, and $2 + (-6) = -4$. Since $-6 + 2$ and $2 + (-6)$ both equal -4, they must equal each other. Therefore, $-6 + 2 = 2 + (-6)$.

c. Verify that the commutative property holds for the sum of -4 and -8.

SOLUTION $-4 + (-8) = -12$, and $-8 + (-4) = -12$. Since $-4 + (-8)$ and $-8 + (-4)$ both equal -12, they must equal each other. Therefore, $-4 + (-8) = -8 + (-4)$.

Subtraction is not commutative, as Example 2 shows.

EXAMPLE 2 Show that subtraction is *not* commutative by showing that $3 - 2 \neq 2 - 3$.

SOLUTION $3 - 2 = 1$, and $2 - 3 = -1$. Since $3 - 2$ and $2 - 3$ do not both equal the same number, $3 - 2 \neq 2 - 3$. By finding one subtraction problem for which the commutative property does not hold, we've shown that subtraction is *not* commutative.

Multiplication Is Commutative Reversing the order of two numbers in a multiplication problem does not change the product (see Example 3). This important property is called the **commutative property of multiplication.**

The commutative property of multiplication

> If a and b represent any real numbers, then
> $$a \cdot b = b \cdot a$$

EXAMPLE 3 **a.** Verify that the commutative property holds for the product of 4 and 5.

SOLUTION $4 \cdot 5 = 20$, and $5 \cdot 4 = 20$. Since $4 \cdot 5$ and $5 \cdot 4$ both equal 20, they must equal each other. Therefore, $4 \cdot 5 = 5 \cdot 4$.

b. Verify that the commutative property holds for the product of -9 and 3.

SOLUTION $-9(3) = -27$, and $3(-9) = -27$. Since $-9(3)$ and $3(-9)$ both equal -27, they must equal each other. Therefore, $-9(3) = 3(-9)$.

Division is not commutative, as Example 4 shows.

EXAMPLE 4

Show that division is *not* commutative by showing that $10 \div 5 \neq 5 \div 10$.

SOLUTION $10 \div 5 = 2$, and $5 \div 10 = \frac{1}{2}$. Since $10 \div 5$ and $5 \div 10$ do not both equal the same number, $10 \div 5 \neq 5 \div 10$. By finding one division problem for which the commutative property does not hold, we've shown that division is *not* commutative.

The Associative Properties

Addition Is Associative   When we are given three numbers to add, we obtain the same answer when we add the first two numbers together first as when we add the last two numbers together first. This property is called the **associative property of addition.** We use grouping symbols such as parentheses ( ), brackets [ ], or braces { } to show which two numbers are to be added together first.

The associative property of addition

> If a, b, and c represent any real numbers, then
> $$(a + b) + c = a + (b + c)$$

EXAMPLE 5

Verify that $(2 + 3) + 4 = 2 + (3 + 4)$.

SOLUTION   On the left side of the equal sign, the parentheses indicate that we must add 2 and 3 first; on the right side of the equal sign, the parentheses indicate that we must add 3 and 4 first.

$$
\begin{array}{c|c}
(2 + 3) + 4 & 2 + (3 + 4) \\
= \quad 5 \quad + 4 & = 2 + \quad 7 \\
= \quad\quad 9 & = \quad\quad 9
\end{array}
$$

Since $(2 + 3) + 4$ and $2 + (3 + 4)$ both equal 9, they must equal each other. Therefore, $(2 + 3) + 4 = 2 + (3 + 4)$, and we've shown that the associative property of addition holds for the sum $2 + 3 + 4$.

Subtraction is not associative, as Example 6 will prove.

EXAMPLE 6

Show that subtraction is *not* associative by showing that $(12 - 6) - 2 \neq 12 - (6 - 2)$.

SOLUTION

$$
\begin{array}{c|c}
(12 - 6) - 2 & 12 - (6 - 2) \\
= \quad 6 \quad - 2 & = 12 - \quad 4 \\
= \quad\quad 4 & = \quad\quad 8
\end{array}
$$

Since $(12 - 6) - 2$ and $12 - (6 - 2)$ do not both equal the same number, $(12 - 6) - 2 \neq 12 - (6 - 2)$. By finding one subtraction problem for which the associative property does not hold, we've shown that subtraction is *not* associative.

Multiplication Is Associative   When we are given three numbers to multiply, we obtain the same answer when we multiply the first two numbers together first as when we multiply the last two numbers together first. This property is called the **associative property of multiplication.** Parentheses, brackets, or braces are used to show which two numbers are to be multiplied together first.

The associative property of multiplication

> If a, b, and c represent any real numbers, then
> $$(a \cdot b) \cdot c = a \cdot (b \cdot c)$$

EXAMPLE 7   Verify that $(3 \cdot 4) \cdot 2 = 3 \cdot (4 \cdot 2)$.

SOLUTION

$$
\begin{array}{ll}
(3 \cdot 4) \cdot 2 & \quad 3 \cdot (4 \cdot 2) \\
= \quad 12 \;\; \cdot 2 & \quad = 3 \cdot \;\; 8 \\
= \quad 24 & \quad = \quad 24
\end{array}
$$

Since $(3 \cdot 4) \cdot 2$ and $3 \cdot (4 \cdot 2)$ both equal 24, they must equal each other. Therefore, $(3 \cdot 4) \cdot 2 = 3 \cdot (4 \cdot 2)$, and we've shown that the associative property of multiplication holds for the product $3 \cdot 4 \cdot 2$.

EXAMPLE 8   Verify that $[(-6)(+2)](-5) = (-6)[(+2)(-5)]$.

SOLUTION

$$
\begin{array}{ll}
[(-6)(+2)](-5) & \quad (-6)[(+2)(-5)] \\
= \quad [-12] \;\; \cdot (-5) & \quad = (-6) \cdot \;\; [-10] \\
= \quad 60 & \quad = \quad 60
\end{array}
$$

Since $[(-6)(+2)](-5)$ and $(-6)[(+2)(-5)]$ both equal 60, they must equal each other. Therefore, $[(-6)(+2)](-5) = (-6)[(+2)(-5)]$, and we've shown that the associative property of multiplication holds for the product $(-6)(+2)(-5)$.

Division is not associative, as a single example will prove.

EXAMPLE 9   Show that division is *not* associative by showing that $(16 \div 4) \div 2 \neq 16 \div (4 \div 2)$.

SOLUTION

$$
\begin{array}{ll}
(16 \div 4) \div 2 & \quad 16 \div (4 \div 2) \\
= \quad 4 \;\; \div 2 & \quad = 16 \div \;\; 2 \\
= \quad 2 & \quad = \quad 8
\end{array}
$$

Since $(16 \div 4) \div 2$ and $16 \div (4 \div 2)$ do not both equal the same number, $(16 \div 4) \div 2 \neq 16 \div (4 \div 2)$. By finding one division problem for which the associative property does not hold, we've shown that division is *not* associative.

Summary of commutative and associative properties

> **1.** The *commutative properties* guarantee that when we have two numbers to add or two numbers to multiply, *reversing the order* of the numbers does not change the answer.
>
> **2.** The *associative properties* guarantee that when we have three numbers to add or three numbers to multiply, we obtain the same answer when we operate on the first two numbers first as when we operate on the last two numbers first.

How to Determine Whether Commutativity or Associativity Has Been Used   In commutativity, the numbers or letters actually exchange places (commute):

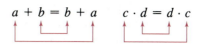

$$a + b = b + a \qquad c \cdot d = d \cdot c$$

In associativity, the numbers or letters stay in their original places, but the grouping is changed:

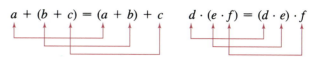

$$a + (b + c) = (a + b) + c \qquad d \cdot (e \cdot f) = (d \cdot e) \cdot f$$

The Distributive Property

Multiplication Is Distributive Over Addition (and Over Subtraction)   The **distributive property** is another axiom—one of the fundamental properties of real numbers that we must accept as true without proof. Whereas each of the properties of real numbers that we've discussed so far was concerned with just one operation, the distributive property is concerned with *two* operations: multiplication *and* addition (or multiplication *and* subtraction).

The distributive property	Multiplication is distributive over addition (and over subtraction). If a, b, and c represent any real numbers, then $$a(b + c) = (ab) + (ac)$$ and $$a(b - c) = (ab) - (ac)$$

EXAMPLE 10   Verify that $5(3 + 7) = (5 \cdot 3) + (5 \cdot 7)$.

SOLUTION

$5(3 + 7)$	$(5 \cdot 3) + (5 \cdot 7)$
$= 5 \ (10)$	$= 15 + 35$
$= 50$	$= 50$

Because $5(3 + 7)$ and $(5 \cdot 3) + (5 \cdot 7)$ both equal 50, they must equal each other. Therefore, $5(3 + 7) = (5 \cdot 3) + (5 \cdot 7)$, and we've verified that the distributive property holds in this example.

EXAMPLE 11   Verify that $8[3 + (-10)] = [8 \cdot 3] + [8 \cdot (-10)]$.

SOLUTION

$8[3 + (-10)]$	$[8 \cdot 3] + [8 \cdot (-10)]$
$= 8 \ [-7]$	$= 24 + [-80]$
$= -56$	$= -56$

Because $8[3 + (-10)]$ and $[8 \cdot 3] + [8 \cdot (-10)]$ both equal -56, they must equal each other. Therefore, $8[3 + (-10)] = [8 \cdot 3] + [8 \cdot (-10)]$, and we've verified that the distributive property holds in this example. The problem could have been written "Verify that $8(3 - 10) = (8 \cdot 3) - (8 \cdot 10)$."

In Example 12 and in the exercises in this section, we include problems that involve the identity and inverse properties as well as the commutative, associative, and distributive properties. We also include a few problems dealing with the multiplication property of zero.

EXAMPLE 12   State whether each of the following is true or false. If the statement is true, give the reason.

SOLUTION

a. $(-7) + 5 = 5 + (-7)$

True; the commutative property of addition (the order of the numbers was changed)

b. $(+6)(-8) = (-8)(+6)$

True; the commutative property of multiplication (the order of the numbers was changed)

c. $[(-3) + 5] + (-2)$
$= (-3) + [5 + (-2)]$

True; the associative property of addition (the grouping was changed)

d. $-6(3 + 5) = (-6 \cdot 3) + (-6 \cdot 5)$

True; multiplication is distributive over addition

e. $-8 + 0 = -8$

True; the additive identity property (0 is the additive identity)

f. $-5 + 5 = 0$

True; the additive inverse property (the additive inverse of -5 is 5)

g. $[(7) \cdot (-4)] \cdot (2)$
$= (7) \cdot [(-4) \cdot (2)]$

True; the associative property of multiplication (the grouping was changed)

h. $(+8) - (-7) = (-7) - (+8)$ *False*

i. $3(4 \cdot 5) = (3 \cdot 4) \cdot (3 \cdot 5)$ *False*

j. $a + (b + c) = (a + b) + c$

True; the associative property of addition (the grouping was changed)

k. $12 \div 6 = 6 \div 12$ *False*

l. $(p \cdot r) \cdot s = p \cdot (r \cdot s)$

True; the associative property of multiplication (the grouping was changed)

m. $3 \times 0 = 3$ *False*

n. $(3 + 5) + 7 = 3 + (7 + 5)$

True; the commutative *and* associative properties of addition

o. $8 \cdot (3 \cdot 6) = (8 \cdot 6) \cdot 3$

True; the commutative *and* associative properties of multiplication

p. $0 \times 6 = 0$

True; the multiplication property of zero

Addition and Multiplication Involving More Than Two Numbers

Because the commutative and associative properties of addition and of multiplication hold for all real numbers, we can add more than two numbers *in any order*, and we can multiply more than two numbers *in any order*.

EXAMPLE 13   Add $-8 + 3 + (-4) + (-6)$.

SOLUTION $-8 + 3 + (-4) + (-6)$

$= [(-8) + (-4) + (-6)] + 3$ Collecting all the negative numbers

$= [-18] + 3$ Adding all the negative numbers first

$= -15$

EXAMPLE 14   Multiply $(-2)(-5)(-3)$.

SOLUTION $(-2)(-5)(-3)$ There are three negative factors

$= (+10)(-3)$

$= -30$ Notice that an *odd* number of negative signs gives a product that is *negative*

EXAMPLE 15 Multiply $(-2)(-5)(-3)(-4)$.

SOLUTION $(-2)(-5)(-3)(-4)$ There are four negative factors

$= (+10)(-3)(-4)$

$= (-30)(-4)$

$= +120$ Notice that an *even* number of negative signs gives a product that is *positive*

We see that when we *multiply* several numbers together, an *odd* number of negative factors gives a product that is *negative* (see Example 14), and an *even* number of negative factors gives a product that is *positive* (see Example 15). Therefore, it is possible to multiply two or more signed numbers as follows.

Multiplying several numerical factors together

Determine the sign of the answer; it will be *positive* if an *even* number of factors are negative, and *negative* if an *odd* number of factors are negative. Then multiply the absolute values of the factors.

Exercises 1.7
Set I

In Exercises 1–36, determine whether each statement is true or false. If the statement is true, give the reason.

1. $7 + 5 = 5 + 7$

2. $9 + 4 = 4 + 9$

3. $(2 + 6) + 3 = 2 + (6 + 3)$

4. $(1 + 8) + 7 = 1 + (8 + 7)$

5. $6 - 2 = 2 - 6$

6. $4 - 7 = 7 - 4$

7. $(a \cdot b) \cdot c = a \cdot (b \cdot c)$

8. $(p \cdot q) \cdot r = p \cdot (q \cdot r)$

9. $8 \div 4 = 4 \div 8$

10. $3 \div 6 = 6 \div 3$

11. $(p)(t) = (t)(p)$

12. $(m)(n) = (n)(m)$

13. $(4) + (-5) = (-5) + (4)$

14. $(-7) + (2) = (2) + (-7)$

15. $5 + (3 + 4) = 5 + (4 + 3)$

16. $6 + (8 + 2) = 6 + (2 + 8)$

17. $6(2 \cdot 3) = (6 \cdot 2) \cdot (6 \cdot 3)$

18. $7(3 + 5) = (7 \cdot 3) + (7 \cdot 5)$

19. $e + f = f + e$

20. $j + k = k + j$

21. $-8 \times 1 = -8$

22. $11 + 0 = 11$

23. $15 + (-15) = 0$

24. $-4 + 4 = 0$

25. $8(5 - 2) = (8 \cdot 5) - (8 \cdot 2)$

26. $9(6 \cdot 4) = (9 \cdot 6) \cdot (9 \cdot 4)$

27. $8 \times 0 = 8$

28. $0 \times (-4) = -4$

29. $9 + (5 + 6) = (9 + 6) + 5$

30. $3 \cdot (8 \cdot 4) = (3 \cdot 4) \cdot 8$

31. $x - 4 = 4 - x$

32. $5 - y = y - 5$

33. $4(a \cdot 6) = (4a)(6)$

34. $m(7 \cdot 5) = (m \cdot 7)(5)$

35. $H + 8 = 8 + H$

36. $4 + P = P + 4$

In Exercises 37–46, complete each statement by using the property indicated.

37. Commutative property:
$$-7 + 3 = \underline{\hspace{2cm}}$$

38. Commutative property:
$$12 + (-5) = \underline{\hspace{2cm}}$$

39. Associative property:
$$4(-3 \cdot 6) = \underline{\hspace{2cm}}$$

40. Associative property:
$$-8[4 \cdot (-2)] = \underline{\hspace{3cm}}$$

41. Distributive property:
$$3(5 - 8) = \underline{\hspace{3cm}}$$

42. Distributive property:
$$6(15 + 2) = \underline{\hspace{3cm}}$$

43. Commutative property:
$$(-4)(3) = \underline{\hspace{3cm}}$$

44. Commutative property:
$$6(-3) = \underline{\hspace{3cm}}$$

45. Associative property:
$$4 + (-3 + 6) = \underline{\hspace{3cm}}$$

46. Associative property:
$$(-3 + 4) + (-2) = \underline{\hspace{3cm}}$$

In Exercises 47–56, perform the indicated operations.

47. $5 + (-2) + 4 + (-8) + (-5)$

48. $-2 + 6 + (-8) + (-12) + 5$

49. $8 + (-3) + (-7) + (-1)$

50. $-2 + (-5) + 6 + (-11)$

51. $(-5)(-4)(-2)$

52. $(-3)(-2)(-8)$

53. $2(-5)(-9)$

54. $(-3)(4)(-2)$

55. $(-2)(-3)(-5)(-4)$

56. $(-4)(-2)(-1)(-7)$

Writing Problems

Express the answers in your own words and in complete sentences.

1. Explain how to determine whether $5 + 8 = 8 + 5$ illustrates the commutative property or the associative property.

2. Explain how to determine whether $(6 \cdot 8) \cdot 2 = 6 \cdot (8 \cdot 2)$ illustrates the commutative property or the associative property.

3. Explain how to determine whether $8 + 0 = 8$ illustrates the additive identity property or the additive inverse property.

Exercises 1.7
Set II

In Exercises 1–36, determine whether each statement is true or false. If the statement is true, give the reason.

1. $5 + 3 = 3 + 5$

2. $(3 + 1) + 5 = 3 + (1 + 5)$

3. $8 - 2 = 2 - 8$

4. $(x \cdot y) \cdot z = x \cdot (y \cdot z)$

5. $10 \div 2 = 2 \div 10$

6. $(3)(-2) = (-2)(3)$

7. $2 + (3 + 4) = 2 + (4 + 3)$

8. $x + y = y + x$

9. $8 + (2 + 5) = (8 + 5) + 2$

10. $a - 2 = 2 - a$

11. $(4 \cdot c)(3) = 4(3 \cdot c)$

12. $3 \div x = x \div 3$

13. $5(-2) = -2(5)$

14. $8 + (3 + 2) = 8 + (2 + 3)$

15. $12 - (6 - 9) = (12 - 6) - 9$

16. $9 \times (-3) = (-3) \times 9$

17. $8(7 \cdot 2) = (8 \cdot 7) \cdot (8 \cdot 2)$

18. $6(5 - 1) = (6 \cdot 5) - (6 \cdot 1)$

19. $3 + (4 + 7) = (3 + 7) + 4$

20. $25 \div 3 = 3 \div 25$

21. $3 + 0 = 3$

22. $0 \times (-13) = -13$

23. $\frac{1}{2} + \left(-\frac{1}{2}\right) = 0$

24. $4 \times (-4) = 0$

25. $3(9 - 4) = (3 \cdot 9) - (3 \cdot 4)$

26. $-3(2 \cdot 4) = (-3 \cdot 2) \cdot (-3 \cdot 4)$

27. $-5 \times 0 = -5$

28. $-16 \times 0 = 0$

29. $18 - 3 = 3 - 18$

30. $5 \cdot (12 \cdot 6) = (5 \cdot 6) \cdot 12$

31. $24 + 16 = 16 + 24$

32. $24 \div (12 \div 2) = (24 \div 12) \div 2$

33. $2 \times (3 \times 4) = (2 \times 3) \times 4$

34. $6 - 15 = 15 - 6$

35. $18 + (3 + 5) = (18 + 3) + 5$

36. $9 - (4 - 12) = (9 - 4) - 12$

In Exercises 37–46, complete each statement by using the property indicated.

37. Commutative property:
$$6 + (-8) = \underline{\hspace{3cm}}$$

38. Commutative property:
$$-6(-4) = \underline{\hspace{3cm}}$$

39. Associative property:
$$4 + [(-3) + 6] = \underline{\hspace{3cm}}$$

40. Associative property:
$$6(-4 \cdot 5) = \underline{\hspace{3cm}}$$

41. Distributive property:
$$-2(8 + 4) = \underline{\hspace{3cm}}$$

42. Distributive property:
$$6(-4 + 8) = \underline{\hspace{3cm}}$$

43. Commutative property:
$$-4 + 3 = \underline{\hspace{3cm}}$$

44. Commutative property:
$$-5(2) = \underline{\hspace{3cm}}$$

45. Associative property:
$$4(-3 \cdot 7) = \underline{\hspace{3cm}}$$

46. Associative property:
$$-1 + [4 + (-2)] = \underline{\hspace{3cm}}$$

In Exercises 47–56, perform the indicated operations.

47. $3 + (-7) + (-6) + 8 + (-9)$

48. $4 + (-6) + 3 + (-2) + (-12)$

49. $9 + (-5) + 7 + (-12)$

50. $-2 + (-8) + (-6) + 11$

51. $4(-5)(-7)$

52. $(-9)(-8)(-1)$

53. $(-5)(-1)(-6)(-2)$

54. $-3(-2)(5)(-4)$

55. $(-5)(-2)(-2)(-2)$

56. $8(-1)(-2)(-3)$

1.8   Finding Powers of Integers and Other Signed Numbers

Because we have learned to multiply signed numbers, we can now consider products in which some number is repeated as a factor.

The shortened notation for a product such as $3 \cdot 3 \cdot 3 \cdot 3$ is 3^4. That is, by definition, $3^4 = 3 \cdot 3 \cdot 3 \cdot 3$. In the expression 3^4, 3 is called the **base**, and 4 is called the **exponent**. The number 4 (the exponent) indicates that 3 (the base) is to be used as a *factor* four times. The entire symbol 3^4 is called an **exponential expression** and is commonly read as "three to the fourth power."

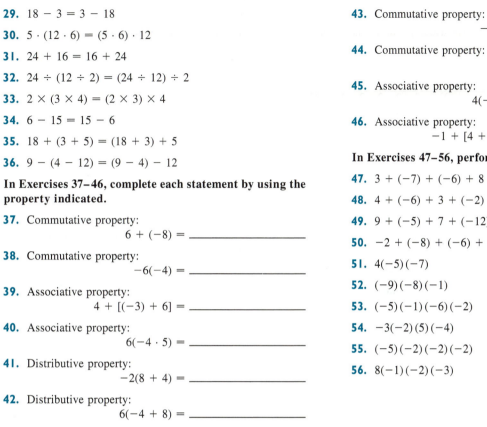

The exponent

Read as "three to the fourth power" ⟶ $3^4 = 81$

The base

	A Word of Caution	$3^4 \neq 3 \cdot 4$
		$3^4 = \underbrace{3 \cdot 3 \cdot 3 \cdot 3}_{\text{Four factors}} = 81$

A Word of Caution   When we write an exponential number, we must be sure that our exponents look like exponents. For example, we must be sure that 3^4 doesn't look like 34.

Powers of Zero

Powers of 0	If a represents any *positive* real number, then
	$$0^a = 0$$

EXAMPLE 7   Examples of powers of zero:

a. $0^2 = 0 \cdot 0 = 0$; or, using the rule above, $0^2 = 0$.
b. $0^5 = 0 \cdot 0 \cdot 0 \cdot 0 \cdot 0 = 0$; or, using the rule above, $0^5 = 0$.

You may find it helpful to memorize the following powers:

$0^2 = 0$	$7^2 = 49$	$0^3 = 0$	$0^4 = 0$
$1^2 = 1$	$8^2 = 64$	$1^3 = 1$	$1^4 = 1$
$2^2 = 4$	$9^2 = 81$	$2^3 = 8$	$2^4 = 16$
$3^2 = 9$	$10^2 = 100$	$3^3 = 27$	$3^4 = 81$
$4^2 = 16$	$11^2 = 121$	$4^3 = 64$	
$5^2 = 25$	$12^2 = 144$	$5^3 = 125$	$2^5 = 32$
$6^2 = 36$	$13^2 = 169$		$2^6 = 64$

Cases where 0 appears as an exponent, such as 5^0, are not discussed at this time.

Exercises *1.8*
Set I

In Exercises 1–8, identify the *base* and the *exponent*, and then find the value of the expression.

1. 3^3 **2.** 2^4 **3.** $(-5)^2$ **4.** $(-6)^3$

5. 7^2 **6.** 3^4 **7.** 0^3 **8.** 0^4

In Exercises 9–48, find the value of each expression.

9. $(-10)^1$ **10.** $(-10)^2$ **11.** 10^3

12. 10^4 **13.** $(-10)^5$ **14.** $(-10)^6$

15. 2^1 **16.** 2^5 **17.** $(-2)^6$

18. $(-2)^7$ **19.** 2^8 **20.** 25^2

21. 40^3 **22.** 0^4 **23.** $(-12)^3$

24. $(-15)^2$ **25.** $(-1)^5$ **26.** $(-1)^7$

27. -2^2 **28.** -3^2 **29.** $(-1)^{99}$

30. $(-1)^{98}$ **31.** $-(-1)^5$ **32.** $-(-1)^6$

33. $-(-9)^2$ **34.** $-(-5)^2$ **35.** 0^8

36. 0^3 **37.** $(12.7)^2$ **38.** $(15.4)^2$

39. $(0.156)^2$ **40.** $(0.087)^2$ **41.** $\left(-\frac{5}{8}\right)^2$

42. $\left(-\frac{2}{3}\right)^4$ **43.** $\left(-\frac{1}{4}\right)^3$ **44.** $\left(-\frac{1}{3}\right)^5$

45. $\left(\frac{3}{16}\right)^1$ **46.** $\left(-\frac{3}{5}\right)^1$ **47.** $(-2.5)^1$

48. $(-0.7)^1$

Writing Problems

Express the answers in your own words and in complete sentences.

1. Explain why $3^4 \neq 12$.

2. Explain why $(-3)^4 \neq -3^4$.

3. Explain why $(-2)^5 = -2^5$.

4. Explain what $(-2)^5$ means.

Exercises 1.8
Set II

In Exercises 1–8, identify the *base* and the *exponent*, and then find the value of the expression.

1. 6^2

2. $(-2)^3$

3. $(-10)^2$

4. -10^2

5. 0^5

6. $(-1)^{35}$

7. $-(-5)^2$

8. $(-1)^{50}$

In Exercises 9–48, find the value of each expression.

9. 10^5

10. 8^2

11. 4^3

12. -3^4

13. $(-3)^4$

14. 20^3

15. 24^2

16. $(-4)^2$

17. -4^2

18. $(-1)^{43}$

19. 0^{23}

20. $-(-8)^2$

21. $(-1)^{132}$

22. 1^{43}

23. $(-2)^5$

24. 30^2

25. 0^{42}

26. $(-7)^2$

27. $(-3)^3$

28. -3^3

29. $-(-3)^3$

30. $-(-3^3)$

31. $-(-1)^7$

32. $-(-2)^4$

33. $-(-2)^6$

34. $-(-4)^2$

35. 0^2

36. 1^3

37. 16^2

38. 0^1

39.  $(0.894)^2$

40. $(0.095)^2$

41. $\left(-\frac{8}{9}\right)^2$

42. $\left(-\frac{3}{4}\right)^4$

43. $\left(-\frac{1}{2}\right)^5$

44. $\left(-\frac{1}{3}\right)^3$

45. $\left(\frac{7}{13}\right)^1$

46. $\left(-\frac{5}{6}\right)^1$

47. $(-3.7)^1$

48. $(0.6)^1$

1.9 Factoring Integers

Tests of Divisibility

We can use the following tests to determine whether a whole number is divisible by 2, 3, or 5.

Divisibility by 2   A whole number is divisible by 2 if its last digit (the unit's digit) is 0, 2, 4, 6, or 8.

Divisibility by 3   A whole number is divisible by 3 if the sum of its digits is divisible by 3.

Divisibility by 5   A whole number is divisible by 5 if its last digit (the unit's digit) is 0 or 5.

 Although there are tests of divisibility by other numbers, those tests are not included in this section.

EXAMPLE 1   Examples of the use of the tests of divisibility:

a. 1 2 , 30 0 , 2,03 4 , and 57 8 are divisible by 2, because the last digit of each number is a 0, 2, 4, 6, or 8.

b. 132 is divisible by 3, because the sum of the digits is $1 + 3 + 2 = 6$, and 6 is divisible by 3.

c. 5,162 is not divisible by 3, because $5 + 1 + 6 + 2 = 14$, and 14 is not divisible by 3.

d. 25 0 and 75 5 are both divisible by 5, because the last digit of each is a 0 or a 5.

An Introduction to Problem-Solving: Making an Organized List

We will discuss a few recognized **problem-solving techniques** in this book. The first of these, the technique of **making an organized list,** is demonstrated in Examples 2 and 3 when we find all the factors of a number. In an *organized* list, the items must be listed in some logical, orderly way.

EXAMPLE 5   Find the prime factorization of 18.

SOLUTION   We can start by factoring 18 into $2 \cdot 9$ and then factor 9 into $3 \cdot 3$ (see below, left), *or* we can start by factoring 18 into $3 \cdot 6$ and then factor 6 into $2 \cdot 3$ (see below, right).

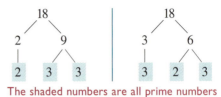

The shaded numbers are all prime numbers

The prime factorizations are $2 \cdot 3 \cdot 3$ and $3 \cdot 2 \cdot 3$. The *factors* are the same; it is only the order that is different. Both factorizations are considered to be *prime factorizations*, because all the factors are prime numbers. If both results had been expressed in exponential form with the smaller base written on the left (that is, as $2 \cdot 3^2$), it would have been obvious that the prime-factored forms were the same.

Prime-factored, exponential form

When a number is written in prime-factored form *and when repeated factors are expressed in exponential form*, we say that the number is in prime-factored, exponential form. We often write the factors with the bases in increasing, numerical order.

In Example 5, the prime-factored, exponential form of 18 is $2 \cdot 3^2$.

An alternative method for finding prime factorizations is demonstrated in Examples 6 and 7.

EXAMPLE 6   Find the prime factorization of each of the following numbers.

a. 24

SOLUTION   We first try to divide 24 by the smallest prime, 2. This number *does* divide exactly into 24 and gives a quotient of 12. We try 2 once again, this time as a divisor of the quotient, 12. Two *does* divide evenly into 12 and gives a quotient of 6. We try 2 yet again, this time as a divisor of the quotient, 6. Two *does* divide evenly into 6 and gives a quotient of 3, which is itself a prime number, so the process ends. (The process ends when each factor is a prime number.) Because we don't know how much space to allow for the division, the work is usually arranged by placing the quotient *under* the number we're dividing into, as follows:

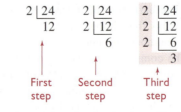

The prime factorization is the product of the numbers that are shaded

First step   Second step   Third step

Therefore, $24 = 2 \cdot 2 \cdot 2 \cdot 3 = 2^3 \cdot 3$, and $2^3 \cdot 3$ is the prime factorization of 24.

b. 20

SOLUTION

Therefore, the prime factorization of 20 is $2 \cdot 2 \cdot 5$, or $2^2 \cdot 5$.

c. 36

S O L U T I O N

$$
\begin{array}{r|r}
2 & 36 \\
2 & 18 \\
3 & 9 \\
& 3
\end{array}
$$

Therefore, the prime factorization of 36 is $2 \cdot 2 \cdot 3 \cdot 3$, or $2^2 \cdot 3^2$.

d. 315

S O L U T I O N Because the units' digit is *odd*, we don't need to try to divide by 2.

$$
\begin{array}{r|r}
3 & 315 \\
3 & 105 \\
5 & 35 \\
& 7
\end{array}
$$

Therefore, the prime factorization of 315 is $3 \cdot 3 \cdot 5 \cdot 7$, or $3^2 \cdot 5 \cdot 7$.

When we're trying to find the prime factors of a number, we do not need to try any prime that has a square greater than that number (see Example 7).

E X A M P L E 7   Find the prime factorization of 131.

S O L U T I O N

Primes in order of size

2	Does not divide 131
3	Does not divide 131
5	Does not divide 131
7	Does not divide 131
11	Does not divide 131
13	Does not divide 131; larger primes need not be tried because $13^2 = 169$, and 169 is greater than 131

Therefore, the prime factorization of 131 is simply 131, since 131 is a prime number.

Exercises    1.9
Set 1

In Exercises 1–16, write "P" if the number is prime and "C" if it is composite, and give the set of all the positive integral factors of the number.

1. 5   **2.** 8

3. 13   **4.** 15

5. 12   **6.** 11

7. 21   **8.** 23

9. 55   **10.** 41

11. 49   **12.** 31

13. 51   **14.** 42

15. 111   **16.** 101

In Exercises 17–26, find all the integral factors of each number.

17. 4

18. 9

19. 10

20. 14

21. 15

22. 16

23. 18

24. 20

25. 21

26. 22

In Exercises 27–42, find the prime factorization of each number.

27. 14   **28.** 15   **29.** 21

30. 22   **31.** 26   **32.** 27

33. 29   **34.** 31   **35.** 32

36. 33   **37.** 34   **38.** 35

39. 84   **40.** 75   **41.** 144

42. 180

43. List any prime numbers greater than 17 and less than 37 that yield a remainder of 1 when divided by 5.

44. List any prime numbers greater than 19 and less than 41 that yield a remainder of 2 when divided by 5.

Writing Problems

Express the answers in your own words and in complete sentences.

1. Explain why 11 is a prime number.

2. Explain why 12 is a composite number.

Exercises   1.9
Set II

In Exercises 1–16, write "P" if the number is prime and "C" if it is composite, and give the set of all the positive integral factors of the number.

1. 17   **2.** 14

3. 18   **4.** 19

5. 61   **6.** 63

7. 81   **8.** 73

9. 39   **10.** 83

11. 100   **12.** 29

13. 129   **14.** 105

15. 89   **16.** 97

In Exercises 17–26, find all the integral factors of each number.

17. 6   **18.** 23

19. 26   **20.** 38

21. 46   **22.** 81

23. 49   **24.** 63

25. 85   **26.** 111

In Exercises 27–42, find the prime factorization of each number.

27. 16   **28.** 18

29. 28   **30.** 30

31. 65   **32.** 78

33. 120   **34.** 112

35. 81   **36.** 49

37. 43   **38.** 72

39. 51   **40.** 38

41. 111   **42.** 88

43. List any prime numbers greater than 13 and less than 29 that yield a remainder of 5 when divided by 7.

44. List any prime numbers less than 41 and greater than 11 that yield a remainder of 1 when divided by 6.

1.10   Finding Roots of Integers and Other Signed Numbers

1.10A   Finding Square Roots of Numbers by Inspection

Just as subtraction is the inverse operation of addition and division is the inverse operation of multiplication, finding *roots* is the inverse operation of raising to powers. Thus, finding the *square root* of a number is the inverse operation of *squaring* a number.

The Principle Square Root   Every positive real number has both a positive and a negative square root; the *positive* square root is called the **principal square root.**

EXAMPLE 1 Determine the square roots of 9.

SOLUTION The number 9 has two square roots: $+3$ and -3.

$+3$ is a square root of 9, because $(+3)^2 = 9$.

-3 is a square root of 9, because $(-3)^2 = 9$.

3 is the *principal* square root of 9, because it is the root that is positive.

The Square Root Symbol The notation for the principal square root of p is $\sqrt{p}$, which is read "the square root of p." We call p itself the radicand. The entire expression $\sqrt{p}$ is called a **radical expression** or, more simply, a **radical**. The parts of a square root are shown below.

The radical sign   The radicand

EXAMPLE 2 Examples of identifying the radicand:

a. $\sqrt{9}$ The radicand is 9.
b. $\sqrt{15}$ The radicand is 15.
c. $\sqrt{\frac{3}{7}}$ The radicand is $\frac{3}{7}$.

Nonnegative Numbers Numbers that are positive or zero—in other words, numbers that are *not negative*—are often called **nonnegative numbers**. That is, if a is a *nonnegative number*, then $a \geq 0$.

Finding the Principal Square Root of a Number When we are asked to find $\sqrt{p}$ (the principal square root of p), we must find some *nonnegative* number whose *square* is p. For example, "Find $\sqrt{9}$" means we must find a nonnegative number whose square is 9. The answer is 3, because $3^2 = 9$. Therefore, $\sqrt{9} = 3$.

The following properties are often used in finding square roots.

PROPERTY 1
The square root rule

> Let a and b represent any nonnegative real numbers.
>
> If $a = b$,   then $\sqrt{a} = \sqrt{b}$

PROPERTY 2
Finding the square root of a perfect square

> If a represents any nonnegative real number, then
>
> $\sqrt{a^2} = a$

Property 1 implies, for example, that since $16 = 4^2$, $\sqrt{16} = \sqrt{4^2}$; and Property 2 implies that $\sqrt{4^2} = 4$ (see Example 3).

Perfect Squares When a number is the square of an integer, we call the number a **perfect square**. Thus, 0, 1, 4, 9, 16, and so on, are perfect squares.

The radicands in this section are perfect squares; therefore, you will find most of the problems in this section easier to do if you have memorized the squares of the whole numbers 0 through 13. When we find a square root by recognizing that the radicand is the square of some number, we say we are finding the square root *by inspection*.

A Word of Caution When $p \geq 0$, $\sqrt{p}$ is *always* positive or zero, because the symbol $\sqrt{p}$ *always* represents the principal square root of p.

EXAMPLE 3 By inspection, find the principal square root of 16; that is, find $\sqrt{16}$.

SOLUTION We know that $16 = 4^2$. Therefore,

$$\sqrt{16} = \sqrt{4^2} = 4$$

Using Property 1
Using Property 2

✓ **Check** $4^2 = 16$

The answer is 4.

❌ **A Word of Caution** Two common errors are shown on the right below:

Correct method	Incorrect method	
$\sqrt{16} = \sqrt{4^2} = 4$	$\sqrt{16} = 4 = 2$	$4 \neq 2$
	$\sqrt{16} = \sqrt{4} = 2$	$\sqrt{16} \neq \sqrt{4}$

❌ **A Word of Caution** The *square* of 16 is 256 (that is, $16^2 = 256$). The *square root* of 16 is 4 (that is, $\sqrt{16} = \sqrt{4^2} = 4$).

EXAMPLE 4 By inspection, find the principal square root of 25; that is, find $\sqrt{25}$.

SOLUTION We know that $25 = 5^2$. Therefore,

$$\sqrt{25} = \sqrt{5^2} = 5$$

Using Property 1
Using Property 2

✓ **Check** $5^2 = 25$

The answer is 5.

EXAMPLE 5 Examples of finding square roots by inspection:

a. $\sqrt{4} = \sqrt{2^2} = 2$ **Check** $2^2 = 4$
b. $\sqrt{36} = \sqrt{6^2} = 6$ **Check** $6^2 = 36$
c. $\sqrt{0} = \sqrt{0^2} = 0$ **Check** $0^2 = 0$
d. $\sqrt{1} = \sqrt{1^2} = 1$ **Check** $1^2 = 1$
e. $\sqrt{0.49} = \sqrt{(0.7)^2} = 0.7$ **Check** $(0.7)^2 = 0.49$
f. $\sqrt{\frac{1}{4}} = \sqrt{\left(\frac{1}{2}\right)^2} = \frac{1}{2}$ **Check** $\left(\frac{1}{2}\right)^2 = \frac{1}{4}$

It is not necessary to show the intermediate step that we have shown in Examples 3, 4, and 5; you can simply write $\sqrt{16} = 4$, $\sqrt{25} = 5$, and so forth.

EXAMPLE 6   Find $-\sqrt{16}$.

SOLUTION   This is a two-step problem. First we find $\sqrt{16}$, and then we find the *additive inverse* of that number. By inspection (or from recalling Example 3), we know that $\sqrt{16}$ is 4. Then

$$-(\sqrt{16}) = -(4) = -4$$

so $-\sqrt{16} = -4$.

Note   Since the square of a real number is never negative, square roots of negative numbers are not real numbers. They are *imaginary numbers*, and we do not discuss imaginary numbers in this book. Therefore, *in this book*, when you're asked to find the square root and the radicand is negative, your answer should be "not a real number." If, in Example 6, you had been asked to find $\sqrt{-16}$, your answer would have been "not a real number," since the radicand is negative. Also, of course, no real number exists whose square is -16. [You can verify that $4^2 = +16$ and that $(-4)^2 = +16$. Also, see Examples 7 and 8.]

EXAMPLE 7   Identify the radicand for $\sqrt{-36}$, and find $\sqrt{-36}$ or write "not a real number."

SOLUTION   The radicand is -36. Since the radicand is negative, we write "not a real number." (Note also that there is no real number whose square is -36.)

EXAMPLE 8   Identify the radicand for $-\sqrt{-25}$, and find $-\sqrt{-25}$ or write "not a real number."

SOLUTION   The radicand is -25. Since the radicand is negative, the answer is "not a real number." (Note that there is no real number whose square is -25; the negative sign in *front* of the radical does not affect the final result.)

Exercises 1.10A
Set I

In each exercise, find the square root by inspection (or write "not a real number"). In Exercises 1–6, identify the radicand before you find the square root.

1. $\sqrt{16}$
2. $\sqrt{25}$
3. $-\sqrt{4}$
4. $-\sqrt{9}$
5. $\sqrt{81}$
6. $\sqrt{36}$
7. $\sqrt{100}$
8. $\sqrt{144}$
9. $-\sqrt{81}$
10. $-\sqrt{121}$

11. $\sqrt{-4}$
12. $\sqrt{-25}$
13. $-\sqrt{-1}$
14. $-\sqrt{-9}$
15. $\sqrt{0.25}$
16. $\sqrt{0.64}$
17. $\sqrt{1.44}$
18. $\sqrt{1.21}$
19. $\sqrt{\frac{1}{16}}$
20. $\sqrt{\frac{1}{121}}$
21. $\sqrt{\frac{9}{25}}$
22. $\sqrt{\frac{16}{81}}$

Writing Problems

Express the answers in your own words and in complete sentences.

1. Explain what $\sqrt{16}$ means.

2. Explain how to check a square root problem.

3. Explain why $\sqrt{49} = 7$.

4. Explain why $\sqrt{81} \neq 9^2$.

5. Explain why $\sqrt{81} \neq \sqrt{9}$.

Exercises 1.10A
Set II

In each exercise, find the square root by inspection (or write "not a real number"). In Exercises 1–6, identify the radicand before you find the square root.

1. $\sqrt{49}$

2. $\sqrt{121}$

3. $-\sqrt{100}$

4. $-\sqrt{144}$

5. $\sqrt{64}$

6. $-\sqrt{36}$

7. $\sqrt{1}$

8. $-\sqrt{25}$

9. $\sqrt{9}$

10. $-\sqrt{169}$

11. $\sqrt{-49}$

12. $-\sqrt{49}$

13. $-\sqrt{-100}$

14. $-\sqrt{-81}$

15. $\sqrt{0.36}$

16. $\sqrt{0.81}$

17. $\sqrt{1.69}$

18. $\sqrt{2.25}$

19. $\sqrt{\frac{1}{25}}$

20. $\sqrt{\frac{1}{144}}$

21. $\sqrt{\frac{4}{49}}$

22. $\sqrt{\frac{36}{121}}$

1.10B Finding Square Roots of Large Integers

Property 3 allows us to find square roots of *products* of numbers and sometimes makes it easier to find square roots of large integers.

PROPERTY 3
The product rule for square roots

> If a and b represent any nonnegative real numbers, then
>
> $$\sqrt{ab} = \sqrt{a}\,\sqrt{b}$$
>
> | The square root of a product | $=$ | the product of the square roots |

EXAMPLE 9

Examples verifying Property 3:

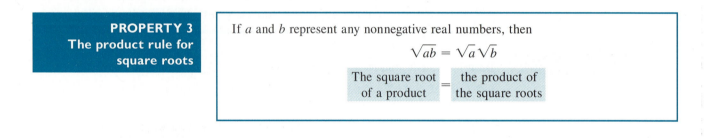

a. $\sqrt{4 \cdot 9} = \sqrt{36} = 6$ and $\sqrt{4} \cdot \sqrt{9} = 2 \cdot 3 = 6$, so $\sqrt{4 \cdot 9} = \sqrt{4} \cdot \sqrt{9}$.

b. $\sqrt{25 \cdot 4} = \sqrt{100} = 10$ and $\sqrt{25} \cdot \sqrt{4} = 5 \cdot 2 = 10$, so $\sqrt{25 \cdot 4} = \sqrt{25} \cdot \sqrt{4}$.

c. $\sqrt{16 \cdot 9} = \sqrt{144} = 12$ and $\sqrt{16} \cdot \sqrt{9} = 4 \cdot 3 = 12$, so $\sqrt{16 \cdot 9} = \sqrt{16} \cdot \sqrt{9}$.

Before we try to find square roots of larger integers, let's observe what happens if we think of Property 2, $\sqrt{a^2} = a$, as meaning the following: When the radicand is expressed as a^2, we can divide the exponent by 2 and drop the radical sign. For example,

$$\sqrt{11^2} = 11^{2/2} = 11^1 = 11$$

$$\sqrt{6^2} = 6^{2/2} = 6^1 = 6$$

$$\sqrt{24^2} = 24^{2/2} = 24^1 = 24$$

and so forth. You can verify that these are all correct answers. Will this procedure work if the exponent of the radicand is an even number greater than 2? Let's try to find $\sqrt{2^6}$ by using this method:

$$\sqrt{2^6} = 2^{6/2} = 2^3 = 8$$

Is this the correct answer? Since $2^6 = 64$, $\sqrt{2^6} = \sqrt{64}$ and $\sqrt{64} = 8$. Yes, the answer is correct. In fact, the following method can always be used.

Finding a square root when all factors of the radicand have even exponents

Express the radicand in prime-factored, exponential form. If the radicand contains more than one base, use Property 3; then divide the exponents of the radicands by 2 and drop the radical signs.

EXAMPLE 10

Examples of finding square roots when all factors of the radicand have even exponents:

a. $\sqrt{3^4} = 3^{4/2} = 3^2$, or 9

b. $\sqrt{7^8} = 7^{8/2} = 7^4$, or 2,401   Use a calculator to evaluate 7^4

              ┌─── Using Property 3

c. $\sqrt{5^2 \cdot 11^2} = \sqrt{5^2} \cdot \sqrt{11^2} = 5^1 \cdot 11^1 = 5 \cdot 11$, or 55

              ┌─── Using Property 3

d. $\sqrt{3^6 \cdot 5^4 \cdot 2^2} = \sqrt{3^6} \cdot \sqrt{5^4} \cdot \sqrt{2^2} = 3^3 \cdot 5^2 \cdot 2^1 = 27 \cdot 25 \cdot 2$, or 1,350

 Note The division shown in parts a and b of Example 10 need not be shown. That is, we can simply write $\sqrt{3^4} = 3^2$, $\sqrt{7^8} = 7^4$, and so forth.

EXAMPLE 11

Find $\sqrt{196}$.

SOLUTION Writing 196 in prime-factored, exponential form, we have $196 = 2^2 \cdot 7^2$. (You can verify that this is correct.)

                        ┌─── Using Property 3

$$\sqrt{196} = \sqrt{2^2 \cdot 7^2} = \sqrt{2^2} \cdot \sqrt{7^2} = 2 \cdot 7 = 14$$

The work can also be thought of as follows:

$$\sqrt{196} = \sqrt{4 \cdot 49} = \sqrt{4} \cdot \sqrt{49} = 2 \cdot 7 = 14$$

✓ **Check** $14^2 = 196$

The answer is 14.

EXAMPLE 12

Find $\sqrt{576}$.

SOLUTION Writing 576 in prime-factored, exponential form, we have $576 = 2^6 \cdot 3^2$. (You can verify that this is correct.)

                        ┌─── Using Property 3

$$\sqrt{576} = \sqrt{2^6 \cdot 3^2} = \sqrt{2^6} \cdot \sqrt{3^2} = 2^3 \cdot 3 = 8 \cdot 3 = 24$$

✓ **Check** $24^2 = 576$

The answer is 24.

An Introduction to Problem-Solving: Guessing-and-Checking

One of the recognized problem-solving techniques is that of **guessing-and-checking** (which is often called the *trial-and-error method*). To use this technique, we simply guess at some answer and check it; if the number we chose doesn't check, we try another number. The method is demonstrated in Examples 13 and 14, where we use it to find square roots of large numbers.

Finding Square Roots of Large Numbers by Guessing-and-Checking (Optional)

It is possible to find square roots of large integers by just selecting various numbers and squaring them. If the square of the number we select equals the radicand, we have found the answer. If not, we can use one of the following properties, which are proved in higher level mathematics classes:

For all positive real numbers a, b, and c,

if $a > b$, then $\sqrt{a} > \sqrt{b}$.

if $a < b$, then $\sqrt{a} < \sqrt{b}$.

if b is between a and c, then $\sqrt{b}$ is between $\sqrt{a}$ and $\sqrt{c}$.

When we use this method in finding square roots (see Examples 13 and 14), we say we are finding the square root by guessing-and-checking.

If it is *given* that the answer is an integer, we can look at the *last digit* of the radicand before deciding which numbers to try; this method is demonstrated in the Alternative Solution for Example 13 and in Example 14.

EXAMPLE 13   Find $\sqrt{289}$ by guessing-and-checking, *given that the answer is an integer.*

SOLUTION   We know that $10^2 = 100$. Because $289 > 100$, $\sqrt{289} > \sqrt{100}$, or $\sqrt{289} > 10$. Therefore, we will try some integers larger than 10.

We try 12: $12^2 = 144$. Because $144 < 289$, 12 is too small.

We try 18: $18^2 = 324$. Because $324 > 289$, 18 is too large.

We try 16: $16^2 = 256$. Because $256 < 289$, 16 is too small.

Because 18 is too large and 16 is too small, we try 17: $17^2 = 289$.

Therefore, $\sqrt{289} = 17$.

ALTERNATIVE SOLUTION $10^2 = 100$ and $20^2 = 400$. Since 289 is between 100 and 400, $\sqrt{289}$ is between $\sqrt{100}$ and $\sqrt{400}$, or $\sqrt{289}$ is an integer between 10 and 20. Therefore, the number *must* be a number from the set {11, 12, 13, 14, 15, 16, 17, 18, 19}. The units' digit of 28**9** is a **9**, and the only digits whose squares *end* in a **9** are 3 and 7 (that is, $3^2 = $ **9** and $7^2 = 4$**9**; the squares of all other digits end in other numbers). The only integers to try are 13 and 17.

We try 13: $13^2 = 169 \neq 289$; the answer is not 13.

We try 17: $17^2 = 289$.

Therefore, $\sqrt{289} = 17$.

EXAMPLE 14   Find $\sqrt{576}$ by guessing-and-checking, *given that the answer is an integer.*

SOLUTION   We note that $20^2 = 400$ and that $30^2 = 900$. Since 576 is between 400 and 900, $\sqrt{576}$ *must* be an integer between 20 and 30. Therefore, the number must be a number from the set {21, 22, 23, 24, 25, 26, 27, 28, 29}. The last digit of 57**6** is a **6**, and the only digits whose squares end in a **6** are 4 and 6 (that is, $4^2 = 1$**6** and $6^2 = 3$**6**; the squares of all other digits end in other numbers). The only integers to try are 24 and 26.

We try 24: $24^2 = 576$.

Therefore, $\sqrt{576} = 24$.

Exercises 1.10B
Set I

Find each of the following square roots. All the answers are integers.

1. $\sqrt{2^8}$ **2.** $\sqrt{7^6}$ **3.** $\sqrt{3^6 \cdot 11^2}$

4. $\sqrt{2^8 \cdot 5^4}$ **5.** $\sqrt{529}$ **6.** $\sqrt{361}$

7. $\sqrt{441}$ **8.** $\sqrt{625}$ **9.** $\sqrt{289}$

10. $\sqrt{324}$ **11.** $\sqrt{729}$ **12.** $\sqrt{1,296}$

Writing Problems

Express the answer in your own words and in complete sentences.

1. Describe the error in the following: $\sqrt{256} = 16^2$.

Exercises 1.10B
Set II

Find each of the following square roots. All the answers are integers.

1. $\sqrt{11^6}$ **2.** $\sqrt{3^8}$

3. $\sqrt{2^4 \cdot 5^2}$ **4.** $\sqrt{7^4 \cdot 3^6}$

5. $\sqrt{400}$ **6.** $\sqrt{484}$

7. $\sqrt{676}$ **8.** $\sqrt{1,024}$

9. $\sqrt{225}$ **10.** $\sqrt{784}$

11. $\sqrt{1,444}$ **12.** $\sqrt{2,809}$

1.10C   Simplifying Square Roots (An Introduction)

To refresh your memory, we'll repeat some facts about rational numbers. A *rational number* is any number that can be expressed in the form a/b, where a and b are integers and $b \neq 0$. When a rational number is expressed in decimal form, the decimal always either terminates or repeats. All natural numbers, all whole numbers, and all integers are rational numbers; all rational numbers are real numbers.

Irrational Numbers   There are infinitely many *real numbers* that are not *rational*. Such numbers are called **irrational numbers**. The decimal form of an irrational number never terminates and never repeats. All irrational numbers are real numbers.

In higher level mathematics courses, we prove that square roots such as $\sqrt{2}$, $\sqrt{3}$, $\sqrt{5}$, and $\sqrt{21}$ are *irrational numbers*; it is not possible to find integers (or even rational numbers) whose squares are 2, 3, 5, or 21. Not *all* numbers that contain square root symbols are irrational; for example, $\sqrt{1}$, $\sqrt{4}$, and $\sqrt{9}$ are *rational* numbers, since $\sqrt{1} = 1$ (1 is a rational number), $\sqrt{4} = 2$ (2 is a rational number), and $\sqrt{9} = 3$ (3 is a rational number).

 Note   No number can be both a rational number and an irrational number.

In later sections, we discuss finding decimal approximations for irrational numbers and graphing irrational numbers on the real number line. In this section, we discuss expressing square roots of natural numbers in simplest radical form.

A number of conditions must be satisfied for a square root to be in *simplest radical form*. The *first* of these conditions is that no factor of the radicand can have an exponent greater than 1 when the radicand is expressed in prime-factored, exponential form. (Other conditions will be given in a later chapter, when we discuss simplifying square roots that involve fractions.)

We can use the following procedure to simplify the square root of a natural number.

Simplifying the square root of a natural number

1. Express the radicand in prime-factored, exponential form.
2. Find the square root of each factor of the radicand as follows:
   a. If the exponent of a factor is an odd number greater than 1, write that factor as the product of two factors—one factor with an exponent of 1 and the other factor with an even exponent. Then proceed with steps b and c.
   b. If the exponent of a factor is an even number, remove that factor from the radical by dividing its exponent by 2 and dropping its radical sign.
   c. If the exponent of a factor is 1, that factor must remain under the radical sign.
3. The simplest radical form of the radical is the product of all the factors found in steps 2b and 2c.

It is customary to write any factors that do *not* contain a radical sign to the *left* of the radical sign.

EXAMPLE 15 Examples of simplifying square roots:

a. $\sqrt{3^5} = \sqrt{3^4 \cdot 3^1} = \sqrt{3^4}\sqrt{3} = 3^2\sqrt{3}$, or $9\sqrt{3}$

——— The exponent (5) is an odd number greater than 1, so we rewrite 3^5 as $3^4 \cdot 3^1$ (one factor has an exponent of 1 and the other factor has an even exponent)

b. $\sqrt{2^7} = \sqrt{2^6 \cdot 2^1} = \sqrt{2^6}\sqrt{2} = 2^3\sqrt{2}$, or $8\sqrt{2}$

——— The exponent (7) is an odd number greater than 1, so we rewrite 2^7 as $2^6 \cdot 2^1$ (one factor has an exponent of 1 and the other factor has an even exponent)

c. $\sqrt{48} = \sqrt{2^4 \cdot 3}$ ←——— The prime factored form of 48
$= \sqrt{2^4}\sqrt{3}$ The exponent (4) is an even number
$= 2^2\sqrt{3}$, or $4\sqrt{3}$

$$\begin{array}{r|l} 2 & 48 \\ 2 & 24 \\ 2 & 12 \\ 2 & 6 \\ & 3 \end{array} \quad 48 = 2^4 \cdot 3$$

d. $\sqrt{75} = \sqrt{3 \cdot 5^2}$ ←——— The prime factored form of 75
$= \sqrt{3}\sqrt{5^2}$
$= \sqrt{3}(5)$
$= 5\sqrt{3}$ Writing the factor that does *not* contain a radical sign to the *left* of the radical

$$\begin{array}{r|l} 3 & 75 \\ 5 & 25 \\ & 5 \end{array} \quad 75 = 3 \cdot 5^2$$

e. $\sqrt{360} = \sqrt{2^3 \cdot 3^2 \cdot 5}$
$= \sqrt{2^2 \cdot 2^1 \cdot 3^2 \cdot 5}$ Rewriting 2^3 as $2^2 \cdot 2^1$
$= \sqrt{2^2 \cdot 3^2} \cdot \sqrt{2 \cdot 5}$
$= 2 \cdot 3\sqrt{2 \cdot 5}$
$= 6\sqrt{10}$

$$\begin{array}{r|l} 2 & 360 \\ 2 & 180 \\ 2 & 90 \\ 3 & 45 \\ 3 & 15 \\ & 5 \end{array} \quad 360 = 2^3 \cdot 3^2 \cdot 5$$

If you see that a radicand contains a factor that is a perfect square, you can simplify the square root by *inspection* (see Example 16).

EXAMPLE 16 Examples of simplifying radicals by inspection:

$\quad$ **a.** $\sqrt{12} = \sqrt{4 \cdot 3}$ $\qquad$ 4 is a factor of 12 and is a perfect square

$\qquad\quad = \sqrt{4} \cdot \sqrt{3}$ $\qquad$ $\sqrt{3}$ cannot be simplified

$\qquad\quad = 2\sqrt{3}$

$\quad$ **b.** $\sqrt{75} = \sqrt{25 \cdot 3}$ $\qquad$ 25 is a factor of 75 and is a perfect square

$\qquad\quad = \sqrt{25} \cdot \sqrt{3}$ $\qquad$ $\sqrt{3}$ cannot be simplified

$\qquad\quad = 5\sqrt{3}$

$\quad$ **c.** $\sqrt{360} = \sqrt{36 \cdot 10}$ $\qquad$ 36 is a factor of 360 and is a perfect square

$\qquad\quad = \sqrt{36} \cdot \sqrt{10}$ $\qquad$ $\sqrt{10}$ cannot be simplified

$\qquad\quad = 6\sqrt{10}$

Note You may not find it necessary to show all the steps that we have shown in Examples 15 and 16.

Exercises 1.10C
Set I

Express each square root in simplest radical form.

1. $\sqrt{2^9}$	**2.** $\sqrt{5^7}$	**3.** $\sqrt{3^3 \cdot 5^6}$	
4. $\sqrt{2^4 \cdot 7^5}$	**5.** $\sqrt{98}$	**6.** $\sqrt{20}$	
7. $\sqrt{18}$	**8.** $\sqrt{45}$	**9.** $\sqrt{8}$	

10. $\sqrt{32}$ $\qquad$ **11.** $\sqrt{44}$ $\qquad$ **12.** $\sqrt{135}$

13. $\sqrt{450}$ $\qquad$ **14.** $\sqrt{192}$ $\qquad$ **15.** $\sqrt{108}$

16. $\sqrt{882}$

Exercises 1.10C
Set II

Express each square root in simplest radical form.

1. $\sqrt{11^5}$	**2.** $\sqrt{3^9}$	**3.** $\sqrt{2^6 \cdot 7^5}$	
4. $\sqrt{5^7 \cdot 3^5}$	**5.** $\sqrt{75}$	**6.** $\sqrt{72}$	
7. $\sqrt{125}$	**8.** $\sqrt{400}$	**9.** $\sqrt{432}$	

10. $\sqrt{200}$ $\qquad$ **11.** $\sqrt{80}$ $\qquad$ **12.** $\sqrt{84}$

13. $\sqrt{243}$ $\qquad$ **14.** $\sqrt{392}$ $\qquad$ **15.** $\sqrt{180}$

16. $\sqrt{675}$

1.10D Finding Square Roots with Calculators and Tables

Square roots of positive numbers can be found or approximated by using a calculator with a square root key $\boxed{\sqrt{x}}$ or, sometimes, by using tables. (See Table I, inside back cover.) We show the keystrokes to use for *some* calculators, but your calculator may require different keystrokes; consult your calculator manual if the keystrokes we show in the examples do not give the correct results.

EXAMPLE 17 Find $\sqrt{710{,}649}$ by using a calculator.

SOLUTION Press the following keys in the order shown:

$\boxed{7}\ \boxed{1}\ \boxed{0}\ \boxed{6}\ \boxed{4}\ \boxed{9}\ \boxed{\sqrt{x}}$

The calculator display shows 843. Therefore, $\sqrt{710{,}649} = 843$. (We cannot find $\sqrt{710{,}649}$ by using Table I.)

As we've mentioned, the decimal form of an irrational number does not terminate and does not repeat. Calculators with a square root key $\boxed{\sqrt{x}}$ give rounded-off *approximations* for square roots that are irrational numbers, and tables of square roots can be used to approximate some square roots (see Examples 18 and 19).

EXAMPLE 18

Approximate $\sqrt{3}$. Use a calculator, rounding off the answer to three decimal places, and also use Table I.

SOLUTION   To approximate $\sqrt{3}$ by using a calculator, press the following keys in the order shown: $\boxed{3}$ $\boxed{\sqrt{x}}$. The display probably shows 1.7320508. (Your calculator may not show the same number of digits.) If we round off this answer to three decimal places, we have

$$\sqrt{3} \approx 1.732*$$

We also can find the approximate value of $\sqrt{3}$ by referring to Table I, which gives the roots rounded off to three decimal places.

Locate 3 in the column headed N.
Read the value of $\sqrt{3}$ to the right of 3 in the column headed $\sqrt{N}$. We see that $\sqrt{3} \approx 1.732$.

N	$\sqrt{N}$
1	1.000
2	1.414
3	1.732
4	2.000
5	2.236

You might verify (by using a calculator) that if you square 1.732, you get a number that is close to 3; if you square 1.7321, you get a number that is a little closer to 3; and if you square 1.7320508, you get a number that is still closer to 3. However, no matter how many decimal places you use for the number you square, you will never get 3 exactly, because $\sqrt{3}$ is an irrational number.

EXAMPLE 19

Approximate $\sqrt{94}$. Use a calculator and round off the answer to three decimal places, and also use Table I.

SOLUTION   To find $\sqrt{94}$ by using a calculator, press the following keys in the order shown: $\boxed{9}$ $\boxed{4}$ $\boxed{\sqrt{x}}$. The display probably shows 9.6953597. When 9.6953597 is rounded off to three decimal places, we get 9.695.

To use Table I, proceed as shown below.

Locate 94 in the column headed N.
Then read that the value of $\sqrt{94} \approx 9.695$ in the column headed $\sqrt{N}$.

N	$\sqrt{N}$
81	9.000
82	9.055
92	9.592
93	9.644
94	9.695
95	9.747
96	9.798

*The symbol $\approx$ (read "is approximately equal to") is used to show that two numbers are *approximately* equal to each other.

There is a paper-and-pencil method for calculating square roots. This method, however, has become obsolete because of the widespread use of calculators; therefore, it is not discussed in this book.

Exercises 1.10D
Set I

In Exercises 1–8, approximate each square root by using a calculator and rounding off the answer to three decimal places or by using Table I, inside the back cover.

1. $\sqrt{13}$ 2. $\sqrt{18}$ 3. $\sqrt{37}$

4. $\sqrt{50}$ 5. $\sqrt{79}$ 6. $\sqrt{60}$

7. $\sqrt{86}$ 8. $\sqrt{92}$

In Exercises 9–12, find each square root by using a calculator.

9. $\sqrt{466,489}$ 10. $\sqrt{674,041}$ 11. $\sqrt{272,484}$

12. $\sqrt{89,401}$

Exercises 1.10D
Set II

In Exercises 1–8, approximate each square root by using a calculator and rounding off the answer to three decimal places or by using Table I, inside the back cover.

1. $\sqrt{31}$ 2. $\sqrt{69}$ 3. $\sqrt{97}$

4. $\sqrt{184}$ 5. $\sqrt{178}$ 6. $\sqrt{145}$

7. $\sqrt{78}$ 8. $\sqrt{125}$

In Exercises 9–12, find each square root by using a calculator.

9. $\sqrt{178,929}$ 10. $\sqrt{373,321}$ 11. $\sqrt{88,804}$

12. $\sqrt{35,344}$

1.10E   Higher Roots; Real Numbers and Their Graphs

Roots other than square roots are called **higher roots**. The parts of the symbol for higher roots are shown below.

The index

The radical sign $\sqrt[n]{p}$ The radicand

When there is no index written, the index is understood to be 2, and the radical will then be a *square root*. Some examples of symbols for higher roots are $\sqrt[3]{8}$, $\sqrt[4]{55}$, and $\sqrt[5]{-32}$; we show how to read and interpret these symbols below.

Principal Roots   *Whenever the radical symbol is used, mathematicians agree that it is to stand for the principal root.*

When the index is an *even* number, we say that the index is **even**. When the index is an *odd* number, we say that the index is **odd**.

Principal higher roots are summarized as follows.

Principal roots

The symbol $\sqrt[n]{p}$ always represents the *principal n*th root of *p*; the index, *n*, must be a natural number greater than 1.

If *p*, the radicand, represents any real number, and if *b* represents some number whose *n*th power is *p* (that is, if $b^n = p$), then

$$\sqrt[n]{p} = b$$

If the radicand is positive, the principal root is positive.

If the radicand is negative:

1. when the index is odd, the principal root is negative;

2. when the index is even, the principal root *is not a real number.*

Reading and Interpreting Symbols for Higher Roots   The symbol $\sqrt[3]{p}$ indicates the *cube* (or *cubic*) *root* of *p*. When the *index* of the radical is a 3, we must find a number whose *cube* is *p*. You will find some cube root problems easier to do if you have memorized the cubes of the first few whole numbers.

EXAMPLE 20   Find $\sqrt[3]{8}$.

SOLUTION   The radicand is positive, so the principal root is positive. We must find a number whose *cube* is 8. That is, we must solve $(?)^3 = 8$. If we have memorized that $2^3 = 8$, then we know that the answer is 2. If we haven't memorized that fact, we must use the guessing-and-checking method: Does $1^3 = 8$? No. Does $2^3 = 8$? Yes. Therefore, $\sqrt[3]{8} = 2$.

EXAMPLE 21   Find $\sqrt[3]{-8}$.

SOLUTION   We must find a number whose *cube* is -8. We know that the principal root will be negative, because the index is odd and the radicand is negative. Because $(-2)^3 = -8$, $\sqrt[3]{-8} = -2$.

The symbol $\sqrt[4]{p}$ indicates the *fourth root* of *p*. When the *index* is a 4, we must find a number whose *fourth power* is *p*.

EXAMPLE 22   Find $\sqrt[4]{16}$.

SOLUTION   The radicand is positive, so the principal root is positive. We must find a positive number whose *fourth* power is 16. Let's use the guessing-and-checking method: Does $1^4 = 16$? No. Does $2^4 = 16$? Yes! Therefore, $\sqrt[4]{16} = 2$.

EXAMPLE 23   Find $\sqrt[4]{-16}$.

SOLUTION   The radicand is negative and the index is even. Therefore, the answer is "not a real number." (There is no *real* number whose fourth power is -16.)

The symbol $\sqrt[5]{p}$ indicates the *fifth root* of *p*, and so forth.

EXAMPLE 24   Find $\sqrt[5]{-1}$.

SOLUTION   We must find a *negative* number whose *fifth* power is -1. Does $(-1)^5 = -1$? Yes. Therefore, $\sqrt[5]{-1} = -1$.

EXAMPLE 25

Examples of finding roots when the radical is preceded by a minus sign:

a. $-\sqrt[4]{16} = -(2) = -2$
b. $-\sqrt[3]{8} = -(2) = -2$
c. $-\sqrt[3]{-8} = -(-2) = 2$
d. $-\sqrt[4]{-16}$ is not a real number, because the radicand is negative and the index is even.

The Real Number System   As we mentioned earlier, any number that can be graphed on the number line is a real number. The relationships among the various sets of numbers are shown in Figure 5. In particular, this figure shows that all natural numbers, all whole numbers, all integers, all rational numbers, and all irrational numbers are real numbers. It also shows that all natural numbers are whole numbers, that all whole numbers are integers, that all integers are rational numbers, that all rational numbers are real numbers, and so forth, and it shows that no number can be both a rational number and an irrational number.

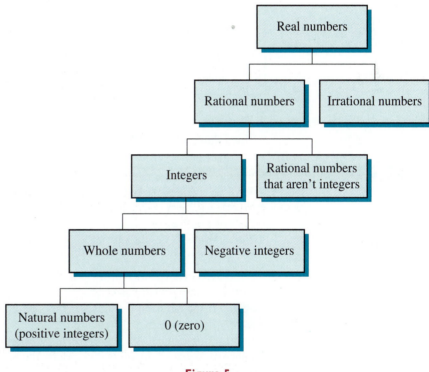

Figure 5

How to Determine Whether a Number with a Radical Sign Is Rational, Irrational, or Not Real   In this book, we will only ask you to decide whether a radical represents a rational number or an irrational number when the radicand is an *integer*.

If p is an integer and if $p \geq 0$, then

$\sqrt{p}$ represents a real, rational number if there is an integer whose square equals p.

$\sqrt{p}$ represents a real, irrational number if there is *not* an integer whose square equals p.

For any $p < 0$, $\sqrt{p}$ represents a number that is *not real*.

By these rules, $\sqrt{16}$ is a real, rational number, because $16 \geq 0$ and there is an integer (4) whose square is 16. Also, $\sqrt{710,649}$ is a real, rational number, because $710,649 \geq 0$ and there is an integer (843) whose square is 710,649 (see Example 17). On the other hand, $\sqrt{3}$ is a real, *irrational* number, because $3 \geq 0$ but there is no integer whose square is 3. Finally, $\sqrt{-4}$, $\sqrt{-0.75}$, and $\sqrt{-\frac{1}{4}}$ are not real, because $-4 < 0$, $-0.75 < 0$, and $-\frac{1}{4} < 0$.

If the index n is *even*, if p is an integer, and if $p \geq 0$, then

> $\sqrt[n]{p}$ represents a real, *rational* number if there is some integer whose nth power equals p, and
>
> $\sqrt[n]{p}$ represents a real, *irrational* number if there is no integer whose nth power equals p.

If the index n is *even* and if $p < 0$, then $\sqrt[n]{p}$ represents a number that is *not real*.

If the index n is *odd* and if p is *any* integer, then

> $\sqrt[n]{p}$ represents a real, *rational* number if there is some integer whose nth power equals p, and
>
> $\sqrt[n]{p}$ represents a real, *irrational* number if there is no integer whose nth power equals p.

According to the above rules, $\sqrt[4]{81}$ is real and rational, because the index (4) is even, 81 is an integer, $81 \geq 0$, and there is an integer (3) whose fourth power is 81. On the other hand, $\sqrt[6]{24}$ is a real, irrational number, because although the index (6) is even, 24 is an integer, and $24 \geq 0$, there is no integer whose sixth power is 24. Also, whereas $\sqrt[4]{-1}$ and $\sqrt[4]{-0.3}$ are not real numbers because the indexes are even and the radicands are negative, $\sqrt[5]{-32}$ is a real, rational number, because the index (5) is odd and there is an integer (-2) whose fifth power is -32. Moreover, $\sqrt[5]{-10}$ is a real, irrational number, because although the index (5) is odd, there is no integer whose fifth power is -10.

Note   If the radicand is a number that is *not* an integer, then it's not always easy to decide whether the square root represents an irrational number or a rational number. For example, $\sqrt{0.04}$ represents a *rational* number, because there is a rational number (0.2) whose square is 0.04. Here, the radicand (0.04) is positive but not an integer. Similarly, $\sqrt{\frac{4}{9}}$ represents a rational number, because there is a rational number $\left(\frac{2}{3}\right)$ whose square is $\frac{4}{9}$. In this book, you will not be asked to determine whether a square root is rational or irrational unless the radicand *is* an integer.

How to Determine Whether a Decimal Number Is Rational or Irrational   If a number in decimal form is a terminating or repeating decimal, it represents a real, rational number. Therefore, 8.37 and $0.2636363\ldots$ represent real, rational numbers. As we have mentioned, when the decimal form of a rational number is a repeating decimal, it is customary to place a bar above the digit or group of digits that repeats; this means that $0.2636363\ldots$ can be written as $0.2\overline{63}$.

If a number in decimal form is a nonterminating, nonrepeating decimal, it represents a real, irrational number. Thus, if we're told that the digits in the decimal number $263.1626892\ldots$ never terminate and never repeat, we know that it is a real, irrational number.

EXAMPLE 26

Name all the numbers from the following list that are (a) natural numbers, (b) integers, (c) rational numbers, (d) irrational numbers, (e) real numbers, and (f) numbers that are *not* real:

$$\sqrt{5},\ \sqrt[3]{17},\ 2.\overline{52},\ 0,\ \tfrac{3}{7},\ 6,\ -4,\ -\sqrt[3]{8},\ \sqrt{-9},\ \text{and } 2.828427125\ldots \text{(never terminates and never repeats)}$$

SOLUTION   Note that $-\sqrt[3]{8} = -2$.

a. There is one natural number in the list: 6.
b. The integers from the list are 0, 6, -4, and $-\sqrt[3]{8}$.
c. The rational numbers are $2.\overline{52}$ (the pair of digits "52" repeats), 0, $\tfrac{3}{7}$, 6, -4, and $-\sqrt[3]{8}$.
d. The irrational numbers are $\sqrt{5}$ (there is no integer whose square is 5), $\sqrt[3]{17}$ (there is no integer whose cube is 17), and $2.828427125\ldots$ (this is a nonterminating, nonrepeating decimal).
e. The real numbers are $\sqrt{5}$, $\sqrt[3]{17}$, $2.\overline{52}$, 0, $\tfrac{3}{7}$, 6, -4, $-\sqrt[3]{8}$, and $2.828427125\ldots$ (remember that all rational numbers and all irrational numbers are real numbers).
f. There is one number that is not real: $\sqrt{-9}$ (the radicand is negative).

Graphing Real Numbers   Recall that we graph a number by placing a dot on the number line above that number. We begin by graphing some positive fractions, mixed numbers, and decimals (see Example 27). In practice, the intended *use* of the graph determines how accurately such numbers must be graphed.

EXAMPLE 27

Graph the numbers $\tfrac{2}{3}$, $3\tfrac{1}{2}$, $\tfrac{31}{5}$, and 7.6 on the number line.

SOLUTION   To graph $\tfrac{2}{3}$, we locate the point that is approximately two-thirds of the way *from* 0 *toward* 1 and place a dot at that point. To graph $3\tfrac{1}{2}$, we locate the point that is approximately halfway between 3 and 4 and place a dot at that point. To graph $\tfrac{31}{5}$, we first convert that number to the mixed number $6\tfrac{1}{5}$ (we could have converted it to decimal form, instead) and then place a dot at the point that is about one-fifth of the way from 6 toward 7. To graph 7.6, we place a dot at the point that is about six-tenths of the way from 7 toward 8. See Figure 6.

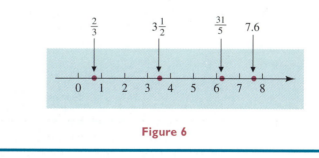

Figure 6

In Example 28, we graph a few negative numbers.

EXAMPLE 28   Graph the numbers -2, $-\frac{1}{2}$, $-2\frac{3}{4}$, and -3.6.

SOLUTION   To graph -2, we place a dot above -2. To graph $-\frac{1}{2}$, we locate the point that is approximately halfway between -1 and 0 and place a dot at that point. To graph $-2\frac{3}{4}$, we place a dot on the point that is three-fourths of the way *from -2 toward -3* (notice that we're moving from *right* to *left*). To graph -3.6, we place a dot on the point that is about six-tenths of the way from -3 toward -4. See Figure 7.

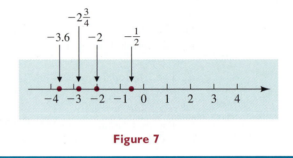

Figure 7

In Example 29, we graph some rational numbers and some irrational numbers (all are real numbers). To graph an *irrational* number, we need to find a decimal approximation for the number. For graphing purposes, the approximation can generally be rounded off to one decimal place.

EXAMPLE 29   Graph the numbers $\sqrt{2}$, $\sqrt[4]{256}$, $-\sqrt{3}$, $\sqrt[3]{-16}$, $\sqrt{5}$, $\sqrt[5]{-1}$, and $\sqrt{38}$.

SOLUTION   We use a calculator or Table I to approximate $\sqrt{2}$, $\sqrt{3}$, $\sqrt{5}$, and $\sqrt{38}$, obtaining the following:

$$\sqrt{2} \approx 1.4, \quad -\sqrt{3} \approx -1.7, \quad \sqrt{5} \approx 2.2, \quad \text{and} \quad \sqrt{38} \approx 6.2$$

$\sqrt[4]{256}$ and $\sqrt[5]{-1}$ are exact: $\sqrt[4]{256} = 4$ (you can verify this) and $\sqrt[5]{-1} = -1$. This leaves $\sqrt[3]{-16}$, and we have not discussed approximating cube roots that are irrational. However, we do know that $(-2)^3 = -8$ and that $(-3)^3 = -27$, so $\sqrt[3]{-16}$ must be between -2 and -3. See Figure 8.

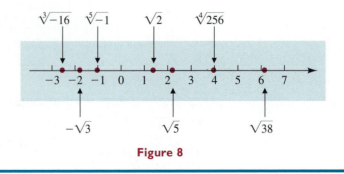

Figure 8

Exercises 1.10E
Set I

In Exercises 1–20, find each of the indicated roots by inspection or by guessing-and-checking or write "not a real number."

1. $\sqrt[3]{64}$
2. $\sqrt[4]{81}$
3. $\sqrt[3]{27}$
4. $\sqrt[3]{125}$
5. $-\sqrt[3]{27}$
6. $-\sqrt{25}$
7. $-\sqrt[4]{1}$
8. $-\sqrt[3]{-32}$
9. $\sqrt[3]{-125}$
10. $\sqrt[3]{-8}$
11. $-\sqrt[4]{-16}$
12. $-\sqrt{-16}$
13. $\sqrt[3]{-1,000}$
14. $\sqrt[3]{-64}$
15. $\sqrt[3]{-1}$
16. $\sqrt[5]{-32}$
17. $\sqrt[6]{729}$
18. $-\sqrt[3]{-216}$
19. $\sqrt{-25}$
20. $\sqrt{-36}$

In Exercises 21 and 22, name all the numbers, if any, from each list that are (a) natural numbers, (b) integers, (c) rational numbers, (d) irrational numbers, (e) real numbers, and (f) numbers that are *not* real.

21. $\sqrt[3]{13}, \frac{1}{2}, -12, \sqrt{-15}, 0.\overline{26}, 0.196732468\ldots$ (never terminates or repeats)

22. $0.67249713\ldots$ (never terminates or repeats), $\sqrt{-27}, 18, \frac{11}{32}, 0.\overline{37}, \sqrt[3]{12}$

23. Graph the numbers $-3, \sqrt{3}, -\frac{9}{4}, 2\frac{2}{5}$, and $\sqrt[3]{-1}$.

24. Graph the numbers $\sqrt{7}, -1\frac{1}{4}, -5, \frac{3}{8}$, and $\sqrt[3]{-27}$.

Writing Problems

Express the answers in your own words and in complete sentences.

1. Explain why $\sqrt{16}$ is rational but $\sqrt{13}$ is irrational.
2. Explain why $\sqrt{-64}$ is not a real number.
3. Explain why $-\sqrt{64}$ is a real number.
4. Explain why $\sqrt[3]{-64}$ is a real number.

Exercises 1.10E
Set II

In Exercises 1–20, find each of the indicated roots by inspection or by guessing-and-checking or write "not a real number."

1. $-\sqrt[3]{-8}$
2. $-\sqrt[3]{-64}$
3. $\sqrt[5]{-1}$
4. $\sqrt[4]{16}$
5. $\sqrt[3]{1,000}$
6. $\sqrt[5]{32}$
7. $\sqrt[3]{216}$
8. $\sqrt[6]{64}$
9. $\sqrt[4]{256}$
10. $\sqrt[5]{243}$
11. $\sqrt[4]{-81}$
12. $-\sqrt[4]{81}$
13. $\sqrt[5]{-100,000}$
14. $-\sqrt[3]{1}$
15. $\sqrt[7]{128}$
16. $\sqrt[4]{625}$
17. $\sqrt[3]{343}$
18. $\sqrt[5]{-243}$
19. $\sqrt[3]{-64}$
20. $-\sqrt[4]{10,000}$

In Exercises 21 and 22, name all the numbers, if any, from each list that are (a) natural numbers, (b) integers, (c) rational numbers, (d) irrational numbers, (e) real numbers, and (f) numbers that are *not* real.

21. $-13, -\frac{51}{22}, 0.\overline{72}, \sqrt{-16}, \sqrt[3]{-125}, 0.26925163\ldots$ (never terminates or repeats)

22. $\sqrt[3]{3}, 1.29715698\ldots$ (never terminates or repeats), $3.\overline{13}, \frac{23}{52}, 45, \sqrt{-100}$

23. Graph the numbers $3\frac{3}{4}, \sqrt{19}, \frac{16}{3}, -1\frac{1}{5}$, and $\sqrt[5]{-1}$.

24. Graph the numbers $\sqrt[4]{81}, -\sqrt{11}, 2\frac{1}{6}, -\frac{1}{3}$, and $\sqrt[3]{-4}$.

1.11   Grouping Symbols; Order of Operations

Earlier in this chapter, we learned how to perform the six operations of arithmetic—addition, subtraction, multiplication, division, raising to powers, and finding roots—on signed numbers. However, most of the problems dealt with only one operation. In this section, we will concentrate on evaluating expressions in which several operations are used.

Grouping Symbols

Operations indicated *within* grouping symbols should be performed before the operations outside the grouping symbols. We've already mentioned that parentheses ( ), brackets [ ], and braces { } can be used as grouping symbols. *All of these grouping symbols* (which must always be used in pairs) *have the same meaning*; different grouping symbols can be used in the same expression.

A fraction bar is also a grouping symbol (see Example 1), as is the bar of a radical sign (see Example 2).

 **Note** We've seen several expressions that contain grouping symbols within which no operation is indicated. In the expression $5(-3)$, for example, no operation is indicated *within* the grouping symbols; that is, there is no operation to perform in "-3."

EXAMPLE 1   Evaluate $\dfrac{-4 + (-2)}{8 - 5}$.

SOLUTION   The fraction bar is a grouping symbol for the addition indicated *above* the bar as well as for the subtraction indicated *below* the bar.

$$\dfrac{-4 + (-2)}{8 - 5} \quad \begin{array}{l} \longleftarrow \text{ We must add } -4 \text{ and } -2 \text{ first} \\ \longleftarrow \text{ We must subtract 5 from 8 next} \end{array}$$

$$= \dfrac{-6}{3}$$

$$= -2$$

EXAMPLE 2   Evaluate $\sqrt{16 + 9}$.

SOLUTION   The bar of the radical sign is a grouping symbol.

$$\sqrt{16 + 9} \quad \text{We must add 16 and 9 } \textit{before} \text{ we take any square roots}$$

$$= \sqrt{25}$$

$$= 5$$

Order of Operations

When two or more operations are indicated in a problem (or when the same operation is indicated more than once) and when there are no grouping symbols to tell us which operation to perform first, what should we do? If the operation symbols in a problem are all addition symbols, we can do the additions in any order; and if they are all multiplication symbols, we can do the multiplications in any order. But what about other cases? We'll consider several problems.

First, let's consider the problem $12 - 6 - 2$. If we do the subtraction on the right first, we have

$$12 - 6 - 2 \overset{?}{=} 12 - (6 - 2) = 12 - 4 = 8$$

but if we do the subtraction on the left first, we have

$$12 - 6 - 2 \overset{?}{=} (12 - 6) - 2 = 6 - 2 = 4$$

Which answer is right? (The correct solution is given in Example 3.)

Next, let's consider the problem $36 \div 6 \div 2$. If we do the division on the left first, we have

$$36 \div 6 \div 2 \overset{?}{=} (36 \div 6) \div 2 = 6 \div 2 = 3$$

whereas if we do the division on the right first, we have

$$36 \div 6 \div 2 \overset{?}{=} 36 \div (6 \div 2) = 36 \div 3 = 12$$

Which answer is correct? (The problem is done correctly in Example 4.)

Finally, let's consider the problem $5 + 4 \cdot 6$. If we do the addition first, we have

$$5 + 4 \cdot 6 \overset{?}{=} (5 + 4) \cdot 6 = 9 \cdot 6 = 54$$

If we do the multiplication first, we have

$$5 + 4 \cdot 6 \overset{?}{=} 5 + (4 \cdot 6) = 5 + 24 = 29$$

Which is right? (The correct solution is given in Example 5.)

So that problems involving more than one operation will have only one correct answer, mathematicians have agreed upon the following order of operations.

Order of operations

1. If there are operations indicated inside grouping symbols, those operations within the grouping symbols should be performed first. A fraction bar and the bar of a radical sign are grouping symbols.

2. The evaluation then proceeds *in this order:*

First: Powers and roots are done.

Next: Multiplication and division are done *in order from left to right.*

Last: Addition and subtraction are done *in order from left to right.*

E X A M P L E 3

Evaluate $12 - 6 - 2$.

S O L U T I O N   There are no grouping symbols, no powers or roots indicated, and no multiplications or divisions indicated. Therefore, we go to the last step and perform the subtractions *in order from left to right.* We have

$$12 - 6 - 2 = (12 - 6) - 2 = 6 - 2 = 4$$

Therefore, the only correct answer for $12 - 6 - 2$ is 4.

The expression $12 - 6 - 2$ must be evaluated by doing the subtractions from left to right because subtraction is not associative. It is possible, however, to think of $12 - 6 - 2$ as the *sum* $12 + (-6) + (-2)$, because *any* subtraction problem can be changed to an addition problem. If the expression is considered as a sum, then the *addition* can be done in *any* order, because addition *is* commutative and associative.

EXAMPLE 4

Evaluate $36 \div 6 \div 2$.

SOLUTION There are no grouping symbols and no powers or roots indicated. We must perform the divisions *in order from left to right*. We have

$$36 \div 6 \div 2 = (36 \div 6) \div 2 = 6 \div 2 = 3$$

Therefore, the only correct answer for $36 \div 6 \div 2$ is 3.

The expression $36 \div 6 \div 2$ must be evaluated by doing the divisions from left to right because division is not associative. However, if the expression is considered as a *product*—that is, as $36 \cdot \frac{1}{6} \cdot \frac{1}{2}$—then the multiplication can be done in any order, because multiplication *is* commutative and associative.

EXAMPLE 5

Evaluate $5 + 4 \cdot 6$.

SOLUTION There are no grouping symbols and no powers or roots indicated. The multiplication must be performed *before* the addition. We then have

$$5 + 4 \cdot 6 = 5 + (4 \cdot 6) = 5 + 24 = 29$$

Therefore, the only correct answer for $5 + 4 \cdot 6$ is 29.

EXAMPLE 6

Evaluate $8 - 6 - 4 + 7$.

SOLUTION If we consider the problem as an addition *and* subtraction problem, the expression must be evaluated from left to right. If, however, the expression is considered as a *sum*, the addition can be done in any order.

— Both methods are correct —

Considered as addition and subtraction and evaluated left to right	Changed to addition and added in any order
$8 - 6 - 4 + 7$	$8 + (-6) + (-4) + (7)$
$= \quad 2 \quad - 4 + 7$	$= 8 + (7) + (-6) + (-4)$
$= \quad -2 \quad + 7$	$= \quad 15 \quad + \quad (-10)$
$= \quad 5$	$= \quad 5$

EXAMPLE 7

Examples of using the correct order of operations:

a. $7 + 3 \cdot 5$ Multiplication must be done before addition

$= 7 + 15$

$= \quad 22$

b. $4^2 + \sqrt{25} - 6$ Powers and roots must be done first

$= 16 + 5 - 6$ Addition must be done next, because addition and subtraction are done left to right

$= \quad 21 \quad - 6$

$= \quad 15$

c. $16 \div 2 \cdot 4$ Division must be done first, because multiplication and division are done left to right

$= \quad 8 \quad \cdot 4$

$= \quad 32$

d. $(-8) \div 2 - (-4)$ Division must be done before subtraction

$= \quad -4 \quad - (-4)$ Subtraction must be changed to addition

$= \quad -4 \quad + \quad 4$

$= \qquad 0$

e. $\sqrt[3]{-8}(-3)^2 - 2(-6)$ Roots and powers must be done first

$= \quad -2(9) \quad - 2(-6)$ Multiplications must be done before subtraction

$= \quad -18 \quad - (-12)$ Subtraction must be changed to addition

$= \quad -18 \quad + \quad 12$

$= \qquad -6$

f. $12\sqrt{25} + 28 \div 4$ There is an understood multiplication symbol between the 12 and the $\sqrt{25}$

$= 12 \cdot 5 + 28 \div 4$ Roots were done first

$= \quad 60 \quad + \quad 7$ Multiplications and divisions were done next

$= \qquad 67$ Addition was done last

g. $5 \cdot (-4) \div 2 \cdot (3 - 8)$ $3 - 8$, the expression *within* the parentheses, must be evaluated first

$= 5 \cdot (-4) \div 2 \cdot (-5)$ The multiplication on the left must be done next

$= \quad -20 \quad \div 2 \cdot (-5)$ The division on the left must be done next

$= \quad -10 \quad \cdot (-5)$

$= \qquad 50$

h. $5 \cdot 3^2$ Powers must be done *before* multiplication

$= 5 \cdot 9$

$= 45$

✖ **A Word of Caution**   Remember that an exponent applies *only* to the immediately preceding symbol. That is, $5 \cdot 3^2 = 5 \cdot 3 \cdot 3$, *not* $15 \cdot 15$.

When grouping symbols appear within other grouping symbols, evaluate the expression inside the *inner* grouping symbols first (see Examples 8 and 9).

EXAMPLE 8   Examples of evaluating expressions that contain grouping symbols within other grouping symbols:

a. $10 - [3 - (2 - 7)]$ $2 - 7$ must be evaluated first:
$2 - 7 = 2 + (-7) = -5$

$= 10 - [3 - (-5)]$ Subtraction must be changed to addition

$= 10 - [3 + (+5)]$

$= 10 - [8]$

$= 2$

b.　　$20 - 2\{5 - [3 - 5(6 - 2)]\}$　　　6 − 2 must be evaluated first

$= 20 - 2\{5 - [3 - 5(4)]\}$　　　Multiplication must be done before subtraction

$= 20 - 2\{5 - [3 - 20]\}$　　　3 − 20 must be evaluated next:
$3 - 20 = 3 + (-20) = -17$

$= 20 - 2\{5 - [-17]\}$　　　Subtraction must be changed to addition

$= 20 - 2\{5 + [+17]\}$　　　The operation within the braces must be performed next

$= 20 - 2\{22\}$　　　Multiplication must be done before subtraction

$= 20 - 44$　　　Subtraction must be changed to addition

$= 20 + (-44)$

$= -24$

A Word of Caution　　A common error in a problem like the one in Example 8b is to start the problem by subtracting 2 from 20, as follows:

$$20 - 2\{5 - [3 - 5(6 - 2)]\} = 18\{5 - [3 - 5(6 - 2)]\}$$

This is incorrect; the expression inside the braces must be multiplied by 2 *before* the subtraction is done. In fact, 2 is *never* subtracted from 20; rather, 2{22}, or 44, is subtracted from 20.

EXAMPLE 9　　Evaluate $27 \div (-3)^2 - 5\left\{6 - \dfrac{8 - 4}{5}\right\}$.

SOLUTION

$27 \div (-3)^2 - 5\left\{6 - \dfrac{8 - 4}{5}\right\}$　　　The innermost grouping symbol here is the fraction bar; we must subtract 4 from 8 first

$= 27 \div (-3)^2 - 5\left\{6 - \dfrac{4}{5}\right\}$　　　Remember from arithmetic that $6 = \dfrac{6}{1} = \dfrac{6 \cdot 5}{1 \cdot 5} = \dfrac{30}{5}$

$= 27 \div (-3)^2 - 5\left\{\dfrac{30}{5} - \dfrac{4}{5}\right\}$

$= 27 \div (-3)^2 - 5\left\{\dfrac{26}{5}\right\}$　　　We must raise to powers before we divide, and $(-3)^2 = 9$

$= 27 \div \quad 9 \quad - \dfrac{5}{1}\left\{\dfrac{26}{5}\right\}$　　　We must divide and multiply before we subtract, and $\dfrac{5}{1}\left\{\dfrac{26}{5}\right\} = \dfrac{5 \cdot 26}{1 \cdot 5} = \dfrac{26 \cdot 5}{1 \cdot 5} = 26$

$= \quad\quad 3 \quad - \quad 26$　　　Subtraction must be changed to addition

$= \quad\quad 3 \quad + (-26)$

$= \quad\quad\quad -23$

EXAMPLE 10　　Evaluate $\sqrt{16} + \sqrt{9}$.

SOLUTION　　We must find the square roots before we add:

$$\sqrt{16} + \sqrt{9} = 4 + 3 = 7$$

Note If you compare this problem with the problem from Example 2, you'll see that $\sqrt{16 + 9} \neq \sqrt{16} + \sqrt{9}$. In general, $\sqrt{a + b} \neq \sqrt{a} + \sqrt{b}$ and $\sqrt{a - b} \neq \sqrt{a} - \sqrt{b}$.

EXAMPLE II Evaluate $\sqrt{13^2 - 12^2}$.

SOLUTION In the expression $\sqrt{13^2 - 12^2}$, the bar of the radical sign acts like a set of grouping symbols; therefore, we must simplify $13^2 - 12^2$ before we take the square root. In order to do this, we must raise to powers first and then subtract.

$$\sqrt{13^2 - 12^2} = \sqrt{169 - 144} = \sqrt{25} = 5$$

A Word of Caution A common error is to write

$$\sqrt{13^2 - 12^2} = 13 - 12$$

We saw in Example 11 that $\sqrt{13^2 - 12^2} = 5$, and $13 - 12 = 1$, not 5. In general, $\sqrt{a^2 + b^2} \neq a + b$, and $\sqrt{a^2 - b^2} \neq a - b$.

Now that we know that multiplication must be done before addition and subtraction, we can restate the distributive properties with fewer parentheses:

$$a(b + c) = ab + ac \qquad \text{and} \qquad a(b - c) = ab - ac$$

We no longer need parentheses around ab and ac.

Order of Operations and Scientific Calculators

(We show keystrokes only for scientific calculators that use algebraic logic.) Most scientific calculators have the correct order of operations built in. Therefore, they will do multiplications and divisions before additions and subtractions, they will work from left to right when doing multiplications and divisions, and so on.

You can tell whether your calculator performs multiplications and divisions before additions and subtractions by trying out simple problems. For example, try the problem $2 + 3 \times 5$ by entering it as follows:

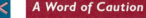

If your calculator display is [⎯⎯⎯⎯⎯⎯⎯⎯ 17] (which *is* the correct answer for $2 + 3 \times 5$), then your calculator does multiplications and divisions before additions and subtractions. If your display is *not* 17, then you must enter the problem differently. Try entering it as $3 \times 5 + 2$ (notice that $2 + 3 \times 5 = 3 \times 5 + 2$, because *addition* is commutative).

You can tell whether your calculator performs multiplications and divisions from left to right by trying the problem $9 \div 3 \times 3$. If you enter the problem as

and see the display [⎯⎯⎯⎯⎯⎯⎯⎯ 9] (which *is* the correct answer for $9 \div 3 \times 3$), then your calculator performs multiplications and divisions from left to right.

To see whether your calculator raises to powers before it multiplies or divides, try the problem 2×3^2. If you enter the problem as

and see the display [⎯⎯⎯⎯⎯⎯⎯⎯ 18] (which *is* the correct answer for 2×3^2), then your calculator raises to powers before it multiplies (or divides).

To see whether your calculator performs additions and subtractions from left to right, try the problems $12 - 6 - 2$ and $15 - 3 + 2$. The answers should be 4 and 14, respectively.

The usual calculator key for raising to powers is $\boxed{y^x}$ or $\boxed{x^y}$. To find 3^5, press $\boxed{3}$ $\boxed{y^x}$ $\boxed{5}$ $\boxed{=}$. The display should be $\boxed{243}$.

We suggest that you rework the problems in the examples in this section with a calculator, being sure to note what keystrokes to use on *your* calculator to obtain the correct answers.

In Examples 12 and 13 we do not show the keystrokes to use for entering the numbers themselves.

EXAMPLE 12   Evaluate $\dfrac{43.6 \times 0.339}{7.42 \times 13.5}$.

SOLUTION   Even if your calculator has the correct order of operations built in, you must be *very* careful in entering this problem. Remember that the fraction bar acts like a set of grouping symbols; that is, the problem could have been written $(43.6 \times 0.339) \div (7.42 \times 13.5)$. Therefore, if your calculator has parentheses, the problem can be entered as follows:

Display
↓
$\boxed{(}$ 43.6 $\boxed{\times}$ .339 $\boxed{)}$ $\boxed{\div}$ $\boxed{(}$ 7.42 $\boxed{\times}$ 13.5 $\boxed{)}$ $\boxed{=}$ 0.147553159

If your calculator doesn't have parentheses but has a memory key, you can find 7.42×13.5 and *store that answer*, finishing the problem as follows:

Display
↓
43.6 $\boxed{\times}$ .339 $\boxed{=}$ $\boxed{\div}$ $\boxed{RCL}$ $\boxed{=}$ 0.147553159

The most efficient way to enter the problem (and the way to enter it if your calculator has neither parentheses nor a memory key) is as follows:

Display
↓
43.6 $\boxed{\times}$ .339 $\boxed{\div}$ 7.42 $\boxed{\div}$ 13.5 $\boxed{=}$ 0.147553159

└— This is ÷ because 13.5 is in the denominator

This method works because $\dfrac{43.6 \times 0.339}{7.42 \times 13.5} = \dfrac{43.6}{1} \times \dfrac{0.339}{1} \times \dfrac{1}{7.42} \times \dfrac{1}{13.5}$.

The answer is approximately 0.147553159.

A Word of Caution   You will get an *incorrect* answer to the problem in Example 12 if your calculator has the correct order of operations built in and you enter the problem as follows:

43.6 $\boxed{\times}$ .339 $\boxed{\div}$ 7.42 $\boxed{\times}$ 13.5 $\boxed{=}$

The calculator interprets this as $43.6 \times 0.339 \div 7.42 \times 13.5$ and, using the correct order of operations, works from left to right. Therefore, the calculator does the problem as if it had been

$$\dfrac{43.6 \times .339}{7.42} \times 13.5, \quad \text{or} \quad \dfrac{43.6 \times .339 \times 13.5}{7.42}$$

The calculator display is $\boxed{26.891563}$, which is *not* the correct answer for $\dfrac{43.6 \times 0.339}{7.42 \times 13.5}$.

EXAMPLE 13   Evaluate $(2.5)^2 \div (5.6 - 11.4)$. Round off the answer to three decimal places.

SOLUTION   On most calculators that use algebraic logic, the problem can be entered as follows:

Display
↓

2.5 $\boxed{x^2}$ $\boxed{\div}$ $\boxed{(}$ 5.6 $\boxed{-}$ 11.4 $\boxed{)}$ $\boxed{=}$ $\boxed{-1.0775862}$

Rounding off the answer to three decimal places, we find that $(2.5)^2 \div (5.6 - 11.4) \approx$ -1.078.

Some calculators have a key for handling common fractions; consult the manual for your calculator for directions on how to use such a key.

Exercises  1.11
Set 1

In Exercises 1–48, evaluate each expression. Be sure to perform the operations in the correct order.

1. $12 - 8 - 6$

2. $15 - 9 - 4$

3. $17 - 11 + 13 - 9$

4. $12 - 8 + 14 - 6$

5. $5 - 8 - 12 - 3 - 14 + 6 - 21$

6. $9 - 11 - 6 - 4 + 15 - 7 - 8$

7. $7 + 2 \cdot 4$

8. $10 + 3 \cdot 6$

9. $9 - 3 \cdot 2$

10. $14 - 8 \cdot 3$

11. $10 \div 2 \cdot 5$

12. $20 \cdot 15 \div 5$

13. $12 \div 6 \div 2$

14. $24 \div 12 \div 6$

15. $(-12) \div 2 \cdot (-3)$

16. $(-18) \div (-3) \cdot (-6)$

17. $8 \cdot 5^2$

18. $6 \cdot 2^4$

19. $(-485)^2 \cdot 0 \cdot (-5)^2$

20. $(-589)^2 \cdot 0 \cdot (-3)^2$

21. $12 \cdot 4 + 16 \div 8$

22. $4 \cdot 3 + 15 \div 5$

23. $28 \div 4 \cdot 2(6)$

24. $48 \div 16 \cdot 2(-8)$

25. $(-2)^2 + (-4)(5) - (-3)^2$

26. $(-5)^2 + (-2)(6) - (-4)^2$

27. $2 \cdot 3 + 3^2 - 4 \cdot 2$

28. $100 \div 5^2 \cdot 6 + 8 \cdot 75$

29. $(10^2)\sqrt{16} + 5(4) - 80$

30. $(5^2)\sqrt{9} + 4(6) - 60$

31. $2 \cdot (-6) \div 3 \cdot (8 - 4)$

32. $5 \cdot (-4) \div 2 \cdot (9 - 4)$

33. $24 - [(-6) + 18]$

34. $17 - [(-9) + 15]$

35. $[12 - (-19)] - 16$

36. $[21 - (-14)] - 29$

37. $[11 - (5 + 8)] - 24$

38. $[16 - (7 + 12)] - 22$

39. $20 - [5 - (7 - 10)]$

40. $16 - [8 - (2 - 7)]$

41. $\dfrac{7 + (-12)}{8 - 3}$

42. $\dfrac{(-14) + (-2)}{9 - 5}$

43. $15 - \{4 - [2 - 3(6 - 4)]\}$

44. $17 - \{6 - [9 - 2(2 - 7)]\}$

45. $32 \div (-2)^3 - 5\left\{7 - \dfrac{6 - 2}{5}\right\}$

46. $36 \div (-3)^2 - 6\left\{4 - \dfrac{9 - 7}{3}\right\}$

47. $\sqrt{3^2 + 4^2}$

48. $\sqrt{13^2 - 5^2}$

In Exercises 49–60, use a calculator to evaluate each expression. Express all answers correct to three decimal places.

49. $\dfrac{63.7 - 14.6}{0.64}$

50. $\dfrac{85.2 - 25.7}{0.32}$

51. $\dfrac{6.79}{3.56 \times 8.623}$

52. $\dfrac{3.45}{1.363 \times 56.8}$

53. $\dfrac{3.79 - 1.062}{45.7 \times 0.245}$

54. $\dfrac{6.78 - 1.479}{0.35 \times 16.2}$

55. $\sqrt{(16.3)^2 - (8.35)^2}$

56. $\sqrt{(23.9)^2 + (38.6)^2}$

57. $(1.5)^2 \div (-2.5) + \sqrt{35}$

58. $(-0.25)^2(-10)^3 + \sqrt{54}$

59. $18.91 - [64.3 - ((8.6)^2 + 14.2)]$

60. $[\sqrt{101.4} - (73.5 - (19.6)^2)] \div 38.2$

Writing Problems

Express the answers in your own words and in complete sentences.

1. Explain what order you would use in doing the problem $8 + 3 \cdot 5$, and explain *why* you would use that order.

2. Explain what order you would use in doing the problem $6 - 12 - 2$, and explain *why* you would use that order.

3. Explain what order you would use in doing the problem $2 \cdot 3^2$, and explain *why* you would use that order.

4. Explain what order you would use in doing the problem $13 - (8 + 2)$, and explain *why* you would use that order.

5. Find and describe the error in $2 + 3 \cdot 5 = 5 \cdot 5 = 25$.

6. Find and describe the error in $45 \div 9 \cdot 5 = 45 \div 45 = 1$.

7. Find and describe the error in $8 - 4 + 1 = 8 - 5 = 3$.

Exercises 1.11
Set II

In Exercises 1–48, evaluate each expression. Be sure to perform the operations in the correct order.

1. $18 - 10 - 5$
2. $13 - 9 + 16 - 8$
3. $26 - 37 + 10 - 15$
4. $15 \div 3 \cdot 5$
5. $7 - 11 - 13 - 2 - 9 + 10 - 32$
6. $-18 \div (-\sqrt{81})$
7. $2 + 5 \cdot 3$
8. $18 \div 6 \div 3$
9. $10 \cdot 15^2 - 4^3$
10. $(-10)^2 \cdot 10 + 0(-20)$
11. $(785)^3(0) + 1^5$
12. $(-5)^2 - (2)(-6) + (-2)^2$
13. $2 + 3(100) \div 25 - 10$
14. $7^2 + \sqrt{64} - 14$
15. $(-12) \div 6 - (-4)$
16. $(10^2)\sqrt{4} - 5 + 36$
17. $15 \cdot 4^2$
18. $8\sqrt{4} - 16 \div 4 \cdot 4$
19. $(863)^4 \cdot 0 \cdot (-23)^2$
20. $\sqrt{64} + \sqrt{36}$
21. $19 + 5 \cdot 10$
22. $\sqrt{64 + 36}$
23. $18 \div 9 \div 3$
24. $24 \div 12 \div 3$
25. $2 \cdot 3^2 + 4 \div 4$
26. $-5^2 - 6^2$
27. $63 - 12 - 38$
28. $35 - 5 \cdot 2$
29. $6\sqrt{9} - 18 \div 6 \cdot 3$
30. $8^2 + 12^2$
31. $4 \cdot (11 - 17) \div 3 \cdot (-2)$
32. $[18 - (-15)] - 13$
33. $33 - [(-16) + 11]$
34. $[22 - (7 + 12)] - 14$
35. $26 - [8 - (6 - 15)]$
36. $\dfrac{(-2) + (-7)}{18 + (-15)}$
37. $23 - \{6 - [5 - 2(8 - 3)]\}$
38. $28 \div (-2)^2 - 6\left\{8 - \dfrac{9 - 4}{3}\right\}$
39. $\sqrt{10^2 - 6^2}$
40. $23 - (45 - 73)$
41. $\dfrac{3 + (-15)}{13 - 7}$
42. $\dfrac{(-8) - (-12)}{2 - 8}$

43. $37 - \{23 - [8 - 9] - 1\}$
44. $13 - \{11 - [1 - 19] - 36\}$
45. $48 \div (-4)^2 - 3\left\{9 - \dfrac{8 - 4}{2}\right\}$
46. $75 \div (-5)^2 - 5\left\{6 - \dfrac{9 - 1}{4}\right\}$
47. $\sqrt{8^2 + 6^2}$
48. $\sqrt{3^2 + 4^2}$

In Exercises 49–60, use a calculator to evaluate each expression. When the answer is not exact, round it off to three decimal places.

49. $\dfrac{76.9 - 13.5}{0.25}$
50. $\dfrac{178.3 - 65.8}{0.75}$
51. $\dfrac{5.69}{23.1 \times 14}$
52. $\dfrac{6,745}{1,145 \times 658}$
53. $\dfrac{7.92 - 2.603}{65.3 \times 0.356}$
54. $\dfrac{8.93 + 0.999}{1.45 \times 27}$
55. $\sqrt{(2.18)^2 + (41.6)^2}$
56. $\sqrt{(1.34)^2 + (4.1)^2}$
57. $(63.1)^2 \div (-3.8) + \sqrt{89}$
58. $\sqrt{(34.5)^2 - (17.8)^2}$
59. $(1.7)^2 \div (-3.2) + \sqrt{43}$
60. $[\sqrt{126.3} - (89.7 - (46.5)^2)] \div 52.6$

1.12

Preparing to Solve Applied Problems

We will begin the formal discussion of solving applied problems by using *algebra* in a later chapter. However, a review of a few general problem-solving techniques for solving applied problems in *arithmetic* might be helpful at this time. Many of these suggestions can be applied to *any* kind of problem.

1. Read the problem carefully. Be sure you understand the problem.

2. Identify what is given and what is being asked for. Draw a picture, if it would be helpful. Ask yourself whether you already have enough information to solve the problem. Do you need more facts? Do you need some special formula?

3. Ask yourself whether you know which operation or operations to use. Will addition solve the problem? Subtraction? Multiplication? Division? If you're not sure, *try something*. Try addition, for example. Do not be afraid to make a mistake at this point.

4. Solve the problem, using the given numbers and facts and the operation(s) you've decided to try.

5. *Check your answer.* This is where you make sure you didn't make a mistake in step 3! Is your answer a reasonable one for the problem you are solving? If not, recheck your calculations. If you still get the same unreasonable answer, then analyze the problem again. Should you have used a different operation? Should you have used a different formula?

6. Be sure you have answered *all* the questions asked in the problem.

In this section, you will not be asked to solve any applied problems. You will, instead, be asked whether enough information is given to allow you to solve the problem and what operation you would use to solve the problem.

EXAMPLE 1 Determine whether enough information has been given to allow you to solve the problem. If so, what operations are needed to solve the problem? If not, what other information do you need?

a. Rachael bought some books for $4.95 each. What was the total cost?

SOLUTION Not enough information is given. We must know *how many* books she bought before we can calculate the total cost.

b. Lil is going to cut a 12-ft length of wire into fourteen pieces of equal length. How long will each piece be?

SOLUTION Enough information is given. We would *divide* to solve the problem.

Exercises 1.12
Set I

In each exercise, do *not* solve the problem! Instead, determine whether enough information has been given for the problem to be solved. If so, what operation is needed to solve the problem? If not, what other information do you need?

1. Koji bought a shirt for $27.99, a pair of slacks for $28.39, and a tie for $15.65. How much did he spend for these items?

2. Eric bought several pencils for 37¢ each. What was the total amount he spent?

3. The members of the math department are going to buy a gift for their student worker; they will share the cost equally. If the gift costs $35, what will each person's share be?

Grouping Symbols **1.11**	( )	Parentheses
	[ ]	Brackets
	{ }	Braces

$$\dfrac{3-2}{5+7} \leftarrow \text{The fraction bar}$$

$$\sqrt{5-1} \ \ulcorner \text{The bar of a radical sign}$$

Order of Operations
1.11

1. If there are any grouping symbols in an expression, that part of the expression within a set of grouping symbols is evaluated first.

2. The evaluation then proceeds in this order:

   First: Powers and roots are done.

   Next: Multiplication and division are done *in order from left to right.*

   Last: Addition and subtraction are done *in order from left to right.*

Solving Applied Problems
1.12

See the suggestions in Section 1.12.

Sections 1.7–1.12   REVIEW EXERCISES   Set I

1. Determine whether each of the following is true or false; if the statement is true, give the reason.

   a. $[(-2) \cdot 3] \cdot 4 = (-2)(3 \cdot 4)$

   b. $5 + (-2) = (-2) + 5$

   c. $5 - (-2) = (-2) - 5$

   d. $a + (b + c) = (a + b) + c$

   e. $(c \cdot d) \cdot e = e \cdot (c \cdot d)$

   f. $5 + (x + 7) = (x + 7) + 5$

   g. $7(12 + 6) = (7 \cdot 12) + (7 \cdot 6)$

2. Find the additive inverse of -23.

3. What is the additive identity?

In Exercises 4–13, perform the indicated operations or write "not a real number."

4. 1^4	**5.** 0^3	**6.** $(-4)^2$
7. -5^2	**8.** $\sqrt{121}$	**9.** $-\sqrt[4]{16}$
10. $\sqrt{-10}$	**11.** $\sqrt[3]{-8}$	**12.** $-\sqrt[5]{-32}$
13. 16^2		

14. a. Find the prime factorization of 270.

   b. List all the positive integral factors of 270.

15. List any prime numbers greater than 19 and less than 31 that yield a remainder of 2 when divided by 7.

16. Name all the numbers from the following list that are **(a)** natural numbers, **(b)** integers, **(c)** rational numbers, **(d)** irrational numbers, **(e)** real numbers, and **(f)** numbers that are *not* real:

   $-\sqrt{5}, -31, 5.3892531 \ldots$ (never terminates or repeats), $\frac{7}{12}, 4.\overline{15}, 8, \sqrt[3]{71}$

17. Use a calculator or Table I to find the decimal approximation of $\sqrt{31}$. (Round off the answer to three decimal places.)

In Exercises 18 and 19, express each square root in simplest radical form.

18. $\sqrt{5^3 \cdot 7^4}$ **19.** $\sqrt{336}$

In Exercises 20–28, evaluate each expression.

20. $18 - 22 - 15 + 6$	**21.** $11 - 7 \cdot 3$
22. $(11 - 7) \cdot 3$	**23.** $15 \div 5 \cdot 3$
24. $36 - [(-7) - 15]$	**25.** $6 - [8 - (3 - 4)]$
26. $(35)^2(0) + 18 \div (-6)$	**27.** $\dfrac{6 + (-14)}{3 - 7}$

28. $\sqrt{10^2 - 8^2}$

29. Complete each statement by using the property indicated.

   a. Distributive Property: $8(-7 + 3) =$ _____

   b. Associative Property: $3(-4 \cdot 6) =$ _____

   c. Commutative Property: $-3 + (-4) =$ _____

30. Determine whether enough information has been given for the following problem to be solved. If so, what operation is needed to solve the problem? If not, what other information do you need?

Michael plans to buy eight books. Some cost $3.95 each and the rest cost $5.95 each. What will the total cost be?

ANSWERS

1a. _____

b. _____

_____

Name

I. Determine whether each of the following is true or false. If the statement is true, give the reason.

c. _____

 a. $6 - (-2) = (-2) - 6$

d. _____

 b. $5 \cdot (2 \cdot 3) = (5 \cdot 2) \cdot 3$

 c. $(-7) + (4 + 3) = [(-7) + (4)] + 3$

e. _____

 d. $-3(4 + 6) = (-3 \cdot 4) + (-3 \cdot 6)$

2. _____

 e. $(-6) \div (-3) = (-3) \div (-6)$

3. _____

2. What is the multiplicative identity?

3. What is the additive inverse of $\frac{2}{3}$?

4. _____

In Exercises 4–12, perform the indicated operations or write "not a real number."

4. $\sqrt{81}$ **5.** 4^3 **6.** 1^3

5. _____

6. _____

7. _____

7. 0^5 **8.** $\sqrt[3]{-125}$ **9.** -11^2

8. _____

9. _____

10. $(-2)^2$ **11.** $-\sqrt[3]{1,000}$ **12.** $\sqrt{-49}$

10. _____

11. _____

12. _____

In Exercises 13–15, express each square root in simplest radical form.

13. $\sqrt{3^{11}}$ **14.** $\sqrt{2^5 \cdot 11^2}$ **15.** $\sqrt{1,352}$

13. _____

14. _____

15. _____

16. a. Find the prime factorization of 240.

16a. _____

 b. List the positive integral factors of 240.

b. _____

_____

17. Complete each statement by using the property indicated.

    a. Commutative property: $7(-3) =$ _____

    b. Associative property: $-4 + (-2 + 6) =$ _____

    c. Distributive property: $4(6 - 2) =$ _____

18. Use a calculator or Table I to find the decimal approximation of $\sqrt{53}$. (Round off the answer to three decimal places.)

19. List any prime numbers greater than 17 and less than 37 that yield a remainder of 4 when divided by 5.

In Exercises 20–28, evaluate each expression.

20. $25 - 8 - 1$ **21.** $48 \div 8 \div 2$ **22.** $8 + 2 \cdot 6$

23. $18 \div 6 \cdot 3$ **24.** $\dfrac{16 + 8}{4 + 2}$ **25.** $5 \cdot 3^2$

26. $\sqrt{17^2 - 15^2}$ **27.** $8\sqrt{16} + 8$ **28.** $12 - [14 - (6 - 8)]$

29. Determine whether enough information has been given for the following problem to be solved. If so, what operation is needed to solve the problem? If not, what other information do you need?

The temperature at 1 P.M. was $5°$ F, and at 11 P.M. it was $-16°$ F. What was the change in temperature?

30. Name all the numbers from the following list that are **(a)** natural numbers, **(b)** integers, **(c)** rational numbers, **(d)** irrational numbers, **(e)** real numbers, and **(f)** numbers that are *not* real:

    $4.35671937\ldots$ (never terminates or repeats), $\sqrt{-169}$, $0.\overline{60}$, $\sqrt[3]{21}$, -53, $-\frac{5}{17}$, 462

17a. _____

b. _____

c. _____

18. _____

19. _____

20. _____

21. _____

22. _____

23. _____

24. _____

25. _____

26. _____

27. _____

28. _____

29. _____

_____

30a. _____

b. _____

c. _____

d. _____

e. _____

f. _____

Chapter I   D I A G N O S T I C   T E S T

The purpose of this test is to see how well you understand the operations with integers and other signed numbers. We recommend that you work this diagnostic test *before* your instructor tests you on this chapter. Allow yourself about 50 minutes to do this test.

   Complete solutions for all the problems on this test, together with section references, are given in the answer section at the end of the book. We suggest that you study the sections referred to for the problems you do incorrectly.

In Problems 1–3, write "true" if the statement is always true; otherwise, write "false."

  1. **a.** The smallest natural number is 1.

   b. $(2 + 3) + 7 = 2 + (3 + 7)$ illustrates the commutative property of addition.

   c. $\sqrt{-25}$ is not a real number.

   d. Division is associative.

  2. **a.** 0 is a real number.

   b. $12 \div 6 = 6 \div 12$

   c. $\sqrt{147}$ is in simplest radical form.

   d. $2(5 \cdot 6) = (2 \cdot 5) \cdot (2 \cdot 6)$

  3. **a.** $\sqrt{16}$ is an irrational number.

   b. Multiplication is distributive over addition.

   c. $9 \cdot 1 = 9$ illustrates the multiplicative identity property.

   d. The additive identity element is 1.

  4. Name all the numbers from the following list that are **(a)** natural numbers, **(b)** integers, **(c)** rational numbers, **(d)** irrational numbers, **(e)** real numbers, and **(f)** numbers that are *not* real:

$$-\sqrt[3]{19}, \ -63, \ 5, \ 0.\overline{38}, \ \tfrac{2}{23},$$
$$0.25375913\ldots\text{(never terminates or repeats)}, \ \sqrt{-121}$$

  5. **a.** Subtract 24 from 5.   **b.** Subtract 8 from -17.

In Problems 6–23, perform the indicated operations or write either "not defined" or "not a real number."

  6. **a.** $4 + (-18)$ **b.** $15 + (-8.62)$

  7. **a.** $-6(0)$ **b.** $-\tfrac{2}{3}\left(-\tfrac{1}{4}\right)$

  8. **a.** $-10 - 5$ **b.** $3 - 22$

  9. **a.** -7^2 **b.** $(-7)^2$

  10. **a.** $3 \div 0$ **b.** $\tfrac{15}{-21}$

  11. **a.** $-\tfrac{5}{8} \div \left(-\tfrac{1}{2}\right)$ **b.** $\tfrac{0}{4.7}$

  12. **a.** $\sqrt{49}$ **b.** $\sqrt[3]{64}$

  13. **a.** $3 - (-12)$ **b.** $0 - 6$

  14. **a.** $2 \cdot 5^2$ **b.** $\sqrt{17^2 - 15^2}$

  15. $23 - 19 - 3 + 11$ 16. $(-6)(-5)(-2)\left(-\tfrac{1}{3}\right)$

  17. $54 \div 9 \cdot 6$ 18. $8 + 9 \cdot 7$

  19. $6 \cdot 4^2 - 4$ 20. $\dfrac{7 - 15}{-2 + 6}$

  21. $5\sqrt{36} - 6(-5)$ 22. $(2^3 - 8)(8^2 - 9^2)$

  23. $\{-10 - [5 + 2(4 - 7)]\} - 3$

  24. **a.** List all the positive integral factors of 54.

   b. Write the prime factorization of 54.

  25. Determine whether enough information has been given for each of the following problems to be solved. If so, what operation is needed to solve the problem? If not, what other information do you need?

   a. At 5 A.M. the temperature in Denver, Colorado, was $-20°$ F. At noon the temperature was 38° F. What was the rise in temperature?

   b. Lupe wrote a check for $37. What was her new checking account balance?

Critical Thinking and Problem-Solving Exercises

Mathematics is much more than simply memorizing a set of rules for solving certain problems; mathematics involves analyzing problems, looking for patterns and relationships, examining data, developing models, and so on. In the Critical Thinking and Problem-Solving Exercises that will appear at the end of each chapter, you are given the opportunity of thinking critically and of devising your own methods of problem-solving—in many cases attacking types of problems different from any you have seen before. Some of the problems are "open-ended" problems—problems that do not have "right" or "wrong" answers; in such problems, analyze the given facts and draw your own conclusions, conclusions that may be different from the conclusions drawn by others.

Methods you might try on these problems include (but are not limited to) guessing-and-checking, making an organized list or a table, looking for patterns, using logic, drawing pictures, making models, and simplifying or working backwards. Your instructor may want you to work in small groups to solve some or all of these problems.

1. Gloria is saving for a new paint job for her car. She was able to save $35 toward the paint job in January, $37 in February (her income was gradually increasing), $39 in March, and $41 in April. If she continues in this way, how much will Gloria save in May? How much will she save in June? If the paint job that she wants costs $299, when will she be able to afford it? Explain your answer.

2. Mary, May, Molly, Mark, Mel, and Merv are sitting in the classroom in front of their professor in an arrangement shown in the figure. Using the following clues, can you decide who is sitting where?

   Molly is next to the professor.

   Mary is between Mark and Molly.

   Mel is next to Mary, and he is closer to the professor than May is.

3. Cherise is thinking of a certain number. Can you guess what number Cherise is thinking about by using the following clues?

   The number is a prime number.

   It is less than 40.

   The sum of its digits is 4.

   The number is greater than 15.

4. By making an organized list, find all the positive factors of 21.

5. The solution of each of the following problems contains an error. Find and describe the errors.

   a. Evaluate $\sqrt{5^2 + 12^2}$.
   $$\sqrt{5^2 + 12^2}$$
   $$= \sqrt{25 + 144}$$
   $$= \sqrt{25} + \sqrt{144}$$
   $$= 5 + 12 = 17$$

   b. Evaluate $2^3 - 3 \cdot 4 \div 2$.
   $$2^3 - 3 \cdot 4 \div 2$$
   $$= 8 - 3 \cdot 4 \div 2$$
   $$= 5 \cdot 4 \div 2$$
   $$= 20 \div 2 = 10$$

Simplifying and Evaluating Algebraic Expressions

CHAPTER 2

n this chapter, we first give some of the basic definitions of algebra. We then discuss using the distributive property in algebra, combining like terms, and removing grouping symbols. We also discuss evaluating algebraic expressions, using formulas, and solving applied problems by using formulas.

2.1   Basic Definitions

We begin this chapter with a number of definitions used in algebra.

Variables   In algebra, we often use letters to represent numbers, and we usually call these letters variables. A **variable** is a letter, object, or symbol that acts as a placeholder for a number that is unknown. The variable may assume different values in a particular problem or discussion.

Constants   A **constant** is an object or symbol that does *not* change its value in a particular problem or discussion. It is usually represented by a number symbol, but it is sometimes represented by one of the first few letters of the alphabet. Thus, in the expression $2x - 7y + 3$, the constants are 2, -7, and 3. (Notice that the negative sign in front of the 7 is considered to be part of the constant.) In the expression $ax + by + c$, it is understood that a, b, and c represent constants and that x and y represent variables.

EXAMPLE 1   List (a) the constants and (b) the variables for $5x + 8y - 3z$.

SOLUTION   It helps to think of this expression in its additive form, $5x + 8y + (-3z)$.

a. The constants are 5, 8, and -3. (Notice that the negative sign in front of the 3 is considered to be part of the constant.)
b. The variables are x, y, and z.

Factors   We mentioned earlier that numbers that are multiplied together to give a product are called the *factors* of that product, and we mentioned that when two symbols (other than two *numbers*) are written next to each other, it is understood that they are to be *multiplied* together.

EXAMPLE 2   Examples of identifying the factors in an expression:

An understood multiplication symbol
a. $2x$ The first factor is 2, and the second factor is x.

Understood multiplication symbols
b. $-7xyz$ The first factor is -7, the second factor is x, the third factor is y, and the fourth factor is z.

Powers of Variables   We have already discussed bases, exponents, and powers of signed numbers, and we know that the expression 3^4 means that 3 is to be used as a factor four times; that is, $3^4 = 3 \cdot 3 \cdot 3 \cdot 3$, or $3^4 = 81$. The *base* is 3 and the *exponent* is 4. In algebra, the base is often a variable, and the exponent is sometimes a variable.

$$\underset{\text{The base}}{\underbrace{}}\, x^{\overset{\text{The exponent}}{5}} = x \cdot x \cdot x \cdot x \cdot x$$

x^5 is read "the fifth power of x" or "x to the fifth power"

Because the exponent is 5, the base (x) is to be used as a factor five times. We say that x^5 is in *exponential form*. Also, just as $2 = 2^1$, $x = x^1$; that is, when there is no exponent, the exponent is understood to be 1.

EXAMPLE 3 Examples of writing repeated factors in exponential form:

a. $xxx = x^3$ x is used as a factor three times

b. $xyxy = xxyy = x^2y^2$

Using the commutative and associative properties of multiplication to rearrange the factors

c. $zxzxyz = xxyzzz = x^2yz^3$

Algebraic Expressions An **algebraic expression** consists of numbers, variables (letters), signs of operation (such as $+$, $-$, etc.), and signs of grouping. (Not all of these need to be present.)

EXAMPLE 4 Examples of algebraic expressions:

a. 8 This algebraic expression consists of a single number

b. z This algebraic expression consists of a single letter (variable)

c. $2x - 3y$

d. $5x^3 - 7x^2 + 4$ **e.** $\dfrac{5s^2 - 2t^2}{\sqrt{3st}}$

f. $\sqrt{b^2 - 4ac}$ **g.** $(x - y)^2 + (t - u)^2$

Terms A **term** of an algebraic expression can consist of one number, one variable, or a *product* of numbers and variables. This means that the only operations that can be indicated within a *term* are multiplication, division, and powers; we include division here because, as we have mentioned, the division $\dfrac{a}{b}$ can be interpreted as the multiplication $a \cdot \dfrac{1}{b}$. If a term is preceded by a minus sign, that sign is *part* of the term. (It is often true that the terms of an algebraic expression are the parts that are separated by addition and subtraction symbols.) Exception: An expression *within grouping symbols* is to be considered as a single unit; it is a *term* or it is a *factor of a term*, even if there are one or more terms *within* the grouping symbols (see Examples 5c, 5d, and 5e).

EXAMPLE 5 Examples of identifying the terms in an algebraic expression:

a. $2s + 5t$

The first term consists of the factors 2 and s

$2s + 5t$ The two terms are separated by an addition symbol

The second term consists of the factors 5 and t

The first term is $2s$, and the second term is $5t$.

b. $3x^2y - 5xy^3 + 7xy$

It is easiest to identify the terms if we think of $3x^2y - 5xy^3 + 7xy$ in its *additive form*—that is, as $3x^2y + (-5xy^3) + 7xy$.

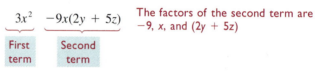

┌── The first term consists of the factors 3, x^2, and y

┌── The second term consists of the factors −5, x, and y^3

The third term consists of the factors 7, x, and y

$$3x^2y + (-5xy^3) + 7xy$$

The first term is $3x^2y$, the second term is $-5xy^3$, and the third term is $7xy$.

c. $(3z + 2x - 1)$

An expression within parentheses is considered as a single unit. Therefore, the algebraic expression $(3z + 2x - 1)$ is considered to be one term.

d. $3x^2 - 9x(2y + 5z)$

An expression within parentheses is considered as a single unit, and here $(2y + 5z)$ is a *factor* of the expression $-9x(2y + 5z)$; therefore, $-9x(2y + 5z)$ consists of *factors* and is considered to be one term.

$$\underbrace{3x^2}_{\substack{\text{First}\\\text{term}}} \quad \underbrace{-9x(2y + 5z)}_{\substack{\text{Second}\\\text{term}}}$$

The factors of the second term are −9, x, and $(2y + 5z)$

The first term is $3x^2$, and the second term is $-9x(2y + 5z)$.

e. $\dfrac{2 - x}{xy} + 5(2x^2 - y)$

The fraction bar is a grouping symbol; also, since the division $\dfrac{a}{b}$ can be interpreted as the multiplication $a \cdot \dfrac{1}{b}$, the division $\dfrac{2 - x}{xy}$ can be interpreted as the multiplication $(2 - x) \cdot \dfrac{1}{x} \cdot \dfrac{1}{y}$. We see, then, that $\dfrac{2 - x}{xy}$ consists of the factors $(2 - x)$, $\dfrac{1}{x}$, and $\dfrac{1}{y}$ and is a single term.

$$\underbrace{\dfrac{2 - x}{xy}}_{\substack{\text{First}\\\text{term}}} + \underbrace{5(2x^2 - y)}_{\substack{\text{Second}\\\text{term}}}$$

The second term consists of the factors 5 and $(2x^2 - y)$

The first term is $\dfrac{2 - x}{xy}$, and the second term is $5(2x^2 - y)$.

Numerical Coefficients When a term consists of a number and one or more variables as *factors*, the factor that is a *number* is said to be the **numerical coefficient** of the term. For example, the numerical coefficient of the term $2\,x$ is 2, and the numerical coefficient of the term $-7\,xyz$ is -7.

When there is *no* number in front of a variable or a product of variables, the numerical coefficient is understood to be 1 if there is a plus sign or no sign in front of the variable(s) (see Example 6c), and −1 if there is a minus sign in front of the variable(s) (see Example 6d).

When a term has a denominator, we determine the numerical coefficient by using the fact that the division $\dfrac{a}{b}$ can be interpreted as the multiplication $\dfrac{1}{b} \cdot a$. For example,

$\dfrac{x}{2}$ can be interpreted as $\dfrac{1}{2} \cdot x$ (see Example 6e), $-\dfrac{z}{5}$ can be interpreted as $-\dfrac{1}{5} \cdot z$ (see Example 6f); furthermore, because $\dfrac{3}{4}xy = \dfrac{3}{4} \cdot \dfrac{x}{1} \cdot \dfrac{y}{1} = \dfrac{3xy}{4}$, $\dfrac{3xy}{4}$ can be interpreted as $\dfrac{3}{4}xy$ (see Example 6g).

EXAMPLE 6

Examples of identifying numerical coefficients:

Term	*Numerical coefficient*
a. $6w$	6
b. $-12xy^2$	-12
c. xy	1

When there is no number in front of a variable, the coefficient is understood to be 1; note that $xy = 1xy$

d. $-s$	-1

When there is a minus sign but no number in front of the variable, the coefficient is understood to be -1; note that $-s = -1s$

e. $\dfrac{x}{2} = \dfrac{1}{2}x$	$\dfrac{1}{2}$
f. $-\dfrac{z}{5} = -\dfrac{1}{5}z$	$-\dfrac{1}{5}$
g. $\dfrac{3xy}{4} = \dfrac{3}{4}xy$	$\dfrac{3}{4}$

Like Terms   Terms that have *equal variable parts* are called **like terms**; only the numerical coefficients can be different. We can also say that if repeated factors are expressed in exponential form (for example, if we write $xxxyy$ as x^3y^2 and xyx as x^2y) and if all variables are written in alphabetical order, then terms that have *identical variable parts* are called **like terms**. Also, terms consisting only of *numbers* are like terms with each other.

EXAMPLE 7

Examples of like terms:

a. $3x$, $4x$, $\frac{1}{2}x$, and $0.7x$ are like terms. They are called "x-terms."
b. $2x^2$, $10xx$, $\frac{3}{4}x^2$, xx, and $2.3x^2$ are like terms. They are called "x^2-terms." (Note that $10xx = 10x^2$ and $xx = x^2$.)
c. $5x^2y$, $2xyx$, yx^2, $\frac{2}{3}x^2y$, and $5.6xxy$ are like terms (note that $x^2y = xyx = yx^2 = xxy$). They are called "x^2y-terms."
d. 5, -3, $\frac{1}{4}$, and 2.6 are like terms. They are called "constant terms."

Unlike Terms   Terms that do *not* have identical variable parts (after repeated factors have been expressed in exponential form and the variables have been written in alphabetical order) are called **unlike terms**.

EXAMPLE 8

Examples of unlike terms:

a. $2x$ and $3y$ are unlike terms.   The variables are different
b. $5x^2$ and $5x$ are unlike terms.   The exponents on the x are different in the two terms

c. $-4y$ and $-4Y$ are unlike terms.

> It is understood that lower-case y and capital Y represent different variables

d. $2xy$ and $2st$ are unlike terms.

> The variable parts are not equal; that is, $xy \neq st$

e. $8x$ and 3 are unlike terms.

> 3 contains *no* variable, but $8x$ contains the variable x

f. $4xyx$ and $10xyy$ are unlike terms.

> $xyx \neq xyy$ (when we write the terms in exponential form, we have $4x^2y$ and $10xy^2$); the variable parts are not identical

---

We will see later that it is possible to *combine* (that is, to write as a single term) only *like terms*.

👉 **Note**   In each term, in addition to writing repeated factors in exponential form and writing variables in alphabetical order, it is customary to write a numerical coefficient (if there is one) to the *left* of any other factors.

Exercises 2.1
Set I

In Exercises 1–4, list **(a)** the different constants and **(b)** the different variables.

1. $2x + 4y + 2$

2. $7s + 3t + 7$

3. $7u - 8v + 2v$

4. $3x - 5y - 2x$

In Exercises 5 and 6, x is either a *factor* of or a *term* of the given expression. Determine which it is.

5. a. $3xyz$ **b.** $3 + x$

    c. $4x(y + 2)$ **d.** $5x$

6. a. $x + 7$ **b.** $7x$

    c. $7xy$ **d.** $x(2 + y)$

In Exercises 7 and 8, $7x$ is either a *factor* of or a *term* of the given expression. Determine which it is.

7. a. $7x + 3$ **b.** $y + 7x$

    c. $7x(3 + y)$ **d.** $7x + (3 + y)$

8. a. $7x(y + 6)$ **b.** $7xy$

    c. $1 + 7x$ **d.** $(y + 6) + 7x$

In Exercises 9–18, **(a)** determine the number of terms and **(b)** write the second term if there is one.

9. $7xy$

10. $5ab$

11. $E - 5F - 3$

12. $R - 2T - 6$

13. $3x^2y + \dfrac{2x + y}{3xy} + 4(3x^2 - y)$

14. $5xy^2 + \dfrac{5x - y}{7xy} + 3(x^2 - 4y)$

15. $5u^2 - 6u(2u + v^2)$

16. $3E^3 - 2E(8E + F^2)$

17. $[(x + y) - (x - y)]$

18. $\{x - [y - (x - y)]\}$

In Exercises 19–26, write the numerical coefficient of the first term.

19. $3x + 7y$ **20.** $4R + 3T$

21. $x^2 - 3xy$ **22.** $x^2 + 5xy$

23. $-b + 4a$ **24.** $-y^2 - x^2$

25. $\dfrac{4xy}{5} + \dfrac{2a}{3}$ **26.** $\dfrac{r}{2} - \dfrac{s}{3} + \dfrac{t}{4}$

In Exercises 27–34, for the set of terms in each exercise, write repeated factors in exponential form and write the variables in alphabetical order whenever necessary; then determine whether or not the terms in that exercise are *all* like terms.

27. $\{7xy^2, 13xyx, 4xyy, -4x^2y\}$

28. $\{9x^2y, 42xyx, 13xy^2, 6yxy\}$

29. $\{7x^5, 12xxxxx, -3x^5\}$

30. $\{9y^3, -4y^3, -23yyy\}$

31. $\{8uv, 12vu, -3uv\}$

32. $\{-6st, 3st, -29ts\}$

33. $\{15x^3, -12x^2, \frac{1}{2}x\}$

34. $\{-14x, 3y, 5z\}$

Writing Problems

Express the answers in your own words and in complete sentences.

1. Explain why $4x$ and $4y$ are not like terms.

2. Explain why $-3x^3$ and $82x^3$ are like terms.

3. Explain why $8x^3$ and $8x^4$ are not like terms.

4. Explain why z is a factor of $18xyz$.

5. Explain why x is not a factor of $3 + x$.

6. Explain why 2 is a term of $z + 2$.

7. Explain why 12 is not a term of $12w$.

Exercises 2.I
Set II

In Exercises 1–4, list (a) the different constants and (b) the different variables.

1. $4x - 7y + 4$

2. $8u - 5v + 3u$

3. $9s + 3t + 9u$

4. $-8x + 2y - 5z$

In Exercises 5 and 6, x is either a *factor* of or a *term* of the given expression. Determine which it is.

5. **a.** xy **b.** $x + y$
   c. $x + 4y$ **d.** $x(y - 3)$

6. **a.** $5y + x$ **b.** $8x$
   c. $x + 8$ **d.** $9x(y - 1)$

In Exercises 7 and 8, $5x$ is either a *factor* of or a *term* of the given expression. Determine which it is.

7. **a.** $7 + 5x$ **b.** $5xy$
   c. $5x(5 + y)$ **d.** $5x + (3 + y)$

8. **a.** $5x - 2$ **b.** $5x(y - 1)$
   c. $5x + y - 1$ **d.** $7 + 5x$

In Exercises 9–18, (a) determine the number of terms and (b) write the second term if there is one.

9. $-2x$

10. $8(-2x)$

11. $3 - 4y + 2z$

12. $3 - (4x + 2z)$

13. $5 - xy$

14. $5xy$

15. $3x^3 + 2(y - 3z)$

16. $8x^2 - 3(x + y) - z$

17. $[a + (b + c)]$

18. $x + y + z$

In Exercises 19–26, write the numerical coefficient of the first term.

19. $18x + 3y$

20. $-18(x + y)$

21. $a - b + c$

22. $-x - y$

23. $-x^2 + 5xy$

24. $-y^2 + 3(x + y)$

25. $-2a^2 - 5ab + b^3$

26. $\dfrac{a}{6} + \dfrac{b}{4} - \dfrac{2c}{3}$

In Exercises 27–34, for the set of terms in each exercise, write repeated factors in exponential form and write the variables in alphabetical order whenever necessary; then determine whether or not the terms in that exercise are *all* like terms.

27. $\{9w^2z, 11wzz, 4w^2z, -5wz^2\}$

28. $\{15, -45, 13, -6\}$

29. $\{-8y^4, 15y^4, -4yyyy\}$

30. $\{17xy, -12xy, 4yx\}$

31. $\{27vw, -11wv, -7vw\}$

32. $\{12xy, 4yz, -5zw\}$

33. $\{-7x^2, 12y^2, \frac{2}{3}x^2\}$

34. $\{16x^2, -13x^4, 8x\}$

2.2 Using the Distributive Property

2.2A Using the Distributive Property to Remove Grouping Symbols

We now discuss removing grouping symbols in multiplication problems in which one factor contains two or more unlike terms; in such problems, we must use the distributive property. Recall that the distributive property is as follows.

Multiplication is distributive over addition and subtraction

$$a(b + c) = ab + ac$$
$$a(b - c) = ab - ac$$

where a, b, and c represent any real numbers.

In arithmetic, when we're simplifying an expression such as $8(3 + 5)$, we can perform the operation inside the grouping symbols first, writing $8(3 + 5) = 8(8) = 64$, or we can apply the distributive property. To do this, we multiply each term inside the grouping symbols by the factor that is outside them and then add the products, as shown in Examples 1 and 2.

EXAMPLE 1   Use the distributive property to remove the parentheses in $8(3 + 5)$; then combine the products.

SOLUTION

$$8(3 + 5) = \overset{\frown}{(8)}(3 + 5)$$

Each term inside the parentheses is multiplied by the factor outside the parentheses; then these products are added

$$= \underset{\text{First product}}{8(3)} + \underset{\text{Second product}}{8(5)} = 24 + 40 = 64$$

Notice that we obtained the same answer as when we added 3 and 5 first.

In Example 2, we show on the left the use of the property $a(b - c) = ab - ac$; on the right, we interpret the subtraction problem as an addition problem and use the property $a(b + c) = ab + ac$. Both methods are acceptable.

EXAMPLE 2   Use the distributive property to remove the parentheses; then combine the products.

a. $2(3 - 7)$

SOLUTION

$$2(3 - 7) = 2(3) - 2(7)$$
$$= 6 - 14$$
$$= -8$$

ALTERNATIVE SOLUTION

$$2(3 - 7) = 2(3 + [-7])$$
$$= 2(3) + 2(-7)$$
$$= 6 + (-14)$$
$$= -8$$

Note also that $2(3 - 7) = 2(-4) = -8$.

b. $-5(2 - 7)$

SOLUTION	ALTERNATIVE SOLUTION
$-5(2 - 7) = -5(2) - (-5)(7)$	$-5(2 - 7) = -5(2 + [-7])$
$= -10 - (-35)$	$= -5(2) + (-5)(-7)$
$= -10 + 35$	$= -10 + 35$
$= 25$	$= 25$

Note also that $-5(2 - 7) = -5(-5) = 25$.

(It is not necessary to show all the steps that we have shown in Example 2.)

When we need to remove grouping symbols that contain more than one term in *algebra*, we very often cannot simplify the expression inside the grouping symbols, and we *must*, therefore, use the distributive property to remove the grouping symbols (see Example 3). (In this section, none of the parentheses contain like terms, and, as we have mentioned, we can combine—that is, write as a single term—only like terms. Therefore, we will not be able to write any of the expressions in this section as a single term.)

EXAMPLE 3   Use the distributive property to remove the parentheses.

a. $2(x + y)$

SOLUTION   Notice that x and y are not *like terms*; therefore, $x + y$ cannot be written as a single term (that is, the two terms cannot be combined).

$$2(x + y) = 2(x) + 2(y)$$
$$= 2x + 2y \qquad \text{2x and 2y are not like terms; they cannot be combined}$$

b. $a(x - y)$

SOLUTION

$$a(x - y) = a(x) - a(y)$$
$$= ax - ay \qquad \text{ax and -ay are not like terms; they cannot be combined}$$

When we use the distributive property, the factor that contains two terms can be on the right, as in $a(b + c)$, or on the left, as in $(b + c)a$. We use the commutative property of multiplication to prove that $(b + c)a = ba + ca$:

PROOF $(b + c)a = a(b + c)$ Using the commutative property of multiplication

$= ab + ac$ Using the distributive property

$= ba + ca$ Using the commutative property of multiplication

Therefore, $(b + c)a = ba + ca$. ■

In Example 4, we show that when we use the distributive property and when the factor that contains two terms is on the left, we multiply each term inside the grouping symbols by the factor on the *right*.

EXAMPLE 4   Use the distributive property to remove the parentheses.

a. $(5x + 3y)(2)$

SOLUTION

$$
\begin{aligned}
(5x + 3y)(2) &= (5x)(2) + (3y)(2) && \text{Using the distributive property} \\
&= (2)(5x) + (2)(3y) && \text{Using the commutative} \\
& && \text{property of multiplication} \\
&= (2 \cdot 5)x + (2 \cdot 3)y && \text{Using the associative} \\
& && \text{property of multiplication} \\
&= \quad 10x \ + \ \ 6y
\end{aligned}
$$

In practice, we usually do not show the steps in which the commutative and associative properties are used. That is, we write only $(5x + 3y)(2) = (5x)(2) + (3y)(2) = 10x + 6y$, or even simply $(5x + 3y)(2) = 10x + 6y$.

b. $(2 + 5y)(-4)$

SOLUTION

$$
\begin{aligned}
(2 + 5y)(-4) &= (2)(-4) + (5y)(-4) && \text{Using the distributive property} \\
&= -8 + (-4)(5y) && \text{Using the commutative property} \\
& && \text{of multiplication} \\
&= -8 + (-4 \cdot 5)y && \text{Using the associative property} \\
& && \text{of multiplication} \\
&= -8 + (-20y) \\
&= -8 - 20y && \text{Can you verify that } -8 - 20y \text{ is} \\
& && \text{equivalent to } -8 + (-20y)?
\end{aligned}
$$

Again, you need not show the steps in which the commutative and associative properties are used.

✖ **A Word of Caution**   In Example 4b, the parentheses around the -4 are *absolutely necessary*. The problem $(2 + 5y) - 4$ is a *subtraction* problem; 4 is to be subtracted from $2 + 5y$. The problem $(2 + 5y)(-4)$ is a *multiplication* problem; $2 + 5y$ is to be multiplied by -4.

✖ **A Word of Caution**   At the end of Example 4b, it would be incorrect to write

$$-8 - 20y = -28y$$

-8 and $-20y$ are not like terms; therefore, $-8 - 20y$ cannot be written as a single term.

Extensions of the Distributive Property

The distributive property can be extended to include any number of terms inside the parentheses. That is,

$$a(b + c - d + e + \cdots) = ab + ac - ad + ae + \cdots$$

and
$$(b + c + d + \cdots)a = ba + ca + da + \cdots$$

EXAMPLE 5 Use the distributive property to remove the grouping symbols.

a. $-4(3x - 7y + 4 - 8z)$

SOLUTION

$$-4(3x - 7y + 4 - 8z) = -4[3x + (-7y) + 4 + (-8z)]$$
$$= -4(3x) + (-4)(-7y) + (-4)(4) + (-4)(-8z)$$
$$= -12x + 28y + (-16) + 32z$$
$$= -12x + 28y - 16 + 32z$$

b. $(3x + 4y - 7)(-2)$

SOLUTION

$$(3x + 4y - 7)(-2) = [3x + 4y + (-7)](-2)$$
$$= 3x(-2) + 4y(-2) + (-7)(-2)$$
$$= -6x + (-8y) + 14$$
$$= -6x - 8y + 14$$

A Word of Caution It is incorrect to write (or think)

$$2(3 \cdot 4) = (2 \cdot 3)(2 \cdot 4)$$

The distributive property applies only when this symbol is a plus or minus sign

We can easily verify that $2(3 \cdot 4) \neq (2 \cdot 3)(2 \cdot 4)$.

$$2(3 \cdot 4) = 2(12) = 24$$

but

$$(2 \cdot 3)(2 \cdot 4) = (6)(8) = 48$$

Therefore, $2(3 \cdot 4) \neq (2 \cdot 3)(2 \cdot 4)$.

Observe that in the expression $2(3 \cdot 4)$, there is only one *term* inside the parentheses. The distributive property can be used *only* when the parentheses contain more than one term.

Exercises 2.2A
Set I

Use the distributive property to remove the parentheses. Do not try to combine any terms.

1. $5(a + 6)$

2. $4(x + 10)$

3. $7(x + y)$

4. $5(m + n)$

5. $3(m - 4)$

6. $3(a - 5)$

7. $4(x - y)$

8. $9(m - n)$

9. $a(6 + x)$

10. $b(7 + y)$

11. $-2(x - 3)$

12. $-3(x - 5)$

13. $(x - 4)(6)$

14. $(3 - 2x)(-5)$

15. $-3(x - 2y + 2)$

16. $-2(x - 3y + 4)$

Writing Problems

Express the answers in your own words and in complete sentences.

2. Explain why $5(z + 3) \neq 5z + 3$.

1. Explain how to remove the parentheses in the expression $8(x + 3)$.

Exercises 2.2A
Set II

Use the distributive property to remove the parentheses. Do not try to combine any terms.

1. $3(x + 4)$ **2.** $2(m - 5)$

3. $x(4 + y)$ **4.** $-3(x - 4)$

5. $(M + N)(3)$ **6.** $(x - y)(-4)$

7. $8(a - b)$ **8.** $(5 - y)(-2x)$

9. $x(y + 2)$ **10.** $(7 + a)(3b)$

11. $-6(x + 2y)$ **12.** $(x + 2y)(-6)$

13. $(y - 2)(4)$ **14.** $(8 - 2y)(-3)$

15. $-7(s - 4t - 5)$

16. $-3(4 - 5u + 7v)$

2.2B Combining Like Terms

We used the distributive property earlier to write a product as a sum of terms; that is, we rewrote the product $a(b + c)$ as the sum of terms $ab + ac$. (We often call this *distributing*, *multiplying out*, or *simplifying*.) We can also use the distributive property to write the sum of terms $ab + ac$ as the product $a(b + c)$. This makes it possible for us to combine *like terms* (see Example 6). We call writing $ab + ac$ as $a(b + c)$ *factoring*; we will discuss factoring in much more detail in a later chapter.

EXAMPLE 6 Combine the like terms in each expression.

a. $3x + 5x$

    SOLUTION $3x$ and $5x$ are like terms.

$$3x + 5x = (3 + 5)x \quad \text{Using the distributive property (factoring)}$$
$$= 8x \quad \text{Simplifying the expression inside the parentheses}$$

b. $7y - 3y$

    SOLUTION $7y$ and $-3y$ are like terms.

$$7y - 3y = (7 - 3)y \quad \text{Using the distributive property (factoring)}$$
$$= 4y \quad \text{Simplifying the expression inside the parentheses}$$

c. $z - 9z$

    SOLUTION Recall that when no numerical coefficient is written, it is understood to be 1. Therefore, the numerical coefficient of z in the first term is understood to be 1.

$$z - 9z = 1z - 9z$$
$$= (1 - 9)z$$
$$= (1 + [-9])z \quad \text{Changing the subtraction to addition}$$
$$= -8z$$

If you look carefully at all the problems in Example 6, you can see that you would get the same answer if you followed the rules given in the box below. No matter which method we use, it is understood that *only like terms can be combined* and that if any term has no written numerical coefficient, the coefficient is understood to be 1.

Combining (adding) like terms	Add the numerical coefficients *of the like terms*; the number so obtained is the numerical coefficient of the sum. The variable part of the sum is the same as the variable part of any *one* of the like terms.

When we say "Add the numerical coefficients," it is understood that if any subtractions are indicated, those subtractions will be mentally changed to additions so that the problem can be treated *as an addition problem*.

In practice, it is neither necessary nor desirable to show all the steps that are shown in Example 6. In Example 6a, we can simply write $3x + 5x = 8x$; in Example 6b, we can write just $7y - 3y = 4y$; and in Example 6c, we can write $z - 9z = -8z$.

When we combine like terms, we often need to change the order in which the terms appear. The commutative and associative properties of addition guarantee that when we do this, the sum remains unchanged.

EXAMPLE 7   Combine the like terms in each expression.

a. $3x - 7x + 10x$

   SOLUTION   Think of this problem as $3x + (-7)x + 10x$.

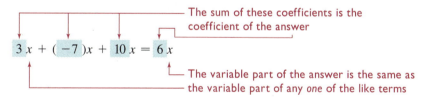

The sum of these coefficients is the coefficient of the answer

$$3 \; x \; + \; (\; -7 \;)x \; + \; 10 \; x \; = \; 6 \; x$$

The variable part of the answer is the same as the variable part of any *one* of the like terms

b. $9x + 5y - 3x$

   SOLUTION   The only terms that can be combined are $9x$ and $-3x$ (they are the only like terms). We can "collect like terms" by rearranging the terms as shown:

$$9x + 5y - 3x$$
$$= (9x - 3x) + 5y \quad \text{Collecting like terms}$$
$$= \quad 6x \quad\; + 5y \quad \text{Combining like terms}$$

c. $12a - 7b - 9a + 4b$

   SOLUTION

$$12a - 7b - 9a + 4b$$
$$= (12a - 9a) + (-7b + 4b) \quad \text{Collecting like terms}$$
$$= \quad 3a \quad\; + \quad (-3b) \quad \text{Combining like terms}$$
$$= 3a - 3b$$

d. $7x - 2y + 9 - 11x + 3 - 4y$

SOLUTION

$$7x - 2y + 9 - 11x + 3 - 4y$$
$$= (7x - 11x) + (-2y - 4y) + (9 + 3) \quad \text{Collecting like terms}$$
$$= -4x + (-6y) + 12 \quad \text{Combining like terms}$$
$$= -4x - 6y + 12$$

e. $-5x^2y + 7xy^2 - 7x^2y$

SOLUTION   Be careful here! The terms $-5x^2y$ and $-7x^2y$ are the only like terms.

$$-5x^2y + 7xy^2 - 7x^2y$$
$$= 7xy^2 + (-5x^2y - 7x^2y) \quad \text{Collecting like terms}$$
$$= 7xy^2 + (-12x^2y) \quad \text{Combining like terms}$$
$$= 7xy^2 - 12x^2y$$

f. $3x^2 - 5x + 5 - 2x^2 + 7x + 11$

SOLUTION

$$3x^2 - 5x + 5 - 2x^2 + 7x + 11$$
$$= (3x^2 - 2x^2) + (-5x + 7x) + (5 + 11) \quad \text{Collecting like terms}$$
$$= x^2 + 2x + 16 \quad \text{Combining like terms}$$

In practice, you need not show all the steps that we have shown in Example 7.
We will consider combining like terms again in a later chapter, after we have discussed exponents in more detail.

Exercises 2.2B
Set I

In each exercise, combine the like terms.

1. $15x - 3x$

2. $12a - 9a$

3. $5a - 12a$

4. $10x - 24x$

5. $2a - 5a + 6a$

6. $3y - 4y + 5y$

7. $5x - 8x + x$

8. $3a - 5a + a$

9. $3x + 2y - 3x$

10. $4a - 2b + 2b$

11. $4y + y - 10y$

12. $3x - x - 5x$

13. $3mn - 5mn + 2mn$

14. $5cd - 8cd + 3cd$

15. $2xy - 5yx + xy$

16. $8mn - 7nm + 3nm$

17. $8x^2y - 2x^2y$

18. $10ab^2 - 3ab^2$

19. $a^2b - 3a^2b$

20. $x^2y^2 - 5x^2y^2$

21. $5ab + 2c - 2ba$

22. $7xy - 3z - 4yx$

23. $5xyz^2 - 2xyz^2 - 4xyz^2$

24. $7a^2bc - 4a^2bc - a^2bc$

25. $5u - 2u + 10v$

26. $8w - 4w + 5v$

27. $8x - 2y - 4x$

28. $9x - 8y + 2x$

29. $7x^2y - 2xy^2 - 4x^2y$

30. $4xy^2 - 5x^2y - 2xy^2$

31. $5x^2 - 3x + 7 - 2x^2 + 8x - 9$

32. $7y^2 + 4y - 6 - 9y^2 - 2y + 7$

33. 12.67 sec + 9.08 sec − 6.73 sec

34. 158.7 ft + 609.5 ft − 421.8 ft − 263.4 ft

Writing Problems

Express the answers in your own words and in complete sentences.

1. Explain why $6x + 10x = 16x$.

2. Explain why $6 + 10x \neq 16x$.

Exercises 2.2B
Set II

In each exercise, combine the like terms.

1. $12a - 4a$

2. $5x - 10x$

3. $18x - 25x$

4. $-4y - 12y$

5. $9z + 14z - 28z$

6. $8x - x + 3x$

7. $3x - 2x + 10x$

8. $7y - 5y - 2y$

9. $7ab + c - 2ba$

10. $8xy - 2xz - 2yz$

11. $6x + x - 9x$

12. $7x + 3x - 10x$

13. $15xy - 3xy + 6xy$

14. $3ab - 2bc + 4ac$

15. $8ab + 2ba - 4ab$

16. $6x - 6 + 4y - 3$

17. $23ab^2 - 17ab^2$

18. $16xy^2 + 5x^2y$

19. $s^3t^2 - 7s^3t^2$

20. $x^2y^4 + 3x^4y^2$

21. $x - 4xy + 3yx$

22. $5a - 4 + 3b - 3$

23. $7xy^2z - 2y^2xz + 5zy^2x$

24. $2x^3 - 2x^2 + 3x - 5x$

25. $5y^2 - 3y^3 + 2y - 4y$

26. $4x - 3y + 7 - 2x + 4 - 6y$

27. $3b - 5a - 9 - 2a + 4 - 5b$

28. $x - 3x^3 + 2x^2 - 5x - 4x^2 + x^3$

29. $y - 2y^2 - 5y^3 - y + 3y^3 - y^2$

30. $a^2b - 11ab + 12ab^2 - 3a^2b + 4ab$

31. $xy^2 + y - 5x^2y + 3xy^2 + x^2y$

32. $4z^2 - 6z + 5 - z^2 + 9z - 10$

33. 37.9 cm − 13.5 cm + 24.8 cm − 19.3 cm

34. 7.2 m − 3.65 m + 8.002 m

2.2C Simplifying an Algebraic Expression

Removing Grouping Symbols

An algebraic expression is not simplified unless all grouping symbols have been removed. The following rules, which result from applications of the distributive property, can be used to remove grouping symbols that contain more than one term. (If there are like terms *within* the grouping symbols, they can be combined *before* these rules are applied.)

1. If a set of grouping symbols containing more than one term is preceded by or followed by a *factor*, use the distributive property (see Examples 8d and 8f).

2. If a set of grouping symbols containing more than one term is *not* preceded by and *not* followed by a factor and is

   a. preceded by no sign at all, drop the grouping symbols (see Examples 8a and 8e).

   b. preceded by a plus sign, drop the grouping symbols if the sign of the first term inside the grouping symbols is an *understood* plus sign (see Example 8b); drop the grouping symbols *and* the plus sign in front of them if the first term inside the grouping symbols has a *written* sign (see Example 8h).

   c. preceded by a minus sign, insert a 1 between the minus sign and the grouping symbol and then use the distributive property (see Examples 8c and 8g).

3. If grouping symbols occur within other grouping symbols, remove the innermost grouping symbols first (see Example 9).

EXAMPLE 8   Remove the grouping symbols in each expression.

a. $(3x - 5) + 2y$

SOLUTION   The parentheses are neither preceded by nor followed by a factor, and they are preceded by no sign at all. (This is an *addition* problem.) Applying Rule 2a, we simply drop the parentheses:

$$(3x - 5) + 2y = 3x - 5 + 2y$$

b. $3x + (4y + 7)$

SOLUTION   The parentheses are preceded by a plus sign and are not followed by a factor; the sign of $4y$ is an *understood* plus sign. (This is an *addition* problem.) Applying Rule 2b, we simply drop the parentheses:

$$3x + (4y + 7) = 3x + 4y + 7$$

c. $-(8 - 6x)$

SOLUTION   The parentheses are neither preceded by nor followed by a factor; they are preceded by a minus sign. (We are to find the *additive inverse* of $8 - 6x$.) Applying Rule 2c, we insert a 1 and then use the distributive property, as follows:

$$-(8 - 6x) = -1[8 + (-6x)] = (-1)(8) + (-1)(-6x) = -8 + 6x$$

Inserting 1 as a *factor* does not change the value of the expression, since $1(8 - 6x) = 8 - 6x$

d. $-2x(4 - 5z)$

SOLUTION   The parentheses are preceded by a factor. (This is a *multiplication* problem.) Using Rule 1, we apply the distributive property:

$$-2x(4 - 5z) = -2x[4 + (-5z)] = (-2x)(4) + (-2x)(-5z) = -8x + 10xz$$

e. $(3x - 5) - y$

SOLUTION   The parentheses are neither preceded by nor followed by a factor. (This is a *subtraction* problem; the y is being subtracted.) Applying Rule 2a, we simply drop the parentheses:

$$(3x - 5) - y = 3x - 5 - y$$

f. $(3x - 5)(-y)$

SOLUTION   The parentheses are followed by a factor. Using Rule 1, we apply the distributive property:

$$(3x - 5)(-y) = [3x + (-5)](-y) = 3x(-y) + (-5)(-y) = -3xy + 5y$$

Note In Example 8e, y is being *subtracted* from $(3x - 5)$, whereas in Example 8f, $(3x - 5)$ is being multiplied by $-y$.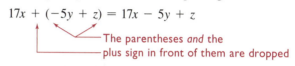

g. $2x - (8 - 6z)$

SOLUTION This is a *subtraction* problem. Applying Rule 2c, we insert a 1 and then use the distributive property, as follows:

$$2x - (8 - 6z) = 2x - 1[8 + (-6z)]$$

$$= 2x + (-1)(8) + (-1)(-6z)$$

$$= 2x - 8 + 6z$$

h. $17x + (-5y + z)$

SOLUTION The parentheses are neither preceded by nor followed by a factor; they are preceded by a plus sign, and the first term inside the parentheses has a written sign. Applying Rule 2b, we drop the parentheses *and* the plus sign in front of them:

$$17x + (-5y + z) = 17x - 5y + z$$

The parentheses *and* the
plus sign in front of them are dropped

When grouping symbols occur within other grouping symbols, it is best to remove the innermost grouping symbols first (see Rule 3 and Example 9).

EXAMPLE 9 Remove the grouping symbols in each expression.

a. $x - [y + (a - b)]$

SOLUTION

$$x - [y + (a - b)] = x - [y + a - b] \qquad \text{Applying Rule 2b}$$

$$= x - 1[y + a + (-b)] \qquad \text{Applying Rule 2c}$$

$$= x + (-1)(y) + (-1)(a) + (-1)(-b) \qquad \text{Using the distributive property}$$

$$= x - y - a + b$$

b. $3 + 2[a - 5(x - 4y)]$

SOLUTION

$$3 + 2[a - 5(x - 4y)]$$

$$= 3 + 2[a - 5x + 20y] \qquad \text{Using the distributive property to remove the parentheses}$$

$$= 3 + 2a - 10x + 40y \qquad \text{Using the distributive property to remove the brackets}$$

c. $(3a - b) - 2\{x - [(y - 2) - z]\}$

SOLUTION

$$(3a - b) - 2\{x - [(y - 2) - z]\}$$

$$= (3a - b) - 2\{x - [y - 2 - z]\} \qquad \text{Using Rule 2a to remove the inner parentheses}$$

$$= (3a - b) - 2\{x - 1[y - 2 - z]\} \qquad \text{Applying Rule 2c}$$

$$= (3a - b) - 2\{x - y + 2 + z\} \qquad \text{Using the distributive property to remove the brackets}$$

$$= 3a - b - 2x + 2y - 4 - 2z \qquad \text{Using Rule 2a to remove the parentheses; using the distributive property to remove the braces}$$

Writing Problems

Express the answers in your own words and in complete sentences.

1. Explain why $8 - (x + y) \neq 8 - x + y$.

2. Explain why $(w + v) - 2 \neq (w + v)(-2)$.

3. Explain why $8 - 6(3a + b) \neq 2(3a + b)$.

Exercises 2.2C
Set II

In Exercises 1–34, remove the grouping symbols.

1. $5 + (x + y)$

2. $-3 + (a - b)$

3. $3 - (x - y)$

4. $-6 - (s + t)$

5. $12 - 2(a + b)$

6. $8 - 2(x - y)$

7. $(a - b) - c$

8. $(s + 3t)(-2u)$

9. $(a - b)(-c)$

10. $(x - y)(-2)$

11. $(8u - 5v)(-7w)$

12. $(x - y) - 2$

13. $(-4x - 3y) - 9z$

14. $(-4x - 3y)(-9z)$

15. $2x - 3(a - b)$

16. $5x - (y + z)$

17. $4(x - 2y) - 3a$

18. $a - (b - c)$

19. $-(x - y) + (2 - a)$

20. $(3 + x) - 3(a - b)$

21. $(a + b)(2) - 6$

22. $2 - 6(a + b)$

23. $x - [a + (y - b)]$

24. $y - [x - (a - b)]$

25. $8 + 4[y - 2(x - b)]$

26. $6 + 7[a - 5(3b - c)]$

27. $15 - 8[x - (y - 3z)]$

28. $14 - 9[a - (b + 4c)]$

29. $x - [-(y - b) + a]$

30. $[8 + 2(a - b) - x] - y$

31. $P - \{x - [y - (4 - z)]\}$

32. $P - \{x - [y - (z - 4)]\}$

33. $-\{-[-(-2 - x) + y]\} - a$

34. $-\{-[-(x - 3) + y] - z\}$

In Exercises 35–58, simplify each algebraic expression.

35. $x + 6(z + 8y + x)$

36. $a - 6(a + 9b - c)$

37. $6x - 3[x - 5(x + 3)]$

38. $11y - 9[y - 3(4 + y)]$

39. $6a - (8a - 3b + 4)$

40. $7x - (4w - 6 + 6y + 13x)$

41. $4x^2 + 2 - (6x + 8 - x^2)$

42. $7y^2 + 6 - (16y^2 + 15 - 4y)$

43. $\sqrt{16z^2}$

44. $\sqrt{36a^8}$

45. $\sqrt{64a^6}$

46. $\sqrt{169x^4y^8}$

47. $\sqrt{x^2y^6}$

48. $\sqrt{400x^4}$

49. $\sqrt{h^8k^4}$

50. $\sqrt{4x^6y^{12}}$

51. $\sqrt{81m^6n^8}$

52. $\sqrt{25a^8b^8c^2}$

53. $\sqrt{144u^{12}v^8}$

54. $\sqrt{x^{14}y^{16}}$

55. $\sqrt{121c^{10}d^6}$

56. $\sqrt{4a^2b^4c^6}$

57. $\sqrt{25r^6s^8t^4}$

58. $\sqrt{121x^{20}y^{30}}$

2.3 Evaluating Algebraic Expressions

In this section, we use our knowledge of the operations on signed numbers to help us find the value of algebraic expressions that contain variables when values for those variables are given.

Evaluating an expression that contains variables

1. Replace each variable by its numerical value, usually enclosing the number within parentheses.
2. Carry out all arithmetic operations, using the correct order of operations.

EXAMPLE 1 Find the value of $3x - 5y$ if $x = 10$ and $y = 4$.

SOLUTION Recall that $3x$ means 3 times x, and $5y$ means 5 times y.

$$3x \quad - 5y$$
$$= 3(10) - 5(4) \quad \text{Replacing each variable with its numerical value}$$
$$= \quad 30 \quad - 20 \quad \text{Carrying out the arithmetic operations}$$
$$= 10$$

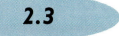

A Word of Caution When we replace a variable with a number, we must enclose the number within parentheses to avoid making the following common errors.

1. Evaluate $3x$ when $x = -2$.

Correct method	*Common errors*
$3x = 3(-2) = -6$	$\cancel{3x = 3 - 2 = 1}$
	$\cancel{3x = 3 - 2 = -6}$

Even though the final answer is correct here, *two* errors have been made! If $x = -2$, $3 - 2$ is not equivalent to $3x$; also, $3 - 2 \neq -6$.

2. Evaluate $4x^2$ when $x = -3$.

Correct method	*Common errors*
$4x^2 = 4(-3)^2 = 4 \cdot 9 = 36$	$\cancel{4x^2 = 4 - 3^2 = 4 - 9 = -5}$
	$\cancel{4x^2 = 4 - 3^2 = 4 + 9 = 13}$
	$\cancel{4x^2 = 4 - 3^2 = 4(9) = 36}$

Even though the final answer is correct here, *two* errors have been made! If $x = -3$, then x^2 is *not* -3^2, it is $(-3)^2$; thus, $4 - 3^2$ is not equivalent to $4x^2$. Furthermore, $4 - 3^2 = 4 - 9$, not $4(9)$.

EXAMPLE 2   Evaluate $3x - 5y$ if $x = -4$ and $y = -6$.

SOLUTION

$$3x \quad - \quad 5y$$

$$= 3(-4) - 5(-6) \quad \text{Replacing } x \text{ with } -4 \text{ and } y \text{ with } -6$$

$$= -12 + 30$$

$$= 18$$

EXAMPLE 3   Find the value of $\dfrac{2a - b}{10c}$ if $a = -1$, $b = 3$, and $c = -2$.

SOLUTION

Remember, this bar is a grouping symbol

$$\frac{2a - b}{10c} = \frac{2(-1) - (3)}{10(-2)} = \frac{-2 - 3}{-20} = \frac{-5}{-20} = +\frac{1 \cdot 5}{4 \cdot 5} = \frac{1}{4}, \text{ or } 0.25$$

EXAMPLE 4   Evaluate $\dfrac{5hgk}{2m}$ if $h = -2$, $g = 3$, $k = -4$, and $m = 6$.

SOLUTION

$$\frac{5hgk}{2m} = \frac{5(-2)(3)(-4)}{2(6)} = \frac{120}{12} = \frac{10 \cdot 12}{1 \cdot 12} = \frac{10}{1} = 10$$

EXAMPLE 5   Find the value of $2a - [b - (3x - 4y)]$ if $a = -3$, $b = 4$, $x = -5$, and $y = 2$.

SOLUTION

$$2a \quad - [b - \quad (3x - 4y)]$$

$$= 2(-3) - \{(4) - [3(-5) - 4(2)]\} \quad \text{Notice that braces and brackets are used to clarify the grouping}$$

$$= 2(-3) - \{4 - [-15 - 8]\}$$

$$= 2(-3) - \{4 - [-23]\}$$

$$= 2(-3) - \{4 + [+23]\} \quad \text{Changing subtraction to addition}$$

$$= \quad -6 \quad - \quad \{+27\}$$

$$= \quad -6 \quad + \quad \{-27\} \quad \text{Changing subtraction to addition}$$

$$= \quad\quad\quad -33$$

EXAMPLE 6   Evaluate $b - \sqrt{b^2 - 4ac}$ when $a = 3$, $b = -7$, and $c = 2$.

SOLUTION

$$b - \sqrt{b^2 - 4ac} \quad \leftarrow \text{This bar is a grouping symbol for } b^2 - 4ac$$

$$= (-7) - \sqrt{(-7)^2 - 4(3)(2)} = (-7) - \sqrt{49 - 24}$$

$$= (-7) - \sqrt{25} = (-7) - 5 = -12$$

> ✖ **A Word of Caution** In Example 6, *two* errors are made if we write
>
> $$b - \sqrt{b^2 - 4ac} = -7 - \sqrt{-7^2 - 4(3)(2)} = -7 - \sqrt{49 - 24}$$
>
> It is incorrect to omit the parentheses around -7 here; furthermore, $-7^2 = -49$, not 49.

EXAMPLE 7

Compare (a) $7 + 3x$ and (b) $10x$ when $x = 5$.

SOLUTION

a. $7 + 3x$ When $x = 5$, $7 + 3x = 7 + 3(5) = 7 + 15 = 22$.
b. $10x$ When $x = 5$, $10x = 10(5) = 50$.

Since $22 \neq 50$, we can see that $7 + 3x \neq 10x$.

Exercises 2.3
Set I

In Exercises 1–28, evaluate each expression when $a = 3$, $b = -5$, $c = -1$, $x = 4$, and $y = -7$.

1. ab

2. xy

3. $a + b$

4. $x + y$

5. $3b$

6. $15b$

7. $9 - 6b$

8. $8 + 7b$

9. b^2

10. $-y^2$

11. $2a - 3b$

12. $3x - 2y$

13. $x - y - 2b$

14. $a - b - 3y$

15. $3b - ab + xy$

16. $4c + ax - by$

17. $x^2 - y^2$

18. $b^2 - c^2$

19. $4 + a(x + y)$

20. $5 - b(a + c)$

21. $2(a - b) - 3c$

22. $3(a - x) - 4b$

23. $3x^2 - 10x + 5$

24. $2y^2 - 7y + 9$

25. $a^2 - 2ab + b^2$

26. $x^2 - 2xy + y^2$

27. $\dfrac{3x}{y + b}$

28. $\dfrac{4a}{c - b}$

In Exercises 29–38, find the value of each expression when $E = -1$, $F = 3$, $G = -5$, $H = -4$, and $K = 0$.

29. $\dfrac{E + F}{EF}$

30. $\dfrac{G + H}{GH}$

31. $\dfrac{(1 + G)^2 - 1}{H}$

32. $\dfrac{1 - (1 + E)^2}{F}$

33. $2E - [F - (3K - H)]$

34. $3H - [K - (4F - E)]$

35. $G - \sqrt{G^2 - 4EH}$

36. $H - \sqrt{H^2 - 4EK}$

🖩 **37.** $\dfrac{\sqrt{2H - 5G}}{0.2F^2}$ (Round off the answer to three decimal places.)

38. $\dfrac{\sqrt{5HG}}{0.5E^2}$

Writing Problems

Express the answer in your own words and in complete sentences.

1. If you're asked to evaluate x^2 when $x = 3$, explain why -3^2 is incorrect notation.

Exercises 2.3
Set II

In Exercises 1–36, evaluate each expression when $a = -2$, $b = 4$, $c = -5$, $x = 3$, and $y = -1$.

1. bc

2. ax

3. $b + c$

4. $a + x$

5. $8y$

6. $-3a$

7. $11 - 3y$

8. $9 - 12a$

9. a^2

10. $-a^2$

11. $4c - 5y$

12. $a - b - 2c$

13. $6x - xy + ab$

14. $c^2 - x^2$

15. $7 - x(a + b)$

16. $3(x - y) - 4c$

17. $b^2 - 4ac$

18. $b^2 - 2bc + c^2$

19. $\dfrac{5b}{x - y}$

20. $\dfrac{a - b}{ab}$

21. $\dfrac{(1 - x)^2 - 1}{y}$

22. $3a - [b - (5x - y)]$

23. $x - \sqrt{x^2 - 4ay}$

24. $\dfrac{b - a}{ab}$

25. $a^2 + 2ab + b^2$

26. $(a + b)^2$

27. $a^2 + b^2$

28. $x^2 - y^2$

29. $(x - y)^2$

30. $x^2 - 2xy + y^2$

31. $\dfrac{a}{b} + \dfrac{c}{x}$

32. $\dfrac{a + c}{b + x}$

33. $\dfrac{a + c}{b}$

34. $\dfrac{a}{b} + \dfrac{c}{b}$

35. $\dfrac{x}{b} - \dfrac{a}{y}$

36. $\dfrac{x - a}{b - y}$

In Exercises 37 and 38, evaluate each expression when $a = -19.32$, $b = 25.73$, and $c = 47.02$. Round off the answers to four decimal places.

37. 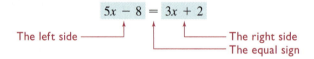 $\dfrac{\sqrt{3b - 4a}}{0.7c^2}$

38. $\dfrac{-b - \sqrt{b^2 - 4ac}}{2a}$

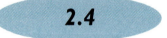

2.4   Using Formulas

In algebra, an **equation** is a statement that two algebraic expressions are equal. The following is an example of an equation:

$$5x - 8 = 3x + 2$$

The left side ——→   ←—— The right side
The equal sign

An equation is made up of three parts:

1. The equal sign ($=$)

2. The expression to the left of the equal sign, called the left side

3. The expression to the right of the equal sign, called the right side

In the equation $5x - 8 = 3x + 2$, the variable x represents an unknown number. Letters other than x may be used in equations to represent unknown numbers.

A **formula** is often an equation that expresses a mathematical or scientific fact. One reason for studying algebra is to prepare for using formulas. People encounter formulas in many real-life situations. In the examples and exercises that follow, we have given the subject areas in which the formulas are used.

To use a formula, we substitute the given values for the variables and simplify the resulting expression. (It is sometimes necessary to do more than this, but in the problems in this section, that will not be the case.) The equal sign is a very important part of the formula, and it must not be dropped or ignored; we write the *full* equation in every step.

EXAMPLE 1   Given the formula $A = \frac{1}{2}bh$, find A when $b = 17$ and $h = 12$.   (Geometry)

SOLUTION   This is the formula for finding the area of a triangle, where A is the area, b is the base, and h is the height, or altitude.

$$A = \frac{1}{2}bh$$

$$A = \frac{1}{2}(17)(12) \quad \text{Substituting 17 for } b \text{ and 12 for } h$$

$$A = 102 \qquad\qquad \text{Notice that we write "} A = \text{" in every step}$$

Therefore, $A = 102$ when $b = 17$ and $h = 12$.

EXAMPLE 2   Given the formula $A = P(1 + rt)$, find A when $P = 1{,}000$, $r = 0.08$, and $t = 1.5$.   (Business)

SOLUTION   This is the formula for finding the amount of money (A) in a savings account at the end of t years, where the interest is simple interest, P is the amount originally invested, and r is the annual interest rate (expressed in decimal form).

$$A = P(1 + rt)$$

$$A = 1{,}000[1 + (0.08)(1.5)] \quad \text{Brackets are used to clarify the grouping}$$

$$A = 1{,}000[1 + 0.12]$$

$$A = 1{,}000[1.12]$$

$$A = 1{,}120$$

Therefore, $A = 1{,}120$ when $P = 1{,}000$, $r = 0.08$, and $t = 1.5$.

EXAMPLE 3   Given the formula $s = \frac{1}{2}gt^2$, find s when $g = 32$ and $t = 5\frac{1}{2}$.   (Physics)

SOLUTION   This is the formula for finding the distance s through which a freely falling object falls in time t, where g is the force due to gravity. When $g = 32$, s must be measured in feet and t must be measured in seconds.

$$s = \frac{1}{2}gt^2$$

$$s = \frac{1}{2}(32)\left(\frac{11}{2}\right)^2 \qquad\qquad 5\frac{1}{2} = \frac{11}{2}$$

$$s = \frac{1}{2}\left(\frac{32}{1}\right)\left(\frac{121}{4}\right)$$

$$s = 484$$

Therefore, $s = 484$ when $g = 32$ and $t = 5\frac{1}{2}$.

EXAMPLE 4   Given the formula $C = \frac{5}{9}(F - 32)$, find C when $F = -13$.   (Science)

SOLUTION   This is the formula for converting temperature from degrees Fahrenheit to degrees Celsius.

$$C = \frac{5}{9}(F - 32)$$

$$C = \frac{5}{9}(-13 - 32)$$

$$C = \frac{5}{9}(-45)$$

$$C = -25$$

Therefore, $C = -25$ when $F = -13$.

Exercises
Set I

Solve the follo
with the probl
mulas from th

1. Water usuall
   ture in degr

2. Tom bought
   tax purposes
   computer be

3. Barbara inv
   simple inter
   the end of 5

Formul

$I = Prt$

$s = \frac{1}{2}gt^2$

$A = P(1$

$A = \pi r^2$

$F = \frac{9}{5}C$

$A = P(1$

$C = \frac{5}{9}(F$

$V = \frac{4}{3}\pi r$

$C = \dfrac{a}{a +}$

$V = C -$

In Exercises 17–22, evaluate each expression when $x = -3$, $y = 5$, and $z = -2$.

17. $4x - y + 2z$

18. $11 - y(x - z)$

19. $3z^2 - 8z + 13$

20. $x^2 + 3xy - y^2$

21. $z - 5[z(x - y) - 2x]$

22. $\dfrac{x^2 - (y - z)^2}{2z - x}$

In Exercises 23–28, solve each problem by using the given formula.

23. $A = \pi r^2$ Find A when $r = 30$ and $\pi \approx 3.14$.

24. $I = Prt$ Find I when $P = 500$, $r = 0.09$, and $t = 1.75$.

25. $C = \frac{5}{9}(F - 32)$ Find C when $F = 59$.

26. $A = P(1 + i)^n$ Find A when $P = 2{,}000$, $n = 4$, and $i = 0.15$.

27. $A = \frac{1}{2}bh$ Find A when $b = 4\frac{1}{2}$ and $h = 1\frac{1}{3}$.

28. $F = \frac{9}{5}C + 32$ Find F when $C = -10$.

In Exercises 29 and 30, solve each problem by using the given formula or by selecting and using one of the formulas listed on the facing page of the inside back cover.

29. If a crowbar is accidentally dropped from the top of a building, how many feet will it have fallen after 3 seconds, assuming that the force of gravity is 32 ft per second²? (When you substitute into the formula $s = \frac{1}{2}gt^2$ and use 32 ft per second² for g, the answer is in feet.)

30. Find the amount of interest earned in three years if $2,500 was invested at 6% simple interest.

17. _____

18. _____

19. _____

20. _____

21. _____

22. _____

23. _____

24. _____

25. _____

26. _____

27. _____

28. _____

29. _____

30. _____

Chapter 2 — DIAGNOSTIC TEST

The purpose of this test is to see how well you understand the basic definitions used in algebra, simplifying algebraic expressions, evaluating algebraic expressions, using formulas, and solving applied problems by using formulas. We recommend that you work this diagnostic test *before* your instructor tests you on this chapter. Allow yourself about 50 minutes.

Complete solutions for all the problems on this test, together with section references, are given in the answer section at the end of the book. We suggest that you study the sections referred to for the problems you do incorrectly.

1. Determine whether $5x$ is a factor of or a term of the expression.

  a. $5xy$ **b.** $5x + y$

  c. $3 + 5x$ **d.** $(3 + 2y) + 5x$

  e. $(3 + 2y)(+5x)$

In Problems 2 and 3, **(a)** determine the number of terms, **(b)** write the second term, and **(c)** write the coefficient of the first term.

2. $6x^3 + \dfrac{7x + 1}{3} - 4y$

3. $-y^4 + 3x - 2z + 4$

In Problems 4 and 5, remove the grouping symbols.

4. $5 - (x - y)$

5. $-2[-4(3c - d) + a] - b$

In Problems 6 and 7, combine like terms.

6. $4x - 3x + 5x$

7. $2a - 5b - 7 - 3b + 4 - 5a$

In Problems 8–11, write each expression in simplest form.

8. $5x - 3(y - x)$

9. $(x - 4)(-5)$

10. $(x - 4) - 5$

11. $3 + 5[3x - (2y - x)]$

In Problems 12–15, evaluate each expression when $a = -4, b = -7, c = -2, x = -6,$ and $y = 5$.

12. $3c - by + cx$

13. $4x - [a - (3c - b)]$

14. $x^2 + 2xy + y^2$

15. $(x + y)^2$

In Problems 16–19, solve each problem by using the given formula.

16. $C = \frac{5}{9}(F - 32)$ Find C when $F = 68$.

17. $A = \pi r^2$ Find A when $\pi \approx 3.14$ and $r = 3$.

18. $A = P(1 + rt)$ Find A when $P = 500, r = 0.10,$ and $t = 3.5$.

19. $V = C - Crt$ Find V when $C = 800, r = 0.06,$ and $t = 10$.

20. Solve the following problem by using the formula $V = \frac{1}{3}\pi r^2 h$, where V is the volume of a right circular cone, r is the radius of its base, h is its altitude, and $\pi \approx 3.14$.

Find the volume of a right circular cone if its radius is 4 ft and its altitude is 3 ft.

Chapters 1 and 2 — CUMULATIVE REVIEW EXERCISES

In Exercises 1–27, perform the indicated operations. If an operation cannot be performed, give a reason.

1. $-16(-2)$ **2.** $-13 + (-8)$

3. $-5 - (-11)$ **4.** $28 \div (-7)$

5. $-15(0)(4)$ **6.** $(-2)^3$

7. $-27 + (10)$ **8.** $\sqrt{16}$

9. $\dfrac{-48}{-12}$ **10.** $0 \div (-6)$

11. $(25)^2$ **12.** $-15(3)$

13. -2^4 **14.** Subtract -9 from -13.

15. $\dfrac{-12}{0}$ **16.** $\sqrt[3]{8}$

17. $\sqrt{64}$ **18.** 0^4

19. $8 - 17$ **20.** $-20 - 9$

21. $24 \div 12 \cdot 2$ **22.** $6 \cdot 2^2$

23. $8 - 6 \cdot 5$ **24.** $64 \div 16 \div 4$

25. $17 - 9 - 2$ **26.** $4 + 17 \times 2$

27. $0 \div 5$

28. Using the formula $C = \frac{5}{9}(F - 32)$, find C when $F = 21\frac{1}{2}$.

In Exercises 29 and 30, write each expression in simplest form.

29. $5x - (z - 2x)$

30. $8 - 3(x - 4 + y)$

In Exercises 31–40, write "true" if the statement is always true. Otherwise, write "false."

31. 1.73 is a rational number.

32. 2/3 is a real number.

33. 1.86235173 ... (never terminates or repeats) is an irrational number.

34. 0 is an irrational number.

35. 0 is a real number.

36. $-\frac{3}{5}$ is the additive inverse of $\frac{3}{5}$.

37. 0 is the multiplicative identity.

38. $52y$ is a term of $6x + 52y$.

39. $4xy$ is a term of $4xyz$.

40. $6s$ is a factor of $6s + 3t$.

Critical Thinking and Problem-Solving Exercises

1. Can you predict what the next four symbols should be for the following list?   A, C, 1, 3, E, G, 5, 7, ____ , ____ ,
____ , ____

2. How many different three-digit numbers are there that are less than 300, that have an even digit in the units' place, and that have a 1, 3, 5, or 8 in the tens' place?

3. Maggie, Loretta, and Brenda are friends. One of them is a chemist, one is a professor of German, and one is an architect. Their last names are Maiolo, Herrin, and Murphy. By using the following clues, can you match the first name, last name, and profession of each of these three women?

> Ms. Murphy is a chemist.
>
> Ms. Maiolo and the German professor had lunch together.
>
> Maggie is not a chemist.
>
> Brenda is an architect.

4. Jenna's 4-year-old son was petting the ducks and goats at a petting zoo; he told Jenna that he had petted 7 animals (ducks and goats only) and he had counted 22 feet on those 7 animals. How many goats did he pet?

5. Suppose your instructor asked you to explain to a student who had missed class what the *terms* of an algebraic expression are. What would your explanation be?

6. The solution of the following problem contains an error. Find and describe the error.
Simplify $x - 2[x - (4 - x) - 8]$.

$$x - 2[x - (4 - x) - 8]$$
$$= x - 2[x - 4 + x - 8]$$
$$= x - 2[2x - 12]$$
$$= x - 4x - 24$$
$$= -3x - 24$$

Solving Equations and Inequalities

CHAPTER 3

The main reason for studying algebra is to equip oneself with the tools necessary for solving problems. Most mathematical problems are solved by the use of equations. In this chapter, we show how to solve simple equations. Methods for solving more difficult equations will be given in later chapters.

3.1 Conditional Equations; Solutions and Solution Sets; Equivalent Equations

As we have mentioned, an equation is a statement that two quantities are equal; it is made up of three parts:

1. The equal sign (=)
2. The expression to the left of the equal sign, called the left side
3. The expression to the right of the equal sign, called the right side

Conditional Equations

An equation may be a *true* statement, such as $7 = 7$; a *false* statement, such as $8 = 3$; or a statement that is sometimes true and sometimes false, such as $x = -2$. An equation that is true for some values of the variable and false for other values is called a **conditional equation**. The equation $x = -2$ is a conditional equation, because it is a true statement if the value of x is -2 and a false statement otherwise.

A Solution of an Equation

A **solution** of an equation is any value of the variable that, when substituted for the variable, makes the two sides of the equation equal. That is, the *solution* makes the equation a true statement.

EXAMPLE 1   Determine whether -2 is a solution of the equation $x = -2$.

SOLUTION   Substituting -2 for x, we have $-2 = -2$, which is a *true* statement. Therefore, -2 is a solution of the equation $x = -2$.

EXAMPLE 2   Determine whether -2 is a solution of the equation $x + 6 = 4$, and determine whether 5 is a solution of that same equation.

SOLUTION   Substituting -2 for x in the equation $x + 6 = 4$, we have

$$(-2) + 6 = 4, \quad \text{or} \quad 4 = 4$$

which is a *true* statement. Therefore, -2 *is* a solution of the equation $x + 6 = 4$.
    Substituting 5 for x, we have

$$(5) + 6 = 4, \quad \text{or} \quad 11 = 4$$

which is a *false* statement. Therefore, 5 is *not* a solution of the equation $x + 6 = 4$.

The Solution Set of an Equation

The **solution set** of an equation is the set of *all* the solutions of the equation. (Recall that we enclose the elements of a set within braces.) If we consider the equation

$x + 6 = 4$ again, we can conclude that -2 is, in fact, the *only* value of x that makes the statement $x + 6 = 4$ true; that is, -2 is the *only* number that we can add to 6 to get a sum of 4. Therefore, the solution set of $x + 6 = 4$ is $\{-2\}$. Similarly, the solution set of the equation $x = -2$ is $\{-2\}$.

Equivalent Equations

Equations that have the same solution set are called **equivalent equations**.

EXAMPLE 3   Examples of equivalent equations:

a. The solution set of the equation $x + 2 = 0$ is $\{-2\}$, because $(-2) + 2 = 0$ is a true statement, and -2 is the only value of x that makes the statement $x + 2 = 0$ true. Since $\{-2\}$ is also the solution set for the equations $x + 6 = 4$ and $x = -2$, the equations $x + 2 = 0$, $x + 6 = 4$, and $x = -2$ are all equivalent equations.

b. The solution set of the equation $x = 4$ is $\{4\}$, because $4 = 4$ is a true statement, and 4 is the only value of x that makes the statement $x = 4$ true. The solution set of the equation $x - 4 = 0$ is $\{4\}$, because $4 - 4 = 0$ is a true statement, and 4 is the only value of x that makes the statement $x - 4 = 0$ true. Therefore, $x = 4$ and $x - 4 = 0$ are equivalent equations.

Exercises 3.1
Set I

1. Is -4 a solution of the equation $x + 2 = 3$?

2. Is 3 a solution of the equation $x - 5 = 2$?

3. Is 4 a solution of the equation $x + 1 = 5$?

4. Is -4 a solution of the equation $x + 4 = 0$?

5. Is -3 in the solution set for $2 + x = -1$?

6. Is -5 in the solution set for $2 + x = -3$?

7. Is 3 in the solution set for $5 + x = 1$?

8. Is 4 in the solution set for $-1 + x = 1$?

Writing Problems

Express the answers in your own words and in complete sentences.

1. Explain why -2 is a solution for the equation $x + 14 = 12$.

2. Explain why 3 is in the solution set for the equation $-2 - x = -5$.

3. Explain why $x + 14 = 12$ and $x + 2 = 0$ are equivalent equations.

Exercises 3.1
Set II

1. Is 5 a solution of the equation $x + 3 = 4$?

2. Is -3 a solution of the equation $x + 5 = 2$?

3. Is -4 a solution of the equation $x + 1 = -3$?

4. Is 4 a solution of the equation $x + 3 = 7$?

5. Is 3 in the solution set for $2 + x = 5$?

6. Is 5 in the solution set for $2 - x = -3$?

7. Is -1 in the solution set for $5 + x = 1$?

8. Is 2 in the solution set for $-1 + x = 1$?

3.2 Solving Equations by Using the Addition and Subtraction Properties of Equality

In this chapter, we solve *first-degree* (or *linear*) *equations* in one variable. These are equations that contain only one variable and that have no exponents on that variable; that is, the exponent on the variable is understood to be 1.

Isolating the Variable   To **isolate the variable** in an equation means to find an equivalent equation that is in the form $x = k$, where k is some number (and where x represents the variable). That is, our goal is to get the variable by itself on one side of the equal sign and a single number (no variables) on the other side; we want the coefficient of the variable to be an understood 1.

We can use a number of properties of equality in finding an equation equivalent to the given equation. We list three of those properties here, and we will list two others in a later section.

The symmetric property of equality	For all real numbers a and b, $$\text{if} \quad a = b, \quad \text{then} \quad b = a$$ *In words*: If the two sides of an equation are interchanged, the new equation is equivalent to the original equation.

The symmetric property permits us to rewrite an equation such as $4 = x$ as $x = 4$. (The equation $x = 4$ is in the *form* $x = k$.)

The addition property of equality	For all real numbers a, b, and c, $$\text{if} \quad a = b, \quad \text{then} \quad a + c = b + c$$ *In words*: If the same number is added to both sides of an equation, the new equation is equivalent to the original equation.

The subtraction property of equality	For all real numbers a, b, and c, $$\text{if} \quad a = b, \quad \text{then} \quad a - c = b - c$$ *In words*: If the same number is subtracted from both sides of an equation, the new equation is equivalent to the original equation.

 Note   It is possible to solve all first-degree equations without using the subtraction property of equality. (This is so because subtraction is the inverse operation of addition.) In fact, we do not show the use of the subtraction property in this book, but some instructors do use it.

 Note   If an equation is in the form $x + a = b$, there are *two terms* on the side of the equation that contains the x, and the coefficient of x is an understood 1.

Solving Equations of the Form $x + a = b$

In this section, we solve equations of the form $x + a = b$ (or $a + x = b$) by reasoning as follows: Because $x + a$ means a is being *added* to x, and because $a + (-a) = 0$, we can isolate x by using the addition property of equality and *adding* $-a$ (the *additive inverse* of a) to both sides of the equation (see Example 1).

Alternatively, because subtraction is the inverse operation of addition, we could use the subtraction property of equality on the equation $x + a = b$ and isolate x by *subtracting* a from both sides of the equation.

When we solve equations by using the properties of equality, we must *always write the complete new equation under the previous equation*. When we use the addition property of equality, the addition can be shown horizontally or vertically; after some practice, many students can do the addition mentally. We will show both horizontal and vertical addition in this section.

We call any solution that we obtain by using the properties of equality an *apparent solution* because, although it *appears* to be a solution of the equation, we may have made errors in obtaining it.

EXAMPLE 1

Solve the equation $2 + x = 10$, and graph its solution on the number line.

SOLUTION Since $2 + x$ means 2 is being added to x, we can isolate x by adding -2 (the additive inverse of 2) to both sides of the equation (or, equivalently, by subtracting 2 from both sides). We first show the addition being done horizontally.

$$2 + x = 10$$

$$(-2) + 2 + x = (-2) + 10 \quad \text{Adding } -2 \text{ to both sides}$$

$$0 \quad + x = 8 \qquad \text{The additive inverse property: } (-2) + 2 = 0$$

$$x = 8 \qquad \text{The additive identity property: } 0 + x = x$$

8 is the apparent solution

Do not omit the equal sign in any step

Notice that the equation $(-2) + 2 + x = (-2) + 10$ was written *under* the equation $2 + x = 10$, that $0 + x = 8$ was written *under* $(-2) + 2 + x = (-2) + 10$, and that $x = 8$ was written *under* $0 + x = 8$.

If we set up the problem vertically, *we must align only like terms in each column:*

2 and -2 are like terms; -2 is written directly under 2

10 and -2 are like terms; -2 is written directly under 10

$$
\begin{aligned}
2 + x &= 10 \\
-2 & -2 \quad \text{Adding } -2 \text{ to both sides (or subtracting 2 from both sides)} \\
0 + x &= 8 \longleftarrow \text{This step is usually not shown} \\
x &= 8 \quad 0 + x = x
\end{aligned}
$$

8 is the apparent solution

Do not omit the equal sign

Note If we made no errors, the solution is 8, and the *solution set* is $\{8\}$; $x = 8$ is not the solution—it is an equation that is equivalent to the equation $2 + x = 10$.

Graph We graph the number 8.

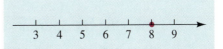

A Word of Caution   A common error is to add 10 or -10 to both sides of the equation in Example 1. We show here that performing either of these additions does *not* lead to a solution of the equation. Although the addition property of equality *does* permit us to add 10 to both sides of the equation, this will *not* isolate x.

Addition shown horizontally

$2 + x = 10$
$10 + 2 + x = 10 + 10$
$12 + x = 20$

Addition shown vertically

$2 + x = 10$
$\underline{10 \qquad 10}$
$12 + x = 20$

— We do not have x by itself on one side

Similarly, the addition property of equality permits us to add -10 to both sides of the equation; however, this will *not* isolate x either.

$2 + x = 10$
$(-10) + 2 + x = (-10) + 10$
$-8 + x = 0$

$2 + x = 10$
$\underline{-10 \qquad -10}$
$-8 + x = 0$

We do not have x by itself on one side

Furthermore, in Example 1, it is incorrect to write

$2 + x = 10 = x = 8$

— An equal sign here implies that $x = 10$, which is *not* true, and also implies that $10 = 8$, which is false

EXAMPLE 2   Solve the equation $8 = H - 4$, and graph its solution on the number line.

SOLUTION   Since $H - 4$ means -4 is being added to H, we can isolate H by adding $+4$ (the additive inverse of -4) to both sides of the equation. Adding $+4$ to both sides gets H by itself on the right side of the equation.

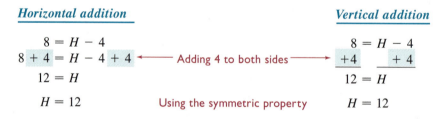

Horizontal addition

$8 = H - 4$
$8 + 4 = H - 4 + 4$ ⟵ Adding 4 to both sides ⟶
$12 = H$

$H = 12$ Using the symmetric property

Vertical addition

$8 = H - 4$
$\underline{+4 \qquad + 4}$
$12 = H$

$H = 12$

Graph   We graph the apparent solution, 12.

Checking the Solution of an Equation

We cannot be sure that an apparent solution is *actually* a solution until we have checked it.

Checking the solution of an equation

1. Replace the variable in the given equation by the apparent solution.
2. Perform the indicated operations on both sides of the equal sign.
3. If the resulting numbers on both sides of the equal sign are the same—that is, if we get a *true* statement—the solution is correct.

EXAMPLE 3

Solve and check $x - 5 = 3$, and graph the solution on the number line.

SOLUTION We isolate x by adding $+5$ (the additive inverse of -5) to both sides of the equation.

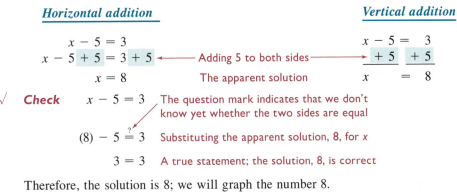

Horizontal addition

$$x - 5 = 3$$
$$x - 5 + 5 = 3 + 5 \quad \longleftarrow \text{Adding 5 to both sides}$$
$$x = 8 \qquad \text{The apparent solution}$$

Vertical addition

$$\begin{array}{rr} x - 5 = & 3 \\ +5 & +5 \\ \hline x = & 8 \end{array}$$

√ **Check** $x - 5 = 3$ The question mark indicates that we don't know yet whether the two sides are equal

$$(8) - 5 \overset{?}{=} 3 \quad \text{Substituting the apparent solution, 8, for } x$$

$$3 = 3 \quad \text{A true statement; the solution, 8, is correct}$$

Therefore, the solution is 8; we will graph the number 8.

Graph

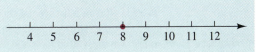

EXAMPLE 4

Solve and check $\frac{1}{4} + x = 3\frac{1}{4}$, and graph the solution on the number line.

SOLUTION We isolate x by adding $-\frac{1}{4}$ (the additive inverse of $\frac{1}{4}$) to both sides of the equation. (We could, instead, subtract $\frac{1}{4}$ from both sides.)

Horizontal addition

$$\frac{1}{4} + x = 3\frac{1}{4}$$
$$\left(-\frac{1}{4}\right) + \frac{1}{4} + x = \left(-\frac{1}{4}\right) + 3\frac{1}{4} \quad \text{Adding } -\frac{1}{4} \text{ to both sides}$$
$$x = 3 \qquad \text{The apparent solution}$$

Vertical addition

$$\begin{array}{rr} \frac{1}{4} + x = & 3\frac{1}{4} \\ -\frac{1}{4} & -\frac{1}{4} \\ \hline x = & 3 \end{array}$$

√ **Check** $\frac{1}{4} + x = 3\frac{1}{4}$

$$\frac{1}{4} + 3 \overset{?}{=} 3\frac{1}{4} \quad \text{Substituting the apparent solution, 3, for } x$$

$$3\frac{1}{4} = 3\frac{1}{4} \quad \text{A true statement}$$

Therefore, the solution is 3, and we will graph the number 3.

Graph

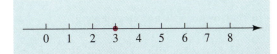

EXAMPLE 5

Solve and check $9.08 = x - 5.47$, and graph the solution on the number line.

SOLUTION We isolate x by adding 5.47 (the additive inverse of -5.47) to both sides of the equation.

Horizontal addition

$$9.08 = x - 5.47$$
$$9.08 + 5.47 = x - 5.47 + 5.47 \quad \text{Adding 5.47 to both sides}$$
$$14.55 = x$$

$$x = 14.55 \quad \text{Using the symmetric property}$$

Vertical addition

$$9.08 = x - 5.47$$
$$\underline{+5.47 \qquad +5.47}$$
$$14.55 = x$$

$$x = 14.55$$

✓ **Check** $\quad 9.08 = x - 5.47$

$$9.08 \overset{?}{=} 14.55 - 5.47 \quad \text{Substituting 14.55 for } x$$

$$9.08 = 9.08 \quad \text{A true statement}$$

Therefore, the solution is 14.55.

Graph

Notice that in all the equations in this section, the coefficient of the variable was always an understood 1 and some number was being added to or subtracted from the variable; we were able to isolate the variable by using the addition property of equality or the subtraction property of equality.

Exercises 3.2
Set I

Solve and check the following equations. For Exercises 1–10, graph the solutions on the number line.

1. $x + 5 = 8$
2. $x + 4 = 9$
3. $x - 3 = 4$
4. $x - 7 = 2$
5. $3 + x = -4$
6. $2 + x = -5$
7. $x + 4 = 21$
8. $x + 15 = 24$
9. $x - 35 = 7$
10. $x - 42 = 9$
11. $9 = x + 5$
12. $11 = x + 8$
13. $12 = x - 11$
14. $14 = x - 15$
15. $-17 + x = 28$
16. $-14 + x = 33$
17. $-28 = -15 + x$
18. $-47 = -18 + x$
19. $x + \frac{1}{2} = 2\frac{1}{2}$
20. $x + \frac{3}{4} = 5\frac{3}{4}$
21. $5.6 + x = 2.8$
22. $3.04 + x = 2.96$
23. $7.84 = x - 3.98$
24. $4.99 = x - 2.08$

Writing Problems

Express the answers in your own words and in complete sentences.

1. Explain how to isolate the variable in order to solve the equation $x - 11 = -3$.

2. Explain how to isolate the variable in order to solve the equation $x + 8 = 2$.

3. Explain why $x + 2 = 8$ and $x + 9 = 15$ are equivalent equations.

Exercises 3.2
Set II

Solve and check the following equations. For Exercises 1–10, graph the solutions on the number line.

1. $x + 7 = 12$

2. $x - 4 = -3$

3. $x - 5 = 8$

4. $3 + x = 1$

5. $6 + x = -9$

6. $-2 + x = 5$

7. $x + 11 = 25$

8. $-5 + x = 0$

9. $x - 18 = 13$

10. $5 + x = 5$

11. $14 = x + 6$

12. $3 = x - 1$

13. $17 = x - 11$

14. $x = 5 - 8$

15. $-21 + x = -41$

16. $-3 + x = 4$

17. $-51 = -37 + x$

18. $-3 + x = -4$

19. $x + 2\frac{1}{4} = 8\frac{1}{4}$

20. $x - 3\frac{1}{2} = 5\frac{1}{2}$

21. $8.4 + x = 6.2$

22. $3.5 + x = 1.07$

23. $5.36 = x - 4.82$

24. $x - 1.35 = -4.2$

3.3 Solving Equations by Using the Multiplication and Division Properties of Equality

3.3A Solving Equations of the Form $ax = b$, $\dfrac{x}{b} = c$, and $\dfrac{ax}{b} = c$

In this section, we solve equations such as $2x = 10$, $\dfrac{x}{3} = -5$, and $\dfrac{2x}{3} = -4$; there is still only one term that contains the variable in each equation, but the numerical coefficient of the variable is no longer a 1. We solve such equations by using one of the following two properties of equality.

The division property of equality	For all real numbers a, b, and c, but $c \neq 0$, $$\text{if} \quad a = b, \quad \text{then} \quad \frac{a}{c} = \frac{b}{c}$$ *In words:* If both sides of an equation are divided by the same nonzero number,* the new equation is equivalent to the original equation. _____ *We cannot divide both sides of an equation by zero, since division by zero is not permitted.

The multiplication property of equality	For all real numbers a, b, and c, but $c \neq 0$, $$\text{if} \quad a = b, \quad \text{then} \quad ac = bc$$ *In words:* If both sides of an equation are multiplied by the same nonzero number,* the new equation is equivalent to the original equation. _____ *If we multiply both sides of an equation by zero, we always get the equation $0 = 0$, which is usually not equivalent to the original equation.

Note If an equation is in the form $ax = b$, $\dfrac{x}{b} = c$, or $\dfrac{ax}{b} = c$, there is just *one* term on the side of the equation that contains the x. That *term* contains two or more *factors*, and the coefficient of x is *not* an understood 1.

Recall that the fundamental property of fractions is $\dfrac{a}{b} = \dfrac{ac}{bc}$. Also recall that the rule for multiplying fractions is $\dfrac{a}{b} \cdot \dfrac{c}{d} = \dfrac{ac}{bd}$. We will be using these properties frequently in the remainder of this chapter.

Solving Equations of the Form $ax = b$

We can solve equations of the form $ax = b$ ($a \neq 0$) by reasoning as follows: Since ax means that a is being *multiplied* by x, and since division is the inverse operation of multiplication, we can isolate x by using the division property of equality and dividing both sides of the equation by a. Dividing ax by a and using the fundamental property of fractions gives

$$\frac{ax}{a} = \frac{x \cdot a}{1 \cdot a} = \frac{x}{1} = x$$

Therefore, the coefficient of x will become 1 (see Example 1, Solution 1).

Alternatively, because the product of a number and its multiplicative inverse (its reciprocal) is 1, we can use the multiplication property of equality and isolate x by multiplying both sides of the equation by $1/a$, the multiplicative inverse (or reciprocal) of a (see Example 1, Solution 2).

EXAMPLE 1 Solve and check the equation $2x = 10$, and graph the solution on the number line.

SOLUTION 1 Using the division property of equality: We isolate x by dividing both sides of the equation by 2, the coefficient of x.

$$2x = 10 \qquad \text{We will divide both sides by 2}$$

$$\frac{2x}{2} = \frac{10}{2} \qquad \text{On the left side, } \frac{2x}{2} = \frac{x \cdot 2}{1 \cdot 2} = \frac{x}{1} = x$$

$$x = 5 \qquad \text{The apparent solution is 5}$$

SOLUTION 2 Using the multiplication property of equality: We isolate x by multiplying both sides of the equation by $\frac{1}{2}$, the multiplicative inverse (or reciprocal) of the coefficient of x.

$$2x = 10 \qquad \text{Recall that } 2 = \tfrac{2}{1} \text{ and that } \left(\tfrac{1}{2}\right)\left(\tfrac{2}{1}\right) = 1$$

$$\left(\tfrac{1}{2}\right)(2)x = \left(\tfrac{1}{2}\right)(10) \qquad \text{On the left side, } \left(\tfrac{1}{2}\right)\left(\tfrac{2}{1}\right)x = 1x = x$$

$$x = 5 \qquad \text{The apparent solution is 5}$$

✓ **Check** $2x = 10$

$$2(5) \overset{?}{=} 10 \qquad \text{Substituting 5 for } x$$

$$10 = 10 \qquad \text{A true statement}$$

Therefore, the solution is 5.

Graph

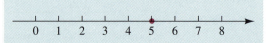

> ✕ **A Word of Caution** Notice the difference between the equations (a) $2x = 10$ and (b) $2 + x = 10$. In (a), 2 and x are *factors*; that is, 2 is multiplied by x. We isolate x by dividing both sides of the equation by 2 (or by multiplying both sides of the equation by $\frac{1}{2}$). In (b), 2 and x are *terms*; that is, 2 is added to x. We isolate x by adding -2 to both sides of the equation (or by subtracting 2 from both sides of the equation).

EXAMPLE 2 Solve and check the equation $9x = -27$, and graph the solution on the number line.

SOLUTION

Using the division property	*Using the multiplication property*
We isolate x by dividing both sides by 9, the coefficient of x	We isolate x by multiplying both sides by $\frac{1}{9}$, the reciprocal of 9
$9x = -27$	$9x = -27$
$\dfrac{9x}{9} = -\dfrac{27}{9}$	$\left(\frac{1}{9}\right)(9x) = \left(\frac{1}{9}\right)(-27)$
$x = -3$	$x = -3$

The apparent solution

✓ **Check** $9x = -27$

$9(-3) \overset{?}{=} -27$ Substituting -3 for x

$-27 = -27$ A true statement

Therefore, the solution is -3.

Graph

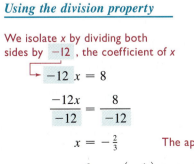

EXAMPLE 3 Solve and check the equation $-12x = 8$, and graph the solution on the number line.

SOLUTION

Using the division property	*Using the multiplication property*
We isolate x by dividing both sides by -12, the coefficient of x	We isolate x by multiplying both sides by $-\frac{1}{12}$, the reciprocal of -12
$-12x = 8$	$-12x = 8$
$\dfrac{-12x}{-12} = \dfrac{8}{-12}$	$\left(-\frac{1}{12}\right)(-12x) = \left(-\frac{1}{12}\right)(8)$
$x = -\frac{2}{3}$	$x = -\frac{2}{3}$

The apparent solution

$\left(\text{You can verify that } \frac{8}{-12} \text{ and } \left(-\frac{1}{12}\right)(8) \text{ both reduce to } -\frac{2}{3}.\right)$

√ **Check**

$$-12x = 8$$

$$-12\left(-\tfrac{2}{3}\right) \overset{?}{=} 8 \quad \text{Substituting } -\tfrac{2}{3} \text{ for } x$$

$$\tfrac{24}{3} \overset{?}{=} 8$$

$$8 = 8 \quad \text{A true statement}$$

Therefore, the solution is $-\tfrac{2}{3}$.

Graph

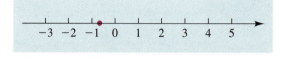

✕ | **A Word of Caution**   In solving the equation $-12x = 8$, a common error is to divide both sides of the equation by 8; this does *not* lead to a solution of the equation, as we see here:

$$\frac{-12x}{8} = \frac{8}{8}, \quad \text{or} \quad \frac{-12x}{8} = 1 \qquad \begin{array}{l}\text{We have not isolated } x; \text{ this}\\ \text{equation has } not \text{ been solved for } x\end{array}$$

Another common error is to divide both sides of the equation by $-12x$. This does *not* lead to a solution of the equation:

$$\frac{-12x}{-12x} = \frac{8}{-12x}, \quad \text{or} \quad 1 = \frac{8}{-12x} \qquad \text{This equation has } not \text{ been solved for } x$$

A third common error is to divide the left side by $-12x$ and the right side by -12; this, of course, is *incorrect*.

$$\frac{-12x}{-12x} = \frac{8}{-12} \qquad \begin{array}{l}\text{It is incorrect to divide one side by}\\ -12x \text{ and the other side by } -12\end{array}$$

For the remainder of this section, we will assume that a and b represent constant real numbers and that $a \ne 0$ and $b \ne 0$. Recall that we can consider the coefficient of $\dfrac{x}{b}$ to be $\dfrac{1}{b}$ and that we can consider the coefficient of $\dfrac{ax}{b}$ to be $\dfrac{a}{b}$. For example, the coefficient of $\dfrac{x}{3}$ is $\dfrac{1}{3}$, and the coefficient of $\dfrac{2x}{5}$ is $\dfrac{2}{5}$.

Solving Equations of the Form $\dfrac{x}{b} = c$

We can solve equations of the form $\dfrac{x}{b} = c$ by using the multiplication property of equality and multiplying both sides of the equation by b, the multiplicative inverse (or reciprocal) of $\dfrac{1}{b}$. Multiplying $\dfrac{x}{b}$ by b and using the fundamental property of fractions gives

$$\frac{b}{1} \cdot \frac{x}{b} = \frac{b \cdot x}{1 \cdot b} = \frac{x \cdot b}{1 \cdot b} = \frac{x}{1} = x$$

Therefore, the coefficient of x will become 1 (see Example 4).

EXAMPLE 4

Solve and check $\dfrac{x}{3} = -5$, and graph the solution on the number line.

SOLUTION The coefficient of x is $\dfrac{1}{3}$; that is, $\dfrac{x}{3} = \dfrac{1}{3}x$.

$$\dfrac{x}{3} = -5 \qquad \text{We will multiply both sides by 3, the reciprocal of } \tfrac{1}{3}$$

$$3\left(\dfrac{x}{3}\right) = 3\,(-5) \qquad \text{On the left side, } \left(\dfrac{3}{1}\right)\left(\dfrac{x}{3}\right) = \dfrac{3 \cdot x}{1 \cdot 3} = \dfrac{x \cdot 3}{1 \cdot 3} = \dfrac{x}{1} = x$$

$$x = -15 \qquad \text{The apparent solution}$$

✓ **Check** $\qquad \dfrac{x}{3} = -5$

$$\dfrac{-15}{3} \overset{?}{=} -5 \qquad \text{Substituting } -15 \text{ for } x$$

$$-5 = -5 \qquad \text{A true statement}$$

The solution is -15.

Graph

EXAMPLE 5

Solve and check $-8 = \dfrac{x}{6}$, and graph the solution on the number line.

SOLUTION The coefficient of x is $\dfrac{1}{6}$; that is, $\dfrac{x}{6} = \dfrac{1}{6}x$.

$$-8 = \dfrac{x}{6} \qquad \text{We will multiply both sides by 6, the reciprocal of } \tfrac{1}{6}$$

$$6\,(-8) = 6\left(\dfrac{x}{6}\right) \qquad \text{On the right side, } \left(\dfrac{6}{1}\right)\left(\dfrac{x}{6}\right) = \dfrac{6 \cdot x}{1 \cdot 6} = \dfrac{x \cdot 6}{1 \cdot 6} = \dfrac{x}{1} = x$$

$$-48 = x \qquad \text{Next, we must use the symmetric property}$$

$$x = -48 \qquad \text{The apparent solution}$$

✓ **Check** $\qquad -8 = \dfrac{x}{6}$

$$-8 \overset{?}{=} \dfrac{-48}{6} \qquad \text{Substituting } -48 \text{ for } x$$

$$-8 = -8 \qquad \text{A true statement}$$

The solution is -48.

Graph

Solving Equations of the Form $\dfrac{ax}{b} = c$

We can solve equations of the form $\dfrac{ax}{b} = c$ by using the multiplication property of equality and multiplying both sides of the equation by $\dfrac{b}{a}$, the reciprocal of $\dfrac{a}{b}$. The product of $\dfrac{a}{b}$ and $\dfrac{b}{a}$ is 1, so the coefficient of x will become 1 (see Example 6, Solution 1).

Because division is the inverse operation of multiplication, it is also possible to *divide* both sides of the equation by $\dfrac{a}{b}$; we don't demonstrate that method in this book.

Equations of the form $\dfrac{ax}{b} = c$ can also be solved in two separate steps. To use the two-step method (which will be demonstrated further in a later chapter), we first "clear fractions" by multiplying both sides of the equation by b; we then use the methods presented earlier in this section to finish solving the equation (see Example 6, Solution 2). We recommend that you become comfortable with both methods.

EXAMPLE 6   Solve and check $\dfrac{2x}{3} = -4$, and graph the solution on the number line.

SOLUTION 1   The coefficient of x is $\dfrac{2}{3}$; that is, $\dfrac{2x}{3} = \dfrac{2}{3}x$.

$$\dfrac{2x}{3} = -4 \qquad \text{We will multiply both sides by } \tfrac{3}{2}, \text{ the reciprocal of } \tfrac{2}{3}$$

$$\left(\dfrac{3}{2}\right)\left(\dfrac{2}{3}\right)x = \left(\dfrac{3}{2}\right)(-4) \qquad \text{On the left side, } \left(\dfrac{3}{2}\right)\left(\dfrac{2x}{3}\right) = \left(\dfrac{3}{2}\right)\left(\dfrac{2}{3}\right)x = \text{1}x = x$$

$$x = -6 \qquad \text{The apparent solution}$$

SOLUTION 2   Using the two-step method, we have

$$\dfrac{2x}{3} = -4 \qquad \text{We will clear fractions by multiplying both sides by 3}$$

$$(3)\left(\dfrac{2x}{3}\right) = (3)(-4) \qquad \text{On the left side, } (3)\left(\dfrac{2x}{3}\right) = \left(\dfrac{3}{1}\right)\left(\dfrac{1}{3}\right)(2x) = \text{1}(2x) = 2x$$

$$2x = 12 \qquad \text{This equation is in the form } ax = b$$

$$\dfrac{2x}{2} = \dfrac{-12}{2} \qquad \text{Dividing both sides by 2} \left(\text{we could, instead, multiply both sides by } \tfrac{1}{2}\right)$$

$$x = -6 \qquad \text{The apparent solution}$$

✓ **Check**
$$\dfrac{2x}{3} = -4$$

$$\dfrac{2(-6)}{3} \overset{?}{=} -4 \qquad \text{Substituting } -6 \text{ for } x$$

$$\dfrac{-12}{3} \overset{?}{=} -4$$

$$-4 = -4 \qquad \text{A true statement}$$

The solution is -6.

Graph

EXAMPLE 7   Solve and check $\dfrac{8x}{5} = -4$, and graph the solution on the number line.

SOLUTION   We show only the two-step method; therefore, we first multiply both sides of the equation by the denominator, 5.

$$\frac{8x}{5} = -4 \qquad \text{We will clear fractions by multiplying both sides by 5}$$

$$5\left(\frac{8x}{5}\right) = 5(-4) \qquad \text{On the left side, } (5)\left(\frac{8x}{5}\right) = \left(\frac{5}{1}\right)\left(\frac{1}{5}\right)(8x) = 1(8x) = 8x$$

$$8\,x = -20 \qquad \text{This equation is in the form } ax = b$$

$$\frac{8x}{8} = \frac{-20}{8} \qquad \text{Dividing both sides by 8 } \left(\text{we could, instead, multiply both sides by } \tfrac{1}{8}\right)$$

$$x = -\tfrac{5}{2} \qquad \text{Recall that } \frac{-20}{8} = -\frac{5\cdot 4}{2\cdot 4} = -\frac{5}{2}$$

✓ **Check** $\dfrac{8x}{5} = -4$

$$\frac{8\left(-\frac{5}{2}\right)}{5} \overset{?}{=} -4 \qquad \text{Substituting } -\tfrac{5}{2} \text{ for } x$$

$$\frac{-20}{5} \overset{?}{=} -4$$

$$-4 = -4 \qquad \text{A true statement}$$

The solution is $-\tfrac{5}{2}$, or $-2\tfrac{1}{2}$.

Graph

Exercises 3.3A
Set I

Solve and check each of the following equations. In Exercises 1–10, graph each solution on the number line.

1. $2x = 8$

2. $3x = 15$

3. $21 = 7x$

4. $42 = 6x$

5. $11x = 33$

6. $12x = 48$

7. $\dfrac{x}{3} = 4$

8. $\dfrac{x}{5} = 3$

9. $\dfrac{x}{5} = -2$

10. $\dfrac{x}{6} = -4$

11. $4 = \dfrac{x}{7}$

12. $3 = \dfrac{x}{8}$

13. $-13 = \dfrac{x}{9}$ 14. $-15 = \dfrac{x}{8}$ 15. $\dfrac{x}{10} = 3.14$ 22. $\dfrac{5x}{7} = -15$ 23. $\dfrac{8x}{3} = 20$ 24. $\dfrac{9x}{5} = 6$

16. $\dfrac{x}{5} = 7.8$ 17. $-12x = 15$ 18. $-8x = 6$ 25. $\dfrac{6x}{5} = 11$ 26. $\dfrac{7x}{5} = 6$ 27. $\dfrac{6x}{7} = -2$

19. $9 = -6x$ 20. $20 = -25x$ 21. $\dfrac{3x}{2} = -6$ 28. $\dfrac{9x}{4} = -6$

Writing Problems

Express the answers in your own words and in complete sentences.

1. Explain how to isolate the variable in order to solve the equation $3x = 5$.

2. Explain how to isolate the variable in order to solve the equation $\dfrac{x}{3} = 5$.

3. Explain why $4x = 60$ and $\dfrac{x}{3} = 5$ are equivalent equations.

Exercises 3.3A
Set II

Solve and check each of the following equations. In Exercises 1–10, graph each solution on the number line.

1. $5x = 35$ 2. $8x = 32$ 3. $24 = 6x$

4. $56 = 7x$ 5. $12x = 36$ 6. $8 = 24x$

7. $\dfrac{x}{7} = 3$ 8. $\dfrac{x}{2} = 13$ 9. $\dfrac{x}{4} = -5$

10. $-5 = \dfrac{x}{6}$ 11. $6 = \dfrac{x}{3}$ 12. $\dfrac{x}{2} = -4$

13. $-12 = \dfrac{x}{6}$ 14. $16 = \dfrac{x}{3}$ 15. $\dfrac{x}{8} = 12.5$

16. $\dfrac{x}{7} = -1$ 17. $-18x = 12$ 18. $6x = -8$

19. $10 = -8x$ 20. $-12 = 9x$ 21. $\dfrac{4x}{5} = -8$

22. $\dfrac{7x}{4} = -14$ 23. $\dfrac{6x}{7} = 20$ 24. $\dfrac{8x}{3} = 6$

25. $\dfrac{5x}{7} = 4$ 26. $\dfrac{3x}{2} = 11$ 27. $\dfrac{16x}{7} = -24$

28. $\dfrac{8x}{3} = -22$

3.3B Solving Equations of the Form $ax + b = c$

We now combine the methods we've learned so far for solving equations.

Solving an equation of the form $ax + b = c$

All numbers on the same side as the variable must be removed.

1. First, remove those numbers being added to or subtracted from the term containing the variable by using the addition (or subtraction) property of equality.
2. Then either
   a. multiply both sides of the equation by the reciprocal of the coefficient of x; or
   b. multiply both sides of the equation by the denominator (if there is one), and then divide both sides of the equation by the coefficient of the variable.

Notice that after we have used step 1, we will use *either* step 2a *or* step 2b. For step 1 in the examples, we will show horizontal addition on the left and vertical addition on the right.

EXAMPLE 8

Solve and check $2x + 3 = 11$, and graph the solution on the number line.

SOLUTION The numbers 2 and 3 must be removed from the side with the x.

Step 1. We need to isolate the variable. Since the 3 is *added* to the term containing the variable, it is removed first, by adding -3 to both sides.

$$2x + 3 = 11$$
$$2x + 3 + (-3) = 11 + (-3)$$
$$2x = 8$$

$$2x + 3 = 11$$
$$\underline{- 3 \quad -3}$$
$$2x = 8$$

Step 2b. Next, since the coefficient of x is 2, we remove it by dividing both sides by 2. (We could, instead, multiply both sides by $\frac{1}{2}$.)

$$\frac{2x}{2} = \frac{8}{2} \qquad \text{Dividing both sides by 2}$$

$$x = 4 \qquad \text{The apparent solution}$$

✓ **Check** $\quad 2x + 3 = 11$

$$2(4) + 3 \stackrel{?}{=} 11 \qquad \text{Substituting 4 for } x$$

$$8 + 3 \stackrel{?}{=} 11$$

$$11 = 11 \qquad \text{A true statement}$$

The solution is 4.

Graph

EXAMPLE 9

Solve the equation $3x - 2 = 11$, and graph its solution on the number line.

SOLUTION

Step 1. We remove the -2 first (by adding 2 to both sides), because -2 is being *added* to the term containing the variable.

$$3x - 2 = 11$$
$$3x - 2 + 2 = 11 + 2$$
$$3x = 13$$

$$3x - 2 = 11$$
$$\underline{+2 \quad +2}$$
$$3x = 13$$

Step 2a. Next, since the coefficient of x is 3, we remove it by multiplying both sides by $\frac{1}{3}$, the reciprocal of 3. (We could, instead, divide both sides by 3.)

$$\left(\tfrac{1}{3}\right)(3x) = \left(\tfrac{1}{3}\right)(13) \qquad \text{On the left side, } \left(\tfrac{1}{3}\right)(3x) = \left(\tfrac{1}{3}\right)\left(\tfrac{3}{1}\right)x = 1x = x$$

$$x = \tfrac{13}{3}, \text{ or } 4\tfrac{1}{3} \qquad \text{The apparent solution}$$

✓ **Check** $\quad 3x = 13$

$$3\left(\tfrac{13}{3}\right) \stackrel{?}{=} 13 \qquad \text{Substituting } \tfrac{13}{3} \text{ for } x$$

$$\tfrac{39}{3} \stackrel{?}{=} 13$$

$$13 = 13 \qquad \text{A true statement}$$

The solution is $\tfrac{13}{3}$, or $4\tfrac{1}{3}$.

Graph

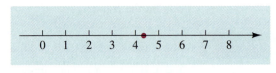

EXAMPLE 10   Solve and check the equation $-12 = 3x + 15$, and graph its solution on the number line.

SOLUTION

Step 1. We remove the 15 first (by adding -15 to both sides), because it is being *added* to the term containing the variable.

$$-12 = 3x + 15$$
$$-12 + (-15) = 3x + 15 + (-15)$$
$$-27 = 3x$$

$$-12 = 3x + 15$$
$$\underline{-15 \qquad -15}$$
$$-27 = 3x$$

Step 2b. $\dfrac{-27}{3} = \dfrac{3x}{3}$ Since the coefficient of x is 3, we remove it by dividing both sides by 3

$-9 = x$ Next, we must use the symmetric property

$x = -9$ The apparent solution

✓ **Check** $-12 = 3x + 15$

$-12 \overset{?}{=} 3(-9) + 15$ Substituting -9 for x

$-12 \overset{?}{=} -27 + 15$

$-12 = -12$ A true statement

The solution is -9.

Graph

EXAMPLE 11   Solve and check the equation $4 = \dfrac{2x}{5} - 6$, and graph its solution on the number line.

SOLUTION

Step 1. We remove the -6 first (by adding 6 to both sides), because -6 is being *added* to the term containing the variable.

$$4 = \dfrac{2x}{5} - 6$$
$$4 + 6 = \dfrac{2x}{5} - 6 + 6$$
$$10 = \dfrac{2x}{5}$$

$$4 = \dfrac{2x}{5} - 6$$
$$\underline{+6 = \qquad + 6}$$
$$10 = \dfrac{2x}{5}$$

Step 2b. The coefficient of x is $\frac{2}{5}$; we can remove it by using the two-step method or by multiplying both sides by $\frac{5}{2}$.

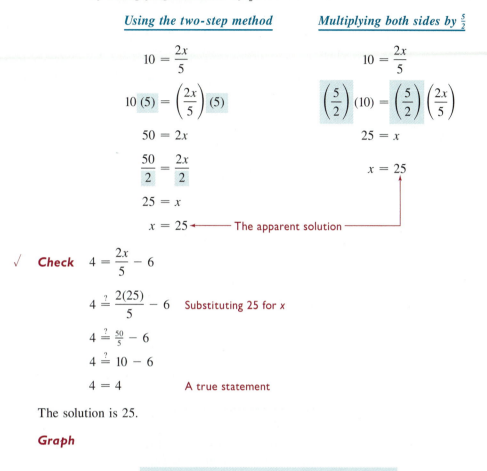

Using the two-step method	*Multiplying both sides by $\frac{5}{2}$*

$$10 = \frac{2x}{5} \qquad\qquad 10 = \frac{2x}{5}$$

$$10\,(5) = \left(\frac{2x}{5}\right)(5) \qquad\qquad \left(\frac{5}{2}\right)(10) = \left(\frac{5}{2}\right)\left(\frac{2x}{5}\right)$$

$$50 = 2x \qquad\qquad 25 = x$$

$$\frac{50}{2} = \frac{2x}{2} \qquad\qquad x = 25$$

$$25 = x$$

$$x = 25 \longleftarrow\text{The apparent solution}\longrightarrow$$

✓ **Check** $4 = \dfrac{2x}{5} - 6$

$$4 \stackrel{?}{=} \frac{2(25)}{5} - 6 \qquad \text{Substituting 25 for } x$$

$$4 \stackrel{?}{=} \tfrac{50}{5} - 6$$

$$4 \stackrel{?}{=} 10 - 6$$

$$4 = 4 \qquad\qquad \text{A true statement}$$

The solution is 25.

Graph

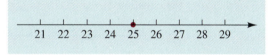

We believe that the two-step method is a little easier for most students. Therefore, we will show only that method for Example 12.

E X A M P L E 12   Solve the equation $\dfrac{3x}{4} + 2 = 11$, and graph its solution on the number line.

S O L U T I O N

Step 1. We remove the 2 first (by adding -2 to both sides), because 2 is being *added* to the term containing the variable.

$$\frac{3x}{4} + 2 = 11 \qquad\qquad\qquad \frac{3x}{4} + 2 = 11$$

$$\frac{3x}{4} + 2 + (-2) = 11 + (-2) \qquad\qquad \underline{ -2 = -2}$$

$$\frac{3x}{4} = 9 \qquad\qquad\qquad \frac{3x}{4} = 9$$

Step 2b. $4\left(\dfrac{3x}{4}\right) = 4\,(9)$ We clear fractions by multiplying both sides by the denominator, 4

$$3x = 36$$ On the left side, $4\left(\dfrac{3x}{4}\right) = \left(\dfrac{4}{1}\right)\left(\dfrac{1}{4}\right)(3x) = 1(3x) = 3x$

$$\dfrac{3x}{3} = \dfrac{36}{3}$$ Dividing both sides by 3

$$x = 12$$ The apparent solution

√ **Check** $\dfrac{3x}{4} + 2 = 11$

$$\dfrac{3(12)}{4} + 2 \overset{?}{=} 11$$ Substituting 12 for x

$$\tfrac{36}{4} + 2 \overset{?}{=} 11$$

$$9 + 2 \overset{?}{=} 11$$

$$11 = 11$$ A true statement

The solution is 12.

Graph

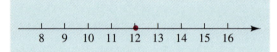

Changing the Signs in an Equation   Multiplying both sides of an equation by -1 is equivalent to changing the sign of *every term* in the equation. For example, consider the equation $5 - x = 7$:

<u>*Changing the sign of every term*</u> | <u>*Multiplying both sides by -1*</u>

$$5 - x = 7$$
$$-5 + x = -7$$

$$5 - x = 7$$
$$(-1)(5 - x) = (-1)7$$
$$-5 + x = -7$$

———— The same equation ————→

Because multiplying both sides of an equation by -1 doesn't change its solution (we're simply using one of the properties of equality), changing the sign of every term of an equation will not change its solution.

If we're solving an equation in which the coefficient of x is negative, we can solve the equation by methods already described (see Examples 13 and 14, Solution 1). Since we're always solving for a *positive x*, however, it may be easier to first make the coefficient of x positive by changing the sign of every term of the equation (see Examples 13 and 14, Solution 2). Both methods are acceptable.

EXAMPLE 13   Solve $5 - x = 7$.

SOLUTION 1   We remove the 5 by adding -5 to both sides.

$$5 - x = 7$$
$$(-5) + 5 - x = (-5) + 7$$
$$-x = 2$$

$$5 - x = \;\; 7$$
$$-5 \qquad -5$$
$$-x = \;\; 2$$

$-x = 2$ has *not* been solved for x

To complete the solution, we will multiply both sides by -1, the reciprocal of -1. (We could, instead, change the signs of all the terms of $-x = 2$.)

$$(-1)(-x) = (-1)(2)$$

$$x = -2 \qquad \text{The apparent solution}$$

Note The equations $x = -2$ and $-x = 2$ are equivalent; however, $-x = 2$ has *not* been solved for x, since the coefficient of $-x$ is understood to be -1, not 1. The equation $x = -2$ *has* been solved for x.

SOLUTION 2 We first change the sign of *every term* of *both* sides of the equation.

$$5 - x = 7 \qquad\qquad\qquad\qquad 5 - x = 7$$

$$-5 + x = -7 \qquad \text{Changing all signs} \qquad -5 + x = -7$$

$$5 + (-5) + x = 5 + (-7) \qquad \text{Adding 5 to both sides} \qquad \underline{+5 \qquad\quad +5}$$

$$x = -2 \qquad \text{The apparent solution} \qquad x = -2$$

Checking will verify that the solution is -2.

EXAMPLE 14 Solve and check $3 - 2x = 9$, and graph the solution on the number line.

SOLUTION 1 Because 3 is being *added* to the term that contains the variable, we remove the 3 by adding -3 to both sides.

$$3 - 2x = 9 \qquad\qquad\qquad 3 - 2x = 9$$

$$-3 + 3 - 2x = -3 + 9 \qquad\qquad \underline{-3 \qquad\quad -3}$$

$$-2x = 6 \qquad\qquad\qquad -2x = 6$$

Since the coefficient of x is -2, we next divide both sides by -2.

$$\frac{-2x}{-2} = \frac{6}{-2} \qquad \text{Dividing both sides by } -2$$

$$x = -3 \qquad \text{The apparent solution}$$

SOLUTION 2 We will first change the sign of *every term* of *both* sides of the equation.

$$3 - 2x = 9 \qquad\qquad\qquad\qquad 3 - 2x = 9$$

$$-3 + 2x = -9 \qquad \text{Changing all signs} \qquad -3 + 2x = -9$$

$$3 + (-3) + 2x = 3 + (-9) \qquad \text{Adding 3 to both sides} \qquad \underline{+3 \qquad\qquad +3}$$

$$2x = -6 \qquad\qquad\qquad\qquad\qquad 2x = -6$$

Since the coefficient of x is 2, we next divide both sides by 2.

$$\frac{2x}{2} = \frac{-6}{2}$$

$$x = -3 \qquad \text{The apparent solution}$$

$\checkmark$ **Check** $\qquad 3 - 2x = 9$

$$3 - 2(-3) \overset{?}{=} 9 \qquad \text{Substituting } -3 \text{ for } x$$

$$3 + 6 \overset{?}{=} 9$$

$$9 = 9 \qquad \text{A true statement}$$

The solution is -3.

Graph

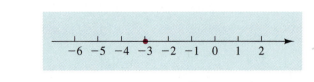

We will look at an alternative way to solve equations that contain fractions in a later chapter.

Exercises 3.3B
Set I

Solve and check the following equations. For Exercises 1–10, graph the solutions on the number line. In Exercises 22 and 24, round off each answer to four decimal places.

1. $4x + 1 = 9$ **2.** $5x + 2 = 12$

3. $6x - 2 = 10$ **4.** $7x - 3 = 4$

5. $2x - 15 = 11$ **6.** $3x - 4 = 14$

7. $4x + 2 = -14$ **8.** $5x + 5 = -10$

9. $14 = 9x - 13$ **10.** $25 = 8x - 15$

11. $12x + 17 = 65$ **12.** $11x + 19 = 41$

13. $8x - 23 = 31$ **14.** $6x - 33 = 29$

15. $14 - 4x = -28$ **16.** $18 - 6x = -44$

17. $8 = 25 - 3x$ **18.** $10 = 27 - 2x$

19. $-73 = 24x + 31$ **20.** $-48 = 36x + 42$

21. $18x - 4.8 = 6$ **22.** $15x - 7.5 = 8$

23. $2.5x - 3.8 = -7.9$ **24.** $3.75x + 0.125 = -0.125$

25. $\dfrac{x}{4} + 6 = 9$ **26.** $\dfrac{x}{5} + 3 = 8$

27. $\dfrac{x}{10} - 5 = 13$ **28.** $\dfrac{x}{20} - 4 = 12$

29. $-14 = \dfrac{x}{6} - 7$ **30.** $-22 = \dfrac{x}{8} - 11$

31. $7 = \dfrac{2x}{5} + 3$ **32.** $9 = \dfrac{3x}{4} + 6$

33. $4 - \dfrac{7x}{5} = 11$ **34.** $3 - \dfrac{2x}{9} = 13$

35. $-24 + \dfrac{5x}{8} = 41$ **36.** $-16 + \dfrac{9x}{4} = 29$

37. $41 = 25 - \dfrac{4x}{5}$ **38.** $54 = 14 - \dfrac{8x}{7}$

Writing Problems

Express the answer in your own words and in complete sentences.

1. Explain what your *first* step would be if you were asked to solve the equation $8x + 5 = 12$. Also explain why you would do that first.

Exercises 3.3B
Set II

Solve and check the following equations. For Exercises 1–10, graph the solutions on the number line.

1. $3x + 2 = 14$ **2.** $5x - 3 = -13$

3. $7x - 4 = 10$ **4.** $8x + 4 = -20$

5. $8x - 12 = 12$ **6.** $6x - 18 = -18$

7. $4x + 11 = -13$ **8.** $6x + 13 = -5$

9. $16 = 5x - 14$ **10.** $14 = 7x - 28$

11. $11x + 19 = 63$ **12.** $8x - 4 = 24$

13. $12x - 17 = 13$ **14.** $11x + 5 = 16$

15. $19 - 10x = -26$ **16.** $8 - 3x = -1$

17. $9 = 32 - 15x$ **18.** $17 = 14 - 3x$

19. $-67 = 18x + 29$ **20.** $-12 = 5x + 3$

21. $5x - 3.6 = 5$ **22.** $4x - 4.26 = 8$

23. $4.1x - 7.4 = -11.5$ **24.** $0.3x + 5.09 = -8.2$

25. $\dfrac{x}{3} + 5 = 7$ **26.** $\dfrac{x}{2} - 4 = -1$

27. $\dfrac{x}{10} - 4 = 11$ **28.** $\dfrac{x}{9} + 3 = 2$

29. $-16 = \dfrac{x}{7} - 9$ **30.** $5 = 3 + \dfrac{x}{2}$

31. $8 = \dfrac{3x}{5} + 4$ **32.** $\dfrac{2x}{4} - 1 = -7$

33. $6 - \dfrac{4x}{5} = 12$ **34.** $8 - \dfrac{2x}{3} = 4$

35. $-19 + \dfrac{7x}{4} = 23$ **36.** $-5 - \dfrac{3x}{4} = 4$

37. $38 = 14 - \dfrac{6x}{11}$ **38.** $5 = 2 - \dfrac{7x}{3}$

3.4   Solving Equations in Which Simplification of Algebraic Expressions Is Necessary

3.4A   Solving Equations in Which the Variable Appears on Both Sides

All the equations solved so far in this chapter have had the variable on only one side of the equation. When the variable appears on both sides of the equation, we start (after combining all like terms) by getting *all* the terms that contain the variable on one side of the equal sign. We do this by using the addition property of equality. (Remember—our goal is to *isolate the variable*, or to get the equation into the form $x = k$.)

It is probably safest to add the *additive inverse* of the *x-term with the smaller coefficient* to both sides of the equation; this makes the coefficient of the sum of the *x*-terms positive (see Example 1, Solution 1). It is not *necessary* to do this, as we show in Example 1, Solution 2; however, it is easy to overlook the negative sign of a negative coefficient and to obtain a wrong answer.

Solving an equation in which the variable appears on both sides	**1.** First combine like terms (if there are any) on each side of the equation. **2.** Remove the term containing the variable from one side of the equation by adding the additive inverse of that term to both sides. **3.** Solve the resulting equation.

EXAMPLE 1   Solve and check $2x - 23 = 6x - 15$, and graph the solution on the number line.

SOLUTION 1   Because $2x$ is the *x*-term with the smaller coefficient, we will add $-2x$ to both sides of the equation; this will remove the *x*-term from the left side of the equation. We'll finish solving the equation by using the methods learned earlier in this chapter. We show horizontal addition first.

$$2x - 23 = 6x - 15$$

$$(-2x) + 2x - 23 = (-2x) + 6x - 15 \qquad \text{Adding } -2x \text{ to both sides}$$

$$-23 = 4x - 15 \qquad \text{Now we must remove } -15 \text{ from the side containing } x$$

$$-23 + 15 = 4x - 15 + 15 \qquad \text{Adding 15 to both sides}$$

$$-8 = 4x \qquad \text{Now we must remove 4 from the side containing } x$$

$$\frac{-8}{4} = \frac{4x}{4} \qquad \text{Dividing both sides by 4}$$

$$-2 = x \qquad \text{Next, we must use the symmetric property}$$

$$x = -2 \qquad \text{The apparent solution}$$

SOLUTION 2   We now solve the same equation by adding $-6x$ to both sides of the equation. This will remove an x-term from the right side of the equation, but it will also give us a negative coefficient for the sum of the x-terms. We will obtain the same solution as before.

$$2x - 23 = 6x - 15$$

$$(-6x) + 2x - 23 = (-6x) + 6x - 15 \qquad \text{Adding } -6x \text{ to both sides}$$

$$-4x - 23 = -15 \qquad \text{Now we must remove } -23 \text{ from the left side}$$

$$-4x - 23 + 23 = -15 + 23 \qquad \text{Adding 23 to both sides}$$

$$-4x = 8 \qquad \text{The coefficient of } x \text{ is } -4$$

$$\frac{-4x}{-4} = \frac{8}{-4} \qquad \text{Dividing both sides by } -4$$

$$x = -2 \qquad \text{The apparent solution}$$

Below, we show the *addition part* of both solutions if vertical addition is used:

Adding $-2x$ *to both sides*	*Adding* $-6x$ *to both sides*

$$
\begin{array}{rcl}
2x - 23 &=& 6x - 15 \\
-2x & & -2x \\
\hline
-23 &=& 4x - 15 \\
+15 & & +15 \\
\hline
-8 &=& 4x
\end{array}
\qquad
\begin{array}{rcl}
2x - 23 &=& 6x - 15 \\
-6x & & -6x \\
\hline
-4x - 23 &=& -15 \\
+23 &=& +23 \\
\hline
-4x & =& 8
\end{array}
$$

We solved these equations above, under Solution 1 and Solution 2.

✓ **Check** $2x - 23 = 6x - 15$

$$2(-2) - 23 \overset{?}{=} 6(-2) - 15 \qquad \text{Substituting } -2 \text{ for } x$$

$$-4 - 23 \overset{?}{=} -12 - 15$$

$$-27 = -27 \qquad \text{A true statement}$$

The solution is -2.

Graph

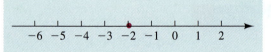

EXAMPLE 2   Solve and check $4x - 5 - x = 13 - 2x - 3$, and graph the solution on the number line.

SOLUTION   We first combine like terms on each side of the equal sign.

$$4x - 5 - x = 13 - 2x - 3$$

$$3x - 5 = -2x + 10 \qquad \text{Like terms have been combined;}$$
$$\text{the smaller coefficient is } -2$$

$$2x + 3x - 5 = 2x + (-2x) + 10$$ Adding $2x$ (the *additive inverse* of $-2x$) to both sides

$$5x - 5 = 10$$ We must now remove -5 from the left side

$$5x - 5 + 5 = 10 + 5$$ Adding 5 (the additive inverse of -5) to both sides

$$5x = 15$$ We must now remove 5 from the left side

$$\frac{5x}{5} = \frac{15}{5}$$ Dividing both sides by 5

$$x = 3$$ The apparent solution

Below, we show the *addition part* of the solution (after like terms have been combined) if vertical addition is used:

$$\begin{array}{rcr} 3x - 5 &=& -2x + 10 \\ +2x & & +2x \\ \hline 5x - 5 &=& 10 \\ + 5 & & + 5 \\ \hline 5x &=& 15 \end{array}$$ This equation was solved above

✓ **Check** $$4x - 5 - x = 13 - 2x - 3$$

$$4(3) - 5 - (3) \stackrel{?}{=} 13 - 2(3) - 3$$ Substituting 3 for x

$$12 - 5 - 3 \stackrel{?}{=} 13 - 6 - 3$$

$$4 = 4$$ A true statement

The solution is 3.

Graph

When we're checking the solution of an equation that has a solution in rounded-off, decimal form, the two sides usually will not be exactly equal to each other, but they should be close (see Example 3).

EXAMPLE 3   Using a calculator, solve and check $8.23 - 4.61x = 3.65$. Round off the answer to three decimal places.

SOLUTION   Let's begin by changing the sign of every term of the equation. This gives the equation $-8.23 + 4.61x = -3.65$.

$$-8.23 + 4.61x = -3.65$$
$$8.23 + (-8.23) + 4.61x = 8.23 + (-3.65)$$
$$4.61x = 4.58$$

$$\begin{array}{l} -8.23 + 4.61x = -3.65 \\ \underline{+8.23 \qquad\qquad +8.23} \\ \qquad\qquad 4.61x = +4.58 \end{array}$$

$$\frac{4.61x}{4.61} = \frac{4.58}{4.61}$$

$$x \approx 0.9934924$$ The calculator display

$$x \approx 0.993$$ Rounding off to three decimal places

√ **Check** $8.23 - 4.61x = 3.65$

$8.23 - 4.61(0.993) \stackrel{?}{=} 3.65$ Substituting 0.993 for x

$8.23 - 4.57773 \stackrel{?}{=} 3.65$

$3.65227 \approx 3.65$

The two sides are not exactly equal to each other, but they are close. The solution is approximately 0.993.

Exercises 3.4A
Set I

Solve and check the following equations. For Exercises 1–10, graph the solutions on the number line. In Exercises 23 and 24, round off each answer to three decimal places.

1. $3x + 14 = 10x$

2. $5x + 56 = 19x$

3. $9x - 21 = 2x$

4. $16x - 3 = 13x$

5. $2x - 7 = x$

6. $5x - 8 = x$

7. $5x = 3x - 4$

8. $7x = 4x - 9$

9. $9 - 2x = x$

10. $8 - 5x = 3x$

11. $3x - 4 = 2x + 5$

12. $5x - 6 = 3x + 6$

13. $6x + 7 = 3 + 8x$

14. $4x + 28 = 7 + x$

15. $7x - 8 = 8 - 9x$

16. $5x - 7 = 7 - 9x$

17. $3x - 7 - x = 15 - 2x - 6$

18. $5x - 2 - x = 4 - 3x - 27$

19. $8x - 13 + 3x = 12 + 5x - 7$

20. $9x - 16 + 6x = 11 + 4x - 5$

21. $7 - 9x - 12 = 3x + 5 - 8x$

22. $13 - 11x - 17 = 5x + 4 - 10x$

23. $7.84 - 1.15x = 2.45$

24. $6.09 - 3.75x = 5.45x$

Writing Problems

Express the answers in your own words and in complete sentences.

1. Explain why $6x + 3 = 4x$ and $2x + 3 = 0$ are equivalent equations.

2. Explain what your first step would be if you were asked to solve the equation $8x + 5 + 2x = 7 + 3x + 6$. Also explain why you would do that first.

Exercises 3.4A
Set II

Solve and check the following equations. For Exercises 1–10, graph the solutions on the number line. In Exercises 23 and 24, round off each answer to three decimal places.

1. $4x + 15 = 19x$

2. $8x - 3 = 5x$

3. $9x - 42 = 7x$

4. $8 + x = 17x$

5. $3x - 10 = x$

6. $4x + 7 = x$

7. $6x = 2x - 8$

8. $9x = 3x + 18$

9. $12 - 5x = x$

10. $21 - 6x = x$

11. $5x - 7 = 4x + 6$

12. $7x + 2 = 8x - 2$

13. $8x + 5 = 14 + 11x$

14. $6x - 2 = 4x - 2$

15. $9x - 13 = 13 - 4x$

16. $-4x + 3 = 3x + 3$

17. $6x - 2 - x = 21 - 3x - 7$

18. $7x + 3 - 2x = 5 - 3x$

19. $4x + 14 + 2x = 12 - 3x - 8$

20. $8x - 3 - 5x = 5 - 2x - 9$

21. $16 - 7x - 4 = 5x + 6 - 4x$

22. $8 + 4x - 3 = 2x + 5 - 7x$

23. $8.42 - 2.35x = 1.25x$

24. $2.67x + 3.4 = -5.33x$

3.4B Solving Equations That Contain Grouping Symbols

When grouping symbols appear in an equation, first we remove them, and then we solve the resulting equation by the methods discussed in the previous sections.

We suggest the following procedure for solving a first-degree equation (an equation in which the exponent on the variable is understood to be 1) in one variable:

Solving a first-degree equation in one variable	**1.** If there are any grouping symbols, remove them.

1. If there are any grouping symbols, remove them.

2. At each step, combine like terms (if there are any) on each side of the equation.

3. If the variable appears on both sides of the equation, rewrite the equation so that all terms containing the variable appear on only one side of the equation; do this by adding the additive inverse of one term that contains the variable to both sides of the equation or by subtracting that term from both sides. (If we add the *additive inverse* of the term with the smaller coefficient to both sides of the equation, the coefficient of the sum of the *x*-terms will be positive.)

4. Remove all the numbers that appear on the same side of the equation as the variable, as follows:

First, remove any numbers being added to (or subtracted from) the term containing the variable by using the addition (or subtraction) property of equality.

Next, complete the solution by multiplying both sides of the equation by the reciprocal of the coefficient of the variable,

or complete the solution by using the two-step method (clearing fractions and then dividing both sides of the equation by the coefficient of the variable).

5. Check the apparent solution in the *original* equation.

Note In step 4 in the box above, we suggest using the addition property first and *then* the multiplication or division property. It is possible to use the multiplication (or even the division) property first, but we believe that the order we recommend in step 4 normally makes solving an equation relatively easy.

EXAMPLE 4 Solve and check the equation $10x - 2(3 + 4x) = 7 - (x - 2)$, and graph its solution on the number line.

SOLUTION

$$10x - 2(3 + 4x) = 7 - (x - 2)$$

$$10x - 6 - 8x = 7 - x + 2 \quad \text{Removing the grouping symbols}$$

Addition shown vertically

$$2x - 6 = -x + 9 \quad \text{Combining like terms; the smaller coefficient is } -1$$

$$x + 2x - 6 = x + (-x) + 9 \quad \text{Adding } x \text{ to both sides to get the } x\text{-terms all on one side}$$

$$3x - 6 = 9 \quad \text{We must now remove } -6 \text{ from the left side}$$

$$3x - 6 + 6 = 9 + 6 \quad \longleftarrow \text{ Adding } +6 \text{ to both sides}$$

$$3x = 15 \quad \text{The coefficient of } x \text{ is } 3$$

$$\frac{3x}{3} = \frac{15}{3} \quad \text{Dividing both sides by 3}$$

$$x = 5 \quad \text{The apparent solution}$$

$$2x - 6 = -x + 9$$
$$\underline{+ x \qquad\qquad + x}$$
$$3x - 6 = \qquad 9$$
$$\underline{+ 6 \qquad\qquad + 6}$$
$$3x \quad = \quad 15$$

√ **Check**

$$10x - 2(3 + 4x) = 7 - (x - 2)$$

$$10(5) - 2(3 + 4 \cdot 5) \stackrel{?}{=} 7 - (5 - 2) \qquad \text{Substituting 5 for } x$$

$$10(5) - 2(3 + 20) \stackrel{?}{=} 7 - (3)$$

$$10(5) - 2(23) \stackrel{?}{=} 7 - 3$$

$$50 - 46 \stackrel{?}{=} 4$$

$$4 = 4 \qquad \text{A true statement}$$

The solution is 5.

Graph

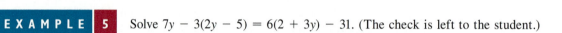

EXAMPLE 5

Solve $7y - 3(2y - 5) = 6(2 + 3y) - 31$. (The check is left to the student.)

SOLUTION

$$7y - 3(2y - 5) = 6(2 + 3y) - 31$$

$$7y - 6y + 15 = 12 + 18y - 31 \qquad \text{Removing grouping symbols}$$

$$y + 15 = 18y - 19 \qquad \begin{array}{l}\text{Combining like terms; the}\\ \text{smaller coefficient is 1}\end{array}$$

$$-y + y + 15 = -y + 18y - 19 \qquad \leftarrow \text{Adding } -y \text{ to both sides}$$

$$15 = 17y - 19 \qquad \begin{array}{l}\text{We must now remove}\\ -19 \text{ from the right side}\end{array}$$

$$15 + 19 = 17y - 19 + 19 \qquad \leftarrow \text{Adding } +19 \text{ to both sides}$$

$$34 = 17y \qquad \begin{array}{l}\text{We must now remove}\\ 17 \text{ from the right side}\end{array}$$

$$\frac{34}{17} = \frac{17y}{17} \qquad \text{Dividing both sides by 17}$$

$$2 = y \qquad \text{Now we must use the symmetric property}$$

$$y = 2 \qquad \text{The apparent solution}$$

Addition shown vertically

$$y + 15 = 18y - 19$$
$$\underline{-y \qquad\quad -y}$$
$$15 = 17y - 19$$
$$\underline{+ 19 \qquad\quad + 19}$$
$$34 = 17y$$

Checking will confirm that the solution is 2.

When the variable appears on both sides of the equation, it is possible to get the variable on one side of the equal sign and the constant on the other side in one step rather than in two steps. This is shown in Example 6, where we add $+15x - 18$ to both sides in one step.

EXAMPLE 6   Solve $5(2 - 3x) - 4 = 5x + [-(2x - 10) + 8]$. (The check is left to the student.)

SOLUTION

$$5(2 - 3x) - 4 = 5x + [-(2x - 10) + 8]$$
$$10 - 15x - 4 = 5x + [-2x + 10 + 8]$$
$$10 - 15x - 4 = 5x + [-2x + 18]$$ Removing grouping symbols
$$10 - 15x - 4 = 5x - 2x + 18$$
$$-15x + 6 = 3x + 18$$
$$15x - 18 + (-15x + 6) = 15x - 18 + (3x + 18)$$ Adding $15x - 18$ to both sides
$$-12 = 18x$$
$$\frac{-12}{18} = \frac{18x}{18}$$ Dividing both sides by 18
$$-\tfrac{2}{3} = x$$ Recall that $\dfrac{-12}{18} = -\dfrac{2 \cdot 6}{3 \cdot 6} = -\dfrac{2}{3}$
$$x = -\tfrac{2}{3}$$ Using the symmetric property of equality

If the addition for solving $-15x + 6 = 3x + 18$ is shown vertically, we have

$$
\begin{array}{rcl}
-15x + 6 = & 3x + 18 \\
\underline{+15x - 18} & \underline{+15x - 18} \\
-12 = & 18x
\end{array}
$$

Adding $+15x - 18$ to both sides

This equation was solved above

Checking will confirm that the solution is $-\tfrac{2}{3}$.

Exercises 3.4B
Set I

Solve and check the following equations. For Exercises 1–10, graph the solutions on the number line.

1. $5x - 3(2 + 3x) = 6$

2. $7x - 2(5 + 4x) = 8$

3. $6x + 2(3 - 8x) = -14$

4. $4x + 5(4 - 5x) = -22$

5. $7x + 5 = 3(3x + 5)$

6. $8x + 6 = 2(7x + 9)$

7. $9 - 4x = 5(9 - 8x)$

8. $10 - 7x = 4(11 - 6x)$

9. $3y - 2(2y - 7) = 2(3 + y) - 4$

10. $4z - 3(5z - 14) = 5(7 + z) - 9$

11. $6(3 - 4x) + 12 = 10x - 2(5 - 3x)$

12. $7(2 - 5x) + 27 = 18x - 3(8 - 4x)$

13. $2(3x - 6) - 3(5x + 4) = 5(7x - 8)$

14. $4(7z - 9) - 7(4z + 3) = 6(9z - 10)$

15. $6(5 - 4h) = 3(4h - 2) - 7(6 + 8h)$

16. $5(3 - 2k) = 8(3k - 4) - 4(1 + 7k)$

17. $2[3 - 5(x - 4)] = 10 - 5x$

18. $3[2 - 4(x - 7)] = 26 - 8x$

19. $3[2h - 6] = 2[2(3 - h) - 5]$

20. $6(3h - 5) = 3[4(1 - h) - 7]$

21. $5(3 - 2x) - 10 = 4x + [-(2x - 5) + 15]$

22. $4(2 - 6x) - 6 = 8x + [-(3x - 11) + 20]$

23. $9 - 3(2x - 7) - 9x = 5x - 2[6x - (4 - x) - 20]$

24. $14 - 2(7 - 4x) - 4x = 8x - 3[2x - (5 - x) - 30]$

25. $-2\{5 - [6 - 3(4 - x)] - 2x\} = 13 - [-(2x - 1)]$

26. $-3\{10 - [7 - 5(4 - x) - 8]\} = 11 - [-(5x - 4)]$

In Exercises 27–30, round off answers to three decimal places.

27. $5.073x - 2.937(8.622 + 7.153x) = 6.208$

28. $21.35 - 27.06x = 34.19(19.22 - 37.81x)$

29. $8.23x - 4.07(6.75x - 5.59) = 3.84(9.18 - x) - 2.67$

30. $11.28(15.93x - 24.66) - 35.42(29.05 - 41.84x)$
$= 22.41(32.56x - 16.29)$

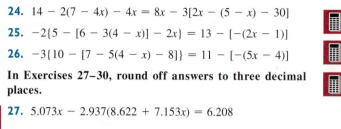

Exercises 3.4B
Set II

Solve and check the following equations. For Exercises 1–10, graph the solutions on the number line.

1. $4x - 5(3 + 2x) = 3$

2. $5x + 3(2 - x) = 8$

3. $8x + 3(4 - 5x) = -16$

4. $9x - 4(x + 3) = 3$

5. $9x + 12 = 2(4x + 5)$

6. $8x + 6 = 5(2x - 4)$

7. $10 - 6x = 4(8 - 7x)$

8. $7 - 3x = 5(7 - 2x)$

9. $2y - 3(5y - 8) = 2(5 + y) - 10$

10. $3x - 5(2x - 3) = 4(2 - x) + 7$

11. $5(6 - 3z) + 18 = -9z - 3(4 - 2z)$

12. $3(3 - 2x) + 5 = 9x - 4(3x - 1)$

13. $3(2z - 6) - 2(6z + 4) = 5(z + 8)$

14. $2(5 - 3x) + 6 = 8x - 5(4 - 2x)$

15. $7(3 - 5h) = 4(3h - 2) - 6(7 + 9h)$

16. $4(2 - x) = 5(3x - 2) - 4(2 + 5x)$

17. $3[2 - 4(k - 6)] = 12 - 6k$

18. $2[2 - 3(x - x)] = 3 - 8x$

19. $4[2x - 5] = 3\{6(7 - x) - 12\}$

20. $5(2x - 3) = 3\{6(1 - x) - 5\} + 10$

21. $5(1 - 2x) - 3 = 4x + [-(2x - 8) + 6]$

22. $20 = 18 - \{-2[3z - 2(z - 1)]\}$

23. $10 - 3(7 - 3x) - 2x = 6x - 5[3x - (4 - x) - 5]$

24. $12 = -\{-3[4z - 2(z - 2)]\}$

25. $-2\{3 - [2 - 4(5 - x) - 7]\} = 12 - [-(3x - 2)]$

26. $6(2 - 3y) - 5 = 5y + [-(2y - 7) + 14]$

In Exercises 27–30, round off answers to three decimal places.

27. $61.25 - 23.04x = 16.19(18.32 - 1.06x)$

28. $7.209x - 4.395(6.281 + 9.154x) = 8.013$

29. $21.82(39.51x - 62.46) - 24.53(50.29 - 48.14x)$
$= 14.28(65.23x - 92.61)$

30. $5.06(18.13x - 4.021) - 6.12(3.062 - 4.31x)$
$= 42.12(31.16x - 10.04)$

3.5 Conditional Equations, Identities, and Equations with No Solution

There are many different kinds of equations. In this section, we discuss three types: conditional equations, identities, and equations with no solution.

Conditional Equations As we mentioned earlier, a *conditional equation* is an equation whose two sides are equal only when certain numbers are substituted for the variable. All equations discussed so far in this chapter have been conditional equations.

Identities If the two sides of an equation are equal when *any* permissible number is substituted for the variable, the equation is called an **identity**. Therefore, an identity has an endless number of solutions.

EXAMPLE 1   Verify that 0, -5, 0.5, and 7 are solutions of the identity $2(5x - 7) = 10x - 14$.

✓   **Check for** 0

$$2(5x - 7) = 10x - 14$$
$$2(5[0] - 7) \overset{?}{=} 10[0] - 14$$
$$2(0 - 7) \overset{?}{=} 0 - 14$$
$$2(-7) \overset{?}{=} -14$$
$$-14 = -14$$

✓   **Check for** -5

$$2(5x - 7) = 10x - 14$$
$$2(5[-5] - 7) \overset{?}{=} 10[-5] - 14$$
$$2(-25 - 7) \overset{?}{=} -50 - 14$$
$$2(-32) \overset{?}{=} -64$$
$$-64 = -64$$

✓   **Check for** 0.5

$$2(5x - 7) = 10x - 14$$
$$2(5[0.5] - 7) \overset{?}{=} 10[0.5] - 14$$
$$2(2.5 - 7) \overset{?}{=} 5 - 14$$
$$2(-4.5) \overset{?}{=} -9$$
$$-9 = -9$$

✓   **Check for** 7

$$2(5x - 7) = 10x - 14$$
$$2(5[7] - 7) \overset{?}{=} 10[7] - 14$$
$$2(35 - 7) \overset{?}{=} 70 - 14$$
$$2(28) \overset{?}{=} 56$$
$$56 = 56$$

The two sides of the equation $2(5x - 7) = 10x - 14$ are equal if *any* real number is substituted for x.

Equations with No Solution   If *no* number will make the two sides of an equation equal, we say that the equation is an **equation with no solution**. (Such an equation is sometimes called a *contradiction* or a *defective equation*.)

EXAMPLE 2   Consider the equation $x + 1 = x + 2$.

Try 0 as a solution: $0 + 1 \neq 0 + 2$.
Try 1 as a solution: $1 + 1 \neq 1 + 2$.
Try 4 as a solution: $4 + 1 \neq 4 + 2$.
Try -6 as a solution: $-6 + 1 \neq -6 + 2$.

Will *any* number work? No. We can think of the problem this way: The equation $x = x$ is true for all numbers x. Because $1 \neq 2$, if we add 1 to the left side and 2 to the right side, we're adding *un*equal numbers to both sides of the equation $x = x$; therefore, $x + 1 \neq x + 2$, so $x + 1 = x + 2$ has no solution.

Usually, we cannot determine whether an equation is a conditional equation, an identity, or an equation with no solution simply by looking at it. Instead, we try to solve the equation by using the methods shown in the preceding sections. In those sections, the equations always reduced to the form $x = k$. In this section, however, three outcomes are possible:

1. If the equation *can* be reduced to the form $x = k$, where k is some real number, the equation is a *conditional equation*.

2. If the variable drops out and the two sides of the equation reduce to the same constant so that we obtain a *true* statement (for example, $0 = 0$), the equation is an *identity*.

3. If the variable drops out and the two sides of the equation reduce to unequal constants so that we obtain a *false* statement (for example, $0 = 3$), the equation is an *equation with no solution*.

E X A M P L E 3

Solve $4x - 2(3 - x) = 12$, or identify the equation as either an identity or an equation with no solution.

S O L U T I O N

$$4x - 2(3 - x) = 12$$

$$4x - 6 + 2x = 12$$

$$6x - 6 = 12 \qquad \text{Combining like terms}$$

$$6x - 6 + 6 = 12 + 6 \quad \longleftarrow \text{Adding 6 to both sides}$$

$$6x = 18$$

$$\frac{6x}{6} = \frac{18}{6} \qquad \text{Dividing both sides by 6}$$

$$x = 3 \qquad \text{This equation is of the form } \boxed{x = k}$$

Vertical addition

$$\begin{array}{r} 6x - 6 = 12 \\ + 6 = +6 \\ \hline 6x \quad = 18 \end{array}$$

Because the equation reduced to the form $\boxed{x = k}$, we know that it is a conditional equation. Checking will confirm that the solution is 3.

E X A M P L E 4

Solve $2(5x - 7) = 10x - 14$, or identify the equation as either an identity or an equation with no solution.

S O L U T I O N

$$2(5x - 7) = 10x - 14$$

$$10x - 14 = 10x - 14 \qquad \text{Removing parentheses}$$

$$-10x + 10x - 14 = -10x + 10x - 14 \quad \text{Adding } -10x \text{ to both sides}$$

$$-14 = -14 \qquad \text{A true statement}$$

Vertical addition

$$\begin{array}{r} 10x - 14 = \quad 10x - 14 \\ -10x \qquad -10x \\ \hline -14 = \qquad -14 \end{array}$$

When we tried to isolate x, all the x's dropped out. Because the two sides of the equation reduced to the same constant (-14) and we obtained a *true statement* ($-14 = -14$), the equation is an identity. (The solution set is the set of *all* real numbers.)

EXAMPLE 5   Solve $3(2x - 5) = 2x + 4(x - 1)$, or identify the equation as either an identity or an equation with no solution.

SOLUTION

$$3(2x - 5) = 2x + 4(x - 1)$$

$6x - 15 = 2x + 4x - 4$ Removing parentheses

$6x - 15 = 6x - 4$ Combining like terms

$-6x + 6x - 15 = -6x + 6x - 4$ ◄ Adding $-6x$ to both sides

$-15 = -4$ A false statement

Vertical addition

$$\begin{array}{rcl} 6x - 15 = & 6x - 4 \\ -6x & -6x \\ \hline -15 = & -4 \end{array}$$

When we tried to isolate x, all the x's dropped out. Because the two sides of the equation reduced to different constants and we obtained a *false statement* $(-15 = -4)$, the equation is an equation with no solution. (The solution set is the empty set, { }.)

Exercises 3.5

Set I

Find the solution of each conditional equation. Identify any equation that is *not* a conditional equation as either an identity or an equation with no solution.

1. $x + 3 = 8$

2. $4 - x = 6$

3. $2x + 5 = 7 + 2x$

4. $10 - 5y = 8 - 5y$

5. $6 + 4x = 4x + 6$

6. $7x + 12 = 12 + 7x$

7. $5x - 2(4 - x) = 6$

8. $8x - 3(5 - x) = 7$

9. $6x - 3(5 + 2x) = -15$

10. $4x - 2(6 + 2x) = -12$

11. $4x - 2(6 + 2x) = -15$

12. $6x - 3(5 + 2x) = -12$

13. $7(2 - 5x) - 32 = 10x - 3(6 + 15x)$

14. $6(3 - 4x) + 10 = 8x - 3(2 - 3x)$

15. $2(2x - 5) - 3(4 - x) = 7x - 20$

16. $3(x - 4) - 5(6 - x) = 2(4x - 21)$

17. $2[3 - 4(5 - x)] = 2(3x - 11)$

18. $3[5 - 2(7 - x)] = 6(x - 7)$

In Exercises 19 and 20, if the equation is conditional, round off the solution to three decimal places.

19. $460.2x - 23.6(19.5x - 51.4) = 1,213.04$

20. $46.2x - 23.6(19.5x - 51.4) = 213.04$

Writing Problems

Express the answers in your own words and in complete sentences.

1. Explain why $x + 7 = 2$ is a conditional equation.

2. Determine whether $3x + 2 = 3(x + 2)$ is a conditional equation, an equation with no solution, or an identity. Then explain why you reached that conclusion.

3. Determine whether $5x + 60 = 5(x + 12)$ is a conditional equation, an equation with no solution, or an identity. Then explain why you reached that conclusion.

Exercises 3.5
Set II

Find the solution of each conditional equation. Identify any equation that is *not* a conditional equation as either an identity or an equation with no solution.

1. $7 - x = 11$

2. $6y - 8 = 3 + 6y$

3. $7x - 4 = 3 + 7x$

4. $3(2 - x) = 5(2x + 1)$

5. $8 - 6x = -6x + 8$

6. $6(2x - 1) = 3(4x + 2) - 12$

7. $4x - 3(2 - x) = 8$

8. $7h - 3(5 - h) = 10$

9. $9x - 3(7 + 3x) = -21$

10. $4(3 - 4x) - 5 = 8(1 - 2x) - 1$

11. $2(x - 4) - (3 + 2x) = 3$

12. $2(7k + 9) - 18 = 14k$

13. $9(3 + 4x) - 17 = 14x + 2(6 + 11x)$

14. $3(2y - 7) - 2(5 - y) = 8y - 31$

15. $8(3x - 2) - (5 + 16x) = 8x - 5$

16. $5(4x - 3) + 6 = 2(3 + 10x)$

17. $5[2 - 3(2 - x)] = 7(2x - 1)$

18. $4(5x - 9) = 3[2 - 4(6 - x)]$

In Exercises 19 and 20, if the equation is conditional, round off the solution to three decimal places.

19. $460.2x - 23.6(19.5x - 51.4) = 213.04$

20. $3.76x - 1.02(5.21x - 10.7) = 21.45$

3.6   Graphing and Solving Inequalities in One Variable

Basic Definitions

An *equation* is a statement that two expressions are *equal*. An **inequality** is a statement that two expressions are *not equal*.

An inequality has three parts:

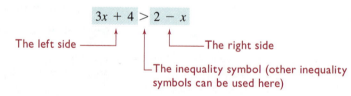

$$3x + 4 > 2 - x$$

The left side

The right side

The inequality symbol (other inequality symbols can be used here)

The following three inequality symbols were introduced in an earlier chapter.

The "unequal to" symbol ($\neq$) $a \neq b$ is read "a is unequal to b."

The "greater than" symbol ($>$) $a > b$ is read "a is greater than b."

The "less than" symbol ($<$) $a < b$ is read "a is less than b."

EXAMPLE   1   Examples of reading the inequalities $\neq$, $>$, and $<$:

a. $5 \neq x - 3$ is read "5 is unequal to x minus 3."

b. $3x - 4 > 7$ is read "$3x$ minus 4 is greater than 7."

c. $2x < 5 - x$ is read "$2x$ is less than 5 minus x."

The "Greater Than or Equal To" Symbol (≥) The inequality $a \geq b$ is read "a is greater than *or* equal to b." This means that if $\begin{cases} either\ a > b \\ or \quad a = b \end{cases}$ is true, then $a \geq b$ is true.

EXAMPLE 2

Examples of the meaning of ≥:

a. $5 \geq 1$ is true because $5 > 1$ is true.

b. $1 \geq 1$ is true because $1 = 1$ is true.

EXAMPLE 3

Examples of reading ≥:

a. $-9 \geq -16$ is read "negative 9 is greater than or equal to negative 16." (The statement $-9 \geq -16$ is true because $-9 > -16$ is true.)

b. $x + 6 \geq 10$ is read "x plus 6 is greater than or equal to 10."

The "Less Than or Equal To" Symbol (≤) The inequality $a \leq b$ is read "a is less than *or* equal to b." This means that if $\begin{cases} either\ a < b \\ or \quad a = b \end{cases}$ is true, then $a \leq b$ is true.

EXAMPLE 4

Examples of the meaning of ≤:

a. $2 \leq 3$ is true because $2 < 3$ is true.

b. $3 \leq 3$ is true because $3 = 3$ is true.

EXAMPLE 5

Examples of reading ≤:

a. $-7 \leq 0$ is read "negative 7 is less than or equal to 0." (The statement $-7 \leq 0$ is true because $-7 < 0$ is true.)

b. $7 \leq 5x - 2$ is read "7 is less than or equal to $5x$ minus 2."

The Solution and the Solution Set of an Inequality

A conditional inequality is an inequality that is true for some values of the variable and false for other values. Examples of conditional inequalities are $3x - 4 > 7$, $2x < 5 - x$, $x + 6 \geq 10$, and $7 \leq 5x - 2$.

A **solution** of a conditional inequality is any number that, when substituted for the variable, makes the inequality a true statement. The **solution set** of an inequality is the set of all numbers that are solutions of the inequality. Whereas the equations we have solved in this chapter have had just one solution (except for identities), an inequality usually has many solutions.

Set-Builder Notation Since we cannot usually *list* all the numbers in the solution set of an inequality, we need to write a *description* of the set, and we often use **set-builder notation** for this description. Set-builder notation is written in the form

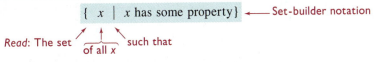

For example, in set-builder notation, the statement "$\{x \mid x > 3\}$" is read "The set of all x such that x is greater than 3."

The Graph of the Solution Set of an Inequality in One Variable

The solution set of an inequality in one variable can be graphed on the number line. We use a *solid circle* to indicate that a particular point *is* in the solution set and a *hollow circle* to indicate that a particular point is *not* in the solution set. In other words, if the inequality is $\leq$ or $\geq$, we graph the end-point as a *solid* circle, and if the inequality is $<$ or $>$, we graph the end-point as a *hollow* circle (see Example 6).

EXAMPLE 6 Graph the solution set of each of the following inequalities:

a. $x \geq 3$

SOLUTION The solution set is the set of all real numbers greater than or equal to 3.

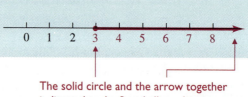

The solid circle and the arrow together indicate that the 3 and all numbers to the *right* of the 3 are solutions

b. $x < 1$

SOLUTION The solution set is the set of all real numbers less than 1.

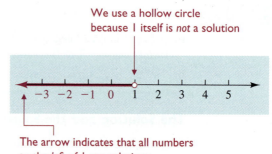

We use a hollow circle because 1 itself is *not* a solution

The arrow indicates that all numbers to the *left* of 1 are solutions

c. $x \neq -2$

SOLUTION The solution set is the set of all real numbers except -2.

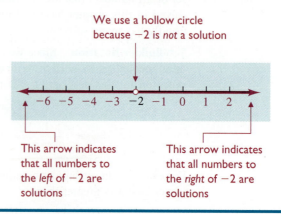

We use a hollow circle because -2 is *not* a solution

This arrow indicates that all numbers to the *left* of -2 are solutions

This arrow indicates that all numbers to the *right* of -2 are solutions

The Sense of an Inequality

The **sense** of an inequality symbol refers to whether the symbol is a *greater than* or a *less than* symbol. For example, the inequalities $1 < 3$ and $5 < 7$ have the *same sense* (both are *less than*), but the inequalities $8 > 5$ and $3 < 6$ have the *opposite* sense (one is *greater than* and the other is *less than*).

$$\left.\begin{array}{l} a > b \\ c > d \end{array}\right\}\begin{array}{l}\text{Same sense} \\ \text{(both are >)}\end{array} \qquad \left.\begin{array}{l} a < b \\ c > d \end{array}\right\}\begin{array}{l}\text{Opposite sense} \\ \text{(one is <, one is >)}\end{array}$$

$$\left.\begin{array}{l} a \le b \\ c \ge d \end{array}\right\}\begin{array}{l}\text{Opposite sense} \\ \text{(one is $\le$, one is $\ge$)}\end{array} \qquad \left.\begin{array}{l} a \le b \\ c \le d \end{array}\right\}\begin{array}{l}\text{Same sense} \\ \text{(both are $\le$)}\end{array}$$

The Properties of Inequalities

Earlier in this chapter, we solved equations by adding the same number to both sides of the equation, multiplying both sides of the equation by the same number, and so on. In Example 7, we're going to examine the effect of adding the same number to both sides of an *inequality*, multiplying both sides of an *inequality* by the same number, and so forth.

EXAMPLE 7

Examine the effect of performing various arithmetic operations on both sides of the inequality $8 < 12$ (note that $8 < 12$ is a *true* statement). (In order to decide whether the sense of the inequality changes, we'll graph the two numbers that result from each arithmetic operation.)

a. Does the sense of the inequality change if we add $+6$ to both sides?

SOLUTION

$$8 < 12$$
$$8 + 6 \; ? \; 12 + 6$$
$$14 \; ? \; 18$$
$$14 < 18$$

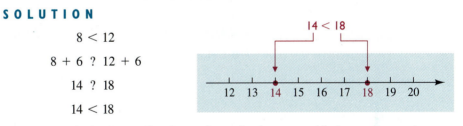

The sense of the inequality is *unchanged* when we add the same number to both sides.

b. Does the sense of the inequality change if we subtract 3 from both sides?

SOLUTION

$$8 < 12$$
$$8 - 3 \; ? \; 12 - 3$$
$$5 \; ? \; 9$$
$$5 < 9$$

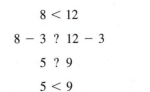
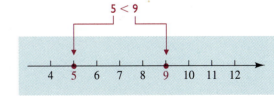

The sense of the inequality is *unchanged* when we subtract the same number from both sides.

c. Does the sense of the inequality change if we multiply both sides by 2 (a *positive* number)?

SOLUTION

$$8 < 12$$
$$8(2) \; ? \; 12(2)$$
$$16 \; ? \; 24$$
$$16 < 24$$

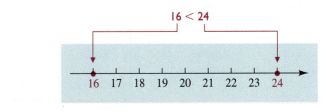

The sense of the inequality is *unchanged* when we multiply both sides by the same *positive* number.

d. Does the sense of the inequality change if we divide both sides by 4 (a *positive* number)?

SOLUTION

$$8 < 12$$

$$\frac{8}{4} \ ? \ \frac{12}{4}$$

$$2 \ ? \ 3$$

$$2 < 3$$

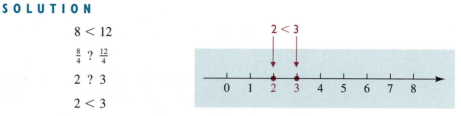

The sense of the inequality is *unchanged* when we divide both sides by the same *positive* number.

e. Does the sense of the inequality change if we multiply both sides by -1 (a *negative* number)?

SOLUTION

$$8 < 12$$

$$8(-1) \ ? \ 12(-1)$$

$$-8 \ ? \ -12$$

$$-8 > -12 \quad \text{The senses are } opposite$$

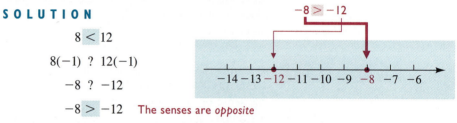

The sense of the inequality is *changed* when we multiply both sides by the same *negative* number.

f. Does the sense of the inequality change if we divide both sides by -2 (a *negative* number)?

SOLUTION

$$8 < 12$$

$$\frac{8}{-2} \ ? \ \frac{12}{-2}$$

$$-4 \ ? \ -6$$

$$-4 > -6 \quad \text{The senses are } opposite$$

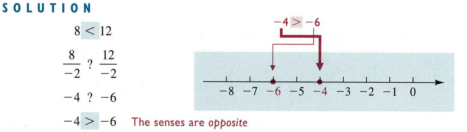

The sense of the inequality is *changed* when we divide both sides by the same *negative* number.

We see that when we multiply or divide both sides of an inequality that contains a $<$, $\leq$, $>$, or $\geq$ symbol by a *negative number*, the *sense* of the inequality changes. We solve simple *inequalities* that contain any one of these inequality symbols by using the following properties of inequalities:

The addition and subtraction properties of inequalities

If an inequality contains one of the symbols $>$, $<$, $\geq$, or $\leq$, the sense of the inequality is unchanged if the same number is added to or subtracted from both sides of the inequality. For example,

if $a < b$, then $a + c < b + c$ and $a - c < b - c$

The senses are the same

where a, b, and c are real numbers. (See Examples 7a and 7b.)

The multiplication and division properties of inequalities	If an inequality contains one of the symbols $>$, $<$, $\geq$, or $\leq$, the sense of the inequality is unchanged if both sides of the inequality are multiplied or divided by the same *positive* number. For example,

$$\text{if } a < b \text{ and } c > 0, \qquad \text{then} \qquad ac < bc \qquad \text{and} \qquad \frac{a}{c} < \frac{b}{c}$$

where a, b, and c are real numbers. (See Examples 7c and 7d.)

However, the sense of the inequality is *changed* if both sides of the inequality are multiplied or divided by the same *negative* number. For example,

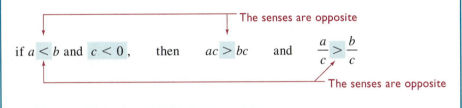

where a, b, and c are real numbers. (See Examples 7e and 7f.)

Solving Inequalities

When we solve an inequality, we must find all the values of the variable that satisfy the inequality. Therefore, in the problems in this book, we want our final inequality to be in one of the forms $x < k$, $x > k$, $x \leq k$, $x \geq k$, or $x \neq k$, where k is some number. To accomplish this, we can use the following procedure:

Solving an inequality that contains $<$, $\leq$, $>$, or $\geq$	Proceed as if you were solving an equation, except that you must *change the sense of the inequality if you multiply or divide both sides of the inequality by a negative number.* Then, if necessary, use the facts that $a < b$ can be rewritten as $b > a$, $a \geq b$ can be rewritten as $b \leq a$, and so on, to get the inequality into one of the forms $x < k, x > k, x \leq k,$ or $x \geq k.$* ——————————————————— *Your instructor may permit you to leave answers in a form such as $k < x$.

Solving an inequality that contains $\neq$	Proceed as if you were solving an equation, except that you must write $\neq$ rather than $=$ in each step.

E X A M P L E   8   Solve $x + 3 < 7$, and graph the solution set on the number line.

S O L U T I O N

$$x + 3 < 7$$
$$x + 3 + (-3) < 7 + (-3) \qquad \text{Adding } -3 \text{ to both sides}$$
$$x < 4$$

This means that if we replace x in the given inequality by *any* number less than 4, we get a true statement. For example, we could replace x by 3 or by -5 (both of these numbers are less than 4):

✓

Check for $x = 3$	**Check for $x = -5$**
$x + 3 < 7$	$x + 3 < 7$
$3 + 3 \overset{?}{<} 7$	$-5 + 3 \overset{?}{<} 7$
$6 < 7$ True	$-2 < 7$ True

The solution set is $\{x \mid x < 4\}$, the set of all x such that x is less than 4.

Graph   We graph the solution set.

We will not show any checks for Examples 9–12.

EXAMPLE 9

Solve $2x - 5 > 1$, and graph the solution set on the number line.

SOLUTION

$$2x - 5 > 1$$
$$2x - 5 + 5 > 1 + 5 \qquad \text{Adding 5 to both sides}$$
$$2x > 6$$
$$\frac{2x}{2} > \frac{6}{2} \qquad \text{Dividing both sides by 2, a } positive \text{ number; the sense is not changed}$$
$$x > 3$$

The solution set is $\{x \mid x > 3\}$, the set of all x such that x is greater than 3.

Graph

EXAMPLE 10

Solve $3x - 2(2x - 7) \le 2(3 + x) - 4$, and graph the solution set on the number line.

SOLUTION 1

$$3x - 2(2x - 7) \le 2(3 + x) - 4$$
$$3x - 4x + 14 \le 6 + 2x - 4 \qquad \text{Removing grouping symbols}$$
$$-x + 14 \le 2x + 2 \qquad \text{Combining like terms}$$
$$-2x + (-x) + 14 \le -2x + 2x + 2 \qquad \text{Adding } -2x \text{ to both sides (also see Solution 2)}$$

$$-3x + 14 \leq 2$$

$$-3x + 14 \boxed{+ (-14)} \leq 2 \boxed{+ (-14)} \qquad \text{Adding } -14 \text{ to both sides}$$

$$-3x \leq -12$$

$$\frac{-3x}{-3} \boxed{\geq} \frac{-12}{-3} \qquad \text{Dividing both sides by } -3, \text{ a } \textit{negative} \\ \text{number; } \textit{the sense is changed}$$

$$x \geq 4$$

SOLUTION 2   It is always possible to avoid having to divide by a negative number; we can do this by moving the variables to whichever side makes the coefficient of the variable positive, as follows:

$$3x - 2(2x - 7) \leq 2(3 + x) - 4$$

$$3x - 4x + 14 \leq 6 + 2x - 4$$

$$-x + 14 \leq 2x + 2 \qquad \text{Combining like terms}$$

$$\boxed{x +} (-x) + 14 \leq \boxed{x +} 2x + 2 \qquad \text{Adding } x \text{ to both sides}$$

$$14 \leq 3x + 2$$

$$14 \boxed{+ (-2)} \leq 3x + 2 \boxed{+ (-2)} \qquad \text{Adding } -2 \text{ to both sides}$$

$$12 \leq 3x$$

$$\frac{12}{3} \leq \frac{3x}{3} \qquad \text{Dividing both sides by 3, a } \textit{positive} \\ \text{number; the sense is not changed}$$

$$4 \leq x$$

$$x \geq 4 \qquad x \geq 4 \text{ has the same meaning as } 4 \leq x$$

The solution set is $\{x \mid x \geq 4\}$, the set of all x such that x is greater than or equal to 4.

Graph

EXAMPLE 11   Solve $\dfrac{8x}{5} - 7 > 3$, and graph the solution set on the number line.

SOLUTION   We must remove the numbers 8, 5, and -7 from the side of the inequality on which the x appears.

$$\frac{8x}{5} - 7 > 3$$

$$\frac{8x}{5} - 7 \boxed{+ 7} > 3 \boxed{+ 7} \qquad \text{Adding 7 to both sides}$$

$$\frac{8x}{5} > 10 \qquad \text{Next, we'll "clear fractions"}$$

Multiplying both sides by 5, a *positive* number; the sense is not changed

$$8x > 50$$

$$\frac{8x}{8} > \frac{50}{8}$$ Dividing both sides by 8, a *positive* number; the sense is not changed

$$x > \tfrac{25}{4} \quad \text{or} \quad x > 6\tfrac{1}{4}$$ Recall that $\dfrac{50}{8} = \dfrac{25 \cdot 2}{4 \cdot 2} = \dfrac{25}{4}$

The solution set is $\left\{ x \mid x > \tfrac{25}{4} \right\}$, the set of all real numbers greater than $\tfrac{25}{4}$.

Graph   For graphing purposes, we think of $\tfrac{25}{4}$ as $6\tfrac{1}{4}$.

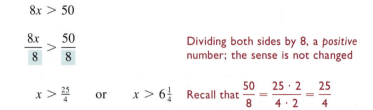

In Example 12, remember that when the inequality contains the $\neq$ symbol, we proceed in exactly the same manner as we do to solve equations, except that we write $\neq$ rather than $=$ in each step.

EXAMPLE 12   Solve $4(x - 3) - 5 \neq 2x - 7$, and graph the solution set on the number line.

SOLUTION

$$4(x - 3) - 5 \neq 2x - 7$$

$$4x - 12 - 5 \neq 2x - 7 \qquad \text{Using the distributive property}$$

$$4x - 17 \neq 2x - 7 \qquad \text{Combining like terms}$$

$$4x - 17 + (-2x + 17) \neq 2x - 7 + (-2x + 17) \qquad \text{Adding $-2x + 17$ to both sides}$$

$$2x \neq 10 \qquad \text{Combining like terms}$$

$$\frac{2x}{2} \neq \frac{10}{2} \qquad \text{Dividing both sides by 2}$$

$$x \neq 5 \qquad \text{Note that we wrote $\neq$ in every step}$$

The solution set is $\{ x \mid x \neq 5 \}$, the set of all x such that x is not equal to 5.

Graph

Exercises 3.6
Set I

Solve each of the following inequalities, and graph each solution set on the number line. In Exercises 23 and 24, round off the decimal approximations to three decimal places.

1. $x - 5 < 2$

2. $x - 4 < 7$

3. $5x + 4 \leq 19$

4. $3x + 5 \leq 14$

5. $6x + 7 > 3 + 8x$

6. $4x + 28 > 7 + x$

7. $2x - 9 > 3(x - 2)$

8. $3x - 11 > 5(x - 1)$

9. $6(3 - 4x) + 12 \geq 10x - 2(5 - 3x)$

10. $7(2 - 5x) + 27 \geq 18x - 3(8 - 4x)$

11. $4(6 - 2x) \neq 5x - 2$

12. $6(2x - 5) + 29 \neq 3x - 7(11 - 4x)$

13. $2[3 - 5(x - 4)] < 10 - 5x$

14. $3[2 - 4(x - 7)] < 26 - 8x$

15. $7(x - 5) - 4x > x - 8$

16. $8(x + 2) \neq 24 - 2(x - 1)$

17. $3x - 5(x + 2) \leq 4x + 8$

18. $5[6 - 2(3 - x)] - 3 < 3x + 4$

19. $\dfrac{7x}{5} - 8 > 3$

20. $\dfrac{6x}{7} - 4 \leq 3$

21. $3 + \dfrac{2x}{9} \leq 5$

22. $9 + \dfrac{4x}{3} < 7$

 23. $12.85x - 15.49 \geq 22.06(9.66x - 12.74)$

 24. $7.12(3.65x - 8.09) + 5.76$
$< 5.18x - 6.92(4.27 - 3.39x)$

Writing Problems

Express the answers in your own words and in complete sentences.

1. Explain why 3 is not a solution for $5x < 15$.

2. Explain what your first step would be if you were asked to solve the inequality $3x < 18$.

3. Explain what your first step would be if you were asked to solve the inequality $3 + x < 18$.

4. Describe the graph of the solution set for $x > 4$.

5. Describe the graph of the solution set for $x \leq -3$.

Exercises 3.6
Set II

Solve each of the following inequalities, and graph each solution set on the number line. In Exercises 23 and 24, round off the decimal approximations to three decimal places.

1. $x + 3 \geq -4$

2. $4 + x > -5$

3. $x + 4 < 7$

4. $5x + 3 < 18$

5. $5x + 7 > 13 + 11x$

6. $7x - 5 \leq 2x - 25$

7. $5x - 6 \leq 3(2 + 3x)$

8. $3x - 2 \geq -2(11 - x)$

9. $5(3 - 2x) + 25 \geq 4x - 6(10 - 3x)$

10. $3(x + 3) - 4 \neq 7x - 3(2 - x)$

11. $2(5 - 3x) \neq 7 - 4x$

12. $6(3 - 2x) - 3(x + 1) < 0$

13. $4[2 - 3(x - 5)] < 2 - 6x$

14. $2(3x - 4) + (x - 1) > 5$

15. $8(x - 3) - (x + 4) < 2x + 2$

16. $3[5 + 3(4 + x)] \le 4(2x - 3)$

17. $5x - 3(3x - 4) \ge 2x + 7$

18. $5 - (3 - x) \ne 3(2 + x)$

19. $\dfrac{4x}{9} - 5 < 2$

20. $3 - \dfrac{2x}{5} > 6$

21. $7 + \dfrac{3x}{5} \le 6$

22. $8 + \dfrac{3x}{7} < 4$

23. $821.4x - 395.2 \ge 604.1(542.8x - 193.7)$

24. $4.01x + 62.1 \le 3.04(5.143 - 6.21x)$

Sections 3.1–3.6   R E V I E W

$$5x - 8 = 3x + 2$$

The left side ——

The right side

The equal sign

Parts of an Equation
3.1

Conditional Equations
3.1

A **conditional equation** is an equation whose two sides are equal only when certain numbers (called *solutions*) are substituted for the variable.

Solution of an Equation
3.1

A **solution** of an equation is a number that, when substituted for the variable, makes the two sides of the equation equal.

Solution Set of an Equation
3.1

The **solution set** of an equation is the set of all the solutions of the equation.

Equivalent Equations
3.1

Equations that have the same solution set are called **equivalent equations**.

Properties of Equality
3.2

For all real numbers a, b, and c:
Symmetry: The equation $a = b$ is equivalent to the equation $b = a$.
Addition: If $a = b$, then $a + c = b + c$. If the same number is added to both sides of an equation, the new equation is equivalent to the original equation.
Subtraction: If $a = b$, then $a - c = b - c$. If the same number is subtracted from both sides of an equation, the new equation is equivalent to the original equation.

3.3

Division: If $a = b$ and if $c \ne 0$, then $\dfrac{a}{c} = \dfrac{b}{c}$. If both sides of an equation are divided by the same nonzero number, the new equation is equivalent to the original equation.
Multiplication: If $a = b$ and if $c \ne 0$, then $ac = bc$. If both sides of an equation are multiplied by the same nonzero number, the new equation is equivalent to the original equation.

To Solve an Equation
3.2, 3.3, 3.4, 3.5

1. If there are any grouping symbols, remove them.

2. In each step, combine like terms (if there are any) on each side of the equation.

3. If the variable appears on both sides of the equation, rewrite the equation so that all terms containing the variable appear on only one side of the equation; do this by adding the additive inverse of one term that contains the variable to both sides of the equation or by subtracting that term from both sides.

4. Remove all numbers that appear on the same side as the variable, as follows:
First, remove any numbers being added to (or subtracted from) the term containing the variable by using the addition (or subtraction) property of equality.
Next, complete the solution by multiplying both sides of the equation by the reciprocal of the coefficient of the variable,
or complete the solution by using the two-step method (clearing fractions and then dividing both sides of the equation by the coefficient of the variable).

5. Three outcomes are possible:

If the equation reduces to the form $x = k$, it is a *conditional equation.*

If the variable drops out and the two sides reduce to the same constant so that we obtain a *true* statement, the equation is an *identity.*

If the variable drops out and the two sides reduce to unequal constants so that we obtain a *false* statement, the equation has *no solution.*

6. If the equation is a conditional equation, check the solution in the *original* equation.

Inequalities
3.6

An **inequality** is a statement that two expressions are not equal.

Set-Builder Notation
3.6

The notation $\{x \mid x \text{ has some property}\}$ is called **set-builder notation**.

Graphing the Solution Set of an Inequality on the Number Line
3.6

To graph the solution set of the inequality $x > c$:

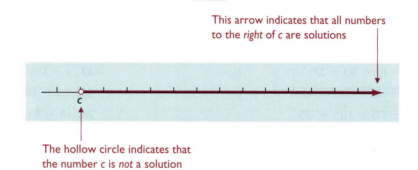

This arrow indicates that all numbers to the *right* of c are solutions

The hollow circle indicates that the number c is *not* a solution

To graph the solution set of the inequality $x \le b$:

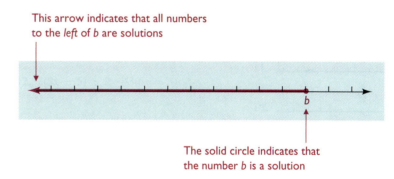

This arrow indicates that all numbers to the *left* of b are solutions

The solid circle indicates that the number b is a solution

The solution sets of other types of inequalities (for example, $x < d$ and $x \ge e$) are graphed using similar procedures.

To Solve an Inequality
3.6

If the inequality contains one of the symbols $<$, $\le$, $>$, or $\ge$, proceed as if you were solving an equation, except that you must *change the sense of the inequality if you multiply or divide both sides of the inequality by a negative number.* Then, if necessary, use the fact that $a < b$ can be rewritten as $b > a$, and so on, to get the inequality into one of the forms $x < k$, $x > k$, $x \le k$, or $x \ge k$. If the inequality contains the symbol $\ne$, proceed as if you were solving an equation, except that you must write $\ne$ rather than $=$ in each step.

15. $4 - 5(x - 7) = 10(4 - x)$ **16.** $2 + 3(4 - x) = 6x - 13$

17. $25 - 2[4(x - 2) - 12x + 4] = 16x + 33$

18. $2\{3[10 - 4(3 - x) + x] - 5\} = 0$

19. $15 - 5[2(8x - 4) - 14x + 8] = 25$

20. $2\{3[2(5 - V) + 4V] - 20\} = 0$

In Exercises 21–26, solve each inequality, and graph the solution set on the number line. (*You* label the points on the number line.)

21. $x + 5 \geq -6$

22. $x - 7 < 2$

23. $3x - 2 > 6 - x$

24. $7x - 2(5 + 4x) \leq 8$

25. $5(6 - 2x) + 3 \geq 2(x + 1) + 11$

26. $3[2 - (4 - x)] + 7 < 4x$

15. _____

16. _____

17. _____

18. _____

19. _____

20. _____

21. _____

22. _____

23. _____

24. _____

25. _____

26. _____

174

Chapter 3 · DIAGNOSTIC TEST

The purpose of this test is to see how well you understand solving simple equations and inequalities. We recommend that you work this diagnostic test *before* your instructor tests you on this chapter. Allow yourself about 50 minutes.

Complete solutions for all the problems on this test, together with section references, are given in the answer section in the back of the book. We suggest that you study the sections referred to for the problems you do incorrectly.

In Problems 1–18, find and check the solution of each conditional equation. Identify any equation that is not a conditional equation as either an identity or an equation with no solution.

1. $x - 5 = 3$ (Graph the solution on the number line.)

2. $5y + 7 = 22$ (Show your check on this problem.)

3. $5z - 7 = 13$

4. $14 + 3x = 7 - 4x$

5. $7x - 4 = 3x + 4(x - 1)$

6. $8 = 4y - 1$

7. $3x - 4 = x + 2(x - 1)$

8. $17 - 3z = -1$

9. $\dfrac{x}{6} = 5.1$

10. $-6 = \dfrac{w}{7}$

11. $\dfrac{x}{4} - 5 = 3$

12. $6x + 1 = 17 - 2x$

13. $3z - 21 + 5z = 4 - 6z + 17$ (Graph the solution on the number line.)

14. $5k - 9(7 - 2k) = 6$ (Show your check on this problem.)

15. $10x - 2(5x - 7) = 14$

16. $2y - 4(3y - 2) = 5(6 + y) - 7$

17. $7(3z + 4) = 14 + 3(7z - 1)$

18. $3[7 - 6(x - 2)] = -3 + 2x$ (Graph the solution on the number line.)

19. Solve the inequality $4x + 5 > -3$, and graph the solution set on the number line.

20. Solve the inequality $5x - 2 \le 10 - x$, and graph the solution set on the number line.

Chapters 1–3 · CUMULATIVE REVIEW EXERCISES

In Exercises 1–4, evaluate each expression or write either "not defined" or "not a real number."

1. $\dfrac{-7}{0}$

2. $\sqrt{8^2 + 6^2}$

3. $7\sqrt{16} - 5(-4)$

4. $25 - \{-16 - [(11 - 7) - 8]\}$

5. Using the formula $V = \frac{4}{3}\pi r^3$, find V when $\pi \approx 3.14$ and $r = 9$.

6. Using the formula $A = P(1 + i)^n$, find A when $P = 3{,}000$, $i = 0.10$, and $n = 2$.

In Exercises 7–12, simplify each algebraic expression.

7. $15x - [9y - (7x - 10y)]$

8. $5(2a - 3b) - 6(4a + 7b)$

9. $8 - 3[2x - (1 - 4x)]$

10. $8x + 3[2 + 7(4 - x)] - 2$

11. $\sqrt{64x^8y^4}$

12. $\sqrt{9x^{10}t^{12}}$

In Exercises 13–19, solve each conditional equation. Identify any equation that is not a conditional equation as either an identity or an equation with no solution.

13. $2x + 1 = 2x + 7$ 14. $3 = 33 - 10x$

15. $10 - 4(2 - 3x) = 2 + 12x$

16. $6 - 4(N - 3) = 2$

17. $\dfrac{C}{7} - 15 = 13$

18. $12(4W - 5) = 9(7W - 8) - 13$

19. $9 - 3(x - 2) = 3(5 - x)$

In Exercises 20–22, solve each inequality.

20. $2x - 5 < 3$

21. $3 - x \ge 4$

22. $5x - 2 \le 8x + 4$

In Exercises 23–30, write "true" if the statement is always true. Otherwise, write "false."

23. $-\frac{3}{4}$ is a real number.

24. $5.\overline{23}$ is an irrational number.

25. -5 is a natural number.

26. All irrational numbers are real numbers.

27. All real numbers are rational numbers.

28. 0 is a real number.

29. $6x$ is a term of $6xy$.

30. $4 + 8 = 8 + 4$ illustrates the commutative property of addition.

Critical Thinking and Problem-Solving Exercises

1. Erica is thinking of a certain number. Can you guess what number Erica is thinking about by using the following four clues?

> The number is a three-digit number.
>
> It is less than 160.
>
> The sum of its digits is 9.
>
> One of its digits is a 2.

2. Discover the pattern and fill in the next three numbers for this set of numbers: 1, 1, 2, 3, 5, 8, ____ , ____ , ____ .

3. If David's daughter, who is in kindergarten, writes the numbers 1 through 100, how many 8's will she be writing altogether?

4. George has a total of 17 pets; some are birds, some are fish, and some are dogs. George's pets have a total of 22 legs, and 12 of his pets cannot fly. How many birds does George have? How many fish? How many dogs?

5. John has six shirts hanging in his closet, one in each of the following six colors: red, tan, taupe, white, blue, and aqua. Using these hints, determine the order in which the shirts are hanging.

> The blue shirt and the taupe shirt are in the two center positions.
>
> The tan shirt and the white shirt are on the two ends.
>
> The red shirt is to the left of the tan shirt and the taupe shirt.
>
> The red shirt is not next to the taupe shirt.

(List the shirts—by color, of course—in order from left to right.)

6. The solution of each of the following equations contains an error. Find and describe the errors.

a. Solve $7 - 4x = -9$.

$$7 - 4x = -9$$
$$-7 + 7 - 4x = -7 - 9$$
$$4x = -16$$
$$\frac{4x}{4} = \frac{-16}{4}$$
$$x = -4$$

b. Solve $2 + 5(x - 7) = x + 3$.

$$2 + 5(x - 7) = x + 3$$
$$7(x - 7) = x + 3$$
$$7x - 49 = x + 3$$
$$-x + 7x - 49 = -x + x + 3$$
$$6x - 49 = 3$$
$$6x - 49 + 49 = 3 + 49$$
$$6x = 52$$
$$\frac{6x}{6} = \frac{52}{6} = \frac{26}{3}$$
$$x = \frac{26}{3}$$

Applications

CHAPTER 4

As we have mentioned, the main reason for studying algebra is to equip oneself with the tools necessary for solving mathematical problems. Most such problems are expressed in words. In this chapter, we present methods for solving some traditional problems that are expressed in words. The skills learned in this chapter can be applied in solving mathematical problems encountered in many fields of study as well as in real-life situations.

4.1 Problem Solving—Translating English Expressions into Algebraic Expressions

Although we can't give you a definite set of rules that will enable you to solve all applied problems, we will suggest a procedure that should help you get started. In this chapter, we discuss several different "types" of applied problems (money problems, mixture problems, distance-rate-time problems, and so forth) in separate sections, because we feel that this technique is most helpful to beginning students. However, the general *method* of attacking applied problems is the same for *all* types of such problems; it is this *method* that you should concentrate on.

Earlier, we gave several suggestions for solving applied problems in arithmetic. We recommend that you review those suggestions at this time, because many of them apply to the solution of applied problems in algebra.

4.1A Key Word Expressions and Their Corresponding Algebraic Operations

When we write an equation that represents the facts given in an applied problem, we call the equation a *mathematical model*. In writing such an equation, it is often helpful to break sentences up into words and phrases—small word expressions. In this section, we show how you can change these words and phrases into *algebraic expressions*. Below is a list of key words and their corresponding algebraic operations or symbols.

+	−	×	÷	=
add	subtract	multiply	divided by	is equal to
sum	difference	times	quotient	equals
plus	minus	product		is
increased by	decreased by	of (in fraction and percent problems)		
more than	diminished by			
	less than			
	subtracted from			

A Word of Caution   Because subtraction is not commutative, care must be taken to put the numbers in an applied subtraction problem in the correct order. For example, the phrases "*m* minus *n*," "the difference of *m* and *n*," and "*m* decreased by *n*" are translated as $m - n$. However, we must *reverse the order* of the variables for the phrases "*m* subtracted from *n*" and "*m* less than *n*"; these are both translated as $n - m$ (see Examples 1d and 1f).

Also notice that "*m* is less than *n*" is *not* a phrase—it is a complete sentence, and it translates as $m < n$.

EXAMPLE 1 Change each of the following English expressions into an algebraic expression.

a. "The sum of A and B"

SOLUTION $A + B$

b. "The product of l and w"

SOLUTION lw

c. "Two decreased by C"

SOLUTION $2 - C$

d. "Two less than C"

SOLUTION $C - 2$ Notice the reversed order of the 2 and the C

e. "Three times the square of x, plus ten"

SOLUTION $3x^2 + 10$

f. "Five subtracted from the quotient of S divided by T"

SOLUTION $\dfrac{S}{T} - 5$ Notice the reversed order of the 5 and the $\dfrac{S}{T}$

Exercises 4.1A
Set I

Translate each phrase or sentence into an algebraic expression.

1. The sum of x and 10   **2.** A added to B

3. Five less than A **4.** B diminished by C

5. The product of 6 and z **6.** A multiplied by B

7. x decreased by 7   **8.** Nine increased by A

9. Four less than 3 times x **10.** Six more than twice x

11. Subtract the product of u and v from x.

12. Subtract x from the product of P and Q.

13. The product of 5 and the square of x

14. The product of 10 and the cube of x

15. The square of the sum of A and B

16. The square of the quotient of A divided by B

17. The sum of x and 7, divided by y

18. T divided by the sum of x and 9

19. The product of x and the difference 6 less than y

20. The product of A and the sum 3 plus B

Exercises 4.1A
Set II

Translate each phrase or sentence into an algebraic expression.

1. Ten added to x **2.** The product of s and t

3. Three less than w **4.** x diminished by 4

5. The sum of u and v **6.** Five increased by x

7. Five decreased by y **8.** Ten plus x

9. Seven more than z **10.** Five times b

11. Twice F, subtracted from 15

12. The quotient of A divided by the sum of C and 10

13. The sum of x and the square of y

14. The quotient of the sum of A and C divided by B

15. The sum of the squares of A and B

16. The square of the sum of x and 4

17. The sum of x and y, divided by z

18. Twice the sum of x and y

19. The product of 7 and the sum of x and y

20. Three less than 5 times y

4.IB Translating English Expressions into Algebraic Expressions

In this section we show that, before we can change a *phrase* or *sentence* into an *algebraic expression*, we must represent *one* unknown number by a *variable* and then represent other unknown numbers (if there are any) *in terms of that same variable*. (In these examples and exercises, there is not enough information given to write an equation or to solve an applied problem.)

Converting a phrase or sentence into an algebraic expression	**1.** Identify which number or numbers are unknown. **2.** Represent *one* of the unknown numbers by a variable, and declare its meaning in a sentence of the form "Let $x = \ldots$." Express any other unknown numbers *in terms of the same variable* with a statement beginning "Then …." (See the Note below this box.) **3.** Convert the phrase or sentence into an algebraic expression, using the variable(s) in place of the unknown number(s).

☞ **Note** When there are two or more unknown quantities, it's important to realize that in step 2, when we write "Let $x = \ldots$," we are making a *choice*. The "Then …" statement means that we no longer have a choice.

For example, if we're told that the width of a rectangle is 4 less than the length, we can write "Let $x =$ the length." Once we've made that choice, however, the width *must* be $x - 4$, and we write "Then $x - 4 =$ the width."

On the other hand, we *can* choose to let x be the width, writing "Let $x =$ the width." Once we've decided to let x be the width, the length *must* be represented by $x + 4$ (or by $4 + x$), and we write "Then $x + 4 =$ the length."

EXAMPLE 2   Change the phrase "twice Albert's salary" into an algebraic expression.

SOLUTION

Step 1. Albert's salary is the unknown number.
Step 2. Let $S =$ Albert's salary.
Step 3. Then $2S$ is the algebraic expression for "twice Albert's salary."

EXAMPLE 3   Change the phrase "the cost of five stamps" into an algebraic expression.

SOLUTION

Step 1. The cost of one stamp is the unknown number.
Step 2. Let $c =$ the cost of one stamp (in cents).
Step 3. Then, since one stamp costs c cents, 5 stamps will cost 5 times c cents, or $5c$ cents. Therefore, $5c$ is the algebraic expression for "the cost of five stamps."

EXAMPLE 4   For the sentence "Mary is 10 years older than Nancy," represent both unknown numbers in terms of the same variable.

SOLUTION I

Step 1. There are two unknown numbers: Mary's age and Nancy's age.
Step 2. Let N = Nancy's age. Then $N + 10$ represents Mary's age, because Mary is 10 years older than Nancy.

SOLUTION 2

Step 1. There are two unknown numbers: Mary's age and Nancy's age.
Step 2. Let M = Mary's age. Then $M - 10$ represents Nancy's age, because Nancy is 10 years younger than Mary.

EXAMPLE 5   For the sentence "The sum of two numbers is 10," represent both unknown numbers in terms of the same variable.

SOLUTION   Let's consider some numbers whose sum is 10. If one number is 3, the other number must be $10 - 3$, or 7. If one number is 6, the other number must be $10 - 6$, or 4. If one number is -8, the other number must be $10 - (-8)$, or 18. (The *sum* of -8 and 18 is 10.) Notice that all of these solutions fit the pattern $10 - x$. Therefore, if one number is x, the other number must be $10 - x$.

Step 1. There are two unknown numbers.
Step 2. Let x = one of the unknown numbers
        Then $10 - x$ = the other unknown number

Notice that $x + (10 - x) = 10$; that is, the sum of the two numbers is 10.

Recall that integers that follow one another in sequence (without interruption) are called *consecutive integers*. Consecutive integers can be represented by variables. If the problem deals with two or more consecutive integers, we can let x equal the first integer. Then the second integer *must* be $x + 1$, the third integer *must* be $x + 2$, and so on.

EXAMPLE 6   For the phrase "the sum of three consecutive integers," represent the *sum* of the three integers in terms of one variable.

SOLUTION

Step 1. There are three unknown integers.
Step 2. Let x = the first integer
        Then $x + 1$ = the second integer
        and $x + 2$ = the third integer
Step 3. Then the *sum* of the three integers is $x + (x + 1) + (x + 2)$.

Adding 2 to any *even* integer gives the next *even* integer; for example, $8 + 2 = 10$, $-16 + 2 = -14$, and so on. Therefore, if a problem deals with consecutive *even* integers, we can let x equal the first even integer. Then the second even integer *must* be $x + 2$, the third even integer *must* be $x + 4$, and so on.

Adding 2 to any *odd* integer gives the next *odd* integer; for example, $9 + 2 = 11$, $-5 + 2 = -3$, and so forth. Therefore, if a problem deals with consecutive *odd* integers, we can let x equal the first odd integer. Then the second odd integer *must* be $x + 2$, the third odd integer *must* be $x + 4$, and so on.

(Notice that if the first integer is x, the second integer is $x + 2$, whether we're dealing with *even* integers or with *odd* integers.)

EXAMPLE 7

For the phrase "the sum of three consecutive odd integers," represent the *sum* of the three integers in terms of one variable.

SOLUTION

Step 1. There are three unknown odd integers.

Step 2. Let $\quad x =$ the first odd integer

Then $x + 2 =$ the second odd integer

and $\quad x + 4 =$ the third odd integer

Step 3. Then the *sum* of the three odd integers is $x + (x + 2) + (x + 4)$.

Some applied problems deal with geometric figures. Several formulas relating to geometric figures are given below. The *perimeter* of a geometric figure is the sum of the lengths of all its sides, or the *distance around* the figure.

Rectangle

Area: $A = lw$

Perimeter: $P = 2l + 2w$

where l is the length and w is the width

Square

Area: $A = s^2$

Perimeter: $P = 4s$

where s is the length of a side

Triangle

Area: $A = \frac{1}{2}bh$

where b is the base and h is the altitude

Perimeter: $P = a + b + c$

where a, b, and c are the lengths of the sides

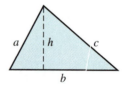

Parallelogram

Area: $A = bh$

where b is the base and h is the altitude

Circle

Area: $A = \pi r^2$

Circumference: $C = 2\pi r$

where $\pi \approx 3.14$ and r is the radius

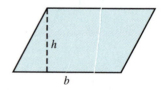

Rectangular solid

Volume: $V = lwh$

Surface area: $S = 2(lw + lh + wh)$

where l is the length, w is the width, and h is the height

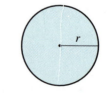

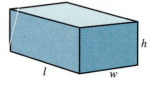

Cube

Volume: $V = e^3$
Surface area: $S = 6e^2$
where e is the length of an edge

Sphere

Volume: $V = \frac{4}{3}\pi r^3$
Surface area: $S = 4\pi r^2$
where $\pi \approx 3.14$ and r is the radius

Right circular cylinder

Volume: $V = \pi r^2 h$
where $\pi \approx 3.14$, r is the radius,
and h is the height

Right circular cone

Volume: $V = \frac{1}{3}\pi r^2 h$
where $\pi \approx 3.14$, r is the radius,
and h is the height

EXAMPLE 8 For the sentence "The height of a right circular cone is 2 ft more than its radius," represent the height and the radius in terms of the same variable.

SOLUTION 1

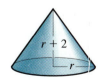

Step 1. There are two unknowns: the height and the radius.
Step 2. Let r = the radius (in feet)
       Then $r + 2$ = the height (in feet)

Because the height is 2 ft more than the radius

SOLUTION 2

Step 1. There are two unknowns: the height and the radius.
Step 2. Let h = the height (in feet)
       Then $h - 2$ = the radius (in feet)

Because the radius is 2 ft less than the height

EXAMPLE 9

If the length of a rectangular solid is 5 yd more than the height and the width is 2 yd less than the height, express the *volume* in terms of the height.

h

$h - 2$ $h + 5$

SOLUTION   The *volume* is the *length* times the *width* times the *height*.

Step 1. There are four unknowns: the height, the length, the width, and the volume.
Step 2. Let h = the height (in yards)
           Then $h + 5$ = the length (in yards)
           and $h - 2$ = the width (in yards)
Step 3. The volume is $(h + 5)(h - 2)h$, or $h(h + 5)(h - 2)$.

👉 **Note**   It would also be possible to let l equal the length and then to express the volume in terms of l or to let w equal the width and then to express the volume in terms of w.

Exercises 4.IB

Set I

In Exercises 1–26, represent one unknown number by a variable, and then convert the phrase or sentence into an algebraic expression.

1. Fred's salary plus seventy-five dollars

2. Jaime's salary minus forty-two dollars

3. Two less than the number of children in Mr. Moore's family

4. The number of players on Jerry's team plus two more players

5. Four times Joyce's age

6. One-fourth of Rene's age

7. Twenty times the cost of a record, increased by eighty-nine cents

8. Seventeen cents less than 5 times the cost of a ballpoint pen

9. One-fifth the cost of a hamburger

10. The length of the building, divided by 8

11. Five times the speed of a car plus 100 miles per hour

12. Twice the speed of a car diminished by 40 miles per hour

13. Ten less than 5 times the square of an unknown number

14. Eight more than 4 times the cube of an unknown number

15. Take the quotient of 7 divided by an unknown number away from 50.

16. Add the quotient of an unknown number divided by 9 to 15.

17. Eleven feet more than twice the length of a rectangle

18. One inch less than the diameter of a circle

19. 2.2 times a certain weight in kilograms

20. 0.62 times a certain distance in kilometers

21. The sum of 8 and an unknown number, divided by the square of that unknown number

22. The sum of the square of an unknown number and 11, divided by the unknown number

23. The sum of 32 and nine-fifths the Celsius temperature

24. Five-ninths times the result of subtracting 32 from the Fahrenheit temperature

25. Six plus the radius of a sphere

26. Five plus the radius of a circle

In Exercises 27–32, represent *one* unknown number by a variable, and express any other unknown numbers in terms of the *same* variable.

27. The length of a rectangle is 12 cm more than its width.

28. The altitude of a triangle is 7 cm less than its base.

29. The combined weight of Walter and Lucy is 320 lb.

30. The combined weight of Teresa and Carlos is 224 lb.

31. The height of a right circular cylinder is 2 times its radius.

32. The radius of a right circular cone is 4 times its height.

33. Express the sum of three consecutive integers in terms of one variable.

34. Express the sum of two consecutive integers in terms of one variable.

35. Express the sum of four consecutive odd integers in terms of one variable.

36. Express the sum of three consecutive even integers in terms of one variable.

37. If the radius of a right circular cylinder is 4 cm less than its height, express the volume of the cylinder in terms of one variable.

38. If the base of a triangle is 5 m less than its altitude, express the area of the triangle in terms of one variable.

39. If the height of a right circular cone is 4 times its radius, express the volume of the cone in terms of one variable.

40. If the radius of a right circular cylinder is 6 times its height, express the volume of the cylinder in terms of one variable.

Exercises *4.IB*
Set II

In Exercises 1–26, represent one unknown number by a variable, and then convert the phrase or sentence into an algebraic expression.

I. The cost of a television set plus thirty-eight dollars

2. Sixteen times the cost of a refrigerator

3. Eight less than the length of a certain bridge

4. One-third of Carol's age

5. Eight times the width of a certain window

6. The cost of seven stamps

7. Three times the cost of a videotape increased by forty-five cents

8. Nathan's age minus 6

9. Two-thirds the cost of a compact disc

10. Nine less than one-fourth of Lyndsey's height

II. Ten times the height of a building, minus 32 m

12. The square of the sum of 8 and an unknown number

13. Five less than the product of 4 and an unknown number

14. One-half the sum of 7 and an unknown number

15. Six more than 3 times the square of an unknown number

16. Eight added to the sum of 10 and an unknown number

17. Three meters less than 3 times the height of the building

18. The speed of the train, divided by 6

19. Twice the result of subtracting an unknown number from 5

20. 8.6 times a certain length in kilometers

21. The square of the result of subtracting 12 from an unknown number

22. The sum of the squares of 3 and an unknown number

23. The square of the sum of 3 and an unknown number

24. The product of 4 and the square of the length of a side of a square

25. Fourteen plus the radius of a circle

26. Five times the radius of a sphere

In Exercises 27–32, represent *one* unknown number by a variable, and express any other unknown numbers in terms of the *same* variable.

27. The sum of two numbers is −22.

28. Henry is three years younger than his sister Claire.

29. The combined age of Esther and Marge is 43.

30. Mrs. Lopez is twenty-nine years older than her daughter Flora.

31. The altitude of a triangle is 8 cm less than its base.

32. The radius of a right circular cylinder is 2 times its height.

33. Express the sum of four consecutive integers in terms of one variable.

34. Express the product of two consecutive integers in terms of one variable.

35. Express the sum of four consecutive even integers in terms of one variable.

36. Express the product of two consecutive odd integers in terms of one variable.

37. If the height of a right circular cone is 6 cm more than its radius, express the volume of the cone in terms of one variable.

38. If the length of a rectangle is 9 yd more than its width, express the perimeter of the rectangle in terms of one variable.

39. If the altitude of a triangle is 3 times its base, express the area of the triangle in terms of one variable.

40. If the height of a right circular cylinder is 5 ft less than its radius, express the volume of the cylinder in terms of one variable.

4.2 Translating Statements into Equations

In this section, we show how to write an equation that represents the facts given in a problem that is expressed in words. We call the steps listed below a *mathematical model*. (We will not yet be *solving* applied problems.)

Translating statements into equations

1. Represent one unknown number by a variable. Represent any *other* unknowns in terms of the same variable.
2. Break up the sentences into key words and phrases.
3. Represent each key word or phrase by an algebraic expression.
4. Arrange the algebraic expressions into an equation that represents the facts given in the problem.

EXAMPLE 1

Translate each of the following sentences into an equation.

a. Fifteen plus twice an unknown number is 37.

SOLUTION

Step 1. Let x = the unknown number. ←——Do not omit this step

Step 2. Fifteen plus twice an unknown number is 37

Step 3.   15   +   2 · x =   37

Step 4. $15 + 2x = 37$ Equation

b. Three times an unknown number is equal to 12 increased by the unknown number.

SOLUTION

Step 1. Let x = the unknown number.

Step 2. Three times an unknown number is equal to 12 increased by the unknown number

Step 3.   3   · x =   12   + x

Step 4. $3x = 12 + x$

c. One-third of an unknown number is 7.

SOLUTION

Step 1. Let x = the unknown number.

Step 2. One-third of an unknown number is 7

Step 3. $\frac{1}{3}$ · x =   7

Step 4. $\frac{1}{3}x = 7$

d. Twice the sum of 6 and an unknown number is equal to 20.

SOLUTION

Step 1. Let x = the unknown number.

Step 2. Twice the sum of 6 and an unknown number is 20

Step 3.   2 · $(6 + x)$ =   20

Step 4. $2(6 + x) = 20$

Note In Example 1d, the parentheses around $6 + x$ are essential, because the entire sum $6 + x$ is to be multiplied by 2, and $2(6 + x) \neq 2 \cdot 6 + x$.

EXAMPLE 2

Write the equation that represents these facts: A piece of wire 63 cm long is to be cut into two pieces. One piece is to be 15 cm longer than the other piece.

SOLUTION 1 (There are *two* unknowns.)

Step 1. Let x = the length of the shorter piece (in centimeters)
Then $x + 15$ = the length of the longer piece (in centimeters)
Step 2. Sum of lengths of two pieces is 63
Steps 3 and 4. $x + (x + 15)$ = 63

SOLUTION 2 (There are *two* unknowns.)

Step 1. Let y = the length of the longer piece (in centimeters)
Then $y - 15$ = the length of the shorter piece (in centimeters)
Step 2. Sum of lengths of two pieces is 63
Steps 3 and 4. $y + (y - 15)$ = 63

EXAMPLE 3

Write an equation that represents these facts: The sum of three consecutive odd integers is 75.

SOLUTION

Step 1. Let x = the first odd integer
Then $x + 2$ = the second odd integer
and $x + 4$ = the third odd integer
Step 2. Sum of three consecutive odd integers is 75
Steps 3 and 4. $x + (x + 2) + (x + 4)$ = 75

EXAMPLE 4

Write an equation that represents these facts: The length of a rectangle is 5 ft more than its width, and its perimeter is 44 ft. (The formula for the perimeter of a rectangle is $P = 2l + 2w$.)

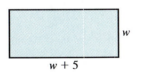

SOLUTION 1

Step 1. Let w = the width of the rectangle (in feet)
Then $w + 5$ = the length of the rectangle (in feet)
Step 2. The perimeter is 44
Steps 3 and 4. $2(w + 5) + 2w$ = 44

SOLUTION 2

Step 1. Let l = the length of the rectangle (in feet)
Then $l - 5$ = the width of the rectangle (in feet)
Step 2. The perimeter is 44
Steps 3 and 4. $2l + 2(l - 5)$ = 44

Exercises 4.2
Set 1

In each exercise, write an equation that represents the given facts. Be sure to state what your variable represents. Do not solve the equation.

1. Thirteen more than twice an unknown number is 25.

2. Twenty-five more than 3 times an unknown number is 34.

3. Five times an unknown number, decreased by 8, is 22.

4. Four times an unknown number, decreased by 5, is 15.

5. Seven minus an unknown number is equal to the unknown number plus 1.

6. Six plus an unknown number is equal to 12 decreased by the unknown number.

7. One-fifth of an unknown number is 4.

8. An unknown number divided by 12 equals 6.

9. When 4 is subtracted from one-half of an unknown number, the result is 6.

10. When 5 is subtracted from one-third of an unknown number, the result is 4.

11. Twice the sum of 5 and an unknown number is equal to 26.

12. Four times the sum of 9 and an unknown number is equal to 18.

13. When the sum of an unknown number and itself is multiplied by 3, the result is 24.

14. Five times the sum of an unknown number and itself is 40.

15. A 75-m rope is to be cut into two pieces. One piece is to be 13 m longer than the other piece.

16. A 46-m wire is to be cut into two pieces. One piece is to be 8 m shorter than the other piece.

17. A 72-cm piece of wire is to be cut into two pieces. One piece is to be twice as long as the other piece.

18. A 72-cm piece of wire is to be cut into two pieces. One piece is to be 3 times as long as the other piece.

19. Michael bought eight more cans of peaches than cans of pears. Altogether, he bought forty-two cans of these two fruits.

20. Kristy bought eleven more cans of peas than cans of corn. Altogether, she bought forty-nine cans of these two vegetables.

21. The sum of four consecutive integers is 106.

22. The sum of three consecutive integers is -72.

23. The length of a rectangle is 4 cm more than its width, and its perimeter is 36 cm.

24. The width of a rectangle is 6 ft less than its length, and its perimeter is 64 ft.

Exercises 4.2
Set II

In each exercise, write an equation that represents the given facts. Be sure to state what your variable represents. Do not solve the equation.

1. Fifteen more than twice an unknown number is 27.

2. Twice the sum of an unknown number and 9 is 46.

3. Four times an unknown number, decreased by 9, is 19.

4. When 5 is subtracted from one-half of an unknown number, the result is 19.

5. Eighteen minus an unknown number is equal to 4 plus the unknown number.

6. When 7 is added to an unknown number, the result is twice that unknown number.

7. Three-eighths of an unknown number is 27.

8. When a number is decreased by 5, the difference is half of the number.

9. When 8 is subtracted from two-thirds of an unknown number, the result is 16.

10. When 3 times an unknown number is subtracted from 20, the result is the unknown number.

11. Three times the result of subtracting an unknown number from 8 is 12.

12. Four times the result of adding an unknown number to itself is 96.

13. When the sum of an unknown number and itself is multiplied by 4, the result is 56.

14. When 6 is subtracted from 5 times an unknown number, the result is the same as when 4 is added to 3 times the unknown number.

15. A 7-yd piece of fabric is to be cut into two pieces. One piece is to be 3 yd longer than the other piece.

16. Brett bought 3 times as many bottles of catsup as jars of mustard. Altogether, he bought twelve containers of these two products.

17. A 42-m rope is to be cut into two pieces. One piece is to be 6 times as long as the other piece.

18. Alice bought 5 times as many skeins (balls) of pink yarn as skeins of white yarn. Altogether, she bought eighteen skeins of these two colors.

19. Teri bought four more cans of car wax than cans of rubbing compound. Altogether, she bought eighteen cans of these two products.

20. A 30-m cable is to be cut into two pieces. One piece is to be 8 m longer than the other.

21. The sum of four consecutive integers is 2.

22. The sum of three consecutive even integers is 78.

23. The length of a rectangle is 19 yd more than its width, and its perimeter is 62 yd.

24. The lengths of all three sides of a triangle are equal, and its perimeter is 42 cm.

4.3   Solving Applied Problems by Using Algebra

Earlier, we showed how to convert the words of an applied problem into an equation. In this section, we show how to solve such problems, some of which lead to inequalities rather than to equations.

In the following examples of solving applied problems, we use the notation Step 1, Step 2, and so forth for the steps you will be *writing*.

Suggestions for solving applied problems

Read   To solve an applied problem, first read it very carefully. *Be sure you understand the problem.* Read it several times, if necessary.

Think   Determine what *type* of problem it is, if possible.* Determine what is unknown. What is being asked for is often found in the last sentence of the problem, which may begin with "What is the..." or "Find the...." Is enough information given so that you *can* solve the problem? Do you need a special formula? What operation(s) must be used? What kind of number is reasonable as the answer? What kind of answer would have to be rejected?

Sketch   Draw a sketch *with labels*, if it might be helpful.

Step 1. Represent one unknown number by a variable, and declare its meaning in a sentence of the form "Let $x = $ ...." Then reread the problem to see how you can represent any other unknown numbers in terms of the same variable, declaring their meaning in sentences that begin "Then ...."

Reread   Reread the sentences, breaking them up into key words and phrases.

Step 2. Translate each key word or phrase into an algebraic expression, and arrange these expressions into an equation or inequality that represents the facts given in the problem.

Step 3. Solve the equation or inequality.

Step 4. Solve for *all* the unknowns asked for in the problem.

Step 5. Check the solution(s) *in the word statement.*

Step 6. State the results clearly.

---

*As we mentioned, we discuss several general types of applied problems, such as money problems, mixture problems, and so on, in this chapter.

EXAMPLE 1 Seven increased by 3 times an unknown number is 13. What is the unknown number?

SOLUTION

Step 1. Let x = the unknown number.

Reread	Seven	increased by	three	times	an unknown number	is	13

Step 2.   7   +   3   · x =   13

Step 3. $7 + 3x = 13$

$-7 + 7 + 3x = -7 + 13$ Adding -7 to both sides

$3x = 6$

$\dfrac{3x}{3} = \dfrac{6}{3}$ Dividing both sides by 3

Step 4. $x = 2$

✓ **Step 5. Check**

Seven	increased by	three	times	an unknown number	is	13

7   +   3   ·   (2)   =   13

$7 + 3(2) \overset{?}{=} 13$ The unknown number is replaced by 2

$7 + 6 \overset{?}{=} 13$

$13 = 13$

Step 6. Therefore, the unknown number is 2.

Note To check an applied problem, you must check the solution in the *word statement*. Any error that may have been made in writing the equation will not be discovered if you simply substitute the solution into the equation.

EXAMPLE 2 Four times an unknown number is equal to twice the sum of 5 and that unknown number. Find the unknown number.

SOLUTION

Step 1.   Let x = the unknown number.

Reread	Four	times	an unknown number	is equal to	twice	the sum of 5 and that unknown number

Step 2.   4   · x =   2 · $(5 + x)$

Step 3. $4x = 2(5 + x)$ The parentheses are essential, since the sum $5 + x$ must be multiplied by 2

$4x = 10 + 2x$ Using the distributive property

$4x + (-2x) = 10 + 2x + (-2x)$ Adding $-2x$ to both sides

$2x = 10$

$\dfrac{2x}{2} = \dfrac{10}{2}$ Dividing both sides by 2

Step 4. $x = 5$

√ **Step 5.**   *Check*   Four times 5 is 20. The sum of 5 and the unknown number (5) is 10, and twice 10 is 20.

Step 6.   Therefore, the unknown number is 5.

EXAMPLE 3

When 7 is subtracted from one-half of an unknown number, the result is 11. What is the unknown number?

SOLUTION   We must be sure 7 goes to the *right* of the subtraction symbol.

Step 1.   Let x = the unknown number.

Reread	When 7 is subtracted from	one-half of an unknown number	the result is	11

Step 2. $\frac{1}{2} \cdot x$ $-$ 7 $=$ 11

Step 3. $\frac{1}{2}x - 7 = 11$

$\frac{1}{2}x - 7 + 7 = 11 + 7$ Adding 7 to both sides

$\frac{1}{2}x = 18$

$2\left(\frac{1}{2}x\right) = 2(18)$ Multiplying both sides by 2

Step 4. $x = 36$

√ **Step 5.**   *Check*   One-half of 36 is 18. When 7 is subtracted from 18, the result is 11.

Step 6.   Therefore, the unknown number is 36.

It is shown in geometry that the sum of the measures of the angles of a triangle is always 180°. Example 4 uses this property.

EXAMPLE 4

If the measure of one angle of a triangle is 73° and the measure of another angle is 54°, what is the measure of the third angle?

SOLUTION

Step 1.   Let x = the measure of the third angle (in degrees).

Reread	The sum of all the angles	equals	180

Step 2. $73 + 54 + x$ $=$ 180

Step 3. $73 + 54 + x = 180$

$127 + x = 180$ Combining like terms

$-127 + 127 + x = -127 + 180$ Adding -127 to both sides

Step 4. $x = 53$

√ **Step 5.**   *Check* $73° + 54° + 53° = 180°$

Step 6.   Therefore, the measure of the third angle is 53°.

EXAMPLE 5

The length of a rectangle is 25 in., and its perimeter is 84 in. Find the width of the rectangle.

SOLUTION

Think The formula for the perimeter of a rectangle is $P = 2l + 2w$.

Step 1. Let w = the width of the rectangle (in inches).

$$2l + 2w \quad = \quad P$$

Reread The length is 25, the width is w | the perimeter is 84

Sketch

Step 2. $2(25) + 2w \quad = \quad 84$

Step 3.
$$2(25) + 2w = 84$$
$$50 + 2w = 84 \qquad \text{Simplifying}$$
$$-50 + 50 + 2w = -50 + 84 \qquad \text{Adding } -50 \text{ to both sides}$$
$$2w = 34$$
$$\frac{2w}{2} = \frac{34}{2} \qquad \text{Dividing both sides by 2}$$

Step 4. $w = 17$

√ **Step 5.** *Check* The perimeter is $2(25 \text{ in.}) + 2(17 \text{ in.}) = 50 \text{ in.} + 34 \text{ in.} = 84 \text{ in.}$

Step 6. Therefore, the width of the rectangle is 17 in.

Example 6 illustrates the solution of a problem that leads to an inequality.

EXAMPLE 6

In an English class, any student needs at least 730 points in order to earn an A, and the final exam is worth 200 points. If Shirley has 545 points just before the final exam, what range of scores will give her an A for the course?

SOLUTION

Step 1. Let x = Shirley's score on the final.

Reread Score before final | plus | score on final | is greater than or equal to | 730

Step 2. $545 \quad + \quad x \quad \geq \quad 730$

Step 3.
$$545 + x \geq 730$$
$$-545 + 545 + x \geq -545 + 730 \qquad \text{Adding } -545 \text{ to both sides}$$

Step 4. $x \geq 185$

√ **Step 5.** *Check* Shirley's score on the final exam can't be greater than 200, since the final exam is worth only 200 points. We will show the check for three different numbers:

Score of 185: $185 + 545 = 730$, and $730 \geq 730$

Score of 192: $192 + 545 = 737$, and $737 \geq 730$

Score of 200: $200 + 545 = 745$, and $745 \geq 730$

Any other numbers between 185 and 200 would also have checked.

Step 6. Therefore, the range of scores that will give Shirley an A for the course is any score greater than or equal to 185 (and less than or equal to 200, since the final exam is worth 200 points).

Exercises 4.3
Set I

In each exercise, set up the problem algebraically and solve. Be sure to state what your variable represents.

In Exercises 1–10, find the unknown number, and check the solution.

1. When twice an unknown number is added to 13, the sum is 25.

2. When 25 is added to 3 times an unknown number, the sum is 34.

3. Five times an unknown number, decreased by 8, is 22.

4. Four times an unknown number, decreased by 5, is 15.

5. Seven minus an unknown number is equal to the unknown number plus 1.

6. Six plus an unknown number is equal to 12 decreased by the unknown number.

7. When 4 is subtracted from one-half of an unknown number, the result is 6.

8. When 5 is subtracted from one-third of an unknown number, the result is 4.

9. Twice the sum of 5 and an unknown number is equal to 26.

10. Four times the sum of 9 and an unknown number is equal to 20.

In Exercises 11–28, solve for the unknowns, and check the solutions.

11. A 36-cm piece of wire is to be cut into two pieces. One of the pieces is to be 10 cm longer than the other piece. Find the length of each piece.

12. A 10-yd piece of fabric is to be cut into two pieces. One of the pieces is to be 2 yd longer than the other piece. Find the length of each piece.

13. Rebecca bought seven more packages of dried apples than packages of dried apricots. If she bought fifteen packages of these two fruits altogether, how many of each did she buy?

14. David bought three more skeins of blue yarn than skeins of green yarn. If he bought thirteen skeins of these two colors altogether, how many of each color did he buy?

15. Find the area of this triangle.

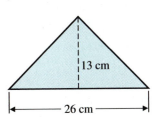

16. Find the volume and the surface area of this rectangular solid.

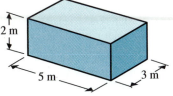

17. Find the approximate volume and surface area of this sphere (round off the answers to two decimal places).

18. Find the approximate area and circumference of this circle (round off the answers to two decimal places).

19. The sum of three consecutive odd integers is 177. Find the integers.

20. The sum of three consecutive even integers is −144. Find the integers.

21. Four times the sum of the first and third of three consecutive integers is 140 more than the second integer. Find the integers.

22. Three times the sum of the first and third of three consecutive integers is 75 more than the second integer. Find the integers.

23. If the measure of one angle of a triangle is 62° and the measure of another angle is 47°, what is the measure of the third angle?

24. If the measure of one angle of a triangle is 18° and the measure of another angle is 37°, what is the measure of the third angle?

25. The length of a rectangle is twice its width, and its perimeter is 102 ft. Find the length and the width.

26. The length of a rectangle is 4 times its width, and its perimeter is 160 cm. Find the length and the width.

27. The width of a rectangle is 5 ft, and its perimeter is 44 ft. Find the length of the rectangle.

28. The width of a rectangle is 7 yd, and its perimeter is 38 yd. Find the length of the rectangle.

In Exercises 29–36, find the unknown number, and check the solution.

29. When twice the sum of 4 and an unknown number is added to the unknown number, the result is the same as when 10 is added to the unknown number.

30. When 6 is subtracted from 5 times an unknown number, the result is the same as when 4 is added to 3 times the unknown number.

31. Three times the sum of 8 and twice an unknown number is equal to 4 times the sum of 3 times the unknown number and 8.

32. Five times the sum of 4 and 6 times an unknown number is equal to 4 times the sum of 10 and twice the unknown number.

33. When 3 times the sum of 4 and an unknown number is subtracted from 10 times the unknown number, the result is equal to 5 times the sum of 9 and twice the unknown number.

34. When twice the sum of 5 and an unknown number is subtracted from 5 times the sum of 6 and twice the unknown number, the result is equal to zero.

35. When 5.75 times the sum of 6.94 and an unknown number is subtracted from 8.66 times the unknown number, the result is equal to 4.69 times the sum of 8.55 and 3.48 times the unknown number. Round off the answer to two decimal places.

36. When 8.23 is subtracted from 4.85 times an unknown number, the result is the same as when 12.62 is added to 5.49 times the unknown number. Round off the answer to two decimal places.

In Exercises 37–40, find all possible solutions for each problem.

37. The sum of an unknown number and 18 is to be at least 5. What is the range of values that the unknown number can have?

38. The sum of an unknown number and 12 is to be at least 47. What is the range of values that the unknown number can have?

39. A rope less than 80 m long is to be cut into two pieces. One piece must be 37 m long. What will the length of the other piece be?

40. A chain more than 25 m long is to be cut into two pieces. One piece must be 8 m long. What will the length of the other piece be?

Exercises 4.3
Set II

In each exercise, set up the problem algebraically and solve. Be sure to state what your variable represents.

In Exercises 1–10, find the unknown number, and check the solution.

1. When 4 times an unknown number is added to 21, the sum is 105.

2. When 7 is added to an unknown number, the result is twice that unknown number.

3. Three times an unknown number, decreased by 12, is 6.

4. When 3 times an unknown number is subtracted from 20, the result is the unknown number.

5. Eighteen minus an unknown number is equal to the unknown number minus 16.

6. One-fifth of an unknown number is 4.

7. When 3 is subtracted from one-third of an unknown number, the result is 7.

8. An unknown number divided by 12 equals 6.

9. Five times the sum of an unknown number and itself is 40.

10. When an unknown number is subtracted from 12, the difference is one-third of the number.

In Exercises 11–28, solve for the unknowns, and check the solutions.

11. A 12-yd piece of fabric is to be cut into two pieces. One of the pieces must be 3 yd longer than the other piece. Find the length of each piece.

12. Susan bought 5 times as many cassette tapes as compact discs. If she purchased twelve of these items altogether, how many of each kind did she buy?

13. Jonathan bought four more packages of unsalted crackers than packages of salted crackers. If he bought sixteen packages of crackers altogether, how many of each kind did he buy?

14. A 42-m piece of cord is to be cut into two pieces. If one piece must be 15 m long, what will the length of the other piece be?

15. Find the area and perimeter of this rectangle.

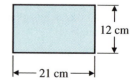
12 cm
21 cm

16. Find the area and perimeter of this square.

14 yd

17. Find the approximate volume of this right circular cone (round off the answer to two decimal places).

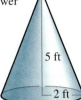

5 ft

2 ft

18. Find the approximate volume of this right circular cylinder (round off the answer to two decimal places).

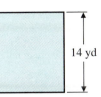

3 m

2 m

19. The sum of three consecutive even integers is −102. Find the integers.

20. The sum of three consecutive odd integers is 27. Find the integers.

21. Twice the sum of the first and third of three consecutive integers is 102 more than the second integer. Find the integers.

22. Three times the sum of the first and third of three consecutive odd integers is 55 more than the second integer. Find the integers.

23. If the measure of one angle of a triangle is 90° and the measure of another angle is 37°, what is the measure of the third angle?

24. If the measures of the three angles of a triangle are equal, find the measure of the angles.

25. The length of a rectangle is 5 times its width, and its perimeter is 204 ft. Find the length and the width.

26. The length of a rectangle is 4 m, and its area is 20 sq. m. Find the width of the rectangle.

27. The length of a rectangle is 7 cm, and its perimeter is 24 cm. Find the width of the rectangle.

28. The base of a triangle is 4 yd, and its area is 38 sq. yd. Find the altitude of the triangle.

In Exercises 29–36, find the unknown number, and check the solution.

29. When 4 times the sum of 5 and an unknown number is added to the unknown number, the result is the same as when 32 is added to the unknown number.

30. Twice the result of subtracting an unknown number from 5 is 8.

31. Twice the sum of 3 and twice an unknown number is equal to 5 times the sum of the unknown number and 1.

32. When an unknown number is subtracted from 11, the result is the same as when the unknown number is added to 3.

33. If the sum of an unknown number and 12 is subtracted from 4 times the unknown number, the result is the unknown number less 4.

34. When 4 times an unknown number is subtracted from 16, the result is twice the sum of 12 and twice the unknown number.

35. When 3.48 times the sum of 9.06 and an unknown number is subtracted from 5.37 times the unknown number, the result is equal to 4.65 times the sum of 2.83 and 8.34 times the unknown number. Round off the answer to two decimal places.

36. Five times the sum of 18 and an unknown number is 2 less than the unknown number.

In Exercises 37–40, find all possible solutions for each problem.

37. The sum of an unknown number and 7 is to be at least 3. What is the range of values that the unknown number can have?

38. Five times an unknown number, plus 7, is greater than 42. What is the range of values that the unknown number can have?

39. A piece of wire less than 35 cm long is to be cut into two pieces. One piece must be 13 cm long. What is the range of values for the length of the other piece?

40. A rope less than 45 m long is to be cut into two pieces. One piece must be 4 times as long as the other piece. What is the range of values for the *shorter* piece?

4.4 Solving Money Problems

In this section, we discuss a type of applied problem commonly referred to as a "coin problem." Not all problems in this section deal with coins, but the method of solving the problems is essentially the same.

4.4A Getting Ready to Solve Money Problems

One important relationship used in solving coin or money problems is the following:

$$
\begin{pmatrix} \text{The value per} \\ \text{item of one} \\ \text{kind of item} \end{pmatrix} \times \begin{pmatrix} \text{The number} \\ \text{of} \\ \text{those items} \end{pmatrix} = \begin{pmatrix} \text{The total value} \\ \text{of that kind} \\ \text{of item} \end{pmatrix}
$$

EXAMPLE 1 Find the number of cents in eight nickels.

SOLUTION
          The value of one nickel (in cents)
          The number of nickels
$5(8) = \boxed{40}$ ← The number of cents

EXAMPLE 2 Find the number of cents in seven quarters.

SOLUTION
          The value of one quarter (in cents)
          The number of quarters
$25(7) = \boxed{175}$ ← The number of cents

EXAMPLE 3 Find the number of cents in x dimes.

SOLUTION
          The value of one dime (in cents)
          The number of dimes
$10(x) = \boxed{10x}$ ← The number of cents

EXAMPLE 4 Find the total value (in cents) of y 35¢ candy bars.

SOLUTION
          The value of one candy bar (in cents)
          The number of candy bars
$35(y) = \boxed{35y}$ ← The total value (in cents)

Another important relationship used in solving coin or money problems is this one:

$$
\begin{pmatrix} \text{The value} \\ \text{of the first} \\ \text{kind of item} \end{pmatrix} + \begin{pmatrix} \text{The value} \\ \text{of the second} \\ \text{kind of item} \end{pmatrix} = \begin{pmatrix} \text{The total value} \\ \text{of both kinds} \\ \text{of items} \end{pmatrix}
$$

The relationship can, of course, be extended to more than two kinds of items. We use this relationship in Examples 5 through 8.

EXAMPLE 5 Find the total value (in dollars) of two adults' movie tickets that cost $5.00 each and four children's tickets that cost $2.50 each.

SOLUTION

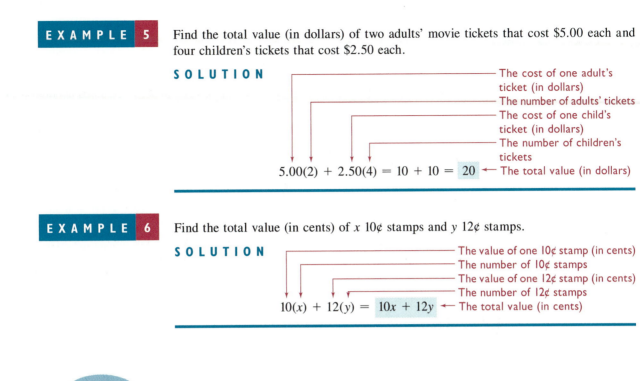

The cost of one adult's ticket (in dollars)
The number of adults' tickets
The cost of one child's ticket (in dollars)
The number of children's tickets

$5.00(2) + 2.50(4) = 10 + 10 = \boxed{20}$ ← The total value (in dollars)

EXAMPLE 6 Find the total value (in cents) of x 10¢ stamps and y 12¢ stamps.

SOLUTION

The value of one 10¢ stamp (in cents)
The number of 10¢ stamps
The value of one 12¢ stamp (in cents)
The number of 12¢ stamps

$10(x) + 12(y) = \boxed{10x + 12y}$ ← The total value (in cents)

Exercises 4.4A
Set I

In each exercise, express the amount requested either as a number or as an algebraic expression.

1. Find the number of cents in seven nickels.

2. Find the number of cents in eleven quarters.

3. Find the number of cents in nine 50¢ pieces.

4. Find the number of cents in twelve dimes.

5. Find the number of cents in x quarters.

6. Find the number of cents in y nickels.

7. Find the total cost of seven adults' tickets and five children's tickets if adults' tickets cost $1.50 each and children's tickets cost 75¢ each.

8. If adults' tickets are $2.50 each and children's tickets are $1.25 each, what will four adults' and nine children's tickets cost?

9. If adults' tickets are $3.50 each and children's tickets are $1.90 each, what will x adults' and y children's tickets cost?

10. Find the total cost of x adults' and y children's tickets if adults' tickets cost $2.75 each and children's tickets cost $1.50 each.

11. Find the total cost of x 25¢ stamps and y 6¢ stamps.

12. Find the total cost of x 4¢ stamps and y 2¢ stamps.

Writing Problems

Express the answers in your own words and in complete sentences.

1. Explain how to find the total value of three dimes and twelve nickels.

2. Explain how to find the total value of five adults' tickets and seven children's tickets if the adults' tickets cost $3.50 each and the children's tickets cost $2.50 each.

3. Suppose Demetrius has x dimes and y nickels. Explain why $10x + 5y$ is *not* the correct algebraic expression for the total number of coins Demetrius has.

Exercises 4.4A
Set II

In each exercise, express the amount requested either as a number or as an algebraic expression.

1. Find the number of cents in nine nickels.

2. Find the number of cents in z dimes.

3. Find the number of cents in fifteen dimes.

4. Find the number of cents in y quarters.

5. Find the number of cents in x 50¢ pieces.

6. Find the number of cents in x dimes and $(x + 4)$ nickels.

7. If adults' tickets cost $4.75 each and children's tickets cost $2.75 each, what will five adults' and three children's tickets cost?

8. If adults' tickets cost $5.25 each and children's tickets cost $3.00 each, what will x adults' tickets and $4x$ children's tickets cost?

9. Find the total cost of x adults' and y children's tickets if adults' tickets cost $4.50 each and children's tickets cost $2.25 each.

10. Find the total cost of x adults' and $(x + 6)$ children's tickets if adults' tickets cost $5.00 each and children's tickets cost $2.50 each.

11. Find the total cost of x 13¢ stamps and y 15¢ stamps.

12. Find the total cost of x 22¢ stamps and $5x$ 18¢ stamps.

4.4B Solving Money Problems

In the examples in this section, we will solve money problems by using the suggestions that were given earlier for solving all applied problems.

EXAMPLE 7   Pete has $115 in his wallet, and all the bills are $10 bills and $5 bills. If his wallet contains seventeen bills altogether, how many $10 bills are there? How many $5 bills are there?

SOLUTION

Think   There are two unknown numbers and, therefore, two choices for the variable. We can let x be the number of $10 bills *or* the number of $5 bills. We'll let x be the number of $10 bills. To answer the question "How many $5 bills are in the wallet?" let's consider some possibilities: Since there are seventeen bills altogether, if there are 3 $10 bills, there will be $17 - 3$, or 14, $5 bills; if there are 15 $10 bills, there will be $17 - 15$, or 2, $5 bills; if there are 8 $10 bills, there will be $17 - 8$, or 9, $5 bills. Notice that all of these solutions fit the pattern $17 - x$. Therefore, if x is the number of $10 bills, there will be $17 - x$ $5 bills.

Step 1. Let $\quad x =$ the number of $10 bills
Then $17 - x =$ the number of $5 bills

Think   This is a money problem. The value of each $10 bill is $10; therefore, the *value* of x $10 bills is $10x$ dollars. The value of each $5 bill is $5; therefore, the *value* of $17 - x$ $5 bills is $5(17 - x)$ dollars.

Reread   Value of $10 bills   plus   value of $5 bills   is $115

Step 2. $\qquad\qquad 10x \qquad + \qquad 5(17 - x) \qquad = \qquad 115$ (in dollars)

Step 3. $10x + 5(17 - x) = 115$

$\qquad\qquad 10x + 85 - 5x = 115$ Using the distributive property

$\qquad\qquad\qquad 5x + 85 = 115$ Combining like terms

$\qquad 5x + 85 + (-85) = 115 + (-85)$ Adding -85 to both sides

$\qquad\qquad\qquad\qquad 5x = 30$

$\qquad\qquad\qquad\qquad \dfrac{5x}{5} = \dfrac{30}{5}$ Dividing both sides by 5

Step 4. $x = 6$ The number of $10 bills

                 $17 - x = 17 - 6 = 11$ The number of $5 bills

✓ **Step 5. Check**   There are 17 bills altogether if there are 6 $10 bills and 11 $5 bills. The value of 6 $10 bills is $60. The value of 11 $5 bills is 11($5), or $55. The sum of these values is $60 + $55, or $115.

Step 6. Therefore, there are six $10 bills and eleven $5 bills.

We solved the problem in Example 7 by letting x be the number of $10 bills. We suggest that you solve the problem by letting x be the number of $5 bills. Your final answer should match the answer in Example 7.

EXAMPLE 8   Dianne has $3.20 in nickels, dimes, and quarters. If she has seven more dimes than quarters and 3 times as many nickels as quarters, how many of each kind of coin does she have?

SOLUTION

Think   This is a money problem; let's express everything in terms of cents. There are three unknown numbers and, therefore, three choices for the variable. We can let N be the number of nickels, D be the number of dimes, or Q be the number of quarters. Because the problem expressed both the number of dimes and the number of nickels in terms of the number of quarters, it will be easiest to let Q be the number of quarters.

Step 1. Let $Q =$ the number of quarters
            Then $Q + 7 =$ the number of dimes   Because there are seven more dimes than quarters

            and $3Q =$ the number of nickels   Because there are 3 times as many nickels as quarters

Think   The value of each quarter is 25¢, so the value of Q quarters is $25Q$ cents. The value of $(Q + 7)$ dimes is $10(Q + 7)$ cents; the value of $(3Q)$ nickels is $5(3Q)$ cents.

Reread	Value of quarters	plus	value of dimes	plus	value of nickels	is	320¢

Step 2. $25Q$ $+$ $10(Q + 7)$ $+$ $5(3Q)$ $=$ 320

Step 3. $25Q + 10(Q + 7) + 5(3Q) = 320$

$$25Q + 10Q + 70 + 15Q = 320 \qquad \text{Simplifying}$$

$$50Q + 70 = 320 \qquad \text{Combining like terms}$$

$$50Q + 70 + (-70) = 320 + (-70) \qquad \text{Adding } -70 \text{ to each side}$$

$$50Q = 250$$

$$\frac{50Q}{50} = \frac{250}{50} \qquad \text{Dividing both sides by 50}$$

Step 4. $Q = 5$ The number of quarters

                       $Q + 7 = 5 + 7 = 12$ The number of dimes

                       $3Q = 3(5) = 15$ The number of nickels

✓ **Step 5. Check**   There are seven more dimes than quarters and three times as many nickels as quarters. The value of 5 quarters is 5(25¢), or 125¢. The value of 12 dimes is 12(10¢), or 120¢. The value of 15 nickels is 15(5¢), or 75¢. The sum of all these values is 125¢ + 120¢ + 75¢, or 320¢, or $3.20.

Step 6. Therefore, there are five quarters, twelve dimes, and fifteen nickels.

Exercises 4.4B
Set I

Set up each problem algebraically, solve, and check. Be sure to state what your variables represent.

1. Bill has thirteen bills in his pocket that have a total value of $95. If these bills consist of $5 bills and $10 bills, how many of each kind are there?

2. Miko has in his wallet eleven bills that have a total value of $85. If the bills consist of $5 bills and $10 bills, how many of each kind are there?

3. Jennifer has twelve coins that have a total value of $2.20. The coins are nickels and quarters. How many of each kind of coin are there?

4. Brian has eighteen coins consisting of nickels and quarters. If the total value of the coins is $2.50, how many of each kind of coin does he have?

5. Derek has $4.00 in nickels, dimes, and quarters. If he has four more quarters than nickels and 3 times as many dimes as nickels, how many of each kind of coin does he have?

6. Staci has $5.50 in nickels, dimes, and quarters. If she has seven more dimes than nickels and twice as many quarters as dimes, how many of each kind of coin does she have?

7. Michael has $2.25 in nickels, dimes, and quarters. If he has three fewer dimes than quarters and as many nickels as the sum of the dimes and quarters, how many of each kind of coin does he have?

8. Muriel has $257 in $10 bills, $5 bills, and $1 bills. If she has five fewer $1 bills than $5 bills and as many $10 bills as the sum of the $5 and $1 bills, how many of each kind of bill does she have?

9. The total receipts for a concert were $19,800 for the 1,080 tickets sold. The promoters sold orchestra seats for $21 each, box seats for $30 each, and balcony seats for $12 each. If there were 5 times as many balcony seats as box seats sold, how many of each kind were sold?

10. The total receipts for a football game were $1,443,200. General admission tickets cost $20 each, reserved seat tickets were $32 each, and box seat tickets were $40 each. If there were twice as many general admission as reserved seat tickets sold and 4 times as many reserved as box seat tickets sold, how many of each kind were sold?

11. Christy spent $3.80 for sixty stamps. She bought only 2¢, 10¢, and 12¢ stamps. If she bought twice as many 10¢ stamps as 12¢ stamps, how many of each kind did she buy?

12. Mark spent $9.80 for 100 stamps. He bought only 6¢, 8¢, and 12¢ stamps. If there were 3 times as many 6¢ stamps as 8¢ stamps, how many of each kind did he buy?

Exercises 4.4B
Set II

Set up each problem algebraically, solve, and check. Be sure to state what your variables represent.

1. Don spent $2.36 for twenty-two stamps. If he bought only 10¢ stamps and 12¢ stamps, how many of each kind did he buy?

2. Karla has fifteen coins with a total value of $2.55. If the coins are only dimes and quarters, how many of each kind of coin does she have?

3. Jill has fifteen coins with a total value of $2.70. If the coins are only dimes and quarters, how many of each kind of coin does she have?

4. Several families went to a movie together. They spent $24.75 for eight tickets. If adults' tickets cost $4.50 each and children's tickets cost $2.25 each, how many of each kind of ticket were purchased?

5. Rachelle has $4.75 in nickels, dimes, and quarters. If she has four more nickels than dimes and twice as many quarters as dimes, how many of each kind of coin does she have?

6. Jason has $4.80 in nickels, dimes, and quarters. If he has three more dimes than quarters and 3 times as many nickels as quarters, how many of each kind of coin does he have?

7. Kevin has $4.75 in nickels, dimes, and quarters. If he has two more dimes than nickels and twice as many quarters as nickels, how many of each kind of coin does he have?

8. Elaine spent $2.88 for fifteen stamps. She bought only 22¢ and 15¢ stamps. How many of each kind did she buy?

9. A class received $233 for selling 200 tickets to the school play. If students' tickets cost $1 each and nonstudents' tickets cost $2 each, how many nonstudent tickets were sold?

10. Heather spent $3.02 for stamps. She bought only 22¢ and 15¢ stamps, and she bought seven more 22¢ stamps than 15¢ stamps. How many of each kind did she buy?

11. Tricia spent $5.72 for twenty-six stamps. She bought only 22¢, 18¢, and 25¢ stamps. If she bought twice as many 22¢ stamps as 18¢ stamps and two more 25¢ stamps than 18¢ stamps, how many of each kind did she buy?

12. The total receipts for a concert were $26,000. Some tickets cost $20 each, some cost $28 each, and the rest cost $36 each. If there were eighty more $28 tickets sold than $36 tickets, and 10 times as many $20 as $36 tickets, how many of each kind were sold?

4.5 Solving Ratio and Rate Problems

Ratios

A **ratio** is used to compare the sizes of two or more *like* quantities. (When we say "like quantities," we mean that we compare inches with inches, pounds with pounds, cubic feet with cubic feet, and so forth.) "The ratio of *a* to *b*" can be written as the division problem $\dfrac{a}{b}$, and *a* and *b* are called the **terms** of the ratio. The terms of the ratio can be any kind of number, except that the denominator cannot be zero.

> ✖ **A Word of Caution** "The ratio of *a* to *b*" is *not* $\dfrac{b}{a}$. It is $\dfrac{a}{b}$. The number that is *before* the word "to" is always in the numerator. The number that is *after* the word "to" is always in the denominator.

We can now give three different meanings to an expression such as $\frac{3}{4}$:

1. 3 of the 4 equal parts into which a unit has been divided (fraction meaning)

2. $3 \div 4$ (division meaning)

3. The ratio of 3 to 4 (comparison—or ratio—meaning)

The meaning chosen depends on how the expression is used.

"The ratio of *a* to *b*" can also be written as $a:b$, but this representation hides the fact that a ratio is a fraction or a rational number. (Notice the word *ratio* in the word *ratio*nal.)

When a ratio is written in fraction form and the denominator is 1, the 1 *must be written*; that is, if the ratio of two numbers is 6 to 1, the ratio must be written as $\frac{6}{1}$, not as 6. Also, if the numerator is greater than the denominator, the fraction *must* be left as an improper fraction; it *cannot* be written as a mixed number.

We usually reduce ratios to lowest terms, and we can use the fundamental property of fractions to reduce ratios. For example, if there are 15 blue shirts and 6 red shirts on a shelf, we might say, "The ratio of blue shirts to red shirts is 15 to 6." In fraction form, this is written $\dfrac{15}{6}$, and $\dfrac{15}{6} = \dfrac{5 \cdot 3}{2 \cdot 3} = \dfrac{5}{2}$; therefore, the ratio of blue shirts to red shirts is *also* 5 to 2, which implies that there are 5 blue shirts for every 2 red shirts. It would be *incorrect* to rewrite the ratio $\frac{5}{2}$ as $2\frac{1}{2}$.

EXAMPLE 1

Marcie has saved $130 and Jan has saved $170. Find **(a)** the ratio of Marcie's savings to Jan's savings and **(b)** the ratio of Jan's savings to Marcie's savings.

SOLUTION

a. The ratio of Marcie's savings to Jan's savings is $\dfrac{\$130}{\$170} = \dfrac{10 \cdot 13}{10 \cdot 17}$, or $\dfrac{13}{17}$.

b. The ratio of Jan's savings to Marcie's savings is $\dfrac{\$170}{\$130} = \dfrac{10 \cdot 17}{10 \cdot 13}$, or $\dfrac{17}{13}$.

The notation $a:b:c$ is usually used when we're comparing three quantities. For example, if the lengths of the sides of a triangle are 14 ft, 21 ft, and 35 ft, we can say that the lengths are in the ratio $14:21:35$. We can also divide *all three numbers* by the largest number that divides exactly into all of them (7) and say that the lengths are in the ratio $2:3:5$.

The key to solving ratio problems is to use the given ratio to help represent the unknown numbers. According to the fundamental property of fractions, $\dfrac{3}{4} = \dfrac{3 \cdot 2}{4 \cdot 2} = \dfrac{3 \cdot 3}{4 \cdot 3} = \dfrac{3 \cdot 4}{4 \cdot 4}$, and so on; in algebra, we can write this fact as $\dfrac{3}{4} = \dfrac{3x}{4x}$, where x represents any real number except 0. Therefore, if we're given some facts about two numbers and are also told, for example, that those numbers are in the ratio of 3 to 5, we can let the smaller number be $3x$ and the larger number be $5x$ (see Examples 2 through 5). This procedure can be generalized as follows:

Representing the unknowns in a ratio problem

> If two unknown numbers are in the ratio of a to b, let the first number be ax and the second number be bx.
>
> If three numbers are in the ratio of a to b to c, let the first number be ax, the second number be bx, and the third number be cx.

 A Word of Caution   In ratio problems, we are not finished when we have found x. We must multiply the value of x by the terms of the ratio in order to find the unknown numbers.

EXAMPLE 2

Two numbers are in the ratio of 3 to 5. Their sum is 32. Find the numbers. (The same problem could have been worded as follows: "Divide 32 into two parts whose ratio is $3:5$," or "Separate 32 into two parts whose ratio is $3:5$.")

SOLUTION

Think   This is a ratio problem.

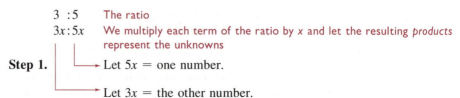

       3 :5   The ratio
      $3x:5x$ We multiply each term of the ratio by x and let the resulting *products* represent the unknowns

Step 1.   Let $5x$ = one number.

           Let $3x$ = the other number.

Reread The sum is 32

Step 2. $3x + 5x = 32$

Step 3. $8x = 32$ Combining like terms

$$\frac{8x}{8} = \frac{32}{8}$$ Dividing both sides by 8

$x = 4$ *Warning*: We are not finished when we have found x; the unknown numbers are $3x$ and $5x$

Step 4. $3x = 3(4) = 12$ The smaller number

$5x = 5(4) = 20$ The larger number

✓ **Step 5. Check**   The ratio of 12 to 20 is $\dfrac{12}{20} = \dfrac{3 \cdot 4}{5 \cdot 4} = \dfrac{3}{5}$, and $12 + 20 = 32$.

Step 6.   Therefore, the numbers are 12 and 20.

EXAMPLE 3   The three sides of a triangle are in the ratio $2:3:4$. The perimeter is 63 ft. Find the lengths of the three sides.

SOLUTION

Think   This is a ratio problem. The formula for the perimeter of a triangle is $P = a + b + c$.

$2\ :3\ :4$ The ratio
$2x:3x:4x$ We multiply each term of the ratio by x and let the resulting *products* represent the unknowns

Step 1.   → Let $4x =$ the number of feet in the longest side.

→ Let $3x =$ the number of feet in the next side.

→ Let $2x =$ the number of feet in the shortest side.

Reread   The perimeter is 63

Step 2. $2x + 3x + 4x = 63$

Step 3. $9x = 63$ Combining like terms

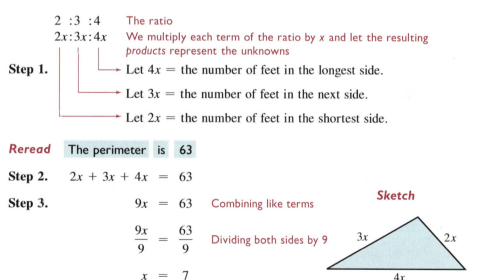

Sketch

$$\frac{9x}{9} = \frac{63}{9}$$ Dividing both sides by 9

$x = 7$

Step 4. $2x = 2(7) = 14$ The number of feet in the shortest side

$3x = 3(7) = 21$ The number of feet in the next side

$4x = 4(7) = 28$ The number of feet in the longest side

✓ **Step 5. Check**   Because $14 \div 7 = 2$, $21 \div 7 = 3$, and $28 \div 7 = 4$, the lengths 14 ft, 21 ft, and 28 ft are in the ratio $2:3:4$. The perimeter is 14 ft + 21 ft + 28 ft, or 63 ft.

Step 6.   Therefore, the lengths of the three sides are 14 ft, 21 ft, and 28 ft.

EXAMPLE 4

Sixty-six hours of a student's week are spent in study, in class, and in work. The times spent in these activities are in the ratio 4:2:5. How many hours are spent in each activity?

SOLUTION

Think   This is a ratio problem. The total hours per week spent on the three activities is 66 hr.

4 :2 :5   The ratio
$4x:2x:5x$ We multiply each term of the ratio by x and let the resulting *products* represent the unknowns

Step 1.   Let $5x$ = the number of hours spent at work.

Let $2x$ = the number of hours spent in class.

Let $4x$ = the number of hours spent in study.

Reread	Sixty-six hours of student's week	are spent in	study,	class,	and	work
Step 2.	66	=	$4x$ +	$2x$	+	$5x$

Step 3. $66 = 4x + 2x + 5x$

$66 = 11x$ Combining like terms

$\dfrac{66}{11} = \dfrac{11x}{11}$ Dividing both sides by 11

$6 = x$

$x = 6$ Using the symmetric property

Step 4. $4x = 4(6) = 24$ The number of hours spent in study

$2x = 2(6) = 12$ The number of hours spent in class

$5x = 5(6) = 30$ The number of hours spent at work

√ **Step 5. Check**   Because $24 \div 6 = 4$, $12 \div 6 = 2$, and $30 \div 6 = 5$, the numbers 24, 12, and 30 are in the ratio 4:2:5. Also, 24 hr + 12 hr + 30 hr = 66 hr.

Step 6. Therefore, the student spends 24 hr per week studying, 12 hr in class, and 30 hr working.

EXAMPLE 5

The length and width of a rectangle are in the ratio of 7 to 5. The perimeter is to be greater than 72. What is the range of values for the *length* of the rectangle?

SOLUTION

Think   This is a ratio problem. The formula for the perimeter of a rectangle is $P = 2l + 2w$.

7 :5   Ratio
$7x:5x$ We multiply each term of the ratio by x and let the resulting *products* represent the unknowns

Step 1.   Let $5x$ = the width.

Let $7x$ = the length.

Sketch

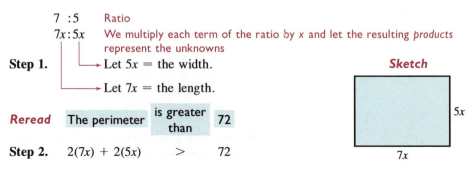

Reread	The perimeter	is greater than	72
Step 2.	$2(7x) + 2(5x)$	>	72

Step 3.	$14x + 10x$	$>$	72	Simplifying
	$24x$	$>$	72	Combining like terms
	$\dfrac{24x}{24}$	$>$	$\dfrac{72}{24}$	Dividing both sides by 24
	x	$>$	3	
Step 4.	$7x$	$>$	$7(3)$	The length is $7x$
	$7x$	$>$	21	The length must be greater than 21

√ **Step 5. Check** We will choose two values (at random) for the length of the rectangle and show the check for these two values. We'll choose lengths of (a) 28 and (b) 47 (notice that both of these are greater than 21).

a. Length = 28: The length was represented by $7x$; therefore, if the length is 28, $7x = 28$, so $x = 4$. Because x is 4, the width ($5x$) must be 5(4), or 20. The perimeter, then, is

$$2(28) + 2(20) = 56 + 40 = 96$$

which is greater than 72.

b. Length = 47: The length was represented by $7x$; therefore, if the length is 47, $7x = 47$, so $x = \frac{47}{7}$. Because x is $\frac{47}{7}$, the width ($5x$) must be $5\left(\frac{47}{7}\right)$, or $\frac{235}{7}$. The perimeter, then, is

$$2(47) + 2\left(\tfrac{235}{7}\right) = 94 + \tfrac{470}{7} = \tfrac{1,128}{7} \approx 161.14$$

which is greater than 72.

Step 6. Therefore, the length must be greater than 21.

Rates

We often find it necessary or desirable to express in fraction form the quotient of one quantity divided by a quantity of a *different* kind. These fractions are technically called *rates*, although many authors do not distinguish between a ratio and a rate. You are undoubtedly familiar with rates such as $\dfrac{\text{miles}}{\text{hour}}$ (miles per hour), $\dfrac{\text{miles}}{\text{gallon}}$ (miles per gallon), $\dfrac{\text{dollars}}{\text{hour}}$ (dollars per hour), and so on.

EXAMPLE 6

A man earns \$103 for every 8 hours he works. Express the rate of dollars to hours as a fraction.

SOLUTION The rate is $\dfrac{103 \text{ dollars}}{8 \text{ hours}}$.

EXAMPLE 7

Gloria can drive about 533 miles on 14 gallons of gasoline. Express the rate of miles to gallons as a fraction.

SOLUTION The rate of miles to gallons is $\dfrac{533 \text{ miles}}{14 \text{ gallons}}$.

When we solve problems involving rates, it is helpful to "cancel like units" whenever possible (see Example 8).

EXAMPLE 8

Mrs. Norton's car uses gasoline at the rate of $16 \dfrac{\text{miles}}{\text{gallon}}$. How many gallons of gasoline will she need in order to drive 464 miles?

SOLUTION

Think   This is a rate problem. Multiplying and "canceling like units" will give

$$\left(\frac{\text{miles}}{\cancel{\text{gallon}}} \right)(\cancel{\text{gallons}}) = \text{miles}$$

Step 1. Let x = the number of gallons of gasoline.

Step 2. $\left(16 \dfrac{\text{miles}}{\cancel{\text{gallon}}} \right)(x \ \cancel{\text{gallons}}) = 464 \text{ miles}$ We cancel "gallons"

We now drop the word "miles"

Step 3. $16x \text{ miles} = 464 \text{ miles}$

$$16x = 464$$

$$\frac{16x}{16} = \frac{464}{16}$$ Dividing both sides by 16

Step 4. $x = 29$

√ **Step 5.** *Check* $\left(16 \dfrac{\text{miles}}{\cancel{\text{gallon}}} \right)(29 \ \cancel{\text{gallons}}) = 464 \text{ miles}$

Step 6. Therefore, Mrs. Norton will need 29 gallons of gasoline.

We will consider ratios and rates again in later chapters.

Exercises 4.5
Set I

In Exercises 1–14, set up each problem algebraically, solve, and check. Be sure to state what your variable represents.

1. Two numbers are in the ratio of 3 to 4. Their sum is 35. Find the numbers.

2. Two numbers are in the ratio of 7 to 3. Their sum is 130. Find the numbers.

3. The length and width of a rectangle are in the ratio of 9 to 4. The perimeter is 78. Find the length and the width.

4. The length and width of a rectangle are in the ratio 7:2. The perimeter is 72. Find the length and the width.

5. The amounts Chuck spends for food, rent, and clothing are in the ratio 4:5:1. If he spends about $850 per month just for these three items, what can he expect to spend for each one of these items in an average month?

6. Each school year, Laura spends about $12,150 for tuition, housing, and food. If the amounts spent just for these three items are in the ratio 4:4:1, how much does she spend in a typical school year for each item?

7. Separate 88 into two parts whose ratio is 6 to 5.

8. Separate 99 into two parts whose ratio is 2 to 7.

9. The three sides of a triangle are in the ratio 4:5:6. The perimeter is 90 ft. Find the lengths of the three sides.

10. The three sides of a triangle are in the ratio 5:6:7. The perimeter is 72 yd. Find the lengths of the three sides.

11. An aunt divided $27,000 among her three nieces in the ratio 2:3:4. How much did each niece receive?

12. A pension fund of $150,000 was invested in stocks, bonds, and real estate in the ratio 5:1:4. How much was invested in stocks? In bonds? In real estate?

13. When Mrs. Mora was mixing the cement for her patio floor, she used a cement-sand-gravel ratio of $1:2\frac{1}{2}:4$. If she used 5 cubic yards of gravel, how much sand did she use? How much cement?

14. In one week's time, a restaurant served 865 dinners. Out of every five dinners served, three were seafood dinners. How many seafood dinners were served during the week?

In Exercises 15–18, set up each problem algebraically and solve.

15. The length and width of a rectangle are in the ratio of 7 to 4. The perimeter is less than 88. What is the range of values for the width?

16. The length and width of a rectangle are in the ratio of 8 to 5. The perimeter is greater than 78. What is the range of values for the length?

17. The three sides of a triangle are in the ratio 2:4:7. The perimeter is greater than 117. What is the range of values for the length of the longest side?

18. The three sides of a triangle are in the ratio 3:5:6. The perimeter is less than 112. What is the range of values for the length of the shortest side?

19. Ruby drove 360 miles in 7 hours. Express the rate of miles to hours as a fraction.

20. Justin bicycled 35 miles in 4 hours. Express the rate of miles to hours as a fraction.

21. Fred can crochet three afghans in 25 days. Express the rate of afghans to days as a fraction.

22. Christopher typed 500 words in 11 minutes. Express the rate of words to minutes as a fraction.

In Exercises 23–26, set up each problem algebraically and solve.

23. If $805 was spent for carpeting that cost $23 \frac{\text{dollars}}{\text{sq. yd}}$, how many square yards were purchased?

24. If Florence earned $432 one week and if her pay was $12 \frac{\text{dollars}}{\text{hour}}$, how many hours did she work?

25. Jianula's car uses gasoline at the rate of $23 \frac{\text{miles}}{\text{gallon}}$. How much gasoline will she use in driving 368 miles?

26. If Frank uses wallpaper that has a rate of coverage of $56 \frac{\text{sq. ft}}{\text{roll}}$, how many rolls will he need to cover 672 sq. ft?

Exercises 4.5
Set II

In Exercises 1–14, set up each problem algebraically, solve, and check. Be sure to state what your variable represents.

1. Two numbers are in the ratio of 3 to 5. Their sum is 40. Find the numbers.

2. The length and width of a rectangle are in the ratio of 7 to 4. The perimeter is 88. Find the length and width.

3. The length and width of a rectangle are in the ratio of 9 to 5. The perimeter is 140. Find the length and the width.

4. A wire that was 135 cm long was cut into two pieces. The ratio of the lengths of the pieces was 4 to 5. Find the length of each piece.

5. The amounts Cameron spends for food, rent, and clothing are in the ratio 4:7:2. If he spends an average of $975 a month for just these three items, how much can he expect to spend for each one of these items in a typical *year*.

6. Marge planted 162 acres of corn and soy beans in the ratio of 5 to 4. How many acres of each did she plant?

7. Separate 126 into two parts whose ratio is 5 to 13.

8. Separate 231 into two parts whose ratio is 17 to 4.

9. The three sides of a triangle are in the ratio 4:5:7. The perimeter is 128. Find the lengths of the three sides.

10. Virginia cut a 65-in. board into two parts whose ratio was 9:4. Find the length of each piece.

11. The gold solder used in a crafts class has gold, silver, and copper in the ratio 5:3:2. How much gold is there in 30 g of this solder?

12. A civil service office tries to employ people without discrimination on the basis of age. If the ratio of people over 40 to those under 40 is 3:4, how many out of 259 people hired would have to be over 40?

13. Fast Eddie's Pro Shop sells four types of bowling balls: types A, B, C, and D. Sales of the balls totaled $3,600 for the year. If the balls sold in the ratio 3:2:1:4, how much of the sales, in dollars, can be attributed to the sale of each type of ball?

14. Susan is making an afghan that requires thirty-two skeins of yarn. The colors brown, beige, and rust are to be used in the ratio 5:8:3. How many skeins of each color should she buy?

In Exercises 15–18, set up each problem algebraically and solve.

15. The length and width of a rectangle are in the ratio of 9 to 4. The perimeter is greater than 156. What is the range of values for the width?

16. The ratio of Ray's age to Patty's age is 3 to 2. The sum of their ages is less than 90 (years). What is the range of values for Ray's age?

Step 4. $x = 15$ The number of pounds of
                                                                    Brand A

                            $50 - x = 50 - 15 = 35$ The number of pounds of
                                                                    Brand B

✓ **Step 5.** *Check*

$$15 \text{ lb @ } \$3.30 = \$\ 49.50$$
$$35 \text{ lb @ } \$3.70 = \underline{\$129.50} \quad 50 \text{ lb @ } \$3.58 = \$179.00$$
$$\text{Total} = \$179.00 \leftarrow \text{Total cost}$$

Step 6. Therefore, the wholesaler must use 15 lb of Brand A and 35 lb of Brand B.

EXAMPLE 2

A 10-lb mixture of walnuts and almonds costs $35.20. If the walnuts cost $4.50 per pound and the almonds cost $3.10 per pound, how many pounds of each kind are in the mixture?

SOLUTION

Think   This is a mixture problem, and this time we're given the total cost of the mixture rather than the unit cost. Because we're *given* the *total cost* of the mixture, we needn't fill in the space for the *unit cost* of the mixture.

Step 1. Let $x =$ the number of pounds of walnuts
            Then $10 - x =$ the number of pounds of almonds

Chart

	Walnuts		Almonds		Mixture
Unit cost	4.50		3.10		/////////
Amount	x	+	$(10 - x)$	=	10
Total cost	$(4.50)x$	+	$3.10(10 - x)$	=	35.20

Step 2. $4.50x + 3.10(10 - x) = 35.20$

Step 3. $45x + 31(10 - x) = 352$ Multiplying both sides by 10
                                                   to remove the decimal points

     $45x + 310 - 31x = 352$ Using the distributive property

            $14x + 310 = 352$

$14x + 310 + (-310) = 352 + (-310)$ Adding -310 to both sides

                $14x = 42$

            $\dfrac{14x}{14} = \dfrac{42}{14}$ Dividing both sides by 14

Step 4. $x = 3$ The number of pounds of
                                                   walnuts

                  $10 - x = 10 - 3 = 7$ The number of pounds of
                                                   almonds

✓ **Step 5.** *Check*

$$3 \text{ lb @ } \$4.50 = \$13.50$$
$$7 \text{ lb @ } \$3.10 = \underline{\$21.70}$$
$$\text{Total} = \$35.20 \quad \text{Total cost}$$

Step 6. Therefore, there are 3 lb of walnuts and 7 lb of almonds.

EXAMPLE 3

Mrs. Reid needs to mix 20 lb of macadamia nuts that cost $8.10 per pound with pecans that cost $5.40 per pound. How many pounds of pecans should she use if the mixture is to cost $6.48 per pound?

SOLUTION

Think This is a mixture problem, and this time we're given the amount of one ingredient but not the total amount of the mixture.

Step 1. Let $x =$ the number of pounds of pecans.

Chart

	Pecans		Macadamia nuts		Mixture
Unit cost	5.40		8.10		6.48
Amount	x	+	20	=	$x + 20$*
Total cost	$(5.40)x$	+	8.10(20)	=	6.48($x + 20$)

Step 2. $5.40x + 8.10(20) = 6.48(x + 20)$

Step 3. $540x + 810(20) = 648(x + 20)$ Multiplying both sides by 100 to remove the decimal points

$540x + 16{,}200 = 648x + 12{,}960$ Simplifying

$-540x + 540x + 16{,}200 = -540x + 648x + 12{,}960$ Adding $-540x$ to both sides

$16{,}200 = 108x + 12{,}960$

$16{,}200 - 12{,}960 = 108x + 12{,}960 - 12{,}960$ Adding $-12{,}960$ to both sides

$3{,}240 = 108x$

$\dfrac{3{,}240}{108} = \dfrac{108x}{108}$ Dividing both sides by 108

Step 4. $x = 30$ The number of pounds of pecans

✓ **Step 5. *Check***

20 lb @ $8.10 = $162   Value of macadamia nuts
30 lb @ $5.40 = $162   Value of pecans
50 lb ← Totals → $324

Price per pound of mixture $= \dfrac{324 \text{ dollars}}{50 \text{ pounds}} = \$6.48$ per pound

Step 6. Therefore, Mrs. Reid should use 30 lb of pecans in the mixture.

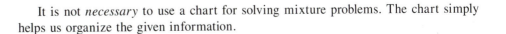

Note Most money problems can be done by using a chart.

It is not *necessary* to use a chart for solving mixture problems. The chart simply helps us organize the given information.

---

*Notice that the *number of pounds* of pecans (x) *plus* the *number of pounds* of macadamia nuts (20) equals the *number of pounds* of the mixture ($x + 20$).

Exercises 4.8
Set I

Set up each problem algebraically, solve, and check. Be sure to state what your variables represent.

1. A merchant makes up a 10-lb mixture of granola and dried apple chunks, and she wants the mixture to cost $3.00 per pound. If the granola costs $2.20 per pound and the apple chunks cost $4.20 per pound, how many pounds of each should she use?

2. Jim wants to make up a 50-lb mixture of macadamia nuts at $8.50 per pound and peanuts at $3.50 per pound. How many pounds of each should he use if the mixture is to cost $5.80 per pound?

3. Alice wants to make up a 30-lb mixture of apples that will be worth 78¢ per pound. One kind of apple costs 95¢ per pound, and the other kind costs 65¢ per pound. How many pounds of each kind should she use?

4. A merchant must make a 60-lb mixture of peanut brittle and English toffee that will be worth $5.60 per pound. If peanut brittle costs $4.20 per pound and English toffee costs $6.60 per pound, how many pounds of each should he use?

5. Margie makes up a 27-lb mixture of gumdrops and caramels that is to be worth $66.00. If the gumdrops cost $2.80 per pound and the caramels cost $2.20 per pound, how many pounds of each should she use?

6. A grocer makes up a 40-lb mixture of two kinds of coffee that is worth $151. If Brand C costs $4.20 per pound and Brand D costs $3.20 per pound, how many pounds of each kind does he use?

7. Dorothy needs a mixture of candy and nuts. How many pounds of candy at $3.00 per pound should she mix with 50 lb of nuts at $2.40 per pound to obtain a mixture worth $2.50 per pound?

8. How many pounds of cashews at $7.60 per pound should be mixed with 30 lb of macadamia nuts at $8.20 per pound in order to obtain a mixture worth $7.96 per pound?

Exercises 4.8
Set II

Set up each problem algebraically, solve, and check. Be sure to state what your variables represent.

1. A 40-lb mixture of peanuts and cashews is to be worth $5.55 per pound. If peanuts cost $2.25 per pound and cashews cost $7.53 per pound, how many pounds of each should be used?

2. A grocer needs to make up a 25-lb mixture of coffee worth $3.52 per pound. If Brand A costs $3.80 per pound and Brand B costs $3.30 per pound, how many pounds of each kind should be used?

3. Walt needs a mixture of walnuts and almonds to be worth $3.92 per pound. If walnuts cost $4.50 per pound and almonds cost $3.50 per pound, how many pounds of each should he use to obtain a 50-lb mixture?

4. Herb made a 25-lb mixture of peanuts and raisins; it was worth $2.88 per pound. If peanuts cost $3.69 per pound and raisins cost $1.44 per pound, how many pounds of each did he use?

5. A 35-lb mixture of two kinds of coffee is worth $118.00. If Brand A costs $3.70 per pound and Brand B costs $3.20 per pound, how many pounds of each kind of coffee were used?

6. Sherma made a 32-lb mixture of nougats and peppermint candies that was worth $70.00. The nougats cost $2.50 per pound and the peppermints cost $2.10 per pound. How many pounds of each did she use?

7. How much Brand A coffee costing $3.95 per pound should be mixed with 25 lb of Brand B coffee costing $3.60 per pound in order to obtain a mixture worth $3.70 per pound?

8. How many pounds of Delicious apples at 95¢ per pound should be mixed with 30 lb of Spartan apples at 90¢ per pound in order to obtain a mixture worth 92¢ per pound?

4.9 Solving Solution Mixture Problems

Another type of mixture problem involves the mixing of liquids. Such problems are often called *solution problems* or *solution mixture problems* because a mixture of two or more liquids is, under certain conditions, a *solution*.* In these problems, it is the *strength* of the solutions that is important, rather than the cost of the solutions.

Before we discuss solution mixture problems, we will discuss chemical solutions in general. A 30% solution of alcohol is a mixture that is 30% pure alcohol and 70% (that is, 100%–30%) water. Examples 1 through 3 demonstrate the use of methods discussed earlier as they apply to solution mixture problems. (The checks for these examples will not be shown, nor will we write Step 1, Step 2, and so on.)

In solution problems, we can find the total amount of the pure substance present in any solution by multiplying the amount of that solution by its *strength*, or *percent*, if the percent is expressed in decimal form (see Example 1).

EXAMPLE 1 How many liters of pure alcohol are there in 2 L of a 30% solution of alcohol?

SOLUTION We must find 30% of 2.
Let x = the number of liters of pure alcohol.

$$x = (0.30)(2) = 0.6 \quad \text{The number of liters of pure alcohol present}$$

EXAMPLE 2 In a 5% glycerin solution, there is 3 mL of pure glycerin present. How many milliliters of solution are there?

SOLUTION We must answer this question: 3 is 5% of what number?
Let x = the number of milliliters of solution.

$$3 = (0.05) \cdot x$$

$$\frac{3}{0.05} = \frac{0.05x}{0.05} \qquad \text{Dividing both sides by 0.05}$$

$$x = 60 \qquad \text{The number of milliliters of solution}$$

EXAMPLE 3 Fifty milliliters of an alcohol solution contains 6 mL of pure alcohol. What percent of the solution is alcohol?

SOLUTION We need to solve this problem: 6 is what percent of 50?
Let x = the fractional part (x must be found and then converted to a percent).

$$6 = x \cdot (50)$$

$$\frac{6}{50} = \frac{50x}{50} \qquad \text{Dividing both sides by 50}$$

$$x = \frac{6}{50} = 0.12 = 12\%$$

Therefore, 12% of the solution is pure alcohol.

Let us now discuss mixtures that involve the mixing of different strengths of solutions. Solution mixture problems can be solved by using a method similar to that used for solving other mixture problems. The general method is similar to the chart method

---

*A solution is a homogeneous mixture of two or more substances.

shown earlier for mixture problems. We use the following relationships rather than a specific formula.

<table>
<tr><td>Three relationships necessary to solve solution mixture problems</td><td>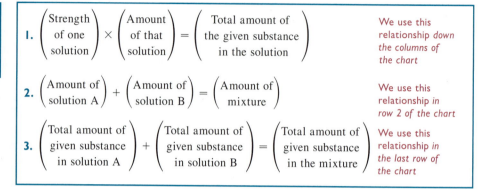</td></tr>
</table>

The chart is set up as follows:

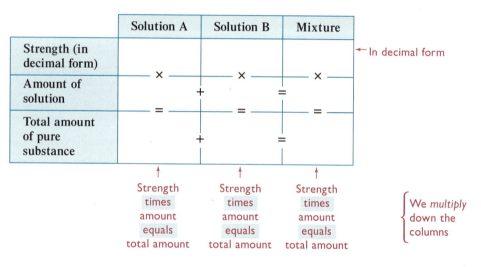

We multiply *down the columns* of the chart to find the total amount of the pure substance present in any one of the solutions. The bottom row of the chart gives the equation.

EXAMPLE 4   How many liters of a 20% alcohol solution must be added to 3 L of a 90% alcohol solution to make an 80% solution?

SOLUTION

Think   This is a solution mixture problem. We'll let x be the amount of the 20% solution.

Step 1. Let x = the number of liters of the 20% solution.

Chart

	20% solution		90% solution		Mixture
Strength (in decimal form)	0.20		0.90		0.80
Amount of solution	x	+	3	=	$x + 3$
Total amount of pure substance	$0.20x$	+	$0.90(3)$	=	$0.80(x + 3)$

Amount of alcohol in 20% solution	+	Amount of alcohol in 90% solution	=	Amount of alcohol in 80% solution

Step 2. $\qquad 0.20x \qquad + \qquad 0.90(3) \qquad = \qquad 0.80(x + 3)$

Step 3. $\qquad 2x + 9(3) = 8(x + 3) \qquad$ Multiplying both sides by 10 to remove the decimal points

$$2x + 27 = 8x + 24 \qquad \text{Simplifying}$$

$$-2x + 2x + 27 = -2x + 8x + 24 \qquad \text{Adding } -2x \text{ to both sides}$$

$$27 = 6x + 24$$

$$27 + (-24) = 6x + 24 + (-24) \qquad \text{Adding } -24 \text{ to both sides}$$

$$3 = 6x$$

$$\frac{3}{6} = \frac{6x}{6} \qquad \text{Dividing both sides by 6}$$

Step 4. $\qquad\qquad x = \dfrac{1}{2} \qquad$ The number of liters of 20% alcohol

√ **Step 5. Check** There is $\left(\frac{1}{2}\text{L}\right)(0.2)$, or 0.1 L of alcohol present in $\frac{1}{2}$ L of a 20% solution, and $(3\text{ L})(0.9)$, or 2.7 L, present in 3 L of a 90% solution. Therefore, there is $(0.1\text{ L} + 2.7\text{ L})$, or 2.8 L, of alcohol going *into* the mixture. The total amount of solution in the mixture will be $(3\text{ L} + \frac{1}{2}\text{ L})$, or 3.5 L, and the *strength* of the solution will be $\dfrac{2.8\text{ L}}{3.5\text{ L}}$, or 0.8, or 80%.

Step 6. Therefore, $\frac{1}{2}$ L of a 20% alcohol solution must be added.

👉 **Note** If we add *water* to an alcohol solution to get a weaker solution, we are adding a *0% solution*. If we add *pure* alcohol to an alcohol solution to get a stronger solution, we are adding a *100% solution* of alcohol.

EXAMPLE 5 How much water should be added to a 10% solution of alcohol to obtain 20 mL of an 8% solution? How much of the 10% solution should be used?

SOLUTION

Think This is a solution mixture problem, and we're adding *water*. The percent of pure alcohol in water is 0%. There is to be 20 mL of solution in the mixture; therefore, if we let x be the number of milliliters of water, there must be $(20 - x)$ mL of the 10% solution.

Step 1. Let $\qquad x =$ the number of milliliters of water to be added
Then $20 - x =$ the number of milliliters of the 10% solution

Chart

	10% solution		Water		Mixture
Strength (in decimal form)	0.10		0.00		0.08
Amount of solution	$20 - x$	+	x	=	20
Total amount of pure substance	$0.10(20 - x)$	+	0	=	$0.08(20)$

Step 2. $0.10(20 - x) + 0 = 0.08(20)$

Step 3. $10(20 - x) = 8(20)$ ← ——— Multiplying both sides by 100 to remove the decimal points

$200 - 10x = 160$ —— Simplifying

$-200 + 200 - 10x = -200 + 160$ —— Adding -200 to both sides

$-10x = -40$

$\dfrac{-10x}{-10} = \dfrac{-40}{-10}$ —— Dividing both sides by -10

Step 4. $x = 4$ —— The number of milliliters of water

$20 - x = 20 - 4 = 16$ —— The number of milliliters of 10% solution

✓ **Step 5. Check** There is $(16 \text{ mL})(0.1)$, or 1.6 mL, of pure alcohol in 16 mL of a 10% solution. Therefore, there is 1.6 mL of pure alcohol going *into* the mixture. The total amount of solution in the mixture is 20 mL, and the *strength* of the mixture is $\dfrac{1.6 \text{ mL}}{20 \text{ mL}}$, or 0.08, or 8%.

Step 6. Therefore, we must add 4 mL of water to 16 mL of the 10% solution.

EXAMPLE 6

How many liters of pure alcohol should be added to 4 L of a 10% solution of alcohol in order to obtain a 25% solution?

SOLUTION

Think This is a solution mixture problem, and we're adding *pure alcohol*; the percent of alcohol in pure alcohol is 100%.

Step 1. Let x = the number of liters of pure alcohol to be added.

Chart

	Pure alcohol		10% solution		Mixture
Strength (in decimal form)	1.00		0.10		0.25
Amount of solution	x	+	4	=	$x + 4$
Total amount of pure substance	$1.00x$	+	$0.10(4)$	=	$0.25(x + 4)$

Step 2. $1.00x + 0.10(4) = 0.25(x + 4)$

Step 3. $100x + 10(4) = 25(x + 4)$ —— Multiplying both sides by 100 to remove the decimal points

$100x + 40 = 25x + 100$ —— Simplifying

$-25x + 100x + 40 = -25x + 25x + 100$ —— Adding $-25x$ to both sides

$75x + 40 = 100$

$75x + 40 - 40 = 100 - 40$ —— Adding -40 to both sides

$75x = 60$

$\dfrac{75x}{75} = \dfrac{60}{75}$ —— Dividing both sides by 75

Step 4. $x = \dfrac{60}{75}$, or 0.8 —— The number of liters of pure alcohol

✓   **Step 5. Check**   There is (obviously) 0.8 L of pure alcohol in 0.8 L of pure alcohol. There is 0.10(4 L), or 0.4 L, of pure alcohol in 4 L of a 10% solution. Therefore, there is (0.8 L + 0.4 L), or 1.2 L, of pure alcohol going *into* the mixture. The total amount of solution in the mixture is (4 L + 0.8 L), or 4.8 L, and the *strength* of the mixture will be $\dfrac{1.2\ L}{4.8\ L}$, or 0.25, or 25%.

Step 6. Therefore, we must add 0.8 L of pure alcohol.

Exercises 4.9
Set I

Set up each problem algebraically, solve, and check. Be sure to state what your variables represent.

1. How many milliliters of pure alcohol are there in 16 mL of a 20% solution of alcohol?

2. How many liters of pure sulfuric acid are there in 2 L of a 20% solution?

3. There is 43 g of sulfuric acid in 500 g of solution. Find the percent of acid in the solution.

4. Five hundred grams of a solution contains 27 g of a drug. Find the strength (in percent form) of the drug.

5. A 40% solution of hydrochloric acid contains 24 mL of pure acid. How many milliliters of solution are there?

6. A 30% solution of potassium chloride contains 15 mL of pure potassium chloride. How many milliliters of solution are there?

7. How many cubic centimeters (cc) of a 20% solution of sulfuric acid must be mixed with 100 cc of a 50% solution to make a 25% solution of sulfuric acid?

8. How many pints of a 2% solution of disinfectant must be mixed with 5 pt of a 12% solution to make a 4% solution of disinfectant?

9. How many milliliters of water must be added to 500 mL of a 40% solution of sodium bromide to reduce it to a 25% solution?

10. How many milliliters of water must be added to 600 mL of a 30% solution of antifreeze to reduce it to a 25% solution?

11. How many liters of pure alcohol must be added to 10 L of a 20% solution of alcohol to make a 50% solution?

12. How many liters of pure alcohol must be mixed with 20 L of a 30% alcohol solution in order to obtain a 44% solution?

Exercises 4.9
Set II

Set up each problem algebraically, solve, and check. Be sure to state what your variables represent.

1. How many liters of pure acetic acid are there in 3 L of a 15% solution of acetic acid?

2. There is 90 mL of pure sulfuric acid in a 15% sulfuric acid solution. How many milliliters of solution are there?

3. There is 66 g of hydrochloric acid in 120 g of solution. What is the percent of hydrochloric acid in the solution?

4. How many milliliters of pure antifreeze are there in 1,600 mL of a 20% solution of antifreeze?

5. A 35% glycerin solution contains 28 L of pure glycerin. How many liters of the glycerin solution are there?

6. There is 48 mL of pure acetic acid in 64 mL of an acetic acid solution. What is the percent of acetic acid in the solution?

7. If 100 gal of a 75% glycerin solution is made up by combining a 30% glycerin solution with a 90% glycerin solution, how much of each solution must be used?

8. How many milliliters of a 25% solution of potassium chloride must be added to 8 mL of pure potassium chloride to obtain a 35% solution?

9. How much water must be added to 60 mL of a 40% alcohol solution to obtain a 24% solution?

10. If 1,600 cc of a 10% dextrose solution is made up by combining a 20% dextrose solution with a 4% dextrose solution, how much of each solution must be used?

11. How many liters of pure alcohol must be added to 15 L of a 60% solution in order to obtain a 70% solution?

12. How many liters of water should be added to 20 L of a 40% solution of hydrochloric acid to reduce it to a 25% solution?

4.10 Miscellaneous Applications

In this section, we provide you with a number of different kinds of applied problems to solve. There may be some money problems, some percent problems, and so forth, but some problems are different from any you have seen so far. All the problems can (and should) be solved by using the algebraic methods that have been covered in this book.

The only way to learn to solve applied problems is to attempt to solve lots of them! In this section, we want you to try your skills at using algebra in problem solving. Therefore, the odd and even problems are not "matched," nor are any examples given. You may need to refer to the formulas listed on the page facing the inside back cover.

Exercises 4.10
Set I

Set up each problem algebraically, solve, and check. Be sure to state what your variables represent.

1. Roy invested some money at 7% simple interest per year. After 3 years, he had earned $756 interest. How much did he invest originally?

2. Forty percent of the members of the Fifty-Plus Club are men. Last Thursday, 20% of the male members of the club and 40% of the female members attended a dinner. If there were 64 people at the members-only dinner, how many members does the club have?

3. The volume of a right circular cylinder is 100π cubic inches, and the radius is 5 in. What is the height (or altitude)?

4. A metal tank is in the shape of a rectangular box, and it contains some water. Right now it is one-third full. If we add 300 cubic inches of water, it will be three-fourths full. What is the volume of the tank?

5. Michael invested $5,000 for 1 year. He invested part of the money at 5% interest (per year) and the rest at 6% interest. If the amount of interest he earned from these investments in one year was $272, how much of the $5,000 did he invest at 5% interest?

6. Cassandra can reach Adam's house from her house in 2 hr if she averages 45 mph. If Bill lives 30 mi further from Cassandra than Adam does, what is the distance between Cassandra's house and Bill's house?

7. If the sum of the measures of two angles is 180°, the angles are said to be **supplementary angles**. If the measure of angle A is four times as great as the measure of its supplement, what is the measure of angle A?

8. Henry gets paid regular wages for the first 40 hours he works, and he gets paid time-and-a-half for overtime. Last week he worked 54 hours and he earned $439.20. What is his regular hourly rate of pay?

9. One number is sixteen less than another number. The sum of the two numbers is 84. Find both numbers.

Exercises 4.10
Set II

Set up each problem algebraically, solve, and check. Be sure to state what your variables represent.

1. Catherine wants to enclose a rectangular area with 900 ft of fencing. If the rectangle is to be twice as long as it is wide, what must the length and width of the rectangle be?

2. One year ago, Brett invested some money at 6% simple interest. If there is now $901 in his account, how much did he invest originally?

3. A metal tank is in the shape of a right circular cylinder, and it contains some water. Right now it is one-fourth full. If we add 80 cubic inches of water, it will be one-third full. What is the volume of the tank?

4. A basket contained only blue balls and pink balls; 30% of those balls were blue. Then 25% of the blue balls and 75% of the pink balls were removed. If 32 balls remained in the basket, how many balls were in the basket originally?

5. The volume of a right circular cone is 240π cubic inches. If the radius of the cone is 6 in., what is the height?

6. Margie is filling an 18-cubic-foot planter with a mixture of potting soil and mushroom compost. She wants to use twice as many cubic feet of potting soil as compost. The potting soil comes in $1\frac{1}{2}$-cubic-foot bags, and the compost comes in 2-cubic-foot bags. How many bags of potting soil should she buy? How many bags of compost?

7. If the sum of the measures of two angles is 90°, the angles are said to be **complementary angles**. If the measure of angle B is one-third as large as the measure of its complement, what is the measure of angle B?

8. At Acme Car Rental, it costs $29.95 per day plus 25¢ per mile to rent a car. If Claire rented a car from Acme for five days and if her total bill was $332.75, how many miles did she drive during those five days?

9. If a 66-oz box of laundry detergent costs $2.29 and a 154-oz box of the same brand of laundry detergent costs $6.79, which size is the better buy? By how much per ounce (round off the answer to the nearest tenth of a cent)?

Sections 4.1–4.10

Method for Solving Applied Problems 4.1–4.10

REVIEW

Read   To solve an applied problem, first read it very carefully. *Be sure you understand the problem*. Read it several times, if necessary.

Think   Determine what *type* of problem it is, if possible. Determine what is unknown. Is there enough information given so that you *can* solve the problem? Do you need a special formula? What operations must be used?

Sketch   Draw a sketch with labels, if it might be helpful.

Step 1. Represent one unknown number by a variable, and declare its meaning in a sentence of the form "Let $x = \ldots$." Then reread the problem to see how you can represent any other unknown numbers in terms of the same variable, declaring their meaning in sentences that begin "Then $\ldots$."

Reread   Reread the sentences, breaking them up into key words and phrases.

Step 2. Translate each key phrase into an algebraic expression, and arrange these expressions into an equation or inequality that represents the facts given in the problem.

Step 3. Solve the equation or inequality.

Step 4. Solve for *all* the unknowns asked for in the problem.

Step 5. Check the solution(s) *in the word statement*.

Step 6. State the results clearly.

(The "chart" method can be used for distance-rate-time problems, for mixture problems, and for solution mixture problems.)

Set up each problem algebraically, solve, and check. Be sure to state what your variables represent.

1. Two numbers are in the ratio of 6 to 7. Their sum is 52. Find the numbers.

2. What is 245% of $450?

3. 77.5 is 31% of what number?

4. Six is what percent of 16?

5. The rent on a $380-a-month apartment was raised 5%. Find the new rent.

6. Solder used for soldering zinc is composed of lead and tin in the ratio 3:5. How many ounces of each are present in 24 oz of solder?

7. The three sides of a triangle are in the ratio 3:4:5. The perimeter is 108. Find the lengths of the three sides of the triangle.

8. Raul has twenty-two bills with a total value of $500. If these bills are $10 bills and $50 bills, how many of each kind are there?

9. Roger bought $2.10 worth of stamps. He bought only 2¢, 10¢, and 12¢ stamps. If there were twice as many 12¢ stamps as 10¢ stamps and twice as many 2¢ stamps as 12¢ stamps, how many of each did he buy?

10. The total receipts for a football game were $492,000. General admission tickets cost $15 each, reserved seat tickets $24 each, and box seat tickets $30 each. If the promoters sold 4,500 more reserved seats than box seats and 4 times as many general admissions as reserved seats, how many of each did they sell?

11. An affirmative action committee at a certain college demanded that 24% of the 1,800 entering freshmen be minority students. If 256 freshman minority students enrolled, how many more minority freshmen would need to be enrolled to satisfy the committee?

12. After sailing downstream for 2 hr, it took a boat 7 hr to return to its starting point. If the speed of the boat in still water was 9 mph, what was the speed of the river?

13. Mrs. Koontz left Downey at 5 A.M., heading toward Washington. Her neighbor Mrs. Fowler left at 6 A.M., also heading toward Washington. By driving 5 mph faster, Mrs. Fowler overtook Mrs. Koontz at 3 P.M. the same day.

    a. What was Mrs. Koontz's average speed?

    b. What was Mrs. Fowler's average speed?

    c. How far had they driven before Mrs. Fowler overtook Mrs. Koontz?

14. A dealer is to make up a 30-lb mixture of nuts costing 85¢ and 95¢ per pound. How many pounds of each must be used in order for the mixture to cost 91¢ per pound?

15. How many cubic centimeters of water must be added to 500 cc of a 25% solution of potassium chloride to reduce it to a 5% solution?

_____

Name

Set up each problem algebraically, solve, and check. Be
sure to state what your variables represent.

1. A woman cut a 5-ft board into two pieces; the lengths of the pieces were in the ratio of
1 to 3. Find the lengths of the pieces *in inches*.

2. What is 175% of $350?

3. Ninety-three is what percent of 124?

4. A 26% solution of hydrochloric acid contains 62.4 mL of pure acid. How many milliliters
of solution are there?

5. There is 45 mL of sulfuric acid in 300 mL of solution. Find the percent of acid in the
solution.

6. Mr. Maiolo's new contract calls for an 8% raise in salary. His present salary is $32,000.
What will his new salary be?

7. A disc jockey plays folk, rock, country-western, and Latino records in the ratio 2:5:3:4.
If he played 126 records in a week, how many records of each type were played?

ANSWERS

1. _____

_____

2. _____

3. _____

4. _____

5. _____

6. _____

7. _____

_____

_____

_____

8. Lee has sixty coins with a total value of $13.35. The coins are dimes, quarters, and 50¢ pieces. If there are 4 times as many dimes as quarters, how many coins of each kind does he have?

9. A 10-lb mixture of nuts and raisins costs $25. If raisins cost $1.90 per pound and nuts $3.40 per pound, how many pounds of each are in the mixture?

10. Cory wants to mix dried figs with dried apricots to make an 8-lb mixture worth $2.70 per pound. If dried figs cost $1.80 per pound and dried apricots cost $4.20 per pound, how many pounds of each should be used?

11. Marsha invested part of $18,000 at 16% and the remainder at 20%. Her total yearly income from these investments was $3,120. How much was invested at each rate? (The interest was *simple* interest.)

12. Drew paddled his kayak downstream for 4 hr. After a lunch break, he paddled back upstream, but it took him 7 hr to get back to his starting point. The speed of the river was 3 mph.

    a. How fast did Drew's kayak move in still water?

    b. How far downstream did he travel?

13. A 100-lb mixture of two different kinds of apples costs $72.50. If one kind of apple costs 90¢ per pound and the other costs 65¢ per pound, how many pounds of each kind were used?

14. How many pounds of Spanish peanuts that cost $2.50 per pound should be mixed with 35 lb of another variety of peanuts that costs $2.60 per pound in order for the mixture to be worth $2.57 per pound?

15. How many cubic centimeters of a 50% phenol solution must be added to 400 cc of a 5% solution to make it a 10% solution?

8. _____

_____

_____

9. _____

_____

10. _____

_____

11. _____

_____

12a. _____

b. _____

13. _____

_____

14. _____

15. _____

234

Chapter 4   DIAGNOSTIC TEST

The purpose of this test is to see how well you understand solving applied problems. We recommend that you work this diagnostic test *before* your instructor tests you on this chapter. Allow yourself about 50 minutes.

Complete solutions for all the problems on this test, together with section references, are given in the answer section in the back of the book. We suggest that you study the sections referred to for the problems you do incorrectly.

Set up each problem algebraically, solve, and check. Be sure to state what your variables represent.

1. A solution contains 28 mL of antifreeze. If it is a 40% solution, how many milliliters of the solution are there?

2. A 20-mL solution of alcohol contains 17 mL of pure alcohol. What is the percent of alcohol in the solution?

3. When 16 is added to 3 times an unknown number, the sum is 37. Find the unknown number.

4. Ellen has twenty-five bills in her purse with a total value of $165. If these bills are all $10 bills and $5 bills, how many of each denomination are there?

5. A 50-lb mixture of granola and dried apricots is to be worth $2.34 per pound. If the granola costs $2.20 per pound and the apricots cost $2.70 per pound, how many pounds of each should be used?

6. Kevin left his home at 7 A.M., driving toward St. George. His brother Jason left their home one hour later, also driving toward St. George. By driving 9 mph faster than Kevin, Jason overtook Kevin at 1 P.M. the same day.

   a. What was Kevin's average speed?

   b. What was Jason's average speed?

   c. How far did they drive before Jason overtook Kevin?

7. How many pounds of nuts that cost $3.50 per pound should be mixed with 30 lb of macadamia nuts that cost $8.00 per pound in order to obtain a mixture that is worth $4.85 per pound?

8. How many milliliters of an 80% solution of alcohol should be added to 140 mL of a 30% solution to make up a 45% solution?

9. The three sides of a triangle are in the ratio $7:9:11$. The perimeter is 135 m. Find the lengths of the three sides.

10. An item that cost a merchant $240 is marked up 30%. What is the selling price?

Chapters 1–4   CUMULATIVE REVIEW EXERCISES

In Exercises 1–3, evaluate each expression or write "not defined."

1. $\frac{0}{0}$

2. $3\sqrt{36} - 4^2(-5)$

3. $46 - 2\{4 - [3(5 - 8) - 10]\}$

4. Using the formula $A = \frac{h}{2}(b + B)$, find A when $h = 5$, $b = 7$, and $B = 11$.

5. Using the formula $S = 4\pi r^2$, find S when $\pi \approx 3.14$ and $r = 5$.

6. List any prime numbers greater than 7 and less than 31 that yield a remainder of 1 when divided by 5.

In Exercises 7 and 8, write each expression in simplest form.

7. $8 + 12\{3 - 2(x + 4)\}$

8. $3x - 2[5x - (-3x + 7)] - 4$

In Exercises 9–14, solve each equation or write either "identity" or "no solution."

9. $32 - 4x = 15$

10. $-11 = \frac{x}{3}$

11. $5w - 12 + 7w = 6 - 8w - 3$

12. $4(2y - 5) = 16 + 3(6y - 2)$

13. $6z - 22 = 2[8 - 4(5z - 1)]$

14. $3 - 5(2x - 3) = 9(2 - x)$

15. What is the additive identity?

16. Is 0 a rational number?

17. Is 17 a real number?

18. Is addition associative?

19. What is the multiplicative identity?

In Exercises 20–25, set up each problem algebraically, solve, and check. Be sure to state what your variables represent.

20. Two numbers are in the ratio of 11 to 9. Their sum is 160. Find the numbers.

21. Patty has fifteen coins with a total value of $1.75. If these coins are all nickels and quarters, how many of each kind are there?

22. Jodi left her home in Lafayette at 6 A.M., heading up the coast. Her husband, Bud, left at 7 A.M., also heading up the coast. By driving 10 mph faster, Bud overtook Jodi at noon. How fast was Bud driving?

23. A merchant wants to make up a 10-lb mixture of two kinds of candy. One kind of candy costs $1.80 per pound, and the other costs $1.20 per pound. How many pounds of each kind of candy should he use if the mixture is to be worth $15.90?

24. How many liters of a 35% solution of alcohol should be mixed with 30 L of a 55% solution to make up a 47% solution?

25. The sum of four consecutive even integers is 108. What are the integers?

Critical Thinking and Problem-Solving Exercises

1. Bruce is looking at a rectangle that is 22 cm long and 28 cm wide, and he wants to construct a *square* that has the same *perimeter* as this rectangle. What would the *area* of the square be?

2. Tina wants to use three-digit code numbers for some articles she is manufacturing. How many different three-digit code numbers can she make up if she uses just the digits 1, 2, 3, and 4, if no digit can be repeated (that is, 224 would not be permitted) and if the *order* in which the digits are used makes a difference (that is, 321 is considered different from 231)?

3. Tess is thinking of a number. Using the following clues, can you guess what the number is? The number is a two-digit prime number that is greater than 50, and the sum of the digits is 10.

4. Ray entered a florist shop to buy some flowers. He saw 4 more purple gladiolas than blue carnations, and the sum of the numbers of purple gladiolas and blue carnations was equal to the total number of pink roses. There were twice as many pink roses as there were yellow daisies. There were one-third as many pink roses as there were white gardenias, and there were 48 white gardenias. Altogether, how many flowers did Ray see?

5. A merchant marked up the price of an item that cost him $240 by 20%. Because the item hadn't sold after a long period of time, he marked down the selling price by 20%, thinking that that would bring the price down to what he had paid for the item. What was wrong with his reasoning? What did the 20% decrease bring the price down to?

Exponents

CHAPTER 5

n this chapter, we discuss the properties of exponents that will be used throughout the remainder of the book. We also continue with our discussion of simplifying algebraic expressions, and we discuss scientific notation. We do not provide proofs of the properties of exponents given in this chapter; either they require proof by mathematical induction, a technique not discussed in this book, or they are true by definition.

5.1   Multiplying Exponential Numbers

The following argument does not *prove* the first property of exponents—the product rule; however, we hope it will convince you that the property is true. Consider the product $x^3 \cdot x^2$. (Notice that x^3 and x^2 are both powers of the *same* base, x.)

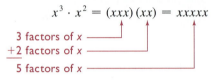

$$x^3 \cdot x^2 = (xxx)(xx) = xxxxx$$

3 factors of x ———
+2 factors of x ———
5 factors of x ———

Now consider the problem x^{3+2}; $x^{3+2} = x^5 = xxxxx$. Because $x^3 \cdot x^2$ and x^{3+2} both equal the same number (both equal $xxxxx$), $x^3 \cdot x^2 = x^{3+2}$.

The general procedure for multiplying exponential numbers *when the bases are the same* is stated in the product rule, which follows.

The product rule

$$x^a \cdot x^b = x^{a+b}$$

In words: To multiply powers of the *same base*, *add* the exponents.

The product rule can be extended to include more factors:

$$x^a x^b x^c \cdots = x^{a+b+c+\cdots}$$

EXAMPLE 1   Examples of using the product rule or of determining that the product rule does not apply:

a. $x^5 \cdot x^2 = x^{5+2} = x^7$

b. $x \cdot x^2 = x^{1+2} = x^3$ Recall that when no exponent is written,
      $x = x^1$ the exponent is understood to be 1

c. $x^3 \cdot x^7 \cdot x^4 = x^{3+7+4} = x^{14}$

d. $x^4 \cdot y^3 \cdot x^2 = x^4 \cdot x^2 \cdot y^3 = x^{4+2} \cdot y^3 = x^6 y^3$

e. $a^x \cdot a^y = a^{x+y}$

f. $x^3 \cdot y^2 = x^3 y^2$ The product rule does not apply in this case, because the bases are different. In fact, $x^3 y^2$ cannot be rewritten with a base appearing only once.

g. $x^2 + x^3$ The product rule does not apply, because the operation is addition rather than multiplication. In fact, $x^2 + x^3$ cannot be rewritten with the base appearing only once.

h. $x^5 + x^5 = 2x^5$ x^5 and x^5 are like terms and *must be combined*

i. $10^3 \cdot 10^4 = 10^{3+4} = 10^7$, or $10,000,000$

j. $2 \cdot 2^3 \cdot 2^2 = 2^{1+3+2} = 2^6 = 64$

✕ **A Word of Caution**	Common errors in using the product rule when the bases are *numbers* (as in Examples 1i and 1j) are shown on the right below:

Correct method	*Incorrect method*
a. $10^3 \cdot 10^4 = 10^{3+4} = 10^7$	$10^3 \cdot 10^4 = 100^{3+4} = 100^7$
b. $2 \cdot 2^3 \cdot 2^2 = 2^{1+3+2} = 2^6$	$2 \cdot 2^3 \cdot 2^2 = 8^{1+3+2} = 8^6$

Remember: $10^3 \cdot 10^4$ means $(10 \cdot 10 \cdot 10) \cdot (10 \cdot 10 \cdot 10 \cdot 10) = 10^7$, *not* 100^7, and $2 \cdot 2^3 \cdot 2^2$ means $(2) \cdot (2 \cdot 2 \cdot 2) \cdot (2 \cdot 2) = 2^6$, *not* 8^6.

> When you multiply powers of the same base, add the exponents but *do not* multiply the bases as well.

As you work the problems in the following exercise set, recall that within each *term* it is customary to write variables in alphabetical order.

Exercises 5.1
Set I

Rewrite each of the following with the base appearing only once whenever possible.

1. $x^3 \cdot x^4$	**2.** $x^2 \cdot x^9$	**3.** $y \cdot y^3$
4. $z \cdot z^4$	**5.** $m^2 \cdot m$	**6.** $a^3 \cdot a$
7. $10^2 \cdot 10^3$	**8.** $10^4 \cdot 10^3$	**9.** $5 \cdot 5^4 \cdot 5^5$
10. $3 \cdot 3^2 \cdot 3^3$	**11.** $x \cdot x^3 \cdot x^4$	**12.** $y \cdot y^5 \cdot y^3$
13. $x^2 y^5$	**14.** $a^3 b^2$	**15.** $3^2 \cdot 5^3$

16. $2^3 \cdot 3^2$	**17.** $a^4 + a^2$	**18.** $x^3 + x^4$
19. $a^x \cdot a^w$	**20.** $x^a \cdot x^b$	**21.** $x^y \cdot y^x$
22. $a^b \cdot b^a$	**23.** $x^2 y^3 x^5$	**24.** $z^3 z^4 w^2$
25. $a^2 b^3 a^5$	**26.** $x^8 y x^4$	**27.** $s^7 + s^4$
28. $t^3 + t^8$	**29.** $7^3 \cdot 7^5$	**30.** $6^4 \cdot 6^{15}$
31. $x^3 + x^3$	**32.** $y^6 + y^6$	**33.** $2^{14} + 2^{14}$
34. $3^{18} + 3^{18}$	**35.** $z^7 + z^7$	**36.** $w^2 + w^2$

 Writing Problems

Express the answers in your own words and in complete sentences.

1. Explain how to multiply x^8 by x^5.

2. Explain why $x^3 + x^2 \neq x^5$.

3. Explain why $3^3 \cdot 3^2 \neq 9^5$.

Exercises 5.1
Set II

Rewrite each of the following with the base appearing only once whenever possible.

1. $x^2 \cdot x^5$	**2.** $y \cdot y^4$	**3.** $u^2 \cdot u$
4. $10^3 \cdot 10^4$	**5.** $3 \cdot 3^2 \cdot 3^2$	**6.** $a \cdot a^3 \cdot a^2$
7. $x^3 y^3$	**8.** $x^a \cdot y^b$	**9.** $7^3 \cdot 7 \cdot 7^5$
10. $3^2 \cdot 4^3$	**11.** $a^4 \cdot a^2$	**12.** $xy^2 x^3$
13. $x^6 \cdot x^9$	**14.** $y^7 \cdot y$	**15.** $8^2 \cdot 3^3$

16. $a^4 \cdot b^8$	**17.** $y^2 + y^{14}$	**18.** $y^2 \cdot y^{14}$
19. $b^x \cdot b^y$	**20.** $b^x \cdot b$	**21.** $s^x \cdot t^y$
22. $s^2 + s^5$	**23.** $a^4 \cdot b^3 \cdot c^2$	**24.** $x + x^4$
25. $x \cdot x^7$	**26.** $xy^3 x^4$	**27.** $x^4 + x^2$
28. $a^5 + a$	**29.** $4^3 \cdot 4^{12}$	**30.** $6^8 + 6^4$
31. $s^5 + s^5$	**32.** $x^6 + y^6$	**33.** $5^{15} + 5^{15}$
34. $2^{28} + 2^{28}$	**35.** $t^4 + t^4$	**36.** $u^2 + u^3$

5.2   Simplifying Products of Factors

In this section, we discuss simplifying a product of factors when there is a *single term* inside each set of grouping symbols. We proceed by following the steps below:

Simplifying a product of factors when each factor contains only one term

1. *Multiply the numerical coefficients.* Remember that the sign of the coefficient of any term is part of the numerical coefficient.
2. *Multiply the variables.* Each variable should appear only once, and the variables are usually written in alphabetical order. Add the exponents on each variable to determine the exponent on that variable. That is, use the product rule $(x^a \cdot x^b = x^{a+b})$.
3. The final answer is the product of the results of steps 1 and 2.

Note   We *cannot* use the distributive property for such problems. The distributive property can be used only when at least one set of grouping symbols contains two or more *terms*; we discuss this further in the next section.

In Example 1, we make use of the associative and commutative properties of multiplication and rearrange the factors so that the numerical coefficients are together, the x's are together, and so on. In practice, this step need not be shown.

EXAMPLE 1   Examples of simplifying products of factors:

a. $(6x)(-2x^3)$

$= (6)(-2)(x \cdot x^3)$ → Step 1: $(6)(-2) = -12$

$= -12x^4$ → Step 2: Using the product rule:
$x \cdot x^3 = x^1 \cdot x^3 = x^{1+3} = x^4$

b. $(-2a^2b)(5a^3b^3)$

$= (-2)(5)(a^2 \cdot a^3)(b \cdot b^3)$ → Step 1: $(-2)(5) = -10$

$= -10a^5b^4$ → Step 2: Using the product rule:
$a^2 \cdot a^3 = a^{2+3} = a^5$ and $b \cdot b^3 = b^{1+3} = b^4$

c. $(-5y^2)(2y)(-4y^3)$

$= (-5)(2)(-4)(y^2yy^3)$

$= 40y^{2+1+3}$

$= 40y^6$

d.

$(-2^4xy^2)(2^3x^2y^3)$

$= (-1)(2^4)(2^3)(xx^2)(y^2y^3)$

$= (-1)(2^{4+3})x^{1+2}y^{2+3}$

$= -2^7x^3y^5$, or $-128x^3y^5$

Alternatively

$(-2^4xy^2)(2^3x^2y^3)$ The exponent 4 applies
only to the 2

$= (-16xy^2)(8x^2y^3)$ $-2^4 = (-1)(2^4)$, or -16

$= (-16)(8)(xx^2)(y^2y^3)$

$= -128x^{1+2}y^{2+3}$

$= -128x^3y^5$

e. $(3xy^2)^3$

$= (3xy^2)(3xy^2)(3xy^2)$

$= (3 \cdot 3 \cdot 3)(xxx)(y^2y^2y^2)$

$= 27x^{1+1+1}y^{2+2+2}$

$= 27x^3y^6$

Recall that the product of an odd number of negative factors is negative and the product of an even number of negative factors is positive. In Example 1, you can determine the sign of the answer by using these properties if you wish. Notice that in Examples 1a, 1b, and 1d we had an odd number of negative factors and the signs of the answers were negative, and in Example 1c we had an even number of negative factors and the sign of the answer was positive.

A term containing exponents is considered *simplified* when each different base appears only once and when the exponent on each base is a single positive integer. For example, $5x^3y$ is simplified, whereas $5xyx^2$ and $5x^{1+2}y$ are not simplified.

Exercises 5.2
Set I

Simplify each of the following products of factors.

1. $(-2a)(4a^2)$
2. $(-3x)(5x^3)$
3. $(-5h^2)(-6h^3)$

4. $(-6k^3)(-8k)$
5. $(-5x^3)^2$
6. $(-7y^4)^2$

7. $(-2a^3)(-4a)(3a^4)$
8. $(-6b^2)(-2b)(-4b^3)$

9. $(-9m)(m^5)(-2m^2)$
10. $(4n^4)(-7n^2)(-n)$

11. $(5x^2)(-7y)$
12. $(-3x^3)(4y)$

13. $(-6m^3n^2)(-4mn^2)$
14. $(-8h^4k)(5h^2k^3)$

15. $(2x^{10}y^2)(-3x^{12}y^7)$
16. $(-2a^2b^5)(-5a^{10}b^{10})$

17. $(3xy^2)^2$
18. $(4x^2y^3)^2$

19. $(5x^4y^5z)(-y^4z^7)$
20. $(-7E^2F^5G^8)(3F^6G^{10})$

21. $(2^3RS^2)(-2^2R^5T^4)$
22. $(-3^2xy)(3^3x^8z^5)$

23. $(-c^2d)(5d^1e^3)(-4c^5e^2)$
24. $(3m^1n^2)(-m^3r^4)(-8n^5r)$

25. $(-2x^2)(-3xy^2z)(-7yz)$
26. $(14y^2)(-5x^3z^4)(-6xyz)$

27. $(3xy)(x^2y^2)(-5x^3y^3)$
28. $(xy)(-2xz)(3yz)$

29. $(-2ab)(5bc)(ac)$
30. $(3xyz)(-2x^2y)(-yz^2)$

31. $(-5x^2y)(2yz^3)(-xz)$
32. $(2a^2b^3)(-3ab^2)(-a^2b^2)$

33. $(-5x^2y^2z)(x^5y^3z)(7xyz^5)$

34. $(-4R^2S^3T^4)(-8RS^3T^4)(-R^6S^1T^5)$

35. $(-3h^2k)(7)(-m^3k^5)(-mh^4)$

36. $(-km^2)(-6n^3)(8)(-k^2m)$

✏️ Writing Problems

Express the answers in your own words and in complete sentences.

1. Explain why $8x^3yx$ is not considered to be in simplest form. What is simplest form for $8x^3yx$?

2. Explain why $(-7x^3)(-5x^2) \neq 35x^6$.

Exercises 5.2
Set II

Simplify each of the following products of factors.

1. $(-3x^2)(8x^3)$
2. $(3a^2)(-2a^4)$
3. $(-7x^3)(-12x^4)$

4. $(-5a^2b)(3ab^2)$
5. $(4x^2)^3$
6. $(3x^2)^2$

7. $(-4b^2)(-2b)(-6b^5)$
8. $(-3t^4)(5t^3)(-10t)$

9. $(-4y^2)(y^7)(-3y)$
10. $(6x^3)(-2x)(7x^4)$

11. $(15x^3)(-3y^2)$
12. $(-12a^4)(-5b)$

13. $(-4c^2d)(-13cd^3)$
14. $(-6e^5f^3)(11e^2f^3)$

15. $(2mn^2)(-4m^2n^2)(2m^3n)$
16. $(5x^2y)(-2xy^2)(-3xy)$

17. $(2m^5n)(-4mn^3)(3m^2n^6)$
18. $(7ab^4)(2a^2b)(-5a^2b^3)$

19. $(8a^4)(-a^4b^3c)$
20. $(3x^7)(-y^3)(4z^2)$

21. $(3x^3)(-3^4x^5)$
22. $(-5^2a^3)(5ab^2)$

23. $(5x^2y)(-3xy^2z)(-2xz^3)$
24. $(-8st^2)(-s^3t)(2t^4)$

25. $(-3ab^3)(-4a^2b^4)(c^5)$
26. $(9x^2y^3)(-2x^4y)(-5xy^3)$

27. $(8ab)(-2ab^4)(-2^4ab^3)$
28. $(-3u^3v^4)(-2^2u^2)(-3^2v^5)$

29. $(-7xy^2)(-3x^2z)(-2xz^3)$
30. $(8ac)(-9b^2c^3)(abc)$

31. $(3s^4)(-5st^3)(-s^2t^4)$
32. $(-2xy^2z)^2$

33. $(2H^2K^4M)(-2^2H^4K^5N)(2^3M^6K^7N)$

34. $(-10x^3y^2z)(-10^2xk^{10}z^1)(-10ky)$

35. $(-2e^1f^2)(-3e^3h)(-6fh)(-5e^2h)$

36. $(-4x^2y)(-7y^2z^2)(-5xy^4z)(-2x^3z^5)$

5.3 Using the Distributive Property; Simplifying Algebraic Expressions

Using the Distributive Property

When we first discussed using the distributive property to remove grouping symbols, the examples and exercises were carefully selected so that the product rule was never needed. In this section, we continue our discussion of simplifying algebraic expressions by combining what we learned earlier in this chapter with our knowledge of the distributive property.

In the preceding section, we learned how to find products of factors when each factor consists of only one term. If one or more factors contains *more* than one term, we must use the distributive property. When we remove grouping symbols by using the distributive property, we must be sure to multiply *each term* inside the grouping symbols by the other factor.

In the first method shown in Examples 1c, 1d, and 1e, we treat subtraction problems as addition problems, and in the alternative method, we use the distributive property in the form $a(b - c) = ab - ac$.

EXAMPLE 1 Examples of using the distributive property to remove parentheses:

a. $x(x^2 + y) = (x)(x^2) + (x)(y)$

$\qquad = x^3 + xy$

b. $3x(4x + x^3y) = (3x)(4x) + (3x)(x^3y)$

$\qquad = 12x^2 + 3x^4y$

c. $-5a(4a^3 - 2a^2b + b^2) = (-5a)(4a^3 + [-2a^2b] + b^2)$

$\qquad = (-5a)(4a^3) + (-5a)(-2a^2b) + (-5a)(b^2)$

$\qquad = -20a^4 + 10a^3b - 5ab^2$

Alternative Method

$-5a(4a^3 - 2a^2b + b^2) = (-5a)(4a^3) - (-5a)(2a^2b) + (-5a)(b^2)$

$\qquad = -20a^4 + 10a^3b - 5ab^2$

d. $(2x - 5)(-4x) = (2x + [-5])(-4x)$

$\qquad = (2x)(-4x) + (-5)(-4x)$

$\qquad = -8x^2 + 20x$

Alternative Method

$(2x - 5)(-4x) = (2x)(-4x) - (5)(-4x)$

$\qquad = -8x^2 + 20x$

e. $(-2x^2 + xy - 5y^2)(-3xy) = (-2x^2 + xy + [-5y^2])(-3xy)$

$\qquad = (-2x^2)(-3xy) + (xy)(-3xy) + (-5y^2)(-3xy)$

$\qquad = 6x^3y - 3x^2y^2 + 15xy^3$

Alternative Method

$(-2x^2 + xy - 5y^2)(-3xy) = (-2x^2)(-3xy) + (xy)(-3xy) - (5y^2)(-3xy)$

$\qquad = 6x^3y - 3x^2y^2 + 15xy^3$

Note In Example 1, it is not necessary to show the intermediate steps that we have shown; you can just write the final answer.

Simplifying Algebraic Expressions

To simplify an algebraic expression, we remove all grouping symbols, simplify each term (each variable can appear only once in a term), and combine all like terms. It is important to write the *complete* expression in every step and, as you are learning, to make only one type of change per step (see Example 2). In Example 2, we treat subtraction problems as addition problems.

EXAMPLE 2 Examples of simplifying algebraic expressions:

a. $5x^2(3x^2 - 2x + 1) - 3x(7x^2 + x - 6)$

$= 5x^2[3x^2 + (-2x) + 1] + (-3x)[7x^2 + x + (-6)]$

$= 5x^2(3x^2) + 5x^2(-2x) + 5x^2(1) + (-3x)(7x^2) + (-3x)(x) + (-3x)(-6)$

$= 15x^4 - 10x^3 + 5x^2 - 21x^3 - 3x^2 + 18x$ Like terms / Simplifying each term

Like terms

$= 15x^4 - 10x^3 - 21x^3 + 5x^2 - 3x^2 + 18x$ Collecting like terms

$= 15x^4 - 31x^3 + 2x^2 + 18x$ Combining like terms

Note It is not customary to show the step in which we *collect* like terms. We will no longer show that step in the examples.

b. $8 - 3[4x^2 - 2x(3x + 7)]$ We must *not* subtract 3 from 8

$= 8 + (-3)[4x^2 + (-2x)(3x + 7)]$ Interpreting subtractions as additions

$= 8 + (-3)[4x^2 + (-2x)(3x) + (-2x)(7)]$ Distributing $-2x$ across $(3x + 7)$

$= 8 + (-3)[4x^2 + (-6x^2) + (-14x)]$ Performing the inner multiplications

Like terms

$= 8 + (-3)[-2x^2 + (-14x)]$ Combining like terms inside the brackets

$= 8 + (-3)(-2x^2) + (-3)(-14x)$ Distributing -3 across $[-2x^2 + (-14x)]$

$= 8 + 6x^2 + 42x$ Simplifying all terms

c. $7b + 2b(3a + 1 - 5b)$

$= 7b + 2b(3a) + 2b(1) + 2b(-5b)$ Using the distributive property

$= 7b + 6ab + 2b - 10b^2$ Simplifying each term

Like terms

$= 9b + 6ab - 10b^2$ Combining like terms

A Word of Caution In Example 4c, it is incorrect to write

$$7b + 2b(3a + 1 - 5b) = 9b(3a + 1 - 5b)$$

Remember: Using the correct order of operations, we must multiply *before* we add.

Raising a Product of Factors to a Power

Next, consider the expression $(xy)^3$.

$$(xy)^3 = (xy)(xy)(xy) = (xxx)(yyy) = x^3 y^3$$

The general procedure for raising a product of factors to a power is as follows.

The rule for raising a product to a power	$(xy)^a = x^a y^a$
	In words: To raise a product of factors to a power, raise *each factor* to that power.

EXAMPLE 2 Examples of using the rule for raising a product to a power:

a. $(xy)^5 = x^5 y^5$

b. $(2x)^3 = 2^3 x^3$ We must raise *numbers* in the parentheses to the power also

✗ A Word of Caution $(2x)^3 \neq 2x^3$. Remember that $2x^3 = 2xxx$ (the exponent applies only to the x), whereas $(2x)^3 = (2x)(2x)(2x) = 2^3 x^3$ (the exponent applies to everything inside the parentheses).

Evaluating Numerical Expressions

EXAMPLE 3 Examples of evaluating numerical expressions:

This exponent applies to everything inside the parentheses (that is, to -2); we must raise to powers *before* we multiply

a. $5(-2)^2 = 5(4) = 20$ $(-2)^2 = (-2)(-2) = 4$

This exponent applies only to the 5

b. $-2 \cdot 5^2 = -2 \cdot 25 = -50$ $5^2 = 25$

This exponent applies to *everything* inside the parentheses [that is, to $(-2 \cdot 5)$]

c. $(-2 \cdot 5)^2 = (-10)^2 = (-10)(-10) = 100$

Alternative Method $(-2 \cdot 5)^2 = (-2)^2 \cdot (5)^2 = 4 \cdot 25 = 100$

Exercises 5.4A
Set I

In Exercises 1–14, remove the parentheses.

I. $(y^2)^5$ **2.** $(N^3)^4$ **3.** $(x^8)^2$ **4.** $(z^4)^7$

5. $(2^4)^2$ **6.** $(3^4)^2$ **7.** $(xy)^5$ **8.** $(ab)^4$

9. $(2c)^6$ **10.** $(3x)^4$ **11.** $(x^4)^7$ **12.** $(v^3)^8$

13. $(10^2)^3$ **14.** $(10^7)^2$

In Exercises 15–20, evaluate each expression.

15. $(-2 \cdot 3)^2$ **16.** $(-3 \cdot 4)^2$

17. $2(-3)^3$ **18.** $3(-2)^3$

19. $-4 \cdot 5^2$ **20.** $-3 \cdot 4^2$

Writing Problems

Express the answer in your own words and in complete sentences.

1. Explain why $(3x)^5 \neq 3x^5$.

Exercises 5.4A
Set II

In Exercises 1–14, remove the parentheses.

1. $(a^3)^4$ 2. $(2x)^4$ 3. $(b^9)^4$ 4. $(x^3)^2$

5. $(2^2)^4$ 6. $(2x)^5$ 7. $(abc)^4$ 8. $(2xy)^5$

9. $(3x)^4$ 10. $(st)^8$ 11. $(u^7)^4$ 12. $(5x)^2$

13. $(10^3)^3$ 14. $(x^6)^4$

In Exercises 15–20, evaluate each expression.

15. $(-3 \cdot 5)^2$ 16. $(-5 \cdot 2)^3$ 17. $5(-2)^5$

18. $-4 \cdot 3^2$ 19. $-6 \cdot 2^2$ 20. $7(-2)^2$

5.4B Exponential Numbers and Division

Note In a later chapter, we will study algebraic fractions in great detail. The fractions found in the remainder of this chapter can be simplified just by using the fundamental property of fractions $\left(\dfrac{a}{b} = \dfrac{ac}{bc}, \text{ if } c \neq 0 \text{ and } b \neq 0 \right)$, the definition of multiplication of fractions $\left(\dfrac{a}{b} \cdot \dfrac{c}{d} = \dfrac{ac}{bd} \right)$, and the properties of exponents that will be developed here.

Dividing Exponential Numbers When the Bases Are the Same

Using the fundamental property of fractions on the division problem $x^5 \div x^3$, or $\dfrac{x^5}{x^3}$, we have

$$\frac{x^5}{x^3} = \frac{xx \cdot xxx}{1 \cdot xxx} = \frac{x^2 \cdot x^3}{1 \cdot x^3} = \frac{x^2}{1} = x^2$$

(We assume $x \neq 0$.)

Now consider x^{5-3}; $x^{5-3} = x^2$. Since $\dfrac{x^5}{x^3}$ and x^{5-3} both equal the same number (x^2), they must equal each other. Therefore, if $x \neq 0$, $\dfrac{x^5}{x^3} = x^{5-3}$.

The general procedure for dividing one exponential number by another *when the bases are the same* is stated in the quotient rule, which follows.

The quotient rule

If $x \neq 0$, $$\frac{x^a}{x^b} = x^{a-b}$$

In words: To divide one exponential number by another *when the bases are the same*, subtract the exponent of the divisor (the exponent in the denominator) from the exponent of the dividend (the exponent in the numerator).

☞ **Note** To see why the restriction $x \neq 0$ is included in the quotient rule, consider the example $\dfrac{0^5}{0^2}$:

$$\frac{0^5}{0^2} = \frac{0 \cdot 0 \cdot 0 \cdot 0 \cdot 0}{0 \cdot 0} = \frac{0}{0}$$

which cannot be determined. For this reason, x cannot be zero in the quotient rule.

EXAMPLE 4 Examples of using the quotient rule or of determining that the quotient rule does not apply (assume $x \neq 0$, $r \neq 0$, and $y \neq 0$):

a. $\dfrac{x^6}{x^2} = x^{6-2} = x^4$

b. $\dfrac{r^{12}}{r^5} = r^{12-5} = r^7$

c. $\dfrac{y^3}{y} = \dfrac{y^3}{y^1} = y^{3-1} = y^2$

d. $\dfrac{10^7}{10^3} = 10^{7-3} = 10^4 = \underbrace{10{,}000}_{\text{4 zeros}}$

e. $\dfrac{x^5}{y^2}$ The quotient rule does not apply when the bases are different. In fact, $\dfrac{x^5}{y^2}$ cannot be rewritten with a base appearing only once.

f. $x^4 - x^2$ The quotient rule does not apply, because the operation is subtraction rather than division. In fact, $x^4 - x^2$ cannot be rewritten with the base appearing only once.

g. $\dfrac{2^a}{2^b} = 2^{a-b}$

In this book, unless otherwise noted, we will assume that none of the variables has a value that makes a denominator zero.

In Example 4, we were careful to have the exponent in the numerator be larger than the exponent in the denominator, so that the quotient always had a positive exponent. When the exponent in the denominator is larger than the exponent in the numerator, the quotient has a negative exponent. Negative exponents are discussed in the next section.

EXAMPLE 5 Examples of dividing exponential expressions when the bases are *not* the same:

a. $\dfrac{-5^2}{(-5)^2} = \dfrac{-1(5)(5)}{(-5)(-5)} = \dfrac{-25}{25} = -1$ The exponent in the numerator applies *only* to the 5

The bases are not the same; in the numerator the base is 5, whereas in the denominator the base is -5. The quotient rule *cannot* be used.

b. $\dfrac{-2^3}{(-2)^3} = \dfrac{-1(2)(2)(2)}{(-2)(-2)(-2)} = \dfrac{-8}{-8} = 1$ The exponent in the numerator applies *only* to the 2

The bases are not the same; in the numerator the base is 2, whereas in the denominator the base is -2. The quotient rule *cannot* be used.

Raising Quotients (or Fractions) to Powers

Consider the problem $\left(\dfrac{x}{y}\right)^3$:

$$\left(\frac{x}{y}\right)^3 = \left(\frac{x}{y}\right)\left(\frac{x}{y}\right)\left(\frac{x}{y}\right) = \frac{xxx}{yyy} = \frac{x^3}{y^3} \qquad \text{Recall that to multiply fractions, we multiply}$$
Recall that to multiply fractions, we multiply the numerators and multiply the denominators

The general procedure for raising a quotient (or fraction) to a power is as follows.

The rule for raising a quotient to a power

If $y \neq 0$,
$$\left(\frac{x}{y}\right)^n = \frac{x^n}{y^n}$$

In words: To raise a quotient (or fraction) to a power, raise both the divisor and the dividend (or both the numerator and the denominator) to that power.

EXAMPLE 6 Examples of raising quotients (or fractions) to powers:

a. $\left(\dfrac{a}{b}\right)^5 = \dfrac{a^5}{b^5}$ Using the rule for raising a quotient to a power

b. $\left(\dfrac{3}{c}\right)^3 = \dfrac{3^3}{c^3} = \dfrac{27}{c^3}$ We must raise *numbers* in parentheses to the power also

The General Property of Exponents

Several of the properties of exponents can be combined into the following general property.

The general property of exponents

$$\left(\frac{x^a y^b}{z^c}\right)^n = \frac{x^{an} y^{bn}}{z^{cn}}$$

In words: When an expression being raised to a power consists of factors only, remove the parentheses by multiplying the exponent of *every factor* by the exponent that is outside the parentheses.

Notice that in the general property of exponents there are no addition or subtraction symbols (that is, x^a, y^b, and z^c are *factors* of the expression within the parentheses), and the exponent of *every* factor within the parentheses must be multiplied by the exponent outside the parentheses; if any factor has no exponent, the exponent of that factor is understood to be 1.

EXAMPLE 7 Examples of simplifying expressions:

a. $(x^2 y^3)^2 = x^{2 \cdot 2} y^{3 \cdot 2} = x^4 y^6$ Each exponent inside the parentheses must be multiplied by 2

The exponents on the 5 and the x are understood to be 1
Those 1's must be multiplied by 3

b. $(5xy^2 z^5)^3 = (5^1 x^1 y^2 z^5)^3 = 5^{1 \cdot 3} x^{1 \cdot 3} y^{2 \cdot 3} z^{5 \cdot 3} = 5^3 x^3 y^6 z^{15} = 125 x^3 y^6 z^{15}$

Note Later, we will simplify expressions such as $(x^2 + y^3)^2$. It is important that you realize that the general property of exponents *cannot* be used in such an expression, because x^2 and y^3 are *terms*, not factors, of $x^2 + y^3$. The general property of exponents applies only when there are no addition or subtraction symbols inside the parentheses.

EXAMPLE 8 Examples of simplifying expressions (assume that none of the factors appearing in any denominator is zero):

a. $\left(\dfrac{x^2 y^5}{z^3}\right)^4 = \dfrac{x^{2\cdot 4} y^{5\cdot 4}}{z^{3\cdot 4}} = \dfrac{x^8 y^{20}}{z^{12}}$

b. $\left(\dfrac{3b^3}{2c^4}\right)^2 = \left(\dfrac{3^1 b^3}{2^1 c^4}\right)^2 = \dfrac{3^{1\cdot 2} b^{3\cdot 2}}{2^{1\cdot 2} c^{4\cdot 2}} = \dfrac{3^2 b^6}{2^2 c^8}$, or $\dfrac{9b^6}{4c^8}$

See the review section for a list of all the properties of exponents.

Exercises 5.4B
Set I

Assume that none of the factors appearing in any denominator is zero. In Exercises 1–20, rewrite each expression with the base appearing only once, if possible.

1. $\dfrac{x^7}{x^2}$ **2.** $\dfrac{y^8}{y^6}$ **3.** $x^4 - x^2$ **4.** $s^8 - s^3$

5. $\dfrac{a^5}{a}$ **6.** $\dfrac{b^7}{b}$ **7.** $\dfrac{10^{11}}{10}$ **8.** $\dfrac{5^6}{5}$

9. $\dfrac{x^8}{y^4}$ **10.** $\dfrac{a^4}{b^3}$ **11.** $\dfrac{a^9}{a^4}$ **12.** $\dfrac{b^{10}}{b^2}$

13. $\dfrac{a^3}{b^2}$ **14.** $\dfrac{x^5}{y^3}$ **15.** $\dfrac{s^5}{s^4}$ **16.** $\dfrac{t^7}{t^6}$

17. $\dfrac{x^{5a}}{x^{3a}}$ **18.** $\dfrac{M^{6x}}{M^{2x}}$ **19.** $\dfrac{y^{6c}}{y^{4c}}$ **20.** $\dfrac{z^{7t}}{z^{5t}}$

In Exercises 21–36, remove the parentheses.

21. $\left(\dfrac{s}{t}\right)^7$ **22.** $\left(\dfrac{x}{y}\right)^9$ **23.** $\left(\dfrac{2}{x}\right)^4$ **24.** $\left(\dfrac{3}{z}\right)^2$

25. $\left(\dfrac{x}{2}\right)^6$ **26.** $\left(\dfrac{c}{5}\right)^3$ **27.** $(a^2 b^3)^2$ **28.** $(x^4 y^5)^3$

29. $(2z^3)^2$ **30.** $(3w^2)^3$ **31.** $\left(\dfrac{xy^4}{z^2}\right)^2$ **32.** $\left(\dfrac{a^3 b}{c^2}\right)^3$

33. $\left(\dfrac{5y^3}{2x^2}\right)^4$ **34.** $\left(\dfrac{6b^4}{7c^2}\right)^3$ **35.** $\left(\dfrac{2^2 x^3}{y^5 z}\right)^3$ **36.** $\left(\dfrac{3^2 a^4}{b^3 c}\right)^2$

In Exercises 37–40, evaluate each expression.

37. $\dfrac{(-4)^2}{-4^2}$ **38.** $\dfrac{-9^2}{(-9)^2}$ **39.** $\dfrac{(-3)^3}{-3^3}$ **40.** $\dfrac{(-4)^3}{-4^3}$

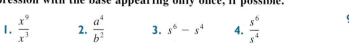

Writing Problems

Express the answers in your own words and in complete sentences.

1. Explain why $\left(\dfrac{x}{2}\right)^5 \neq \dfrac{x^5}{2}$.

2. Explain why $\left(\dfrac{2x^3}{y^2}\right)^3 \neq \dfrac{2x^9}{y^6}$.

Exercises 5.4B
Set II

Assume that none of the factors appearing in any denominator is zero. In Exercises 1–20, rewrite each expression with the base appearing only once, if possible.

1. $\dfrac{x^9}{x^3}$ **2.** $\dfrac{a^4}{b^2}$ **3.** $s^6 - s^4$ **4.** $\dfrac{s^6}{s^4}$

5. $\dfrac{y^7}{y}$ **6.** $\dfrac{3^5}{3}$ **7.** $\dfrac{8^6}{8^2}$ **8.** $\dfrac{15^9}{15^4}$

9. $\dfrac{a^9}{b^4}$ **10.** $\dfrac{a^6}{a^4}$ **11.** $\dfrac{g^9}{g^2}$ **12.** $\dfrac{x^{12}}{b^2}$

13. $\dfrac{m^4}{n^2}$ **14.** $\dfrac{x^5}{x}$ **15.** $\dfrac{c^6}{c^5}$ **16.** $\dfrac{d^7}{f^5}$ **30.** $(8xy^3)^2$ **31.** $\left(\dfrac{x^3y}{z^2}\right)^3$ **32.** $\left(\dfrac{s^4t^2}{u^3}\right)^3$

17. $\dfrac{y^{7e}}{y^{5e}}$ **18.** $\dfrac{z^{8a}}{z^{3a}}$ **19.** $\dfrac{H^{4n}}{H^{2n}}$ **20.** $z^{4a} - z^{2a}$ **33.** $\left(\dfrac{3a^5}{2x^2}\right)^4$ **34.** $\left(\dfrac{5}{2t^5}\right)^2$ **35.** $\left(\dfrac{2^2x^5}{yz^2}\right)^4$

In Exercises 21–36, remove the parentheses.

36. $\left(\dfrac{3^2a^4}{b^2c}\right)^3$

21. $\left(\dfrac{u}{v}\right)^4$ **22.** $\left(\dfrac{8}{c}\right)^2$ **23.** $\left(\dfrac{3}{a}\right)^3$

In Exercises 37–40, evaluate each expression.

24. $\left(\dfrac{x}{2}\right)^5$ **25.** $\left(\dfrac{t}{5}\right)^2$ **26.** $\left(\dfrac{a}{c}\right)^7$ **37.** $\dfrac{-2^4}{(-2)^4}$ **38.** $\dfrac{-7^2}{(-7)^2}$ **39.** $\dfrac{-3^3}{(-3)^3}$ **40.** $\dfrac{(-2)^5}{-2^5}$

27. $(h^3k^4)^2$ **28.** $(2x^4y^2)^3$ **29.** $(5a^3)^2$

5.5 Zero and Negative Exponents

5.5A Definitions of Zero and Negative Exponents

The Zero Exponent

To understand the definition of x^0, let's consider dividing one exponential number by another when the bases are the same *and* when the exponents are equal. In particular, let's consider the problem $x^4 \div x^4$, or $\dfrac{x^4}{x^4}$, where $x \neq 0$.

$$x^4 \div x^4 = \frac{x^4}{x^4} = 1 \qquad \text{\color{red}{Any nonzero number divided by itself is 1}}$$

If we apply the quotient rule to the problem $\dfrac{x^4}{x^4}$, we have $\dfrac{x^4}{x^4} = x^{4-4} = x^0$. Since $\dfrac{x^4}{x^4} = 1$ and since applying the quotient rule gives $\dfrac{x^4}{x^4} = x^0$, to be consistent, x^0 has to equal 1. In fact, mathematicians *define* x^0 to be 1 if $x \neq 0$ (see the zero exponent rule).

The zero exponent rule

> *By definition:* If $x \neq 0$, then $\qquad x^0 = 1$
>
> *In words:* Any nonzero number raised to the zero power equals 1.

 Note To see why we included the statement $x \neq 0$ in the zero exponent rule, let's apply the quotient rule to the problem $\frac{0}{0}$:

$$\frac{0}{0} = \frac{0^1}{0^1} = 0^{1-1} = 0^0$$

But $\frac{0}{0}$ is undefined. Therefore, 0^0 is also undefined.

EXAMPLE 1 Examples of zero as an exponent (assume $a \neq 0$ and $x \neq 0$):

  a. $10^0 = 1$ Using the zero exponent rule
  b. $a^0 = 1$ Using the zero exponent rule
  c. $(6x)^0 = 1$ The exponent applies to everything inside the parentheses
  d. $6x^0 = 6 \cdot 1 = 6$ The exponent applies only to the x

Negative Exponents

To understand the definition of x^{-n}, let's consider dividing one exponential number by another when the bases are the same *and* when the exponent of the divisor is *greater than* the exponent of the dividend. In particular, let's consider the problem $x^3 \div x^5$, or $\dfrac{x^3}{x^5}$.

$$\frac{x^3}{x^5} = \frac{1 \cdot xxx}{xx \cdot xxx} = \frac{1 \cdot x^3}{x^2 \cdot x^3} = \frac{1}{x^2} \qquad \text{\textcolor{red}{Using the fundamental property of fractions}}$$

However, if we use the quotient rule, we have

$$\frac{x^3}{x^5} = x^{3-5} = x^{-2}$$

Since $\dfrac{x^3}{x^5} = \dfrac{1}{x^2}$ and since applying the quotient rule gives $\dfrac{x^3}{x^5} = x^{-2}$, to be consistent, x^{-2} has to equal $\dfrac{1}{x^2}$.

Also, according to the product rule and the zero exponent rule,

$$x^2 \cdot x^{-2} = x^{2+(-2)} = x^0 = 1$$

Then, since the product of x^2 and x^{-2} is 1, x^{-2} must be the multiplicative inverse of x^2. Again, then, we *want* x^{-2} to equal $\dfrac{1}{x^2}$. In fact, mathematicians *define* x^{-n} to equal $\dfrac{1}{x^n}$ when $x \neq 0$.

The first negative exponent rule

By definition: If $x \neq 0$, then $\qquad x^{-n} = \dfrac{1}{x^n}$

Another way of seeing that the definitions of the zero exponent and negative exponents are reasonable is to examine the following pattern of powers:

Notice that on the left side of the equal sign the exponent on 10 decreases by 1 as we move down the column

Notice that on the right side of the equal sign each number is divided by 10 to get the number below it

$$10^4 = 10,000$$

$$10^3 = 1,000$$

$$10^2 = 100$$

$$10^1 = 10$$

$$10^0 = 1$$

$$10^{-1} = \frac{1}{10}$$

$$10^{-2} = \frac{1}{10^2}$$

$$10^{-3} = \frac{1}{10^3}$$

$$10,000 \div 10 = 1,000$$

$$1,000 \div 10 = 100$$

$$100 \div 10 = 10$$

$$10 \div 10 = 1$$

$$1 \div 10 = \frac{1}{10}$$

$$\frac{1}{10} \div 10 = \frac{1}{100} = \frac{1}{10^2}$$

$$\frac{1}{10^2} \div 10 = \frac{1}{1,000} = \frac{1}{10^3}$$

Numerical Bases with Negative Exponents A number with a negative exponent is not *necessarily* negative, although it *can* be negative (see Example 2 and the Word of Caution that follows it).

EXAMPLE 2

Examples of evaluating expressions with numerical bases and negative exponents:

a. $5^{-1} = \frac{1}{5}$ — Notice that $\frac{1}{5}$ is a positive number

b. $3^{-2} = \frac{1}{3^2}$, or $\frac{1}{9}$ — Notice that $\frac{1}{9}$ is a positive number

c. $(-10)^{-4} = \frac{1}{(-10)^4} = \frac{1}{10,000}$ — The exponent applies to everything inside the parentheses

Notice that $\frac{1}{10,000}$ is a positive number

d. $(-2)^{-5} = \frac{1}{(-2)^5} = \frac{1}{-32}$, or $-\frac{1}{32}$ — The exponent applies to everything inside the parentheses

When negative exponents are removed, the results *can* be negative

e. $-10^{-4} = (-1)(10^{-4}) = \frac{-1}{1} \cdot \frac{1}{10^4} = \frac{-1 \cdot 1}{1 \cdot 10^4} = \frac{-1}{10,000}$, or $-\frac{1}{10,000}$

The exponent applies only to the 10

✖ **A Word of Caution** One common error is to think that a number with a negative exponent is always a negative number (see parts a, b, and c, below), and another is to think that $-x = \frac{1}{x}$ (see part d, below).

Correct method	*Incorrect method*
a. $3^{-2} = \frac{1}{3^2}$, or $\frac{1}{9}$	$3^{-2} = -6$ or $3^{-2} = -9$
b. $10^{-1} = \frac{1}{10}$	$10^{-1} = -10$
c. $10^{-2} = \frac{1}{100}$	$10^{-2} = -100$
d. $-4 = -4$ (-4 cannot be rewritten)	$-4 = \frac{1}{4}$

Removing Negative Exponents Examples 3 through 6 demonstrate removing negative exponents.

EXAMPLE 3

Examples of removing negative exponents:

a. $x^{-5} = \frac{1}{x^5}$ — Using the first negative exponent rule

b. $7^{-2} = \frac{1}{7^2}$, or $\frac{1}{49}$ — Using the first negative exponent rule

c. $3^0 a^{-3} = 1 \cdot \frac{1}{a^3} = \frac{1}{a^3}$ — Using the zero exponent rule and the first negative exponent rule

d. $6z^{-5} = 6(z^{-5}) = \left(\frac{6}{1}\right)\left(\frac{1}{z^5}\right) = \frac{6 \cdot 1}{1 \cdot z^5} = \frac{6}{z^5}$ — The exponent applies only to the z

e. $-3x^{-2} = -3(x^{-2}) = -\left(\frac{3}{1}\right)\left(\frac{1}{x^2}\right) = -\frac{3 \cdot 1}{1 \cdot x^2} = -\frac{3}{x^2}$ — The exponent applies only to the x

f. $(-3x)^{-2} = \frac{1}{(-3x)^2} = \frac{1}{(-3)^2 x^2} = \frac{1}{9x^2}$ — The exponent applies to everything inside the parentheses

The next two properties can be proved (although we don't provide the proofs); they follow from the first negative exponent rule, which is $x^{-n} = \dfrac{1}{x^n}$.

The second negative exponent rule

If $x \neq 0$, then $\qquad\qquad\qquad \dfrac{1}{x^{-n}} = x^n$

The third negative exponent rule

If $x \neq 0$ and $y \neq 0$, then $\qquad \left(\dfrac{x}{y}\right)^{-n} = \left(\dfrac{y}{x}\right)^n$

The second and third negative exponent rules can both be used to remove negative exponents (see Examples 4 and 5).

EXAMPLE 4   Examples of using the second negative exponent rule to remove negative exponents:

a. $\dfrac{1}{x^{-4}} = x^4$ Using the second negative exponent rule

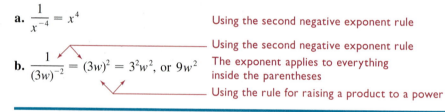

b. $\dfrac{1}{(3w)^{-2}} = (3w)^2 = 3^2 w^2$, or $9w^2$

       Using the second negative exponent rule

The exponent applies to everything inside the parentheses

Using the rule for raising a product to a power

A Word of Caution Notice that $\dfrac{1}{(3w)^{-2}} \neq \dfrac{1}{3w^{-2}}$. Using the second negative exponent rule, we see that

$$\frac{1}{3w^{-2}} = \frac{1}{3} \cdot \frac{1}{w^{-2}} = \frac{1}{3} \cdot w^2 = \frac{1}{3} \cdot \frac{w^2}{1} = \frac{1 \cdot w^2}{3 \cdot 1} = \frac{w^2}{3}, \quad \text{whereas} \quad \frac{1}{(3w)^{-2}} = 9w^2$$

EXAMPLE 5   Examples of using the third negative exponent rule to remove negative exponents:

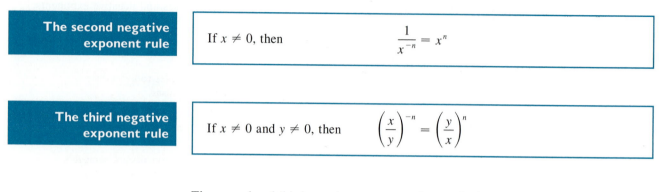

a. $\left(\dfrac{a}{b}\right)^{-3} = \left(\dfrac{b}{a}\right)^3 = \dfrac{b^3}{a^3}$

       Using the third negative exponent rule

Using the rule for raising a quotient to a power

b. $\left(-\dfrac{x}{2}\right)^{-5} = \left(-\dfrac{2}{x}\right)^5 = -\dfrac{2^5}{x^5}$, or $-\dfrac{32}{x^5}$

       Using the third negative exponent rule

Using the fact that a negative number raised to an odd power is negative

Note If a single number or variable appears in the numerator or in the denominator of a fraction, that number can still be considered a factor of the numerator or denominator. For example, $\dfrac{5}{x} = \dfrac{1 \cdot 5}{1 \cdot x}$; therefore, we can say that 5 is a factor of the numerator and x is a factor of the denominator of $\dfrac{5}{x}$ (see Examples 6c, 6d, and 6e).

EXAMPLE 6   Examples of removing negative exponents in products of factors:

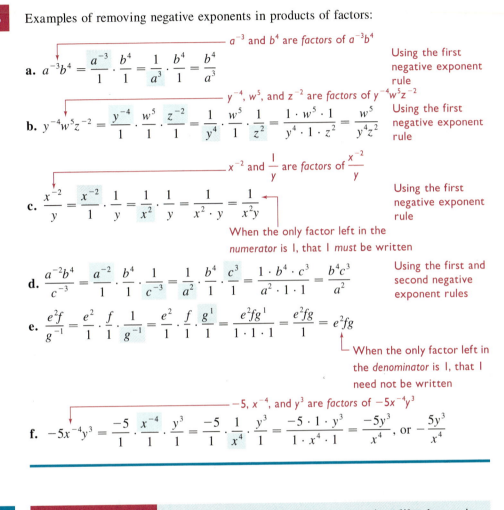

a^{-3} and b^4 are *factors* of $a^{-3}b^4$

a. $a^{-3}b^4 = \dfrac{a^{-3}}{1} \cdot \dfrac{b^4}{1} = \dfrac{1}{a^3} \cdot \dfrac{b^4}{1} = \dfrac{b^4}{a^3}$

Using the first negative exponent rule

y^{-4}, w^5, and z^{-2} are *factors* of $y^{-4}w^5z^{-2}$

b. $y^{-4}w^5z^{-2} = \dfrac{y^{-4}}{1} \cdot \dfrac{w^5}{1} \cdot \dfrac{z^{-2}}{1} = \dfrac{1}{y^4} \cdot \dfrac{w^5}{1} \cdot \dfrac{1}{z^2} = \dfrac{1 \cdot w^5 \cdot 1}{y^4 \cdot 1 \cdot z^2} = \dfrac{w^5}{y^4z^2}$

Using the first negative exponent rule

x^{-2} and $\dfrac{1}{y}$ are *factors* of $\dfrac{x^{-2}}{y}$

c. $\dfrac{x^{-2}}{y} = \dfrac{x^{-2}}{1} \cdot \dfrac{1}{y} = \dfrac{1}{x^2} \cdot \dfrac{1}{y} = \dfrac{1}{x^2 \cdot y} = \dfrac{1}{x^2y}$

Using the first negative exponent rule

When the only factor left in the numerator is 1, that 1 *must* be written

d. $\dfrac{a^{-2}b^4}{c^{-3}} = \dfrac{a^{-2}}{1} \cdot \dfrac{b^4}{1} \cdot \dfrac{1}{c^{-3}} = \dfrac{1}{a^2} \cdot \dfrac{b^4}{1} \cdot \dfrac{c^3}{1} = \dfrac{1 \cdot b^4 \cdot c^3}{a^2 \cdot 1 \cdot 1} = \dfrac{b^4c^3}{a^2}$

Using the first and second negative rules

e. $\dfrac{e^2f}{g^{-1}} = \dfrac{e^2}{1} \cdot \dfrac{f}{1} \cdot \dfrac{1}{g^{-1}} = \dfrac{e^2}{1} \cdot \dfrac{f}{1} \cdot \dfrac{g^1}{1} = \dfrac{e^2fg^1}{1 \cdot 1 \cdot 1} = \dfrac{e^2fg}{1} = e^2fg$

When the only factor left in the *denominator* is 1, that 1 need not be written

-5, x^{-4}, and y^3 are *factors* of $-5x^{-4}y^3$

f. $-5x^{-4}y^3 = \dfrac{-5}{1} \cdot \dfrac{x^{-4}}{1} \cdot \dfrac{y^3}{1} = \dfrac{-5}{1} \cdot \dfrac{1}{x^4} \cdot \dfrac{y^3}{1} = \dfrac{-5 \cdot 1 \cdot y^3}{1 \cdot x^4 \cdot 1} = \dfrac{-5y^3}{x^4}$, or $-\dfrac{5y^3}{x^4}$

A Word of Caution   A common error in simplifying expressions like the one in Example 6c is to omit the *numerator* and write the final answer as x^2y. This is incorrect, since $\dfrac{1}{x^2y} \neq x^2y$.

A Word of Caution   A common error in simplifying expressions like the one in Example 6f is to confuse the expression $-5x^{-4}y^3$ with the expression $5^{-1}x^{-4}y$ and, in simplifying $-5x^{-4}y^3$, to move the 5 to the denominator; however, $-5 \neq \frac{1}{5}$. It *is* true that $5^{-1}x^{-4}y^3 = \dfrac{y^3}{5x^4}$, but the problem in Example 6f is *not* $5^{-1}x^{-4}y^3$, it is $-5x^{-4}y^3$.

Rewriting Expressions Without Denominators

The first and second negative exponent rules can be used to remove denominators; it may be necessary to *insert* negative exponents in such problems (see Example 7).

simplify algebraic expressions. Most of the expressions in this section can be simplified in any of several different ways. For example, an expression such as $\dfrac{x^3}{x^5}$ can be simplified in several ways, two of which are shown below:

$$\frac{x^3}{x^5} = x^{3-5} = x^{-2} = \frac{1}{x^2} \qquad \text{Using the quotient rule and the first negative exponent rule}$$

and

$$\frac{x^3}{x^5} = \frac{x^3}{1} \cdot \frac{1}{x^5} = \frac{1}{x^{-3}} \cdot \frac{1}{x^5} = \frac{1 \cdot 1}{x^{-3} \cdot x^5} = \frac{1}{x^{-3+5}} = \frac{1}{x^2} \qquad \text{Using the second negative exponent rule and the product rule}$$

In the remaining examples, we demonstrate just one way of simplifying each algebraic expression.

EXAMPLE 8   Examples of applying the properties of exponents:

a. $a^4 \cdot a^{-3} = a^{4+(-3)} = a^1 = a$ Using the product rule

b. $x^{-5} \cdot x^2 = x^{-5+2} = x^{-3} = \dfrac{1}{x^3}$ Using the product rule and the first negative exponent rule

c. $(y^{-2})^{-1} = y^{(-2)(-1)} = y^2$ Using the power rule

d. $(x^2)^{-4} = x^{2(-4)} = x^{-8} = \dfrac{1}{x^8}$ Using the power rule and the first negative exponent rule

e. $\dfrac{y^{-2}}{y^{-6}} = y^{(-2)-(-6)} = y^{-2+6} = y^4$ Using the quotient rule

f. $\dfrac{z^{-4}}{z^{-2}} = z^{(-4)-(-2)} = z^{-4+2} = z^{-2} = \dfrac{1}{z^2}$ Using the quotient rule and the first negative exponent rule

g. $h^3 h^0 h^{-2} = h^{3+0+(-2)} = h^1 = h$ Using the product rule

h. $\left(\dfrac{a}{2bc^3}\right)^0 = 1$ Using the zero exponent rule

---

The general property of exponents also applies to expressions with zero and negative exponents (see Example 9).

EXAMPLE 9   Examples of using the general property of exponents when there are zero and negative exponents:

a. $(x^3 y^{-1})^5 = x^{3 \cdot 5} y^{(-1)(5)} = x^{15} y^{-5} = \dfrac{x^{15}}{1} \cdot \dfrac{1}{y^5} = \dfrac{x^{15} \cdot 1}{1 \cdot y^5} = \dfrac{x^{15}}{y^5}$

The properties of exponents apply to numerical bases as well as to variable bases

b. $\left(\dfrac{2a^{-3} b^2}{c^5}\right)^3 = \dfrac{2^{1 \cdot 3} a^{-3 \cdot 3} b^{2 \cdot 3}}{c^{5 \cdot 3}} = \dfrac{2^3 a^{-9} b^6}{c^{15}} = \dfrac{8b^6}{a^9 c^{15}}$ Using the general property of exponents and the first negative exponent rule

c. $\left(\dfrac{3^2 c^{-4}}{d^3}\right)^{-1} = \dfrac{3^{2(-1)} c^{(-4)(-1)}}{d^{3(-1)}} = \dfrac{3^{-2} c^4}{d^{-3}} = \dfrac{c^4 d^3}{3^2}, \text{ or } \dfrac{c^4 d^3}{9}$ Using the general property of exponents and the first and second negative exponent rules

d. $(x^{-3} y^2)^{-2} = x^{(-3)(-2)} y^{2(-2)} = x^6 y^{-4} = \dfrac{x^6}{y^4}$ Using the general property of exponents and the first negative exponent rule

e. $(-3x)^{-2} = (-3)^{-2}x^{-2} = \dfrac{1}{(-3)^2} \cdot \dfrac{1}{x^2} = \dfrac{1}{9} \cdot \dfrac{1}{x^2} = \dfrac{1}{9x^2}$ We saw $(-3x)^{-2}$ simplified a different way in Example 3f

Simplifying 5^0 first

f. $(\,5^0\,h^{-2})^{-3} = (\,1\,h^{-2})^{-3} = (h^{-2})^{-3} = h^{(-2)(-3)} = h^6$

When the bases are numerical, we often write the final result without exponents unless the base and/or the exponent are very large (see Example 10).

EXAMPLE 10   Examples of using the properties of exponents with numerical bases:

a. $5^5 \cdot 5^{-3} = 5^{5+(-3)} = 5^2$, or 25   Using the product rule

b. $(2^3)^{-1} = 2^{3(-1)} = 2^{-3} = \dfrac{1}{2^3}$, or $\dfrac{1}{8}$ Using the power rule and the first negative exponent rule

c. $\dfrac{5^0}{5^2} = 5^{0-2} = 5^{-2} = \dfrac{1}{5^2}$, or $\dfrac{1}{25}$ Using the quotient rule and the first negative exponent rule

d. $\dfrac{7^2}{7^{-3}} = 7^{2-(-3)} = 7^{2+3} = 7^5$, or 16,807   Using the quotient rule (and a calculator)

Using the product rule and the quotient rule

Using the power rule

Using the first negative exponent rule

e. $\left(\dfrac{10^{-2} \cdot 10^5}{10^4}\right)^3 = (10^{-2+5-4})^3 = (10^{-1})^3 = 10^{-1(3)} = 10^{-3} = \dfrac{1}{10^3}$, or $\dfrac{1}{1,000}$

Note   The problem in Example 10e can also be done by using the general property of exponents first, as follows:

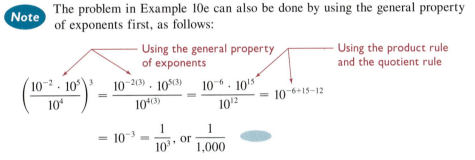

Using the general property of exponents   Using the product rule and the quotient rule

$$\left(\dfrac{10^{-2} \cdot 10^5}{10^4}\right)^3 = \dfrac{10^{-2(3)} \cdot 10^{5(3)}}{10^{4(3)}} = \dfrac{10^{-6} \cdot 10^{15}}{10^{12}} = 10^{-6+15-12}$$

$$= 10^{-3} = \dfrac{1}{10^3}, \text{ or } \dfrac{1}{1,000}$$

A Word of Caution   Common errors in using the properties of exponents are shown on the right below:

Correct method

a. $5^5 \cdot 5^{-3} = 5^{5+(-3)} = 5^2$, or 25

b. $10^{-2} \cdot 10^5 = 10^{-2+5} = 10^3$, or 1,000

Incorrect method

$5^5 \cdot 5^{-3} = 25^{5+(-3)} = 25^2$, or 625

$10^{-2} \cdot 10^5 = 100^{-2+5} = 100^3$, or 1,000,000

Remember: When you multiply powers of the same base, add the exponents but *do not* multiply the bases as well.

As we've mentioned, we will simplify more algebraic expressions that contain fractions in a later chapter.

Exercises 5.5B
Set I

Assume that none of the factors appearing in any denominator is zero.

In Exercises 1–28, simplify each expression; write each answer using only positive exponents.

1. $x^{-3} \cdot x^4$
2. $y^6 \cdot y^{-2}$
3. $(x^2)^{-4}$
4. $(z^3)^{-2}$
5. $(a^{-2})^3$
6. $(b^{-5})^2$
7. $(y^{-3})^{-4}$
8. $(s^{-2})^{-7}$
9. $(m^{-2}n)^4$
10. $(p^{-3}r)^5$
11. $(x^{-2}y^3)^{-4}$
12. $(w^{-3}z^4)^{-2}$
13. $\dfrac{x^6}{x^9}$
14. $\dfrac{t^4}{t^8}$
15. $\dfrac{y^{-2}}{y^5}$
16. $\dfrac{z^{-2}}{z^2}$
17. $\dfrac{d^4}{d^{-5}}$
18. $\dfrac{c^3}{c^{-4}}$
19. $\left(\dfrac{M^{-2}}{N^3}\right)^4$
20. $\left(\dfrac{R^5}{S^{-4}}\right)^3$
21. $x^{3m} \cdot x^{-m}$
22. $y^{-2n} \cdot y^{5n}$
23. $(x^{3b})^{-2}$
24. $(y^{2a})^{-3}$
25. $\dfrac{x^{2a}}{x^{-5a}}$
26. $\dfrac{a^{3x}}{a^{-5x}}$
27. $\dfrac{b^{-2n}}{b^{3n}}$
28. $\dfrac{y^{-3m}}{y^{2m}}$

In Exercises 29–44, evaluate each expression.

29. $10^3 \cdot 10^{-2}$
30. $2^{-3} \cdot 2^2$
31. $10^4 \cdot 10^{-2}$
32. $3^{-2} \cdot 3^3$
33. $(10^2)^{-1}$
34. $(2^{-3})^2$
35. $5^0 \cdot 7^2$
36. $4^3 \cdot 2^0$
37. $\dfrac{10^2}{10^{-5}}$
38. $\dfrac{2^3}{2^{-2}}$
39. $\dfrac{10^0}{10^2}$
40. $\dfrac{5^2}{5^0}$
41. $\dfrac{10^{-3} \cdot 10^2}{10^5}$
42. $\dfrac{2^3 \cdot 2^{-4}}{2^2}$
43. $\dfrac{3^4}{3^2 \cdot 3^{-8}}$
44. $\dfrac{7^3}{7^{-5} \cdot 7^4}$

Writing Problems

Express the answers in your own words and in complete sentences.

1. Explain why $3x^0y^4$ is not in simplest form.

2. Explain why x^{-2} is not in simplest form.

Exercises 5.5B
Set II

Assume that none of the factors appearing in any denominator is zero.

In Exercises 1–28, simplify each expression; write each answer using only positive exponents.

1. $y^{-7} \cdot y^9$
2. $x^0 \cdot y^{-4}$
3. $(a^3)^{-5}$
4. $(x^2)^{-4}$
5. $(x^{-4})^5$
6. $(y^{-3})^{-7}$
7. $(z^{-2})^{-4}$
8. $(a^3)^{-6}$
9. $(x^{-3}y)^5$
10. $(t^{-2}u^3)^4$
11. $(c^{-4}d^5)^{-3}$
12. $(a^{-2}b^4)^{-3}$
13. $\dfrac{a^4}{a^6}$
14. $\dfrac{b^3}{b^7}$
15. $\dfrac{x^{-3}}{x^5}$
16. $\dfrac{y^{-4}}{y^4}$
17. $\dfrac{c^5}{c^{-4}}$
18. $\dfrac{d^4}{d^{-4}}$
19. $\left(\dfrac{R^4}{S^{-2}}\right)^3$
20. $\left(\dfrac{x^4}{2y^{-3}}\right)^{-3}$
21. $x^{2n} \cdot x^{-n}$
22. $a^{-3m} \cdot b^{3m}$
23. $(k^{-2c})^2$
24. $(3^4xy^{2n})^0$
25. $\dfrac{y^{4n}}{y^{-3n}}$
26. $\dfrac{z^{-2m}}{z^{-4m}}$
27. $\dfrac{x^{-2a}}{x^{5a}}$
28. $\dfrac{b^{-4m}}{b^{-2m}}$

In Exercises 29–44, evaluate each expression.

29. $10^{-4} \cdot 10^3$
30. $11^4 \cdot 11^{-7}$
31. $3^5 \cdot 3^{-3}$
32. $5^{-2} \cdot 7^0$
33. $(2^{-4})^2$
34. $(3^3)^{-1}$
35. $3^4 \cdot 5^0$
36. $3^{-3} \cdot 3^{-2}$
37. $\dfrac{10^3}{10^{-2}}$
38. $\dfrac{13^0}{13^{-2}}$
39. $\dfrac{5^0}{5^4}$
40. $\dfrac{7^{-2}}{7^{-5}}$
41. $\dfrac{10^{-2} \cdot 10^3}{10^4}$
42. $\dfrac{3^5 \cdot 3^{-3}}{3^4}$
43. $\dfrac{2^3}{2^{-6} \cdot 2^4}$
44. $\dfrac{11^2 \cdot 11^0}{11^5 \cdot 11^{-4}}$

5.6 Scientific Notation

Now that we have discussed zero and negative exponents, we can introduce *scientific notation*, a notation that is used in many sciences. Scientific notation gives us a shorter way of writing very large numbers, such as Avogadro's number ($\approx 602{,}000{,}000{,}000{,}000{,}000{,}000{,}000$, used in chemistry), and very small numbers, such as Boltzmann's constant ($0.000\,000\,000\,000\,000\,138$, used in physics) (see Example 7). We can also use scientific notation to simplify the arithmetic in a problem such as $\dfrac{30{,}000{,}000 \times 0.0005}{0.000\,000\,6 \times 80{,}000}$ (see Example 8). Scientific notation also gives us a way of indicating a certain accuracy for measurements and other rounded-off numbers, although we do not discuss that topic in this book.

Scientific calculators usually express answers that are very large or very small in scientific notation (see Examples 9 and 10). Also, you should be aware that many calculators will give the *wrong answer* if very large or very small numbers are entered *unless* the numbers are entered in scientific notation (see Example 11).

Multiplying a Decimal Number by a Power of Ten   Before we discuss converting a number to scientific notation and the uses of scientific notation, we briefly review multiplying a decimal number by a power of 10.

Recall from arithmetic that to multiply a decimal number by 10^n, we move the decimal point n places to the right if $n > 0$ and $|n|$ places to the left if $n < 0$ (see Example 1).

EXAMPLE 1   Examples of multiplying a decimal number by a power of 10:

a. $8.021 \times 10^4 = 8\,0{,}210. = 80{,}210$ The exponent is 4, and $4 > 0$; the decimal point is moved 4 places to the *right*

b. $6.03 \times 10^{-3} = 0.006\,03.$ The exponent is -3, $-3 < 0$, and $|-3| = 3$; the decimal point is moved 3 places to the *left*

Scientific Notation

A positive number written in scientific notation is written in the form $\boxed{a \times 10^n}$, where a is a number greater than or equal to 1 but less than 10, and n is an integer. For example, 8.021×10^4 is correctly written in scientific notation, because 8.021 is a number between 1 and 10, and 10^4 is a power of 10. Since a is to be greater than or equal to 1 but less than 10, it must have *exactly one nonzero digit to the left of its decimal point.*

When we convert a number to scientific notation, we are using the facts that multiplying a number by 1 doesn't change its value and that $10^n \times 10^{-n} = 1$; therefore, when we convert a number to scientific notation, we do not change its value. Consider the following examples.

EXAMPLE 2   Convert 0.0712 to scientific notation.

SOLUTION   When 0.0712 is written in scientific notation, the decimal point must be between the 7 and the 1; that is, the decimal point must be moved two places to the right. Because the decimal point needs to be moved two places to the right, we'll multiply 0.0712 by $10^2 \times 10^{-2}$.

This product is 1

$$0.0712 = 0.0712 \times 10^2 \times 10^{-2} = (0.0712 \times 10^2) \times 10^{-2} = \underline{7.12} \times 10^{-2}$$

This number is in scientific notation

EXAMPLE 3   Convert 27,400 to scientific notation.

SOLUTION   When 27,400 is written in scientific notation, the decimal point must be between the 2 and the 7; that is, the decimal point must be moved four places to the left. Because the decimal point needs to be moved four places to the left, we'll multiply 27,400 by $10^{-4} \times 10^4$.

$$27{,}400 = 27{,}400 \times \overbrace{10^{-4} \times 10^4}^{\text{This product is 1}} = \underbrace{(27{,}400 \times 10^{-4})}_{} \times 10^4 = \underbrace{2.74}_{} \times 10^4$$

This number is in scientific notation

An alternative method is demonstrated in Example 4.

EXAMPLE 4   Convert each of the following numbers to scientific notation. That is, write each number in the form $a \times 10^n$, where n is an integer and where a has exactly one nonzero digit to the left of its decimal point.

a. 61,200

SOLUTION   Because there is no decimal point, we place a caret ($_\wedge$) just to the right of the last digit of the number; we then have

$$61{,}200_\wedge$$

(We will *drop* the caret in the final answer; it just helps us decide what exponent to use on the 10.) Next, we find a by placing a decimal point in the number so that there is exactly one nonzero digit to the left of the decimal point, giving

$$6.1200_\wedge \quad \text{This number is } a; \text{ that is, } a = 6.1200_\wedge$$

Because the caret is to the *right* of the decimal point (that is, because 61,200 > 10), n is *positive*, and because there are four digits between the decimal point and the caret, the absolute value of n is 4.

Therefore, $61{,}200 = 6.1200 \times 10^4$.

b. 0.032

SOLUTION   We replace the decimal point with a caret, and we find a by placing a (new) decimal point so that there is exactly one nonzero digit to its left; we then have

$$0_\wedge 03.2 \quad \text{This number is } a; \text{ that is, } a = 0_\wedge 03.2$$

Because the caret is to the *left* of the decimal point (that is, because 0.032 < 1), n is *negative*, and because there are two digits between the decimal point and the caret, the absolute value of n is 2, so $n = -2$.

Therefore, $0.032 = 3.2 \times 10^{-2}$.

The method shown in Example 4 is generalized as follows.

Converting a positive number to scientific notation	**Step 1.** Finding a
	    a. Replace the decimal point with a caret ($_\wedge$). (If there is no decimal point, place the caret just to the right of the last digit of the number.)
	    b. Place a decimal point in the number so that there is exactly one nonzero digit to the left of the decimal point. (The number thus written is a.)
	Step 2. Finding the correct power of 10
	    a. The number of digits separating the caret and the decimal point in step 1 gives the absolute value (the numerical value) of the exponent of 10. If the decimal point and the caret coincide (that is, if both fall in the same position), the exponent is zero.
	    b. The sign of the exponent of 10 is *positive* if the caret is to the right of the decimal point and *negative* if the caret is to the left of the decimal point.
	
	The number in scientific notation is the product of the two numbers found in steps 1 and 2. (We drop the caret in the final answer.)

This method implies that if the number to be converted to scientific notation is greater than or equal to 10, the exponent of 10 will be positive; if the number is less than 1, the exponent of 10 will be negative. If the number is between 1 and 10 (or equal to 1), the exponent of 10 will be zero.

EXAMPLE 5

Examples of converting decimal numbers to scientific notation:

Decimal notation	*Finding a*	*Scientific notation*
a. 0.00753	$0_\wedge 007.53 \times 10^?$	7.53×10^{-3}

In the middle column, we replaced the decimal point with a caret and placed a decimal point just after the 7. The caret is *three* digits to the *left* of the decimal point; therefore, the exponent that will replace the question mark is -3, as shown in the third column.

b. 86,100,000	$8.6100000_\wedge \times 10^?$	8.61×10^7
c. 2.49	$2_\wedge 49 \times 10^?$	2.49×10^0
d. 6,410	$6.410_\wedge \times 10^?$	6.41×10^3
e. 0.0003015	$0_\wedge 0003.015 \times 10^?$	3.015×10^{-4}

a, the number between 1 and 10 ⎯⎯ The power of 10

Notice that in parts a and e, the caret is to the left of the decimal point and the exponent on the 10 is negative, and in parts b and d, the caret is to the right of the decimal point and the exponent on the 10 is positive. In part c, the decimal point and the caret coincide and the exponent on the 10 is zero.

It is sometimes necessary to convert a number such as 672.3×10^2 or 0.0791×10^{-3} to scientific notation or to decimal notation, as shown in Example 6.

EXAMPLE 6

Convert each number to scientific notation and then to decimal notation.

a. 672.3×10^2

SOLUTION Note that 672.3×10^2 is *not* in scientific notation, because 672.3 does not have exactly one nonzero digit to the left of the decimal point.

$$672.3 = 6.723 \times 10^2$$

Therefore,

$$
\begin{aligned}
672.3 \times 10^2 &= (6.723 \times 10^2) \times 10^2 \\
&= 6.723 \times (10^2 \times 10^2) \\
&= 6.723 \times 10^4 \qquad \text{Scientific notation} \\
&= 67{,}230 \qquad\qquad \text{Decimal notation}
\end{aligned}
$$

b. 0.0791×10^{-3}

SOLUTION

$$0.0791 = 7.91 \times 10^{-2}$$

Therefore,

$$
\begin{aligned}
0.0791 \times 10^{-3} &= (7.91 \times 10^{-2}) \times 10^{-3} \\
&= 7.91 \times (10^{-2} \times 10^{-3}) \\
&= 7.91 \times 10^{-5} \qquad \text{Scientific notation} \\
&= 0.0000791 \qquad \text{Decimal notation}
\end{aligned}
$$

EXAMPLE 7

Express (a) Avogadro's number and (b) Boltzmann's constant in scientific notation.

SOLUTION

a. Avogadro's number is approximately 602,000,000,000,000,000,000,000.

The caret is to the right of the decimal point; there are 23 digits between the caret and the decimal point

$$
\begin{aligned}
602{,}000{,}000{,}000{,}000{,}000{,}000{,}000 &= 6.020\,000\,000\,000\,000\,000\,000\,00_\wedge \times 10^{23} \\
&= 6.02 \times 10^{23}
\end{aligned}
$$

b. Boltzmann's constant is $0.000\,000\,000\,000\,000\,138$.

The caret is to the left of the decimal point; there are 16 digits between the caret and the decimal point

$$
\begin{aligned}
0.000\,000\,000\,000\,000\,138 &= 0_\wedge 000\,000\,000\,000\,000\,1.38 \times 10^{-16} \\
&= 1.38 \times 10^{-16}
\end{aligned}
$$

EXAMPLE 8

Use scientific notation in solving this problem: $\dfrac{30{,}000{,}000 \times 0.0005}{0.000\,000\,6 \times 80{,}000}$.

SOLUTION

$$
\frac{30{,}000{,}000 \times 0.0005}{0.000\,000\,6 \times 80{,}000} = \frac{(3 \times 10^7) \times (5 \times 10^{-4})}{(6 \times 10^{-7}) \times (8 \times 10^4)}
$$

Writing each factor in scientific notation

$$
= \frac{(\overset{1}{\cancel{3}} \times 5) \times (10^7 \times 10^{-4})}{(\underset{2}{\cancel{6}} \times 8) \times (10^{-7} \times 10^4)}
$$

Collecting the powers of 10

$$= \frac{5}{16} \times \frac{10^3}{10^{-3}}$$

Simplifying what's in the parentheses *and* writing the problem as a product of two fractions

$$= 0.3125 \times (10^3 \times 10^3)$$

Writing 5/16 in decimal form and writing $1/10^{-3}$ as 10^3

Converting 0.3125 to scientific notation

$$= (3.125 \times 10^{-1}) \times 10^6$$

$$= 3.125 \times 10^5$$

The answer in scientific notation

$$= 312,500$$

The answer in decimal notation

Calculators and Scientific Notation

Scientific Notation in Calculator Displays   As we mentioned, when a scientific calculator is used and answers are very large or very small, the calculator display will probably be in scientific notation. On the calculator, however, numbers in scientific notation are displayed in a different (and possibly misleading) way. The calculator display $\boxed{5.06 \ ^{03}}$ does *not* mean 5.06 to the third power. It means 5.06×10^3. (Calculator displays of $\boxed{5.06 \ 03}$ and $\boxed{5.06 \ E \ 3}$ also mean 5.06×10^3.)

EXAMPLE 9    Find 80,000,000 × 300,000, using a scientific calculator.

SOLUTION   The display probably shows $\boxed{2.4 \ ^{13}}$, $\boxed{2.4 \ 13}$, or $\boxed{2.4 \ E \ 13}$. These displays all mean 2.4×10^{13}.

EXAMPLE 10   Find 0.0000008 ÷ 400, using a scientific calculator.

SOLUTION   The display probably shows $\boxed{2. \ ^{-09}}$, $\boxed{2. \ -09}$, or $\boxed{2. \ E \ -09}$. These displays all mean 2.0×10^{-9}.

Entering Numbers into a Calculator in Scientific Notation   Very large and very small numbers must be entered into most scientific calculators in scientific notation. The keystrokes used to indicate operations vary, of course, with the brand of calculator. We will assume, in Example 11, that the key for indicating that you're entering a number in scientific notation is marked $\boxed{EE}$, but it could be marked $\boxed{EXP}$. You must consult your calculator manual for details about *your* calculator.

EXAMPLE 11   Given that the formula for simple interest is $I = Prt$, use a calculator to find the (simple) interest for one year on $4,100,000,000,000 (this is the approximate amount of the national debt in 1993) if the interest rate is $4\frac{3}{4}\%$.

SOLUTION   We'll use $t = 1$, $r = 0.0475$ ($0.0475 = 4\frac{3}{4}\%$ in decimal form), and $P = 4,100,000,000,000$. This last number is too large to enter into most calculators without using scientific notation. Changing 4,100,000,000,000 to scientific notation, we have

$$4{,}100{,}000{,}000{,}000 = 4.100\,000\,000\,000_\wedge \times 10^{12} = 4.1 \times 10^{12}$$

Therefore, $I = (4.1 \times 10^{12})(0.0475)(1)$.

Keystrokes   *Display*

$\boxed{4}$ $\boxed{.}$ $\boxed{1}$ $\boxed{EE}$ $\boxed{1}$ $\boxed{2}$ $\boxed{\times}$ $\boxed{.}$ $\boxed{0}$ $\boxed{4}$ $\boxed{7}$ $\boxed{5}$ $\boxed{=}$ $\boxed{1.9475 \quad 11}$

This calculator display means 1.9475×10^{11}, or 194,750,000,000. Therefore, at a $4\frac{3}{4}\%$ interest rate, the interest on the national debt for *one year* is about $194,750,000,000.

A Word of Caution   In Example 11, watch your calculator display carefully if you enter the value for P as

Your calculator may ignore the last few zeros; if so, you will probably get an *incorrect* answer.

EXAMPLE 12   Use a scientific calculator, entering the numbers in scientific notation, to find

$$\frac{(3 \times 10^7) \times (5 \times 10^{-4})}{(6 \times 10^{-7}) \times (8 \times 10^4)}.$$

SOLUTION   This is the problem we solved in Example 8. Using the correct order of operations, we must treat the problem as if it were

$$\frac{3 \times 10^7}{1} \times \frac{5 \times 10^{-4}}{1} \times \frac{1}{6 \times 10^{-7}} \times \frac{1}{8 \times 10^4}$$

$$= (3 \times 10^7) \times (5 \times 10^{-4}) \div (6 \times 10^{-7}) \div (8 \times 10^4)$$

Keystrokes	Display

| 3 | EE | 7 | × | 5 | EE | 4 | +/− | ÷ | 6 | EE | 7 | +/− |
| ÷ | 8 | EE | 4 | = |

| 3.125   05 |

This calculator display means 3.125×10^5, or 312,500, which is the answer we obtained in Example 8.

The even-numbered applied problems at the end of the following exercise set are not "matched" to the odd-numbered problems.

Exercises   5.6
Set I

In Exercises 1–6, express each number in decimal notation.

1. 8.06×10^3 **2.** 3.14×10^4

3. 1.32×10^{-3} **4.** 8.2×10^{-4}

5. 5.26×10^0 **6.** 9.11×10^0

In Exercises 7–18, express each number in scientific notation.

7. 35,300   **8.** 825,000

9. 0.00312   **10.** 0.000145

11. 8.97   **12.** 2.497

13. 0.815   **14.** 0.274

15. 0.0002   **16.** 0.006

17. 45   **18.** 12

In Exercises 19 and 20, use scientific notation to solve the problem.

19. $\dfrac{5,000,000 \times 0.000003}{0.0006 \times 20,000}$

20. $\dfrac{0.0000004 \times 700,000,000}{32,000,000 \times 0.00005}$

In Exercises 21–24, perform the indicated operations with a scientific calculator, and express each answer correctly in scientific notation.

21. $860,000 \times 630,000$ **22.** $0.0000009 \div 3,000$

23. $\sqrt{0.00000081}$ **24.** $\sqrt{0.00000225}$

In Exercises 25–28, use a scientific calculator.

25. By definition (in chemistry and physics), 1 mole of any substance contains approximately 6.02×10^{23} molecules. How many molecules will 700 moles of hydrogen contain? (Express the answer in scientific notation.)

26. The indebtedness of one of the developing nations is $120,000,000,000. If the interest rate on this debt is 7.6% simple interest per year, what is the interest on this debt for one year?

27. If the spacecraft Voyager traveled 4,400,000,000 miles in 12 years, what was its average speed in miles per hour? (Assume that each year has 365 days, and round off the answer to the nearest mile per hour.)

28. The speed of light is about 186,000 miles per second. How many miles does light travel in one day? (Express the answer in scientific notation and in decimal notation.)

Writing Problems

Express the answer in your own words and in complete sentences.

1. Explain why 23.5×10^4 is not correctly expressed in scientific notation.

Exercises 5.6
Set II

In Exercises 1–6, express each number in decimal notation.

1. 5.23×10^4 **2.** 6.34×10^2

3. 7.12×10^{-4} **4.** 2.3×10^{-3}

5. 7.32×10^0 **6.** 3.71×10^0

In Exercises 7–18, express each number in scientific notation.

7. 87,600   **8.** 25,000,000

9. 0.00631   **10.** 0.00614

11. 3.69   **12.** 3.9

13. 0.153   **14.** 0.456

15. 0.0052   **16.** 0.00003

17. 28   **18.** 2.9

In Exercises 19 and 20, use scientific notation to solve the problem.

19. $\dfrac{80,000,000 \times 0.00006}{0.00016 \times 300,000}$

20. $\dfrac{0.0000034 \times 600,000,000}{17,000,000 \times 0.00003}$

In Exercises 21–24, perform the indicated operations with a scientific calculator, and express each answer correctly in scientific notation.

21. $60,000 \times 730,000,000$ **22.** $0.00000012 \div 4,000$

23. $\sqrt{0.00000036}$ **24.** $\sqrt{0.00000289}$

In Exercises 25–28, use a scientific calculator.

25. The indebtedness of one of the developing nations is $52,000,000,000. If the interest rate on this debt is 7.6% simple interest per year, what is the interest on this debt for two years?

26. The planet Neptune is about 2,700,000,000 miles from the earth. What is this distance in kilometers? (One mile ≈ 1.61 km.)

27. A unit used in measuring the length of light waves is the Ångström. One micron is a millionth of a meter, and one Ångström is one ten-thousandth of a micron. One Ångström is what part of a meter? (Express the answer in scientific notation.)

28. The speed of light is about 186,000 miles per second. How long will it take light to reach us from the planet Neptune? (Assume that Neptune is about 2,700,000,000 miles from the earth, and round off the answer to the nearest hour.)

Sections 5.1–5.6   REVIEW

Properties of Exponents

Section	Property	Description
5.1	$x^a \cdot x^b = x^{a+b}$	The product rule
5.4	$(x^a)^b = x^{ab}$	The power rule
5.4	$(xy)^a = x^a y^a$	The rule for raising a product to a power
5.4	$\dfrac{x^a}{x^b} = x^{a-b}, \quad \text{if } x \neq 0$	The quotient rule
5.4	$\left(\dfrac{x}{y}\right)^n = \dfrac{x^n}{y^n}, \quad \text{if } y \neq 0$	The rule for raising a quotient to a power
5.4	$\left(\dfrac{x^a y^b}{z^c}\right)^n = \dfrac{x^{an} y^{bn}}{z^{cn}}, \quad \text{if } z \neq 0$	The general property of exponents

5.5 $x^0 = 1$, if $x \neq 0$ The zero exponent rule

5.5 $x^{-n} = \dfrac{1}{x^n}$, if $x \neq 0$ The first negative exponent rule

5.5 $\dfrac{1}{x^{-n}} = x^n$, if $x \neq 0$ The second negative exponent rule

5.5 $\left(\dfrac{x}{y}\right)^{-n} = \left(\dfrac{y}{x}\right)^{n}$, if $y \neq 0$ and $x \neq 0$ The third negative exponent rule

Simplifying a Product of Factors When Each Factor Has Only One Term
5.2

1. *Multiply the numerical coefficients.* The sign of the coefficient of any term is *part of* the numerical coefficient.

2. *Multiply the variables.* Each variable should appear only once, and the variables are usually written in alphabetical order. Use the product rule to determine the exponent on each variable.

3. The final answer is the product of the results of steps 1 and 2.

Simplified Form of Expressions with Exponents
5.2

A term with exponents is considered *simplified* when each different base appears only once and when the exponent on each base is a single positive integer.

Simplifying a Product of Factors When One Factor Has More Than One Term
5.3

Use the distributive property:

$$a(b + c) = ab + ac$$

Simplifying an Algebraic Expression
5.1, 5.2, 5.3, 5.4, and 5.5

Remove all grouping symbols, simplify each term, and combine all like terms.

Scientific Notation
5.6

A number is correctly expressed in **scientific notation** if it is in the form $a \times 10^n$, where a is greater than or equal to 1 but less than 10 and n is any integer.

Sections 5.1–5.6   **REVIEW EXERCISES**   **Set I**

Assume that none of the factors appearing in any denominator is zero.

In Exercises 1–34, write each expression in simplest form.

1. $m^2 m^3$
2. yy^7
3. $2 \cdot 2^2$
4. $x^{2y} x^{5y}$
5. $10 \cdot 10^y$
6. $a^5 \cdot a^{-3}$
7. $5x(x^2 + 7)$
8. $(5x^2 y + 3x - 1)(-2x)$
9. $(-5ef^2)(-f^7 g^3)(-10e^4 g^2)(-2ef)$
10. $(3x)^3$
11. $(x^2 y^3)^4$
12. $(n^{-5})^{-3}$
13. $(p^{-3})^5$
14. $(2c)^{-4}$
15. $(k^{-7})^0$
16. $(s^4 t^{-1})^{-3}$
17. $(-2x^4)^2$
18. $(-5a^{-4})^3$
19. $(-10)^{-3}$
20. $\dfrac{r^7}{r^5}$
21. $\dfrac{x^{-4}}{x^5}$
22. $\left(\dfrac{c}{d}\right)^4$
23. $\dfrac{m^0}{m^{-3}}$
24. $x^4 + x^2$
25. $\dfrac{n^{-6}}{n^0}$
26. $\left(\dfrac{x^2 y^3}{2z^4}\right)^5$
27. $\left(\dfrac{a^{-4}}{b^3 c^0}\right)^{-5}$

28. $x^{-5} \cdot x^{-3}$
29. $\left(\dfrac{r^{-6}}{s^5 t^{-3}}\right)^4$
30. $\left(\dfrac{3x^4 y^{-3}}{z^3 w^{-5}}\right)^0$
31. $(5a^3 b^{-4})^{-2}$
32. $(9x^8 y^3)(-7x^4 y^5)$
33. $2x(3x^2 - x) - (3x^2 - 4)$
34. $2m^2 n(3mn^2 - 2n) - 5mn^2(2m - 3m^2 n)$

In Exercises 35–39, evaluate each expression.

35. 4^{-2}
36. $(10^{-2})^2$
37. $\dfrac{2^0}{2^{-3}}$
38. $4^0 \cdot 3^2$
39. $\dfrac{(-8)^2}{-8^2}$

In Exercises 40–42, write each expression without a denominator, using negative exponents whenever necessary.

40. $\dfrac{a^3}{b^2}$
41. $\dfrac{m^2}{n^{-3}}$
42. $\dfrac{u^{-4} v^3}{10^2 w^{-5}}$

43. Express 45,300 in scientific notation.
44. Express 0.03156 in scientific notation.

Name

Assume that none of the factors appearing in any denominator is zero.

In Exercises 1–34, write each expression in simplest form.

1. $x^4 y^2 x^6$

2. $3 \cdot 3^7$

3. $a^2 b^7$

4. $m^x m^y$

5. $x^{-7} \cdot x^4$

6. $4^2 \cdot 4^4$

7. $3x(2x - 1)$

8. $9a^2 b(ab + 1)$

9. $(3x^4 - x^3 - x^2)(-2x)$

10. $7 + 4(3x^4 - 5 - x^2)$

11. $(-x^2)(3xy^4)(5y^3)$

12. $(-9w^2 z^3)(7z^4)(-wx^2 z^4)$

13. $(6e^3)^0$

14. $(2xy^2)^5$

15. $(m^{-2})^4$

16. $(m^{-1} n^2)^{-3}$

17. $c^{-5} \cdot d^0$

18. $(-2t^{-3})^2$

19. $(3a^{-2} b)^{-4}$

20. $m^{3x} \cdot m^{-x}$

21. $\dfrac{y^6}{y^2}$

ANSWERS

1. _____

2. _____

3. _____

4. _____

5. _____

6. _____

7. _____

8. _____

9. _____

10. _____

11. _____

12. _____

13. _____

14. _____

15. _____

16. _____

17. _____

18. _____

19. _____

20. _____

21. _____

22. $\left(\dfrac{y}{2}\right)^5$

23. $\dfrac{x^{-2}}{x^2}$

24. $x^5 - x^2$

25. $\left(\dfrac{x}{2}\right)^{-4}$

26. x^3y^5

27. $\left(\dfrac{r^{-3}}{s^2t^{-2}}\right)^2$

28. $\left(\dfrac{3p}{m^2n^{-1}}\right)^{-3}$

29. $\left(\dfrac{5a^2}{b^4}\right)^{-2}$

30. $\left(\dfrac{18x^{-3}y}{z^2w^{-3}}\right)^0$

31. $\left(\dfrac{x^8y}{2z}\right)^2$

32. $(2c^{-4}d^2)^{-2}$

33. $-5x^{-3}$

34. $10k(4k^2 - 3k - 5) - 2k(5k^2 - 4k + 10)$

In Exercises 35–39, evaluate each expression.

35. 5^{-2}

36. $(2^{-3})^2$

37. $\dfrac{10^{-2}}{10^0}$

38. $\dfrac{(-11)^2}{-11^2}$

39. $2^4 \cdot 3^0$

In Exercises 40–42, write each expression without a denominator, using negative exponents whenever necessary.

40. $\dfrac{z}{y^{-2}}$

41. $\dfrac{h^2}{k^2}$

42. $\dfrac{u^{-3}v^2}{5^2w}$

43. Express 0.00297 in scientific notation.

44. Express 120,000,000 in scientific notation.

22. _____

23. _____

24. _____

25. _____

26. _____

27. _____

28. _____

29. _____

30. _____

31. _____

32. _____

33. _____

34. _____

35. _____

36. _____

37. _____

38. _____

39. _____

40. _____

41. _____

42. _____

43. _____

44. _____

Chapter 5   DIAGNOSTIC TEST

The purpose of this test is to see how well you understand working with exponents. We recommend that you work this diagnostic test *before* your instructor tests you on this chapter. Allow yourself about 50 minutes.

Complete solutions for all the problems on this test, together with section references, are given in the answer section in the back of the book. We suggest that you study the sections referred to for the problems you do incorrectly.

In Problems 1–17, simplify each expression. Assume that none of the factors appearing in any denominator is zero.

1. $(-3xy)(5x^3y)(-2xy^4)$
2. $2xy^2(x^2 - 3y - 4)$
3. $x^3 \cdot x^4$
4. $(x^2)^3$
5. $\dfrac{x^5}{x^2}$
6. x^{-4}
7. x^2y^{-3}
8. $\dfrac{a^{-3}}{b}$
9. $\dfrac{x^{5a}}{x^{3a}}$
10. $(4^{3x})^0$
11. $(x^2y^4)^3$
12. $(a^{-3}b)^2$
13. $\left(\dfrac{p^3}{q^2}\right)^2$
14. $\left(\dfrac{x}{y^2}\right)^{-3}$
15. $\left(\dfrac{3x^{-2}}{y^{-3}}\right)^{-1}$
16. $3h(2k^2 - 5h) - h(2h - 3k^2)$
17. $x(x^2 + 2x + 4) - 2(x^2 + 2x + 4)$

In Problems 18–23, evaluate each expression.

18. $2^3 \cdot 2^2$
19. $10^{-4} \cdot 10^2$
20. 5^{-2}
21. $(2^{-3})^2$
22. $\dfrac{10^{-3}}{10^{-4}}$
23. $(5^0)^2$

24. Write the expression $\dfrac{a^3}{b}$ without a denominator, using negative exponents if necessary.

25. Write each number in scientific notation.
   a. 1,326   b. 0.527

Chapters 1–5   CUMULATIVE REVIEW EXERCISES

In Exercises 1–6, write "true" if the statement is always true; otherwise, write "false."

1. The additive identity is 1.
2. The additive inverse of $\frac{8}{5}$ is $-\frac{8}{5}$.
3. Subtraction is commutative.
4. Division is associative.
5. $3{,}870{,}000 = 3.87 \times 10^{-6}$ in scientific notation.
6. $3x$ is a factor of the expression $3x + 7$.

In Exercises 7–11, evaluate each expression, or write "not defined."

7. $\dfrac{14 - 23}{-8 + 11}$
8. $\sqrt{13^2 - 5^2}$
9. $6(-3) - 4\sqrt{36}$
10. $\{-12 - [7 + (3 - 9)]\} - 15$
11. $\dfrac{0}{-11}$

In Exercises 12 and 13, solve each problem by using the given formula.

12. $C = \frac{5}{9}(F - 32)$. Find C when $F = -13$.
13. $A = P(1 + rt)$. Find A when $P = 1{,}200$, $r = 0.15$, $t = 4$.
14. Solve and check the equation $3x - 7(2 - x) = 8x - 22$.

In Exercises 15–18, simplify each expression.

15. $(-5p)(4p^3)$
16. $(-4h^1j^5)(-j^3k^4)(-5hk)(-10j^2h^7)$
17. $7 - 4(3xy - 3)$
18. $3xy^2(8x - 2xy) - 5xy(10xy - 3xy^2)$

In Exercises 19–22, set up each problem algebraically, solve, and check.

19. Several families went to a movie together. They spent $16.25 for nine tickets. If adults' tickets cost $2.50 and children's tickets cost $1.25, how many of each kind of ticket was purchased?

20. When twice the sum of 11 and an unknown number is subtracted from 6 times the sum of 8 and twice the unknown number, the result is 6. What is the unknown number?

21. A rectangle is 7 cm longer than it is wide. If its perimeter is 66 cm, what is its width? What is its length?

22. A boat cruised downstream for 5 hr before stopping for the night. The next day, it took 7 hr for the boat to get back to its starting point. If the speed of the river both days was 4 mph, find the speed of the boat in still water.

Critical Thinking and Problem-Solving Exercises

1. Dominique was thinking about numbers one day, and she made the following observations:

$$1 + 3 = 2^2$$

$$1 + 3 + 5 = 3^2$$

$$1 + 3 + 5 + 7 = 4^2$$

Verify that her observations are correct, and *without doing the actual addition*, predict what the answer for $1 + 3 + 5 + 7 + 9$ is. Then check your prediction by doing the addition.

2. Kent learned that he could buy liquid laundry detergent and fabric softener on sale if he bought them by the case. Each case of laundry detergent contains 4 bottles, and each case of fabric softener holds 6 bottles. Kent wants to buy a total of 60 bottles of these two items. If he does buy at least one case of each item, what are the possible combinations of cases of laundry detergent and fabric softener that will give him a total of 60 bottles?

3. Four friends, Monica, Alan, Bobbie, and Tony, went to a restaurant. They each ordered one item, and all the items were different. One ordered a soft drink, one a cup of coffee, one a piece of pie, and one a donut. Neither Alan nor Tony ordered pie or donuts, Monica didn't order a cup of coffee or donuts, and Alan ordered a soft drink. What did each person order?

4. Dick, David, and Larry wanted to attend a baseball game, and in checking to see whether they had enough money for tickets, hot dogs, and so on, they discovered that Dick and David together had $67, Dick and Larry together had $70, and David and Larry together had $73. How much did each man have individually?

5. Suppose your instructor asked you to explain to a student who had missed class *why $x^{-n} = \dfrac{1}{x^n}$*. What would your explanation be?

6. Each of the following solutions contains an error. Find and describe the error.

a. Evaluate $2^2 \cdot 2^3$.
Solution:　　　　　　$2^2 \cdot 2^3 = 4^5$

b. Simplify $2^3 \cdot 3^2$.
Solution:　　　　　　$2^3 \cdot 3^2 = 6^5$

c. Simplify -5^{-2}.
Solution:　　　　　　$-5^{-2} = \frac{1}{25}$

d. Simplify $(2x^2y^4)^3$.
Solution:　　　　$(2x^2y^4)^3 = 2x^6y^{12}$

Polynomials

CHAPTER 6

n this chapter, we look in detail at a particular type of algebraic expression called a *polynomial*. Polynomials have the same importance in algebra that whole numbers have in arithmetic. Just as much of the work in arithmetic involves operations with whole numbers, much of the work in algebra involves operations with polynomials.

Because a polynomial is a special kind of algebraic expression, we have already discussed some of the work with polynomials. In this chapter, we review and extend these concepts.

6.1   Basic Definitions

Because of its importance, we repeat here the definition of a *term* of an algebraic expression. A *term* can consist of one number, one variable, or a *product* of numbers and variables; each plus or minus sign is part of the term that follows it. (An expression within grouping symbols is considered a single term.)

Recall that a term containing exponents is considered *simplified* when each different base appears only once and when the exponent on each base is a single *positive* number.

Polynomials

A **polynomial in one variable** is an algebraic expression that, in simplified form, contains only terms of the form ax^n, where a (the coefficient) represents any real number, x represents any variable, and n represents any whole number. For example, $8x^4 + 3x^2 - 3x$ is a polynomial in x, because each of the *terms* is of the form ax^n.

A polynomial with only one term is called a **monomial**, a polynomial with two unlike terms is called a **binomial**, and a polynomial with three unlike terms is called a **trinomial**. We will use the general term **polynomial** for polynomials with four or more terms.

EXAMPLE 1   Examples of algebraic expressions that are polynomials in one variable:

a. $-2x$ This polynomial is a monomial in x

b. $7y + 3y^3$ This polynomial is a binomial in y

c. 8 This polynomial is a monomial; it is of the form $8x^0$, because $8x^0 = 8 \cdot 1 = 8$

d. $z^7 - 2z^2 + 3$ This polynomial is a trinomial in z; 3 is called the *constant term*

e. $x^3 + 3x^2 + 3x + 1$ This is a polynomial in x; 1 is called the *constant term*

f. $\dfrac{1}{x^{-5}}$ This is a polynomial because, in simplified form, it becomes x^5

If an algebraic expression in simplified form contains terms with negative (or fractional*) exponents *on the variables* or if it contains simplified terms with variables in a denominator or under a radical sign, then the algebraic expression is *not* a polynomial.

---

*Fractional exponents (that is, exponents that are not integers) are not discussed in this book.

EXAMPLE 2 Examples of algebraic expressions that are *not* polynomials:

a. $3x^{-5}$ This is *not* a polynomial because it has a negative exponent on a variable and, in simplified form, would have a variable in the denominator

b. $\dfrac{1}{x+2}$ This is *not* a polynomial because it has a variable in the denominator

c. $\sqrt{3+x}$ This is *not* a polynomial because the variable is under a radical sign

An algebraic expression with two variables is a **polynomial in two variables** if, in its simplified form, (1) it contains no negative or fractional exponents on the variables, (2) no variables are in denominators, and (3) no variables are under radical signs.

EXAMPLE 3 Examples of algebraic expressions that are polynomials in two variables:

a. $xy^3\sqrt{7}$ This polynomial is a monomial; note that *constants* can be under radical signs

b. $-4xy + \frac{1}{2}x^2y$ This polynomial is a binomial; note that *constants* can be in denominators

c. $x^3y^2 - 2x + 3y^2 - 1$ This is a polynomial in x and y

d. $7uv^4 - 5u^2v + 2u$ Polynomials can contain any variables; this is a polynomial in u and v

The Degree of a Term of a Polynomial   If a polynomial contains only one variable, then the **degree of any term** of that polynomial is the exponent on the variable in that term. If a polynomial contains more than one variable, the degree of any term of that polynomial is the *sum* of the exponents on the variables in that term.

EXAMPLE 4 Examples of finding the degree of a term:

a. $5x^3$ Third degree

b. $6x^2y$ Third degree because $6x^2y = 6x^2y^1$ $\quad 2 + 1 = 3$

c. 14 Zero degree because $14 = 14x^0$

d. 2^5x^3 Third degree because we consider only exponents on *variables*

e. $-2u^3vw^2$ Sixth degree because $-2u^3vw^2 = -2u^3v^1w^2$ $\quad 3 + 1 + 2 = 6$

The Degree of a Polynomial   The **degree of a polynomial** is defined to be the degree of its highest-degree term. Therefore, to find the degree of a polynomial, we first find the degree of each of its terms. The *largest* of these numbers will be the degree of the polynomial.*

EXAMPLE 5 Examples of finding the degree of a polynomial:

a. $9x^3 - 7x + 5$

        Zero-degree term
        First-degree term
        Third-degree term ← Highest-degree term

Therefore, $9x^3 - 7x + 5$ is a third-degree polynomial.

*Mathematicians define the zero polynomial, 0, as having no degree.

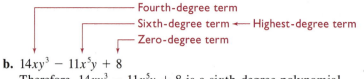

b. $14xy^3 - 11x^5y + 8$

Therefore, $14xy^3 - 11x^5y + 8$ is a sixth-degree polynomial.

c. $6^5 a^2bc^3 + 12ab^6c^2$ Exponents on constants do not affect the degree

This is a ninth-degree polynomial, because the term with the highest degree $(12ab^6c^2)$ is of degree 9.

Descending and Ascending Powers   If the exponents on one variable get smaller as we read the terms from left to right, the polynomial is arranged in **descending powers** of that variable. If the exponents on one variable get larger as we read the terms from left to right, the polynomial is arranged in **ascending powers** of that variable. For example,

The exponents get smaller as we read from left to right

$$8x^3 - 3x^2 + 5x^1 + 7$$

$7 = 7x^0$

Therefore, $8x^3 - 3x^2 + 5x + 7$ is arranged in descending powers of x.

EXAMPLE 6   Arrange $5 - 2x^2 + 4x$ in descending powers of x.

SOLUTION $-2x^2 + 4x + 5$

Your instructor may prefer that you write polynomials in one variable in descending powers of that variable.

A polynomial with more than one variable can be arranged in descending or ascending powers of any *one* of its variables (see Example 7).

EXAMPLE 7   Arrange $3x^3y - 5xy + 2x^2y^2 - 10$ (a) in descending powers of x and then (b) in descending powers of y.

SOLUTION

a. $3x^3y + 2x^2y^2 - 5xy - 10$ Arranged in descending powers of x

b. $2x^2y^2 + 3x^3y - 5xy - 10$ Arranged in descending powers of y

Since y is to the same power in both terms, we write the higher-degree *term* first

Leading Coefficients   The **leading coefficient** of a polynomial is defined to be the numerical coefficient of its highest-degree term.

EXAMPLE 8   Examples of naming the leading coefficients for the polynomials from Example 5:

a. The leading coefficient of $9x^3 - 7x + 5$ is 9.   The highest-degree term is $9x^3$

b. The leading coefficient of $14xy^3 - 11x^5y + 8$ is -11.   The highest-degree term is $-11x^5y$

c. The leading coefficient of $6^5a^2bc^3 + 12ab^6c^2$ is 12.   The highest-degree term is $12ab^6c^2$

Exercises
Set I

In Exercises 1–16, if the expression is a polynomial, **(a)** find the degree of the first term and **(b)** find the degree of the polynomial. If it is *not* a polynomial, write "not a polynomial."

1. $3x + 2x^2$

2. $4y + 5y^3$

3. $\dfrac{1}{x} - 3x^2 + 3$

4. $8x^4 - \dfrac{3}{2x} - 2x$

5. $3xy^2 + x^3y^3 - 3x^2y - y^3$

6. $6mn^2 + 8m^3 - 12m^2n - n^3$

7. $xy + 3x^4 + 5$

8. $xy + 5x^3 + 2$

9. $x^{-2} + 5x^{-1} + 4$

10. $y^{-3} + y^{-2} + 6$

11. $\sqrt{x - 3} + x^3$

12. $y^2 + \sqrt{2 + x}$

13. $x^2\sqrt{5} - xy + 3$

14. $x^3\sqrt{6} - xy + 2$

15. $\dfrac{3}{2x^2 - 3x + 1}$

16. $\dfrac{8}{3x^3 - 2x - 5}$

In Exercises 17–20, write each polynomial in descending powers of the indicated variable, and find the leading coefficient.

17. $7x^3 - 4x - 5 + 8x^5$; powers of x

18. $10 - 3y^5 + 4y^2 - 2y^3$; powers of y

19. $8xy^2 + xy^3 - 4x^2y$; powers of y

20. $3x^3y + x^4y^2 - 3xy^3$; powers of x

Writing Problems

Express the answers in your own words and in complete sentences.

1. Explain why $\sqrt{x + 3}$ is not a polynomial.

2. Explain why $\dfrac{4}{x - 1}$ is not a polynomial.

3. Explain why $3^4x^5 + 7x^2 - 1$ is a fifth-degree polynomial.

Exercises 6.1
Set II

In Exercises 1–16, if the expression is a polynomial, **(a)** find the degree of the first term and **(b)** find the degree of the polynomial. If it is *not* a polynomial, write "not a polynomial."

1. $8x + 5x^3$

2. $4y^3 + \dfrac{5}{4x}$

3. $7x^3 + \dfrac{9}{2x} - 3x^2$

4. $5x^{-3} - 2x + 4$

5. $x^3y - 8x^4y^3 + xy - y^5$

6. $8xyz^2 + 3x^2y - z^3$

7. $xy + 8x^3 + 5$

8. $7uv + 8$

9. $3x^{-4} + 2x^{-2} + 6$

10. $\dfrac{1}{2x^2 - 5x}$

11. $x^2 - \sqrt{2x - 5}$

12. $x^2\sqrt{7} - 3 + 4x$

13. $5 - xy + x^2\sqrt{2}$

14. $\dfrac{5}{2x^4 + 3x^2 - 3}$

15. $\dfrac{2}{6x^2 - 5x + 8}$

16. $\sqrt{9x + 5} - x^2$

In Exercises 17–20, write each polynomial in descending powers of the indicated variable, and find the leading coefficient.

17. $17a - 15a^3 + a^{10} - 4a^5$; powers of a

18. $3x^2y + 8x^3 + y^3 - xy^5$; powers of y

19. $9x^2y^4 - 8x^5y^2 + 3x^4 - 3$; powers of x

20. $5st - 9rs^2t - rt^2 - 3rs$; powers of r

6.2 Simplifying, Adding, and Subtracting Polynomials

Simplifying Polynomials

We simplify a polynomial the same way we simplify an algebraic expression: We remove all grouping symbols, simplify each term, and combine all like terms.

Adding Polynomials

Polynomials can be added horizontally by removing the grouping symbols and combining like terms. (In most *addition* problems, this means that we can simply "drop the grouping symbols.")

It is often helpful to underline like terms with the same kind of marking (perhaps using single underlining for the x^2-terms, double underlining for the x-terms, and so forth) before we combine them (see Example 1).

EXAMPLE 1 Examples of adding polynomials, with answers written in descending powers of x:

a. $(3x^2 + 5x - 4) + (2x + 5) + (x^3 - 4x^2 + x)$

$= 3x^2 + 5x - 4 + 2x + 5 + x^3 - 4x^2 + x$ Removing the parentheses

$= x^3 - x^2 + 8x + 1$ Combining like terms and writing the answer in descending powers of x

b. $(5x^3y^2 - 3x^2y^2 + 4xy^3) + (4x^2y^2 - 2xy^2) + (-7x^3y^2 + 6xy^2 - 3xy^3)$

$= 5x^3y^2 - 3x^2y^2 + 4xy^3 + 4x^2y^2 - 2xy^2 - 7x^3y^2 + 6xy^2 - 3xy^3$ Removing the parentheses

$= -2x^3y^2 + x^2y^2 + xy^3 + 4xy^2$

c. $4x^2y + 8xyx + 6yx^2 - 3xy^2$ The underlined terms *are* like terms, since $xyx = x^2y$ and $yx^2 = x^2y$

$= 4x^2y + 8x^2y + 6x^2y - 3xy^2$

$= 18x^2y - 3xy^2$

In vertical addition, which is sometimes desirable, it is important to have all like terms lined up vertically.

Adding polynomials vertically

> **1.** Arrange the polynomials under one another *so that like terms are in the same vertical column.*
>
> **2.** Find the sum of the terms in each vertical column.

EXAMPLE 2 Add $(3x^2 + 2x - 1)$, $(2x + 5)$, and $(4x^3 + 7x^2 - 6)$ vertically.

SOLUTION

$$
\begin{array}{r}
3x^2 + 2x - 1 \\
2x + 5 \\
4x^3 + \ 7x^2 \qquad - 6 \\
\hline
4x^3 + 10x^2 + 4x - 2
\end{array}
$$

EXAMPLE 3

Add $(8x^2y - 3xy^2 + xy - 2)$, $(4xy^2 - 7xy)$, and $(-5x^2y + 9)$ vertically.

SOLUTION

$$
\begin{array}{r}
8x^2y - 3xy^2 + xy - 2 \\
4xy^2 - 7xy \\
-5x^2y \qquad\qquad\quad + 9 \\
\hline
3x^2y + xy^2 - 6xy + 7
\end{array}
$$

Subtracting Polynomials

We subtract polynomials the same way we subtract signed numbers: We *add* the additive inverse of the polynomial being subtracted to the polynomial we're subtracting *from*. We can find the additive inverse of a polynomial by changing *all* the signs of the polynomial (this is the method we show) *or* by multiplying the polynomial by -1.

Subtracting polynomials

> If the problem is to be set up *vertically*, place the polynomial being subtracted *under* the other polynomial, *with like terms in the same vertical column.*
>
> For horizontal *or* vertical subtraction:
>
> **1.** Change the subtraction symbol to an addition symbol *and* find the *additive inverse* of the polynomial *being subtracted.*
>
> **2.** Find the sum of the resulting polynomials.

EXAMPLE 4

Subtract the following polynomials as indicated.

This is the polynomial being subtracted

a. $(-4x^3 + 8x^2 - 2x - 3) - (4x^3 - 7x^2 + 6x + 5)$

SOLUTION

The sign of $4x^3$ is understood to be positive

$(-4x^3 + 8x^2 - 2x - 3) - (+4x^3 - 7x^2 + 6x + 5)$

Changing the subtraction symbol to an addition symbol *and*

$= (-4x^3 + 8x^2 - 2x - 3) + (-4x^3 + 7x^2 - 6x - 5)$ finding the additive inverse of the polynomial being subtracted

$= -4x^3 + 8x^2 - 2x - 3 - 4x^3 + 7x^2 - 6x - 5$ Removing the grouping symbols

$= -8x^3 + 15x^2 - 8x - 8$ Combining like terms

b. Subtract $(-4x^2y + 10xy^2 + 9xy - 7)$ from $(11x^2y - 8xy^2 + 7xy + 2)$.

This is the polynomial being subtracted

SOLUTION

$(11x^2y - 8xy^2 + 7xy + 2) - (-4x^2y + 10xy^2 + 9xy - 7)$

Changing subtraction to addition

$= (11x^2y - 8xy^2 + 7xy + 2) + (+4x^2y - 10xy^2 - 9xy + 7)$

Removing the grouping symbols

$= 11x^2y - 8xy^2 + 7xy + 2 + 4x^2y - 10xy^2 - 9xy + 7$

Combining like terms

$= 15x^2y - 18xy^2 - 2xy + 9$

c. Subtract $(2x^2 - 5x + 3)$ from the sum of $(8x^2 - 6x - 1)$ and $(4x^2 + 7x - 9)$.

S O L U T I O N

$$[(8x^2 - 6x - 1) + (4x^2 + 7x - 9)] - (2x^2 - 5x + 3)$$

$$= [8x^2 - 6x - 1 + 4x^2 + 7x - 9] + (-2x^2 + 5x - 3) \qquad \text{Changing subtraction to addition}$$

$$= [12x^2 + x - 10] + (-2x^2 + 5x - 3) \qquad \text{Combining like terms inside the brackets}$$

$$= 12x^2 + x - 10 - 2x^2 + 5x - 3 \qquad \text{Removing the grouping symbols}$$

$$= 10x^2 + 6x - 13 \qquad \text{Combining like terms}$$

d. $(x^2 + 5) - [(x^2 - 3) + (2x^2 - 1)]$

S O L U T I O N

$$(x^2 + 5) - [(x^2 - 3) + (2x^2 - 1)]$$

$$= (x^2 + 5) - [x^2 - 3 + 2x^2 - 1] \qquad \text{Removing the innermost parentheses}$$

$$= (x^2 + 5) - [3x^2 - 4] \qquad \text{Combining like terms inside the brackets}$$

$$= (x^2 + 5) + [-3x^2 + 4] \qquad \text{Changing subtraction to addition}$$

$$= x^2 + 5 - 3x^2 + 4 \qquad \text{Removing the grouping symbols}$$

$$= -2x^2 + 9 \qquad \text{Combining like terms}$$

In Examples 5 and 6, we show vertical subtraction. Because subtraction in long division problems is always done vertically, it is essential that you know how to subtract vertically.

EXAMPLE 5 Subtract the lower polynomial from the one above it: $5x^2 + 3x - 6$
                                                                        $\underline{3x^2 - 5x + 2}$

S O L U T I O N

$$\begin{array}{r} 5x^2 + 3x - 6 \\ -(3x^2 - 5x + 2) \end{array} \rightarrow \begin{array}{r} 5x^2 + 3x - 6 \\ +(-3x^2 + 5x - 2) \\ \hline 2x^2 + 8x - 8 \end{array}$$

We change the sign of *each* term in the polynomial being subtracted; then we *add* the resulting terms

EXAMPLE 6 Subtract $(3x^2 - 2x - 7)$ from $(x^3 - 2x - 5)$ vertically.

S O L U T I O N

$$\begin{array}{r} x^3 \qquad\quad - 2x - 5 \\ -(3x^2 - 2x - 7) \end{array} \rightarrow \begin{array}{r} x^3 \qquad\qquad - 2x - 5 \\ +(-3x^2 + 2x + 7) \\ \hline x^3 \quad -3x^2 \qquad\quad + 2 \end{array}$$

We change signs and *add*

 Note   Your instructor may require you to change the signs *mentally* and may not allow you to show the sign changes.

Exercises 6.2

Set I

Perform the indicated operations, and simplify the results.

1. $(2m^2 - m + 4) + (3m^2 + m - 5)$

2. $(5n^2 + 8n - 7) + (6n^2 - 6n + 10)$

3. $(2x^3 - 4) + (4x^2 + 8x) + (-9x + 7)$

4. $(5 + 8z^2) + (4 - 7z) + (z^2 + 7z)$

5. $(3x^2 + 4x - 10) - (5x^2 - 3x + 7)$

6. $(2a^2 - 3a + 9) - (3a^2 + 4a - 5)$

7. Subtract $(-5b^2 + 4b + 8)$ from $(8b^2 + 2b - 14)$.

8. Subtract $(-8c^2 - 9c + 6)$ from $(11c^2 - 4c + 7)$.

9. $(6a - 5a^2 + 6) + (4a^2 + 6 - 3a)$

10. $(2b + 7b^2 - 5) + (4b^2 - 2b + 8)$

11. Subtract $(5a + 3a^2 - 4)$ from $(4a^2 + 6 - 3a)$.

12. Subtract $(2b + b^2 - 7)$ from $(8 + 3b^2 - 7b)$.

13. Add: $\begin{array}{l} 17a^3 \quad\;\; + 4a - 9 \\ \underline{ 8a^2 - 6a + 9} \end{array}$

14. Add: $\begin{array}{l} \quad\;\; - b^3 + 5b^2 - 8 \\ \underline{-20b^4 + 2b^3 \qquad\;\; + 7} \end{array}$

15. Add: $\begin{array}{l} 14x^2y^3 - 11xy^2 + 8xy \\ -9x^2y^3 + \;\,6xy^2 - 3xy \\ \underline{\;\;7x^2y^3 - \;\,4xy^2 - 5xy} \end{array}$

16. Add: $\begin{array}{l} 12a^2b - \;\,8ab^2 + \;\,6ab \\ -7a^2b + 11ab^2 - \;\,3ab \\ \underline{\;\;4a^2b - \;\;\;\,ab^2 - 13ab} \end{array}$

In Exercises 17–20, subtract the lower polynomial from the one above it.

17. Subtract: $\begin{array}{l} 15x^3 - 4x^2 \qquad\; + 12 \\ \underline{\;\;8x^3 \qquad\quad\; + 9x - \;\,5} \end{array}$

18. Subtract: $\begin{array}{l} \qquad\quad - 14y^2 + \;\,6y - 24 \\ \underline{7y^3 + 14y^2 - 13y} \end{array}$

19. Subtract: $\begin{array}{l} 10a^2b - 6ab + 5ab^2 \\ \underline{\;\;3a^2b + 6ab - 7ab^2} \end{array}$

20. Subtract: $\begin{array}{l} 14m^3n^2 - 9m^2n^2 - 6mn \\ \underline{-8m^3n^2 - 5m^2n^2 + 3mn} \end{array}$

21. $(7m^8 - 4m^4) + (4m^4 + m^5) + (8m^8 - m^5)$

22. $(8h - 4h^6) + (5h^7 + 3h^6) + (9h - 5h^7)$

23. $(6r^3t + 14r^2t - 11) + (19 - 8r^2t + r^3t) + (8 - 6r^2t)$

24. $(13m^2n^2 + 4mn + 23) + (17 + 4mn - 9m^2n^2) + (-29 - 8mn)$

25. $(7x^2y^2 - 3x^2y + xy + 7) - (3x^2y^2 - 5xy + 4 + 7x^2y)$

26. $(4x^2y^2 + x^2y - 5xy - 4) - (9 - 5x^2y^2 - xy + 3x^2y)$

27. $(x^2 + 4) - [(x^2 - 5) - (3x^2 + 1)]$

28. $(3x^2 - 2) - [(4 - x^2) - (2x^2 - 1)]$

29. Subtract $(2x^2 - 4x + 3)$ from the sum of $(5x^2 - 2x + 1)$ and $(-4x^2 + 6x - 8)$.

30. Subtract $(6y^2 + 3y - 4)$ from the sum of $(-2y^2 + y - 9)$ and $(8y^2 - 2y + 5)$.

31. Subtract the sum of $(x^3y + 3xy^2 - 4)$ and $(2x^3y - xy^2 + 5)$ from the sum of $(5 + xy^2 + x^3y)$ and $(-6 - 3xy^2 + 4x^3y)$.

32. Subtract the sum of $(2m^2n - 4mn^2 + 6)$ and $(-3m^2n + 5mn^2 - 4)$ from the sum of $(5 + m^2n - mn^2)$ and $(3 + 4m^2n + 2mn^2)$.

33. $(7.239x^2 - 4.028x + 6.205) + (-2.846x^2 + 8.096x + 5.307)$

34. $(29.62x^2 + 35.78x - 19.80) + (7.908x^2 - 29.63x - 32.84)$

Writing Problems

Express the answer in your own words and in complete sentences.

1. Explain why we must put a set of parentheses around the polynomial being subtracted when we subtract one polynomial from another.

Exercises 6.2
Set II

Perform the indicated operations, and simplify the results.

1. $(2x^2 - 3x + 1) + (4x^2 + 5x - 3)$

2. $(5z + 7) - (7z^2 - 8)$

3. $(8x^3 - 4x^2) + (x^2 - 3x) + (8 - 2x^2)$

4. $(3x^2 + 5x) + (2x^3 - 6x^2) - (3 - 5x)$

5. $(9x^2 - 2x + 3) - (12x^2 - 8x - 9)$

6. $(3 - x^3 + 5x) - (-8x - x^2 + 8x)$

7. Subtract $(8y^3 - 3y)$ from $(y^2 - 3y + 12)$.

8. Subtract $(-3x^3 + 2x^2 - 3x + 2)$ from $(1 - x - x^2 - x^3)$.

9. $(9c + 2c^2 - 8) + (3 - 17c - 8c^2)$

10. $(8x^3 - x + 2) - (x^2 - x + 2)$

11. Subtract $(9 - z + z^2)$ from $(-3z^2 - z + 9)$.

12. Subtract $(2x - 7x^2 + 3x^3)$ from $(-x + 9x^2 - 1 + x^3)$.

13. Add: $\begin{array}{r} 8x^3 \quad\quad\;\; + 3x - 7 \\ \underline{5x^2 - 5x + 7} \end{array}$

14. Add: $\begin{array}{r} 4x^3 + 7x^2 - 5x + 4 \\ \underline{2x^3 - 5x^2 + 5x - 6} \end{array}$

15. Add: $\begin{array}{r} 3y^4 - 2y^3 + 4y + 10 \\ -5y^4 + 2y^3 + 4y - 6 \\ \underline{7y^4 \quad\quad\;\; - 6y - 8} \end{array}$

16. Add: $\begin{array}{r} 18x^2y - 3xy^2 + 4xy \\ -6x^2y + 8xy^2 - 9xy \\ \underline{-4x^2y - 9xy^2 + xy} \end{array}$

In Exercises 17–20, subtract the lower polynomial from the one above it.

17. Subtract: $\begin{array}{r} 10x^3 - 5x^2 + 6x - 1 \\ \underline{-2x^3 + 3x^2 + 9x - 5} \end{array}$

18. Subtract: $\begin{array}{r} 5z^3 \quad\quad\;\; - 7z + 8 \\ \underline{8z^3 - 10z^2 + 7z} \end{array}$

19. Subtract: $\begin{array}{r} 7x^2y^2 - 8xy^2 + xy - 6 \\ \underline{2x^2y^2 - 6xy^2 - 5xy + 9} \end{array}$

20. Subtract: $\begin{array}{r} -3x^2 + 3x \\ \underline{5x^2 - 2x + 6} \end{array}$

21. $(8x^6 - 3x^4) + (x - 7x^4) + (3x^4 - x)$

22. $(5x^2 - 3x + 1) - (5x^2 - 3x + 1)$

23. $(7x^2y + 4xy^2 - 5) + (8xy^2 - 7x^2y + xy) + (-2 - xy^2)$

24. $(5x^2y - 4x + 3y^2) + (-yx^2 + 6x - y^2)$

25. $(9x^4y - x^2y^2 - 5 + 8xy^3) - (3x^2y^2 - 4xy^3 + 7 - 6x^3)$

26. $(3y^2z - 4x^2y^2 + 5) - (5x^2y^2 - 4y^2z)$

27. $(3x^2 + 5) - [(x^2 - 3) - (2x^2 + 8)]$

28. $(x^2 - 3xy + 4) - [x^2 - 3xy - (6 + x^2)]$

29. Subtract $(3x^2 - 7x - 1)$ from the sum of $(x^2 - x + 2)$ and $(x^2 - 3x - 6)$.

30. Subtract $(2y^2 + 3y - 9)$ from the sum of $(y - y^2 + 2)$ and $(5y - y^2 + 7)$.

31. Subtract the sum of $(xy^2 - 2x^2y - 2)$ and $(3x^2y - 4xy^2 - 6)$ from the sum of $(3 - x^2y)$ and $(2xy^2 - 5x^2y - 8)$.

32. Subtract $(-3m^2n^2 + 2mn - 7)$ from the sum of $(6m^2n^2 - 8mn + 9)$ and $(-10m^2n^2 + 18mn - 11)$.

33. $(5.416x - 34.54x^2 + 7.806) +$ $(51.75x^2 - 1.644x - 9.444)$

34. $(5.886x^2 - 3.009x + 7.966) -$ $[4.961x^2 - 54.51x - (7.864 - 1.394x^2)]$

6.3   Multiplying Polynomials

When we multiplied $-2x^3$ by $6x$ in the Exponents chapter, we were really multiplying a monomial by a monomial [recall that $(6x)(-2x^3) = -12x^4$], and when we multiplied $x^2 + y$ by x in that same chapter [recall that $x(x^2 + y) = x^3 + xy$], we were really multiplying a polynomial with more than one term by a monomial. There will be some problems such as these in the next exercise set.

6.3A Products of Two Binomials

Since we often need to find the product of two binomials, it is helpful to be able to find their product by inspection (that is, without writing anything down except the answer). First, however, we show the step-by-step procedure for multiplying two binomials (see Example 1); when we multiply two binomials, it is necessary to use the distributive property *more than once*.

EXAMPLE 1   Multiply $(3x + 2)(4y + 5)$, and simplify the result.

SOLUTION 1   We first treat $3x + 2$ as if it were a single number.

Step 1. $(3x + 2)(4y + 5) = (3x + 2)(4y) + (3x + 2)(5)$

Step 2. We now use the distributive property again on $(3x + 2)(4y)$ and on $(3x + 2)(5)$:

$$= (3x)(4y) + (2)(4y) + (3x)(5) + (2)(5) \qquad \text{This step need not be shown}$$

Step 3.
$$= 12xy + 8y + 15x + 10$$

SOLUTION 2   This time we treat $4y + 5$ as if it were a single number.

Step 1. $(3x + 2)(4y + 5) = (3x)(4y + 5) + (2)(4y + 5)$

Step 2. We use the distributive property again on $(3x)(4y + 5)$ and on $(2)(4y + 5)$:

$$= (3x)(4y) + (3x)(5) + (2)(4y) + (2)(5) \qquad \text{This step need not be shown}$$

Step 3.
$$= 12xy + 15x + 8y + 10$$

Notice that the answer obtained by using Solution 2 is equal to the answer obtained by using Solution 1; only the order in which the terms are listed is different.

Let's agree on some terminology so that we can more easily discuss finding products of two binomials by inspection.

Consider the product from Example 1 again: Because $3x + 2$ and $4y + 5$ are binomials, we call the $3x$ and the $4y$ the *first terms* of the binomials and the 2 and the 5 the *last terms* of the binomials.

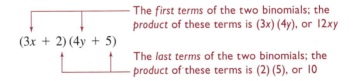

The *first terms* of the two binomials; the product of these terms is $(3x)(4y)$, or $12xy$

$(3x + 2)(4y + 5)$

The *last terms* of the two binomials; the product of these terms is $(2)(5)$, or 10

Because the $3x$ and the 5 are the "outside" terms, when we multiply them together we call their product the *outer product*. Because the 2 and the $4y$ are the "inside" terms, we call their product the *inner product*.

The *outer product*: $(3x)(5) = 15x$

$(3x + 2)(4y + 5)$

The *inner product*: $(2)(4y) = 8y$

Observe from Solution 2 of Example 1 that in the answer for $(3x + 2)(4y + 5)$, the first term was $12xy$, the product of the two *first terms*; the second term was $15x$, the *outer product*; the third term was $8y$, the *inner product*; and the last term was 10, the product of the two *last terms*.

The *FOIL* Method In general, the product of two binomials is a sum of four terms. The first term is the product of the two *F*irst terms, the second term is the *O*uter product, the third term is the *I*nner product, and the fourth term is the product of the two *L*ast terms. We do not *have* to write the terms in this order, but if we do, we can use the letters *F*, *O*, *I*, *L* as a memory device and say that we're using the *FOIL* method to find the product of two binomials by inspection. You must remember, however, that the four numbers so found are *terms* and must be added together. It might be even better to think of the procedure as the $F + O + I + L$ method. We show the FOIL method under Solution 2 in Examples 2 through 6.

EXAMPLE 2

Multiply $(2x - 5)(6y - 1)$, and simplify the result.

SOLUTION 1 We will think of this problem as $[2x + (-5)][6y + (-1)]$.

$$(2x - 5)(6y - 1)$$

$$= [2x + (-5)][6y + (-1)] \qquad \text{Thinking of each factor as a } sum \text{ of terms}$$

$$= [2x + (-5)](6y) + [2x + (-5)](-1) \qquad \text{Using the distributive property}$$

$$= (2x)(6y) + (-5)(6y) + (2x)(-1) + (-5)(-1) \qquad \text{Using the distributive property again}$$

$$= 12xy - 30y - 2x + 5$$

$$= 12xy - 2x - 30y + 5 \qquad \text{Expressing the answer in descending powers of } x$$

SOLUTION 2 Using the FOIL method, we have

First	+	Outer	+	Inner	+	Last
$(2x - 5)(6y - 1)$		$(2x - 5)(6y - 1)$		$(2x - 5)(6y - 1)$		$(2x - 5)(6y - 1)$
$12xy$	+	$(-2x)$	+	$(-30y)$	+	5

The product, then, is $12xy - 2x - 30y + 5$.

 Note It is not necessary (or desirable) to show all the steps that we show in the examples.

EXAMPLE 3

Multiply $(2x^2 + 7)(4x + 2)$, and simplify the result.

SOLUTION 1

$$(2x^2 + 7)(4x + 2)$$

$$= (2x^2 + 7)(4x) + (2x^2 + 7)(2)$$

$$= (2x^2)(4x) + (7)(4x) + (2x^2)(2) + (7)(2)$$

$$= 8x^3 + 28x + 4x^2 + 14$$

$$= 8x^3 + 4x^2 + 28x + 14 \qquad \text{Expressing the answer in descending powers of } x$$

SOLUTION 2 Using the FOIL method, we have

First	+	Outer	+	Inner	+	Last
$(2x^2 + 7)(4x + 2)$		$(2x^2 + 7)(4x + 2)$		$(2x^2 + 7)(4x + 2)$		$(2x^2 + 7)(4x + 2)$
$8x^3$	+	$4x^2$	+	$28x$	+	14

The product, then, is $8x^3 + 4x^2 + 28x + 14$.

When the FOIL method is used in a multiplication problem that is of the form $(ax + by)(cx + dy)$, the inner and outer products will always be *like terms* and they must be combined (see Example 4, Solution 2).

EXAMPLE 4

Multiply $(x + 2)(x - 5)$, and simplify the result.

SOLUTION 1

$(x + 2)(x - 5)$

$= (x + 2)[x + (-5)]$ — Thinking of $x - 5$ as the sum $x + (-5)$

$= (x + 2)(x) + (x + 2)(-5)$ — Using the distributive property

$= (x)(x) + (2)(x) + (x)(-5) + (2)(-5)$ — Using the distributive property again

$= x^2 + 2x - 5x - 10$

$= x^2 - 3x - 10$ — Combining like terms

SOLUTION 2 Using the FOIL method, we have

First	+	Outer	+	Inner	+	Last
$(x + 2)(x - 5)$		$(x + 2)(x - 5)$		$(x + 2)(x - 5)$		$(x + 2)(x - 5)$
x^2	+	$(-5x)$	+	$2x$	+	(-10)
x^2	+			$(-3x)$	+	(-10)

The product is $x^2 - 3x - 10$.

In fact, when the FOIL method is used on a problem that is of the form $(ax + by)(cx + dy)$, there will be two terms in the product if the sum of the inner and outer products is zero; otherwise, there will be three terms in the product. The product can be found quickly by using the following procedure:

Multiplying when the product is of the form $(ax + by)(cx + dy)$

1. The *first term* of the product is the product of the first terms of the binomials.
2. The *middle term* of the product is the *sum* of the inner and outer *products*.
3. The *last term* of the product is the product of the last terms of the binomials.

 **Note** Plus or minus signs between the *terms* of the product are essential.

EXAMPLE 5

Multiply $(x - 3)(x - 4)$, and simplify the result.

SOLUTION 1 $(x - 3)(x - 4) = (x - 3)[x + (-4)]$

$= (x - 3)(x) + (x - 3)(-4)$

$= x^2 - 3x + (-4x) + 12$

$= x^2 - 7x + 12$

SOLUTION 2 The problem is of the form $(ax + by)(cx + dy)$; therefore, we can use the rules given in the box above. We think of the problem as

$$[x + (-3)][x + (-4)]$$

Step 1. *F* $(x)(x)$, or x^2 The product of the *First* terms

　　　　+

Step 2. *O* $(x)(-4)$, or $-4x$ The *Outer* product

　　　　+

　　　　I $(-3)(x)$, or $-3x$ The *Inner* product

　　　　+ $-7x$ The *sum* of the outer and inner products

Step 3. *L* $(-3)(-4)$, or 12 The product of the *Last* terms

The product, then, is $x^2 + (-7x) + 12$, or $x^2 - 7x + 12$.

EXAMPLE 6 Simplify $(3x + 2)(4x - 5)$.

SOLUTION 1

$$
\begin{aligned}
(3x + 2)(4x - 5) &= (3x + 2)[4x + (-5)] \\
&= (3x + 2)(4x) + (3x + 2)(-5) \\
&= (3x)(4x) + (2)(4x) + (3x)(-5) + (2)(-5) \\
&= 12x^2 + 8x + (-15x) + (-10) \\
&= 12x^2 + (-7x) - 10 \\
&= 12x^2 - 7x - 10
\end{aligned}
$$

SOLUTION 2 $(3x + 2)[4x + (-5)]$

Step 1. *F* $(3x)(4x)$, or $12x^2$ The product of the *First* terms

　　　　+

Step 2. *O* $(3x)(-5)$, or $-15x$ The *Outer* product

　　　　+

　　　　I $(2)(4x)$, or $8x$ The *Inner* product

　　　　+ $-7x$ The *sum* of the outer and inner products

Step 3. *L* $(2)(-5)$, or -10 The product of the *Last* terms

The product is $12x^2 + (-7x) + (-10)$, or $12x^2 - 7x - 10$.

In Examples 7 and 8, we show the short method only.

EXAMPLE 7 Simplify $(5x - 4y)(6x + 7y)$.

SOLUTION $[5x + (-4y)](6x + 7y)$

Step 1. *F* $(5x)(6x)$, or $30x^2$ The product of the *First* terms

　　　　+

Step 2. *O* $(5x)(7y)$, or $35xy$ The *Outer* product

　　　　+

　　　　I $(-4y)(6x)$, or $-24xy$ The *Inner* product

　　　　+ $11xy$ The *sum* of the outer and inner products

Step 3. *L* $(-4y)(7y)$, or $-28y^2$ The product of the *Last* terms

The product is $30x^2 + 11xy + (-28y^2)$, or $30x^2 + 11xy - 28y^2$.

EXAMPLE 8   Simplify $(3x - 8y)^2$.

SOLUTION   Because raising to a power is repeated multiplication, we have

$$(3x - 8y)^2 = (3x - 8y)(3x - 8y)$$

Step 1. F $(3x)(3x)$, or $9x^2$ The product of the *First* terms
             +
Step 2. O $(3x)(-8y)$, or $-24xy$ The *Outer* product
             +
             I $(-8y)(3x)$, or $-24xy$ The *Inner* product
             + $-48xy$ The *sum* of the outer and inner products
Step 3. L $(-8y)(-8y)$, or $64y^2$ The product of the *Last* terms

The final answer is $9x^2 - 48xy + 64y^2$.

✖ **A Word of Caution**   A common error is to write

$$(3x - 8y)^2 = 9x^2 - 64y^2 \qquad \text{or} \qquad (3x - 8y)^2 = 9x^2 + 64y^2$$

Example 8 shows that neither of these answers is correct. In a later section, we discuss a short method for correctly simplifying $(3x - 8y)^2$.

In Example 9, the problem is not of the form $(ax + by)(cx + dy)$. We will use the FOIL method.

EXAMPLE 9   Multiply $(2x^2 - 5x)(3x + 7)$, and simplify the result.

SOLUTION

F $(2x^2)(3x)$, or $6x^3$ The product of the *First* terms
+
O $(2x^2)(7)$, or $14x^2$ The *Outer* product
+
I $(-5x)(3x)$, or $-15x^2$ The *Inner* product
+
L $(-5x)(7)$, or $-35x$ The product of the *Last* terms

The product is $6x^3 + 14x^2 + (-15x^2) + (-35x)$. The outer and inner products are like terms; they must be combined.
Therefore, $(2x^2 - 5x)(3x + 7) = 6x^3 - x^2 - 35x$.

Exercises 6.3A
Set I

Simplify—but be careful! Some problems are products of monomials, some are products of a monomial and a polynomial, some are products of two binomials, and some are polynomials raised to a power.

1. $(5x + 2)(3y + 4)^*$

2. $(7x + 6)(2y + 1)$

3. $(8x - 7)(5y - 3)$

4. $(8y - 3)(y - 4)$

5. $(5x^2 - 2y)(3x + y)$

6. $(2s + 3t)(s^2 - 4t)$

7. $(x + 3)(x - 2)$

8. $(a - 4)(a + 3)$

9. $(y + 8)(y - 9)$

10. $(z - 3)(z + 10)$

11. $(7x^3y)(-2y^2)(4x)$

12. $(-4ab^2)(3a^4)(2b^3)$

13. $2x(3x^2 + 7)$

14. $3y(4y^3 - 2)$

15. $(x + 1)(x + 4)$

16. $(x + 3)(x + 1)$

35. $(b - 4)^2$

36. $(b - 6)^2$

17. $(a + 5)(a + 2)$

18. $(a + 7)(a + 1)$

37. $(3x + 1)(x + 2)$

38. $(2x + 3)(x + 2)$

19. $(5a^2 - b^2)(2a)$

20. $(7c^2 - d^2)(3c)$

39. $(2x + 4)(4x - 3)$

40. $(3x + 5)(2x - 1)$

21. $(x + 8)(x - 8)$

22. $(z + 10)(z - 10)$

41. $(4x - 6)(5x - 2)$

42. $(5x - 4)(3x - 4)$

23. $(m - 4)(m + 2)$

24. $(n - 3)(n + 7)$

43. $(2x + 5)^2$

44. $(3x + 4)^2$

25. $(y + 9)(y - 12)$

26. $(z - 5)(z - 11)$

45. $(3x^2 - 2)(2x + 1)$

46. $(8a^2 + 1)(3a - 2)$

27. $(x + y)(-ab)$

28. $(x - y)(ab)$

47. $(4x - y)(2x + 7y)$

48. $(3x - 2y)(4x + 5y)$

29. $(x + y)(a - b)$

30. $(x - y)(a + b)$

49. $(7x - 10y)(7x - 10y)$

50. $(4u - 9v)(4u - 9v)$

31. $(4x)^2$

32. $(3x)^2$

51. $(3a + 2b)^2$

52. $(2x - 6y)^2$

33. $(4 + x)^2$

34. $(3 + x)^2$

53. $(4c - 3d)(4c + 3d)$

54. $(5e + 2f)(5e - 2f)$

 ## Writing Problems

Express the answer in your own words and in complete sentences.

I. Explain what *FOIL* stands for.

Exercises 6.3A
Set II

Simplify—but be careful! Some problems are products of monomials, some are products of a monomial and a polynomial, some are products of two binomials, and some are polynomials raised to a power.

1. $(6x + 1)(8y + 5)$

2. $(3y - 2)(x + 3)$

3. $(4y - 7)(2x - 6)$

4. $(3x)(x + y)$

5. $(3m^2 + 2n)(m - n)$

6. $(3m + 2n)(m - n)$

7. $(a + 2)(a - 5)$

8. $(a + 2)(a)(-5)$

9. $(x - 4)(x + 7)$

10. $(x^2 - 4)(x + 7)$

11. $(8x)(2y^2)(-3x)$

12. $(8x)(2y^2 - 3x)$

13. $3m(4m^3 - 5)$

14. $3m(4m^3)(-5)$

15. $(x + 2)(x + 3)$

16. $(y - 1)(y - 1)$

17. $(m + 3)(m + 5)$

18. $(m - 3)(m + 5)$

19. $(2x^2 - y^2)(12x)$

20. $(2x - y)^2$

21. $(t + 7)(t - 7)$

22. $(s + 8)(s + 8)$

23. $(h - 6)(h + 3)$

24. $(h + 6)(h - 3)$

25. $(w + 7)(w - 8)$

26. $(w - 7)(w + 8)$

27. $(a + b)(-cd)$

28. $(a + b)(a - b)$

29. $(a + b)(c - d)$

30. $(a + b)^2$

31. $(7x)^2$

32. $(2 + y)^2$

33. $(7 + x)^2$

34. $(2y)^2$

35. $(y - 5)^2$

36. $(-5y)^2$

37. $(5x + 1)(x + 3)$

38. $(2a + 5b)(a + b)$

39. $(3c + 2)(c + 1)$

40. $(4x^2 + 1)(x + 2)$

41. $(y - 5)(y - 5)$

42. $(4x - 7)^2$

43. $(a + 4)^2$

44. $(5x - 3)^2$

45. $(5x^2 - 3)(2x + 3)$

46. $(x - 1)(x^2 + 2)$

47. $(11x + 10y)(3x - 4y)$

48. $(10x - 7y)(8x + 9y)$

49. $(2x - 3y)(2x + 3y)$

50. $(2x - 3y)^2$

51. $(5m + 2n)^2$

52. $(2y + 5z)^2$

53. $(2y + 5z)(2y - 5z)$

54. $(3x + 4)(2x - 5)$

6.3B   Multiplying a Polynomial by a Polynomial

In this section, we discuss multiplying polynomials when one or both polynomials contain more than two terms, and we discuss vertical multiplication of polynomials.

EXAMPLE 10 Multiply $(3x^2 - 4x + 6)(5x - 2)$.

SOLUTION 1 Using the distributive property, we first multiply each term of the factor on the right by $(3x^2 - 4x + 6)$, and we treat $(5x - 2)$ as $[5x + (-2)]$; then we must use the distributive property *again* on the two resulting products.

$$(3x^2 - 4x + 6)[5x + (-2)] = (3x^2 - 4x + 6)(5x) + (3x^2 - 4x + 6)(-2)$$
$$= 15x^3 - 20x^2 + 30x - 6x^2 + 8x - 12$$
$$= 15x^3 - 26x^2 + 38x - 12$$

SOLUTION 2 Using the distributive property, we first multiply each term of the factor on the left by $(5x - 2)$; then we will need to use the distributive property again on the three resulting products.

$$(3x^2 - 4x + 6)(5x - 2) = 3x^2(5x - 2) - 4x(5x - 2) + 6(5x - 2)$$
$$= 15x^3 - 6x^2 - 20x^2 + 8x + 30x - 12$$
$$= 15x^3 - 26x^2 + 38x - 12$$

We can arrange the multiplication vertically if, as we're multiplying, we're careful to arrange like terms in the same vertical column. (This is necessary because we will be *adding* the like terms in each column.) It might help to compare this procedure with the procedure used in arithmetic for multiplying whole numbers.

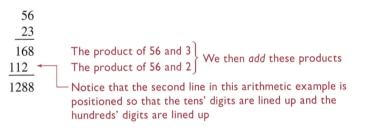

In Example 11, we repeat the problem from Example 10, but we set up the problem vertically. Vertical multiplication of polynomials can be done from right to left (as in arithmetic) *or* from left to right.

EXAMPLE 11 Multiply $(3x^2 - 4x + 6)(5x - 2)$, setting up the problem vertically.

SOLUTION We usually place the polynomial that contains more terms above the other polynomial.

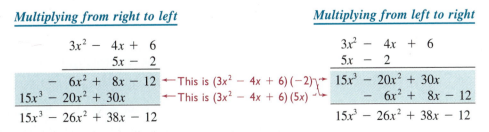

We *add* the polynomials that are shaded (notice that like terms are in the same column); remember that when we *add* like terms, the exponents *do not change*.

EXAMPLE 12

Multiply $(2x^2 - 5x)(3x + 7)$, setting up the problem vertically. (This is the same problem as in Example 9.)

SOLUTION

Multiplying from right to left

$$
\begin{array}{r}
2x^2 - 5x \\
3x + 7 \\
\hline
14x^2 - 35x \\
6x^3 - 15x^2 \\
\hline
6x^3 - x^2 - 35x
\end{array}
$$

← This is $(2x^2 - 5x)(7)$

← This is $(2x^2 - 5x)(3x)$

Multiplying from left to right

$$
\begin{array}{r}
2x^2 - 5x \\
3x + 7 \\
\hline
6x^3 - 15x^2 \\
14x^2 - 35x \\
\hline
6x^3 - x^2 - 35x
\end{array}
$$

EXAMPLE 13

Multiply $(2m + 3m^2 - 5)(2 + m^2 - 3m)$.

SOLUTION The multiplication is simplified by first arranging the polynomials in descending powers of m.

Multiplying from right to left

$$
\begin{array}{r}
3m^2 + 2m - 5 \\
m^2 - 3m + 2 \\
\hline
6m^2 + 4m - 10 \\
-9m^3 - 6m^2 + 15m \\
3m^4 + 2m^3 - 5m^2 \\
\hline
3m^4 - 7m^3 - 5m^2 + 19m - 10
\end{array}
$$

Multiplying from left to right

$$
\begin{array}{r}
3m^2 + 2m - 5 \\
m^2 - 3m + 2 \\
\hline
3m^4 + 2m^3 - 5m^2 \\
-9m^3 - 6m^2 + 15m \\
6m^2 + 4m - 10 \\
\hline
3m^4 - 7m^3 - 5m^2 + 19m - 10
\end{array}
$$

If the multiplication is done horizontally, we have

$$(2m + 3m^2 - 5)(2 + m^2 - 3m)$$
$$= (2m + 3m^2 - 5)(2) + (2m + 3m^2 - 5)(m^2) + (2m + 3m^2 - 5)(-3m)$$
$$= 4m + 6m^2 - 10 + 2m^3 + 3m^4 - 5m^2 + (-6m^2) + (-9m^3) + 15m$$
$$= 3m^4 - 7m^3 - 5m^2 + 19m - 10$$

EXAMPLE 14

Multiply $(a^2 + 3a + 9)(a - 3)$.

SOLUTION

Multiplying from right to left

$$
\begin{array}{r}
a^2 + 3a + 9 \\
a - 3 \\
\hline
-3a^2 - 9a - 27 \\
a^3 + 3a^2 + 9a \\
\hline
a^3 - 27
\end{array}
$$

Multiplying from left to right

$$
\begin{array}{r}
a^2 + 3a + 9 \\
a - 3 \\
\hline
a^3 + 3a^2 + 9a \\
-3a^2 - 9a - 27 \\
\hline
a^3 - 27
\end{array}
$$

If the multiplication is done horizontally, we have

$$(a^2 + 3a + 9)(a - 3)$$
$$= (a^2 + 3a + 9)(a) + (a^2 + 3a + 9)(-3)$$
$$= a^3 + 3a^2 + 9a + (-3a^2) + (-9a) + (-27)$$
$$= a^3 - 27$$

Missing Terms Consider the polynomial $ax^3 + bx + c$. Because there is no x^2 term written, we say that we have a *missing term*; that is, the coefficient of x^2 is an understood zero. When we multiply and divide polynomials, it is usually desirable to write any missing terms with a coefficient of zero. That is, we would write $x^3 + x - 1$ as $x^3 + 0x^2 + x - 1$ (see Example 15).

EXAMPLE 15

Multiply $(x^3 - 1 + x)(2x^2 + 2 - x)$.

SOLUTION We write both polynomials in descending powers of x.

Multiplying from right to left | *Multiplying from left to right*

Note that $0x^2$ was inserted to save a place for the terms that arise in the multiplication

$$
\begin{array}{r}
x^3 + 0x^2 + x - 1 \\
2x^2 - x + 2 \\
\hline
2x^3 + 0x^2 + 2x - 2 \\
- x^4 + 0x^3 - x^2 + x \\
2x^5 + 0x^4 + 2x^3 - 2x^2 \\
\hline
2x^5 - x^4 + 4x^3 - 3x^2 + 3x - 2
\end{array}
$$

$$
\begin{array}{r}
x^3 + 0x^2 + x - 1 \\
2x^2 - x + 2 \\
\hline
2x^5 + 0x^4 + 2x^3 - 2x^2 \\
- x^4 + 0x^3 - x^2 + x \\
2x^3 + 0x^2 + 2x - 2 \\
\hline
2x^5 - x^4 + 4x^3 - 3x^2 + 3x - 2
\end{array}
$$

If the multiplication is done horizontally, it is not necessary to make any allowance for missing terms or to write the polynomials in descending powers.

$$(x^3 - 1 + x)(2x^2 + 2 - x)$$

$$= (x^3 - 1 + x)(2x^2) + (x^3 - 1 + x)(2) + (x^3 - 1 + x)(-x)$$

$$= 2x^5 - 2x^2 + 2x^3 + 2x^3 - 2 + 2x - x^4 + x - x^2$$

$$= 2x^5 - x^4 + 4x^3 - 3x^2 + 3x - 2$$

Powers of Polynomials

We can raise any polynomial to any power by using repeated multiplication. [In Example 8, we found $(3x - 8y)^2$ by using repeated multiplication.]

EXAMPLE 16

Simplify $(a - b)^3$.

SOLUTION To simplify $(a - b)^3$, we must remove the grouping symbols; therefore, we must use $a - b$ as a factor three times.

$$(a - b)^3 = (a - b)(a - b)(a - b)$$

First find $(a - b)^2$ Then multiply that product by $(a - b)$

$$
\begin{array}{r}
a - b \\
a - b \\
\hline
- ab + b^2 \\
a^2 - ab \\
\hline
a^2 - 2ab + b^2
\end{array}
$$

$$
\begin{array}{r}
a^2 - 2ab + b^2 \\
a - b \\
\hline
- a^2b + 2ab^2 - b^3 \\
a^3 - 2a^2b + ab^2 \\
\hline
a^3 - 3a^2b + 3ab^2 - b^3
\end{array}
$$

Therefore, $(a - b)^3 = a^3 - 3a^2b + 3ab^2 - b^3$.

> ✕ **A Word of Caution** A common error in squaring a binomial is shown below:
>
> $$(x + 3)^2 = x^2 + 9$$
>
> However,
>
> $$(x + 3)^2 \neq x^2 + 3^2$$
>
>   ↑
>   └──── *x* and 3 are *terms* of *x* + 3
>
> To verify that $(x + 3)^2 \neq x^2 + 3^2$, let's let $x = 4$. Then $(x + 3)^2 = (4 + 3)^2 = 7^2 = 49$, but $x^2 + 3^2 = 4^2 + 3^2 = 16 + 9 = 25$, and $25 \neq 49$. *The square of a sum does not equal the sum of the squares.*
>
> **Correct method**
> _____
>
> $$(x + 3)^2 = (x + 3)(x + 3)$$
> $$= (x + 3)x + (x + 3)(3) \quad \text{Using the distributive property}$$
> $$= x^2 + 3x + 3x + 9 \quad \text{Using the distributive property again}$$
> $$= x^2 + 6x + 9 \quad \text{Combining like terms}$$
>
>   ↑
>   └ This is the term that is often left out
>
> Again, if *x* is 4, $x^2 + 6x + 9 = 4^2 + 6(4) + 9 = 16 + 24 + 9 = 49$.
>
> It *is* true that the square of a *product* equals the product of the squares; that is,
>
> $$(3x)^2 = 3^2 x^2 \quad \text{Using the rule for raising a product to a power}$$
>
>   ↑
> └──── 3 and *x* are factors of 3*x*

In a later section, we will show a special method for squaring a binomial.

Exercises   6.3B
Set I
_____

Simplify each expression.

1. $(x - 3)(2x^2 + x - 1)$

2. $(x - 2)(3x^2 + x - 1)$

3. $(x^2 + x + 1)(x^2 + x + 1)$

4. $(x^2 - x - 1)(x^2 - x - 1)$

5. $(4z)(z^2 - 4z + 16)$

6. $(-5a)(a^2 + 5a + 25)$

7. $(z + 4)(z^2 - 4z + 16)$

8. $(a - 5)(a^2 + 5a + 25)$

9. $(4 - 3z^3 + z^2 - 5z)(4 - z)$

10. $(3 + 2v^2 - v^3 + 4v)(2 - v)$

11. $(-3x^2y + xy^2 - 4y^3)(-2xy)$

12. $(-4xy^2 - x^2y + 3x^3)(-3xy)$

13. $(2x^2 - x + 4)(x^2 - 5x - 3)$

14. $(6a^2 + 2a - 3)(2a^2 - 3a + 5)$

15. $(7 - 2y + 3y^2)(8y + 4y^2 - 3)$

16. $(x - 4 + 6x^2)(9 - 3x + x^2)$

17. $(5x - 2)^2$

18. $(2x - 5)^2$

19. $(x + y)^2(x - y)^2$

20. $(x - 2)^2(x + 2)^2$

21. $(x + 2)^3$

22. $(x + 3)^3$

23. $(x^2 + 2x - 3)^2$

24. $(y^2 - 4y - 5)^2$

25. $(x + 3)^4$

26. $(x + 2)^4$

Writing Problems

Express the answers in your own words and in complete sentences.

1. Explain why $(x + 5)^2 \neq x^2 + 5^2$.

2. Explain why $(x + 8)^3 \neq x^3 + 8^3$.

Exercises 6.3B
Set II

Simplify each expression.

1. $(x - 1)(5x^2 + x - 1)$

2. $(x + 1)(2x^2 + x + 1)$

3. $(x^2 + 2x + 1)(x^2 + 2x + 1)$

4. $(x^2 - 2x + 1)(x^2 - 2x - 1)$

5. $(3x)(x^2 - 3x + 9)$

6. $(-2x)(4 + 2x + x^2)$

7. $(3 + x)(x^2 - 3x + 9)$

8. $(2 - x)(4 + 2x + x^2)$

9. $(2 - 2x^3 + x^2 - 3x)(3 - x)$

10. $(x + 3x^3 + 4)(5 - x)$

11. $(3xy^2 - 5x^2y + 4)(-2x^2y)$

12. $(a^3 - 3a^2b + 3ab^2 - b^3)(-5a^2b)$

13. $(7 + 2x^3 + 3x)(4 - x)$

14. $(4 + a^4 + 3a^2 - 2a)(a + 3)$

15. $(5y^2 - 2y - 6)(2y^2 - 4y + 3)$

16. $(2 + 3x^2 - 4x)(6x - 7 + 2x^2)$

17. $(3x + 4)^2$

18. $(x^2 - 2x + 1)(x^2 - 2x + 1)$

19. $(x - 1)^2(x + 1)^2$

20. $(y + 2)^2(y - 2)^2$

21. $(x + 1)^3$

22. $(2w - 3)^3$

23. $(x^2 + 4x + 4)^2$

24. $(z^2 - 3z - 4)^2$

25. $(x + 1)^4$

26. $(x - 1)^4$

6.4 Special Products

Some products are especially important and have special formulas. You *must* learn these formulas and how to use them so that you will be able to do the factoring problems that are so important in all of higher mathematics.

6.4A The Product of the Sum and Difference of Two Terms

Because $a + b$ is the *sum of two terms* and $a - b$ is the *difference of two terms*, we call the product $(a + b)(a - b)$ the *product of the sum and difference of two terms*.

PROPERTY 1

$$(a + b)(a - b) = a^2 - b^2$$

In words: The product of the sum and difference of two terms equals the square of the first term minus the square of the second term.

PROOF $(a + b)(a - b) = (a + b)[a + (-b)]$

$$= (a + b)(a) + (a + b)(-b) \quad \text{Using the distributive property}$$

$$= a^2 + ba - ab - b^2 \quad \text{Then } ba - ab = ab - ab = 0$$

$$= a^2 - b^2 \quad \blacksquare$$

When we use Property 1 to find a product, we say that we are finding the product *by inspection*.

EXAMPLE 1   Examples of finding the product of the sum and difference of two terms:

a. $(x + 2)(x - 2) = (x)^2 - (2)^2 = x^2 - 4$
b. $(2x + 3y)(2x - 3y) = (2x)^2 - (3y)^2 = 4x^2 - 9y^2$
c. $(10x^2 - 7y^3)(10x^2 + 7y^3) = (10x^2)^2 - (7y^3)^2 = 100x^4 - 49y^6$
d. $(5a^3b^2 + 6cd^4)(5a^3b^2 - 6cd^4) = (5a^3b^2)^2 - (6cd^4)^2 = 25a^6b^4 - 36c^2d^8$

Notice in Example 1 that (1) all the answers have two terms, (2) all the answers contain a minus sign, and (3) the absolute values of both terms in each of the answers are perfect squares.

Exercises  6.4A
Set I

Find the products by inspection.

1. $(x + 3)(x - 3)$
2. $(z + 4)(z - 4)$
3. $(w - 6)(w + 6)$
4. $(y - 5)(y + 5)$
5. $(5a + 4)(5a - 4)$
6. $(6a - 5)(6a + 5)$
7. $(2u + 5v)(2u - 5v)$
8. $(3m - 7n)(3m + 7n)$
9. $(4b - 9c)(4b + 9c)$
10. $(7a - 8b)(7a + 8b)$
11. $(2x^2 - 9)(2x^2 + 9)$
12. $(10y^2 - 3)(10y^2 + 3)$
13. $(1 + 8z^3)(1 - 8z^3)$
14. $(9v^4 - 1)(9v^4 + 1)$
15. $(5xy + z)(5xy - z)$
16. $(10ab + c)(10ab - c)$
17. $(7mn + 2rs)(7mn - 2rs)$
18. $(8hk + 5ef)(8hk - 5ef)$

Exercises 6.4A
Set II

Find the products by inspection.

1. $(h - 7)(h + 7)$
2. $(u + 9)(u - 9)$
3. $(3m + 5)(3m - 5)$
4. $(2x - 3y)(2x + 3y)$
5. $(6a - 7b)(6a + 7b)$
6. $(3x + 7)(3x - 7)$
7. $(8x - 2y)(8x + 2y)$
8. $(5x - 12y)(5x + 12y)$
9. $(7a - 9b)(7a + 9b)$
10. $(13c + d)(13c - d)$
11. $(4h^2 - 5)(4h^2 + 5)$
12. $(8x^2 + 1)(8x^2 - 1)$
13. $(1 + 9k^3)(1 - 9k^3)$
14. $(3x^2y - 8xy^2)(3x^2y + 8xy^2)$
15. $(4w + 3xy)(4w - 3xy)$
16. $(9ab - 2c)(9ab + 2c)$
17. $(5uv - 8ef)(5uv + 8ef)$
18. $(9st + 4z^2)(9st - 4z^2)$

6.4B The Square of a Binomial

Properties 2 and 3 help us quickly square a binomial (see Example 2).

PROPERTY 2
The square of a binomial sum

$$(a + b)^2 = a^2 + 2ab + b^2$$

PROOF
$$
\begin{aligned}
(a + b)^2 &= (a + b)\,(a + b) \\
&= (a + b)\,(a) + (a + b)\,(b) \\
&= a^2 + ba + ab + b^2 \qquad \text{Then } ba + ab = ab + ab = 2ab \\
&= a^2 + 2ab + b^2 \quad \blacksquare
\end{aligned}
$$

PROPERTY 3
The square of a binomial difference

$$(a - b)^2 = a^2 - 2ab + b^2$$

PROOF
$$
\begin{aligned}
(a - b)^2 &= (a - b)\,(a - b) \\
&= (a - b)\,[a + (-b)] \\
&= (a - b)\,(a) + (a - b)\,(-b) \\
&= a^2 - ba - ab + b^2 \qquad \text{Then } -ba - ab = -ab - ab = -2ab \\
&= a^2 - 2ab + b^2 \quad \blacksquare
\end{aligned}
$$

Properties 2 and 3 can be combined and stated in words as follows:

Squaring a binomial

1. The *first term* of the result is the square of the first term of the binomial.
2. The *middle term* of the result is *twice* the product of the two terms of the binomial.
3. The *last term* of the result is the square of the last term of the binomial.

We can use Property 3 to square a binomial difference, or we can treat the difference like a sum (see Examples 2b and 2c).

Although the square of a binomial *can* be found by using the methods learned in the section on multiplying binomials, you are strongly urged to use Property 2 or Property 3 in solving such problems.

EXAMPLE 2 Examples of squaring binomials:

a. $(m + n)^2 = (m)^2 + 2(m)\,(n) + (n)^2 = m^2 + 2mn + n^2$ Using Property 2

b. $(a - 3)^2 = (a)^2 - 2(a)\,(3) + 3^2 = a^2 - 6a + 9$ Using Property 3

   or $(a - 3)^2 = [a + (-3)]^2 = (a)^2 + 2(a)\,(-3) + (-3)^2 = a^2 - 6a + 9$

   └── Treating the difference like a sum and using Property 2

c. $(2x - 5)^2 = (2x)^2 - 2(2x)\,(5) + 5^2 = 4x^2 - 20x + 25$ Using Property 3

   or $(2x - 5)^2 = [2x + (-5)]^2 = (2x)^2 + 2(2x)\,(-5) + (-5)^2 = 4x^2 - 20x + 25$

   └── Treating the difference like a sum and using Property 2

> **A Word of Caution**   A common error in squaring a sum of terms is to confuse squaring a sum of terms with squaring a product of factors.
> It *is* true that $(ab)^2 = a^2 b^2$ (note that a and b are *factors* of ab). However, $(a + b)^2 \neq a^2 + b^2$; that is, $(a + b)^2$ cannot be found by finding the sum of the squares of a and b.
>
> | *Correct method* | *Incorrect method* |
> |---|---|
> | $(a + b)^2 = (a + b)(a + b)$ | $(a + b)^2 = a^2 + b^2$ |
> | $\quad\quad\quad = a^2 + 2ab + b^2$ | a and b are *terms* of $a + b$ |
>
> When squaring a binomial, do not forget this middle term

When we square a *monomial*, the answer has one term; when we square a *binomial*, the answer has *three* terms.

Exercises   6.4B
Set I

Simplify each expression.

1. $(x - 1)^2$ **2.** $(x - 5)^2$ **3.** $(x + 3)^2$

4. $(x + 4)^2$ **5.** $(4x - 1)^2$ **6.** $(7x - 1)^2$

7. $(12x + 1)^2$ **8.** $(11x + 1)^2$ **9.** $(2s + 4t)^2$

10. $(3u + 7v)^2$ **11.** $(5x - 3y)^2$ **12.** $(4x - 7y)^2$

13. $(3x + 2z)^2$ **14.** $(2x + 7s)^2$ **15.** $(7x - 8y)^2$

16. $(9x - 7y)^2$

Writing Problems

Express the answer in your own words and in complete sentences.

1. Explain how you would simplify $(2x + 5)^2$.

Exercises   6.4B
Set II

Simplify each expression.

1. $(x + 1)^2$ **2.** $(x - 12)^2$ **3.** $(x + 15)^2$

4. $(x - 9)^2$ **5.** $(9x - 1)^2$ **6.** $(8x + 1)^2$

7. $(15y + 1)^2$ **8.** $(10z - 2)^2$ **9.** $(6x + 3y)^2$

10. $(5x + 12u)^2$ **11.** $(10x - 3t)^2$ **12.** $(12x + 5y)^2$

13. $(11x + 2y)^2$ **14.** $(6x - 5a)^2$ **15.** $(9x - 5z)^2$

16. $(3s - 10t)^2$

6.5 Dividing Polynomials

6.5A Dividing a Polynomial by a Monomial

Dividing a Monomial by a Monomial

When we divided x^5 by x^3 and x^6 by x^4 earlier, we were really dividing a monomial by a monomial, and we did those divisions by using the quotient rule.

If the monomials have numerical coefficients and/or contain more than one variable, we can use the definition of multiplication in reverse $\left(\dfrac{ac}{bd} = \dfrac{a}{b} \cdot \dfrac{c}{d}\right)$ and rewrite the problem as a *product* of fractions. Then we can reduce any numerical coefficients and simplify any fractions that contain variables by using the quotient rule (see Example 1).

EXAMPLE 1 Examples of dividing a monomial by a monomial:

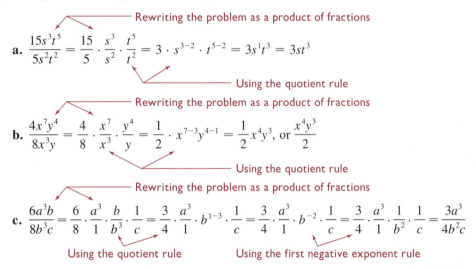

Note You do not need to show all the steps that we have shown in Example 1.

Note The divisions in Example 1 could have been done by using the fundamental property of fractions. For example,

$$\frac{15s^3t^5}{5s^2t^2} = \frac{3st^3(5s^2t^2)}{1(5s^2t^2)} = \frac{3st^3}{1} = 3st^3, \quad \frac{4x^7y^4}{8x^3y} = \frac{x^4y^3(4x^3y)}{2(4x^3y)} = \frac{x^4y^3}{2},$$

and

$$\frac{6a^3b}{8b^3c} = \frac{3a^2(2b)}{4b^2c(2b)} = \frac{3a^3}{4b^2c}$$

In a later chapter, we will discuss still other methods of reducing fractions.

Dividing a Polynomial by a Monomial

Next, we consider division problems in which the *dividend* has more than one term and the *divisor* has only one term. That is, we consider problems in which the *divisor* is a monomial. (Recall that in a division problem, the *dividend* is the number we're dividing *into* and the *divisor* is the number we're dividing *by*.)

We can change a division problem that is in the form $\dfrac{a+b}{c}$ to the form $\dfrac{a}{c} + \dfrac{b}{c}$. Let's verify this with an example from arithmetic:

$$\frac{6+8}{2} = \frac{14}{2} = 7 \qquad \bigg| \qquad \frac{6}{2} + \frac{8}{2} = 3 + 4 = 7$$

We see that we obtain the same result (7) for both problems; therefore, $\dfrac{6+8}{2} = \dfrac{6}{2} + \dfrac{8}{2}$. The following rule for dividing a polynomial by a monomial is based on the fact that we can write $\dfrac{a+b}{c}$ as $\dfrac{a}{c} + \dfrac{b}{c}$ and $\dfrac{a-b}{c}$ as $\dfrac{a}{c} - \dfrac{b}{c}$.

Dividing a polynomial by a monomial

> Divide *each* term of the polynomial by the monomial, and simplify each term. The quotient is the sum of the simplified terms.

PROOF THAT $\dfrac{a+b}{c} = \dfrac{a}{c} + \dfrac{b}{c}$

$$\frac{a+b}{c} = (a+b)\left(\frac{1}{c}\right) \qquad \text{Writing the division problem as a multiplication problem}$$

$$= a\left(\frac{1}{c}\right) + b\left(\frac{1}{c}\right) \qquad \text{Using the distributive property}$$

$$= \frac{a}{c} + \frac{b}{c}$$

(The proof that $\dfrac{a-b}{c} = \dfrac{a}{c} - \dfrac{b}{c}$ is similar.) ∎

EXAMPLE 2 Examples of dividing a polynomial by a monomial:

a. $\dfrac{4x+2}{2} = \dfrac{4x}{2} + \dfrac{2}{2} = \dfrac{4}{2} \cdot \dfrac{x}{1} + 1 = 2x + 1$

b. $\dfrac{4x^3 - 6x^2}{2x} = \dfrac{4x^3}{2x} - \dfrac{6x^2}{2x} = \dfrac{4}{2} \cdot \dfrac{x^3}{x} - \dfrac{6}{2} \cdot \dfrac{x^2}{x} = 2x^{3-1} - 3x^{2-1} = 2x^2 - 3x$

c. $\dfrac{9x^3 - 6x^2 + 12x}{3x} = \dfrac{9x^3}{3x} - \dfrac{6x^2}{3x} + \dfrac{12x}{3x} = \dfrac{9}{3} \cdot \dfrac{x^3}{x} - \dfrac{6}{3} \cdot \dfrac{x^2}{x} + \dfrac{12}{3} \cdot \dfrac{x}{x}$

$$= 3x^{3-1} - 2x^{2-1} + 4x^{1-1} = 3x^2 - 2x^1 + 4x^0$$

$$= 3x^2 - 2x + 4(1) = 3x^2 - 2x + 4$$

d. $\dfrac{4x^4 - 8x^3 + 16}{-4x} = \dfrac{4x^4 + (-8x^3) + 16}{-4x} = \dfrac{4x^4}{-4x} + \dfrac{-8x^3}{-4x} + \dfrac{16}{-4x}$

$$= \frac{4}{-4} \cdot \frac{x^4}{x} + \frac{-8}{-4} \cdot \frac{x^3}{x} + \frac{16}{-4} \cdot \frac{1}{x}$$

$$= -1 \cdot x^{4-1} + 2 \cdot x^{3-1} + (-4)\left(\frac{1}{x}\right)$$

$$= -x^3 + 2x^2 - \frac{4}{x}$$

e. $\dfrac{15x^4y^2 + 20y^3z - 10xz^2}{5xyz} = \dfrac{15x^4y^2 + 20y^3z + (-10xz^2)}{5xyz}$

$$= \frac{15x^4y^2}{5xyz} + \frac{20y^3z}{5xyz} + \frac{-10xz^2}{5xyz}$$

$$= \frac{15}{5} \cdot \frac{x^4}{x} \cdot \frac{y^2}{y} \cdot \frac{1}{z} + \frac{20}{5} \cdot \frac{1}{x} \cdot \frac{y^3}{y} \cdot \frac{z}{z} + \frac{-10}{5} \cdot \frac{x}{x} \cdot \frac{1}{y} \cdot \frac{z^2}{z}$$

$$= 3 \cdot x^{4-1} \cdot y^{2-1} \cdot \frac{1}{z} + 4 \cdot \frac{1}{x} \cdot y^{3-1} \cdot z^{1-1}$$

$$- 2 \cdot x^{1-1} \cdot \frac{1}{y} \cdot z^{2-1}$$

$$= 3x^3y^1 \cdot \frac{1}{z} + 4 \cdot \frac{1}{x} \cdot y^2z^0 - 2x^0 \cdot \frac{1}{y} \cdot z^1$$

$$= \frac{3x^3y}{z} + \frac{4y^2}{x} - \frac{2z}{y}$$

f. $\dfrac{4a^2bc^2 - 6ab^2c^2 + 12bc}{-6abc} = \dfrac{4a^2bc^2 + (-6ab^2c^2) + 12bc}{-6abc}$

$$= \frac{4a^2bc^2}{-6abc} + \frac{-6ab^2c^2}{-6abc} + \frac{12bc}{-6abc}$$

$$= \frac{4}{-6} \cdot \frac{a^2}{a} \cdot \frac{b}{b} \cdot \frac{c^2}{c} + \frac{-6}{-6} \cdot \frac{a}{a} \cdot \frac{b^2}{b} \cdot \frac{c^2}{c} + \frac{12}{-6} \cdot \frac{1}{a} \cdot \frac{b}{b} \cdot \frac{c}{c}$$

$$= -\frac{2}{3} \cdot a^{2-1} \cdot 1 \cdot c^{2-1} + 1 \cdot 1 \cdot b^{2-1} \cdot c^{2-1}$$

$$+ (-2) \cdot \frac{1}{a} \cdot 1 \cdot 1$$

$$= -\frac{2}{3}a^1c^1 + b^1c^1 - 2 \cdot \frac{1}{a} = -\frac{2}{3}ac + bc - \frac{2}{a}$$

☞ **Note** You do not need to show all the steps that we have shown in Example 2.

Exercises 6.5A
Set I

Perform the indicated divisions; express all results in simplest form.

1. $\dfrac{6x^5}{3x^2}$

2. $\dfrac{8y^8}{2y^4}$

3. $\dfrac{4a^2bc^2}{12ac}$

4. $\dfrac{5x^3yz^2}{15xz}$

5. $\dfrac{-5x^3y^2}{-10x^2y}$

6. $\dfrac{-7s^2t^5}{-21st^3}$

7. $\dfrac{8u^5v^2w}{-12uv^3w^4}$

8. $\dfrac{14c^4d^3e^2}{-21c^2de^5}$

9. $\dfrac{3x + 6}{3}$

10. $\dfrac{10x + 15}{5}$

11. $\dfrac{4 + 8x}{4}$

12. $\dfrac{5 - 10x}{5}$

13. $\dfrac{6x - 8y}{2}$

14. $\dfrac{5x - 10y}{5}$

15. $\dfrac{2x^2 + 3x}{x}$

16. $\dfrac{4y^2 - 3y}{y}$

17. $\dfrac{15x^3 - 5x^2}{5x^2}$

18. $\dfrac{12y^4 - 6y^2}{6y^2}$

19. $\dfrac{3a^2b - ab}{ab}$ **20.** $\dfrac{5mn^2 - mn}{mn}$ **21.** $\dfrac{8x^7 + 4x^5 - 12x^3}{4x^2}$

28. $\dfrac{21m^2n^5 - 35m^3n^2 - 14m^2n^2}{7m^2n^2}$

22. $\dfrac{6y^6 + 18y^4 - 12y^3}{6y^2}$ **23.** $\dfrac{5x^5 - 4x^3 + 10x^2}{-5x^2}$

29. $\dfrac{5x^3 - 8x^2 + 3}{-5x^2}$ **30.** $\dfrac{7y^3 - 9y^2 + 8}{-7y^2}$

24. $\dfrac{7y^4 - 5y^3 + 14y^2}{-7y^2}$ **25.** $\dfrac{-15x^2y^2z^2 - 30xyz}{-5xyz}$

31. $\dfrac{15x^3y^2 - 30xy^3 + 20xy}{15x^2y^2}$ **32.** $\dfrac{28m^2n^3 - 35m^3n^2 - 14mn}{14m^2n^2}$

26. $\dfrac{-24a^2b^2c^2 - 16abc}{-8abc}$ **27.** $\dfrac{13x^4y^2 - 26x^2y^3 + 39x^2y^2}{13x^2y^2}$

Writing Problems

Express the answers in your own words and in complete sentences.

1. Find and describe the error in $\dfrac{\overset{2}{\cancel{4}} + 3x}{\underset{1}{\cancel{2}}} = 2 + 3x$.

2. Find and describe the error in $\dfrac{6z + \overset{1}{\cancel{5}}}{\underset{1}{\cancel{5}}} = 6z + 1$.

Exercises   6.5A
Set II

Perform the indicated divisions; express all results in simplest form.

1. $\dfrac{10b^6}{5b^2}$ **2.** $\dfrac{5d^7}{2d^4}$ **3.** $\dfrac{3xy^3z^2}{12xy}$

21. $\dfrac{42w^8 + 14w^4 - 28w^3}{7w^2}$ **22.** $\dfrac{15x^7 + 20x^4 - 35x^3}{5x^3}$

4. $\dfrac{3a^2bc^3}{15ac}$ **5.** $\dfrac{-6x^3y^3}{-12xy^2}$ **6.** $\dfrac{-14a^3b^5}{-21ab^2}$

23. $\dfrac{8x^4 - 4x^2 + 12x^3}{-8x^2}$ **24.** $\dfrac{9y^2 - 3y^3 + 18y^5}{-9y^2}$

25. $\dfrac{-16a^3b^2c^3 - 32abc}{-16abc}$ **26.** $\dfrac{-18s^2t^3u^2 - 9stu^2}{-9stu}$

7. $\dfrac{9s^4t^2u}{-12st^3u^4}$ **8.** $\dfrac{-35a^4b^3c^2}{21a^2bc^4}$ **9.** $\dfrac{9x + 12}{3}$

27. $\dfrac{12a^5b^3 - 24a^3b^3 + 48a^2b^2}{12a^2b^2}$

10. $\dfrac{15y + 20}{5}$ **11.** $\dfrac{10 + 20x}{10}$ **12.** $\dfrac{8 - 10y}{8}$

28. $\dfrac{15x^5y^2 - 25x^2y^3 - 10x^2y^2}{15x^2y^2}$

29. $\dfrac{6x^3 - 8x^2 + 3}{-6x^2}$

13. $\dfrac{5x - 15y}{5}$ **14.** $\dfrac{4x - 6}{2}$ **15.** $\dfrac{3x^2 - 6x}{3x}$

30. $\dfrac{9y^4 - 9y^3 + 6}{6y^2}$

16. $\dfrac{8x^3 - 10x}{4x}$ **17.** $\dfrac{8x^3 - 4x^2}{4x^2}$ **18.** $\dfrac{15x^3 - 30x}{15x}$

31. $\dfrac{13x^3y^2 - 26xy^3 + 39xy}{13x^2y^2}$

19. $\dfrac{6ab^2 - ab}{ab}$ **20.** $\dfrac{12xy^2 - 6x^2y}{9xy}$

32. $\dfrac{8a^3b^2c - 4a^2bc - 10ac}{4abc}$

6.5B Dividing a Polynomial by a Polynomial

The method used to divide one polynomial (the *dividend*) by a polynomial (the *divisor*) with two or more terms is similar to the method used to divide one whole number by another (using long division) in arithmetic. In fact, it may help to review a long division problem from arithmetic. Let's consider the problem $684 \div 5$.

> We first divide 6 by 5; this gives a quotient of 1, and this 1 is written directly above the 6

$$
\begin{array}{r}
136 \ \text{R} \ 4 \\
5\overline{)684} \\
\underline{5} \\
18 \\
\underline{15} \\
34 \\
\underline{30} \\
4
\end{array}
$$

This is 1×5; we next subtract 5 from 6, getting 1, and bring down the 8 from the dividend

We next divide 18 by 5, getting 3; this 3 is written directly above the 8
This is 3×5; we next subtract 15 from 18, getting 3, and bring down 4

We next divide 34 by 5, getting 6; this 6 is written directly above the 4
This is 6×5; we next subtract 30 from 34, getting 4

The remainder is 4

The long-division algorithm* for dividing polynomials will be demonstrated step by step; it can be summarized as follows. (Study hint: As you read the steps in the summary, follow them in Example 3.)

| **Dividing one polynomial by another** | 1. Arrange the divisor and the dividend in descending powers of one variable. In the *dividend*, write any *missing terms* with a coefficient of zero. |

Dividing one polynomial by another

  1. Arrange the divisor and the dividend in descending powers of one variable. In the *dividend*, write any *missing terms* with a coefficient of zero.

  2. Find the first term of the quotient by dividing the first term of the dividend by the first term of the divisor.

  3. Multiply the *entire* divisor by the first term of the quotient. Place this product under the dividend, lining up like terms.

  4. Subtract the product found in step 3 from the dividend, and bring down at least one term from the dividend; we'll call these terms the remainder. If the degree of the remainder is greater than or equal to the degree of the divisor, continue with steps 5 through 8.

  5. Find the next term of the quotient by dividing the first term of the remainder by the first term of the divisor.

  6. Multiply the entire divisor by the term found in step 5, placing this product under the remainder found in step 4 and lining up like terms.

  7. Subtract the product found in step 6 from the polynomial above it, bringing down at least one more term from the dividend.

  8. Repeat steps 5 through 7 until the remainder is 0 *or* until the degree of the remainder is less than the degree of the divisor.

  9. Check your answer: Divisor $\times$ quotient $+$ remainder $=$ dividend.

 Note   In steps 4 and 7, it is acceptable to "bring down" *all* the remaining terms of the dividend (after subtracting) rather than just one term.

When the remainder is not zero, it can be preceded by an R (for Remainder) and written to the right of the quotient. An alternative method is to write the remainder in fraction form; in this case, the *numerator* of the fraction is the *remainder*, the *denominator* of the fraction is the *divisor*, and this fraction must be *added to* the quotient.

*An *algorithm* is a special method or procedure for solving a certain kind of problem.

EXAMPLE 3   Divide $(5x^2 + 6x^3 - 5 - 4x)$ by $(3 + 2x)$.

SOLUTION   The degree of the divisor is 1.

Step 1. We first arrange the divisor and the dividend in descending powers of x.

$$2x + 3 \overline{)6x^3 + 5x^2 - 4x - 5}$$

Step 2. To find the first term of the quotient, we divide the first term of the dividend by the first term of the divisor.

$$\begin{array}{r} 3x^2 \\ 2x + 3 \overline{)6x^3 + 5x^2 - 4x - 5}\end{array}$$

The first term of the quotient is $\dfrac{6x^3}{2x} = \dfrac{6}{2} \cdot \dfrac{x^3}{x} = 3x^2$

Step 3. We multiply the *entire* divisor by the first term of the quotient.

$$\begin{array}{r} 3x^2 \\ 2x + 3 \overline{)6x^3 + 5x^2 - 4x - 5} \\ 6x^3 + 9x^2 \end{array}$$

Notice that like terms are lined up vertically
This is $3x^2(2x + 3)$

Step 4. We subtract the product found in step 3 from the dividend *and* we bring down a term of the dividend.

$$\begin{array}{r} 3x^2 \\ 2x + 3 \overline{)6x^3 + 5x^2 - 4x - 5} \\ \ominus \quad \ominus \\ 6x^3 + 9x^2 \\ \hline -4x^2 - 4x \end{array}$$

We subtract (change signs and add) and bring down a term of the dividend
The remainder → $-4x^2 - 4x$ ← The degree of the remainder is 2

The division must be continued, because the degree of the remainder is greater than the degree of the divisor.

Step 5. To find the second term of the quotient, we divide the first term of the remainder by the first term of the divisor.

$$\begin{array}{r} 3x^2 - 2x \\ 2x + 3 \overline{)6x^3 + 5x^2 - 4x - 5} \\ \ominus \quad \ominus \\ 6x^3 + 9x^2 \\ \hline -4x^2 - 4x \end{array}$$

The second term of the quotient is $\dfrac{-4x^2}{2x} = \dfrac{-4}{2} \cdot \dfrac{x^2}{x} = -2x$

Step 6. We multiply the entire divisor by the *second term* of the quotient.

$$\begin{array}{r} 3x^2 - 2x \\ 2x + 3 \overline{)6x^3 + 5x^2 - 4x - 5} \\ \ominus \quad \ominus \\ 6x^3 + 9x^2 \\ \hline -4x^2 - 4x \\ -4x^2 - 6x \end{array}$$

Notice that like terms are lined up vertically
This is $(-2x)(2x + 3)$

Step 7. We subtract the product found in step 6 from the polynomial above it, also bringing down one term of the dividend.

$$\begin{array}{r} 3x^2 - 2x \\ 2x + 3 \overline{)6x^3 + 5x^2 - 4x - 5} \\ \ominus \quad \ominus \\ 6x^3 + 9x^2 \\ \hline -4x^2 - 4x \\ \oplus \quad \oplus \\ -4x^2 - 6x \\ \hline 2x - 5 \end{array}$$

We will subtract and bring down the next term of the dividend
The remainder → $2x - 5$ The degree is 1

Step 8. The degree of the remainder *equals* the degree of the divisor; therefore, we must repeat steps 5, 6, and 7.

The third term of the quotient is $\dfrac{2x}{2x} = 1$

$$
\begin{array}{r}
3x^2 - 2x + 1 \\
2x + 3 \overline{)\ 6x^3 + 5x^2 - 4x - 5} \\
\ominus \quad \ominus \\
6x^3 + 9x^2 \\
-4x^2 - 4x \\
\oplus \quad \oplus \\
-4x^2 - 6x \\
2x - 5 \\
\ominus \quad \ominus \\
2x + 3 \quad \longleftarrow \text{This is } 1(2x + 3) \\
\text{The remainder} \longrightarrow -8 \longleftarrow \text{The degree is 0}
\end{array}
$$

The division is now finished, because the degree of the remainder is less than the degree of the divisor.

✓ **Step 9. Check**   (Divisor) × (Quotient) + (Remainder) = Dividend

$$(2x + 3)(3x^2 - 2x + 1) + (-8) \overset{?}{=} 6x^3 + 5x^2 - 4x - 5$$

$$(6x^3 + 5x^2 - 4x + 3) + (-8) \overset{?}{=} 6x^3 + 5x^2 - 4x - 5$$

$$6x^3 + 5x^2 - 4x - 5 = 6x^3 + 5x^2 - 4x - 5$$

Answer $3x^2 - 2x + 1 \text{ R } -8$, *or* $3x^2 - 2x + 1 + \dfrac{-8}{2x + 3}$, *or*

$$3x^2 - 2x + 1 - \dfrac{8}{2x + 3}$$

Note   It is important to check the answers to division problems; most errors in division seem to occur in the *subtraction* stage of the division.

EXAMPLE 4   Divide $(-3x + x^2 - 10)$ by $(2 + x)$.

SOLUTION

Step 1. We first arrange the divisor and the dividend in descending powers of x.

$$x + 2 \overline{)x^2 - 3x - 10}$$

The first term of the quotient is $\dfrac{x^2}{x} = x$

Step 2.
$$
\begin{array}{r}
x \\
x + 2 \overline{)\ x^2 - 3x - 10}
\end{array}
$$

Step 3.
$$
\begin{array}{r}
x \\
x + 2 \overline{)x^2 - 3x - 10} \\
x^2 + 2x
\end{array}
$$
Multiplying

This is $x(x + 2)$

Step 4.
$$
\begin{array}{r}
x \\
x + 2 \overline{)\ x^2 - 3x - 10} \\
\ominus \quad \ominus \\
x^2 + 2x \\
-5x - 10
\end{array}
$$
Changing signs and adding

Bringing down the next term

In Exercises 31 and 32, solve each inequality.

31. $6x - 5 < 13$ **32.** $5(3 + 5x) \leq 30x - 5$

In Exercises 33–36, set up each problem algebraically, solve, and check. Be sure to state what your variables represent.

33. Susan worked twenty-two problems correctly on a math test that had twenty-five problems. Find her percent score.

34. A business paid $125 for an item. What was the selling price of the item if it was then marked up 40%?

35. A 10-lb mixture of walnuts and almonds is to be worth $3.52 per pound. If walnuts cost $4.50 per pound and almonds cost $3.10 per pound, how many pounds of each should be used?

36. The three sides of a triangle are in the ratio $3:5:7$. The perimeter is 75. Find the lengths of the three sides.

Critical Thinking and Problem-Solving Exercises

1. Opal arranged three coins into a small triangle. Then she formed a second (equilateral) triangle by adding one more row of coins, using three additional coins in the third row, so that there were six coins used altogether. She formed a third triangle by adding another row, using a total of ten coins. If she keeps on in this way, how many coins will she add for the last row of the fourth triangle, and how many coins will be used altogether? How many coins will be in the last row of the sixth triangle?

2. At noon, Lori finished a batch of cookies for her grandchildren; by 2 P.M., half of the cookies had been eaten; by 3 P.M., one-third of the remaining cookies had been eaten; by 4 P.M., one-fourth of the remaining cookies were gone; by 5 P.M., one-fifth of the remaining cookies had disappeared; by 6 P.M., one-sixth of the remaining cookies were gone and there were only ten cookies left. How many cookies did Lori bake?

3. The solution of each of the following problems contains an error. Find and describe each error.

    a. Simplify $(x + 4)^2$.
$$(x + 4)^2 = x^2 + 16$$

    b. Simplify $2x(3x^2 - 4x) + 5x^2(x - 2)$.
$$2x(3x^2 - 4x) + 5x^2(x - 2)$$
$$= 6x^3 - 8x^2 + 5x^3 - 10x^2$$
$$= 11x^6 - 18x^4$$

    c. Divide, using long division: $(x^3 - 7x + 12) \div (x - 3)$.

$$
\begin{array}{r}
x^2 \quad - \quad 4 \\
x - 3 \overline{) x^3 - 7x + 12} \\
\ominus \quad \oplus \\
x^3 - 3x \\
\hline
- 4x + 12 \\
\oplus \quad \ominus \\
- 4x + 12 \\
\hline
0
\end{array}
$$

    d. The sum of 4 times a number and 12 is 6 less than twice the number. Find the number. Let $x =$ the number.
$$4x + 12 = 6 - 2x$$
$$6x + 12 = 6$$
$$6x = -6$$
$$x = -1$$

4. Suppose you knew that the answer to a multiplication problem was $x^2 - 3x - 10$, and suppose also that you knew that one of the factors was $x + 2$. Can you find the other factor? If so, describe the method you used and name the other factor.

Factoring

C H A P T E R 7

n this chapter, we discuss factoring polynomials, solving those equations that can be solved by factoring, and solving applied problems that lead to such equations. Factoring is an important topic in algebra.

7.1 Factoring Polynomials That Contain a Common Factor

Factoring a sum of terms means rewriting it, if possible, *as a single term* that is a *product of factors*. When we *factor* a polynomial, we are "undoing" multiplication. For example, $3x + 6$ (which is a sum of terms) factors into $3(x + 2)$, which is a product of factors; observe that $3(x + 2)$ has just *one term*. [You might verify that $3(x + 2) = 3x + 6$.] It is essential that the techniques of factoring be mastered, because factoring is used a great deal in the remainder of this course and in higher-level mathematics courses.

7.1A Factoring Polynomials That Contain a Common Monomial Factor

Finding the Greatest Common Factor (GCF) of Several Integers

The **greatest common factor (GCF)** of two or more integers is the largest integer that is a factor of *all* those integers; that is, it is the *largest* number that divides exactly into all the numbers.

   For example, let's consider the numbers 24 and 16. The set of factors of 24 is the set $\{1, 2, 3, 4, 6, 8, 12, 24\}$, and the set of factors of 16 is the set $\{1, 2, 4, 8, 16\}$. The *common factors* (the numbers that are *common* to *both* sets) are 1, 2, 4, and 8, and the largest of *these* numbers is 8. Therefore, 8 is the *greatest common factor*, or GCF, of 24 and 16; that is, 8 is the *largest* number that divides exactly into both 24 and 16.

   In practice, we do not write all the factors of the numbers to find the GCF; if the numbers are small, we might find the GCF "by inspection." (You probably knew before we listed all the factors of 24 and 16 that the *largest* number that divides exactly into both numbers is 8.) If the numbers are large and/or have a large number of factors, we can use the following prime-factoring method: We first write the numbers in prime-factored, exponential form. Then we write down, as a *factor* of the GCF, each base that is common to *all* the numbers, and we raise each of these bases to the *lowest* power to which it occurs in any of the numbers. The GCF is the product of these factors (see Example 1).

EXAMPLE 1   Find the GCF of each set of numbers:

a. 24 and 16

> **SOLUTION**   We first write 24 and 16 in prime-factored, exponential form:
>
> $$24 = 2^3 \cdot 3 \qquad 16 = 2^4$$
>
> The exponents are 3 and 4; the *smaller* exponent is 3
>
> The common base
>
> The only *base* common to both numbers is 2; this *base* will be a factor of the GCF. (The base 3 is *not* common to both numbers.) The *smaller* exponent on the base 2 is 3; therefore, the GCF is 2^3, or 8.

b. 6 and 35

S O L U T I O N   We write the numbers in prime-factored, exponential form:

$$6 = 2 \cdot 3 \qquad 35 = 5 \cdot 7$$

There are *no* common bases; therefore, the GCF is 1.

Another way to see that the GCF is 1 is to write all the positive factors of both numbers: The set of all the positive factors of 6 is { 1 , 2, 3, 6}, and the set of all the positive factors of 35 is { 1 , 5, 7, 35}. The only number common to *both* sets is 1, so 1 is the *largest* factor common to both sets; that is, the GCF is 1.

c. 252, 189, and 126

S O L U T I O N   We first write the numbers in prime-factored, exponential form:

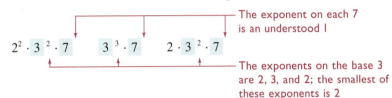

The exponent on each 7 is an understood 1

The exponents on the base 3 are 2, 3, and 2; the smallest of these exponents is 2

The bases common to *all* the numbers are 3 and 7; these bases will be factors of the GCF. (The base 2 is *not* common to all three numbers.) Because the smallest exponent on the base 3 is 2, we raise 3 to the second power so that one factor of the GCF is 3^2. Because the smallest exponent on 7 is 1, we raise 7 to the first power so that another factor of the GCF is 7^1. The GCF is the product of 3^2 and 7^1, which is $3^2 \cdot 7^1$, or $3^2 \cdot 7$, or 63.

Finding the Greatest Common Factor (GCF) of a Polynomial

The **greatest common factor (GCF)** of a polynomial is the largest expression that is a factor of all the terms in the polynomial. We find the GCF as follows.

Finding the greatest common factor (GCF)	1. Write each numerical coefficient in prime-factored form. Repeated factors must be expressed in exponential form.
	2. Write down each different base, numerical or variable, that is common to *all* the terms.
	3. Raise each of the bases in step 2 to the *lowest* power to which it occurs in any of the terms.
	4. The *greatest common factor* (GCF) is the *product* of all the factors found in step 3. It may be positive or negative.*

*See Example 10, second solution, for a case in which it is practical to have a negative GCF.

E X A M P L E   2   Find the GCF of $15x^3 + 9x$.

S O L U T I O N   We might find by inspection that the GCF of 15 and 9 is 3, or, writing 15 and 9 in prime-factored, exponential form, we have

$$15x^3 + 9x = 3^1 \cdot 5^1 \cdot x^3 + 3^2 \cdot x^1$$

The *bases* common to both terms are 3 and *x*. The smaller exponent on the 3 is 1, and the smaller exponent on the *x* is 1. Therefore, we raise 3 to the first power (this gives 3^1), and we raise *x* to the first power (this gives x^1). The GCF is the *product* of 3^1 and x^1, or $3x$. (The GCF could also be $-3x$.)

Factoring Polynomials That Have a Common Factor

Earlier, we used the distributive property to rewrite a product of factors as a sum of terms; that is, $a(b + c) = ab + ac$. In this section, we use the distributive property to rewrite a *sum of terms* as a *product of factors*; that is, we want to "undo" a multiplication, whenever possible. (It is not *always* possible to rewrite a sum of terms as a product of factors, since a sum of terms is not always factorable.) When we use the distributive property to get

$$ab + ac = a(b + c)$$

we say we are *factoring out* the greatest common factor. Notice that the right side of the equation contains *only one term*.

To factor a polynomial that contains a factor common to all the terms of the polynomial, we can follow the suggestions in the following box. They are based on the fact that we're asking ourselves a question such as "If $5x^3 + 9x$ had been the *answer to a multiplication problem*, what would the original factors have been?"

Factoring out the GCF

I. Combine like terms, if there are any.

2. Find the GCF of all the terms. It will often, but not always, be a monomial.

3. Find the *polynomial factor** by dividing each term of the polynomial being factored by the GCF. *The polynomial factor will always have as many terms as the expression in step 1.* It should contain only *integer* coefficients.

4. The factored form of the given polynomial is the product of the factors found in steps 2 and 3.

5. Check the result by using the distributive property to remove the parentheses; the resulting product should be the polynomial from step 1.

---

*We will call this factor the *polynomial factor* because it will be a polynomial and will *always* have more than one term.

EXAMPLE 3   Factor $15x^3 + 9x$.

SOLUTION   We found in Example 2 that the GCF is $3x$. To find the polynomial factor, we divide $15x^3$ by $3x$ and then divide $9x$ by $3x$.

$$15x^3 + 9x$$

$$= 3x(\quad + \quad) \qquad \text{The polynomial factor should contain as many terms as the original expression}$$

$$= 3x(5x^2 + 3)$$

This term is $\dfrac{9x}{3x} = \dfrac{9}{3} \cdot \dfrac{x}{x} = 3 \cdot 1 = 3$

This term is $\dfrac{15x^3}{3x} = \dfrac{15}{3} \cdot \dfrac{x^3}{x} = 5x^{3-1} = 5x^2$

Therefore, the factors of $15x^3 + 9x$ are $3x$ and $(5x^2 + 3)$.

$$15x^3 + 9x = \underset{\text{The GCF}}{3x} \; (\underset{\text{The polynomial factor}}{5x^2 + 3})$$

✓ **Check** $3x(5x^2 + 3) = (3x)(5x^2) + (3x)(3) = 15x^3 + 9x$

Therefore, $15x^3 + 9x = 3x(5x^2 + 3)$. Notice that the answer, $3x(5x^2 + 3)$, contains *only one term*, and the expression within the parentheses cannot be factored.

> ✕ **A Word of Caution** Two errors commonly made in factoring $15x^3 + 9x$ are shown below:
>
> $$15x^3 + 9x = 3x \cdot 5x^2 + 3$$
> $$15x^3 + 9x = (3x) \cdot 5x^2 + 3$$
>
> Neither answer is correct. There *must* be parentheses around $5x^2 + 3$. Note that
>
> $$3x \cdot 5x^2 + 3 = 15x^3 + 3 \neq 15x^3 + 9x$$
> $$(3x) \cdot 5x^2 + 3 = 15x^3 + 3 \neq 15x^3 + 9x$$

EXAMPLE 4 Factor $6x + 4$, or write "not factorable."

SOLUTION We might find by inspection that the GCF of 6 and 4 is 2, or, writing 6 and 4 in prime-factored, exponential form, we have $6 = 2 \cdot 3$ and $4 = 2^2$.

$$6x + 4 = 2^1 \cdot 3 \cdot x + 2^2$$

The only common base is 2, and the smaller exponent on the 2 is 1. Therefore, the GCF is 2^1, or 2.

$$6x + 4 = 2(3x + 2)$$

- The GCF
- The polynomial factor
- This term of the polynomial factor is $\dfrac{4}{2} = 2$
- This term of the polynomial factor is $\dfrac{6x}{2} = \dfrac{6}{2} \cdot x = 3x$

Therefore, the factors of $6x + 4$ are 2 and $(3x + 2)$.

✓ **Check** $2(3x + 2) = 6x + 4$
The answer, $2(3x + 2)$, contains *only one term*, and the polynomial factor has no common factor left in it.

In this book, when we say to *factor a polynomial*, we mean to factor the polynomial so that all the *constants* are integers. When we do this, we say we are factoring *over the integers*.

EXAMPLE 5 Factor $5x + 2y$.

SOLUTION Although 5 and x are factors of the *first term* of $5x + 2y$, neither one is a factor of the *second term*. In fact, $5x$ and $2y$ have no common integral factors other than ± 1. While it is true that

$$5x + 2y = 5\left(x + \tfrac{2}{5}y\right)$$

and that

$$5x + 2y = 2\left(\tfrac{5}{2}x + y\right)$$

the parentheses in both cases contain constants that are *not integers*. Therefore, $5x + 2y$ is *not factorable* over the integers.

Prime (or Irreducible) Polynomials

It is important to realize that most polynomials *cannot* be factored over the integers. A polynomial is said to be **prime** (or **irreducible**) over the integers if it cannot be expressed as a product of polynomials of lower degree such that all the constants in the factors are integers. Thus, in Example 5, the polynomial $5x + 2y$ is a *prime polynomial*. (When a polynomial is prime, *no one* can factor it; you must realize that the meaning of "The polynomial is prime" is different from the meaning of "I can't factor the polynomial.")

The Uniqueness of Factorization

We know that a composite number can be expressed in one and only one prime-factored form, except for the order in which the factors are written. A similar property holds for the factorization of polynomials.

> Factorization of polynomials over the integers is *unique*; that is, when a polynomial has been factored completely over the integers, the answer can be expressed in one and only one prime-factored form, except for the order in which the factors are written and the signs of the factors of the product.

EXAMPLE 6   Factor $24x^3 - 16x^5$, or write "not factorable."

SOLUTION   We first write 24 and 16 in prime-factored, exponential form:

$$24x^3 - 16x^5 = 2^3 \cdot 3 \cdot x^3 - 2^4 \cdot x^5$$

The common bases are 2 and x. The lower power of the base 2 is 3, and the lower power of the base x is 3. Hence, the GCF is $2^3 \cdot x^3$, or $8x^3$.

The GCF
The polynomial factor

$$24x^3 - 16x^5 = 8x^3(3 - 2x^2)$$

This term of the polynomial factor is

$$\frac{-16x^5}{8x^3} = \frac{-16}{8} \cdot \frac{x^5}{x^3} = -2x^{5-3} = -2x^2$$

This term of the polynomial factor is $\dfrac{24x^3}{8x^3} = \dfrac{24}{8} \cdot \dfrac{x^3}{x^3} = 3$

Therefore, the factors of $24x^3 - 16x^5$ are $8x^3$ and $(3 - 2x^2)$.

√ **Check** $8x^3(3 - 2x^2) = (8x^3)(3) + (8x^3)(-2x^2) = 24x^3 - 16x^5$

The answer, $8x^3(3 - 2x^2)$, contains *only one term*, and the polynomial factor, $3 - 2x^2$, has no common factor left in it.

The factoring *could* also be done by using $-8x^3$ as the GCF:

$$24x^3 - 16x^5 = -8x^3(-3 + 2x^2) = -8x^3(2x^2 - 3)$$

√ **Check** $-8x^3(2x^2 - 3) = -8x^3(2x^2) + (-8x^3)(-3) = -16x^5 + 24x^3$
$$= 24x^3 - 16x^5$$

Because factorization of polynomials is unique, the following answers are all acceptable (they differ only in the order in which the factors or the terms of the factors are written or in the signs of both factors):

$$8x^3(3 - 2x^2), \quad (3 - 2x^2)(8x^3), \quad -8x^3(-3 + 2x^2), \quad -8x^3(2x^2 - 3),$$
$$(-3 + 2x^2)(-8x^3), \quad (2x^2 - 3)(-8x^3)$$

A Word of Caution In Example 6, it is incorrect to write

$$24x^3 - 16x^5 = (-3 + 2x^2) - 8x^3$$ Without parentheses around it,
$-8x^3$ is a *term*, not a *factor*

or $\quad 24x^3 - 16x^5 = (2x^2 - 3) - 8x^3$

Both equal $-3 + 2x^2 - 8x^3$, *not* $24x^3 - 16x^5$.

EXAMPLE 7

Factor $2x^4 + 4x^3 - 8x^2$, or write "not factorable."

SOLUTION We might find by inspection that the GCF of 2, 4, and -8 is 2, or, writing the coefficients in prime-factored, exponential form, we have

$$2x^4 + 4x^3 - 8x^2 = 2^1 x^4 + 2^2 x^3 - 2^3 x^2$$

The common bases are 2 and x. The lowest power of the base 2 is 1, and the lowest power of the base x is 2. Therefore, the GCF is $2^1 \cdot x^2$, or $2x^2$.

The GCF
The polynomial factor
$$2x^4 + 4x^3 - 8x^2 = 2x^2(x^2 + 2x - 4)$$

This term is $\dfrac{-8x^2}{2x^2} = \dfrac{-8}{2} \cdot \dfrac{x^2}{x^2} = -4$

This term is $\dfrac{4x^3}{2x^2} = \dfrac{4}{2} \cdot x^{3-2} = 2x$

This term is $\dfrac{2x^4}{2x^2} = \dfrac{2}{2} \cdot \dfrac{x^4}{x^2} = 1 \cdot x^{4-2} = x^2$

√ **Check** $\quad 2x^2(x^2 + 2x - 4) = 2x^2(x^2) + 2x^2(2x) + 2x^2(-4) = 2x^4 + 4x^3 - 8x^2$

Therefore, $2x^4 + 4x^3 - 8x^2 = 2x^2(x^2 + 2x - 4)$.

If an expression has been factored, it contains only one term. However, even if an expression contains only one term, it still may not be factored *completely*. Any expression inside grouping symbols should always be examined carefully to see whether it can be factored further.

A Word of Caution Suppose we had factored the trinomial in Example 7 as follows:

$$2x^4 + 4x^3 - 8x^2 = 2(x^4 + 2x^3 - 4x^2)$$

This *is* in factored form, since it contains only one term, but it is not factored *completely*, because $x^4 + 2x^3 - 4x^2$ can still be factored. Similarly, if the trinomial is factored as

$$2x^4 + 4x^3 - 8x^2 = x^2(2x^2 + 4x - 8)$$

the factoring is not complete, since $2x^2 + 4x - 8$ can still be factored. In other words, $x^4 + 2x^3 - 4x^2$ and $2x^2 + 4x - 8$ are not *prime* polynomials.

EXAMPLE 8   Factor $6a^3b^3 - 8a^2b^2 + 10a^3b$, or write "not factorable."

SOLUTION   We might find by inspection that the GCF of 6, -8, and 10 is 2, or, writing the coefficients in prime-factored, exponential form, we have

$$6a^3b^3 - 8a^2b^2 + 10a^3b = 2^1 \cdot 3^1 a^3 b^3 - 2^3 a^2 b^2 + 2^1 \cdot 5^1 a^3 b$$

The GCF is $2^1 \cdot a^2 \cdot b^1$, or $2a^2b$.

$$6a^3b^3 - 8a^2b^2 + 10a^3b = 2a^2b(3ab^2 - 4b + 5a)$$

This term is $\dfrac{10a^3b}{2a^2b} = \dfrac{10}{2} \cdot \dfrac{a^3}{a^2} \cdot \dfrac{b}{b} = 5a$

This term is $\dfrac{-8a^2b^2}{2a^2b} = \dfrac{-8}{2} \cdot \dfrac{a^2}{a^2} \cdot \dfrac{b^2}{b} = -4b$

This term is $\dfrac{6a^3b^3}{2a^2b} = \dfrac{6}{2} \cdot \dfrac{a^3}{a^2} \cdot \dfrac{b^3}{b} = 3ab^2$

✓ **Check** $2a^2b(3ab^2 - 4b + 5a) = 2a^2b(3ab^2) + 2a^2b(-4b) + 2a^2b(5a)$

$$= 6a^3b^3 - 8a^2b^2 + 10a^3b$$

Therefore, $6a^3b^3 - 8a^2b^2 + 10a^3b = 2a^2b(3ab^2 - 4b + 5a)$.

EXAMPLE 9   Factor $3xy - 6y^2 - 3y$, or write "not factorable."

SOLUTION   Writing the coefficients in prime-factored, exponential form, we get

$$3xy - 6y^2 - 3y = 3xy - 2 \cdot 3y^2 - 3y \qquad \text{The GCF is } 3y$$

$$3xy - 6y^2 - 3y = 3y(x - 2y - 1)$$

This term is $\dfrac{-3y}{3y} = \dfrac{-3}{3} \cdot \dfrac{y}{y} = -1$

This term is $\dfrac{-6y^2}{3y} = \dfrac{-6}{3} \cdot \dfrac{y^2}{y} = -2y$

This term is $\dfrac{3xy}{3y} = \dfrac{3}{3} \cdot x \cdot \dfrac{y}{y} = x$

✓ **Check** $3y(x - 2y - 1) = 3y(x) + 3y(-2y) + 3y(-1) = 3xy - 6y^2 - 3y$

Therefore, $3xy - 6y^2 - 3y = 3y(x - 2y - 1)$.

EXAMPLE 10   Factor $-15x^2y^3 - 20xy^4 + 25x^3y^2$, or write "not factorable."

SOLUTION   Writing the coefficients in prime-factored, exponential form, we have

$$-15x^2y^3 - 20xy^4 + 25x^3y^2 = -3 \cdot 5x^2y^3 - 2^2 \cdot 5xy^4 + 5^2x^3y^2$$

First Solution   Using $5xy^2$ as the GCF, we have

The GCF

The polynomial factor

$$-15x^2y^3 - 20xy^4 + 25x^3y^2 = 5xy^2(-3xy - 4y^2 + 5x^2)$$

This term is $\dfrac{25x^3y^2}{5xy^2} = \dfrac{25}{5} \cdot \dfrac{x^3}{x} \cdot \dfrac{y^2}{y^2} = 5x^2$

This term is $\dfrac{-20xy^4}{5xy^2} = \dfrac{-20}{5} \cdot \dfrac{x}{x} \cdot \dfrac{y^4}{y^2} = -4y^2$

This term is $\dfrac{-15x^2y^3}{5xy^2} = \dfrac{-15}{5} \cdot \dfrac{x^2}{x} \cdot \dfrac{y^3}{y^2} = -3xy$

✓ **Check** $5xy^2(-3xy - 4y^2 + 5x^2) = 5xy^2(-3xy) + 5xy^2(-4y^2) + 5xy^2(5x^2)$

$$= -15x^2y^3 - 20xy^4 + 25x^3y^2$$

Therefore, one solution is $-15x^2y^3 - 20xy^4 + 25x^3y^2 = 5xy^2(-3xy - 4y^2 + 5x^2)$.

Second Solution   Using $-5xy^2$ as the GCF, we have

The GCF

The polynomial factor

$$-15x^2y^3 - 20xy^4 + 25x^3y^2 = -5xy^2(3xy + 4y^2 - 5x^2)$$

This term is $\dfrac{25x^3y^2}{-5xy^2} =$

$\dfrac{25}{-5} \cdot \dfrac{x^3}{x} \cdot \dfrac{y^2}{y^2} = -5x^2$

This term is $\dfrac{-20xy^4}{-5xy^2} = \dfrac{-20}{-5} \cdot \dfrac{x}{x} \cdot \dfrac{y^4}{y^2} = 4y^2$

This term is $\dfrac{-15x^2y^3}{-5xy^2} = \dfrac{-15}{-5} \cdot \dfrac{x^2}{x} \cdot \dfrac{y^3}{y^2} = 3xy$

✓ **Check** $-5xy^2(3xy + 4y^2 - 5x^2) = -5xy^2(3xy) + (-5xy^2)(4y^2) + (-5xy^2)(-5x^2)$

$$= -15x^2y^3 - 20xy^4 + 25x^3y^2$$

Therefore, *another* solution is $-15x^2y^3 - 20xy^4 + 25x^3y^2 = -5xy^2(3xy + 4y^2 - 5x^2)$. Both factored forms are equally correct.

Exercises 7.1A
Set 1

In Exercises 1–8, find the greatest common factor (GCF) of each set of numbers.

1. 12, 24   **2.** 24, 40   **3.** 21, 10

4. 16, 9   **5.** 32, 24, 40   **6.** 18, 6, 30

7. 18, 12, 48   **8.** 16, 12, 48

In Exercises 9–46, factor each polynomial completely and check, or write "not factorable."

9. $12x + 8$ **10.** $6x + 9$ **11.** $5a - 8$

12. $7b - 2$ **13.** $2x + 8$ **14.** $3x + 9$

15. $5a - 10$ **16.** $7b - 14$ **17.** $6y - 3$

18. $15z - 5$ **19.** $9x^2 + 3x$ **20.** $8y^2 - 4y$

21. $10a^3 - 25a^2$ **22.** $27b^2 - 18b^4$ **23.** $21w^2 - 20z^2$

24. $15x^3 - 16y^3$ **25.** $2a^2b + 4ab^2$ **26.** $3mn^2 + 6m^2n$

27. $12c^3d^2 - 18c^2d^3$ **28.** $15ab^3 - 45a^2b^4$

29. $4x^3 - 12x - 24x^2$ **30.** $18y - 6y^2 - 30y^3$

31. $4x^4 - 7x^2 + 1$ **32.** $3x^2 + 2x - 2$

33. $8x^3 - 6x^2 + 2x$ **34.** $9y^4 + 6y^3 - 3y^2$

35. $24a^4 + 8a^2 - 40$ **36.** $45b^3 - 15b^4 - 30$

37. $-14x^8y^9 + 42x^5y^4 - 28xy^3$

38. $-21u^7v^8 - 63uv^5 + 35u^2v^5$

39. $15h^2k - 8hk^2 + 9st$

40. $10uv^3 + 5u^2v - 4wz$

41. $-44a^{14}b^7 - 33a^{10}b^5 + 22a^{11}b^4$

42. $-26e^8f^6 + 13e^{10}f^8 - 39e^{12}f^5$

43. $18u^{10}v^5 + 24 - 14u^{10}v^6$

44. $30a^3b^4 - 15 + 45a^8b^7$

45. $18x^3y^4 - 12y^2z^3 - 48x^4y^3$

46. $32m^5n^7 - 24m^8p^9 - 40m^3n^6$

Writing Problems

Express the answers in your own words and in complete sentences.

1. Explain what the GCF of two or more numbers *is*. (Don't explain how to find it.)

2. Explain why the GCF of $6x^3y + 9xy^2$ is *not* 3.

Exercises 7.1A
Set II

In Exercises 1–8, find the greatest common factor (GCF) of each set of numbers.

1. 10, 25   **2.** 28, 8   **3.** 36, 15   **4.** 24, 35

5. 25, 10, 15   **6.** 16, 24, 40   **7.** 15, 18, 12   **8.** 16, 8, 12

In Exercises 9–46, factor each polynomial completely and check, or write "not factorable."

9. $6h + 9$ **10.** $12x - 5$ **11.** $2x - 7$

12. $8k - 12$ **13.** $6h + 18$ **14.** $8x + 9$

15. $8k - 16$ **16.** $6 - 12x$ **17.** $8k - 4$

18. $3x - 22$ **19.** $4a^2 + 2a$ **20.** $10x + 25$

21. $8x^3 - 12x^2$ **22.** $4y - 12y^2$ **23.** $9w^2 + 16z^2$

24. $14a^2 - 21a^3$ **25.** $5xy^2 + 10x^2y$ **26.** $x^2 + 4$

27. $15a^2b^3 - 12ab^2$ **28.** $12x^2y + 9xy^2$

29. $16z - 8z^3 - 12z^2$ **30.** $12x^3 - 5y + 3x$

31. $8x^4 - 3x^2 + 5$ **32.** $12x^3 - 28x^2 - 8x$

33. $12x^4 - 6x^3 + 3x^2$ **34.** $4y^2 - 8y^4 + 16y^5$

35. $42x^5 - 6x^4 + 12x^3$ **36.** $-7a^2 + 14a^3 - 35a^4$

37. $-12ab^2 + 9a^2b - 36ab$ **38.** $20z^5 - 30z^3 - 10z^2$

39. $12h^2k - 18hk^2 - 35mp$ **40.** $30e^3f + 18 - 12ef^2$

41. $10m^2n - 21mn^3 - 13mn$

42. $-16u^3v^2 + 24uv^3 - 40v^4w^2$

43. $25x^2y^3 + 10 - 15x^3y^2$

44. $-5xy^2 + 7x - 3y$

45. $15a^3y^4 - 18ay^3 + 12a^2y$

46. $x^4 + 16$

7.1B Factoring Polynomials That Contain a Common Binomial Factor

Sometimes an expression has a GCF that is not a monomial. Such an expression can still be factored, using the procedure given in the last subsection. This type of factoring is used in *factoring by grouping* and also in higher-level mathematics courses.

EXAMPLE 11   Factor $a(x + y) + b(x + y)$.

SOLUTION   This expression contains two terms, and therefore it is *not* in factored form. The common factor, $x + y$, is not a monomial; it is a binomial.

The greatest common factor is $(x + y)$

$$a(\,x + y\,) + b(\,x + y\,) = (x + y)(\,a + b\,)$$

The GCF

This term is $\dfrac{b(x + y)}{x + y} = b$

This term is $\dfrac{a(x + y)}{x + y} = a$

Therefore, $a(x + y) + b(x + y) = (x + y)(a + b)$.

✓ **Check** $(x + y)(a + b) = (x + y)a + (x + y)b = a(x + y) + b(x + y)$

The answer, $(x + y)(a + b)$, contains only one term, so it is in factored form.

EXAMPLE 12

Factor $b(a - 1) + (a - 1)$.

SOLUTION   The expression contains two terms. The common binomial factor is $a - 1$.

The greatest common factor is $(a - 1)$

$$b(\,a - 1\,) + (\,a - 1\,) = (a - 1)(\,b + 1\,)$$

The GCF

This term is $\dfrac{(a - 1)}{a - 1} = 1$

This term is $\dfrac{b(a - 1)}{a - 1} = b$

Notice that the expression $(a - 1)(b + 1)$ contains just one term.

✓ **Check** $(a - 1)(b + 1) = (a - 1)(b) + (a - 1)(1) = b(a - 1) + (a - 1)$

Therefore, $b(a - 1) + (a - 1) = (a - 1)(b + 1)$.

It is possible to have a common monomial factor *and* a common binomial factor in the same expression (see Example 13).

EXAMPLE 13

Factor $6s^3t^2(x - y) + 8s^2t^3(x - y)$, or write "not factorable."

SOLUTION   Writing the coefficients in prime-factored, exponential form gives us

$$6s^3t^2(x - y) + 8s^2t^3(x - y) = 2 \cdot 3s^3t^2(x - y) + 2^3s^2t^3(x - y)$$

The expression contains two terms. The greatest common *monomial* factor is $2s^2t^2$, and the greatest common *binomial* factor is $(x - y)$. The GCF, then, is the product of both of these factors, or $2s^2t^2(x - y)$.

The GCF

$$2 \cdot 3s^3t^2(x - y) + 2^3s^2t^3(x - y) = 2s^2t^2(x - y)(\,3s + 4t\,)$$

This term is $\dfrac{2^3s^2t^3(x - y)}{2s^2t^2(x - y)}$

This term is $\dfrac{2 \cdot 3s^3t^2(x - y)}{2s^2t^2(x - y)}$

Notice that the expression $2s^2t^2(x - y)(3s + 4t)$ contains *just one term*.

✓ **Check** $2s^2t^2(x - y)(3s + 4t) = [2s^2t^2(x - y)](3s) + [2s^2t^2(x - y)](4t)$
$$= [2s^2t^2(3s)(x - y)] + [2s^2t^2(4t)(x - y)]$$
$$= 6s^3t^2(x - y) + 8s^2t^3(x - y)$$

Therefore, $6s^3t^2(x - y) + 8s^2t^3(x - y) = 2s^2t^2(x - y)(3s + 4t)$.

Exercises 7.1B
Set I

Factor each polynomial completely, or write "not factorable."

1. $c(s + t) + b(s + t)$ $(s+t)(c+b)$ **2.** $a(b + c) + d(b + c)$

3. $x(a - b) + 5(a - b)$ $(a-b)(x+5)$ **4.** $y(s - t) + 7(s - t)$

5. $x(u + v) - 3(u + v)$ **6.** $s(t + u) - 2(t + u)$

7. $8(x - y) - a(x - y)$ **8.** $7(a - b) - c(a - b)$

9. $4(s - t) + u(s - t) - v(s - t)$

10. $a(x - y) + 5(x - y) - b(x - y)$.

11. $3x^2y^3(a + b) + 9xy^2(a + b)$

12. $10ab^3(s + t^2) + 15a^2b(s + t^2)$

13. $4u^3v^5(x - y) + 6u^4v^4(x - y)$

14. $10x^2y^5(3a - 2b) + 15x^5y^3(3a - 2b)$

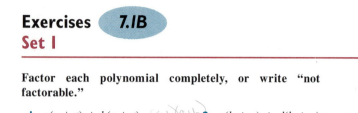

Writing Problems

Express the answer in your own words and in complete sentences.

1. Explain why the GCF of $10xy^2(a + b) + 5y^3(a + b)$ is *not* $(a + b)$.

Exercises 7.1B
Set II

Factor each polynomial completely, or write "not factorable."

1. $x(f + g) + s(f + g)$ **2.** $y(u - v) - z(u - v)$

3. $u(x - y) + 4(x - y)$ **4.** $d(a - b) - 3(a - b)$

5. $a(b + c) - 9(b + c)$ **6.** $x(y + z) - w(y + z)$

7. $9(a - b) - c(a - b)$ **8.** $y(x - z) - 9(x - z)$

9. $8(x - y) + z(x - y) - w(x - y)$

10. $f(a + b) - g(a + b) + (a + b)$

11. $8x^3y^2(c - d) + 4x^2y^3(c - d)$

12. $14a^3b^3(x - 2) + 7ab^2(x - 2)$

13. $12x^4y^3(a + b) + 8x^3y^2(a + b)$

14. $9a^2b^4(u - v) - 12a^3b^7(u - v)$

7.2 Factoring the Difference of Two Squares

Any polynomial that can be expressed in the form $a^2 - b^2$ is called a **difference of two squares**.

Because $(a + b)(a - b) = a^2 - b^2$ (we proved this earlier), $a^2 - b^2$ *factors into* $(a + b)(a - b)$. This is so, of course, because factoring is "undoing" multiplication. That is, in factoring $a^2 - b^2$, we are asking ourselves the question "If $a^2 - b^2$ had been the *answer* to a multiplication problem, what would the original factors have been?" Recall that the *answer* to a multiplication problem that is in the form $(a + b)(a - b)$ always has two terms and always contains a minus sign; furthermore, the absolute values of the two terms are always *perfect squares* (that is, the answer is always a *difference of two squares*—it is always in the form $a^2 - b^2$).

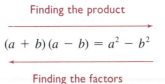

$$(a + b)(a - b) = a^2 - b^2$$

Finding the product

Finding the factors

Factoring the difference of two squares

1. Write (or think of) the expression to be factored in the form $a^2 - b^2$.

2. Make the following blank outline for the factors:

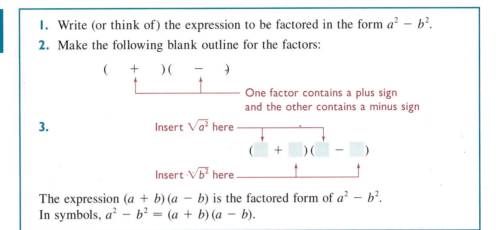

$$(\quad + \quad)(\quad - \quad)$$

One factor contains a plus sign and the other contains a minus sign

3. Insert $\sqrt{a^2}$ here

$$(\quad + \quad)(\quad - \quad)$$

Insert $\sqrt{b^2}$ here

The expression $(a + b)(a - b)$ is the factored form of $a^2 - b^2$.
In symbols, $a^2 - b^2 = (a + b)(a - b)$.

Recall that $25x^2$ is a *perfect square* because 25 is the *square of an integer* and the exponent on x is *even*, and $100a^4b^6$ is a perfect square because 100 is the square of an integer and the exponents on a and b are even. Also recall that $\sqrt{25x^2} = 5x$ and $\sqrt{100a^4b^6} = 10a^2b^3$. We will make use of facts such as these when we factor a difference of two squares.

EXAMPLE 1 Factor $x^2 - 4$, or write "not factorable."

SOLUTION

Step 1. $x^2 - 4 = (x)^2 - (2)^2$, which is in the form $a^2 - b^2$ (that is, it is a *difference of two squares*).

Step 2. $x^2 - 4 = (\quad + \quad)(\quad - \quad)$ Making a blank outline

Step 3. $x^2 - 4 = (x + 2)(x - 2)$ $\sqrt{x^2}$ $\sqrt{4}$

✓ **Check** $(x + 2)(x - 2) = (x)^2 - (2)^2 = x^2 - 4$

Therefore, $x^2 - 4 = (x + 2)(x - 2)$.

EXAMPLE 2 Factor $25y^4 - 9z^2$, or write "not factorable."

SOLUTION

Step 1. $25y^4 - 9z^2 = (5y^2)^2 - (3z)^2$, which is in the form $a^2 - b^2$.

Step 2. $25y^4 - 9z^2 = (\quad + \quad)(\quad - \quad)$

Step 3. $25y^4 - 9z^2 = (5y^2 + 3z)(5y^2 - 3z)$ $\sqrt{25y^4}$ $\sqrt{9z^2}$

✓ **Check** $(5y^2 + 3z)(5y^2 - 3z) = (5y^2)^2 - (3z)^2 = 25y^4 - 9z^2$

Therefore, $25y^4 - 9z^2 = (5y^2 + 3z)(5y^2 - 3z)$.

EXAMPLE 3   Factor $49a^6b^2 - 81c^4d^8$, or write "not factorable."

SOLUTION

Step 1. $49a^6b^2 - 81c^4d^8 = (7a^3b)^2 - (9c^2d^4)^2$, which is in the form $a^2 - b^2$.

Step 2. $49a^6b^2 - 81c^4d^8 = (\quad + \quad)(\quad - \quad)$

Step 3. $49a^6b^2 - 81c^4d^8 = (7a^3b + 9c^2d^4)(7a^3b - 9c^2d^4)$
$\sqrt{49a^6b^2}$
$\sqrt{81c^4d^8}$

√ **Check** $(7a^3b + 9c^2d^4)(7a^3b - 9c^2d^4) = (7a^3b)^2 - (9c^2d^4)^2 = 49a^6b^2 - 81c^4d^8$

Therefore, $49a^6b^2 - 81c^4d^8 = (7a^3b + 9c^2d^4)(7a^3b - 9c^2d^4)$.

✕ **A Word of Caution**   A common error is to think that

$$a^2 + b^2 = (a + b)(a + b)$$

This is not so, however, because $(a + b)(a + b) = a^2 + 2ab + b^2$, not $a^2 + b^2$. Another common error is to think that

$$a^2 + b^2 = (a + b)(a - b)$$

This is not so, however, because $(a + b)(a - b) = a^2 - b^2$. Still another error is to think that

$$a^2 + b^2 = (a - b)(a - b)$$

But this cannot be so, because $(a - b)(a - b) = a^2 - 2ab + b^2$.

In fact, a *sum* of two squares (that is, a polynomial that can be expressed in the form $a^2 + b^2$) is *not factorable* over the integers. (Exception: If the exponents on the variables are multiples of 4 or 6, the polynomial *may* be factorable; however, we will not discuss the methods of such factoring in this book.)

EXAMPLE 4   Factor $x^2 + 4$, or write "not factorable."

SOLUTION   There is no common monomial factor. Because $x^2 + 4$ is a *sum of two squares*, we conclude that it is *not factorable* over the integers.

We should begin every factoring problem by looking for and factoring out any factor common to every term. To be sure that a polynomial has been factored *completely*, we should always check each factor to see whether it can still be factored. Then we try to factor the polynomial in the parentheses (see Examples 5 and 6).

✕ **A Word of Caution**   When we have factored out a common factor and then factored further, we must be sure to include *every* factor in the final answer.

EXAMPLE 5   Factor $16x^4 - 4x^2$, or write "not factorable."

SOLUTION $16x^4 - 4x^2$ is a difference of two squares; however, there *is* a common monomial factor: $4x^2$. Therefore, we first factor out $4x^2$.

$$16x^4 - 4x^2 = 4x^2\underbrace{(4x^2 - 1)}$$ ← This has been factored, but not completely
↑ A difference of two squares

$$= 4x^2(2x + 1)(2x - 1)$$

√ **Check** $4x^2(2x + 1)(2x - 1) = 4x^2(4x^2 - 1) = 16x^4 - 4x^2$

Therefore, $16x^4 - 4x^2 = 4x^2(2x + 1)(2x - 1)$.

EXAMPLE 6 Factor $81s^2 - 144s^2t^2$, or write "not factorable."

SOLUTION $81s^2 - 144s^2t^2$ is a difference of two squares; however, there *is* a common monomial factor: $9s^2$. Therefore, we first factor out $9s^2$.

$$81s^2 - 144s^2t^2 = 9s^2\underbrace{(9 - 16t^2)}$$ ← This has been factored, but not completely
↑ A difference of two squares

$$= 9s^2(3 + 4t)(3 - 4t)$$

√ **Check** $9s^2(3 + 4t)(3 - 4t) = 9s^2(9 - 16t^2) = 81s^2 - 144s^2t^2$

Therefore, $81s^2 - 144s^2t^2 = 9s^2(3 + 4t)(3 - 4t)$.

Exercises 7.2
Set I

Factor each polynomial completely and check, or write "not factorable."

1. $m^2 - n^2$
2. $u^2 - v^2$
3. $x^2 - 9$
4. $x^2 - 25$
5. $a^2 - 1$
6. $1 - b^2$
7. $4c^2 - 1$
8. $16d^2 - 1$
9. $16x^2 - 9y^2$
10. $25a^2 - 4b^2$
11. $9h^2 - 10k^2$
12. $16e^2 - 15f^2$
13. $4x^4 - 2x$
14. $9a^4 - 3a$
15. $16x^2 + 1$
16. $25y^2 + 1$
17. $49u^4 - 36v^4$
18. $81m^6 - 100n^4$
19. $x^6 - a^4$
20. $b^2 - y^6$
21. $2x^2 - 18$
22. $3x^2 - 12$
23. $2x^2 + 9$
24. $5y^2 + 2$
25. $a^2b^2 - c^2d^2$
26. $m^2n^2 - r^2s^2$
27. $49 - 25w^2z^2$
28. $36 - 25u^2v^2$
29. $4h^4k^4 - 1$
30. $9x^4y^4 - 1$
31. $81a^4b^6 - 16m^2n^8$
32. $49c^8d^4 - 100e^6f^2$
33. $49x^4y^2 - 7x^2$
34. $25x^2y^4 - 5y^2$
35. $5x^3 - 45xy^2$
36. $11r^3 - 44rs^2$
37. $225x^2s^2 - 100x^2t^2$
38. $64y^2u^2 - 144y^2v^2$

Writing Problems

Express the answer in your own words and in complete sentences.

1. Explain why $a^2 - b^2 \neq (a - b)^2$.

EXAMPLE 1

Examples of partially filled-in outlines for factoring:

a. $x^2 + 6x + 8 = (\,x\qquad)(\,x\qquad)$

b. $z^2 - z - 6 = (\,z\qquad)(\,z\qquad)$ } Partially filled-in outlines

c. $m^2 + 7m + 12 = (\,m\qquad)(\,m\qquad)$

We continue our discussion of the product $(x + 2)(x + 5)$.

The *product* of 2 and 5 is 10

$$(x + 2\,)(x + 5\,) = x^2 + 7\,x + 10$$

The *sum* of 2 and 5 is 7

When we're working backwards from multiplication and are trying to factor $x^2 + 7x + 10$, we ask ourselves the question "If $x^2 + 7x + 10$ had been the answer to a multiplication problem, what would the original factors have been?" We can begin factoring $x^2 + 7x + 10$ by making the blank outline and filling in the first term of each binomial. Because the sign of the *last* term of $x^2 + 7x + 10$ is a plus sign, the rule of signs indicates that the signs of the two last terms of the binomials will be the same. Because the sign of the *middle* term of $x^2 + 7x + 10$ is positive, both signs will be positive, and we can add those signs to the blank outline.

The *product* must be +10

$$x^2 + 7x + 10 = (x + \quad)(x + \quad)$$

The *sum* must be +7

To fill in the *last terms* of the binomials, we must find two numbers whose *product* is 10 and whose *sum* is 7. The pairs of positive factors of 10 are $(1)(10)$ and $(2)(5)$. The *sum* of 1 and 10 is 11, not 7, but the sum of 2 and 5 *is* 7. Therefore, we have

$$x^2 + 7x + 10 = (x + 2)(x + 5) \qquad or \qquad (x + 5)(x + 2)$$

The order in which the two factors are listed is unimportant; both $(x + 2)(x + 5)$ and $(x + 5)(x + 2)$ are correct factored forms of $x^2 + 7x + 10$.

The general procedure for factoring a trinomial that is in the form $x^2 + bx + c$ is summarized as follows:

Factoring a trinomial that can be expressed in the form $x^2 + bx + c$

1. Factor out the GCF, if there is one.

2. Arrange the trinomial in descending powers of the variable.

3. Make a blank outline. If the rule of signs indicates that the signs of the two last terms of the binomials will be the same, you can add those signs to the outline.

4. The first term inside each set of parentheses will be the square root of the first term of the trinomial. If the third term of the trinomial contains a variable, the square root of that variable must be a *factor* of the second term of each binomial.

5. To find the last term of each binomial:

　a. List (mentally, at least) *all* pairs of integral factors of the coefficient of the last term of the trinomial.

　b. *Select the particular pair of factors that has a sum equal to the coefficient of the middle term of the trinomial.* If no such pair of factors exists, the trinomial is not factorable.

6. Check the result by multiplying the binomials together. The resulting product should be the trinomial from step 2. Be sure neither of the factors can be factored further.

👉 **Note** *Alternative Step 5:* It is also possible to consider only the *positive* factors of the coefficient of the last term of the trinomial and to put the signs in *last*. In this case, when the rule of signs tells us that the signs are going to be the same, we select those factors whose *sum* equals the absolute value of the coefficient of the last term of the trinomial, and when the rule of signs tells us that the signs are going to be *different*, we select those factors whose *difference* equals the absolute value of the coefficient of the middle term of the trinomial. We demonstrate this method at the end of Example 4.

EXAMPLE 2 Factor $x^2 + 5x + 6$.

SOLUTION

Steps 1 and 2. There is no common factor. The trinomial is in descending powers of x.

Step 3. The last sign of $x^2 + 5x + 6$ is a plus sign; the rule of signs for factoring trinomials tells us that the signs in the binomials will be *the same*. The middle term of $x^2 + 5x + 6$ is positive; therefore, both signs will be plus signs.

$$(\quad + \quad)(\quad + \quad) \leftarrow \text{The outline}$$

Step 4. The first term in each binomial is $\sqrt{x^2} = x$:

$$(x + \quad)(x + \quad) \leftarrow \text{The partially filled-in outline}$$

Step 5. We must find a pair of numbers whose *product* is 6 (6 is the last term of the trinomial) and whose *sum* is 5 (5 is the coefficient of the middle term of the trinomial). Pairs of positive factors of 6 are (1)(6) and (2)(3). We must select the pair whose *sum* is 5. This is the pair (2)(3), and we have

$$x^2 + 5x + 6 = (x + 2)(x + 3) \qquad \text{or} \qquad (x + 3)(x + 2)$$

Notice that $(x + 2)(x + 3)$ contains only *one term*; it *is* in factored form. If we had used (1)(6), we would have had ~~(x + 1)(x + 6)~~, but

$$(x + 1)(x + 6) = x^2 + 7x + 6$$

└ An incorrect middle term

which is not the polynomial we're factoring.

┌ A correct middle term

✓ **Step 6. Check** $(x + 2)(x + 3) = x^2 + 5x + 6$

Therefore, $x^2 + 5x + 6 = (x + 2)(x + 3)$ or $(x + 3)(x + 2)$.

✖ **A Word of Caution** Three common errors in factoring $x^2 + 5x + 6$ are shown below:

$$x^2 + 5x + 6 = (x + 2) \cdot x + 3$$
$$x^2 + 5x + 6 = x + 2 \cdot (x + 3)$$
$$x^2 + 5x + 6 = x + 2 \cdot x + 3$$

Note that *both sets of parentheses are essential.*

$$(x + 2) \cdot x + 3 = x^2 + 2x + 3 \neq x^2 + 5x + 6$$
$$x + 2 \cdot (x + 3) = x + 2x + 6 = 3x + 6 \neq x^2 + 5x + 6$$
$$x + 2 \cdot x + 3 = x + 2x + 3 = 3x + 3 \neq x^2 + 5x + 6$$

EXAMPLE 3

Factor $m^2 - 9m + 8$, or write "not factorable."

SOLUTION

Steps 1 and 2. There is no common factor. The trinomial is in descending powers.

Step 3. $m^2 - 9m + 8$

———— A plus sign here tells us that the signs in the binomials will be the same

———— A minus sign here tells us that both signs will be minus signs

$(\ -\ )(\ -\ )$ ← The outline

Step 4. The first terms of the binomial factors must be m.

$(m -\quad)(m -\quad)$ ← The partially filled-in outline

Step 5. The pairs of negative factors of 8 are $(-1)(-8)$ and $(-2)(-4)$. We must select the pair whose sum is -9. This is the pair -1 and -8. (The *signs* of -1 and -8 are already in the outline.) Therefore,

$$m^2 - 9m + 8 = (m - 1)(m - 8) = (m - 8)(m - 1)$$

✓ **Step 6. Check** $(m - 1)(m - 8) = m^2 - 9m + 8$

Therefore, $m^2 - 9m + 8 = (m - 1)(m - 8)$ or $(m - 8)(m - 1)$.

✗ **A Word of Caution** Although $(m - 2)(m - 4)$ gives us the correct *first* term and the correct *third* term for the problem in Example 3, the *middle term* is incorrect:

$$(m - 2)(m - 4) = m^2 - 6m + 8 \neq m^2 - 9m + 8$$

EXAMPLE 4

Factor $4a - 12 + a^2$, or write "not factorable."

SOLUTION

Step 1. There is no common factor.

Step 2. $a^2 + 4a - 12$ The polynomial is now in descending powers of a

———— A minus sign here tells us that the signs in the binomials will be different

Step 3. $(\quad)(\quad)$ ← The outline

Step 4. The first terms of the binomial factors must be a.

$(a\quad)(a\quad)$ ← The partially filled-in outline

Step 5. The pairs of factors of -12 are $(1)(-12)$, $(-1)(12)$, $(2)(-6)$, $(-2)(6)$, $(3)(-4)$, and $(-3)(4)$. We must select the pair whose sum is $+4$; this is the pair -2 and 6. Therefore,

$$a^2 + 4a - 12 = (a - 2)(a + 6) = (a + 6)(a - 2)$$

Alternatively, we can consider only the pairs of *positive* factors of 12: $(1)(12)$, $(2)(6)$, and $(3)(4)$. Because the rule of signs tells us that the signs will be *different*, we select the pair of factors whose *difference* equals $|+4|$ (the absolute value of the coefficient of the middle term of the trinomial). This is the pair $(2)(6)$; we insert these numbers into the blank outline *before* we insert any signs:

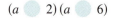

Now we insert the signs *so that the sum of the inner and outer products is equal to* $+4a$, the middle term of the trinomial. This gives

$$(a - 2)(a + 6)$$

✓ **Step 6. Check** $(a - 2)(a + 6) = a^2 + 4a - 12 = 4a - 12 + a^2$

Therefore, $4a - 12 + a^2 = (a - 2)(a + 6)$ or $(a + 6)(a - 2)$.

EXAMPLE 5

Factor $a^2 - 4a - 12$, or write "not factorable."

SOLUTION

$$a^2 - 4a - 12$$

A minus sign here tells us that the signs in the binomials will be different

See Example 4, Step 5, for pairs of factors of -12. We must select the pair whose sum is -4; the pair is 2 and -6. This gives

$$a^2 - 4a - 12 = (a + 2)(a - 6) = (a - 6)(a + 2)$$

✓ **Check** $(a + 2)(a - 6) = a^2 - 4a - 12$

Therefore, $a^2 - 4a - 12 = (a + 2)(a - 6)$ or $(a - 6)(a + 2)$.

EXAMPLE 6

Factor $3x^2 - 24x - 60$, or write "not factorable."

SOLUTION This polynomial is *not* in the form $x^2 + bx + c$; however, let's factor out the GCF, 3:

$$3x^2 - 24x - 60 = 3(x^2 - 8x - 20)$$

$x^2 - 8x - 20$ *is in the form* $x^2 + bx + c$

Although $3(x^2 - 8x - 20)$ has only one term and *has* been factored, it's not completely factored. We now try to factor $x^2 - 8x - 20$.

$$x^2 - 8x - 20$$

A minus sign here tells us that the signs in the binomials will be different

The pairs of factors of -20 are $(1)(-20)$, $(-1)(20)$, $(2)(-10)$, $(-2)(10)$, $(4)(-5)$, and $(-4)(5)$. The pair whose sum is -8 is the pair 2 and -10. Therefore,

$$x^2 - 8x - 20 = (x + 2)(x - 10)$$

and $$3x^2 - 24x - 60 = 3(x + 2)(x - 10)$$

✓ **Check** $3(x + 2)(x - 10) = 3(x^2 - 8x - 20) = 3x^2 - 24x - 60$

We must be sure to include 3 as a factor of the answer. Therefore, $3x^2 - 24x - 60 = 3(x + 2)(x - 10)$, or $3(x - 10)(x + 2)$.

EXAMPLE 7

Factor $x^4 + 9x^3 + 20x^2$, or write "not factorable."

SOLUTION There *is* a common factor. The GCF is x^2; we factor it out first. *We must then be sure to include x^2 as a factor of the answer.*

$$x^4 + 9x^3 + 20x^2 = x^2(x^2 + 9x + 20)$$

The expression $x^2(x^2 + 9x + 20)$ has been factored; however, the polynomial factor is still factorable:

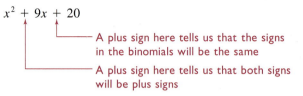

$$x^2 + 9x + 20$$

A plus sign here tells us that the signs in the binomials will be the same

A plus sign here tells us that both signs will be plus signs

The pairs of positive factors of 20 are $(1)(20)$, $(2)(10)$, and $(4)(5)$. The pair whose sum is 9 is the pair 4 and 5. This gives

$$x^4 + 9x^3 + 20x^2 = x^2(x^2 + 9x + 20) = x^2(x + 4)(x + 5)$$

√ **Check** $x^2(x + 4)(x + 5) = x^2(x^2 + 9x + 20) = x^4 + 9x^3 + 20x^2$

Therefore, $x^4 + 9x^3 + 20x^2 = x^2(x + 4)(x + 5)$ or $x^2(x + 5)(x + 4)$.

EXAMPLE 8   Factor $x^2 - 11xy + 24y^2$, or write "not factorable."

SOLUTION   There is no common factor.

$$x^2 - 11xy + 24y^2$$

A plus sign here tells us that both signs will be the same

A minus sign here tells us that they will both be minus signs

Because the third term contains the variable y^2, we must be sure to have $\sqrt{y^2} = y$ as a *factor* of the second term of each binomial.

$$\sqrt{y^2} = y$$
$$(x - \quad y)(x - \quad y) \leftarrow \text{The partially filled-in outline}$$

The pairs of negative factors of 24 are $(-1)(-24)$, $(-2)(-12)$, $(-3)(-8)$, and $(-4)(-6)$. The pair whose sum is -11 is the pair -3 and -8. (The *signs* of -3 and -8 are already in the outline.) Therefore,

$$x^2 - 11xy + 24y^2 = (x - 3y)(x - 8y)$$

√ **Check** $(x - 3y)(x - 8y) = x^2 - 11xy + 24y^2$

Therefore, $x^2 - 11xy + 24y^2 = (x - 3y)(x - 8y)$ or $(x - 8y)(x - 3y)$.

EXAMPLE 9   Factor $x^2 - 5x + 3$, or write "not factorable."

SOLUTION   There is no common factor.

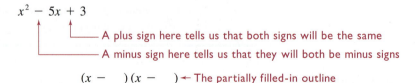

$$x^2 - 5x + 3$$

A plus sign here tells us that both signs will be the same

A minus sign here tells us that they will both be minus signs

$$(x - \quad)(x - \quad) \leftarrow \text{The partially filled-in outline}$$

We must find a pair of negative integers whose product is 3 and whose sum is -5. There is no such pair of numbers, since $(-1)(-3) = 3$ but $(-1) + (-3) \neq -5$. Therefore, $x^2 - 5x + 3$ is *not factorable*.

Exercises 7.3A
Set I

Factor each polynomial completely and check, or write "not factorable."

1. $x^2 + 6x + 8$
2. $x^2 + 9x + 8$
3. $x^2 + 5x + 4$
4. $x^2 + 4x + 4$
5. $k^2 + 7k + 6$
6. $k^2 + 5k + 6$
7. $7u + u^2 + 10$
8. $11u + u^2 + 10$
9. $y^2 - 2y + 8$
10. $y^2 - 7y + 8$
11. $b^2 - 9b + 14$
12. $b^2 - 15b + 14$
13. $z^2 - 9z + 20$
14. $z^2 - 12z + 20$
15. $18 + x^2 - 11x$
16. $18 + x^2 - 9x$
17. $x^2 + 9x - 10$
18. $y^2 - 3y - 10$
19. $z^2 - z - 6$
20. $m^2 + 5m - 6$
21. $5x^2 + 10x$
22. $8y^3 + 4y^2$
23. $x^2 + 4x - 5$
24. $y^2 + 6y - 7$

25. $x^2 + 100$
26. $z^2 + 1$
27. $z^5 + 9z^4 - 10z^3$
28. $x^4 + 7x^3 - 8x^2$
29. $t^2 + 11t - 30$
30. $m^2 - 17m - 30$
31. $u^4 + 12u^2 - 64$
32. $v^4 - 30v^2 - 64$
33. $16 + v^2 - 8v$
34. $16 + v^2 - 10v$
35. $b^2 - 11bd - 60d^2$
36. $c^2 + 17cx - 60x^2$
37. $r^2 - 13rs - 48s^2$
38. $s^2 + 22st - 48t^2$
39. $x^4 + 2x^3 - 35x^2$
40. $x^4 + 2x^3 - 48x^2$
41. $14x^2 - 15x + x^3$
42. $8x^3 - 9x^2 + x^4$
43. $3x^2 + 6x - 24$
44. $5x^2 + 15x - 50$
45. $12 + 4x^2 - 16x$
46. $24 + 2x^2 - 14x$
47. $x^4 + 6x^3 + x^2$
48. $y^4 + 5y^3 + y^2$
49. $x^4 - 9$
50. $a^4 - 25$

Writing Problems

Express the answers in your own words and in complete sentences.

1. Explain why $x^2 + 7x + 6$ factors into $(x + 1)(x + 6)$.

2. Explain why $x^2 + x + 1$ is not factorable.

3. Explain why $x^2 - 5x + 6 \neq (x + 1)(x - 6)$.

4. Explain why $x^2 - 5x - 6 \neq (x - 2)(x - 3)$.

Exercises 7.3A
Set II

Factor each polynomial completely and check, or write "not factorable."

1. $m^2 + 7m + 12$
2. $x^2 + 9x + 20$
3. $h^2 + 11h + 18$
4. $x^2 + 13x + 12$
5. $a^2 + 10a + 16$
6. $x^2 + x + 1$
7. $10k + k^2 + 24$
8. $21y + y^2 + 20$
9. $x^2 - x + 12$
10. $x^2 - 7x + 12$
11. $n^2 - 9n + 18$
12. $x^2 - 5x + 4$
13. $z^2 - 8z + 16$
14. $y^2 - 2y + 9$
15. $15 + x^2 - 8x$
16. $13n + n^2 - 14$
17. $w^2 + 2w - 24$
18. $3z - 18 + z^2$
19. $r^2 - 9r - 20$
20. $y^2 - 8y - 20$
21. $3x^2 + 12x$
22. $c^2 + 121$
23. $x^2 - 4x - 45$
24. $a^4 + a^3 + a^2$

25. $a^2 + 25$
26. $5a^2 + 25$
27. $z^5 - 5z^4 - 6z^3$
28. $x^4 + 5x^3 + 5x^2$
29. $n^2 - 10n - 24$
30. $x^2 + 4x - 45$
31. $v^4 - 18v^2 + 45$
32. $u^4 - 4u^3 - 21u^2$
33. $36 + y^2 - 15y$
34. $u^4 - 21 - 4u^2$
35. $w^2 - 8wz - 48z^2$
36. $x^2 - 15xy + 54y^2$
37. $p^2 - 9pt - 52t^2$
38. $a^2 + 10ab + 16b^2$
39. $s^4 + 3s^3 - 28s^2$
40. $s^2 - 3st + 28t^2$
41. $y^3 - 30y^2 + y^4$
42. $y^2 - 11yz + 30z^2$
43. $4x^2 - 8x - 32$
44. $6x^2 + 6 + 6x$
45. $-6x + 2x^2 - 20$
46. $-5x + 5x^3$
47. $t^4 + 5t^3 + t^2$
48. $x^2 - 5x - 24$
49. $x^2 - 121$
50. $y^2 + 121$

7.3B Factoring a Trinomial of the Form $ax^2 + bx + c$

A trinomial of the form $ax^2 + bx + c$ (or $ax^2 + bxy + cy^2$) is often called the *general trinomial*, and the method we now show for factoring it might be called the *trial method* or the guessing-and-checking method. When we try to factor such a trinomial, we *assume* that it will factor into a product of two binomials. (We may decide later that the trinomial does *not* factor.)

The following method can be used for factoring any trinomial that can be expressed in the form $ax^2 + bx + c$.

Factoring a trinomial that can be expressed in the form $ax^2 + bx + c$	**1.** Factor out the GCF, if there is one. **2.** Arrange the trinomial in descending (or ascending) powers of one variable. (If the leading coefficient is negative, either arrange the trinomial in ascending powers or factor out -1 before proceeding.) **3. a.** If the leading coefficient of the trinomial is 1, factor by using the techniques given in the preceding section. **b.** If the leading coefficient of the trinomial is *not* 1, proceed with steps 4–7. **4.** Make a blank outline, and fill in the *variable parts* of each binomial. (In deciding what the variable parts of each term of the binomial should be, use the method given in the preceding section.) If the rule of signs indicates that the signs of the last terms of the binomials will both be positive or both be negative, you may put those signs in the outline. **5.** List (mentally, at least) all pairs of factors of the coefficient of the first term of the trinomial and of the last term of the trinomial. **6.** By guessing and checking, select the pairs of factors (if they exist) from step 5 that make the sum of the inner and outer products of the binomials equal to the middle term of the trinomial. If no such pairs exist, the trinomial is *not factorable*. **7.** Check the result by multiplying the binomials together; the resulting product should be the trinomial from step 2. Check to be sure that none of the factors can be factored further.

EXAMPLE 10   Factor $2x^2 + 7x + 5$, or write "not factorable."

SOLUTION   There is no common factor; the trinomial is already in descending powers of x. The signs of the last terms of the binomial factors will both be plus signs.

$$(\quad + \quad)(\quad + \quad)$$

The first term of each binomial factor must contain an x (as a factor) in order to give the x^2 in $2x^2$, the first term of the trinomial.

$$(\; x + \quad)(\; x + \quad)$$

The only factors of 2 are 1 and 2. We put these numbers in the outline.

$$(\; 1x + \quad)(\; 2x + \quad)$$

We have two ways to try the factors of 5:

$$(1x + 5)(2x + 1) = 2x^2 + 11x + 5$$

└──── An incorrect middle term

or

The correct factored form

$$(1x + 1)(2x + 5) = 2x^2 + 7x + 5$$

The correct middle term

The sum of the inner and outer products must equal the middle term of the trinomial. We found the correct pair by guessing-and-checking.

✓ **Check** $(x + 1)(2x + 5) = 2x^2 + 7x + 5$

Therefore, $2x^2 + 7x + 5 = (x + 1)(2x + 5)$ or $(2x + 5)(x + 1)$.

EXAMPLE 11

Factor $13x + 5x^2 + 6$, or write "not factorable."

SOLUTION There is no common factor.

$$13x + 5x^2 + 6 = 5x^2 + 13x + 6 \leftarrow \text{Arranging in descending powers of } x$$

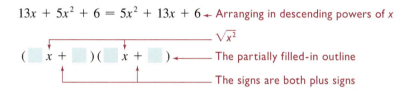
$\sqrt{x^2}$

$(\ \ x + \ \ )(\ \ x + \ \ ) \leftarrow$ The partially filled-in outline

The signs are both plus signs

Next, we list all pairs of factors of the leading coefficient and of the last term of the trinomial.

Factors of the leading coefficient	*Factors of the last term*
$5 = 1 \cdot 5$	$6 = 1 \cdot 6 = 2 \cdot 3$

There are two pairs of factors to try for 6: (1)(6) and (2)(3). Therefore, we must try the following combinations:

$$(5x + 1)(x + 6)$$
$$(5x + 6)(x + 1)$$
$$(5x + 2)(x + 3)$$
$$(5x + 3)(x + 2)$$

In all these combinations, the product of the two first terms is $5x^2$, and the product of the two last terms is 6. However, only in the last combination is the sum of the inner and outer products equal to $13x$.

✓ **Check** $(5x + 3)(x + 2) = 5x^2 + 13x + 6 = 13x + 5x^2 + 6$

Therefore, $13x + 5x^2 + 6 = (5x + 3)(x + 2)$ or $(x + 2)(5x + 3)$.

When the leading coefficient or the coefficient of the third term is a prime number, it is probably easiest to insert the factors of that number first (see Example 12).

EXAMPLE 12

Factor $4z^2 - 3z - 7$, or write "not factorable."

SOLUTION There is no common factor, and the polynomial is in descending powers of z. The signs will be different. Because 7 is a prime number, we will fill in the factors of 7 first.

$(\ \ z \ \ 7)(\ \ z \ \ 1) \leftarrow$ The partially filled-in outline

Now we must insert the factors of 4 (either 1 and 4 or 2 and 2) and the signs so that the sum of the inner and outer products is $-3z$. We must consider the following combinations:

$(2z + 7)(2z - 1)$ $(2z - 7)(2z + 1)$

$(1z + 7)(4z - 1)$ $(1z - 7)(4z + 1)$

$(4z + 7)(1z - 1)$ $(4z - 7)(1z + 1)$ ←This is the only combination in which the sum of the inner and outer products is $-3z$

√ **Check** $(4z - 7)(z + 1) = 4z^2 - 3z - 7$

Therefore, $4z^2 - 3z - 7 = (4z - 7)(z + 1)$ or $(z + 1)(4z - 7)$.

EXAMPLE 13

Factor $5x^2 - x + 2$, or write "not factorable."

SOLUTION These are the only combinations we need to consider:

$(5x - 1)(x - 2)$ The sum of the outer and inner products is $-11x$

$(5x - 2)(x - 1)$ The sum of the outer and inner products is $-7x$

In both of these combinations, the product of the two first terms is $5x^2$ and the product of the two last terms is $+2$. However, in neither case is the sum of the inner and outer products $-x$. Therefore, $5x^2 - x + 2$ is *not factorable*.

EXAMPLE 14

Factor $19xy - 15y^2 - 6x^2$, or write "not factorable."

SOLUTION Because the coefficients of x^2 and y^2 are both negative, we factor out -1 before proceeding.

$$19xy - 15y^2 - 6x^2 = -6x^2 + 19xy - 15y^2 = -1(6x^2 - 19xy + 15y^2)$$

We now factor $6x^2 - 19xy + 15y^2$.

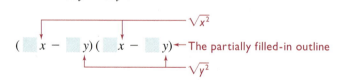

$(\quad x - \quad y)(\quad x - \quad y)$ ←The partially filled-in outline

These are the combinations we must consider:

$(6x - 15y)(1x - 1y)$ $(6x - 1y)(1x - 15y)$

$(6x - 3y)(1x - 5y)$ $(6x - 5y)(1x - 3y)$

$(2x - 15y)(3x - 1y)$ $(2x - 1y)(3x - 15y)$

$(2x - 5y)(3x - 3y)$ $(2x - 3y)(3x - 5y)$ ← This is the only combination in which the sum of the inner and outer products is $-19xy$

When we check the result, we must remember that we had factored out -1 earlier, and we must multiply $(2x - 3y)(3x - 5y)$ *by this -1*.

√ **Check** $-1(2x - 3y)(3x - 5y) = -1(6x^2 - 19xy + 15y^2)$

$$= -6x^2 + 19xy - 15y^2 = 19xy - 15y^2 - 6x^2$$

Therefore, $19xy - 15y^2 - 6x^2 = -(2x - 3y)(3x - 5y)$. Other acceptable answers include $(3y - 2x)(3x - 5y)$ and $(2x - 3y)(5y - 3x)$.

EXAMPLE 15

Factor $10u^2 - 12u + 8u^3$, or write "not factorable."

SOLUTION   We first arrange the trinomial in descending powers of u and factor out the GCF.

$$10u^2 - 12u + 8u^3 = 8u^3 + 10u^2 - 12u = 2u(4u^2 + 5u - 6)$$

We now try to factor the polynomial in parentheses, $4u^2 + 5u - 6$.

$$(\quad u \bigcirc \blacksquare\ )(\quad u \bigcirc \blacksquare\ ) \leftarrow \text{The partially filled-in outline}$$

The combinations to consider:

$(2u + 6)(2u - 1)$	$(2u - 6)(2u + 1)$
$(2u + 2)(2u - 3)$	$(2u - 2)(2u + 3)$
$(4u + 1)(1u - 6)$	$(4u - 1)(1u + 6)$
$(4u + 6)(1u - 1)$	$(4u - 6)(1u + 1)$
$(4u + 2)(1u - 3)$	$(4u - 2)(1u + 3)$
$(4u + 3)(1u - 2)$	$(4u - 3)(1u + 2)$ $\leftarrow$ This is the only combination in which the sum of the inner and outer products is $+5u$

If our factoring was correct,

$$4u^2 + 5u - 6 = (4u - 3)(u + 2)$$

and

$$10u^2 - 12u + 8u^3 = 2u(4u - 3)(u + 2)$$

✓ **Check** $2u(4u - 3)(u + 2) = 2u(4u^2 + 5u - 6) = 8u^3 + 10u^2 - 12u$

$$= 10u^2 - 12u + 8u^3$$

Therefore, $10u^2 - 12u + 8u^3 = 2u(4u - 3)(u + 2)$ or $2u(u + 2)(4u - 3)$.

Note   It is not necessary to write down all possible combinations as we did in Examples 11 through 15; in practice, we usually try out each combination as we write it down, rather than listing all possible combinations first.

Hint: If the original trinomial does not have a common factor, then neither *binomial factor* can have a common factor. This fact can help eliminate some combinations from consideration (see Example 16).

EXAMPLE 16

Factor $12a^2 + 7ab - 10b^2$, or write "not factorable."

SOLUTION   There is no common factor. The outline is as follows:

$$(\quad a \bigcirc \blacksquare\ b)(\quad a \bigcirc \blacksquare\ b)$$

If we first try 3 and 4 as the factors of 12, and 2 and 5 as the factors of 10, we have these combinations to consider:

★ $(3a + 5b)(4a - 2b)$	★ $(3a - 5b)(4a + 2b)$
$(3a + 2b)(4a - 5b)$	$(3a - 2b)(4a + 5b)$ $\leftarrow$ In this combination, the sum of the inner and outer products is $+7ab$

The starred combinations both have a common factor of 2 in one of the binomials. Since the original trinomial had no common factor, these combinations need not be given any consideration.

✓ **Check** $(3a - 2b)(4a + 5b) = 12a^2 + 7ab - 10b^2$

Therefore, $12a^2 + 7ab - 10b^2 = (3a - 2b)(4a + 5b)$.

Sometimes, the first term is a constant and the last term contains the variable. In this case, we proceed in almost the same way as when the first term contains the variable (see Example 17).

EXAMPLE 17 Factor $8 + 2x - x^2$, or write "not factorable."

SOLUTION If we arranged this trinomial in descending powers of x, the first term would be *negative*. The expression is easier to factor if we leave it in ascending powers of x. The outline is as follows:

$$(\;\blacksquare\;\bullet\;x)(\;\blacksquare\;\bullet\;x)$$

If we first try 2 and 4 as the factors of 8, we have this combination:

$$(2\;\bullet\;x)(4\;\bullet\;x)$$

If we put the signs in as

$$\overline{(2 - x)(4 + x)}$$

we do *not* get the correct middle term; we get $-2x$, not $+2x$. Therefore, we'll try

$$(2 + x)(4 - x)$$

✓ **Check** $(2 + x)(4 - x) = 8 + 2x - x^2$

Therefore, $8 + 2x - x^2 = (2 + x)(4 - x)$ or $(4 - x)(2 + x)$.

When the first and third terms of the trinomials are perfect squares, the trinomial *may* be the square of a binomial.

EXAMPLE 18 Factor $9x^2 - 12xy + 4y^2$, or write "not factorable."

SOLUTION Because $9x^2 = (3x)^2$ and $4y^2 = (2y)^2$ (that is, the first and third terms are perfect squares), it is possible that $9x^2 - 12xy + 4y^2$ is the square of a binomial. Since the sign of the middle term of the trinomial is negative, we'll try $(3x - 2y)^2$.

✓ **Check** $(3x - 2y)^2 = (3x)^2 - 2(3x)(2y) + (2y)^2 = 9x^2 - 12xy + 4y^2$

Therefore, $9x^2 - 12xy + 4y^2 = (3x - 2y)^2$.

 Note $9x^2 - 12xy + 4y^2$ could also have been factored by using the trial method; if this had been done, the answer would have been $(3x - 2y)(3x - 2y)$, which is also an acceptable answer.

Note If the answer to each of the following four questions is "yes," the trinomial *is definitely* the square of a binomial:

1. Are the first and third terms both positive? (If they are both *negative*, factor out -1 before proceeding.)

2. Is the first term a perfect square?

3. Is the third term a perfect square?

4. Is the absolute value of the middle term *twice* the product of the square root of the first term and the square root of the third term?

If the answer to any of these questions is "no," the trinomial is not the square of a binomial, but it might be factorable by using the trial method.

You might want to begin the factoring of any trinomial by asking yourself whether it is the square of a binomial.

Exercises 7.3B
Set I

Factor each polynomial completely and check, or write "not factorable."

1. $3x^2 + 7x + 2$
2. $3x^2 + 5x + 2$
3. $5x^2 + 7x + 2$
4. $5x^2 + 11x + 2$
5. $7x + 4x^2 + 3$
6. $13x + 4x^2 + 3$
7. $5x^2 + 20x + 4$
8. $5x^2 + 11x + 4$
9. $5a^2 - 16a + 3$
10. $5m^2 - 8m + 3$
11. $3b^2 - 22b + 7$
12. $3u^2 - 10u + 7$
13. $5z^2 - 36z + 7$
14. $5z^2 - 12z + 7$
15. $14n - 5 + 3n^2$
16. $2k - 7 + 5k^2$
17. $9x^2 - 49$
18. $16y^2 - 1$
19. $7x^2 + 23xy + 6y^2$
20. $7a^2 + 43ab + 6b^2$
21. $7h^2 - 11hk + 4k^2$
22. $7h^2 - 16hk + 4k^2$

23. $3t^2 + 19tz - 6z^2$
24. $3w^2 - 11wx - 6x^2$
25. $49x^2 - 42x + 9$
26. $25x^2 - 20x + 4$
27. $18u^3 + 39u^2 - 15u$
28. $12y^3 + 14y^2 - 10y$
29. $4x^2 + 4x + 4$
30. $6x^2 + 6x + 18$
31. $7x^2 - 49$
32. $5x^2 - 25$
33. $6 - 17v + 5v^2$
34. $6 - 11v + 5v^2$
35. $20x + 3x^2 + 12$
36. $13x + 3x^2 + 12$
37. $45x^2 - 120x + 80$
38. $64x^2 - 96x + 36$
39. $8x^2 - 2x - 15$
40. $8x^2 - 19x - 15$
41. $6y^2 - 19y + 10$
42. $6y^2 - 23y + 10$
43. $9a^2 + 24a - 20$
44. $9b^2 + 3b - 20$
45. $6e^4 - 7e^2 - 20$
46. $10f^4 - 29f^2 - 21$
47. $x^2 + 144$
48. $y^2 + 9$

Exercises 7.3B
Set II

Factor each polynomial completely and check, or write "not factorable."

1. $3y^2 + 16y + 5$
2. $2z^2 + 9z + 7$
3. $4a^2 + 9a + 5$
4. $3b^2 + b + 4$
5. $13x + 6x^2 + 2$
6. $2 + 6x^2 + 7x$
7. $2x^2 + 6x + 3$
8. $6x^2 + 15x + 9$
9. $7a^2 - 22a + 3$
10. $7a^2 - 20a - 3$
11. $2x^2 - 15x + 7$
12. $7 + 2x^2 + 9x$
13. $3t^2 - 16t + 5$
14. $3t^2 - 16t - 5$
15. $13x - 7 + 2x^2$
16. $9x + 2x^2 - 7$
17. $25x^2 - 16$
18. $16x^2 + 25$
19. $5x^2 + 13x + 6$
20. $10a^3 + 14a^2 - 12a$
21. $5x^2 - 17x + 6$
22. $5x^2 - 13x - 6$

23. $5x^2 + 30x + 6$
24. $5x^2 + 11x + 6$
25. $25s^2 - 30s + 9$
26. $9x^2 - 37x + 4$
27. $14m^3 - 20m^2 + 6m$
28. $3e^2 - 20e - 7$
29. $5s^2 - 5s + 5$
30. $5h^2 - 8h + 3$
31. $6x^2 - 36$
32. $36x^2 - 1$
33. $12 - 17u - 5u^2$
34. $8ab + 5b^2 + 3a^2$
35. $15x + 2x^2 + 18$
36. $7x^2 + 2x - 5$
37. $70k^2 - 24k + 2$
38. $6s^2 - 17st - 5t^2$
39. $12x^2 + 7x - 10$
40. $8x^2 - 6x - 9$
41. $5x^2 - 16x + 12$
42. $6t^2 - 5t - 21$
43. $4m^2 - 16m + 7$
44. $14x^2 - 4x - 10$
45. $8v^4 - 14v^2 - 15$
46. $8x^3 - 8x^2 + 8x$
47. $z^2 + 4$
48. $2x^2 - 72$

Sections 7.1–7.3

REVIEW

Greatest Common Factor
7.1

The **greatest common factor (GCF)** of two integers is the largest integer that is a factor of both integers.

Methods of Factoring
7.1

1. If there is a common factor, find the greatest common factor (GCF) and factor it out.

7.2

2. A difference of two squares is always factorable:

$$a^2 - b^2 = (a + b)(a - b)$$

7.3A

3. To factor a trinomial that can be expressed in the form $x^2 + bx + c$, see Section 7.3A.

7.3B

4. Use guessing-and-checking to try to factor a trinomial that can be expressed in the form $ax^2 + bx + c$.

Sections 7.1–7.3   **REVIEW EXERCISES   Set I**

Factor each polynomial completely and check, or write "not factorable."

1. $8x - 4$

2. $5 + 7a$

3. $m^2 - 4$

4. $25 + n^2$

5. $x^2 + 10x + 21$

6. $y^2 - 10y + 15$

7. $2u^2 + 4u$

8. $3b - 6b^2 + 12b^3$

9. $z^2 - 7z - 18$

10. $7x - 30 - x^2$

11. $4x^2 - 25x + 6$

12. $4x^2 + 20x + 25$

13. $9k^2 - 144$

14. $100 + 81p^2$

15. $8 - 2a^2$

16. $10c^2 - 44c + 16$

17. $15u^2v - 3uv$

18. $5ab^2 - 10a^2b - 5ab$

19. $10x^2 - xy - 24y^2$

20. $8x^5y^2 - 12x^2y^4 - 16x^2y^2$

21. $3y^2 - 300$

22. $t^2 - 72 - t$

23. $5u^2 + 2ut - 7t^2$

24. $x^2 + x + 8$

25. $7x^2 + 8x + 1$

26. $6x^2 + 48x + 90$

27. $x^2y^2 + 9$

28. $s^2 - 16s + 63$

29. $3(u + v) - t(u + v)$

30. $3st - 8uv$

31. $4m^2 + 23m - 6$

32. $v(w - 5) - 8(w - 5)$

_____

Name

Factor each polynomial completely and check, or write "not factorable."

1. $6 - 30u$

2. $36 - m^2$

3. $8x + 5y$

4. $x^2 + 9x + 18$

5. $12v^2 - 8v$

6. $z^2 + 2z - 35$

7. $4e^2 - 27e + 18$

8. $x^2 - 8x + 8$

9. $64 - 49p^2$

10. $48 - 3a^2$

11. $16x^2 - 8x + 1$

12. $3x^2 + 6xy + 9y^2$

13. $3a^2b - 15ab^2 - 3ab$

14. $24hk^2 - 6hk$

15. $40x^2 + 10xy - 50y^2$

16. $21u^2 + 13uv - 20v^2$

ANSWERS

1. _____

2. _____

3. _____

4. _____

5. _____

6. _____

7. _____

8. _____

9. _____

10. _____

11. _____

12. _____

13. _____

14. _____

15. _____

16. _____

17. $8m^3n^2 - 14mn^4 - 6mn^2$

18. $4x^2 + 8x + 8$

19. $28x^2 - 13xy - 6y^2$

20. $21ab^2 - 7ab$

21. $x^2 - x - 56$

22. $u^2 + 45 - 14u$

23. $y^2 + y + 9$

24. $11x^2 - 78xy + 7y^2$

25. $10x^2 - 10x - 60$

26. $11x^2 - 12x + 1$

27. $u(v + 6) + w(v + 6)$

28. $16 + a^2b^2$

29. $6c^2 - 23c - 4$

30. $5(x + y) - z(x + y)$

31. $2y^2 - 128$

32. $5xy + 10uv$

17. _____

18. _____

19. _____

20. _____

21. _____

22. _____

23. _____

24. _____

25. _____

26. _____

27. _____

28. _____

29. _____

30. _____

31. _____

32. _____

7.4

Factoring by Grouping

If a polynomial contains four terms, we can sometimes factor it by first separating its terms into two groups and factoring each group separately. Since we will still have more than one term at this point, the expression will not yet be factored. However, if we then see that each of the groups has a common factor, we will be able to factor the polynomial.

The procedure that follows assumes that any factor common to all *four* terms has already been factored out.

Factoring an expression of four terms by grouping two and two	**1.** Arrange the four terms into two groups of two terms each so that each group of two terms is factorable.
	2. Factor each *group*. You will now have two terms. *The expression will not yet be factored.*
	3. If the two terms from step 2 now have a common binomial factor, factor out that binomial factor, and the polynomial is factored. If the two terms from step 2 do *not* have a common binomial factor, try a different arrangement of the original four terms; if you cannot arrange the terms of the original polynomial so that there is a common binomial factor in step 2, you cannot factor the polynomial.*

*Some polynomials that contain four terms can be grouped so that a difference of two squares is formed. However, we do not discuss such factoring in this book.

EXAMPLE 1 Factor $ax + ay + bx + by$.

SOLUTION

The GCF $= a$ ⎯⎯⎯   ⎯⎯⎯ The GCF $= b$

Step 1. $ax + ay + bx + by$

This polynomial is not in factored form because it has two terms

Step 2. $= a(\,x + y\,) + b(\,x + y\,)$ $(x + y)$ is the GCF of these two terms

Step 3. $= (x + y)(a + b)$

The GCF

This term is $\dfrac{b(x + y)}{(x + y)} = b$

This term is $\dfrac{a(x + y)}{(x + y)} = a$

✓ **Check** $(x + y)(a + b) = (x + y)a + (x + y)b = ax + ay + bx + by$

Therefore, $ax + ay + bx + by = (x + y)(a + b)$ or $(a + b)(x + y)$.

✕ **A Word of Caution**   Be careful not to make the following common error:

$$\underset{\substack{\\ \text{No sign here}}}{ax + ay + bx + by = a(x + y) \quad b(x + y)}$$

$$= (x + y)(a + b)$$

Even though the final answer is correct, *two* errors have been made, since

$$ax + ay + bx + by \neq a(x + y) \quad b(x + y)$$

and $$a(x + y) \quad b(x + y) \neq (x + y)(a + b)$$

It is often possible to group terms differently and still be able to factor the expression (see Examples 2 and 3). *The same factors are obtained no matter what grouping is used*, because factorization over the integers is unique.

EXAMPLE 2   Factor $ab - b + a - 1$.

SOLUTION

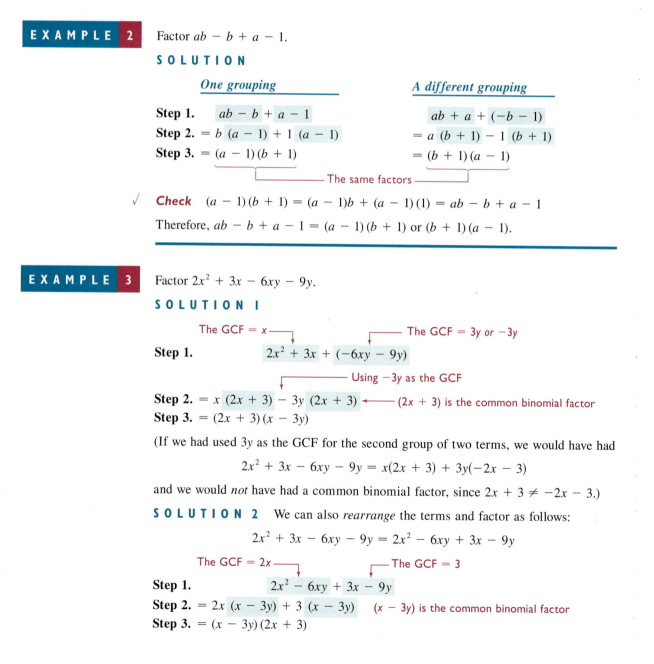

One grouping	*A different grouping*
Step 1. $ab - b + a - 1$	$ab + a + (-b - 1)$
Step 2. $= b\,(a - 1) + 1\,(a - 1)$	$= a\,(b + 1) - 1\,(b + 1)$
Step 3. $= (a - 1)(b + 1)$	$= (b + 1)(a - 1)$

—————— The same factors ——————

✓ **Check** $(a - 1)(b + 1) = (a - 1)b + (a - 1)(1) = ab - b + a - 1$

Therefore, $ab - b + a - 1 = (a - 1)(b + 1)$ or $(b + 1)(a - 1)$.

EXAMPLE 3   Factor $2x^2 + 3x - 6xy - 9y$.

SOLUTION I

The GCF $= x$ ⟶   ⟵ The GCF $= 3y$ or $-3y$

Step 1. $2x^2 + 3x + (-6xy - 9y)$

Using $-3y$ as the GCF

Step 2. $= x\,(2x + 3) - 3y\,(2x + 3)$ ⟵ $(2x + 3)$ is the common binomial factor
Step 3. $= (2x + 3)(x - 3y)$

(If we had used $3y$ as the GCF for the second group of two terms, we would have had

$$2x^2 + 3x - 6xy - 9y = x(2x + 3) + 3y(-2x - 3)$$

and we would *not* have had a common binomial factor, since $2x + 3 \neq -2x - 3$.)

SOLUTION 2   We can also *rearrange* the terms and factor as follows:

$$2x^2 + 3x - 6xy - 9y = 2x^2 - 6xy + 3x - 9y$$

The GCF $= 2x$ ⟶   ⟵ The GCF $= 3$

Step 1. $2x^2 - 6xy + 3x - 9y$

Step 2. $= 2x\,(x - 3y) + 3\,(x - 3y)$ $(x - 3y)$ is the common binomial factor
Step 3. $= (x - 3y)(2x + 3)$

$\checkmark$ **Check** $(2x + 3)(x - 3y) = (2x + 3)(x) + (2x + 3)(-3y) = 2x^2 + 3x - 6xy - 9y$

Therefore, $2x^2 + 3x - 6xy - 9y = (2x + 3)(x - 3y)$, or $(x - 3y)(2x + 3)$.

In Example 4, one group of two terms does not have a common factor. Rather, it factors because it is a difference of two squares.

EXAMPLE 4 Factor $a^2 - b^2 + 3a - 3b$.

SOLUTION

A difference of two squares

3 is a common factor

Step 1. $a^2 - b^2 + 3a - 3b$

$a - b$ is the common binomial factor

Step 2. $= (a + b)(a - b) + 3(a - b)$

Step 3. $= (a - b)[(a + b) + 3]$

This term is $\dfrac{3(a - b)}{(a - b)}$

This term is $\dfrac{(a + b)(a - b)}{(a - b)}$

$\checkmark$ **Check** $(a - b)(a + b + 3) = (a - b)a + (a - b)b + (a - b)(3)$

$$= a^2 - ab + ab - b^2 + 3a - 3b$$

$$= a^2 - b^2 + 3a - 3b$$

Therefore, $a^2 - b^2 + 3a - 3b = (a - b)(a + b + 3)$.

✗ **A Word of Caution**   An expression is not factored until it has been written as a single term that is a product of factors. To see this, let us consider Example 1 again.

$$ax + ay + bx + by$$

$$= a(x + y) + b(x + y)$$ This expression is *not* in factored form because it has two terms

First term   Second term

$$= (x + y)(a + b)$$ The factored form of $ax + ay + bx + by$

Single term

Expressions with more than four terms may also be factored by grouping; however, in this book we consider factoring only expressions of four terms by grouping, and we consider only those problems in which there is a common binomial factor in step 2.

Exercises **7.4**
Set I

Factor each polynomial completely and check, or write "not factorable."

1. $am + bm + an + bn$

2. $cu + cv + du + dv$

3. $st + 4t + 3su + 12u$

4. $5cz + 5z + 7ct + 7t$

5. $3xr - 6yr + 4x - 8y$

6. $4ms - 6mt + 10ns - 15nt$

7. $mx - nx - my + ny$

8. $ah - ak - bh + bk$

9. $xy + x - y - 1$

10. $ad - d + a - 1$

11. $3a^2 - 6ab + 2a - 4b$

12. $2h^2 - 6hk + 5h - 15k$

13. $6e^2 - 2ef - 9e + 3f$

14. $8m^2 - 4mn - 6m + 3n$

15. $h^2 - k^2 + 2h + 2k$

16. $x^2 - y^2 + 4x + 4y$

17. $x^3 + 3x^2 - 4x + 12$

18. $a^3 + 5a^2 - 2a + 10$

19. $a^3 - 2a^2 - 4a + 8$

20. $x^3 - 3x^2 - 9x + 27$

21. $10xy - 15y + 8x - 12$

22. $35 - 42m - 18mn + 15n$

23. $a^2 - 4 + ab - 2b$

24. $x^2 - 25 - xy + 5y$

Writing Problems

Express the answer in your own words and in complete sentences.

1. Find and describe two errors in the following:

$$ac + bc + ad + bd = c(a + b)d(a + b) = (a + b)(c + d)$$

Exercises **7.4**
Set II

Factor each polynomial completely and check, or write "not factorable."

1. $cz + dz + cw + dw$

2. $sx + sy + tx + ty$

3. $xz + 5z + 2sx + 10s$

4. $4cz - 10c + 6bz - 15b$

5. $4ab - 12ac + 3b - 9c$

6. $5yz - 3xz + 15xy - 9x^2$

7. $hw - kw - hz + kz$

8. $bx - by - cx + cy$

9. $ef + f - e - 1$

10. $ef + f - e + 1$

11. $2s^2 - 6st + 5s - 15t$

12. $6cd + 4c - 15de - 10e$

13. $6a^2 - 3ab - 14a + 7b$

14. $10x^2 - 5xy - 4x + 2y$

15. $x^2 - y^2 + 5x - 5y$

16. $s^2 - t^2 + 2s + 2t$

17. $y^3 - 2y^2 - 4y - 8$

18. $a^3 - 3a^2 - 5a + 15$

19. $u^3 - 3u^2 - 9u + 27$

20. $v^3 + 2v^2 - 25v - 50$

21. $12xy - 8y + 15x - 10$

22. $x^3 - 2x^2 + 9x - 18$

23. $x^2 - 9 + xy + 3y$

24. $y^2 - 16 - xy - 4x$

7.5

The Master Product Method for Factoring Trinomials (Optional)

The master product method, which makes use of factoring by grouping, is a method for factoring trinomials that eliminates some of the guesswork. It can also be used to determine whether or not a trinomial is factorable. The procedure that follows assumes that any common factors have been factored out.

Factoring the general trinomial by the master product method	Arrange the trinomial in descending powers of one variable: $$ax^2 + bxy + cy^2$$ 1. Find the master product (MP) by multiplying the first and last coefficients of the trinomial being factored (MP $= a \cdot c$). 2. Write the pairs of factors of the master product (MP). 3. Choose the pair of factors whose sum is the coefficient of the middle term (b). 4. Rewrite the given trinomial, replacing the middle term with a sum of two terms whose coefficients are the pair of factors found in step 3. 5. Factor the expression in step 4 by grouping. 6. Check the result by multiplying the binomial factors to see if their product is the given trinomial. Check to be sure that neither factor can be factored further.

EXAMPLE 1   Factor $7x + 2x^2 + 5$ by the master product method.

SOLUTION

$$7x + 2x^2 + 5 = 2x^2 + 7x + 5 \longleftarrow \text{Arranging in descending powers of } x$$

Step 1. The master product (MP) $= (2)(+5) = 10$; the middle coefficient is $+7$.

Step 2.
$$10 = (-1)(-10) = (1)(10)$$
$$= (-2)(-5) = (2)(5)$$

Step 3. The pair whose sum is $+7$ is $(2)(5)$.

Step 4. $2x^2 + 7x + 5 = 2x^2 + 2x + 5x + 5$ Replacing $7x$ with $2x + 5x$

Step 5. $= 2x(x + 1) + 5(x + 1)$ Factoring by grouping

$$= (x + 1)(2x + 5)$$

√ **Step 6. Check** $(x + 1)(2x + 5) = 2x^2 + 7x + 5 = 7x + 2x^2 + 5$

Therefore, $7x + 2x^2 + 5 = (x + 1)(2x + 5)$ or $(2x + 5)(x + 1)$. You can verify that if, in step 4, we had written the polynomial as $2x^2 + 5x + 2x + 5$, the final factored form would have been the same.

EXAMPLE 2   Factor $4z^2 - 3z - 7$ by the master product method.

SOLUTION

Step 1. The master product (MP) $= (4)(-7) = -28$; the middle coefficient is -3.

Step 2. $-28 = (-1)(28) = (1)(-28)$

$= (-2)(14) = (2)(-14)$

Step 3. $= (-4)(7) = (4)(-7) \longleftarrow$ This is the pair whose sum is -3

Step 4. $4z^2 - 3z - 7 = 4z^2 + 4z - 7z - 7$ Replacing $-3z$ with $4z - 7z$

Step 5. $= 4z(z + 1) - 7(z + 1)$ Factoring by grouping

                             $= (z + 1)(4z - 7)$

√ Step 6. **Check** $(z + 1)(4z - 7) = 4z^2 - 3z - 7$

Therefore, $4z^2 - 3z - 7 = (z + 1)(4z - 7)$.

EXAMPLE 3

Factor $12a^2 + 5a - 10$, or write "not factorable."

SOLUTION

Step 1. The master product (MP) $= (12)(-10) = -120$; the middle coefficient is $+5$.

Step 2.

$$-120 = (-1)(120) = (1)(-120)$$
$$= (-2)(60) = (2)(-60)$$
$$= (-3)(40) = (3)(-40)$$
$$= (-4)(30) = (4)(-30)$$
$$= (-5)(24) = (5)(-24)$$
$$= (-6)(20) = (6)(-20)$$
$$= (-8)(15) = (8)(-15)$$
$$= (-10)(12) = (10)(-12)$$

None of the sums of these pairs is $+5$. Therefore, the trinomial is *not factorable*.

☞ **Note** The master product method can also be used with trinomials of the form $x^2 + bx + c$; however, we think the method presented earlier is shorter and simpler for trinomials of that type.

Exercises   7.5

Do Exercises 1–30 of Exercises 7.3B using the master product method.

7.6   Factoring Completely; Selecting the Method of Factoring

Factoring Completely

We've emphasized the importance of **factoring completely** since the beginning of this chapter. When there is a common factor, be sure to look for and factor out the *greatest* common factor. Once an expression has been factored, be sure to look carefully at *each* polynomial factor to see if that factor can be factored further. If you do these things correctly, the polynomial will be factored completely.

Selecting the Method of Factoring

The following procedure can be used to select the correct method for completely factoring a polynomial.

First, check for a common factor, no matter how many terms the polynomial has.

When there is a common factor, be sure to find the *greatest* common factor and factor it out (Section 7.1). Then examine the polynomial factor to see whether it can be factored further, asking yourself the following questions.

If the polynomial to be factored has two terms:

  1. Is it a difference of two squares? (Section 7.2B)
  2. Is it a sum of two squares? (If so, it is *not* factorable.)

If the polynomial to be factored has three terms:

  1. Can it be expressed in the form $x^2 + bx + c$? (Section 7.3A)
  2. Can it be expressed in the form $ax^2 + bx + c$? (Section 7.3B)
  3. Is the trinomial the square of a binomial? (Section 7.3B)

If the polynomial to be factored has four terms, can it be factored by grouping? (Section 7.4)

Check to see whether any factor can still be factored. When the expression is completely factored, the same factors are obtained no matter what method is used.

Check the result by multiplying the factors together.

☞ **Note** Always remember that there are two steps to checking a factoring problem: *Accuracy* is checked by multiplication, and *completeness* is checked by inspection.

EXAMPLE 1 Factor $27x^2 - 12y^2$, or write "not factorable."

SOLUTION

$$27x^2 - 12y^2 \qquad \text{3 is the GCF}$$
$$\text{This factor can still be factored}$$
$$= 3(\; 9x^2 - 4y^2 \;) \longleftarrow (9x^2 - 4y^2) = (3x + 2y)(3x - 2y)$$
$$= 3(3x + 2y)(3x - 2y)$$

√ **Check** $3(3x + 2y)(3x - 2y) = 3(9x^2 - 4y^2) = 27x^2 - 12y^2$

Therefore, $27x^2 - 12y^2 = 3(3x + 2y)(3x - 2y)$.

EXAMPLE 2 Factor $3x^3 - 27x + 5x^2 - 45$, or write "not factorable."

SOLUTION This polynomial has four terms; we'll try to factor by grouping.

$$\underbrace{3x^3 - 27x} + \underbrace{5x^2 - 45}$$
$$= 3x(x^2 - 9) + 5(x^2 - 9)$$
$$\text{This factor can still be factored}$$
$$= (\; x^2 - 9 \;)(3x + 5) \longleftarrow (x^2 - 9) = (x + 3)(x - 3)$$
$$= (x + 3)(x - 3)(3x + 5)$$

√ **Check** $(x + 3)(x - 3)(3x + 5) = [(x + 3)(x - 3)](3x + 5)$
$$= [x^2 - 9](3x + 5)$$
$$= [x^2 - 9](3x) + [x^2 - 9](5)$$
$$= 3x^3 - 27x + 5x^2 - 45$$

Therefore, $3x^3 - 27x + 5x^2 - 45 = (x + 3)(x - 3)(3x + 5)$.

EXAMPLE 3   Factor $5x^2 + 10x - 40$, or write "not factorable."

SOLUTION

$$5x^2 + 10x - 40 \qquad \text{5 is the GCF}$$

—— This factor can still be factored

$$= 5(\;x^2 + 2x - 8\;) \leftarrow (x^2 + 2x - 8) = (x - 2)(x + 4)$$

$$= 5(x - 2)(x + 4)$$

√ **Check** $5(x - 2)(x + 4) = 5(x^2 + 2x - 8) = 5x^2 + 10x - 40$

Therefore, $5x^2 + 10x - 40 = 5(x - 2)(x + 4)$.

EXAMPLE 4   Factor $a^4 - b^4$, or write "not factorable."

SOLUTION   This polynomial has two terms and is a difference of two squares.

$$a^4 - b^4$$

—— This factor can still be factored

$$= (a^2 + b^2)(\;a^2 - b^2\;) \leftarrow (a^2 - b^2) = (a + b)(a - b)$$

$$= (a^2 + b^2)(a + b)(a - b)$$

√ **Check** $(a^2 + b^2)(a + b)(a - b) = (a^2 + b^2)(a^2 - b^2) = a^4 - b^4$

Therefore, $a^4 - b^4 = (a^2 + b^2)(a + b)(a - b)$.

Exercises 7.6
Set I

Factor each polynomial completely and check, or write "not factorable."

1. $2x^2 - 8y^2$

2. $3x^2 - 27y^2$

3. $5a^4 - 20b^2$

4. $6m^2 - 54n^4$

5. $x^4 - y^4$

6. $a^4 - 16$

7. $4v^2 + 14v - 8$

8. $6v^2 - 27v - 15$

9. $8z^2 - 12z - 8$

10. $18z^2 - 21z - 9$

11. $4x^2 - 100$

12. $9x^2 - 36$

13. $12x^2 + 10x - 8$

14. $45x^2 - 6x - 24$

15. $ab^2 - 2ab + a$

16. $au^2 - 2au + a$

17. $x^4 - 81$

18. $16y^8 - z^4$

19. $16x^2 + 16$

20. $25b^2 + 100$

21. $2u^3 + 2u^2v - 12uv^2$

22. $3m^3 - 3m^2n - 36mn^2$

23. $8h^3 - 20h^2k + 12hk^2$

24. $15h^2k - 35hk^2 + 10k^3$

25. $a^5b^2 - 4a^3b^4$

26. $x^2y^4 - 100x^4y^2$

27. $2ax^2 - 8a^3y^2$

28. $3b^2x^4 - 12b^2y^2$

29. $12 + 4x - 3x^2 - x^3$

30. $45 - 9z - 5z^2 + z^3$

31. $6my - 4nz + 15mz - 5zn$

32. $10xy + 5mn - 6xy - mn$

33. $x^4 - 8x^2 + 16$

34. $y^4 - 18y^2 + 81$

35. $x^8 - 1$

36. $a^8 - b^8$

37. $x^2 - a^2 - 4x + 4a$

38. $m^2 - 25 + mn - 5n$

39. $6ac - 6bd + 6bc - 6ad$

40. $10cy - 6cz + 5dy - 3dz$

Writing Problems

Express the answers in your own words and in complete sentences.

1. Describe the two kinds of checks that you should perform on every factoring problem.

2. Describe the error in the answer to the following problem:

   Factor $4x^2 - 4x - 8$ completely.

   Answer: $(4x + 4)(x - 2)$

Exercises 7.6
Set II

Factor each polynomial completely and check, or write "not factorable."

1. $3a^2 - 75b^2$
2. $7c^2 - 63b^2$
3. $4h^4 - 36b^2$
4. $9x^4 - 36y^2$
5. $x^4 - 16y^4$
6. $m^4 - 1$
7. $10x^2 + 25x - 15$
8. $6x^2 + 15x - 9$
9. $10y^2 + 14y - 12$
10. $6x^2 + 6x + 12$
11. $16x^2 - 36$
12. $25a^2 - 100$
13. $30w^2 + 27w - 21$
14. $8x^2 + 22x - 6$
15. $h^2k - 4hk + 4k$
16. $x^2y + 4xy + 4$
17. $81c^4 - 16$
18. $81c^2 + 16$
19. $5x^2 + 20$
20. $4m^3n^3 - mn^5$
21. $5wz^2 + 5w^2z - 10w^3$
22. $2t^2r^4 - 18t^4$
23. $12x^2y - 42xy^2 + 36y^3$
24. $xy + 3y - 4x - 12$
25. $6x^3y^2 - 12xy^4$
26. $5x^4y + 20x^2y^3$
27. $3bx^2 - 12b^3y^2$
28. $9x^2 + 36$
29. $45 + 9b - 5b^2 - b^3$
30. $12 + 4x - 3x^2 - x^3$
31. $3xy + 2xz - 8xw + 3xz$
32. $8wx + 5xy - 4yz - 11yz$
33. $x^4 - 2x^2 + 1$
34. $x^3 + 3x^2 - 25x - 75$
35. $y^8 - 1$
36. $x^2 + 1$
37. $a^2 - 4b^2 + 2a + 4b$
38. $10ac + 10ad - 5bc - 5bd$
39. $6ef + 3gf - 12eh - 9gh$
40. $x^2 - 9y^2 + 2x - 6y$

7.7   Solving Equations by Factoring

Factoring has many applications. In this section, we use factoring to solve equations.

A **polynomial equation** is an equation that has a polynomial on both sides of the equal sign; the polynomial on one side of the equal sign can be the zero polynomial, 0. The *degree* of the equation equals the degree of the highest-degree *term* in the equation.

Polynomial equations with a first-degree term as the highest-degree term are called **first-degree**, or **linear, equations**. (All the equations that have been solved so far have been first-degree equations.) Polynomial equations with a second-degree term as the highest-degree term are called **second-degree**, or **quadratic, equations.**

EXAMPLE   I   Examples of polynomial equations:

a. $-5x = 3$ A linear (or first-degree) equation in one variable

b. $2x^2 + 7 = -4x$ A quadratic (or second-degree) equation in one variable.

We are now ready to solve quadratic and higher-degree equations. One method of solving such equations is based on the zero-factor property, which is stated without proof.

The zero-factor property

If the product of two factors is zero, then one or both of the factors must be zero. That is,

$$\text{if } a \cdot b = 0, \text{ then } \begin{cases} a = 0 \\ \text{or } b = 0 \\ \text{or both } a = 0 \text{ and } b = 0 \end{cases}$$

where a and b represent any real numbers.

Note This means, for example, that if $6x = 0$, then $x = 0$, because $6 \neq 0$.

The zero-factor property can be extended to include more than two factors; that is, if a product of factors is zero, at least one of the factors must be zero. We use the zero-factor property in solving higher-degree equations. The method is summarized below.

Solving an equation by factoring

1. Move all nonzero terms to one side of the equal sign by adding the same expression to both sides. *Only zero must remain on the other side.* Then arrange the polynomial in descending powers.
2. Factor the polynomial completely.
3. Using the zero-factor property, set each factor equal to zero.* (Any factors that are nonzero constants *cannot* equal zero; see Example 3.)
4. Solve each resulting first-degree equation.
5. Check all apparent solutions in the original equation.

---

*If any of the factors that contain variables are *not* first-degree polynomials, we cannot solve the equation at this time.

In Examples 2 through 4, we already have zero on one side of the equal sign, and the polynomial has already been factored. Therefore, we proceed with step 3.

EXAMPLE 2   Solve $(x - 1)(x - 2) = 0$.

SOLUTION   Since $(x - 1)(x - 2) = 0$,

Step 3. $(x - 1) = 0$ *or* $(x - 2) = 0$ Using the zero-factor property

Step 4. If $x - 1 = 0$ If $x - 2 = 0$

then $x - 1 + 1 = 0 + 1$ then $x - 2 + 2 = 0 + 2$

$x = 1$ $x = 2$

✓ **Step 5.**   *Check for $x = 1$*   *Check for $x = 2$*

$(x - 1)(x - 2) = 0$ $(x - 1)(x - 2) = 0$

$(1 - 1)(1 - 2) \overset{?}{=} 0$ $(2 - 1)(2 - 2) \overset{?}{=} 0$

$(0)(-1) \overset{?}{=} 0$ $(1)(0) \overset{?}{=} 0$

$0 = 0$ $0 = 0$

Therefore, 1 and 2 are the solutions for the equation $(x - 1)(x - 2) = 0$.

EXAMPLE 3   Solve $3(x + 2)(x - 1) = 0$.

SOLUTION   Since $3(x + 2)(x - 1) = 0$,

			Using the zero-factor property
Step 3. $3 \neq 0$	$x + 2 = 0$	$x - 1 = 0$	
Step 4.	$x + 2 - 2 = 0 - 2$	$x - 1 + 1 = 0 + 1$	
	$x = -2$	$x = 1$	

Step 5. We leave the checking of the solutions -2 and 1 to you.

EXAMPLE 4   Solve $2x(x - 3)(x + 4) = 0$.

SOLUTION   Since $2x(x - 3)(x + 4) = 0$,

			Using the zero-factor property
Step 3. $2x = 0$	$x - 3 = 0$	$x + 4 = 0$	
Step 4. $\dfrac{2x}{2} = \dfrac{0}{2}$	$x - 3 + 3 = 0 + 3$	$x + 4 - 4 = 0 - 4$	
$x = 0$	$x = 3$	$x = -4$	

Step 5. Checking will verify that the solutions are 0, 3, and -4.

✖ **A Word of Caution**   In Example 3, don't make the mistake of thinking that 3 is a solution of the equation because 3 is one of the factors of the polynomial. We can write $3 \neq 0$ (as we did in the example), we can *divide* both sides of the equation by 3, or we can ignore the 3; 3 is *not* a solution of the equation. Similarly, in Example 4, 2 is *not* a solution of the equation, even though 2 is one of the factors of the polynomial.

In Example 4, we could have considered the 2 and the x as *separate* factors and simply written $x = 0$ rather than $2x = 0$.

EXAMPLE 5   Solve $x^2 - 9x = 0$.

SOLUTION   We have zero on one side of the equal sign. We proceed with step 2.

$$x^2 - 9x = 0$$

Step 2. $\quad x(x - 9) = 0$ Factoring the polynomial

		Using the zero-factor property
Step 3.	$x = 0$	$x - 9 = 0$
Step 4.		$x - 9 + 9 = 0 + 9$
	$x = 0$	$x = 9$

√ **Step 5.**

Check for $x = 0$	**Check for $x = 9$**
$x^2 - 9x = 0$	$x^2 - 9x = 0$
$0^2 - 9(0) \overset{?}{=} 0$	$9^2 - 9(9) \overset{?}{=} 0$
$0 - 0 \overset{?}{=} 0$	$81 - 81 \overset{?}{=} 0$
$0 = 0$	$0 = 0$

The solutions are 0 and 9.

EXAMPLE 6   Solve $6x^2 = 5 - 7x$.

SOLUTION

$$6x^2 = 5 - 7x$$

Step 1. $\quad 6x^2 + 7x - 5 = 5 - 7x + 7x - 5$ Adding $7x - 5$ to both sides

$$6x^2 + 7x - 5 = 0$$

Step 2. $\quad (2x - 1)(3x + 5) = 0$ Factoring the polynomial

Step 3. $\qquad\qquad 2x - 1 = 0 \qquad\qquad\qquad 3x + 5 = 0$ Setting each factor

Step 4. $\quad 2x - 1 + 1 = 0 + 1 \qquad 3x + 5 - 5 = 0 - 5$ equal to zero (using the zero-factor property)

$$2x = 1 \qquad\qquad\qquad 3x = -5$$

$$\frac{2x}{2} = \frac{1}{2} \qquad\qquad\qquad \frac{3x}{3} = \frac{-5}{3}$$

$$x = \tfrac{1}{2} \qquad\qquad\qquad x = -\tfrac{5}{3}$$

✓ **Step 5.**   **Check for $x = \tfrac{1}{2}$**   **Check for $x = -\tfrac{5}{3}$**

$$6x^2 = 5 - 7x \qquad\qquad\qquad 6x^2 = 5 - 7x$$

$$6\left(\tfrac{1}{2}\right)^2 \overset{?}{=} 5 - 7\left(\tfrac{1}{2}\right) \qquad 6\left(-\tfrac{5}{3}\right)^2 \overset{?}{=} 5 - 7\left(-\tfrac{5}{3}\right)$$

$$6\left(\tfrac{1}{4}\right) \overset{?}{=} 5 - \tfrac{7}{2} \qquad\qquad 6\left(\tfrac{25}{9}\right) \overset{?}{=} 5 + \tfrac{35}{3}$$

$$\frac{6}{4} \overset{?}{=} \frac{10 - 7}{2} \qquad\qquad \frac{2 \cdot 25}{3} \overset{?}{=} \frac{15 + 35}{3}$$

$$\tfrac{3}{2} = \tfrac{3}{2} \qquad\qquad\qquad \tfrac{50}{3} = \tfrac{50}{3}$$

The solutions are $\tfrac{1}{2}$ and $-\tfrac{5}{3}$.

✕   **A Word of Caution**   The product must equal zero, or no conclusions can be drawn about the factors.

$\qquad$ Suppose $(x - 1)(x - 3) = 8$. ⟵——————— No conclusion can be drawn because the product $\neq 0$

A common error is to think that

$$\text{if} \qquad (x - 1)(x - 3) = 8$$

$$\text{then} \quad x - 1 = 8 \quad\Big|\quad x - 3 = 8$$

$$x = 9 \quad\Big|\quad x = 11$$

This "solution" is incorrect, because

$$\text{if} \quad x = 9 \qquad\qquad\qquad or \quad \text{if} \quad x = 11$$

$$\text{then} \quad (x - 1)(x - 3) \qquad\qquad \text{then} \quad (x - 1)(x - 3)$$

$$= (9 - 1)(9 - 3) \qquad\qquad\qquad = (11 - 1)(11 - 3)$$

$$= 8 \cdot 6 \qquad\qquad\qquad\qquad\quad = 10 \cdot 8$$

$$= 48 \qquad\qquad\qquad\qquad\qquad = 80$$

$$\neq 8 \qquad\qquad\qquad\qquad\qquad \neq 8$$

The correct solutions are found as follows:

$$(x - 1)(x - 3) = 8$$

$$x^2 - 4x + 3 = 8$$

Step 1. $x^2 - 4x + 3 - 8 = 8 - 8$ Adding -8 to both sides

$$x^2 - 4x - 5 = 0$$

Step 2. $(x - 5)(x + 1) = 0$ Factoring the polynomial

Setting each factor

Step 3. $x - 5 = 0$ $x + 1 = 0$ equal to zero

Step 4. $x - 5 + 5 = 0 + 5$ $x + 1 - 1 = 0 - 1$

$$x = 5 \qquad\qquad x = -1$$

Step 5. Checking will verify that 5 and -1 are the solutions.

E X A M P L E 7

Solve $(x - 5)(x + 4) = -14$.

S O L U T I O N We first remove the parentheses.

$$(x - 5)(x + 4) = -14$$

$$x^2 - x - 20 = -14 \qquad\qquad \text{Removing the parentheses}$$

Step 1. $x^2 - x - 20 + 14 = -14 + 14$ Adding 14 to both sides

$$x^2 - x - 6 = 0$$

Step 2. $(x + 2)(x - 3) = 0$ Factoring the polynomial

Setting each factor

Step 3. $x + 2 = 0$ $x - 3 = 0$ equal to zero

Step 4. $x + 2 - 2 = 0 - 2$ $x - 3 + 3 = 0 + 3$

$$x = -2 \qquad\qquad x = 3$$

✓ **Step 5.**   **Check for $x = -2$**   **Check for $x = 3$**

$$(x - 5)(x + 4) = -14 \qquad (x - 5)(x + 4) = -14$$

$$(-2 - 5)(-2 + 4) \overset{?}{=} -14 \qquad (3 - 5)(3 + 4) \overset{?}{=} -14$$

$$(-7)(2) \overset{?}{=} -14 \qquad\qquad (-2)(7) \overset{?}{=} -14$$

$$-14 = -14 \qquad\qquad\qquad -14 = -14$$

The solutions are -2 and 3.

E X A M P L E 8

Solve $3x^3 = 4x - x^2$.

S O L U T I O N

$$3x^3 = 4x - x^2$$

Step 1. $3x^3 + x^2 - 4x = 4x - x^2 + x^2 - 4x$ Adding $x^2 - 4x$ to both sides

$$3x^3 + x^2 - 4x = 0$$

Step 2. $x(3x^2 + x - 4) = 0$ Factoring out the GCF

$$x(x - 1)(3x + 4) = 0 \qquad\qquad \text{Factoring the polynomial factor}$$

Setting each factor equal to zero

Step 3. $x = 0$	$x - 1 = 0$	$3x + 4 = 0$
Step 4.	$x - 1 + 1 = 0 + 1$	$3x + 4 - 4 = 0 - 4$
		$3x = -4$
		$\dfrac{3x}{3} = \dfrac{-4}{3}$
$x = 0$	$x = 1$	$x = -\frac{4}{3}$

✓ **Step 5.** **Check for $x = 0$** | **Check for $x = 1$** | **Check for $x = -\frac{4}{3}$**

$3x^3 = 4x - x^2$

$3(0)^3 \overset{?}{=} 4(0) - 0^2$

$3(0) \overset{?}{=} 0 - 0$

$0 = 0$

$3x^3 = 4x - x^2$

$3(1)^3 \overset{?}{=} 4(1) - 1^2$

$3(1) \overset{?}{=} 4 - 1$

$3 = 3$

$3x^3 = 4x - x^2$

$3\left(-\frac{4}{3}\right)^3 \overset{?}{=} 4\left(-\frac{4}{3}\right) - \left(-\frac{4}{3}\right)^2$

$3\left(-\frac{64}{27}\right) \overset{?}{=} -\frac{16}{3} - \frac{16}{9}$

$-\frac{64}{9} \overset{?}{=} -\frac{48}{9} - \frac{16}{9}$

$-\frac{64}{9} = -\frac{64}{9}$

The solutions are 0, 1, and $-\frac{4}{3}$.

A Word of Caution Pay close attention to whether you're asked to *factor* a polynomial or *solve* a polynomial equation! If you're asked to *factor* $x^2 - 9x$, the correct answer is "$x(x - 9)$," *not* "$x = 0$ or $x = 9$." (You can't attach "$= 0$" to $x^2 - 9x$.) On the other hand, if you're asked to *solve the equation* $x^2 - 9x = 0$, the correct answer is "The solutions are 0 and 9," *not* "$x(x - 9)$"; $x(x - 9)$ is not the solution for an equation.

Exercises 7.7

Set I

Solve and check each of the following equations.

1. $(x - 5)(x + 4) = 0$
2. $(x + 7)(x - 2) = 0$

3. $3x(x - 4) = 0$
4. $5x(x + 6) = 0$

5. $(x + 10)(2x - 3) = 0$
6. $(x - 8)(3x + 2) = 0$

7. $x^2 + 9x + 8 = 0$
8. $x^2 + 6x + 8 = 0$

9. $x^2 - x - 12 = 0$
10. $x^2 + x - 12 = 0$

11. $x^2 = 64$
12. $x^2 = 144$

13. $6x^2 - 10x = 0$
14. $6y^2 - 21y = 0$

15. $24w = 4w^2$
16. $20m = 5m^2$

17. $5a^2 = 16a - 3$
18. $3z^2 = 22z - 7$

19. $3u^2 = 2u + 5$
20. $5k^2 = 34k + 7$

21. $(x - 2)(x - 3) = 2$
22. $(x - 3)(x - 5) = 3$

23. $x(x - 4) = 12$
24. $x(x - 2) = 15$

25. $4x(2x - 1)(3x + 7) = 0$
26. $5x(4x - 3)(7x - 6) = 0$

27. $2x^3 + x^2 = 3x$
28. $4x^3 = 10x - 18x^2$

29. $2a^3 - 10a^2 = 0$
30. $4b^3 - 24b^2 = 0$

Writing Problems

Express the answers in your own words and in complete sentences.

1. Explain what is wrong with the following statement:

If $xy = 8$, then $x = 8$ or $y = 8$.

2. Explain what is wrong with the following statement:

If $(x + 2)(x + 3) = 6$, then $x + 2 = 6$ or $x + 3 = 6$.

3. Explain how to check the solutions of a polynomial equation.

Exercises 7.7
Set II

Solve and check each of the following equations.

1. $(x + 3)(x - 5) = 0$

2. $(y - 8)(y + 9) = 0$

3. $2y(y - 7) = 0$

4. $7x(x + 4) = 0$

5. $(z - 6)(3z + 2) = 0$

6. $(5x - 10)(4x + 5) = 0$

7. $a^2 + 8a + 12 = 0$

8. $x^2 + 2x = 48$

9. $m^2 + 3m - 18 = 0$

10. $2x^2 + 11x + 15 = 0$

11. $w^2 - 24 = 5w$

12. $x^2 = 4x + 5$

13. $5h^2 - 20h = 0$

14. $k^3 = 5k^2$

15. $12t = 6t^2$

16. $x = 10 - 2x^2$

17. $3n^2 = 7n + 6$

18. $(x - 4)(x + 2) = -9$

19. $13x + 3 = -4x^2$

20. $y(3y - 2)(y + 5) = 0$

21. $(y - 3)(y - 6) = -2$

22. $10x^2 + 11x = 6$

23. $u(u - 9) = -14$

24. $(x - 2)(x + 4) = 7$

25. $2x(3x - 2)(5x + 9) = 0$

26. $8y(2y^2 + 5y - 3) = 0$

27. $21x^2 + 60x = 18x^3$

28. $12y^2 = 4y^3 + 5y$

29. $3a^2 + 18a = 0$

30. $(x - 5)(x + 6) = -10$

7.8   Solving Applied Problems by Factoring

To refresh your memory, we repeat here the main suggestions for solving applied problems. In this section, most of the equations we need to solve (see step 3) will be second-degree (quadratic) equations.

The method for solving applied problems

Read   To solve an applied problem, first read it very carefully.

Think   Determine what *type* of problem it is, if possible. Determine what is unknown. Do you need a special formula?

Sketch   Draw a sketch *with labels*, if it might be helpful.

Step 1. Represent one unknown number by a variable, and declare its meaning in a sentence of the form "Let $x = \ldots$." Then express any other unknown numbers in terms of the same variable, declaring their meaning in sentences that begin "Then $\ldots$."

Reread   Reread the entire problem, breaking it up into key words and phrases.

Step 2. Translate each key word or phrase into an algebraic expression, and arrange these expressions into an equation or inequality that represents the facts given in the problem.

Step 3. Solve the equation or inequality.

Step 4. Solve for *all* the unknowns asked for in the problem.

Step 5. Check the solution(s) *in the word statement.*

Step 6. State the results clearly.

EXAMPLE 1   The difference of two numbers is 3. Their product is 10. What are the two numbers?

SOLUTION

Step 1. Let x = the smaller number   Since the difference is 3, the larger number
Then $x + 3$ = the larger number   must be $x + 3$ if the smaller number is x

Reread   The product   is   10

Step 2. $x(x + 3) = 10$ Remember: It is incorrect to say now that
$x = 10$ or $x + 3 = 10$

Step 3. $x^2 + 3x = 10$ Since this is a quadratic equation, we must
now get 0 on one side

$x^2 + 3x - 10 = 10 - 10$ Adding -10 to both sides

$x^2 + 3x - 10 = 0$

$(x - 2)(x + 5) = 0$ Factoring the polynomial

Setting each factor
equal to 0

$x - 2 = 0$ | $x + 5 = 0$

$x - 2 + 2 = 0 + 2$ | $x + 5 - 5 = 0 - 5$

Step 4. $x = 2$ | $x = -5$ The smaller numbers

$x + 3 = 2 + 3 = 5$ | $x + 3 = -5 + 3 = -2$ The larger
numbers

✓ **Step 5. Check for 2 and 5**   The difference is $5 - 2 = 3$; the product is $5 \cdot 2 = 10$.
Check for -5 and -2   The difference is $-2 - (-5) = -2 + 5 = 3$; the
product is $(-2)(-5) = 10$.

Step 6. Therefore, the numbers 2 and 5 are one solution, and the numbers -5 and -2
are another solution.

A Word of Caution   In Example 1, a common error is to state that the *required
numbers* are 2 and -5, once we have found that $x = 2$ or
$x = -5$. *This is not correct*. The difference of 2 and -5 is *not* 3, and the product
of 2 and -5 is *not* 10. Rather, 2 and -5 are possible values for *the smaller number*; we must then find the larger numbers.

EXAMPLE 2   Find two consecutive integers whose product is 19 more than their sum.

SOLUTION

Step 1. Let x = the smaller integer
Then $x + 1$ = the larger integer

Reread   The product   is   19   more than   the sum

Step 2. $x(x + 1)$ $=$ 19 $+$ $x + (x + 1)$

Step 3. $x^2 + x = 19 + 2x + 1$

$x^2 + x = 2x + 20$

$x^2 + x - 2x - 20 = 2x + 20 - 2x - 20$ Adding $-2x - 20$ to both sides

$x^2 - x - 20 = 0$

$(x + 4)(x - 5) = 0$ Factoring the polynomial

$$x + 4 = 0 \qquad\qquad x - 5 = 0$$

Setting each factor equal to 0

$$x + 4 - 4 = 0 - 4 \qquad\qquad x - 5 + 5 = 0 + 5$$

Step 4. $\qquad x = -4 \qquad\qquad x = 5 \qquad$ The smaller integers

$$x + 1 = -4 + 1 = -3 \qquad\qquad x + 1 = 5 + 1 = 6 \qquad$$ The larger integers

√ **Step 5.** *Check for* −4 *and* −3 Their sum is −7. Their product is 12. 12 is 19 more than −7.

Check for 5 *and* 6 Their sum is 11. Their product is 30. 30 is 19 more than 11.

Step 6. Therefore, the numbers −4 and −3 are one solution, and the numbers 5 and 6 are another solution.

EXAMPLE 3

Find three consecutive odd integers such that the product of the first two is 21 more than 6 times the third.

SOLUTION

Step 1. Let $\qquad x =$ the first odd integer

Then $x + 2 =$ the second odd integer

and $\quad x + 4 =$ the third odd integer

Reread | The product of the first two integers | is | 21 | more than | 6 times the third

Step 2. $\qquad x(x + 2) \quad = \quad 21 \quad + \quad 6(x + 4)$

Step 3. $\qquad x^2 + 2x = 21 + 6x + 24$

$$x^2 + 2x = 6x + 45$$

$$x^2 + 2x - 6x - 45 = 6x + 45 - 6x - 45 \qquad$$ Adding $-6x - 45$ to both sides

$$x^2 - 4x - 45 = 0$$

$$(x + 5)(x - 9) = 0 \qquad\qquad$$ Factoring the polynomial

$$x + 5 = 0 \qquad\qquad x - 9 = 0 \qquad$$ Setting each factor equal to 0

$$x + 5 - 5 = 0 - 5 \qquad\qquad x - 9 + 9 = 0 + 9$$

Step 4. $\qquad x = -5 \qquad\qquad x = 9 \qquad$ The first integers

$$x + 2 = -5 + 2 = -3 \qquad\qquad x + 2 = 9 + 2 = 11 \qquad$$ The second integers

$$x + 4 = -5 + 4 = -1 \qquad\qquad x + 4 = 9 + 4 = 13 \qquad$$ The third integers

√ **Step 5.** $\quad$ *Check for* −5, −3, −1 $\qquad$ *Check for* 9, 11, 13

$$(-5)(-3) \stackrel{?}{=} 21 + 6(-1) \qquad (9)(11) \stackrel{?}{=} 21 + 6(13)$$

$$15 \stackrel{?}{=} 21 - 6 \qquad\qquad 99 \stackrel{?}{=} 21 + 78$$

$$15 = 15 \qquad\qquad\qquad 99 = 99$$

Step 6. Therefore, the integers −5, −3, and −1 are one solution, and the integers 9, 11, and 13 are another solution.

In solving applied problems about geometric figures, you may find it helpful to make a drawing of the figure and write the given information on the figure. Lengths of sides of geometric figures cannot be negative or zero; therefore, any value of the variable that would make a length negative or zero must be rejected.

EXAMPLE 4

One square has a side 3 ft longer than the side of a second square. If the area of the larger square is 4 times as great as the area of the smaller square, find the length of a side of each square.

SOLUTION

Formula Needed $A = s^2$, where A = the area and s = the length of a side of the square.

Step 1. Let $\quad x$ = the length of a side of the smaller square (in feet)

Then $x + 3$ = the length of a side of the larger square (in feet)

Sketch

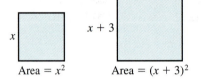

Area = x^2 $\qquad$ Area = $(x + 3)^2$

Reread

The area of the larger square	is	4 times as great as	the area of the smaller square

Step 2. $\qquad (x + 3)^2 \qquad = \qquad 4 \cdot \qquad x^2$

Step 3. $\qquad (x + 3)^2 = 4x^2$

$x^2 + 6x + 9 = 4x^2$

$-4x^2 + x^2 + 6x + 9 = -4x^2 + 4x^2 \qquad$ Adding $-4x^2$ to both sides

$-3x^2 + 6x + 9 = 0$

$-3(x^2 - 2x - 3) = 0 \qquad$ Factoring out -3

$-3(x + 1)(x - 3) = 0 \qquad$ Factoring the polynomial factor

Setting each factor equal to 0

$-3 \neq 0$	$x + 1 = 0$	$x - 3 = 0$
	$x + 1 - 1 = 0 - 1$	$x - 3 + 3 = 0 + 3$

Step 4. $\qquad\qquad\qquad x = -1 \qquad\qquad x = 3$ (ft) $\quad$ Length of side of smaller square

-1 must be rejected, since a length cannot be negative

$x + 3 = 6$ (ft) $\quad$ Length of side of larger square

✓ **Step 5. Check** The area of the smaller square is $(3 \text{ ft})^2 = 9 \text{ ft}^2$. The area of the larger square is $(6 \text{ ft})^2 = 36 \text{ ft}^2$. The area of the larger square (36 sq. ft) is 4 times as great as the area of the smaller square (9 sq. ft).

Step 6. Therefore, the length of a side of the smaller square is 3 ft, and the length of a side of the larger square is 6 ft.

EXAMPLE 5

The width of a rectangle is 5 cm less than its length. Its area is 10 more (numerically*) than its perimeter. What are the dimensions of the rectangle?

SOLUTION

Formulas Needed $A = lw$ and $P = 2l + 2w$, where A = the area, l = the length, w = the width, and P = the perimeter.

---

*We say that the area is "numerically" 10 more than the perimeter because the area is measured in *square centimeters*, whereas the perimeter is measured in *centimeters*.

Step 1. Let $l =$ the length (in centimeters)
Then $l - 5 =$ the width (in centimeters)
Notice that the area is $l(l - 5)$ and the perimeter is $2l + 2(l - 5)$.

Reread | The area | is | 10 | more than | the perimeter | | *Sketch*

Step 2. $l(l - 5)$ $=$ 10 $+$ $2l + 2(l - 5)$

Step 3. $l(l - 5) = 10 + 2l + 2(l - 5)$

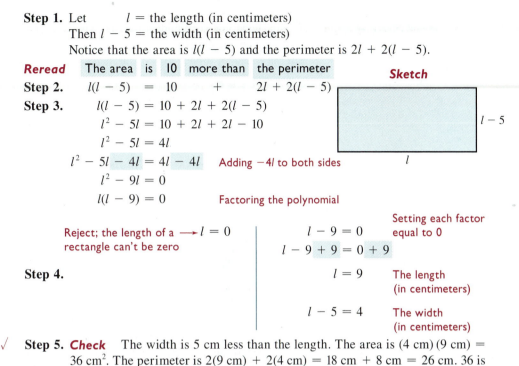

$l^2 - 5l = 10 + 2l + 2l - 10$

$l^2 - 5l = 4l$

$l^2 - 5l - 4l = 4l - 4l$ Adding $-4l$ to both sides

$l^2 - 9l = 0$

$l(l - 9) = 0$ Factoring the polynomial

Reject; the length of a $\longrightarrow l = 0$ | $l - 9 = 0$ Setting each factor equal to 0
rectangle can't be zero | $l - 9 + 9 = 0 + 9$

Step 4. $l = 9$ The length (in centimeters)

     $l - 5 = 4$ The width (in centimeters)

✓ **Step 5. Check** The width is 5 cm less than the length. The area is $(4\text{ cm})(9\text{ cm}) = 36\text{ cm}^2$. The perimeter is $2(9\text{ cm}) + 2(4\text{ cm}) = 18\text{ cm} + 8\text{ cm} = 26\text{ cm}$. 36 is 10 more than 26.

Step 6. Therefore, the rectangle has a length of 9 cm and a width of 4 cm.

EXAMPLE 6

The length of a rectangular solid is 2 cm more than its width. The height is 3 cm, and the volume is 72 cc (cubic centimeters). Find the width and the length.

SOLUTION

Formula Needed $V = lwh$, where $V =$ the volume, $l =$ the length, $w =$ the width, and $h =$ the height of the box.

Step 1. Let $w =$ the width (in centimeters)
Then $w + 2 =$ the length (in centimeters)   *Sketch*

Think The volume is $(w + 2)(w)(3)$.

Reread | The volume | is | 72

Step 2. $(w + 2)(w)(3) = 72$

Step 3. $3w^2 + 6w = 72$

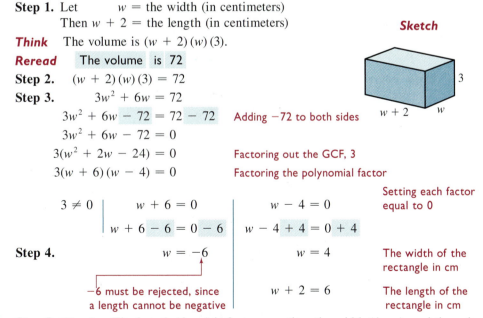

$3w^2 + 6w - 72 = 72 - 72$ Adding -72 to both sides

$3w^2 + 6w - 72 = 0$

$3(w^2 + 2w - 24) = 0$ Factoring out the GCF, 3

$3(w + 6)(w - 4) = 0$ Factoring the polynomial factor

$3 \neq 0$ | $w + 6 = 0$ | $w - 4 = 0$ Setting each factor equal to 0

   $w + 6 - 6 = 0 - 6$ | $w - 4 + 4 = 0 + 4$

Step 4. $w = -6$ $w = 4$ The width of the rectangle in cm

-6 must be rejected, since | $w + 2 = 6$ The length of the
a length cannot be negative | rectangle in cm

✓ **Step 5. Check** The length (6 cm) is 2 cm more than the width (4 cm), and the volume is $(6\text{ cm})(4\text{ cm})(3\text{ cm}) = 72\text{ cm}^3$.

Step 6. Therefore, the width of the solid is 4 cm, and the length is 6 cm.

EXAMPLE 7   A 3-in.-wide mat surrounds a picture. The area of the *picture itself* is 96 sq. in. If the width of the outside of the mat is three times the length of the outside of the mat, what are the (outside) dimensions of the mat?

SOLUTION

Formula Needed $A = lw$, where A = the area, l = the length, and w = the width of a rectangle.

Step 1. Let m = the length of the outside of the mat (in inches)
    Then $3m$ = the width of the outside of the mat (in inches)

Sketch

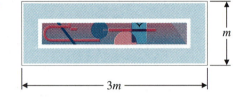

Think   To find the width of the *picture*, we must subtract 3 in. (the width of the *mat*) *twice* from the width of the *outside* of the mat. Similarly, to find the length of the *picture*, we must subtract 3 in. (the width of the *mat*) *twice* from the length of the *outside* of the mat. Therefore, we have

The width of the *picture* (in inches) is $3m - 3 - 3$, or $3m - 6$.

The length of the *picture* (in inches) is $m - 3 - 3$, or $m - 6$.

The area of the *picture* (in inches) is $(3m - 6)(m - 6)$.

Reread   The area of the picture ⎢ is ⎢ 96

Step 2. $(3m - 6)(m - 6) = 96$

Step 3. $(3m - 6)(m - 6) = 96$

$3m^2 - 24m + 36 = 96$

$3m^2 - 24m + 36 - 96 = 96 - 96$ Adding -96 to both sides

$3m^2 - 24m - 60 = 0$

$3(m^2 - 8m - 20) = 0$ Factoring out 3

$3(m + 2)(m - 10) = 0$ Factoring the polynomial factor

Setting each factor equal to 0

$3 \neq 0$ ⎢ $m + 2 = 0$ ⎢ $m - 10 = 0$

⎢ $m + 2 - 2 = 0 - 2$ ⎢ $m - 10 + 10 = 0 + 10$

Step 4. $m = -2$ ⎢ $m = 10$ The length of the outside of the mat in in.

-2 must be rejected, since a length cannot be negative

$3m = 30$ The width of the outside of the mat in in.

√ **Step 5.** **Check**   The outside of the mat is three times as wide as it is long. The length of the *picture* is $(10 - 3 - 3)$ in., or 4 in. The width of the *picture* is $(30 - 3 - 3)$ in., or 24 in. The area of the picture is (4 in.)(24 in.), or 96 sq. in.

Step 6. Therefore, the outside dimensions of the mat are 10 in. by 30 in.

Exercises 7.8

Set I

Set up each problem algebraically, solve, and check. Be sure to state what your variables represent.

1. The difference of two numbers is 5. Their product is 14. Find the numbers.

2. The difference of two numbers is 6. Their product is 27. Find the numbers.

3. The sum of two numbers is 12. Their product is 35. Find the numbers.

4. The sum of two numbers is −4. Their product is −12. Find the numbers.

5. Find three consecutive integers such that the product of the first two plus the product of the last two is 8.

6. Find three consecutive integers such that the product of the first two plus the product of the first and third is 14.

7. Find three consecutive even integers such that twice the product of the first two is 16 more than the product of the last two.

8. Find three consecutive odd integers such that twice the product of the last two is 91 more than the product of the first two.

9. The length of a rectangle is 5 ft more than its width. Its area is 84 sq. ft. What are its dimensions?

10. The width of a rectangle is 3 ft less than its length. Its area is 28 sq. ft. What are its dimensions?

11. One square has a side 3 cm shorter than the side of a second square. The area of the larger square is 4 times as great as the area of the smaller square. Find the length of a side of each square.

12. One square has a side 4 ft longer than the side of a second square. The area of the larger square is 9 times as great as the area of the smaller square. Find the length of a side of each square.

13. The width of a rectangle is 4 yd less than its length. Its area is 17 more (numerically) than its perimeter. What are the dimensions of the rectangle?

14. The area of a square is twice its perimeter (numerically). What is the length of a side?

15. The length of a rectangular solid is 4 cm more than its height. Its width is 5 cm, and its volume is 225 cc. Find its length and its height.

16. The length of a rectangular solid is 2 in. more than its width. Its height is 6 in., and its volume is 378 cu. in. Find its length and its width.

17. A 2-in.-wide mat surrounds a picture. The area of the *picture itself* is 216 sq. in. If the width of the outside of the mat is four times the length of the outside of the mat, what are the (outside) dimensions of the mat? (See the figure.)

18. A 4-ft-wide exposed aggregate walk surrounds a rectangular garden. (Exposed aggregate is a type of cement.) The area of the *garden itself* is 192 sq. ft. If the length of the outside of the walk is twice the width of the outside of the walk, what are the outside dimensions of the walk? (See the figure.)

19. A 2-in.-wide mat surrounds a picture. The area of the *picture itself* is 144 sq. in. If the width of the outside of the mat is 10 in. more than the length of the outside of the mat, what are the (outside) dimensions of the mat? What are the dimensions of the picture? (See the figure.)

20. A 1-in.-wide border surrounds the printing on a page. The area of the *printing itself* is 135 sq. in. If the length of the outside of the border is 6 in. more than the width of the outside of the border, what are the (outside) dimensions of the paper that the printing is on? What are the dimensions of the printing itself? (See the figure.)

Exercises 7.8
Set II

Set up each problem algebraically, solve, and check. Be sure to state what your variables represent.

1. The difference of two numbers is 12. Their product is 28. Find the numbers.

2. The difference of two numbers is 19. Their product is −84. Find the numbers.

3. The sum of two numbers is 10. Their product is −24. Find the numbers.

4. Find two consecutive numbers whose product is 1 less than their sum.

5. Find three consecutive integers such that the product of the first two minus the third is 7.

6. Find three consecutive integers such that the product of the first and third minus the second is 41.

7. Find three consecutive odd integers such that twice the product of the first two is 7 more than the product of the last two.

8. Bruce's vegetable garden is now square. If he forms a rectangle by increasing the length of one side by 3 ft and the length of the adjacent side by 6 ft, the area of the rectangle will be 3 times as great as the area of the square. What is the size of the vegetable garden now?

9. The length of a rectangle is 8 m more than its width. Its area is 48 sq. m. What are its dimensions?

10. The width of a rectangular box equals the length of the side of a certain cube. The length of the box is 3 cm more than its width, and the height of the box is 1 cm less than its width. The volume of the box is 9 cc more than the volume of the cube. Find the dimensions of the cube and of the box.

11. One square has a side 2 km shorter than the side of a second square. The area of the larger square is 9 times as great as the area of the smaller square. Find the length of a side of each square.

12. Find three consecutive integers such that the product of the first two is 53 more than the sum of the last two.

13. The width of a rectangle is 2 in. less than its length. Its area is 4 more (numerically) than its perimeter. What are the dimensions of the rectangle?

14. The length of a rectangular solid is 8 in. more than its width. Its height is 2 in., and its volume is 96 cu. in. Find its length and its width.

15. The length of a rectangular solid is 5 cm more than its height. Its width is 3 cm, and its volume is 18 cc. Find its length and its height.

16. A pieced quilt measuring 5 ft by 6 ft is to be surrounded by a border of uniform width. How wide should the border be if the area of the border is to be 4 sq. ft less than the area of the pieced quilt? (See the figure.)

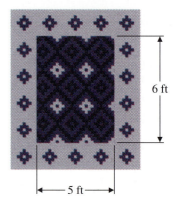

17. A 3-in.-wide mat surrounds a picture. The area of the *picture itself* is 176 sq. in. If the length of the outside of the mat is twice its width, what are the dimensions of the outside of the mat? What are the dimensions of the picture? (See the figure.)

18. A rectangular piece of cardboard is three times as long as it is wide. A 3-cm square is to be cut from each corner, and the sides will be turned up to form a box with an open top. (See the figure.) The volume of the box is to be 540 cm³. Find the dimensions of the original piece of cardboard, and find the dimensions of the box.

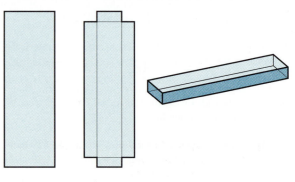

19. A 3-in.-wide mat surrounds a picture. The area of the *picture itself* is 80 sq. in. (see the figure). If the width of the outside of the mat is 2 in. more than the length of the outside of the mat, what are the (outside) dimensions of the mat? What are the dimensions of the picture?

20. The length of a rectangular solid is 8 cm. Its width is 2 cm more than its height, and its volume is 280 cc. Find its height and its width.

Figure for Exercise 19

Sections 7.4–7.8 REVIEW

Factoring by Grouping
7.4

If a polynomial contains four terms, it can sometimes be factored by grouping.

Master Product Method
7.5

The master product method can be used to factor the general trinomial when the leading coefficient is not equal to 1.

Factoring Completely
7.6

First, check for a common factor; if there is one, find the GCF and factor it out.
If the polynomial to be factored has two terms:

1. Is it a difference of two squares? (Section 7.2B)

2. Is it a sum of two squares? (If so, it is *not* factorable.)

If the polynomial to be factored has three terms:

1. Can it be expressed in the form $x^2 + bx + c$? (Section 7.3A)

2. Can it be expressed in the form $ax^2 + bx + c$? (Section 7.3B)

3. Is the trinomial the square of a binomial? (Section 7.3B)

If the polynomial to be factored has four terms, can it be factored by grouping? (Section 7.4)
Check to see whether any factor already obtained can be factored further.
Check the result by multiplying the factors together.

To Solve an Equation by Factoring
7.7

1. Move all nonzero terms to one side of the equal sign by adding the same expression to both sides. Only zero must remain on the other side. Then arrange the polynomial in descending powers.

2. Factor the polynomial.

3. Using the zero-factor property, set each factor equal to zero.

4. Solve each resulting first-degree equation.

5. Check all apparent solutions in the *original* equation.

A Word of Caution Pay close attention to whether you're asked to *factor* a polynomial or *solve* a polynomial equation.

Applied Problems Solved by Factoring
7.8

Many applied problems lead to second-degree (quadratic) equations that can be solved by factoring.

19. $5t^2 = 40t$

20. $12m^2 = 10 - 7m$

21. $5h(h - 11)(h + 3) = 0$

22. $30x^3 = 87x^2 + 63x$

23. $14x^2 = 26x + 4$

24. $32x^2 + 96x + 72 = 0$

In Exercises 25–30, set up each problem algebraically, solve, and check. Be sure to state what your variables represent.

25. The difference of two numbers is 6. Their product is 72. Find the numbers.

26. One side of a square is 8 cm longer than the side of a second square. The area of the larger square is 9 times as great as the area of the smaller square. Find the length of a side of each square.

27. The length of a rectangle is 7 yd more than its width. Its area is 4 more (numerically) than its perimeter. Find the dimensions of the rectangle.

28. Find two consecutive integers whose product is 10 less than 4 times their sum.

29. Find three consecutive even integers such that twice the product of the first two is 10 more than the third integer.

30. The width of a rectangular solid is 3 cm more than its height. Its length is 18 cm and its volume is 324 cc. Find its height and its width.

19. _____

20. _____

21. _____

22. _____

23. _____

24. _____

25. _____

_____

26. _____

_____

27. _____

_____

28. _____

_____

29. _____

_____

_____

30. _____

_____

Chapter 7   DIAGNOSTIC TEST

The purpose of this test is to see how well you understand factoring. We recommend that you work this diagnostic test *before* your instructor tests you on this chapter. Allow yourself about 50 minutes.

Complete solutions for all the problems on this test, together with section references, are given in the answer section in the back of this book. We suggest that you study the sections referred to for the problems you do incorrectly.

In Problems 1–12, factor each polynomial completely and check, or write "not factorable."

1. $8x + 12$

2. $5x^3 - 35x^2$

3. $25x^2 - 121y^2$

4. $8x^2 + 7$

5. $5a^2 - 180$

6. $z^2 + 9z + 8$

7. $m^2 + 5m - 6$

8. $11x^2 - 18x + 7$

9. $x^2 + 7x - 6$

10. $4y^2 + 19y - 5$

11. $5n - mn - 5 + m$

12. $6h^2k - 8hk^2 + 2k^3$

In Problems 13–18, solve and check each equation.

13. $3x^2 - 12x = 0$

14. $x^2 + 20 = 12x$

15. $3x^3 = x^2 + 10x$

16. $(x - 7)(x + 6) = -22$

17. $3x^2 = 75$

18. $2x^2 - 5x = 3$

In Problems 19 and 20, set up each problem algebraically, solve, and check. Be sure to state what your variables represent.

19. Find three consecutive odd integers such that twice the product of the first two minus the product of the first and third is 49.

20. The length of a rectangle is 3 ft more than its width. Its area is 28 sq. ft. Find its width and its length.

Chapters 1–7   CUMULATIVE REVIEW EXERCISES

1. Evaluate $24 \div 2\sqrt{16} - 3^2 \cdot 5$.

2. Evaluate $\sqrt{x^2 + y^2}$ if $x = -5$ and $y = -12$.

3. Using the formula $C = \dfrac{a}{a + 12} \cdot A$, find C if $a = 8$ and $A = 35$.

In Exercises 4 and 5, simplify each expression, and write the answers using only positive exponents.

4. $(3y^{-3}z^2)^{-2}$

5. $\left(\dfrac{15x^2}{10x^3}\right)^3$

6. Write the following in scientific notation.
   a. 57,300,000   b. 0.00351

In Exercises 7–10, perform the indicated operations and simplify.

7. $(2x^2 + 5x - 3) - (-4x^2 + 8x + 10) + (6x^2 + 3x - 8)$

8. $(y - 4)(3y^2 - 2y + 5)$

9. $\dfrac{12a^2 - 3a}{3a}$

10. $(8x^2 - 2x - 17) \div (2x - 3)$

In Exercises 11–16, factor each polynomial completely and check, or write "not factorable."

11. $36x^2 - 18x$

12. $1 - 36t^2$

13. $x^2 - 8x + 15$

14. $7x^2 - x - 5$

15. $5k^2 - 34k - 7$

16. $3n^2 - 2n - 5$

In Exercises 17–20, solve and check each equation.

17. $x(5x + 9) = 2$

18. $5(3x + 2) = 2(1 - x) - 3$

19. $2(x^2 + 6) = 11x$

20. $5(2b - 4) = 8(3b + 5) - 4(9 + 6b)$

In Exercises 21–25, set up each problem algebraically, solve, and check.

21. The sum of two consecutive odd integers is -22. Find the integers.

22. The length and width of a rectangle are in the ratio of 3 to 2. The area is 150 sq. m. Find the width and the length.

23. How many pounds of cashews at $7.50 per pound should be mixed with 20 lb of peanuts at $3.50 per pound if the mixture is to be worth $5.90 per pound?

24. A chemist added 15 L of a 30% solution to a 10% solution to obtain a 15% solution. How much of the 10% solution was there before the 30% solution was added?

25. Margaret has $2.60 in nickels, dimes, and quarters. If she has 1 more dime than quarters and 3 times as many nickels as quarters, how many of each kind of coin does she have?

Critical Thinking and Problem-Solving Exercises

1. A container is in the shape of a right circular cylinder, and its radius is 5 cm. At present, it is one-fourth full of water. If 300 cc of water is added, it will be one-third full. What is the volume of the container?

2. If you are to receive change of 28¢ in some combination of quarters, dimes, nickels, and pennies, how many different combinations are possible? (For example, 2 dimes, 1 nickel, and 3 pennies counts as one combination.)

3. Eric and Maria are saving money. Maria adds $250 to her savings account every month, and Eric adds $100 to his savings account every month. Right now, Maria has $1,000 in her account, and Eric has $800 in his. How long will it be before Maria has exactly twice as much in her account as Eric has in his account?

4. Wing, Merwin, and Tom are friends. One of them is a journalist, one is a professor of physics, and one is a musician. Their last names are Jones, Smith, and White. By using the following clues, can you match the first name, last name, and profession of each of these three men?

    Mr. White is a journalist.

    Mr. Jones played chess with the physics professor.

    Tom is a musician.

    Wing is not a journalist.

5. Scotty is thinking of a number. If he subtracts 5 from the number and multiplies that new number by the sum of 3 and the original number, the product is 33. Can you find the number Scotty is thinking of? (*Hint*: There are two answers.)

6. The solution of each of the following problems contains an error. Find and describe each error.

   a. Solve $x^2 - 3x = 4$.

   $$x^2 - 3x = 4$$
   $$x(x - 3) = 4$$
   $$x = 4 \qquad x - 3 = 4$$
   $$x = 7$$

   b. Solve $4x - 2(x - 3) < 3(x + 3)$.

   $$4x - 2(x - 3) < 3(x + 3)$$
   $$4x - 2x + 6 < 3x + 9$$
   $$2x + 6 < 3x + 9$$
   $$-x + 6 < 9$$
   $$-x < 3$$
   $$x < -3$$

   c. Solve $x^2 + 3x - 4 = 0$.

   Solution: $(x + 4)(x - 1)$

Rational Expressions

CHAPTER 8

In this chapter, we define rational expressions, and we discuss how to perform operations on rational expressions and how to solve those equations and applied problems that involve them. A knowledge of the different methods of factoring is *essential* to your work with rational expressions.

8.1   Basic Definitions

Rational Expressions

A **rational expression** (also called a **fraction**, or sometimes an **algebraic fraction**) is an algebraic expression of the form $\dfrac{P}{Q}$, where P and Q are polynomials and $Q \neq 0$. We call P the **numerator** and Q the **denominator** of the rational expression.* We also call P and Q the **terms** of the rational expression.

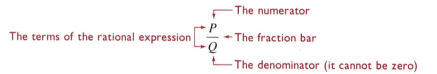

The terms of the rational expression ⌐ $\dfrac{P}{Q}$ — The numerator, — The fraction bar, — The denominator (it cannot be zero)

Note Until we discuss solving rational equations, it is understood that none of the variables has a value that would make a denominator zero.

A Word of Caution If you are accustomed to writing fractions with a slanted bar (/), *you are strongly urged to break the habit!* We make this recommendation because it is incorrect to write an answer that should be $\dfrac{x-5}{x+3}$, for example, as $x - 5/x + 3$. Notice that $x - 5/x + 3 = x - \dfrac{5}{x} + 3$ (division must be done before addition and subtraction). Then, since $\dfrac{x-5}{x+3} \neq x - \dfrac{5}{x} + 3$, $\dfrac{x-5}{x+3} \neq x - 5/x + 3$. If you do use a slanted bar for rational expressions, be sure to put parentheses around any numerator or denominator that contains more than one term.

If a slanted bar is used, parentheses are necessary for denominators with more than one *factor*, also. In this case, a common error is to write an answer that should be $\dfrac{1}{2x}$ as $1/2x$. Note, however, that $1/2x = \dfrac{1}{2}x$ (the division must be done first because it's on the *left*). Then, since $\dfrac{1}{2x} \neq \dfrac{1}{2}x$, $\dfrac{1}{2x} \neq 1/2x$.

---

*We usually refer to expressions in which the numerator and denominator are natural numbers as *fractions* and to expressions in which the numerator and denominator are polynomials (other than natural numbers) as *rational expressions.*

Equivalent Rational Expressions and the Fundamental Property of Rational Expressions

Just as equivalent fractions are fractions that have the same value, **equivalent rational expressions** are rational expressions that have the same value. The fundamental property of fractions holds even when the numerator and denominator are polynomials; we then call the property the **fundamental property of rational expressions.** It allows us to reduce rational expressions and also to "build" rational expressions in order to add and subtract them.

The fundamental property of rational expressions

> If P, Q, and C are polynomials, and if $Q \neq 0$ and $C \neq 0$, then
>
> $$\frac{P \cdot C}{Q \cdot C} = \frac{P}{Q}$$

The fundamental property of rational expressions permits us to do the following:

1. Multiply both numerator and denominator by the same nonzero number; that is, if $Q \neq 0$ and $C \neq 0$, then $\dfrac{P}{Q} = \dfrac{P \cdot C}{Q \cdot C}$. For example, if $x \neq 0$,

$$\frac{4}{7} = \frac{4 \cdot x}{7 \cdot x} = \frac{4x}{7x}$$

2. Divide both numerator and denominator by the same nonzero number; that is, if $Q \neq 0$ and $C \neq 0$, then $\dfrac{P \cdot C}{Q \cdot C} = \dfrac{P}{Q}$. Another way of expressing this is as follows: If $D \neq 0$, $\dfrac{P \div D}{Q \div D} = \dfrac{P}{Q}$; for example, if $x \neq 0$,

$$\frac{4x}{7x} = \frac{4x \div x}{7x \div x} = \frac{4}{7}$$

✖ **A Word of Caution** We do *not* get a rational expression equivalent to the one we started with if we *add* the same number to or *subtract* the same number from both the numerator and the denominator. For example, $\dfrac{2}{3} \neq \dfrac{2 + 4}{3 + 4}$ even though $\dfrac{2}{3} = \dfrac{2 \cdot 4}{3 \cdot 4}$, and $\dfrac{6}{9} \neq \dfrac{6 - 3}{9 - 3}$ even though $\dfrac{6}{9} = \dfrac{6 \div 3}{9 \div 3}$.

EXAMPLE 1   Determine whether the following pairs of rational expressions are equivalent.

SOLUTION

a. $\dfrac{1}{2}, \dfrac{1 + x}{2 + x}$ Not equivalent; we can't get the second rational expression from the first by multiplying or dividing both 1 and 2 by the same number.

b. $\dfrac{3}{x}, \dfrac{6}{2x}$ Equivalent; if we multiply both 3 and x by 2, we get 6 and $2x$.

c. $\dfrac{8x}{12y}, \dfrac{2x}{3y}$ Equivalent; if we divide both $8x$ and $12y$ by 4, we get $2x$ and $3y$.

d. $\dfrac{3 + x}{4 + x}, \dfrac{3}{4}$ Not equivalent; we can't get the second rational expression from the first by multiplying or dividing both $(3 + x)$ and $(4 + x)$ by the same number.

Exercises 8.1
Set I

Determine whether each pair of rational expressions is equivalent.

1. $\dfrac{x}{2y}, \dfrac{5x}{10y}$

2. $\dfrac{a}{7b}, \dfrac{2a}{14b}$

3. $\dfrac{x}{2y}, \dfrac{x+5}{2y+5}$

4. $\dfrac{3c}{5d}, \dfrac{3c+8}{5d+8}$

5. $\dfrac{6(x+1)}{12(3x-2)}, \dfrac{x+1}{2(3x-2)}$

6. $\dfrac{9(x+5)}{27(2x-7)}, \dfrac{x+5}{3(2x-7)}$

7. $\dfrac{s-t}{t-s}, \dfrac{t-s}{s-t}$

8. $\dfrac{w-5}{5-w}, \dfrac{5-w}{w-5}$

Writing Problems

Express the answers in your own words and in complete sentences.

1. Explain one way of determining whether two rational expressions are equivalent.

2. Explain why $\dfrac{1}{2+x} \neq 1/2 + x$.

Exercises 8.1
Set II

Determine whether each pair of rational expressions is equivalent.

1. $\dfrac{a}{5b}, \dfrac{10a}{50b}$

2. $\dfrac{x}{3y}, \dfrac{x+4}{3y+4}$

3. $\dfrac{x}{4y}, \dfrac{x+2}{4y+2}$

4. $\dfrac{7x}{3y}, \dfrac{7x(3+z)}{3y(3+z)}$

5. $\dfrac{5(2x-7)}{15(2x-5)}, \dfrac{2x-7}{3(2x-5)}$

6. $\dfrac{4(x+3)}{16(2x-5)}, \dfrac{4x+3}{32x-5}$

7. $\dfrac{w-3t}{3t-w}, \dfrac{3t-w}{w-3t}$

8. $\dfrac{s+5}{s-5}, \dfrac{s-5}{s+5}$

8.2 Reducing Rational Expressions to Lowest Terms

> A rational expression is in **lowest terms** if the greatest common factor (GCF) of its numerator and denominator is 1.

In this section and in the remainder of the book, it is understood that all answers that are rational expressions are to be reduced to lowest terms unless otherwise indicated.

If the numerator and denominator of a rational expression do not have a common factor other than 1, we cannot reduce the expression.

If the numerator and denominator of a rational expression do have a common factor, we can use the fundamental property of rational expressions and divide both the numerator and the denominator by any common factors; the new rational expression will be equivalent to the original one. We can reduce the rational expression to lowest terms in one step by dividing both numerator and denominator by their GCF.

In Arithmetic   Because 12 and 18 have a common factor, we can reduce the fraction $\frac{12}{18}$ as follows:

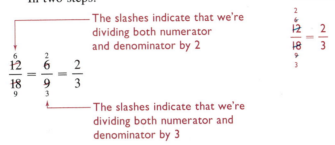

2 and 3 are both factors of the numerator

$$\frac{12}{18} = \frac{2 \cdot 2 \cdot 3}{2 \cdot 3 \cdot 3} = \frac{2}{3} \leftarrow \text{Numerator and denominator were both divided by } 2 \cdot 3$$

2 and 3 are both factors of the denominator

In practice, the work is often done in one of the following ways:

In two steps:

The slashes indicate that we're dividing both numerator and denominator by 2

$$\frac{\overset{6}{\cancel{12}}}{\underset{9}{\cancel{18}}} = \frac{\overset{2}{\cancel{6}}}{\underset{3}{\cancel{9}}} = \frac{2}{3}$$

The slashes indicate that we're dividing both numerator and denominator by 3

This work is often shown as follows:

$$\frac{\overset{2}{\overset{6}{\cancel{\cancel{12}}}}}{\underset{9}{\underset{3}{\cancel{\cancel{18}}}}} = \frac{2}{3}$$

In one step: We determine that the GCF of 12 and 18 is 6. Then

$$\frac{\overset{2}{\cancel{12}}}{\underset{3}{\cancel{18}}} = \frac{2}{3} \qquad \text{The slashes indicate that we're dividing both numerator and denominator by 6}$$

In Algebra   A rational expression can be reduced whenever the numerator and denominator have a common *factor*, and several different methods can be used to reduce a rational expression to lowest terms.

The fundamental property of rational expressions $\left(\dfrac{P \cdot C}{Q \cdot C} = \dfrac{P}{Q} \right)$ permits us to divide both numerator and denominator of a rational expression by any nonzero expression. Therefore, we can reduce a rational expression by finding the greatest common factor (GCF) of the numerator and denominator (after factoring them both completely) and dividing both numerator and denominator by their GCF (see Example 1).

EXAMPLE 1   Reduce each of the following rational expressions to lowest terms by finding the GCF of the numerator and denominator.

a. $\dfrac{4x^2y}{2xy}$

SOLUTION   The GCF of $4x^2y$ and $2xy$ is $2xy$.

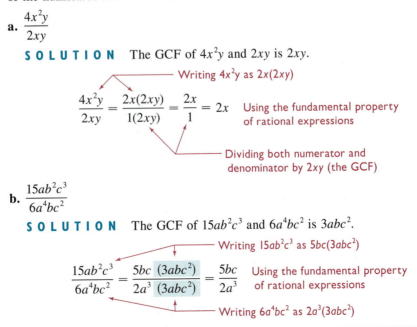

Writing $4x^2y$ as $2x(2xy)$

$$\frac{4x^2y}{2xy} = \frac{2x(2xy)}{1(2xy)} = \frac{2x}{1} = 2x \qquad \begin{array}{l}\text{Using the fundamental property} \\ \text{of rational expressions}\end{array}$$

Dividing both numerator and denominator by $2xy$ (the GCF)

b. $\dfrac{15ab^2c^3}{6a^4bc^2}$

SOLUTION   The GCF of $15ab^2c^3$ and $6a^4bc^2$ is $3abc^2$.

Writing $15ab^2c^3$ as $5bc(3abc^2)$

$$\frac{15ab^2c^3}{6a^4bc^2} = \frac{5bc\,(3abc^2)}{2a^3\,(3abc^2)} = \frac{5bc}{2a^3} \qquad \begin{array}{l}\text{Using the fundamental property} \\ \text{of rational expressions}\end{array}$$

Writing $6a^4bc^2$ as $2a^3(3abc^2)$

We can also reduce a rational expression to lowest terms by dividing both numerator and denominator by *any* factor common to both and then checking for any *other* common factors (see Example 2).

EXAMPLE 2

Examples of reducing rational expressions by dividing both numerator and denominator by *any* common factors:

Dividing both numerator and denominator by 2

Dividing both numerator and denominator by y

a. $\dfrac{4x^2y}{2xy} = \dfrac{\overset{2 \cdot x \cdot 1}{4x^2y}}{\underset{1 \cdot 1 \cdot 1}{2xy}} = \dfrac{2x}{1} = 2x$

Dividing both numerator and denominator by x

Dividing both numerator and denominator by 3

Dividing both numerator and denominator by b

b. $\dfrac{15ab^2c^3}{6a^4bc^2} = \dfrac{\overset{5 \cdot 1 \cdot b \cdot c}{15ab^2c^3}}{\underset{2 \cdot a^3 \cdot 1 \cdot 1}{6a^4bc^2}} = \dfrac{5bc}{2a^3}$

Dividing both numerator and denominator by c^2

Dividing both numerator and denominator by a

c. $\dfrac{z}{2z} = \dfrac{\overset{1}{z}}{\underset{1}{2z}} = \dfrac{1}{2}$

Dividing both numerator and denominator by z

Note   A factor of 1 will always remain in the numerator and denominator after they have been divided by factors common to both.

Furthermore, when we divided a monomial by a monomial in an earlier chapter, we were actually reducing a rational expression, and this method gives us yet another way of reducing rational expressions: We can reduce a rational expression to lowest terms by first rewriting it as a product of two or more rational expressions, then using the quotient rule $\left(\dfrac{x^a}{x^b} = x^{a-b}\right)$, and finally simplifying the result (see Example 3).

EXAMPLE 3

Examples of reducing rational expressions by using the quotient rule:

a. $\dfrac{4x^2y}{2xy} = \dfrac{\overset{2}{4}}{\underset{1}{2}} \cdot \dfrac{x^2}{x^1} \cdot \dfrac{y^1}{y^1} = 2 \cdot x^{2-1} \cdot y^{1-1} = 2x^1y^0 = 2x(1) = 2x$

These steps need not be shown

b. $\dfrac{15ab^2c^3}{6a^4bc^2} = \dfrac{\overset{5}{15}}{\underset{2}{6}} \cdot \dfrac{a^1}{a^4} \cdot \dfrac{b^2}{b^1} \cdot \dfrac{c^3}{c^2} = \dfrac{5}{2} \cdot a^{1-4} \cdot b^{2-1} \cdot c^{3-2} = \dfrac{5}{2}a^{-3}b^1c^1 = \dfrac{5bc}{2a^3}$

c. $\dfrac{8x^3}{2x} = \dfrac{\overset{4}{8}}{\underset{1}{2}} \cdot \dfrac{x^3}{x^1} = 4 \cdot x^{3-1} = 4x^2$

d. $\dfrac{6a^3b^4}{9ab^2} = \dfrac{\overset{2}{6}}{\underset{3}{9}} \cdot \dfrac{a^3}{a^1} \cdot \dfrac{b^4}{b^2} = \dfrac{2}{3} \cdot a^{3-1} \cdot b^{4-2} = \dfrac{2}{3}a^2b^2 \text{ or } \dfrac{2a^2b^2}{3}$

Recall that

$$-(b - a) = -1(b - a) = -b + a = a - b$$

Therefore, $a - b$ can always be substituted for $-(b - a)$, and $-(b - a)$ can always be substituted for $a - b$. We often use these facts when we're working with rational expressions. (See the second solution for Example 4e.)

In Example 4, the numerators and denominators generally contain two or more terms. We cannot attempt to reduce such rational expressions until we have *factored* the numerators and denominators. Remember, we can divide numerator and denominator only by common *factors*.

EXAMPLE 4 Examples of reducing rational expressions:

a. $\dfrac{3x - 9}{x^2 - 9} = \dfrac{3(\overset{1}{\cancel{x - 3}})}{(x + 3)(\underset{1}{\cancel{x - 3}})} = \dfrac{3}{x + 3}$ Dividing both numerator and denominator by $(x - 3)$, the only common *factor*

Notice that $\dfrac{3}{x + 3}$ *cannot be reduced*, since 3 is not a *factor* of the denominator.

b. $\dfrac{x^2 - 4x - 5}{x^2 + 5x + 4} = \dfrac{(\overset{1}{\cancel{x + 1}})(x - 5)}{(\underset{1}{\cancel{x + 1}})(x + 4)} = \dfrac{x - 5}{x + 4}$ Dividing both numerator and denominator by $(x + 1)$, the GCF

c. $\dfrac{3x^2 - 5xy - 2y^2}{6x^3y + 2x^2y^2} = \dfrac{(x - 2y)(\overset{1}{\cancel{3x + y}})}{2x^2y(\underset{1}{\cancel{3x + y}})} = \dfrac{x - 2y}{2x^2y}$ Dividing both numerator and denominator by $(3x + y)$, the GCF

Notice that $x - y$ and $y - x$ are the additive inverses of each other

$(-1)(x - y) = (y - x)$

d. $\dfrac{x - y}{y - x} = \dfrac{(-1)(x - y)}{(-1)(y - x)} = \dfrac{(\overset{1}{\cancel{y - x}})}{(-1)(\underset{1}{\cancel{y - x}})} = \dfrac{1}{-1} = -1$ Dividing both numerator and denominator by $(y - x)$, the GCF

Multiplying both numerator and denominator by -1

Note This *result* is quite important. It permits us to replace any rational expression of the *form* $\dfrac{x - y}{y - x}$ with -1; that is, any rational expression in which the numerator and denominator are the additive inverses of each other can be replaced with -1.

e. $\dfrac{2b^2 + ab - 3a^2}{4a^2 - 9ab + 5b^2} = \dfrac{(b - a)(2b + 3a)}{(a - b)(4a - 5b)}$

$(-1)(b - a) = (a - b)$

Dividing both numerator and denominator by $(a - b)$, the GCF

$= \dfrac{(-1)(b - a)(2b + 3a)}{(-1)(a - b)(4a - 5b)} = \dfrac{(\overset{1}{\cancel{a - b}})(2b + 3a)}{(-1)(\underset{1}{\cancel{a - b}})(4a - 5b)}$

Multiplying both numerator and denominator by -1

$= \dfrac{(2b + 3a)}{(-1)(4a - 5b)} = \dfrac{2b + 3a}{5b - 4a}$ Removing the parentheses *or*

$$\frac{(b-a)(2b+3a)}{(a-b)(4a-5b)} = \frac{(b-a)(2b+3a)}{-(b-a)(4a-5b)}$$ Dividing both numerator and denominator by $(b-a)$, the GCF

Substituting $-(b-a)$ for $a-b$

$$= \frac{2b+3a}{-(4a-5b)}$$

$$= \frac{2b+3a}{5b-4a}$$ Removing the parentheses

or, if we use the result of part d,

$$\frac{(b-a)(2b+3a)}{(a-b)(4a-5b)} = \frac{b-a}{a-b} \cdot \frac{2b+3a}{4a-5b}$$ $\frac{b-a}{a-b}$ is of the form $\frac{x-y}{y-x}$

$$= -1\left(\frac{2b+3a}{4a-5b}\right)$$ Replacing $\frac{b-a}{a-b}$ with -1

$$= -\frac{2b+3a}{4a-5b} *$$

f. $\dfrac{x+3}{x+6}$

This rational expression cannot be reduced, since neither x nor 3 is a *factor* of the numerator or the denominator.

g. $\dfrac{x+y}{x}$ x is not a factor of the numerator

This rational expression cannot be reduced.

A Word of Caution   A common error made in reducing fractions is to forget that the expression that the numerator and denominator are divided by must be a *factor* of both (see Examples 4f and 4g).

3 is a *term*, not a *factor*, of the numerator

$$\frac{\cancel{3}+2}{\cancel{3}} = 2$$ Incorrect reduction

The above reduction is incorrect because

$$\frac{3+2}{3} = \frac{5}{3} \neq 2$$

In Example 5, we use the method shown in Example 3 to simplify some algebraic expressions that are not rational expressions. (They are not rational expressions because the numerators and denominators aren't polynomials.) The *answers* are rational expressions.

---

*We will see in the next section that $-\dfrac{2b+3a}{4a-5b}$ is equivalent to $\dfrac{2b+3a}{5b-4a}$.

EXAMPLE 5   Examples of simplifying algebraic expressions:

a. $\dfrac{12x^{-2}}{4x^{-3}} = \dfrac{\overset{3}{\cancel{12}}}{\underset{1}{\cancel{4}}} \cdot \dfrac{x^{-2}}{x^{-3}} = 3 \cdot x^{-2-(-3)} = 3 \cdot x^{-2+3} = 3 \cdot x^1 = 3x$

b. $\dfrac{5a^4b^{-3}}{10a^{-2}b^{-4}} = \dfrac{\overset{1}{\cancel{5}}}{\underset{2}{\cancel{10}}} \cdot \dfrac{a^4}{a^{-2}} \cdot \dfrac{b^{-3}}{b^{-4}} = \dfrac{1}{2} \cdot a^{4-(-2)} \cdot b^{-3-(-4)}$

$\qquad\qquad = \dfrac{1}{2} \cdot a^{4+2} \cdot b^{-3+4} = \dfrac{1}{2} \cdot a^6 \cdot b^1 = \dfrac{1}{2}a^6b, \text{ or } \dfrac{a^6b}{2}$

c. $\left(\dfrac{9xy^{-2}}{15x^3y^{-4}}\right)^2 = \left(\dfrac{\overset{3}{\cancel{9}}}{\underset{5}{\cancel{15}}} \cdot \dfrac{x^1}{x^3} \cdot \dfrac{y^{-2}}{y^{-4}}\right)^2 = \left(\dfrac{3}{5} \cdot x^{1-3} \cdot y^{-2-(-4)}\right)^2 = \left(\dfrac{3}{5} \cdot x^{-2} \cdot y^{-2+4}\right)^2$

$\qquad\qquad = \left(\dfrac{3}{5} \cdot \dfrac{1}{x^2} \cdot y^2\right)^2 = \left(\dfrac{3y^2}{5x^2}\right)^2 = \dfrac{3^2y^{2\cdot2}}{5^2x^{2\cdot2}} = \dfrac{9y^4}{25x^4}$

Exercises 8.2
Set I

In Exercises 1–52, reduce each rational expression to lowest terms.

1. $\dfrac{2x}{6x^3}$

2. $\dfrac{3y^3}{9y^4}$

3. $\dfrac{6ab^2}{3ab}$

4. $\dfrac{10m^2n}{5mn}$

5. $\dfrac{9x^4}{6x^6}$

6. $\dfrac{9y^3}{6y^7}$

7. $\dfrac{10x^4}{15x^3}$

8. $\dfrac{15y^5}{9y^2}$

9. $\dfrac{12h^4k^3}{8h^2k}$

10. $\dfrac{16a^5b^3}{12ab^2}$

11. $\dfrac{15x^3yz^4}{18xy^3}$

12. $\dfrac{9a^2b^3c^4}{21bc^6}$

13. $\dfrac{24st^4u^2}{15s^2tu^3}$

14. $\dfrac{18u^2v^5w}{8vw^2}$

15. $\dfrac{21x^3y^5}{14yz^3}$

16. $\dfrac{25a^2b}{35ab^2}$

17. $\dfrac{5x - 10}{x - 2}$

18. $\dfrac{3x + 12}{x + 4}$

19. $\dfrac{7x - 21}{15x^2 - 45x}$

20. $\dfrac{12y - 18}{4y^3 - 6y^2}$

21. $\dfrac{6x^2y}{30x^3y - 18xy^2}$

22. $\dfrac{9ab^2}{18a^3b^2 - 36ab^3}$

23. $\dfrac{8x^2y + 12xy^2}{12x^3y + 18x^2y^2}$

24. $\dfrac{18s^3t + 30s^2t^3}{24s^2t^2 + 40st^4}$

25. $-\dfrac{5x - 6}{6 - 5x}$

26. $\dfrac{4 - 3z}{3z - 4}$

27. $\dfrac{5x^2 + 30x}{10x^2 - 40x}$

28. $\dfrac{4x^3 - 4x^2}{12x^2 - 12x}$

29. $\dfrac{2 + 4}{4}$

30. $\dfrac{3 + 9}{3}$

31. $\dfrac{5 + x}{5}$

32. $\dfrac{x + 8}{8}$

33. $\dfrac{x^2 - 1}{x + 1}$

34. $\dfrac{x^2 - 4}{x - 2}$

35. $\dfrac{6x^2 - x - 2}{10x^2 + 3x - 1}$

36. $\dfrac{8x^2 - 10x - 3}{12x^2 + 11x + 2}$

37. $\dfrac{x^2 - y^2}{(x + y)^2}$

38. $\dfrac{a^2 - 9b^2}{(a - 3b)^2}$

39. $\dfrac{2y^2 + xy - 6x^2}{3x^2 + xy - 2y^2}$

40. $\dfrac{10y^2 + 11xy - 6x^2}{4x^2 - 4xy - 15y^2}$

41. $\dfrac{8x^2 - 2y^2}{2ax - ay + 2bx - by}$

42. $\dfrac{3x^2 - 12y^2}{ax + 2by + 2ay + bx}$

43. $\dfrac{(-1)(z - 8)}{8 - z}$

44. $\dfrac{x - 12}{(-1)(12 - x)}$

45. $\dfrac{(-1)(a - 2b)(b - a)}{(2a + b)(a - b)}$

46. $\dfrac{8(n - 2m)}{(-1)(3n + m)(2m - n)}$

47. $\dfrac{9 - 16x^2}{16x^2 - 24x + 9}$

48. $\dfrac{25 - 9x^2}{9x^2 - 30x + 25}$

49. $\dfrac{10 + x - 3x^2}{2x^2 + x - 10}$

50. $\dfrac{12 - 19x + 5x^2}{3x^2 - 5x - 12}$

51. $\dfrac{18 - 3x - 3x^2}{6x^2 + 6x - 36}$

52. $\dfrac{16 + 4x - 2x^2}{8x^2 - 16x - 64}$

In Exercises 53–76, simplify each algebraic expression.

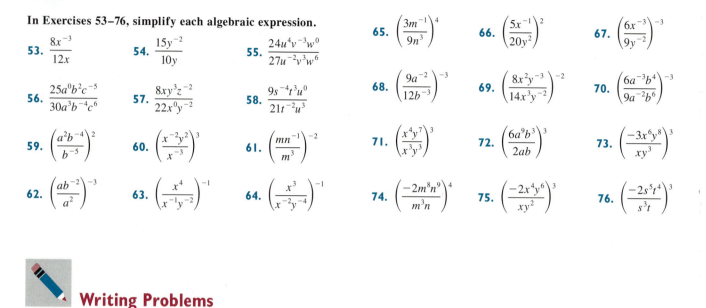

53. $\dfrac{8x^{-3}}{12x}$

54. $\dfrac{15y^{-2}}{10y}$

55. $\dfrac{24u^4v^{-3}w^0}{27u^{-2}v^3w^6}$

56. $\dfrac{25a^0b^2c^{-5}}{30a^3b^{-4}c^6}$

57. $\dfrac{8xy^3z^{-2}}{22x^0y^{-2}}$

58. $\dfrac{9s^{-4}t^3u^0}{21t^{-2}u^3}$

59. $\left(\dfrac{a^2b^{-4}}{b^{-5}}\right)^2$

60. $\left(\dfrac{x^{-2}y^2}{x^{-3}}\right)^3$

61. $\left(\dfrac{mn^{-1}}{m^3}\right)^{-2}$

62. $\left(\dfrac{ab^{-2}}{a^2}\right)^{-3}$

63. $\left(\dfrac{x^4}{x^{-1}y^{-2}}\right)^{-1}$

64. $\left(\dfrac{x^3}{x^{-2}y^{-4}}\right)^{-1}$

65. $\left(\dfrac{3m^{-1}}{9n^3}\right)^4$

66. $\left(\dfrac{5x^{-1}}{20y^2}\right)^2$

67. $\left(\dfrac{6x^{-3}}{9y^{-2}}\right)^{-3}$

68. $\left(\dfrac{9a^{-2}}{12b^{-3}}\right)^{-3}$

69. $\left(\dfrac{8x^2y^{-3}}{14x^3y^{-2}}\right)^{-2}$

70. $\left(\dfrac{6a^{-3}b^4}{9a^{-2}b^6}\right)^{-3}$

71. $\left(\dfrac{x^4y^7}{x^3y^3}\right)^3$

72. $\left(\dfrac{6a^9b^3}{2ab}\right)^3$

73. $\left(\dfrac{-3x^6y^8}{xy^3}\right)^3$

74. $\left(\dfrac{-2m^8n^9}{m^3n}\right)^4$

75. $\left(\dfrac{-2x^4y^6}{xy^2}\right)^3$

76. $\left(\dfrac{-2s^5t^4}{s^3t}\right)^3$

Writing Problems

Express the answer in your own words and in complete sentences.

1. Describe one method of reducing a rational expression to lowest terms.

Exercises 8.2
Set II

In Exercises 1–52, reduce each rational expression to lowest terms.

1. $\dfrac{12y^3}{16y^5}$

2. $\dfrac{2ab}{8a^2b^3}$

3. $\dfrac{8mn^3}{4n^2}$

4. $\dfrac{15x^3y}{10xy^2}$

5. $\dfrac{12b^6}{8b^3}$

6. $\dfrac{12x^3}{10x}$

7. $\dfrac{18z^6}{10z^4}$

8. $\dfrac{24x^4}{9x^2}$

9. $\dfrac{25a^3b^4}{15ab^3}$

10. $\dfrac{18a^8}{12a^4}$

11. $\dfrac{20h^2j^3k}{35h^4jk^3}$

12. $\dfrac{35kl^3m^4}{28k^4l^3m^6}$

13. $\dfrac{15m^2n^2q}{25m^3nq^4}$

14. $\dfrac{14xy^3z^4}{12x^2y^4z}$

15. $\dfrac{8s^3t^4}{12s^4t^3}$

16. $\dfrac{14x^5y^3}{21xy^2}$

17. $\dfrac{8x-12}{2x-3}$

18. $\dfrac{8x+5}{5+8x}$

19. $\dfrac{9y-27}{12y^2-36y}$

20. $\dfrac{12x}{4x^3-6x^2}$

21. $\dfrac{5s^2t}{30s^4t^2-20s^2t}$

22. $\dfrac{9xy^2-18x^2y}{18x^3y^2-36xy^3}$

23. $\dfrac{27a^2b+18ab^3}{36a^3b+24a^2b^3}$

24. $\dfrac{10x^2+5xy}{12xy-6y^2}$

25. $-\dfrac{7x-1}{1-7x}$

26. $-\dfrac{3x+4}{4+3x}$

27. $\dfrac{6x^2+9x}{12x^2+9x}$

28. $\dfrac{x^2-9}{3x+9}$

29. $\dfrac{5+10}{10}$

30. $\dfrac{5+10x}{10}$

31. $\dfrac{7+x}{7}$

32. $\dfrac{21+3x}{3}$

33. $\dfrac{x^2-16}{x+4}$

34. $\dfrac{x-1}{x^2-1}$

35. $\dfrac{4x^2-9x+2}{4x^2+7x-2}$

36. $\dfrac{x^2-9}{x^2+5x+6}$

37. $\dfrac{x^2-25}{(x+5)^2}$

38. $\dfrac{x^2-16}{x^2-x-12}$

39. $\dfrac{3y^2+5xy-2x^2}{3x^2-8xy-3y^2}$

40. $\dfrac{x^2+x-20}{x^2+2x-15}$

41. $\dfrac{8x^2-18y^2}{2ax-3ay+2bx-3by}$

42. $\dfrac{x^2-11x+30}{x^2-9x+20}$

43. $\dfrac{(-1)(a-5)}{5-a}$

44. $\dfrac{12a^3b+6a^2b^2}{18a^2b^2+9ab^3}$

45. $\dfrac{(-1)(x-y)(3y-x)}{(3y+x)(y-x)}$

46. $\dfrac{15m^2n^2-15mn^3}{10m^2-10mn}$

47. $\dfrac{4-25x^2}{(5x-2)^2}$

48. $\dfrac{16x^2-y^2}{y^2-8xy+16x^2}$

49. $\dfrac{8+10y-12y^2}{2y^2-3y-20}$

50. $\dfrac{60a^2-110a+30}{15a+5a^2-10a^3}$

51. $\dfrac{36-3x-3x^2}{9x^2+9x-108}$

52. $\dfrac{6x^3-54x}{108x^2-9x^3-9x^4}$

In Exercises 53–76, simplify each algebraic expression.

53. $\dfrac{16h^{-2}}{10h}$

54. $\dfrac{18x^{-3}}{10x}$

55. $\dfrac{9a^{-3}b^{-2}c^4}{12a^{-1}b^{-4}c^0}$

56. $\dfrac{15c^{-4}d^{-1}}{20c^{-2}d^{-4}}$

57. $\dfrac{24m^{-1}np^3}{18m^{-3}n^4p}$

58. $\left(\dfrac{35x^0y^{-1}}{42x^{-4}y^{-3}}\right)^2$

59. $\left(\dfrac{6a^{-3}b^3}{15a^2b^{-2}}\right)^2$

60. $\dfrac{15a^{-5}}{18a^{-2}b^{-3}}$

61. $\left(\dfrac{m^{-1}n^3}{m}\right)^{-2}$

62. $\dfrac{9a^2b^{-3}}{15a^{-5}b^2}$

63. $\left(\dfrac{x^2}{x^{-3}y^{-2}}\right)^{-1}$

64. $\left(\dfrac{ab^{-3}}{2b^{-2}}\right)^{-3}$

65. $\left(\dfrac{15R^4S}{21RS^{-2}}\right)^3$

66. $\dfrac{8x^{-5}y^{-3}}{18x^{-2}y^{-1}}$

67. $\left(\dfrac{14a^{-3}b^2}{42a^{-4}}\right)^3$

68. $\left(\dfrac{2x^2y^{-3}}{x^{-3}y}\right)^{-4}$

69. $\left(\dfrac{8x^{-4}y}{12xy^{-3}}\right)^{-4}$

70. $\left(\dfrac{12c^{-3}}{9d^{-4}}\right)^{-3}$

71. $\left(\dfrac{36a^3b^{-2}}{20b^3c^{-1}}\right)^{-3}$

72. $\left(\dfrac{9x^{-3}b^4}{6x^{-2}b^6}\right)^0$

73. $\left(\dfrac{-6s^3t^{-2}}{s^{-4}t^5u^4}\right)^3$

74. $\left(\dfrac{-4x^3y^2}{xy^0z^4}\right)^4$

75. $\left(\dfrac{de^2f^{-4}}{-2d^{-2}}\right)^3$

76. $\dfrac{12a^0b^0}{21a^{-2}b^3c}$

8.3 Multiplying and Dividing Rational Expressions

Multiplying Rational Expressions

Multiplication of rational expressions is defined as follows:

$$\frac{P}{Q} \cdot \frac{R}{S} = \frac{P \cdot R}{Q \cdot S}$$

where P, Q, R, and S are polynomials and $Q \neq 0$ and $S \neq 0$.

In practice, however, we can often reduce the resulting rational expression. Therefore, we give the following suggestions for multiplying rational expressions:

Multiplying rational expressions

1. Factor any numerators or denominators that contain more than one term.

2. Divide the numerators and denominators by all factors common to both. (The common factors can be, but do not have to be, in the same rational expression.)

3. The numerator of the product is the product of all the factors remaining in the numerators, and the denominator of the product is the product of all the factors remaining in the denominators. If the only factor remaining in the numerator is 1, that 1 must be *written*.

We can write the product as a single rational expression either *before* we divide by the common factors (see Examples 1 and 2a–c) or *after* we divide by the common factors (see Examples 2d–e).

EXAMPLE 1 Multiply $\frac{4}{9} \cdot \frac{3}{8}$.

SOLUTION

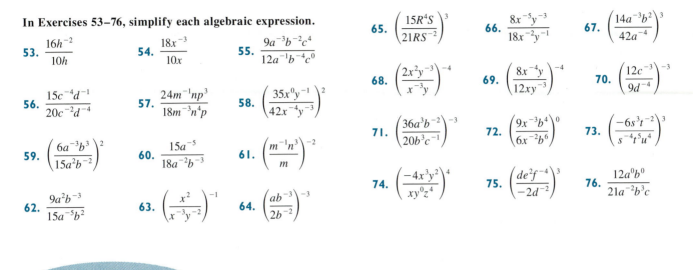

Writing 4 as $2 \cdot 2$

$$\frac{4}{9} \cdot \frac{3}{8} = \frac{2 \cdot 2 \cdot 3}{3 \cdot 3 \cdot 2 \cdot 2 \cdot 2} = \frac{1(2 \cdot 2 \cdot 3)}{2 \cdot 3(2 \cdot 2 \cdot 3)} = \frac{1}{6}$$ Dividing both numerator and denominator by $2 \cdot 2 \cdot 3$

Writing 9 as $3 \cdot 3$ and 8 as $2 \cdot 2 \cdot 2$

EXAMPLE 2

Examples of multiplying rational expressions:

a. $\dfrac{1}{m^2} \cdot \dfrac{m}{5} = \dfrac{1 \cdot \overset{1}{\cancel{m}}}{\underset{m}{\cancel{m^2}} \cdot 5} = \dfrac{1}{5m}$ Dividing both numerator and denominator by m

b. $\dfrac{2y^3}{3x^2} \cdot \dfrac{12x}{5y^2} = \dfrac{2\overset{y}{\cancel{y^3}} \cdot \overset{4 \cdot 1}{\cancel{12x}}}{\underset{1 \cdot x}{\cancel{3x^2}} \cdot 5\cancel{y^2}} = \dfrac{8y}{5x}$ Dividing both numerator and denominator by 3, x, and y^2

Dividing both numerator and denominator by 2, x, and $(x-3)$

c. $\dfrac{x}{2x-6} \cdot \dfrac{4x-12}{x^2} = \dfrac{x}{2(x-3)} \cdot \dfrac{4(x-3)}{x^2} = \dfrac{\overset{1}{\cancel{x}} \cdot \overset{2}{\cancel{4}}(\cancel{x-3})^{\,1}}{\underset{1}{\cancel{2}}(\cancel{x-3}) \cdot \underset{x}{\cancel{x^2}}} = \dfrac{2}{x}$

Dividing both numerator and denominator by $2x$ and $(x+2)$

d. $\dfrac{x+2}{6x^2} \cdot \dfrac{8x}{x^2-x-6} = \dfrac{(\cancel{x+2})^{\,1}}{\underset{3x}{\cancel{6x^2}}} \cdot \dfrac{\overset{4}{\cancel{8x}}}{(\cancel{x+2})(x-3)} = \dfrac{4}{3x(x-3)}$

Dividing both numerator and denominator by $5xy$ and $(x+y)$

e. $\dfrac{10xy^3}{x^2-y^2} \cdot \dfrac{2x^2+xy-y^2}{15x^2y} = \dfrac{\overset{2 \; y^2}{\cancel{10xy^3}}}{(\cancel{x+y})(x-y)} \cdot \dfrac{(\cancel{x+y})(2x-y)}{\underset{3x}{\cancel{15x^2y}}}$

$\qquad\qquad = \dfrac{2y^2(2x-y)}{3x(x-y)}$

Dividing Rational Expressions

Recall that, in a division problem, the number we're dividing *by* is called the *divisor*, and the number we're dividing *into* is called the *dividend*; also recall from arithmetic that to divide one rational number by another we multiply the dividend by the multiplicative inverse (the reciprocal) of the divisor.

In Arithmetic

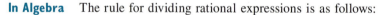

The multiplicative inverse of the divisor

$$\dfrac{3}{5} \div \dfrac{4}{7} = \dfrac{3}{5} \cdot \dfrac{7}{4} = \dfrac{3 \cdot 7}{5 \cdot 4} = \dfrac{21}{20}$$

The dividend ⎦   ⎣ The divisor

In Algebra   The rule for dividing rational expressions is as follows:

Dividing rational expressions

Multiply the dividend by the multiplicative inverse (the reciprocal) of the divisor:

$$\dfrac{P}{Q} \div \dfrac{S}{T} = \dfrac{P}{Q} \cdot \dfrac{T}{S}$$

where P, Q, S, and T are polynomials and $Q \neq 0$, $S \neq 0$, and $T \neq 0$. Reduce the resulting rational expression, if possible.

P R O O F   In both arithmetic and algebra, a division problem is equivalent to a rational expression $\left(\text{that is, } P \div Q = \dfrac{P}{Q}\right)$. For example, $6 \div 2 = \dfrac{6}{2}$, and $(3x^2+5) \div (5x-1) = \dfrac{3x^2+5}{5x-1}$.

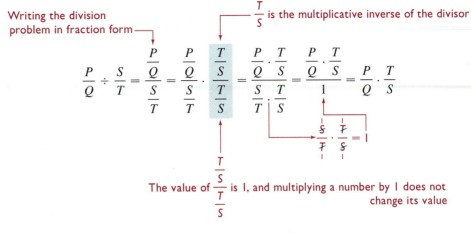

Therefore, $\dfrac{P}{Q} \div \dfrac{S}{T} = \dfrac{P}{Q} \cdot \dfrac{T}{S}$. ■

EXAMPLE 3 Examples of dividing rational expressions:

Changing the division problem to a multiplication problem

a. $\dfrac{4}{3x} \div \dfrac{12}{x^3} = \dfrac{4}{3x} \cdot \dfrac{x^3}{12} = \dfrac{4x^3}{3 \cdot 12x} = \dfrac{\overset{1}{4}}{3 \cdot \underset{3}{12}} \cdot \dfrac{x^3}{x} = \dfrac{1}{9}x^{3-1} = \dfrac{1}{9}x^2$, or $\dfrac{x^2}{9}$

These steps need not be shown

The problem can also be done as follows:

$$\dfrac{4}{3x} \div \dfrac{12}{x^3} = \dfrac{\overset{1}{4}}{3x} \cdot \dfrac{\overset{x^2}{x^3}}{\underset{3}{12}} = \dfrac{x^2}{9}$$

b. $\dfrac{4r^3}{9s^2} \div \dfrac{8r^2s^4}{15rs} = \dfrac{4r^3}{9s^2} \cdot \dfrac{15rs}{8r^2s^4} = \dfrac{4 \cdot 15r^4s}{9 \cdot 8r^2s^6} = \dfrac{\overset{1}{4} \cdot \overset{5}{15}}{\underset{2}{8} \cdot \underset{3}{9}} \cdot \dfrac{r^4}{r^2} \cdot \dfrac{s^1}{s^6} = \dfrac{5}{6}r^{4-2}s^{1-6}$

These steps need not be shown

$$= \dfrac{5}{6}r^2s^{-5} = \dfrac{5r^2}{6s^5}$$

The problem can also be done as follows:

$$\dfrac{4r^3}{9s^2} \div \dfrac{8r^2s^4}{15rs} = \dfrac{\overset{r}{4r^3}}{\underset{3\,s}{9s^2}} \cdot \dfrac{\overset{5}{15rs}^1}{\underset{2}{8r^2s^4}} = \dfrac{5r^2}{6s^5}$$

c. $\dfrac{y^2 - x^2}{4xy - 2y^2} \div \dfrac{2x - 2y}{2x^2 + xy - y^2} = \dfrac{y^2 - x^2}{4xy - 2y^2} \cdot \dfrac{2x^2 + xy - y^2}{2x - 2y}$ ← This expression must be *factored* before it can be reduced

$$= \dfrac{(y - x)(y + x)}{2y(2x \overset{1}{\cancel{-}\, y})} \cdot \dfrac{(x + y)(2x \overset{1}{\cancel{-}\, y})}{2(x - y)}$$

$$= \dfrac{(y - x)(x + y)^2}{(x - y)(4y)}$$

$$= \boxed{\dfrac{y - x}{x - y}} \cdot \dfrac{(x + y)^2}{4y} = \boxed{-1} \cdot \dfrac{(x + y)^2}{4y} = -\dfrac{(x + y)^2}{4y}$$

Replacing $\dfrac{y - x}{x - y}$ with -1

d. $\dfrac{3y^3 - 3y^2}{16y^5 + 8y^4} \div \dfrac{3y^2 + 6y - 9}{4y + 12} = \dfrac{3y^2(y - 1)}{8y^4(2y + 1)} \cdot \dfrac{4(y + 3)}{3(y^2 + 2y - 3)}$

$$= \dfrac{\overset{1}{\cancel{3}}\overset{1}{\cancel{y^2}}(\cancel{y - 1})}{\underset{2\;y^2}{\cancel{8}\cancel{y^4}(2y + 1)}} \cdot \dfrac{\overset{1}{\cancel{4}}(\cancel{y + 3})}{\underset{1}{\cancel{3}}(\underset{1}{\cancel{y + 3}})(\underset{1}{\cancel{y - 1}})}$$

$$= \dfrac{1}{2y^2(2y + 1)}$$

> **A Word of Caution** $2y^2(2y + 1)$ is not an acceptable answer for Example 3d. The 1 in the numerator cannot be omitted.

Order of Operations

The usual order of operations applies when we're multiplying and dividing rational expressions. If division is interpreted as *and changed to* multiplication, then the multiplication can, of course, be done in any order (see Example 4).

EXAMPLE 4   Perform the indicated operations.

a. $\dfrac{3x}{5y} \div \dfrac{7x^2}{6y^3} \div \dfrac{4y}{9z}$

SOLUTION I   Working from left to right, we have

$$\dfrac{3x}{5y} \div \dfrac{7x^2}{6y^3} \div \dfrac{4y}{9z} = \left(\dfrac{3x}{5y} \div \dfrac{7x^2}{6y^3}\right) \div \dfrac{4y}{9z} = \left(\dfrac{\overset{1}{\cancel{3x}}}{\cancel{5y}} \cdot \dfrac{6\overset{y^2}{\cancel{y^3}}}{7\underset{x}{\cancel{x^2}}}\right) \div \dfrac{4y}{9z} = \dfrac{18\overset{9y}{\cancel{y^2}}}{35x} \cdot \dfrac{9z}{\underset{2}{\cancel{4y}}} = \dfrac{81yz}{70x}$$

SOLUTION 2   Changing all divisions to multiplication yields

$$\dfrac{3x}{5y} \div \dfrac{7x^2}{6y^3} \div \dfrac{4y}{9z} = \dfrac{\overset{1}{\cancel{3x}}}{\cancel{5y}} \cdot \dfrac{6\overset{3y^2z}{\cancel{y^3}}}{7\underset{x}{\cancel{x^2}}} \cdot \dfrac{9z}{\underset{2}{\cancel{4y}}} = \dfrac{81yz}{70x}$$

b. $\dfrac{7x^2}{3y} \div \dfrac{5x}{2y^3} \cdot \dfrac{3y}{4x}$

SOLUTION I   Working from left to right, we have

$$\dfrac{7x^2}{3y} \div \dfrac{5x}{2y^3} \cdot \dfrac{3y}{4x} = \left(\dfrac{7x^2}{3y} \div \dfrac{5x}{2y^3}\right) \cdot \dfrac{3y}{4x} = \left(\dfrac{7\overset{x}{\cancel{x^2}}}{\underset{1}{\cancel{3y}}} \cdot \dfrac{2\overset{y^2}{\cancel{y^3}}}{\underset{1}{\cancel{5x}}}\right) \cdot \dfrac{3y}{4x} = \dfrac{\overset{7}{\cancel{14}}xy^2}{\underset{5}{\cancel{15}}} \cdot \dfrac{\overset{1}{\cancel{3y}}}{\underset{2}{\cancel{4x}}} = \dfrac{7y^3}{10}$$

SOLUTION 2   Changing all divisions to multiplication yields

$$\dfrac{7x^2}{3y} \div \dfrac{5x}{2y^3} \cdot \dfrac{3y}{4x} = \dfrac{7\overset{x}{\cancel{x^2}}}{\underset{1}{\cancel{3y}}} \cdot \dfrac{2\overset{1}{\cancel{y^3}}}{\underset{1}{\cancel{5x}}} \cdot \dfrac{\overset{1}{\cancel{3y}}}{\underset{2}{\cancel{4x}}} = \dfrac{7y^3}{10}$$

The Three Signs of a Rational Expression

Every rational expression has three signs associated with it, even if those signs are not written: the sign of the rational expression itself, the sign of the numerator, and the sign of the denominator. Consider the rational expression $\frac{3}{4}$:

The sign of the numerator

The sign of the rational expression $\rightarrow +\dfrac{+3}{+4}$

The sign of the denominator

Recall that $\dfrac{-1}{-1} = 1$ and that $(-1)(-1) = 1$, and remember that the multiplicative identity property guarantees that when we multiply a number by 1, the value of the number is unchanged. Therefore, we can multiply a rational expression by $\dfrac{-1}{-1}$ or by $(-1)(-1)$ without changing its value. Recall, too, that $\dfrac{-1}{1} = -1$ and that $\dfrac{1}{-1} = -1$; therefore, we can replace -1 with $\dfrac{-1}{1}$ or with $\dfrac{1}{-1}$, so

$$(-1)\left(\dfrac{-1}{1}\right) = 1 \quad \text{and} \quad (-1)\left(\dfrac{1}{-1}\right) = 1$$

Let's see what happens to the three signs of the rational expression $\frac{3}{4}$ as we multiply it by $\dfrac{-1}{-1}$, by $(-1)\left(\dfrac{-1}{1}\right)$, and by $(-1)\left(\dfrac{1}{-1}\right)$, all of which equal 1.

$$+\dfrac{+3}{+4} = + \left(\dfrac{-1}{-1}\right)\left(\dfrac{3}{4}\right) = +\dfrac{(-1)(3)}{(-1)(4)} = +\dfrac{-3}{-4}$$

Multiplying by $\dfrac{-1}{-1}$ leaves the value unchanged

The sign of the numerator and the sign of the denominator are different from $+\dfrac{+3}{+4}$; that is, *two* of the three signs are different

$$+\dfrac{+3}{+4} = + (-1)\left(\dfrac{-1}{1}\right)\left(\dfrac{3}{4}\right) = -\dfrac{(-1)(3)}{(1)(4)} = -\dfrac{-3}{+4}$$

Multiplying by $(-1)\left(\dfrac{-1}{1}\right)$ leaves the value unchanged

The sign of the rational expression and the sign of the numerator are different from $+\dfrac{+3}{+4}$; that is, *two* of the three signs are different

$$+\dfrac{+3}{+4} = + (-1)\left(\dfrac{1}{-1}\right)\left(\dfrac{3}{4}\right) = -\dfrac{(1)(3)}{(-1)(4)} = -\dfrac{+3}{-4}$$

Multiplying by $(-1)\left(\dfrac{1}{-1}\right)$ leaves the value unchanged

The sign of the rational expression and the sign of the denominator are different from $+\dfrac{+3}{+4}$; that is, *two* of the three signs are different

Let's see what happens to the three signs of the rational expression $\dfrac{-2}{3}$ as we multiply it by $\dfrac{-1}{-1}$, by $(-1)\left(\dfrac{-1}{1}\right)$, and by $(-1)\left(\dfrac{1}{-1}\right)$, all of which equal 1.

$$+\dfrac{-2}{+3} = + \left(\dfrac{-1}{-1}\right)\left(\dfrac{-2}{3}\right) = +\dfrac{(-1)(-2)}{(-1)(3)} = +\dfrac{+2}{-3}$$

The sign of the numerator and the sign of the denominator are different from $+\dfrac{-2}{+3}$; that is, *two* of the three signs are different

$$+\dfrac{-2}{+3} = + (-1)\left(\dfrac{-1}{1}\right)\left(\dfrac{-2}{3}\right) = -\dfrac{(-1)(-2)}{(1)(3)} = -\dfrac{+2}{+3}$$

The sign of the rational expression and the sign of the numerator are different from $+\dfrac{-2}{+3}$; that is, *two* of the three signs are different

$$+\dfrac{-2}{+3} = + (-1)\left(\dfrac{1}{-1}\right)\left(\dfrac{-2}{3}\right) = -\dfrac{(1)(-2)}{(-1)(3)} = -\dfrac{-2}{-3}$$

The sign of the rational expression and the sign of the denominator are different from $+\dfrac{-2}{+3}$; that is, *two* of the three signs are different

We see, then, that changing two of the three signs of a rational expression does not change the value of that rational expression.

The rule of signs for rational expressions

> If any *two* of the three signs of a rational expression are changed, the value of the rational expression is unchanged. Rational expressions obtained in this way are *equivalent rational expressions*.

This rule of signs is sometimes helpful when we are reducing rational expressions or when we are performing operations, such as addition, multiplication, and so forth, on rational expressions. Furthermore, the rule of signs enables us to write an answer such as $-\dfrac{2b + a}{4a - 5}$ in an alternative form. Changing the sign of the rational expression and the sign of the denominator yields

$$-\frac{2b + a}{4a - 5} = + \frac{2b + a}{-(4a - 5)} = \frac{2b + a}{5 - 4a}$$

$$\left(\text{We mentioned in the preceding section that } -\frac{2b + a}{4a - 5} \text{ is equivalent to } \frac{2b + a}{5 - 4a}. \right)$$

Note The rule of signs for rational expressions implies that when a rational expression contains a negative factor, the negative sign of that factor can be written in front of the numerator *or* in front of the rational expression *or* in front of the denominator. For example,

$$\frac{-2}{3x} = -\frac{2}{3x} = \frac{2}{-3x}$$

We can even think of the rule of signs as permitting us to *move* a negative sign from the numerator to in front of the rational expression, from the numerator to the denominator, and so on. (However, most authors and instructors prefer that the negative sign *not* be in the *denominator*.)

When the numerator and denominator each contain just one term, many instructors believe that the rational expression should not contain more than one negative sign; for example, many instructors insist that $\dfrac{-5x}{-y}$ be rewritten as $\dfrac{5x}{y}$, and that $-\dfrac{-4z}{-3w}$ be rewritten as $-\dfrac{4z}{3w}$ or as $\dfrac{-4z}{3w}$.

EXAMPLE 5 Find the missing term in each expression.

a. $-\dfrac{-3x}{2y} = \dfrac{?}{2y}$

SOLUTION Because the signs of the *denominators* are the same in both rational expressions (both are understood to be +) and the signs of the *rational expressions* are different, the signs of the *numerators* must be different. Therefore,

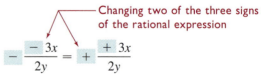
Changing two of the three signs of the rational expression

$$-\frac{-\ 3x}{2y} = +\frac{+\ 3x}{2y}$$

The missing term is $3x$.

b. $\dfrac{x}{-5} = \dfrac{-x}{?}$

SOLUTION Because the signs of the *rational expressions* are the same in both rational expressions (both are understood to be +) and the signs of the *numerators* are different, the signs of the *denominators* must be different. Therefore,

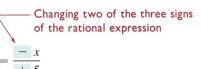

The missing term is 5.

c. $-\dfrac{30}{3x - 5} = \dfrac{30}{?}$

S O L U T I O N   The signs of the numerators are the same (both are understood to be +) and the signs of the rational expressions are different, so the signs of the denominators must also be different.

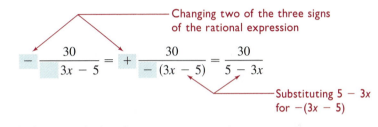

Therefore, the missing term is $5 - 3x$.

d. $\dfrac{a - 1}{-2} = -\dfrac{?}{-2}$

S O L U T I O N   The signs of the denominators are the same (both are −) and the signs of the rational expressions are different, so the signs of the numerators must also be different.

$$\frac{a - 1}{-2} = -\frac{-(a - 1)}{-2} = -\frac{1 - a}{-2}$$

Changing two of the three signs of the rational expression

Substituting $1 - a$ for $-(a - 1)$

The missing term is $1 - a$.

e. $\dfrac{x - y}{(a - b)(u + v)} = \dfrac{?}{(b - a)(u + v)}$

S O L U T I O N   The signs of the rational expressions are both understood to be +, and the signs of the denominators are different because $b - a$ is the *additive inverse* of $a - b$. The signs of the numerators must, therefore, be different.

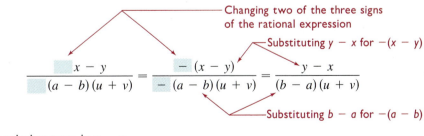

The missing term is $y - x$.

The problems in Examples 5b and 5e can also be done by applying the fundamental property of rational expressions and multiplying both numerator and denominator of each expression by -1.

Exercises 8.3
Set I

In Exercises 1–40, perform the indicated operations.

1. $\frac{5}{6} \div \frac{5}{3}$

2. $\frac{3}{8} \div \frac{21}{12}$

3. $\frac{4a^3}{5b^2} \cdot \frac{10b}{8a^2}$

4. $\frac{6d^2}{8c} \cdot \frac{12c^2}{9d^3}$

5. $\frac{3x^2}{16} \div \frac{x}{8}$

6. $\frac{4y^3}{7} \div \frac{8y^2}{21}$

7. $\frac{3x^4y^2z}{18xy} \cdot \frac{15z}{x^3yz^2}$

8. $\frac{21a^2b^5}{4b^2c^2} \cdot \frac{6c^3}{7a^2b^3c}$

9. $\frac{x}{x+2} \cdot \frac{5x+10}{x^3}$

10. $\frac{b}{b+5} \cdot \frac{3b+15}{b^4}$

11. $\frac{y-2}{y} \cdot \frac{6}{3y-6}$

12. $\frac{a-1}{a} \cdot \frac{8}{4a-4}$

13. $\frac{b^3}{a+3} \div \frac{4b^2}{2a+6}$

14. $\frac{m^4}{m+2} \div \frac{6m^3}{3m+6}$

15. $\frac{5s-15}{30s} \div \frac{s-3}{45s^2}$

16. $\frac{3n-6}{15n} \div \frac{n-2}{20n^2}$

17. $\frac{a+4}{a-4} \div \frac{a^2+8a+16}{a^2-16}$

18. $\frac{x+8}{x-8} \div \frac{x^2+16x+64}{x^2-64}$

19. $\frac{5}{z+4} \cdot \frac{z^2-16}{(z-4)^2}$

20. $\frac{7}{x+3} \cdot \frac{x^2-9}{(x-3)^2}$

21. $\frac{3a-3b}{4c+4d} \cdot \frac{2c+2d}{b-a}$

22. $\frac{7x-7y}{8a+8b} \cdot \frac{4a+4b}{y-x}$

23. $\frac{4x-8}{4} \cdot \frac{x+2}{x^2-4}$

24. $\frac{5x-20}{5} \cdot \frac{x+4}{x^2-16}$

25. $\frac{4a+4b}{ab^2} \div \frac{3a+3b}{a^2b}$

26. $\frac{5x-5y}{x^2y} \div \frac{4x-4y}{xy^2}$

27. $\frac{a^2-9b^2}{a^2-6ab+9b^2} \div \frac{a+3b}{a-3b}$

28. $\frac{x^2-y^2}{x^2-2xy+y^2} \div \frac{x+y}{x-y}$

29. $\frac{x-y}{9x+9y} \div \frac{x^2-y^2}{3x^2+6xy+3y^2}$

30. $\frac{x-5y}{8x+40y} \div \frac{x^2-25y^2}{4x^2+40xy+100y^2}$

31. $\frac{a^2-9b^2}{6a^2-36ab+54b^2} \div \frac{a+3b}{2a-6b}$

32. $\frac{x^2-y^2}{8x^2-16xy+8y^2} \div \frac{x+y}{4x-4y}$

33. $\frac{2x^3y+2x^2y^2}{6x} \div \frac{x^2y^2-xy^3}{y-x}$

34. $\frac{5st^4+5s^2t^3}{10st^3} \div \frac{s^2t^2-s^3t}{s-t}$

35. $\frac{5x^2}{12y^3} \div \frac{10y}{7x^2} \div \frac{8x}{3y}$

36. $\frac{40s^3}{9t^2} \div \frac{12t}{17s^3} \div \frac{34s^4}{3t^2}$

37. $\frac{5x^3}{7xy} \div \frac{2y}{3x^4} \cdot \frac{14xy^3}{9y^2}$

38. $\frac{14a^3b}{5b^4} \div \frac{4b^3}{7a^4} \cdot \frac{8ab^3}{49ab}$

39. $\frac{x-3}{x+2y} \div \frac{x-2y}{x^2-4y^2} \cdot \frac{x}{xy-3y}$

40. $\frac{x-5}{x+5y} \div \frac{x-5y}{x^2-25y^2} \cdot \frac{y}{x^2-5x}$

In Exercises 41–52, find each missing term.

41. $-\frac{5}{6} = \frac{?}{6}$

42. $-\frac{8}{9} = \frac{?}{9}$

43. $\frac{5}{-y} = \frac{-5}{?}$

44. $\frac{2}{-x} = \frac{?}{x}$

45. $\frac{2-x}{-9} = \frac{?}{9}$

46. $\frac{5-y}{-2} = \frac{?}{2}$

47. $\frac{6-y}{5} = \frac{y-6}{?}$

48. $\frac{8-x}{7} = \frac{x-8}{?}$

49. $-\frac{4}{x-5} = \frac{4}{?}$

50. $-\frac{3}{x-4} = \frac{3}{?}$

51. $\frac{a-b}{(c+d)(5-x)} = \frac{?}{(c+d)(x-5)}$

52. $\frac{x-y}{(u+v)(c-2)} = \frac{?}{(u+v)(2-c)}$

✏️ Writing Problems

Express the answer in your own words and in complete sentences.

1. Correctly using the terms *divisor* and *dividend*, explain how to divide one rational expression by another.

Exercises 8.3
Set II

In Exercises 1–40, perform the indicated operations.

1. $\frac{2}{3} \div \frac{1}{2}$

2. $\frac{5}{8} \div \frac{5}{14}$

3. $\frac{5x^3}{4y^2} \cdot \frac{8y}{10x^2}$

4. $\frac{8y^2}{12x} \cdot \frac{6x^2}{4y^3}$

5. $\frac{5a^2}{15} \div \frac{a}{3}$

6. $\frac{5x^3}{6} \div \frac{8x^2}{20}$

7. $\frac{6a^3b^2c^3}{24a^2b} \cdot \frac{18b}{a^2bc^2}$

8. $\frac{18x^2y^3}{10x^3y^2} \cdot \frac{5z^3}{9x^2y^2z}$

9. $\frac{s}{s+3} \cdot \frac{4s+12}{s^4}$

10. $\frac{x+8}{x+4} \cdot \frac{4x+16}{x^3}$

11. $\frac{x+7}{x} \cdot \frac{2}{3x+21}$

12. $\frac{y-3}{12y+12} \cdot \frac{9}{4y-12}$

13. $\frac{x^3}{y+5} \div \frac{4x^2}{3y+15}$

14. $\frac{x^5}{x-3} \div \frac{2x^7}{2x-6}$

15. $\frac{4a-16}{12a} \div \frac{a-4}{48a^3}$

16. $\frac{12b-6}{12b} \div \frac{2b-1}{24b^2}$

17. $\frac{x+1}{x-1} \div \frac{x^2+2x+1}{x^2-1}$

18. $\frac{x^2+x-2}{x-1} \div \frac{x^2+5x+6}{x^2}$

19. $\frac{9}{z+3} \cdot \frac{z^2-9}{(z-3)^2}$

20. $\frac{z^3}{z+4} \div \frac{z-1}{z^2+3z-4}$

21. $\frac{5x-5y}{3x+3y} \cdot \frac{4x+4y}{y-x}$

22. $\frac{3-y}{y+1} \cdot \frac{4+5y+y^2}{y^2+y-12}$

23. $\frac{2x-10}{8} \cdot \frac{x+5}{x^2-25}$

24. $\frac{x^2+10x+25}{x^2-25} \div \frac{5-x}{x+5}$

25. $\frac{5x+5y}{xy^3} \div \frac{8x+8y}{x^3y}$

26. $\frac{12f+16}{15f} \div \frac{6f^3+8f^2}{20f^4}$

27. $\frac{s^2-4t^2}{s^2-4st+4t^2} \div \frac{s+2t}{s-2t}$

28. $\frac{x^3-5x}{9x} \div \frac{4x^3-20x}{12x^2}$

29. $\frac{a-4b}{a+4b} \div \frac{a^2-16b^2}{2a^2+16ab+32b^2}$

30. $\frac{2b^2c-2bc^2}{b+c} \div \frac{4bc^2-4b^2c}{4b+4c}$

31. $\frac{x^2-25y^2}{3x^2-30xy+75y^2} \div \frac{x+5y}{4x-20y}$

32. $\frac{2x-1}{y} \div \frac{2x^2+x-1}{y^2}$

33. $\frac{8x^2y^2+16xy^3}{4y} \div \frac{2x^3y-2x^2y^2}{y-x}$

34. $\frac{a+2}{a+1} \cdot \frac{a^2+a}{a^2-4}$

35. $\frac{6s^2}{5t^3} \div \frac{15t}{11s^2} \div \frac{22s}{9t}$

36. $\frac{8m^3}{9n^2} \div \frac{12n}{17m^3} \cdot \frac{3m^4}{34n^2}$

37. $\frac{9u^3}{8uv} \div \frac{6v}{7u^4} \cdot \frac{4uv^3}{21v^2}$

38. $\frac{9a^3b}{11b^4} \div \frac{18b^3}{55a^4} \div \frac{5ab^3}{21ab}$

39. $\frac{x-7}{x-5y} \div \frac{x+5y}{x^2-25y^2} \cdot \frac{x}{xy-7y}$

40. $\frac{3s-5}{3s+5t} \div \frac{3s-5t}{9s^2-25t^2} \cdot \frac{t}{9st-15t}$

In Exercises 41–52, find each missing term.

41. $-\frac{4}{3} = \frac{?}{3}$

42. $-\frac{4}{3} = \frac{?}{-3}$

43. $\frac{8}{-b} = \frac{-8}{?}$

44. $\frac{9}{-a} = \frac{?}{a}$

45. $\frac{1-y}{-5} = \frac{?}{5}$

46. $\frac{5-y}{-2} = \frac{?}{2}$

47. $\frac{8-a}{3} = \frac{a-8}{?}$

48. $\frac{5-x}{a-b} = \frac{x-5}{?}$

49. $-\frac{7}{y-2} = \frac{7}{?}$

50. $-\frac{a-2}{x-3} = \frac{2-a}{?}$

51. $\frac{u-v}{(a+3)(7-x)} = \frac{?}{(a+3)(x-7)}$

52. $\frac{3-a}{(x+1)(y-3)} = \frac{?}{(3-y)(1+x)}$

8.4　Adding and Subtracting Rational Expressions That Have the Same Denominator

In Arithmetic　Recall from arithmetic that when we add two or more fractions that have the same denominator, the numerator of the sum is the sum of the numerators and the denominator of the sum is the same as any *one* of the denominators. Recall also that when we subtract one fraction from another, the numerator of the difference is the difference of the numerators and the denominator of the difference is the same as any *one* of the denominators. That is,

$$\frac{a}{c} + \frac{b}{c} = \frac{a+b}{c} \qquad \text{and} \qquad \frac{a}{c} - \frac{b}{c} = \frac{a-b}{c}$$

EXAMPLE 1　Examples of adding and subtracting fractions that have the same denominator in arithmetic:

a. $\dfrac{1}{9} + \dfrac{2}{9} + \dfrac{4}{9} = \dfrac{1+2+4}{9} = \dfrac{7}{9}$

$\frac{3}{9}$ reduces to $\frac{1}{3}$

b. $\dfrac{5}{9} - \dfrac{2}{9} = \dfrac{5-2}{9} = \dfrac{3}{9} = \dfrac{1}{3}$

In Algebra　In algebra, rational expressions can be added by using the following procedure:

Adding rational expressions that have the same denominator

1. The numerator of the sum is the sum of the numerators, and the denominator of the sum is the same as any *one* of the denominators. That is, if P, Q, and R are polynomials and if $Q \neq 0$, then

$$\frac{P}{Q} + \frac{R}{Q} = \frac{P+R}{Q}$$

2. The resulting rational expression should be reduced, if possible.

PROOF　First recall that $\dfrac{P}{Q}$ is equivalent to $P\left(\dfrac{1}{Q}\right)$ and $\dfrac{R}{Q}$ is equivalent to $R\left(\dfrac{1}{Q}\right)$. Therefore,

Factoring out $\dfrac{1}{Q}$

$$\frac{P}{Q} + \frac{R}{Q} = P\left(\frac{1}{Q}\right) + R\left(\frac{1}{Q}\right) = \frac{1}{Q}(P+R) = \frac{1}{Q} \cdot \frac{P+R}{1} = \frac{P+R}{Q} \quad\blacksquare$$

EXAMPLE 2　Examples of adding rational expressions that have the same denominator:

a. $\dfrac{2}{x} + \dfrac{5}{x} = \dfrac{2+5}{x} = \dfrac{7}{x}$

Factoring the numerator and the denominator

b. $\dfrac{4}{6+3x} + \dfrac{2x}{6+3x} = \dfrac{4+2x}{6+3x} = \dfrac{2(2+x)}{3(2+x)} = \dfrac{2}{3}$

Factoring -3 from $15 - 3d$

c. $\dfrac{15}{d-5} + \dfrac{-3d}{d-5} = \dfrac{15 + (-3d)}{d-5} = \dfrac{15-3d}{d-5} = \dfrac{-3(\cancel{d}-5)^{1}}{\cancel{d-5}_{1}} = -3$

 Note   Example 2c can also be finished as follows:

$$\frac{15-3d}{d-5} = \frac{3(5-d)}{d-5} = 3\left(\frac{5-d}{d-5}\right) = 3(-1) = -3$$

Any subtraction problem can be changed to an addition problem by using the definition of subtraction, $a - b = a + (-b)$, even when the problem involves rational expressions. Subtraction of rational expressions can also be done by using the following procedure:

Subtracting rational expressions that have the same denominator	**1.** The numerator of the difference is the difference of the numerators, and the denominator of the difference is the same as any *one* of the denominators. That is, if P, Q, and R are polynomials and if $Q \neq 0$, then $$\frac{P}{Q} - \frac{R}{Q} = \frac{P-R}{Q}$$ **2.** The resulting rational expression should be reduced, if possible.

EXAMPLE 3   Examples of subtracting rational expressions that have the same denominator:

a. $\dfrac{3}{4a} - \dfrac{5}{4a} = \dfrac{3}{4a} + \left(-\dfrac{5}{4a}\right) = \dfrac{3}{4a} + \left(\dfrac{-5}{4a}\right) = \dfrac{3+(-5)}{4a} = \dfrac{-2}{4a} = -\dfrac{1}{2a}$ Changing subtraction to addition

or $\dfrac{3}{4a} - \dfrac{5}{4a} = \dfrac{3-5}{4a} = \dfrac{-2}{4a} = -\dfrac{1}{2a}$ Using the procedure in the box

b. $\dfrac{7}{x-2} - \dfrac{4}{x-2} = \dfrac{7}{x-2} + \left(-\dfrac{4}{x-2}\right) = \dfrac{7}{x-2} + \left(\dfrac{-4}{x-2}\right) = \dfrac{7+(-4)}{x-2}$

$$= \frac{3}{x-2}$$

or $\dfrac{7}{x-2} - \dfrac{4}{x-2} = \dfrac{7-4}{x-2} = \dfrac{3}{x-2}$

c. $\dfrac{4x}{2x-y} - \dfrac{2y}{2x-y} = \dfrac{4x-2y}{2x-y} = \dfrac{2(\cancel{2x-y})^{1}}{\cancel{2x-y}_{1}} = 2$ (Showing the second method only)

✖ **A Word of Caution**   In Example 3a, $-1/2a$ is an *incorrect* answer, and in Example 3b, $3/x - 2$ is an *incorrect* answer. That is,

$$-1/2a = -\frac{1}{2}a, \quad \text{not } -\frac{1}{2a}; \qquad \text{and} \quad 3/x - 2 = \frac{3}{x} - 2, \quad \text{not } \frac{3}{x-2}$$

In a subtraction problem, if the numerator of the rational expression being subtracted contains more than one term, we *must* put parentheses around that numerator when we rewrite the problem as a single rational expression (see Example 4).

EXAMPLE 4   Subtract $\dfrac{5}{2x + 3} - \dfrac{x + 1}{2x + 3}$.

SOLUTION

$$\frac{5}{2x + 3} - \frac{x + 1}{2x + 3} = \frac{5 - (x + 1)}{2x + 3} = \frac{5 - x - 1}{2x + 3} = \frac{4 - x}{2x + 3}$$

A Word of Caution   It is *incorrect* to do Example 4 this way:

$$\frac{5}{2x + 3} - \frac{x + 1}{2x + 3} = \frac{5 - x + 1}{2x + 3} = \frac{6 - x}{2x + 3}$$

Remember: Parentheses are always needed when an expression with more than one term is being subtracted. See Example 4 for the correct solution.

Denominators That Are Additive Inverses   When two denominators are not identical but are the *additive inverses* of each other, we can make the rational expressions have the same denominator by changing the signs of the numerator and denominator of one of the rational expressions (see Example 5).

EXAMPLE 5   Add $\dfrac{9}{x - 2} + \dfrac{5}{2 - x}$.

SOLUTION $x - 2$ and $2 - x$ are the additive inverses of each other.

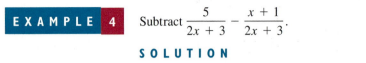

Changing the sign of the
numerator and the sign of the denominator

$$\frac{9}{x - 2} + \frac{5}{2 - x} = \frac{9}{x - 2} + \frac{-5}{-(2 - x)} = \frac{9}{x - 2} + \frac{-5}{x - 2} = \frac{9 - 5}{x - 2} = \frac{4}{x - 2}$$

In subtraction problems in which the denominators are the additive inverses of each other, we can change the sign of the denominator and the sign of the rational expression being subtracted (see Example 6).

EXAMPLE 6   Subtract $\dfrac{8}{y - 5} - \dfrac{3}{5 - y}$.

SOLUTION

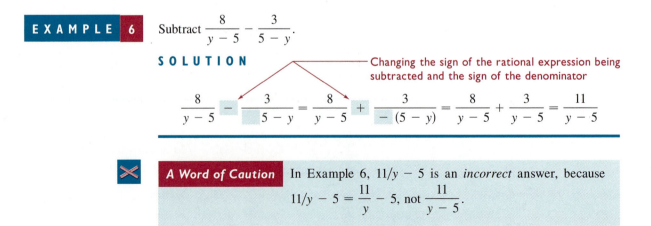

Changing the sign of the rational expression being
subtracted and the sign of the denominator

$$\frac{8}{y - 5} - \frac{3}{5 - y} = \frac{8}{y - 5} + \frac{3}{-(5 - y)} = \frac{8}{y - 5} + \frac{3}{y - 5} = \frac{11}{y - 5}$$

A Word of Caution   In Example 6, $11/y - 5$ is an *incorrect* answer, because

$$11/y - 5 = \frac{11}{y} - 5, \text{ not } \frac{11}{y - 5}.$$

✕ **A Word of Caution**   A common error is to confuse *addition of rational expressions* with *solving equations* and to multiply both rational expressions by the same number. This is incorrect, since multiplying an expression by any number other than 1 changes the *value* of the expression.

Correct method

$$\frac{1}{3x} + \frac{4}{3x} = \frac{5}{3x}$$

Incorrect method

$$\frac{1}{3x} + \frac{4}{3x} = (3x)\left(\frac{1}{3x}\right) + (3x)\left(\frac{4}{3x}\right) = 1 + 4 = 5$$

Exercises 8.4
Set I

Perform the indicated operations.

1. $\dfrac{7}{a} + \dfrac{2}{a}$

2. $\dfrac{5}{b} + \dfrac{2}{b}$

3. $\dfrac{6}{x - y} - \dfrac{2}{x - y}$

4. $\dfrac{7}{m + n} - \dfrac{1}{m + n}$

5. $\dfrac{2}{3a} + \dfrac{4}{3a}$

6. $\dfrac{8}{5z} + \dfrac{2}{5z}$

7. $\dfrac{2y}{y + 1} + \dfrac{2}{y + 1}$

8. $\dfrac{10x}{2x + 3} + \dfrac{15}{2x + 3}$

9. $\dfrac{3}{x + 3} + \dfrac{x}{x + 3}$

10. $\dfrac{5}{y + 5} + \dfrac{y}{y + 5}$

11. $\dfrac{3x}{x - 4} - \dfrac{12}{x - 4}$

12. $\dfrac{7x}{x - 2} - \dfrac{14}{x - 2}$

13. $\dfrac{x - 3}{y - 2} - \dfrac{x + 5}{y - 2}$

14. $\dfrac{z - 4}{a - b} - \dfrac{z + 3}{a - b}$

15. $\dfrac{a + 2}{2a + 1} - \dfrac{1 - a}{2a + 1}$

16. $\dfrac{6x - 1}{3x - 2} - \dfrac{3x + 1}{3x - 2}$

17. $\dfrac{4x - 1}{2x + 3} + \dfrac{5x - 3}{2x + 3}$

18. $\dfrac{8x + 5}{3x - 2} + \dfrac{4x + 3}{3x - 2}$

19. $\dfrac{9 + 7x}{7x - 9} - \dfrac{3x - 2}{7x - 9}$

20. $\dfrac{4 + 5x}{5x - 4} - \dfrac{6x - 5}{5x - 4}$

21. $\dfrac{-x}{x - 2} - \dfrac{2}{2 - x}$

22. $\dfrac{-b}{2a - b} - \dfrac{2a}{b - 2a}$

23. $\dfrac{-15w}{1 - 5w} - \dfrac{3}{5w - 1}$

24. $\dfrac{-35}{6w - 7} - \dfrac{30w}{7 - 6w}$

25. $\dfrac{7z}{8z - 4} + \dfrac{6 - 5z}{4 - 8z}$

26. $\dfrac{5x}{9x - 3} + \dfrac{4 - 7x}{3 - 9x}$

27. $\dfrac{31 - 8x}{12 - 8x} - \dfrac{5 - 16x}{8x - 12}$

28. $\dfrac{13 - 30w}{15 - 10w} - \dfrac{10w + 17}{10w - 15}$

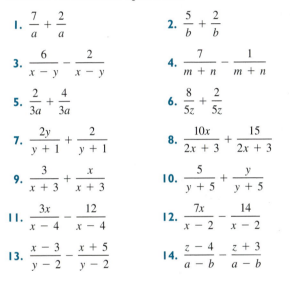

Writing Problems

Express the answer in your own words and in complete sentences.

1. Explain why $\dfrac{5x}{x + 3} - \dfrac{2x + 1}{x + 3} \neq \dfrac{3x + 1}{x + 3}$.

Exercises 8.4
Set II

Perform the indicated operations.

1. $\dfrac{8}{y} + \dfrac{7}{y}$

2. $\dfrac{6}{x} - \dfrac{2}{x}$

3. $\dfrac{8}{x - y} - \dfrac{5}{x - y}$

4. $\dfrac{x + 1}{a + b} + \dfrac{x - 1}{a + b}$

5. $\dfrac{7}{5x} + \dfrac{8}{5x}$

6. $\dfrac{15}{7x} - \dfrac{1}{7x}$

7. $\dfrac{4x}{x + 3} + \dfrac{12}{x + 3}$

8. $\dfrac{2x + 1}{x + 4} - \dfrac{x - 3}{x + 4}$

9. $\dfrac{3}{2a + 3} + \dfrac{2a}{2a + 3}$

10. $\dfrac{9x}{3x - y} - \dfrac{3y}{3x - y}$

19. $\dfrac{8 + 3x}{3x - 8} - \dfrac{4x - 5}{3x - 8}$

20. $\dfrac{18 + 7x}{3x - 7} - \dfrac{5x - 6}{3x - 7}$

11. $\dfrac{5a}{a - 2b} - \dfrac{10b}{a - 2b}$

12. $\dfrac{8}{z - 4} - \dfrac{2z}{z - 4}$

21. $\dfrac{-s}{s - 5} - \dfrac{5}{5 - s}$

22. $\dfrac{5}{2x + 3} - \dfrac{x + 4}{2x + 3}$

13. $\dfrac{a - 5}{x - 4} - \dfrac{a + 3}{x - 4}$

14. $\dfrac{4x + y}{2x + y} + \dfrac{2x + 2y}{2x + y}$

23. $\dfrac{-3m}{5 - m} - \dfrac{15}{m - 5}$

24. $\dfrac{2x}{3x - 8} + \dfrac{5x - 8}{8 - 3x}$

15. $\dfrac{7x - 1}{3x - 1} - \dfrac{x + 1}{3x - 1}$

16. $\dfrac{b}{b - 2a} + \dfrac{2a}{2a - b}$

25. $\dfrac{8x}{12x - 15} + \dfrac{25 - 12x}{15 - 12x}$

26. $\dfrac{5x - 2}{4x - 5} - \dfrac{x + 3}{4x - 5}$

17. $\dfrac{12x - 5}{9x + 2} + \dfrac{7x - 4}{9x + 2}$

18. $\dfrac{11x + 3}{6x - 1} + \dfrac{11x + 3}{6x - 1}$

27. $\dfrac{11x - 9y}{8x - 6y} - \dfrac{13x - 9y}{6y - 8x}$

28. $\dfrac{18y - 21}{10y - 16} - \dfrac{3y + 3}{10y - 16}$

8.5 The Least Common Multiple (LCM) and the Least Common Denominator (LCD)

The Multiples of a Number

The **multiples of a number** are the numbers that result from multiplying that number by the natural numbers (see Example 1).

EXAMPLE 1 Examples of finding the multiples of a number:

a. The *multiples* of 12 are 12, 24, 36, 48, 60, 72, 84, 96, 108, 120, and so forth. They are found as follows:

$$12 \cdot 1 = 12 \qquad 12 \cdot 2 = 24 \qquad 12 \cdot 3 = 36$$
$$12 \cdot 4 = 48 \qquad 12 \cdot 5 = 60 \qquad 12 \cdot 6 = 72$$
$$12 \cdot 7 = 84 \qquad 12 \cdot 8 = 96 \qquad 12 \cdot 9 = 108$$

and so forth.

b. The *multiples* of 15 are 15, 30, 45, 60, 75, 90, 105, 120, and so forth. They are found as follows:

$$15 \cdot 1 = 15 \qquad 15 \cdot 2 = 30$$
$$15 \cdot 3 = 45 \qquad 15 \cdot 4 = 60$$
$$15 \cdot 5 = 75 \qquad 15 \cdot 6 = 90$$

and so forth.

The Least Common Multiple (LCM)

The **least common multiple (LCM)** of two or more numbers is the *smallest* number that is a multiple of all of the numbers (see Example 2).

EXAMPLE 2 Find the least common multiple (LCM) of 12 and 15.

SOLUTION From Example 1, we have

Multiples of 12: $\{12, 24, 36, 48, 60, 72, 84, 96, 108, 120, \ldots\}$

Multiples of 15: $\{15, 30, 45, 60, 75, 90, 105, 120, \ldots\}$

Common multiples are 60, 120 (these two are shaded), 180, 240, 300, and so forth. (There are *infinitely* many common multiples.) The *smallest* of these is 60. Therefore, 60 is the *least common multiple (LCM)* of 12 and 15.

In practice, we usually don't find the LCM of numbers by listing their multiples; instead, we use the method described in the box below for finding the LCM of two or more polynomials.

The least common multiple of two or more *polynomials* is the *smallest* polynomial that is exactly divisible by each of the given polynomials. We can find the LCM by using the method in the following box.

Finding the least common multiple of polynomials	**1.** Factor each polynomial completely. Repeated factors must be expressed in exponential form.
	2. Write down each different base that appears in any of the factorizations.
	3. Raise each base to the *highest power* to which it occurs in *any* of the factorizations.
	4. The least common multiple (LCM) is the product of all the factors found in step 3.

EXAMPLE 3 Find the LCM of $12b^3$ and $15b^2$.

SOLUTION

Step 1. $12b^3 = 2^2 \cdot 3 \cdot b^3$; $15b^2 = 3 \cdot 5 \cdot b^2$ The polynomials in factored form
Step 2. $2, 3, 5, b$ Writing all the different bases
Step 3. $2^2, 3^1, 5^1, b^3$ Raising each base to the *highest* power to
                                            which it occurs in any of the factorizations
Step 4. The LCM is the *product* of all the expressions from step 3, so the LCM $= 2^2 \cdot 3^1 \cdot 5 \cdot b^3$, or $60b^3$.

Note Example 3 confirms that the LCM of 12 and 15 is 60.

The Least Common Denominator (LCD)

The **least common denominator (LCD)** of two or more rational expressions is the *least common multiple* of their denominators.

EXAMPLE 4 Find the LCD for the terms in $\dfrac{3}{2} + \dfrac{4}{y}$.

SOLUTION

Step 1. The denominators cannot be factored.
Step 2. 2 and y are the different bases.
Step 3. $2^1, y^1$ The highest power of each base
Step 4. The LCD is $2^1 \cdot y^1$, or $2y$.

EXAMPLE 5 Find the LCD for the terms in $\dfrac{2}{x} + \dfrac{5}{x^2}$.

SOLUTION

Step 1. The denominators are already factored.
Step 2. x (x is the only base.)
Step 3. x^2 The highest power of x in either denominator is x^2
Step 4. The LCD is x^2.

EXAMPLE 6 Find the LCD for the terms in $\dfrac{7}{18x^2y} + \dfrac{5}{8xy^4}$.

SOLUTION

Step 1. $2 \cdot 3^2 \cdot x^2 \cdot y; \quad 2^3 \cdot x \cdot y^4$ The denominators in factored form
Step 2. $2, 3, x, y$ All the different bases
Step 3. $2^3, 3^2, x^2, y^4$ The highest powers
Step 4. The LCD is $2^3 \cdot 3^2 \cdot x^2 \cdot y^4$, or $72x^2y^4$.

EXAMPLE 7 Find the LCD for the terms in $\dfrac{2}{x} + \dfrac{x}{x+2}$.

SOLUTION

Step 1. The denominators cannot be factored.
Step 2. $x, (x+2)$
Step 3. $x^1, (x+2)^1$
Step 4. The LCD is $x(x+2)$.

EXAMPLE 8 Find the LCD for the terms in $\dfrac{16}{a^2b} + \dfrac{a-2}{2a(a-b)} - \dfrac{b+1}{4b^3(a-b)}$.

SOLUTION

Step 1. $a^2b; \quad 2a(a-b); \quad 2^2b^3(a-b)$
Step 2. $2, a, b, (a-b)$ All the different bases
Step 3. $2^2, a^2, b^3, (a-b)^1$ The highest powers
Step 4. The LCD is $2^2a^2b^3(a-b)^1$, or $4a^2b^3(a-b)$.

EXAMPLE 9 Find the LCD for the terms in $\dfrac{8}{3x-3} - \dfrac{5}{x^2+2x+1}$.

SOLUTION

Step 1. $3x - 3 = 3(x-1); \quad x^2 + 2x + 1 = (x+1)^2$
Step 2. $3, (x-1), (x+1)$
Step 3. $3^1, (x-1)^1, (x+1)^2$
Step 4. The LCD is $3(x-1)(x+1)^2$.

EXAMPLE 10 Find the LCD for the terms in $\dfrac{2x-3}{x^2+10x+25} - \dfrac{14}{4x^2+20x} + \dfrac{4x-3}{x^2+2x-15}$.

SOLUTION

Step 1. $x^2 + 10x + 25 = (x+5)^2$
$\quad\quad 4x^2 + 20x = 4x(x+5) = 2^2x(x+5)$
$\quad\quad x^2 + 2x - 15 = (x+5)(x-3)$
Step 2. $2, x, (x+5), (x-3)$
Step 3. $2^2, x, (x+5)^2, (x-3)$
Step 4. The LCD is $4x(x+5)^2(x-3)$.

Exercises  8.5
Set I

In Exercises 1–8, find the LCM of the given pairs of polynomials.

1. 12, 4   **2.** 3, 9   **3.** 3, x **4.** 2y, y

5. 5x, 2x **6.** 3z, 4z **7.** 15x^2, 12x^3y **8.** 18a^2bc, 16ab^2

In Exercises 9–22, find the LCD. Do *not* add the rational expressions.

9. $\dfrac{7}{12u^3v^2} - \dfrac{11}{18uv^3}$

10. $\dfrac{13}{50x^3y^4} - \dfrac{17}{20x^2y^5}$

11. $\dfrac{4}{a} + \dfrac{a}{a+3}$

12. $\dfrac{5}{b} + \dfrac{b}{b-5}$

13. $\dfrac{x}{2x+4} - \dfrac{5}{4x}$

14. $\dfrac{4}{3x} + \dfrac{2x}{3x+6}$

15. $\dfrac{3}{4z^2} + \dfrac{2z}{z^2+2z+1} - \dfrac{4z}{z+1}$

16. $\dfrac{2x}{x^2-2x+1} - \dfrac{5}{x-1} + \dfrac{11}{12x^3}$

17. $\dfrac{x-4}{x^2+3x+2} + \dfrac{3x+1}{x^2+2x+1}$

18. $\dfrac{2x+3}{x^2-x-12} + \dfrac{x-4}{x^2+6x+9}$

19. $\dfrac{x^2+1}{12x^3+24x^2} - \dfrac{4x+3}{x^2-4x+4} + \dfrac{1}{x^2-4}$

20. $\dfrac{2y+5}{y^2+6y+9} - \dfrac{7y}{y^2-9} - \dfrac{11}{8y^2-24y}$

21. $\dfrac{3x+1}{6x^2+x-2} + \dfrac{x^2+1}{9x^3+12x^2+4x} + \dfrac{5x^2-1}{4x^2-4x+1}$

22. $\dfrac{5x+1}{10x^2+13x-3} + \dfrac{3x^2-1}{4x^3+12x^2+9x} + \dfrac{x-4}{25x^2-10x+1}$

 ## Writing Problems

Express the answer in your own words and in complete sentences.

1. Explain what the LCM of two or more polynomials is. (Do *not* explain how to find it.)

Exercises 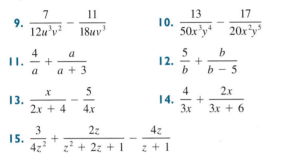 8.5
Set II

In Exercises 1–8, find the LCM of the given pairs of polynomials.

1. 21, 24   **2.** 2, 5   **3.** 2, y **4.** 4y^2, 6y

5. 2x, 3x **6.** 9x^2y, 30y^2z **7.** 8y^4, 12x^2y **8.** 5x^3, 10x^5

In Exercises 9–22, find the LCD. Do *not* add the rational expressions.

9. $\dfrac{5}{9x^2y} - \dfrac{7}{6xy^2}$

10. $\dfrac{5}{2+x^2} + \dfrac{9}{2x^2}$

11. $\dfrac{2}{x} + \dfrac{x}{x+5}$

12. $\dfrac{x+1}{4} - \dfrac{x+3}{12}$

13. $\dfrac{a}{3a+6} - \dfrac{5}{3a}$

14. $\dfrac{9y}{x^2-y^2} + \dfrac{6x}{x^2+2xy+y^2}$

15. $\dfrac{3}{2x} + \dfrac{x}{x^2+4x+4} - \dfrac{1}{x+2}$

16. $\dfrac{11}{x^2+16x+64} - \dfrac{7}{2x^2-128} + \dfrac{1}{4x-32}$

17. $\dfrac{x-5}{x^2+4x+3} + \dfrac{5x-3}{x^2+7x+12}$

18. $\dfrac{5}{8z^3} - \dfrac{8z}{z^2-4} + \dfrac{5z}{9z+18}$

19. $\dfrac{x^2+3}{9x^3+45x^2} - \dfrac{5x+2}{x^2-10x+25} + \dfrac{1}{x^2-25}$

20. $\dfrac{a^2+1}{3a^2+5a-2} - \dfrac{1}{3a^2+6a} + \dfrac{5a}{18a^4-6a^3}$

21. $\dfrac{5x+1}{4x^2-11x-3} + \dfrac{x+3}{16x^3+8x^2+x} + \dfrac{3x^2}{x^2-6x+9}$

22. $\dfrac{6x^2}{6x^2+5x-1} + \dfrac{3}{30x^3-5x^2} + \dfrac{5+x}{10x^4+10x^3}$

8.6   Adding and Subtracting Rational Expressions That Have Different Denominators

Because the rule for adding rational expressions is $\dfrac{P}{Q} + \dfrac{R}{Q} = \dfrac{P + R}{Q}$, we can add and subtract rational expressions only when the *denominators are the same*. To add (or subtract) rational expressions that *don't* have the same denominator, we first find the least common denominator (the LCD) and then convert each rational expression to an equivalent rational expression that has the LCD as its denominator. When we do this, we often say we are *building* rational expressions.

To convert a rational expression to one with the LCD as its denominator, we can use the fundamental property of rational expressions $\left(\dfrac{P}{Q} = \dfrac{P \cdot C}{Q \cdot C} \right)$. Using this property is equivalent to multiplying an expression by 1, as the following argument shows:

The identity property of multiplication ($a = a \cdot 1$) guarantees that multiplying a number by 1 does not change its value

$$\frac{P}{Q} = \frac{P}{Q} \cdot 1 = \frac{P}{Q} \cdot \frac{C}{C} = \frac{P \cdot C}{Q \cdot C}$$

Substituting $\dfrac{C}{C}$ for 1, since $\dfrac{C}{C} = 1$

In Example 1, we compare using the fundamental property of rational expressions with multiplying a rational expression by a rational expression whose value is 1.

EXAMPLE 1   Examples comparing the methods of building rational expressions:

a. Convert $\frac{5}{6}$ to a fraction with a denominator of 12.

Using the fundamental property of rational expressions

$$\frac{5}{6} = \frac{5 \cdot 2}{6 \cdot 2} = \frac{10}{12}$$

Multiplying by 1

$$\frac{5}{6} = \frac{5}{6} \cdot \frac{2}{2} = \frac{5 \cdot 2}{6 \cdot 2} = \frac{10}{12}$$

b. Convert $\dfrac{x}{x + 2}$ to a rational expression with a denominator of $3y(x + 2)$.

Using the fundamental property of rational expressions

$$\frac{x}{x + 2} = \frac{x \, (3y)}{(x + 2) \, (3y)} = \frac{3xy}{3y(x + 2)}$$

Multiplying by 1

$$\frac{x}{x + 2} = \frac{x}{x + 2} \cdot \frac{3y}{3y} = \frac{3xy}{3y(x + 2)}$$

c. Convert $\dfrac{2}{x}$ to a rational expression with a denominator of $x(x + 2)$, or $x^2 + 2x$.

Using the fundamental property of rational expressions

$$\frac{2}{x} = \frac{2 \cdot (x + 2)}{x \cdot (x + 2)} = \frac{2x + 4}{x^2 + 2x}$$

Multiplying by 1

$$\frac{2}{x} = \frac{2}{x} \cdot \frac{x + 2}{x + 2} = \frac{2(x + 2)}{x(x + 2)} = \frac{2x + 4}{x^2 + 2x}$$

The procedure given in the following box can be used for adding and subtracting rational expressions that do not have the same denominator.*

Adding or subtracting rational expressions that do not have the same denominator

1. Find the LCD.

2. Convert each rational expression to an equivalent rational expression that has the LCD as its denominator, as follows:

    a. To find the factor to multiply by, divide the LCD by the denominator.

    b. Multiply the numerator and denominator by the quotient from step 2a. In practice, we compare the denominator we're *building* with the factored form of the LCD; then we multiply both numerator and denominator (of the rational expression we're building) by any missing factors.

3. Add or subtract the resulting rational expressions.

4. Reduce the resulting sum or difference, if possible.

Example 2 reviews adding fractions with unlike denominators in arithmetic.

EXAMPLE 2   Add $\frac{1}{2} + \frac{2}{3} + \frac{3}{4}$.

SOLUTION   Because $4 = 2^2$, the LCD is $2^2 \cdot 3$, or 12.

$$\frac{1}{2} = \frac{1 \cdot 6}{2 \cdot 6} = \frac{6}{12}$$ Multiplying numerator and denominator by 6 because $12 \div 2 = 6$

$$\frac{2}{3} = \frac{2 \cdot 4}{3 \cdot 4} = \frac{8}{12}$$ Multiplying numerator and denominator by 4 because $12 \div 3 = 4$

$$+\frac{3}{4} = +\frac{3 \cdot 3}{4 \cdot 3} = +\frac{9}{12}$$ Multiplying numerator and denominator by 3 because $12 \div 4 = 3$

$$\frac{23}{12}$$ Adding the numerators

EXAMPLE 3   Add $\dfrac{2}{x} + \dfrac{5}{x^2}$.

SOLUTION

Step 1. The LCD is x^2.   We found the LCD in a previous example

Step 2. $\dfrac{2}{x} = \dfrac{2 \cdot x}{x \cdot x} = \dfrac{2x}{x^2}$ Multiplying numerator and denominator by x because $x^2 \div x = x$

$$+\frac{5}{x^2} = +\frac{5}{x^2} = +\frac{5}{x^2}$$

Step 3. $\dfrac{2}{x} + \dfrac{5}{x^2} = \dfrac{2x}{x^2} + \dfrac{5}{x^2} = \dfrac{2x + 5}{x^2}$ The terms of the numerator are not like terms

Step 4. The rational expression cannot be reduced.

---

*In step 2b in the rules in the box (and in Examples 2 through 9), we suggest building a rational expression by using the fundamental property of rational expressions. As we have mentioned, it is equally acceptable to build a rational expression by multiplying it by another rational expression whose value is 1.

A Word of Caution A common error is to reduce rational expressions right after converting them to equivalent rational expressions with
the LCD as the denominator (see step 2). The addition then cannot be done, because the rational expressions no longer have the same denominator; we simply
get an addition problem identical to the one we started with.

This is the problem we started with

$$\frac{2}{x} + \frac{5}{x^2} = \frac{2\overset{1}{x}}{\underset{x}{x^2}} + \frac{5}{x^2} = \frac{2}{x} + \frac{5}{x^2}$$

We cannot add these rational expressions

EXAMPLE 4

Add $\dfrac{7}{18x^2y} + \dfrac{5}{8xy^4}$.

SOLUTION

Step 1. The LCD is $72x^2y^4$.   We found the LCD in a previous example

$$\frac{\text{LCD}}{\text{denominator}} = \frac{72x^2y^4}{18x^2y} = 4y^3$$

Step 2. $\dfrac{7}{18x^2y} = \dfrac{7\,(4y^3)}{18x^2y\,(4y^3)} = \dfrac{28y^3}{72x^2y^4}$

$\dfrac{5}{8xy^4} = \dfrac{5\,(9x)}{8xy^4\,(9x)} = \dfrac{45x}{72x^2y^4}$

$$\frac{\text{LCD}}{\text{denominator}} = \frac{72x^2y^4}{8xy^4} = 9x$$

Step 3. $\dfrac{7}{18x^2y} + \dfrac{5}{8xy^4} = \dfrac{28y^3}{72x^2y^4} + \dfrac{45x}{72x^2y^4} = \dfrac{28y^3 + 45x}{72x^2y^4}$

Step 4. The rational expression cannot be reduced.

In applying step 2, if the multiplication involves any expression containing more
than one term, we *must* put parentheses around that expression (see Example 5).

EXAMPLE 5

Add $\dfrac{2}{x} + \dfrac{x}{x + 2}$.

SOLUTION

Step 1. The LCD is $x(x + 2)$.   We found the LCD in a previous example

$$\frac{\text{LCD}}{\text{denominator}} = \frac{x(x + 2)}{x} = x + 2$$

Step 2. $\dfrac{2}{x} = \dfrac{2\,(x + 2)}{x\,(x + 2)} = \dfrac{2x + 4}{x(x + 2)}$

$\dfrac{x}{x + 2} = \dfrac{x \cdot x}{(x + 2) \cdot x} = \dfrac{x^2}{x(x + 2)}$

Note the parentheses

$$\frac{\text{LCD}}{\text{denominator}} = \frac{x(x + 2)}{(x + 2)} = x$$

Step 3. $\dfrac{2}{x} + \dfrac{x}{x + 2} = \dfrac{2x + 4}{x(x + 2)} + \dfrac{x^2}{x(x + 2)} = \dfrac{2x + 4 + x^2}{x(x + 2)}$

Step 4. $\dfrac{2x + 4 + x^2}{x(x + 2)} = \dfrac{x^2 + 2x + 4}{x(x + 2)}$ $x^2 + 2x + 4$ does not factor

The rational expression cannot be reduced.

EXAMPLE 6   Subtract $3 - \dfrac{2a}{a+2}$.

SOLUTION

Step 1. The LCD is $a + 2$.

Step 2. $3 = \dfrac{3\,(a+2)}{1\,(a+2)} = \dfrac{3a+6}{a+2}$ $\dfrac{\text{LCD}}{\text{denominator}} = \dfrac{a+2}{1} = a+2$

$\dfrac{2a}{a+2} = \dfrac{2a}{a+2} = \dfrac{2a}{a+2}$

Step 3. $3 - \dfrac{2a}{a+2} = \dfrac{3a+6}{a+2} - \dfrac{2a}{a+2} = \dfrac{3a+6-2a}{a+2} = \dfrac{a+6}{a+2}$

Step 4. The rational expression cannot be reduced.

EXAMPLE 7   Subtract $\dfrac{z-1}{z+3} - \dfrac{z+3}{z-1}$.

SOLUTION   We cannot divide the first numerator and the second denominator by $z - 1$ or the first denominator and the second numerator by $z + 3$, because this is not a multiplication problem.

Step 1. The LCD is $(z+3)(z-1)$.

Step 2. $\dfrac{z-1}{z+3} = \dfrac{(z-1)\,(z-1)}{(z+3)\,(z-1)}$

$\dfrac{z+3}{z-1} = \dfrac{(z+3)\,(z+3)}{(z-1)\,(z+3)}$ — Note the parentheses

Step 3. $\dfrac{z-1}{z+3} - \dfrac{z+3}{z-1} = \dfrac{(z-1)\,(z-1)}{(z+3)\,(z-1)} - \dfrac{(z+3)\,(z+3)}{(z-1)\,(z+3)}$

$= \dfrac{z^2-2z+1}{(z+3)(z-1)} - \dfrac{z^2+6z+9}{(z+3)(z-1)}$

These parentheses are essential

$= \dfrac{(z^2-2z+1)-(z^2+6z+9)}{(z+3)(z-1)}$

$= \dfrac{z^2-2z+1-z^2-6z-9}{(z+3)(z-1)}$

$= \dfrac{-8z-8}{(z+3)(z-1)} = \dfrac{-8(z+1)}{(z+3)(z-1)}$

Step 4. The rational expression cannot be reduced. (Note: Other forms of the answer are also acceptable.)

A Word of Caution   An error frequently made in Example 7 is shown below:

$$\dfrac{\overset{1}{\cancel{z-1}}}{\underset{1}{\cancel{z+3}}} - \dfrac{\overset{1}{\cancel{z+3}}}{\underset{1}{\cancel{z-1}}} = 1 - 1 = 0$$

It is incorrect to divide the numerator of one rational expression and the denominator of a *different* rational expression by a common factor when adding or subtracting; such division can be done *only* when we multiply rational expressions.

EXAMPLE 8   Subtract $\dfrac{x - 1}{2x^2 + 11x + 12} - \dfrac{x - 2}{2x^2 + x - 3}$.

SOLUTION

Step 1. In order to find the LCD, we must factor each denominator.

$$2x^2 + 11x + 12 = (2x + 3)(x + 4)$$

$$2x^2 + x - 3 = (2x + 3)(x - 1)$$

The LCD is $(2x + 3)(x + 4)(x - 1)$.

Step 2. $\dfrac{x - 1}{2x^2 + 11x + 12} = \dfrac{x - 1}{(2x + 3)(x + 4)} = \dfrac{(x - 1)\,(x - 1)}{(2x + 3)(x + 4)\,(x - 1)}$

$\dfrac{x - 2}{2x^2 + x - 3} = \dfrac{x - 2}{(2x + 3)(x - 1)} = \dfrac{(x - 2)\,(x + 4)}{(2x + 3)(x - 1)\,(x + 4)}$

Step 3. $\dfrac{x - 1}{2x^2 + 11x + 12} - \dfrac{x - 2}{2x^2 + x - 3}$

$= \dfrac{(x - 1)(x - 1)}{(2x + 3)(x + 4)(x - 1)} - \dfrac{(x - 2)(x + 4)}{(2x + 3)(x - 1)(x + 4)}$

$= \dfrac{x^2 - 2x + 1}{(2x + 3)(x + 4)(x - 1)} - \dfrac{x^2 + 2x - 8}{(2x + 3)(x + 4)(x - 1)}$

$= \dfrac{x^2 - 2x + 1 - (x^2 + 2x - 8)}{(2x + 3)(x + 4)(x - 1)}$ The parentheses are essential

$= \dfrac{x^2 - 2x + 1 - x^2 - 2x + 8}{(2x + 3)(x + 4)(x - 1)}$

$= \dfrac{-4x + 9}{(2x + 3)(x + 4)(x - 1)}$

Step 4. The rational expression cannot be reduced.

✗ A Word of Caution It is incorrect to *multiply* all the rational expressions by the LCD; we change the *value* of an expression if we multiply it by any number *except* 1. Example 8 *cannot* be done as follows:

$(2x + 3)(x + 4)(x - 1) \cdot \dfrac{x - 1}{(2x + 3)(x + 4)} - (2x + 3)(x + 4)(x - 1) \cdot \dfrac{x - 2}{(2x + 3)(x - 1)}$

$= (x - 1)(x - 1) - (x + 4)(x - 2) = x^2 - 2x + 1 - (x^2 + 2x - 8)$

$= -4x + 9$

Order of Operations When we're working with rational expressions, the usual order of operations applies. Therefore, we must do all multiplications and divisions (in order from left to right) before we do any additions or subtractions (see Example 9).

EXAMPLE 9   Perform the indicated operations: $\dfrac{x + 5}{x - 2} - \dfrac{x + 3}{x^2 + 9x + 20} \div \dfrac{x - 4}{x^2 - 16}$.

SOLUTION   The division must be done *before* the subtraction.

$$\frac{x+5}{x-2}-\frac{x+3}{x^2+9x+20}\div\frac{x-4}{x^2-16}=\frac{x+5}{x-2}-\frac{x+3}{x^2+9x+20}\cdot\frac{x^2-16}{x-4}$$ Changing division to multiplication

$$=\frac{x+5}{x-2}-\frac{x+3}{(x+4)(x+5)}\cdot\frac{(x+4)(x-4)}{x-4}$$

$$=\frac{x+5}{x-2}-\frac{x+3}{x+5}$$ The LCD is $(x-2)(x+5)$

Using the fundamental property of rational expressions

$$=\frac{(x+5)(x+5)}{(x-2)(x+5)}-\frac{(x+3)(x-2)}{(x+5)(x-2)}$$

$$=\frac{x^2+10x+25-(x^2+x-6)}{(x-2)(x+5)}$$

$$=\frac{x^2+10x+25-x^2-x+6}{(x-2)(x+5)}$$

$$=\frac{9x+31}{(x-2)(x+5)}$$

Exercises 8.6
Set I

Perform the indicated operations.

1. $\frac{3}{a^2}+\frac{2}{a^3}$

2. $\frac{5}{u}+\frac{4}{u^3}$

3. $\frac{1}{2}+\frac{3}{x}-\frac{5}{x^2}$

4. $\frac{2}{3}-\frac{1}{y}+\frac{4}{y^2}$

5. $\frac{2}{27x^3y^2}-\frac{5}{18xy^5}$

6. $\frac{5}{48a^2b}-\frac{7}{18ab^4}$

7. $5+\frac{2}{x}$

8. $3+\frac{4}{y}$

9. $\frac{5}{4y^2}+\frac{9}{6y}$

10. $\frac{10}{5x}+\frac{3}{4x^2}$

11. $3x-\frac{3}{x}$

12. $4y-\frac{5}{y}$

13. $\frac{3}{a}+\frac{a}{a+3}$

14. $\frac{5}{b}+\frac{b}{b-5}$

15. $\frac{x}{2x+4}+\frac{-5}{4x}$

16. $\frac{4}{3x}+\frac{-2x}{3x+6}$

17. $x+\frac{2}{x}-\frac{3}{x-2}$

18. $m+\frac{3}{m}-\frac{2}{m-4}$

19. $\frac{a+b}{b}+\frac{b}{a-b}$

20. $\frac{y+x}{x}+\frac{x}{y-x}$

21. $\frac{3}{2e-2}-\frac{2}{3e-3}$

22. $\frac{2}{3f+6}-\frac{1}{5f+10}$

23. $\frac{3}{m-2}-\frac{5}{2-m}$

24. $\frac{7}{n-5}-\frac{2}{5-n}$

25. $\frac{x+1}{x-1}-\frac{x-1}{x+1}$

26. $\frac{x+4}{x-4}-\frac{x-4}{x+4}$

27. $\frac{y}{x^2-xy}+\frac{x}{y^2-xy}$

28. $\frac{b}{ab-a^2}+\frac{a}{ab-b^2}$

29. $\frac{2x}{x-3}-\frac{2x}{x+3}+\frac{36}{x^2-9}$

30. $\frac{x}{x+4}-\frac{x}{x-4}-\frac{32}{x^2-16}$

31. $\frac{x}{x^2+4x+4}+\frac{1}{x+2}$

32. $\frac{2x}{x^2-2x+1}-\frac{5}{x-1}$

33. $\frac{2x+3}{x^2+4x-5}-\frac{x+5}{2x^2+x-3}$

34. $\frac{3x+2}{x^2+6x-7}-\frac{x+7}{3x^2-x-2}$

35. $\frac{x^2}{x^2+8x+16}-\frac{x+1}{x^2-16}$

36. $\dfrac{x^2}{x^2 + 6x + 9} - \dfrac{x + 2}{x^2 - 9}$

37. $\dfrac{x + 1}{2x^3 + 3x^2 - 5x} - \dfrac{x - 1}{2x^3 + 7x^2 + 5x}$

38. $\dfrac{x + 1}{3x^3 + x^2 - 4x} - \dfrac{x - 1}{3x^3 + 7x^2 + 4x}$

39. $\dfrac{5}{8x + 6} - \dfrac{5}{8x} + \dfrac{4}{4x^2 + 3x}$

40. $\dfrac{4}{9x + 15} - \dfrac{4}{9x} + \dfrac{3}{3x^2 + 5x}$

41. $\dfrac{8}{18x - 45} - \dfrac{1}{24x^2 - 60x} - \dfrac{7}{12x^2}$

42. $\dfrac{7}{50x - 10} - \dfrac{1}{20x^2 - 4x} - \dfrac{9}{8x^2}$

43. $\dfrac{3}{8x^2} - \dfrac{x + 3}{4x^2 - 8x} \div \dfrac{x^2 + 2x - 3}{x^2 - 2x}$

44. $\dfrac{8}{9x^2} - \dfrac{x + 5}{3x^2 - 12x} \div \dfrac{x^2 + 8x + 15}{x^2 - 4x}$

45. $\dfrac{x + 3}{x - 5} - \dfrac{x - 2}{x^2 - x - 12} \div \dfrac{x + 7}{x^2 + 3x - 28}$

46. $\dfrac{x + 4}{x - 2} - \dfrac{x - 6}{x^2 + x - 12} \div \dfrac{x + 5}{x^2 + 2x - 15}$

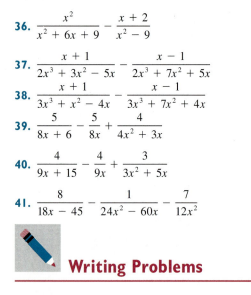

Writing Problems

Express the answers in your own words and in complete sentences.

1. Explain why multiplying a rational expression by $\dfrac{x + 3}{x + 3}$ does not change its value.

2. Explain how to build $\dfrac{x}{x - 4}$ to a rational expression with a denominator of $x^2 - 7x + 12$.

3. Find and describe the error in the following:

$$\dfrac{x + 3}{x - 5} + \dfrac{x + 2}{x + 1}$$

$$= (x - 5)(x + 1)\left(\dfrac{x + 3}{x - 5}\right) + (x - 5)(x + 1)\left(\dfrac{x + 2}{x + 1}\right)$$

$$= x^2 + 4x + 3 + x^2 - 3x - 10 = 2x^2 + x - 7$$

4. Find and describe the error in the following:

$$\dfrac{\overset{1}{x + 1}}{\underset{1}{x - 2}} + \dfrac{\overset{1}{x - 2}}{\underset{1}{x + 1}} = 1 + 1 = 2$$

Exercises   8.6
Set II

Perform the indicated operations.

1. $\dfrac{3}{x} + \dfrac{4}{x^2}$

2. $\dfrac{5}{2x^3y} + \dfrac{3}{4xy^2}$

3. $\dfrac{1}{3} - \dfrac{1}{a} + \dfrac{2}{a^2}$

4. $\dfrac{1}{6} - \dfrac{1}{k} + \dfrac{3}{k^2}$

5. $\dfrac{9}{50x^4y^3} - \dfrac{8}{75xy^5}$

6. $\dfrac{5}{36bc^2} - \dfrac{7}{90ab^3c}$

7. $2 + \dfrac{4}{z}$

8. $m - \dfrac{3}{m} + \dfrac{2}{m + 4}$

9. $\dfrac{5}{6xy^2} + \dfrac{7}{8x^2y}$

10. $\dfrac{z + 1}{z + 2} - \dfrac{z - 1}{z - 2}$

11. $4m - \dfrac{5}{m}$

12. $\dfrac{2}{x + 5} + \dfrac{3}{x - 5}$

13. $\dfrac{2}{x} + \dfrac{x}{x + 4}$

14. $\dfrac{2}{x + 5} - \dfrac{3}{x - 5}$

15. $\dfrac{x}{3x + 6} + \dfrac{-3}{2x}$

16. $\dfrac{x + 5}{x - 2} - \dfrac{x - 2}{x + 5}$

17. $a + \dfrac{3}{a} - \dfrac{2}{a - 2}$

18. $\dfrac{2}{a + 3} - \dfrac{4}{a - 1}$

19. $\dfrac{a - b}{b} + \dfrac{b}{a + b}$

20. $\dfrac{x + 2}{x - 3} - \dfrac{x + 3}{x - 2}$

21. $\dfrac{7}{4x - 4} - \dfrac{4}{7x - 7}$

22. $\dfrac{2x + 3}{x + 8} + \dfrac{x + 8}{2x + 3}$

23. $\dfrac{5}{y - 7} - \dfrac{8}{7 - y}$

24. $\dfrac{x - y}{x} - \dfrac{x}{x + y}$

25. $\dfrac{x - 1}{x + 2} - \dfrac{x - 2}{x + 1}$

26. $\dfrac{5}{x - 3} - \dfrac{4}{3 - x}$

27. $\dfrac{c}{d^2 - cd} + \dfrac{d}{c^2 - cd}$

28. $\dfrac{x - 5}{x + 5} - \dfrac{x + 5}{x - 5}$

29. $\dfrac{x}{x - 1} - \dfrac{x}{x + 1} + \dfrac{2}{x^2 - 1}$

30. $\dfrac{x + 2}{x^2 + x - 2} + \dfrac{3}{x^2 - 1}$

31. $\dfrac{x}{x^2 + 10x + 25} + \dfrac{1}{x + 5}$

32. $\dfrac{b}{ab - a^2} - \dfrac{a}{b^2 - ab}$

33. $\dfrac{4x + 3}{x^2 + 3x - 4} - \dfrac{x + 4}{4x^2 - x - 3}$

34. $\dfrac{a^2}{4a^2 + 12a + 9} - \dfrac{a + 3}{4a^2 - 9}$

35. $\dfrac{x^2}{x^2 + 12x + 36} - \dfrac{x + 3}{x^2 - 36}$

36. $\dfrac{a^2}{9a^2 - 25} - \dfrac{a + 1}{9a^2 - 30a + 25}$

37. $\dfrac{x + 1}{3x^3 + x^2 - 4x} - \dfrac{x - 1}{3x^3 + 7x^2 + 4x}$

38. $\dfrac{5}{3x^4 - 27x^2} + \dfrac{2}{x^3 + 6x^2 + 9x} - \dfrac{3}{6x - 2x^2}$

39. $\dfrac{7}{6x + 15} - \dfrac{5}{6x} + \dfrac{3}{2x^2 + 5x}$

40. $\dfrac{2}{15x + 5} - \dfrac{3}{20x} + \dfrac{5}{3x^2 + x}$

41. $\dfrac{5}{12x + 18} - \dfrac{1}{30x^2 + 45x} - \dfrac{7}{18x^2}$

42. $\dfrac{5}{14x^2 - 14x} - \dfrac{5}{21x^2} - \dfrac{6}{35x - 35}$

43. $\dfrac{5}{6x^2} - \dfrac{x + 3}{5x^2 - 15x} \div \dfrac{x^2 + 7x + 10}{x^2 + 2x - 15}$

44. $\dfrac{3x - 1}{2x^3} - \dfrac{3x - 1}{3x^2 - 7x + 2} \cdot \dfrac{x^2 - 3x + 2}{4x^2 - 3x - 1}$

45. $\dfrac{x - 5}{x + 3} - \dfrac{x + 2}{x^2 + 7x + 12} \div \dfrac{3x^2 + 6x}{x^2 + 3x - 4}$

46. $\dfrac{x^2 + 1}{x^2 - 1} - \dfrac{2x^2 + x - 1}{x - 1} \div \dfrac{2x^2 - 5x + 2}{x - 2}$

8.7   Simplifying Complex Fractions

A **complex fraction** is an algebraic expression that contains one or more rational expressions in its numerator and/or in its denominator. The following are examples of complex fractions:

$$\dfrac{\dfrac{2}{x}}{3}, \quad \dfrac{a}{\dfrac{1}{c}}, \quad \dfrac{\dfrac{3}{z}}{\dfrac{5}{z}}, \quad \dfrac{\dfrac{3}{x} - \dfrac{2}{y}}{\dfrac{5}{x} + \dfrac{3}{y}}$$

The parts of a complex fraction are as follows:

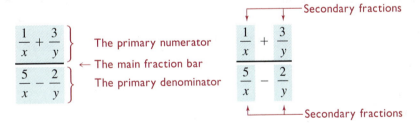

$\left.\dfrac{\dfrac{1}{x} + \dfrac{3}{y}}{\dfrac{5}{x} - \dfrac{2}{y}}\right\}$ → The primary numerator
← The main fraction bar
→ The primary denominator

Secondary fractions → $\dfrac{\dfrac{1}{x} + \dfrac{3}{y}}{\dfrac{5}{x} - \dfrac{2}{y}}$ ← Secondary fractions

Simplifying complex fractions

Method 1: Multiply the complex fraction by a rational expression in which both the numerator and the denominator equal the LCD of the secondary fractions (the value of this rational expression will be 1); then simplify the results.

Method 2: First simplify the primary numerator and the primary denominator of the complex fraction separately; then divide the simplified numerator by the simplified denominator.

Note that in some of the following examples, the solution can be found more easily by method 1 than by method 2; in others, the opposite is true.

EXAMPLE 1   Simplify $\dfrac{\frac{1}{2} + \frac{3}{4}}{\frac{5}{6} - \frac{2}{3}}$.

SOLUTION 1   To use method 1, we first determine that the LCD of the secondary denominators 2, 4, 6, and 3 is 12.

$$\frac{12}{12} \cdot \frac{\frac{1}{2} + \frac{3}{4}}{\frac{5}{6} - \frac{2}{3}} = \frac{12\left(\frac{1}{2} + \frac{3}{4}\right)}{12\left(\frac{5}{6} - \frac{2}{3}\right)} = \frac{\frac{12}{1}\left(\frac{1}{2}\right) + \frac{12}{1}\left(\frac{3}{4}\right)}{\frac{12}{1}\left(\frac{5}{6}\right) - \frac{12}{1}\left(\frac{2}{3}\right)} = \frac{6 + 9}{10 - 8} = \frac{15}{2} = 7\frac{1}{2}$$

└─ Multiplying by 1

SOLUTION 2   Using method 2, we have

$$\frac{\frac{1}{2} + \frac{3}{4}}{\frac{5}{6} - \frac{2}{3}} = \left(\frac{1}{2} + \frac{3}{4}\right) \div \left(\frac{5}{6} - \frac{2}{3}\right) = \left(\frac{2}{4} + \frac{3}{4}\right) \div \left(\frac{5}{6} - \frac{4}{6}\right) = \frac{5}{4} \div \frac{1}{6} = \frac{5}{\underset{2}{4}} \cdot \frac{\overset{3}{6}}{1} = \frac{15}{2} = 7\frac{1}{2}$$

✖ **A Word of Caution**   An error frequently made in Example 1 is shown below:

$$\left(\tfrac{1}{2} + \tfrac{3}{4}\right) \div \left(\tfrac{5}{6} - \tfrac{2}{3}\right) = \left(\tfrac{1}{2} + \tfrac{3}{4}\right) \times \left(\tfrac{6}{5} - \tfrac{3}{2}\right)$$

This is incorrect, since the multiplicative inverse of $\left(\frac{5}{6} - \frac{2}{3}\right)$ is not $\left(\frac{6}{5} - \frac{3}{2}\right)$. [You might verify that $\left(\frac{5}{6} - \frac{2}{3}\right)\left(\frac{6}{5} - \frac{3}{2}\right) \neq 1$.]

EXAMPLE 2   Simplify $\dfrac{\frac{4b^2}{9a^2}}{\frac{8b}{3a^3}}$. ←── The main fraction bar

SOLUTION 1   To use method 1, we first determine that the LCD of the secondary denominators $9a^2$ and $3a^3$ is $9a^3$.

$$\frac{9a^3}{9a^3}\left(\frac{\frac{4b^2}{9a^2}}{\frac{8b}{3a^3}}\right) = \frac{\frac{9a^3}{1}\left(\frac{4b^2}{9a^2}\right)}{\frac{9a^3}{1}\left(\frac{8b}{3a^3}\right)} = \frac{4ab^2}{24b} = \frac{ab}{6}$$

└─ The value of this fraction is 1

SOLUTION 2   Using method 2, we have

$$\frac{\frac{4b^2}{9a^2}}{\frac{8b}{3a^3}} = \frac{4b^2}{9a^2} \div \frac{8b}{3a^3} = \frac{\overset{b}{4b^2}}{\underset{3}{9a^2}} \cdot \frac{\overset{a}{3a^3}}{\underset{2}{8b}} = \frac{ab}{6}$$

EXAMPLE 3   Simplify $\dfrac{\frac{2}{x} - \frac{3}{x^2}}{5 + \frac{1}{x}}$.

SOLUTION 1   To use method 1, we first determine that the LCD of the secondary denominators x and x^2 is x^2.

$$\frac{x^2}{x^2}\left(\frac{\frac{2}{x} - \frac{3}{x^2}}{5 + \frac{1}{x}}\right) = \frac{\frac{x^2}{1}\left(\frac{2}{x}\right) - \frac{x^2}{1}\left(\frac{3}{x^2}\right)}{\frac{x^2}{1}\left(\frac{5}{1}\right) + \frac{x^2}{1}\left(\frac{1}{x}\right)} = \frac{2x - 3}{5x^2 + x}$$

SOLUTION 2 Using method 2, we have

$$\frac{\dfrac{2}{x} - \dfrac{3}{x^2}}{5 + \dfrac{1}{x}} = \left(\frac{2}{x} - \frac{3}{x^2}\right) \div \left(5 + \frac{1}{x}\right) = \left(\frac{2x}{x^2} - \frac{3}{x^2}\right) \div \left(\frac{5x}{x} + \frac{1}{x}\right)$$

$$= \frac{2x - 3}{x^2} \div \frac{5x + 1}{x} = \frac{2x - 3}{x^2} \cdot \frac{\overset{1}{\cancel{x}}}{5x + 1} = \frac{2x - 3}{5x^2 + x}$$

EXAMPLE 4   Simplify $\dfrac{1 + \dfrac{2}{a}}{1 - \dfrac{4}{a^2}}$.

SOLUTION 1   The LCD of the secondary denominators a and a^2 is a^2.

$$\frac{a^2}{a^2}\left(\frac{1 + \dfrac{2}{a}}{1 - \dfrac{4}{a^2}}\right) = \frac{\dfrac{a^2}{1}\left(\dfrac{1}{1}\right) + \dfrac{a^2}{1}\left(\dfrac{2}{a}\right)}{\dfrac{a^2}{1}\left(\dfrac{1}{1}\right) - \dfrac{a^2}{1}\left(\dfrac{4}{a^2}\right)} = \frac{a^2 + 2a}{a^2 - 4} = \frac{a(a + 2)}{(a + 2)(a - 2)} = \frac{a}{a - 2}$$

SOLUTION 2   Using method 2, we have

$$\frac{1 + \dfrac{2}{a}}{1 - \dfrac{4}{a^2}} = \left(1 + \frac{2}{a}\right) \div \left(1 - \frac{4}{a^2}\right) = \left(\frac{a}{a} + \frac{2}{a}\right) \div \left(\frac{a^2}{a^2} - \frac{4}{a^2}\right)$$

$$= \frac{a + 2}{a} \div \frac{a^2 - 4}{a^2} = \frac{a + 2}{a} \cdot \frac{a^2}{a^2 - 4} = \frac{\overset{1}{\cancel{a + 2}}}{\underset{1}{\cancel{a}}} \cdot \frac{\overset{a}{\cancel{a^2}}}{(\cancel{a + 2})(a - 2)} = \frac{a}{a - 2}$$

EXAMPLE 5   Simplify $\dfrac{\dfrac{2}{x + 3} + \dfrac{1}{x}}{\dfrac{3}{x + 3} - \dfrac{2}{x}}$.

SOLUTION 1   The LCD of the secondary denominators is $x(x + 3)$.

$$\frac{x(x + 3)}{x(x + 3)}\left(\frac{\dfrac{2}{x + 3} + \dfrac{1}{x}}{\dfrac{3}{x + 3} - \dfrac{2}{x}}\right) = \frac{\dfrac{x(x + 3)}{1}\left(\dfrac{2}{x + 3}\right) + \dfrac{x(x + 3)}{1}\left(\dfrac{1}{x}\right)}{\dfrac{x(x + 3)}{1}\left(\dfrac{3}{x + 3}\right) - \dfrac{x(x + 3)}{1}\left(\dfrac{2}{x}\right)}$$

$$= \frac{2x + (x + 3)}{3x - (x + 3)(2)} = \frac{2x + x + 3}{3x - 2x - 6} = \frac{3x + 3}{x - 6}$$

(We mentally factor the numerator and see that the answer is not reducible.)

SOLUTION 2   Using method 2, we have

$$\frac{\dfrac{2}{x + 3} + \dfrac{1}{x}}{\dfrac{3}{x + 3} - \dfrac{2}{x}} = \left(\frac{2}{x + 3} + \frac{1}{x}\right) \div \left(\frac{3}{x + 3} - \frac{2}{x}\right)$$

$$= \left(\frac{2 \, x}{x \, (x + 3)} + \frac{1 \, (x + 3)}{x \, (x + 3)}\right) \div \left(\frac{3 \, x}{x \, (x + 3)} - \frac{2 \, (x + 3)}{x \, (x + 3)}\right)$$

$$= \left(\frac{2x + (x + 3)}{x(x + 3)}\right) \div \left(\frac{3x - 2(x + 3)}{x(x + 3)}\right)$$

$$= \left(\frac{2x + x + 3}{x(x + 3)}\right) \div \left(\frac{3x - 2x - 6}{x(x + 3)}\right) = \left(\frac{3x + 3}{x(x + 3)}\right) \div \left(\frac{x - 6}{x(x + 3)}\right)$$

$$= \left(\frac{3x + 3}{\underset{1}{\cancel{x(x + 3)}}}\right)\left(\frac{\overset{1}{\cancel{x(x + 3)}}}{x - 6}\right) = \frac{3x + 3}{x - 6}$$

Exercises 8.7
Set I

Simplify each complex fraction.

1. $\dfrac{\frac{3}{4}}{\frac{5}{6}}$
2. $\dfrac{\frac{3}{5}}{\frac{3}{4}}$
3. $\dfrac{\frac{2}{3}}{\frac{4}{9}}$
4. $\dfrac{\frac{5}{6}}{\frac{5}{9}}$

5. $\dfrac{\frac{3}{4} - \frac{1}{2}}{\frac{5}{8} + \frac{1}{4}}$
6. $\dfrac{\frac{5}{6} - \frac{1}{3}}{\frac{2}{9} + \frac{1}{6}}$
7. $\dfrac{\frac{3}{5} + 2}{2 - \frac{3}{4}}$
8. $\dfrac{\frac{3}{16} + 5}{6 - \frac{7}{8}}$

9. $\dfrac{\frac{5x^3}{3y^4}}{\frac{10x}{9y}}$
10. $\dfrac{\frac{8a^4}{5b}}{\frac{4a^3}{15b^2}}$
11. $\dfrac{\frac{18cd^2}{5a^3b}}{\frac{12cd^2}{15ab^2}}$
12. $\dfrac{\frac{8x^2y}{7z^3}}{\frac{12xy^2}{21z^5}}$

13. $\dfrac{\frac{x+3}{5}}{\frac{2x+6}{10}}$
14. $\dfrac{\frac{a-4}{3}}{\frac{2a-8}{9}}$
15. $\dfrac{\frac{x+2}{2x}}{\frac{x+1}{4x^2}}$

16. $\dfrac{\frac{x-3}{3x^2}}{\frac{x-9}{9x}}$
17. $\dfrac{\frac{a}{b} + 1}{\frac{a}{b} - 1}$
18. $\dfrac{2 + \frac{x}{y}}{2 - \frac{x}{y}}$

19. $\dfrac{\frac{1}{x} + x}{\frac{1}{x} - x}$
20. $\dfrac{a - \frac{4}{a}}{a + \frac{4}{a}}$
21. $\dfrac{\frac{c}{d} + 2}{\frac{c^2}{d^2} - 4}$

22. $\dfrac{\frac{x^2}{y^2} - 1}{\frac{x}{y} - 1}$
23. $\dfrac{x + \frac{x}{y}}{1 + \frac{1}{y}}$
24. $\dfrac{1 - \frac{1}{b}}{3 - \frac{3}{b}}$

25. $\dfrac{\frac{1}{x^2} - \frac{1}{y^2}}{\frac{1}{x} + \frac{1}{y}}$
26. $\dfrac{\frac{1}{a^2} - \frac{1}{4}}{\frac{1}{a} - \frac{1}{2}}$
27. $\dfrac{\frac{2}{x} - \frac{4}{x^2}}{\frac{1}{x} - \frac{2}{x^2}}$

28. $\dfrac{\frac{2}{y^2} + \frac{1}{y}}{\frac{8}{y^2} + \frac{4}{y}}$

29. $\dfrac{\frac{x}{x+1} + \frac{4}{3x}}{\frac{x}{x+1} - \frac{3}{x}}$

30. $\dfrac{\frac{4x}{4x+1} + \frac{1}{2x}}{\frac{2}{4x+1} + \frac{2}{x}}$

31. $\dfrac{\frac{1}{4x} + \frac{x}{x-6}}{\frac{x-1}{x} - \frac{3}{x-6}}$

32. $\dfrac{\frac{x+1}{y} + \frac{1}{x-1}}{\frac{1}{x-1} + \frac{1}{2y}}$

Writing Problems

Express the answer in your own words and in complete sentences.

1. Find and describe the error in the following:

Exercises 8.7
Set II

Simplify each complex fraction.

1. $\dfrac{\frac{5}{6}}{\frac{2}{9}}$
2. $\dfrac{\frac{3}{5}}{\frac{9}{10}}$
3. $\dfrac{\frac{5}{6}}{\frac{10}{18}}$
4. $\dfrac{\frac{8}{9}}{\frac{8}{27}}$

5. $\dfrac{\frac{7}{9} - \frac{1}{3}}{\frac{5}{2} - \frac{1}{4}}$
6. $\dfrac{\frac{5}{6} - \frac{1}{3}}{\frac{3}{2} - \frac{1}{4}}$
7. $\dfrac{\frac{3}{4} + 2}{3 - \frac{1}{2}}$
8. $\dfrac{\frac{1}{8} + 3}{4 - \frac{1}{2}}$

9. $\dfrac{\frac{3a^3}{5b^2}}{\frac{6a^2}{10b^3}}$
10. $\dfrac{\frac{4x^4}{5y^3}}{\frac{8x^2}{10y^4}}$
11. $\dfrac{\frac{16bc^2}{5a^2d}}{\frac{4ac^4}{10b^2d^3}}$
12. $\dfrac{\frac{9x^2y^3}{11wz^2}}{\frac{18xy^4}{22w^3z}}$

13. $\dfrac{\frac{x-2}{4}}{\frac{3x-6}{12}}$
14. $\dfrac{\frac{a}{b} - 2}{2 + \frac{a}{b}}$
15. $\dfrac{\frac{x+3}{3x}}{\frac{x+1}{6x^2}}$
16. $\dfrac{\frac{1}{x+2}}{\frac{4}{3x+6}}$

17. $\dfrac{\frac{4}{x} + 1}{\frac{4}{x} - 1}$
18. $\dfrac{2}{1 + \frac{1}{x}}$
19. $\dfrac{\frac{3}{a} - a}{\frac{3}{a} + a}$

20. $\dfrac{1 - \dfrac{1}{a^2}}{\dfrac{1}{a} - \dfrac{1}{a^2}}$ **21.** $\dfrac{\dfrac{x^2}{y^2} - 4}{\dfrac{x}{y} + 2}$ **22.** $\dfrac{\dfrac{1}{a} + \dfrac{1}{b}}{\dfrac{1}{ab}}$ **29.** $\dfrac{\dfrac{x}{x+3} + \dfrac{3}{2x}}{\dfrac{x}{x+3} - \dfrac{2}{x}}$ **30.** $\dfrac{\dfrac{1}{6x} + \dfrac{x}{x-3}}{\dfrac{x-2}{x} - \dfrac{2}{x-3}}$

23. $\dfrac{2 + \dfrac{2}{b}}{a + \dfrac{a}{2}}$ **24.** $\dfrac{\dfrac{1}{x^2} - 9}{\dfrac{1}{x} + 3}$ **25.** $\dfrac{\dfrac{1}{x^2} - \dfrac{4}{y^2}}{\dfrac{1}{x} + \dfrac{2}{y}}$ **31.** $\dfrac{\dfrac{3x}{3x+1} + \dfrac{1}{3x}}{\dfrac{3}{3x+1} + \dfrac{3}{x}}$ **32.** $\dfrac{\dfrac{x+4}{y} + \dfrac{1}{x-4}}{\dfrac{4}{x-4} + \dfrac{1}{4y}}$

26. $\dfrac{\dfrac{x+1}{6x^2}}{\dfrac{x+2}{2x}}$ **27.** $\dfrac{\dfrac{1}{y^2} - \dfrac{1}{9}}{\dfrac{1}{y} + \dfrac{1}{3}}$ **28.** $\dfrac{\dfrac{a^2}{9} - \dfrac{1}{b^2}}{\dfrac{a}{3} + \dfrac{1}{b}}$

Sections 8.1–8.7

REVIEW

Rational Expressions
8.1

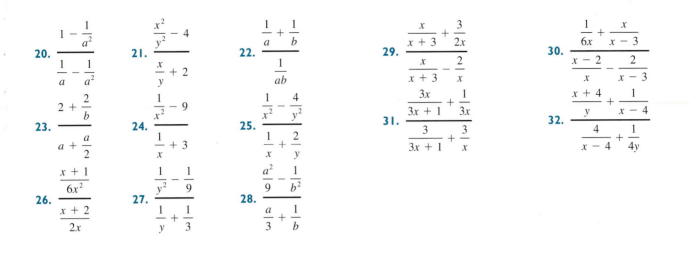

The terms of the rational expression → $\dfrac{P}{Q}$
— The numerator
← The fraction bar
— The denominator (cannot be zero)

where P and Q are polynomials and $Q \neq 0$.

Equivalent Rational Expressions
8.1

Equivalent rational expressions are rational expressions that have the same value. The fundamental property of rational expressions states that $\dfrac{P \cdot C}{Q \cdot C} = \dfrac{P}{Q}$, where P, Q, and C are polynomials and where $Q \neq 0$ and $C \neq 0$.

To Reduce a Rational Expression to Lowest Terms
8.2

1. Factor the numerator and denominator completely and find their GCF.
2. Divide both numerator and denominator by their GCF.

(We can, instead, divide both the numerator and the denominator by any factor common to both and then check for other common factors.)

To Multiply Rational Expressions
8.3

1. Factor any numerators or denominators that contain more than one term.
2. Divide the numerators and denominators by all factors common to both.
3. The numerator of the product is the product of all the factors remaining in the numerators, and the denominator of the product is the product of all the factors remaining in the denominators. If the only factor remaining in the numerator is 1, that 1 must be *written*.

$$\frac{P}{Q} \cdot \frac{R}{S} = \frac{P \cdot R}{Q \cdot S}, \quad Q \neq 0, S \neq 0$$

To Divide Rational Expressions
8.3

Multiply the dividend by the multiplicative inverse (the reciprocal) of the divisor.

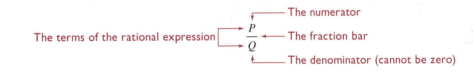

$$\frac{P}{Q} \div \frac{S}{T} = \frac{P}{Q} \cdot \frac{T}{S}, \quad Q \neq 0, T \neq 0, S \neq 0$$

The dividend ↑   ↑ The divisor

Reduce the resulting rational expression, if possible.

The Three Signs of a Rational Expression
8.3

Every rational expression has three signs associated with it: the sign of the entire rational expression, the sign of the numerator, and the sign of the denominator. If any two of the three signs of a rational expression are changed, the value of the rational expression is unchanged.

To Add or Subtract Rational Expressions That Have the Same Denominator **8.4**	1. The numerator of the sum is the sum of the numerators, and the denominator of the sum is the same as any *one* of the denominators. That is, if P, Q, and R are polynomials and if $Q \neq 0$, then $$\frac{P}{Q} + \frac{R}{Q} = \frac{P + R}{Q} \quad \text{and} \quad \frac{P}{Q} - \frac{R}{Q} = \frac{P - R}{Q}$$ 2. The resulting rational expression should be reduced, if possible.
To Find the LCM of Two or More Polynomials **8.5**	1. Factor each polynomial completely. Express repeated factors in exponential form. 2. Write down each different base that appears in any of the factorizations. 3. Raise each base to the highest power to which it occurs in any of the factorizations. 4. The **least common multiple (LCM)** is the product of all the factors found in step 3.
To Find the LCD **8.5**	The **least common denominator (LCD)** of two or more rational expressions is the LCM of their denominators.
To Add or Subtract Rational Expressions That Do Not Have the Same Denominator **8.6**	1. Find the LCD. 2. Convert each rational expression to an equivalent rational expression that has the LCD as its denominator. 3. Add or subtract the resulting rational expressions. 4. Reduce the resulting rational expression, if possible.
Complex Fractions **8.7**	A **complex fraction** is an algebraic expression that contains a rational expression in its numerator and/or its denominator.
To Simplify Complex Fractions **8.7**	*Method 1:* Multiply the complex fraction by a rational expression in which both the numerator and the denominator equal the LCD of the secondary fractions; then simplify the results. *Method 2:* First simplify the primary numerator and the primary denominator of the complex fraction; then divide the simplified numerator by the simplified denominator.

Sections 8.1–8.7   R E V I E W   E X E R C I S E S   Set I

In Exercises 1 and 2, determine whether the pairs of rational expressions are equivalent.

1. $\dfrac{x + 1}{2x + 2}, \dfrac{1}{2}$

2. $\dfrac{4x + 5}{3x + 5}, \dfrac{4x}{3x}$

In Exercises 3–5, find each missing term.

3. $\dfrac{x - 6}{7} = -\dfrac{?}{-7}$

4. $\dfrac{2}{a - b} = -\dfrac{?}{b - a}$

5. $\dfrac{3 - x}{4} = \dfrac{x - 3}{?}$

In Exercises 6–12, reduce each rational expression to lowest terms.

6. $\dfrac{6ab^4}{3a^3b^3}$

7. $\dfrac{2 + 4m}{2}$

8. $\dfrac{3 - 6n}{3}$

9. $\dfrac{a^2 - 4}{a + 2}$

10. $\dfrac{x - 5}{x^2 - 3x - 10}$

11. $\dfrac{a - b}{ax + ay - bx - by}$

12. $\dfrac{42x^4y^5}{35x^8y^2}$

13. Simplify $\left(\dfrac{6a^4b^{-2}}{8a^{-1}b^3}\right)^{-2}$

In Exercises 14–28, perform the indicated operations.

14. $\dfrac{8}{k} + \dfrac{2}{k}$

15. $5 - \dfrac{3}{2x}$

16. $\dfrac{4m}{2m - 3} - \dfrac{2m + 3}{2m - 3}$

17. $\dfrac{-5a^2}{3b} \div \dfrac{10a}{9b^2}$

18. $\dfrac{21x^3}{4y^2} \div \dfrac{-7x}{8y^4}$

19. $\dfrac{3x + 6}{6} \cdot \dfrac{2x^2}{4x + 8}$

20. $\dfrac{y - 3}{2} - \dfrac{y + 4}{3}$

21. $\dfrac{a - 2}{a - 1} + \dfrac{a + 1}{a + 2}$

22. $\dfrac{k + 3}{k - 2} - \dfrac{k + 2}{k - 3}$

23. $\dfrac{2x^2 - 6x}{x + 2} \div \dfrac{x}{4x + 8}$

24. $\dfrac{4z^2}{z - 5} \div \dfrac{z}{2z - 10}$

25. $\dfrac{3x + 4}{2x^2 + 7x + 6} - \dfrac{x + 2}{3x^2 + 10x + 8}$

26. $\dfrac{5x - 20}{3x^3 + 24x^2 + 48x} - \dfrac{x + 4}{5x^2 - 80}$

27. $\dfrac{x + 6}{2x^2 - 11x - 6} + \dfrac{x - 6}{2x^2 + 13x + 6}$

28. $\dfrac{3x + 2}{6x^2 - 13x + 5} - \dfrac{7x - 1}{3x^2 - 2x - 5}$

In Exercises 29–32, simplify each complex fraction.

29. $\dfrac{\dfrac{5k^2}{3m^2}}{\dfrac{10k}{9m}}$

30. $\dfrac{3 - \dfrac{a}{b}}{2 + \dfrac{a}{b}}$

31. $\dfrac{\dfrac{a^2}{b^2} - 1}{\dfrac{a}{b} - 1}$

32. $\dfrac{1 - \dfrac{16}{x^2}}{1 - \dfrac{4}{x}}$

Name _____

In Exercises 1 and 2, determine whether the pairs of rational expressions are equivalent.

1. $\dfrac{5x}{3y}, \dfrac{5x+1}{3y+1}$

2. $\dfrac{3x(x-1)}{2x(x-1)}, \dfrac{3x}{2x}$

In Exercises 3–5, find each missing term.

3. $\dfrac{6}{x-3} = \dfrac{-6}{?}$

4. $\dfrac{4-x}{7} = -\dfrac{?}{7}$

5. $\dfrac{x-3}{(2+x)(5-2x)} = \dfrac{?}{(x+2)(2x-5)}$

In Exercises 6–12, reduce each rational expression to lowest terms.

6. $\dfrac{6ab^3}{2ab}$

7. $\dfrac{x+2}{x^2-2x-8}$

8. $\dfrac{9-y^2}{3-y}$

9. $\dfrac{x+3}{x^2-x-12}$

10. $\dfrac{42x^4y^5}{35x^8y^2}$

11. $\dfrac{4z^3+4z^2-24z}{2z^2+4z-6}$

12. $\dfrac{6k^3-12k^2-18k}{3k^2+3k-36}$

13. Simplify $\left(\dfrac{6x^2y^{-4}}{9x^{-1}y^2}\right)^{-2}$.

In Exercises 14–28, perform the indicated operations.

14. $\dfrac{3}{x} + \dfrac{5}{x}$

15. $4 - \dfrac{3}{2x}$

ANSWERS

1. _____
2. _____
3. _____
4. _____
5. _____
6. _____
7. _____
8. _____
9. _____
10. _____
11. _____
12. _____
13. _____
14. _____
15. _____

16. $\dfrac{2a}{3a + 1} - \dfrac{3a - 1}{3a + 1}$

17. $\dfrac{x + 1}{2} - \dfrac{x - 3}{5}$

18. $\dfrac{y - 2}{y + 1} - \dfrac{y - 1}{y + 2}$

19. $\dfrac{15x^3}{4y^2} \div \dfrac{5x^2}{8y}$

20. $\dfrac{4x - 4}{2} \cdot \dfrac{6x^2}{3x - 3}$

21. $\dfrac{z^2 + 3z + 2}{z^2 + 2z + 1} \div \dfrac{z^2 + 2z - 3}{z^2 - 1}$

22. $\dfrac{5x - 5}{10} \cdot \dfrac{4x^3}{2x - 2}$

23. $\dfrac{x + 4}{5} - \dfrac{x - 2}{3}$

24. $\dfrac{x^2 + 2x - 3}{2x^2 - x - 1} \cdot \dfrac{6x + 3}{3x^2 + 11x + 6}$

25. $\dfrac{15x + 3}{50x^2 - 20x + 2} - \dfrac{4}{75x^2 - 3}$

26. $\dfrac{5}{6x^2 + 5x - 1} - \dfrac{x + 1}{36x^2 - 12x + 1}$

27. $\dfrac{5x - 1}{10x^2 + 37x + 7} + \dfrac{11x + 2}{10x^2 + 33x - 7}$

28. $\dfrac{x - 3}{2x^2 + 11x + 15} - \dfrac{x + 3}{2x^2 - x - 15}$

16. _____

17. _____

18. _____

19. _____

20. _____

21. _____

22. _____

23. _____

24. _____

25. _____

26. _____

27. _____

28. _____

29. _____

30. _____

31. _____

32. _____

In Exercises 29–32, simplify each complex fraction.

29. $\dfrac{\dfrac{10a^2b}{12a^4b^3}}{\dfrac{5ab^2}{16a^2b^3}}$

30. $\dfrac{2 + \dfrac{a}{b}}{\dfrac{a}{b} - 2}$

31. $\dfrac{\dfrac{1}{x^2} - \dfrac{9}{y^2}}{\dfrac{1}{x} - \dfrac{3}{y}}$

32. $\dfrac{\dfrac{x}{y} + 2}{\dfrac{x}{y} - 2}$

Sections 8.1–8.7 DIAGNOSTIC TEST

The purpose of this test is to see how well you understand operations involving rational expressions. If you will be tested on Sections 8.1–8.7, we recommend that you work this diagnostic test *before* your instructor tests you on this material. Allow yourself about 50 minutes to do this test.

Complete solutions for all the problems on this test, together with section references, are given in the answer section in the back of the book. We suggest that you study the sections referred to for the problems you do incorrectly.

1. Determine whether $\dfrac{3x + 2}{5y + 2}$ and $\dfrac{3x}{5y}$ are equivalent rational expressions.

In Problems 2–4, reduce each rational expression to lowest terms.

2. $\dfrac{12a^6b^3}{8a^2b^5}$

3. $\dfrac{5x + 1}{25x^2 + 10x + 1}$

4. $\dfrac{x^2 - 4y^2}{2x^2 + 3xy - 2y^2}$

5. Find each missing term.

   a. $\dfrac{5}{-3x} = \dfrac{-5}{?}$ b. $\dfrac{17}{x - 2} = \dfrac{?}{2 - x}$

6. Find the least common multiple (LCM) of $6x^3y$ and $15x^2y^2$.

In Problems 7–18, perform the indicated operations. Reduce all answers to lowest terms.

7. $\dfrac{x}{3y^2} \cdot \dfrac{6xy}{5y^3}$

8. $\dfrac{x}{x^2 + 6x + 8} \cdot \dfrac{3x + 12}{12x^2 + 24x}$

9. $\dfrac{4}{x - 5} \div \dfrac{8}{x^2 - 5x}$

10. $\dfrac{7y}{7y + y^2} \div \dfrac{1 - y^2}{7 + 8y + y^2}$

11. $\dfrac{8}{3y^2} - \dfrac{2}{3y^2}$

12. $\dfrac{8x}{x + 7} + \dfrac{3}{x + 7}$

13. $\dfrac{8x}{4x - 1} - \dfrac{2}{4x - 1}$

14. $\dfrac{3x}{x - 1} + \dfrac{4}{x + 3}$

15. $\dfrac{x - 3}{x + 5} + \dfrac{x + 5}{x - 3}$

16. $\dfrac{x + 14}{x^2 + 7x + 12} - \dfrac{x + 5}{x^2 + 5x + 6}$

17. $\dfrac{x}{x - 4} - \dfrac{x}{x + 4} - \dfrac{32}{x^2 - 16}$

18. $\dfrac{x + 7}{3x^2 - 22x + 7} - \dfrac{x - 7}{3x^2 + 20x - 7}$

In Problems 19 and 20, simplify the complex fractions.

19. $\dfrac{\dfrac{4x^3}{9y^4}}{\dfrac{2x}{27y^2}}$

20. $\dfrac{\dfrac{6}{x} - \dfrac{4}{x^2}}{5 + \dfrac{2}{x}}$

8.8 Solving Rational Equations

A **rational equation** is an equation that contains one or more rational expressions.

8.8A Excluded Values of Rational Expressions

Before we discuss solving rational equations, we must discuss **excluded values** of rational expressions. Because we cannot permit a denominator to be zero, any value of the variable that makes a denominator zero *must be excluded*. We find excluded values as follows.

Finding excluded values	If there are variables in the denominator, set the denominator equal to zero and solve the resulting equation; the solutions of that equation are the values of the variable that must be excluded.
	If there are no variables in the denominator, no values of the variable need be excluded.

EXAMPLE 1   Examples of finding excluded values for rational expressions:

a. $\dfrac{x}{5}$ There are no variables in the denominator.
Therefore, no values of the variable need be excluded.

b. $\dfrac{4x + 1}{x + 2}$ We set $x + 2$ equal to zero and solve:

$$x + 2 = 0$$

$$x = -2$$

Since the denominator would be zero if x were replaced with -2, -2 is an excluded value.

c. $\dfrac{x^2 + 1}{x^2 + 3x - 4}$ We set $x^2 + 3x - 4$ equal to zero and solve:

$$x^2 + 3x - 4 = 0$$

$$(x + 4)(x - 1) = 0$$

$x + 4 = 0$	$x - 1 = 0$
$x + 4 - 4 = 0 - 4$	$x - 1 + 1 = 0 + 1$
$x = -4$	$x = 1$

Therefore, -4 and 1 are excluded values.

d. $\dfrac{7}{x}$ We set x equal to zero and solve: $x = 0$. Therefore, 0 is an excluded value.

e. $\dfrac{1}{2}$ There are no variables in the denominator. Therefore, no values of the variable need be excluded.

Exercises
Set I

Determine what value (or values) of the variable must be excluded.

1. $\dfrac{3x + 4}{x - 2}$ **2.** $\dfrac{5 - 4x}{x + 3}$ **3.** $\dfrac{x}{10}$ **4.** $\dfrac{y}{20}$

5. $\dfrac{x - 4}{3x^2 - 6x}$ **6.** $\dfrac{3x + 2}{4x^2 - 12x}$

7. $\dfrac{3 + x}{x^2 - x - 2}$ **8.** $\dfrac{x - 5}{x^2 + x - 12}$

Exercises 8.8A
Set II

Determine what value (or values) of the variable must be excluded.

1. $\dfrac{6x + 4}{x - 7}$ **2.** $\dfrac{x + 8}{x}$ **3.** $\dfrac{x}{7}$ **4.** $\dfrac{7}{x}$

5. $\dfrac{x - 7}{5x^2 - 10x}$ **6.** $\dfrac{5x - 3}{x^2 - 2x - 24}$

7. $\dfrac{8 + x}{x^2 - 6x - 27}$ **8.** $\dfrac{x + 2}{x^2 + 6x + 9}$

8.8B Solving Rational Equations That Simplify to First-Degree Equations

In this section, after we remove denominators and grouping symbols from the rational equation, the equation that remains will be a *first-degree equation* (a polynomial equation whose highest-degree term is first-degree).

The techniques for solving rational equations differ greatly from the techniques for performing operations, such as addition, subtraction, and so forth, on rational expressions.

In an earlier chapter, we "cleared fractions" (that is, we removed the denominators) in equations that contained only *one* denominator by multiplying both sides of the equation by that denominator (the denominator was always a *number*, never a variable). If an equation has *more than one* denominator, we can clear fractions by multiplying both sides of the equation by the least common denominator (LCD).

Extraneous Roots   When we clear fractions by multiplying both sides of the equation by the LCD, the new equation will be equivalent to the original equation *as long as the expression we're multiplying by (the LCD) is not zero*. (Recall that equivalent equations are equations that have exactly the same solution set.) If any denominator contains a variable, the LCD will also contain a variable. Therefore, we must *exclude* any values of the variable that make the LCD zero; otherwise, we would be multiplying both sides of the equation by zero.

After we've cleared fractions and solved the new equation, some (or all) of the *apparent* solutions to the new equation may be *excluded values*. These solutions, which are called **extraneous roots**, are *not* solutions of the original equation, and they must be rejected. (Any value of the variable that, when checked, gives a false statement—such as $3 = 0$—is also an extraneous root and must be rejected. We will see roots of this kind when we solve radical equations in a later chapter.)

If we check an apparent solution that happens to be an excluded value in the original equation, we will always have zeros in denominators; this indicates to us that that solution must be rejected.

Solving Rational Equations

We can solve rational equations by using the addition, subtraction, multiplication, and division properties of equality, as outlined below.

Solving a rational equation that simplifies to a first-degree equation	1. Find the LCD, and find all excluded values.
	2. Remove denominators by multiplying *both sides of the equation* (that is, by multiplying *every term*) by the LCD.
	3. **a.** Remove all grouping symbols. **b.** Collect and combine like terms on each side of the equal sign.
	4. **a.** Move all the terms that contain the variable to one side of the equal sign and all other terms to the other side. **b.** Divide both sides of the equation by the coefficient of the variable, or multiply both sides of the equation by the reciprocal of the coefficient of the variable.
	5. Reject any apparent solutions that are excluded values.
	6. Check any other apparent solutions in the original equation.

EXAMPLE 2

Solve $\dfrac{x}{2} + \dfrac{x}{3} = 5$.

SOLUTION

Step 1.   The LCD is 6. There are no excluded values.

Step 2.

$$6\left(\frac{x}{2} + \frac{x}{3}\right) = 6\,(5)$$

We must use the distributive property on the left side

$$6\left(\frac{x}{2}\right) + 6\left(\frac{x}{3}\right) = 6\,(5)$$

Therefore, *each term of the equation* is multiplied by the LCD, 6

$$\frac{\overset{3}{6}}{1}\left(\frac{x}{\underset{1}{2}}\right) + \frac{\overset{2}{6}}{1}\left(\frac{x}{\underset{1}{3}}\right) = \frac{6}{1}\left(\frac{5}{1}\right)$$

Step 3a. $3x + 2x = 30$

Step 3b. $5x = 30$

Step 4. $\dfrac{5x}{5} = \dfrac{30}{5}$ or $\left(\tfrac{1}{5}\right)(5x) = \left(\tfrac{1}{5}\right)(30)$

$$x = 6 \qquad\qquad\qquad x = 6$$

Step 5.   Step 5 does not apply, as there were no excluded values.

✓ **Step 6.**   **Check**

$$\frac{x}{2} + \frac{x}{3} = 5$$

$$\tfrac{6}{2} + \tfrac{6}{3} \overset{?}{=} 5$$

$$3 + 2 \overset{?}{=} 5$$

$$5 = 5$$

The solution is 6.

EXAMPLE 3

Solve $\dfrac{x-4}{2} - \dfrac{x}{5} = \dfrac{1}{10}$.

SOLUTION

Step 1.   The LCD is 10. There are no excluded values.

Step 2.

$$10\left(\frac{x-4}{2} - \frac{x}{5}\right) = 10\left(\frac{1}{10}\right)$$

Multiplying both sides by the LCD

$$\frac{\overset{5}{10}}{1}\left(\frac{x-4}{\underset{1}{2}}\right) - \frac{\overset{2}{10}}{1}\left(\frac{x}{\underset{1}{5}}\right) = \frac{\overset{1}{10}}{1}\left(\frac{1}{\underset{1}{10}}\right)$$

Using the distributive property on the left side

$$5(x-4) - 2x = 1$$

Step 3a. $5x - 20 - 2x = 1$

Step 3b. $3x - 20 = 1$

Step 4a. $3x - 20 + 20 = 1 + 20$

$$3x = 21$$

Step 4b. $\dfrac{3x}{3} = \dfrac{21}{3}$ or $\left(\dfrac{1}{3}\right)(3x) = \left(\dfrac{1}{3}\right)(21)$

$$x = 7 \qquad\qquad\qquad x = 7$$

Step 5. Step 5 does not apply, as there were no excluded values.

✓ **Step 6.** *Check*

$$\frac{x-4}{2} - \frac{x}{5} = \frac{1}{10}$$

$$\frac{7-4}{2} - \frac{7}{5} \stackrel{?}{=} \frac{1}{10}$$

$$\frac{3}{2} - \frac{7}{5} \stackrel{?}{=} \frac{1}{10}$$

$$\frac{15}{10} - \frac{14}{10} \stackrel{?}{=} \frac{1}{10}$$

$$\frac{1}{10} = \frac{1}{10}$$

The solution is 7.

 Note There could not have been any extraneous roots in Examples 2 and 3, because the LCD did not contain a variable.

EXAMPLE 4 Solve $\dfrac{x}{x-3} = \dfrac{3}{x-3} + 4$.

SOLUTION

Step 1. The LCD is $x - 3$. Because 3 makes the denominators $x - 3$ zero, 3 is an excluded value.

Step 2. $(x - 3)\left(\dfrac{x}{x-3}\right) = (x - 3)\left(\dfrac{3}{x-3} + 4\right)$

$$\frac{\overset{1}{(x - 3)}}{1} \cdot \frac{x}{\underset{1}{(x - 3)}} = \frac{\overset{1}{(x - 3)}}{1} \cdot \frac{3}{\underset{1}{(x - 3)}} + \frac{(x - 3)}{1} \cdot \frac{4}{1}$$

$$x = 3 + 4(x - 3)$$

Using the distributive property on the right side

Step 3a. $x = 3 + 4x - 12$

Step 3b. $x = 4x - 9$

Step 4a. $x + (-4x) = 4x - 9 + (-4x)$

$$-3x = -9$$

Step 4b. $\dfrac{-3x}{-3} = \dfrac{-9}{-3}$ or $\left(-\dfrac{1}{3}\right)(-3x) = \left(-\dfrac{1}{3}\right)(-9)$

$$x = 3 \qquad\qquad\qquad x = 3$$

Step 5. Since 3 is an excluded value, it must be rejected. Because 3 was the only apparent solution, the given equation has *no solution*.

 Note If we had failed to look for excluded values, the check would have shown that there was no solution.

✓ **Check**

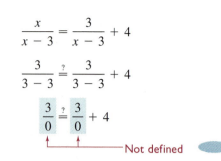

$$\frac{x}{x-3} = \frac{3}{x-3} + 4$$

$$\frac{3}{3-3} \stackrel{?}{=} \frac{3}{3-3} + 4$$

$$\frac{3}{0} \stackrel{?}{=} \frac{3}{0} + 4$$

Not defined

Note It is also possible to solve rational equations as follows:

1. Perform the indicated operations on each side of the equal sign. (You should now have exactly two rational expressions, one on each side of the equal sign.)

2. Find the LCD of the resulting rational expressions, and change both of them to rational expressions with the LCD as the denominator. The equation should now be in the form $\dfrac{P}{Q} = \dfrac{R}{Q}$.

3. Use the fact that if $\dfrac{P}{Q} = \dfrac{R}{Q}$, then $P = R$. (If two rational expressions are equal and if their denominators are equal, then the two numerators must be equal. In other words, set the two numerators equal to each other.)

We do not demonstrate this method.

Exercises 8.8B
Set I

Solve each equation.

1. $\dfrac{x}{3} + \dfrac{x}{4} = 7$

2. $\dfrac{x}{5} + \dfrac{x}{3} = 8$

3. $\dfrac{10}{c} = 2$

4. $\dfrac{6}{z} = 4$

5. $\dfrac{a}{2} - \dfrac{a}{5} = 6$

6. $\dfrac{b}{3} - \dfrac{b}{7} = 12$

7. $\dfrac{9}{2x} = 3$

8. $\dfrac{14}{3x} = 7$

9. $\dfrac{M-2}{5} + \dfrac{M}{3} = \dfrac{1}{5}$

10. $\dfrac{y+2}{4} + \dfrac{y}{5} = \dfrac{1}{4}$

11. $\dfrac{7}{x+4} = \dfrac{3}{x}$

12. $\dfrac{5}{x+6} = \dfrac{2}{x}$

13. $\dfrac{3x}{x-2} = 5$

14. $\dfrac{4}{x+3} = \dfrac{2}{x}$

15. $\dfrac{3x-1}{x-2} = 3 + \dfrac{2x+1}{x-2}$

16. $\dfrac{5x-2}{x-4} = 7 + \dfrac{4x+2}{x-4}$

17. $\dfrac{x}{x^2+1} = \dfrac{2}{1+2x}$

18. $\dfrac{3x}{3x^2+2} = \dfrac{1}{x+1}$

19. $\dfrac{2x-1}{3} + \dfrac{3x}{4} = \dfrac{5}{6}$

20. $\dfrac{3z-2}{4} + \dfrac{3z}{8} = \dfrac{3}{4}$

21. $\dfrac{2(m-3)}{5} - \dfrac{3(m+2)}{2} = \dfrac{7}{10}$

22. $\dfrac{5(x-4)}{6} - \dfrac{2(x+4)}{9} = \dfrac{5}{18}$

23. $\dfrac{x}{x-2} = \dfrac{2}{x-2} + 5$

24. $\dfrac{x}{x+5} = 4 - \dfrac{5}{x+5}$

25. $\dfrac{4}{x^2+x-6} + \dfrac{3}{x^2+5x+6} = \dfrac{23}{6x^2-24}$

26. $\dfrac{5}{x^2+3x-4} + \dfrac{2}{x^2+5x+4} = \dfrac{17}{6x^2-6}$

27. $\dfrac{7}{9x^2-81} - \dfrac{5}{x^2+4x+3} = \dfrac{-10}{63x^2-126x-189}$

28. $\dfrac{5}{2x^2-50} - \dfrac{3}{x^2+4x-5} = \dfrac{11}{8x^2-48x+40}$

Writing Problems

Express the answers in your own words and in complete sentences.

1. Explain how the method used in solving a rational equation differs from the method used in adding two rational expressions.

2. Explain what is meant by an extraneous root.

Exercises 8.8B
Set II

Solve each equation.

1. $\dfrac{x}{5} + \dfrac{x}{2} = 7$

2. $\dfrac{x}{6} + \dfrac{x-6}{3} = 2$

3. $\dfrac{8}{z} = 4$

4. $\dfrac{15}{x-1} = 3$

5. $\dfrac{a}{4} - \dfrac{a}{3} = 1$

6. $\dfrac{x}{3} - \dfrac{x}{5} = 2$

7. $\dfrac{12}{5x} = 2$

8. $\dfrac{3}{x} + \dfrac{1}{2} = \dfrac{6}{x}$

9. $\dfrac{y-1}{2} + \dfrac{y}{5} = \dfrac{3}{10}$

10. $\dfrac{24}{x} = \dfrac{5}{2} + \dfrac{4}{x}$

11. $\dfrac{6}{y-2} = \dfrac{3}{y}$

12. $\dfrac{5}{x-3} = \dfrac{2}{x}$

13. $\dfrac{6x}{x-2} = 8$

14. $\dfrac{3}{x+7} + \dfrac{17}{2x} = \dfrac{1}{x}$

15. $\dfrac{7x-2}{x-1} = 5 + \dfrac{6x-1}{x-1}$

16. $\dfrac{3}{x+1} - \dfrac{2}{x} = \dfrac{5}{2x}$

17. $\dfrac{x}{3x^2+5} = \dfrac{1}{1+3x}$

18. $\dfrac{4x-1}{x-6} = 2 + \dfrac{3x+5}{x-6}$

19. $\dfrac{2z-4}{3} + \dfrac{3z}{2} = \dfrac{5}{6}$

20. $\dfrac{5}{1+x} - \dfrac{5}{x} = \dfrac{1}{x}$

21. $\dfrac{2(m-1)}{5} - \dfrac{3(m+1)}{2} = \dfrac{3}{10}$

22. $\dfrac{5x}{5x^2+2} = \dfrac{1}{x+2}$

23. $\dfrac{y}{y-3} + 10 = \dfrac{3}{y-3}$

24. $\dfrac{x}{x+3} = 2 + \dfrac{3}{x+3}$

25. $\dfrac{5}{x^2+x-12} + \dfrac{7}{x^2+7x+12} = \dfrac{21}{4x^2-36}$

26. $\dfrac{3}{x^2-5x+4} + \dfrac{2}{x^2-16} = \dfrac{5}{5x^2+15x-20}$

27. $\dfrac{3}{5x^2-20} - \dfrac{4}{x^2+3x+2} = \dfrac{-50}{25x^2-25x-50}$

28. $\dfrac{3}{x^2-3x-10} - \dfrac{5}{3x^2-75} = \dfrac{-31}{18x^2+126x+180}$

8.8C Solving Proportions

One special kind of rational equation is called a **proportion**. A proportion is a statement that two ratios or two rates are equal. That is, a proportion is an equation of the form $\dfrac{P}{Q} = \dfrac{S}{T}$, where P, Q, S, and T are polynomials and $Q \neq 0$ and $T \neq 0$. (P, Q, S, and T are called the *terms* of the proportion.) When we multiply both sides of the equation $\dfrac{P}{Q} = \dfrac{S}{T}$ by QT (the LCD), we obtain the new equation $PT = QS$. [Notice that $\dfrac{P}{Q} = \dfrac{S}{T}$ is equivalent to $QT\left(\dfrac{P}{Q}\right) = QT\left(\dfrac{S}{T}\right)$, which is equivalent to $PT = QS$.] When we rewrite $\dfrac{P}{Q} = \dfrac{S}{T}$ as $PT = QS$, we often say we are "cross-multiplying."

The cross-multiplication property

$$\text{If } \frac{P}{Q} = \frac{S}{T}, \quad \text{then} \quad PT = QS.$$

We can cross-multiply as the first step in solving a proportion; however, we must be sure to check for excluded values before we cross-multiply.

EXAMPLE 5   Solve the proportion $\dfrac{4}{7} = \dfrac{3}{z}$ for z.

SOLUTION   We can use the multiplication property of equality, or we can cross-multiply. Zero is an *excluded value*, since $z = 0$ would make a denominator zero.

Method 1

$$\frac{4}{7} = \frac{3}{z} \qquad \text{The LCD is } 7z$$

$$(\overset{1}{7}z)\left(\frac{4}{\underset{1}{7}}\right) = (7\overset{1}{z})\left(\frac{3}{\underset{1}{z}}\right) \qquad \begin{array}{l}\text{Multiplying both}\\\text{sides by } 7z\end{array}$$

Method 2

$$\frac{4}{7} \quad \frac{3}{z}$$

$$4z = 3(7) \qquad \text{Cross-multiplying}$$

Continuing, for either method, we have

$$4z = 21$$

$$\frac{4z}{4} = \frac{21}{4} \qquad \begin{array}{l}\text{Dividing both sides by 4}\\\left(\text{or multiplying both sides by } \tfrac{1}{4}\right)\end{array}$$

$$z = \frac{21}{4}$$

$\frac{21}{4}$ is not an excluded value, and checking will verify that $\frac{21}{4}$ is the solution.

EXAMPLE 6   Solve $\dfrac{3}{t} = -4$.

SOLUTION   This equation *can* be treated like a proportion (see method 2). Zero is an excluded value, since the denominator would be zero if $t = 0$.

Method 1

$$\frac{3}{t} = -4$$

$$\overset{1}{t}\left(\frac{3}{\underset{1}{t}}\right) = t(-4) \qquad \begin{array}{l}\text{Multiplying both sides}\\\text{by } t, \text{ the LCD}\end{array}$$

$$3 = -4t$$

Method 2

$$\frac{3}{t} = \frac{-4}{1} \qquad \begin{array}{l}\text{Rewriting } -4 \text{ as } \dfrac{-4}{1},\\\text{to get a proportion}\end{array}$$

$$3 \cdot 1 = t(-4) \qquad \text{Cross-multiplying}$$

$$3 = -4t$$

Continuing, for either method, and dividing both sides by -4 $\left(\text{or multiplying both sides by } -\tfrac{1}{4}\right)$, we have

$$t = \frac{3}{-4} = -\frac{3}{4}$$

√ **Check** $-\frac{3}{4}$ is not an excluded value, since 0 was the *only* excluded value.

$$\frac{3}{t} = -4$$

$$\frac{3}{-\frac{3}{4}} \stackrel{?}{=} -4 \quad \text{Remember: } \frac{3}{-\frac{3}{4}} = 3 \div \left(-\frac{3}{4}\right) = 3\left(-\frac{4}{3}\right) = -4$$

$$-4 = -4$$

Therefore, the solution is $-\frac{3}{4}$.

EXAMPLE 7

Solve $\dfrac{9x}{x-3} = 6$.

SOLUTION This equation *can* be treated like a proportion (see method 2). Because the denominator is zero if $x = 3$, 3 is an excluded value.

Method 1	*Method 2*
$\dfrac{9x}{x-3} = 6$ The LCD is $x-3$	$\dfrac{9x}{x-3} = \dfrac{6}{1}$ Writing 6 as $\dfrac{6}{1}$ makes the equation a proportion
$(x-3)\left(\dfrac{9x}{x-3}\right) = (x-3)(6)$	$9x \cdot 1 = 6(x-3)$ Cross-multiplying
$9x = 6(x-3)$	$9x = 6(x-3)$

Continuing, for either method, we have

$$9x = 6x - 18 \qquad \text{Removing the parentheses}$$

$$9x - 6x = 6x - 18 - 6x \qquad \text{Adding } -6x \text{ to both sides}$$

$$3x = -18$$

$$\frac{3x}{3} = \frac{-18}{3} \qquad \text{Dividing both sides by 3}$$

$$x = -6$$

√ **Check** -6 is not an excluded value, since 3 was the *only* excluded value.

$$\frac{9x}{x-3} = 6$$

$$\frac{9(-6)}{-6-3} \stackrel{?}{=} 6$$

$$\frac{-54}{-9} \stackrel{?}{=} 6$$

$$6 = 6$$

The solution is -6.

EXAMPLE 8   Solve $\dfrac{x + 1}{x - 2} = \dfrac{7}{4}$.

SOLUTION   Since $x = 2$ will make a denominator zero, 2 is an excluded value.

$$\frac{x + 1}{x - 2} = \frac{7}{4} \qquad \text{This is a proportion}$$

$$4(x + 1) = 7(x - 2) \qquad \substack{\text{Cross-multiplying } or \text{ multiplying} \\ \text{both sides by the LCD}}$$

$$4x + 4 = 7x - 14 \qquad \text{Removing the parentheses}$$

$$4x + 4 \;\boxed{- 7x - 4} = 7x - 14 \;\boxed{- 7x - 4} \qquad \text{Adding } -7x - 4 \text{ to both sides}$$

$$-3x = -18$$

$$\frac{-3x}{-3} = \frac{-18}{-3} \qquad \text{Dividing both sides by } -3$$

$$x = 6$$

✓ **Check**   6 is not an excluded value, since 2 was the *only* excluded value.

$$\frac{x + 1}{x - 2} = \frac{7}{4}$$

$$\frac{6 + 1}{6 - 2} \overset{?}{=} \frac{7}{4}$$

$$\frac{7}{4} = \frac{7}{4}$$

The solution is 6.

The terms of a proportion can be any kind of number, except that no denominator can be zero. In some problems, it is especially easy to *finish* solving the equation by multiplying both sides of the equation by the multiplicative inverse of the coefficient of the variable; we show this method in Examples 9 and 10. In the alternative method shown in Examples 9 and 10, we simplify a complex fraction before cross-multiplying. (The checking will not be shown in Examples 9, 10, and 11.)

EXAMPLE 9   Solve for P: $\dfrac{P}{3} = \dfrac{\frac{5}{6}}{5}$.

SOLUTION   No values of the variable need be excluded.

$$\frac{P}{3} = \frac{\frac{5}{6}}{5}$$

$$5P = (\overset{1}{\cancel{3}})\left(\frac{5}{\underset{2}{\cancel{6}}}\right) \qquad \text{Cross-multiplying}$$

$$5P = \frac{5}{2}$$

$$\frac{1}{5}\,(\overset{1}{\cancel{5}}P) = \left(\frac{1}{5}\right)\left(\frac{\overset{1}{\cancel{5}}}{2}\right) \qquad \substack{\text{Multiplying both} \\ \text{sides by } \frac{1}{5}}$$

$$P = \frac{1}{2}$$

Alternative method

Simplifying $\dfrac{\frac{5}{6}}{5}$, we have

$$\frac{\frac{5}{6}}{5} = \frac{5}{6} \div \frac{5}{1} = \frac{5}{6} \cdot \frac{1}{\underset{1}{\cancel{5}}} = \frac{1}{6}$$

Then

$$\frac{P}{3} = \frac{1}{6} \qquad \text{Substituting } \frac{1}{6} \text{ for } \frac{\frac{5}{6}}{5}$$

$$6P = (3)(1) \qquad \text{Cross-multiplying}$$

$$6P = 3$$

$$\frac{\overset{1}{\cancel{6}}P}{\underset{1}{\cancel{6}}} = \frac{\overset{1}{\cancel{3}}}{\underset{2}{\cancel{6}}} \qquad \substack{\text{Dividing both} \\ \text{sides by } 6}$$

$$P = \frac{1}{2}$$

Checking will verify that the solution is $\frac{1}{2}$.

EXAMPLE 10

Solve $\dfrac{x}{4} = \dfrac{3\frac{1}{2}}{5\frac{1}{4}}$.

SOLUTION   There are no excluded values.

$$\frac{x}{4} = \frac{3\frac{1}{2}}{5\frac{1}{4}}$$

$$\left(5\tfrac{1}{4}\right)x = \left(3\tfrac{1}{2}\right)(4) \qquad \text{Cross-multiplying}$$

$$\tfrac{21}{4}x = \left(\tfrac{7}{2}\right)\overset{2}{(\cancel{4})} \qquad \begin{array}{l}\text{Writing mixed numbers} \\ \text{as improper fractions}\end{array}$$

$$\left(\tfrac{4}{21}\right)\left(\tfrac{21}{4}\right)x = \overset{}{\underset{3}{\left(\tfrac{4}{\cancel{21}}\right)}}\left(\tfrac{\cancel{14}}{1}\right) \qquad \begin{array}{l}\text{Multiplying both} \\ \text{sides by } \tfrac{4}{21}\end{array}$$

$$x = \tfrac{8}{3} \text{ or } 2\tfrac{2}{3}$$

(The problem could also have been finished by using the two-step procedure.)

Checking will verify that the solution is $\frac{8}{3}$, or $2\frac{2}{3}$.

Alternative method

Simplifying $\dfrac{3\frac{1}{2}}{5\frac{1}{4}}$, we have

$$\frac{3\frac{1}{2}}{5\frac{1}{4}} = \frac{\frac{7}{2}}{\frac{21}{4}} = \frac{7}{2} \div \frac{21}{4} = \frac{\overset{1}{\cancel{7}}}{\underset{1}{\cancel{2}}} \cdot \frac{\overset{2}{\cancel{4}}}{\underset{3}{\cancel{21}}} = \frac{2}{3}$$

Then

$$\frac{x}{4} = \frac{2}{3} \qquad \text{Substituting } \tfrac{2}{3} \text{ for } \tfrac{3\frac{1}{2}}{5\frac{1}{4}}$$

$$3x = 4(2) \qquad \text{Cross-multiplying}$$

$$\frac{3x}{3} = \frac{8}{3}, \text{ or } 2\frac{2}{3}$$

EXAMPLE 11

Solve for B: $\dfrac{0.24}{2.7} = \dfrac{4}{B}$.

SOLUTION

$$\frac{0.24}{2.7} = \frac{4}{B}$$

$$\frac{24}{270} = \frac{4}{B} \qquad \begin{array}{l}\text{Multiplying both numerator and denominator of } \tfrac{0.24}{2.7} \\ \text{by 100 to eliminate the decimal points}\end{array}$$

$$24B = 4(270) \qquad \text{Cross-multiplying}$$

$$\frac{24B}{24} = \frac{\overset{1}{\cancel{4}}(\overset{45}{\cancel{270}})}{\underset{\underset{1}{6}}{\cancel{24}}} \qquad \begin{array}{l}\text{We do } not \text{ multiply 270 by 4 } before \text{ we reduce because} \\ \text{the arithmetic is easier if we don't}\end{array}$$

$$B = 45$$

Checking will verify that the solution is 45.

> ✖ **A Word of Caution**   Because a proportion can be solved by cross-multiplying, a common error is to try to cross-multiply when finding a *sum* or *product* of rational expressions. *It is incorrect to cross-multiply when adding or multiplying fractions or rational expressions.*

Incorrect application of cross-multiplication	*Correct application of cross-multiplication*	*Incorrect application of cross-multiplication*
This is a *sum* $$\frac{3}{4} \xcancel{+} \frac{2}{3} = 9 + 8$$	This is an *equation* $$\text{If } \frac{16}{6} = \frac{8}{3},$$	This is a *product* $$\frac{16}{6} \xcancel{\cdot} \frac{8}{3} = \frac{16 \cdot 3}{6 \cdot 8}$$
Correct sum:	then $16 \cdot 3 = 6 \cdot 8$	Correct product:
$$\frac{3}{4} + \frac{2}{3} = \frac{9}{12} + \frac{8}{12} = \frac{17}{12}$$	$$48 = 48$$	$$\frac{16}{6} \cdot \frac{8}{3} = \frac{\overset{8}{\cancel{16}} \cdot 8}{\underset{3}{\cancel{6}} \cdot 3} = \frac{64}{9}$$

Exercises
Set I

Solve each equation.

1. $\dfrac{x}{14} = \dfrac{-3}{7}$ **2.** $\dfrac{x}{12} = \dfrac{-5}{6}$ **3.** $\dfrac{x}{4} = \dfrac{2}{3}$

4. $\dfrac{x}{5} = \dfrac{6}{4}$ **5.** $\dfrac{8}{x} = \dfrac{4}{5}$ **6.** $\dfrac{10}{x} = \dfrac{15}{4}$

7. $\dfrac{4}{7} = \dfrac{x}{21}$ **8.** $\dfrac{15}{12} = \dfrac{x}{9}$ **9.** $\dfrac{100}{x} = \dfrac{40}{30}$

10. $\dfrac{144}{36} = \dfrac{96}{x}$ **11.** $\dfrac{x+1}{x-1} = \dfrac{3}{2}$ **12.** $\dfrac{x+1}{5} = \dfrac{x-1}{3}$

13. $\dfrac{2x+7}{9} = \dfrac{2x+3}{5}$ **14.** $\dfrac{2x+7}{3x+10} = \dfrac{3}{4}$

15. $\dfrac{5x-10}{10} = \dfrac{3x-5}{7}$ **16.** $\dfrac{8x-2}{3x+4} = \dfrac{3}{2}$

17. $\dfrac{\frac{3}{4}}{6} = \dfrac{P}{16}$ **18.** $\dfrac{\frac{2}{5}}{4} = \dfrac{P}{25}$

19. $\dfrac{A}{9} = \dfrac{3\frac{1}{3}}{5}$ **20.** $\dfrac{A}{8} = \dfrac{2\frac{1}{4}}{18}$

21. $\dfrac{7.7}{B} = \dfrac{3.5}{5}$ **22.** $\dfrac{6.8}{B} = \dfrac{17}{57.4}$

23. $\dfrac{P}{100} = \dfrac{\frac{3}{2}}{15}$ **24.** $\dfrac{P}{100} = \dfrac{\frac{7}{5}}{35}$

25. $\dfrac{12\frac{1}{2}}{100} = \dfrac{A}{48}$ **26.** $\dfrac{16\frac{2}{3}}{100} = \dfrac{9}{B}$

Exercises 8.8C
Set II

Solve each equation.

1. $\dfrac{x}{15} = \dfrac{-2}{5}$ **2.** $\dfrac{x}{-18} = \dfrac{7}{-6}$ **3.** $\dfrac{x}{15} = \dfrac{6}{5}$

4. $\dfrac{x+2}{5} = \dfrac{5-x}{8}$ **5.** $\dfrac{81}{x} = \dfrac{9}{5}$ **6.** $\dfrac{4}{13} = \dfrac{16}{x}$

7. $\dfrac{18}{28} = \dfrac{x}{14}$ **8.** $\dfrac{x}{100} = \dfrac{75}{125}$ **9.** $\dfrac{26}{x} = \dfrac{39}{14}$

10. $\dfrac{x}{18} = \dfrac{24}{30}$ **11.** $\dfrac{x+5}{x-5} = \dfrac{27}{10}$ **12.** $\dfrac{x+8}{x-8} = 5$

13. $\dfrac{4x+3}{2} = \dfrac{4x+1}{3}$ **14.** $\dfrac{3x-7}{x-5} = 2$

15. $\dfrac{2x+5}{3} = \dfrac{3x-1}{2}$ **16.** $\dfrac{15}{22} = \dfrac{x}{33}$

17. $\dfrac{\frac{2}{3}}{4} = \dfrac{P}{12}$ **18.** $\dfrac{\frac{5}{6}}{\frac{1}{6}} = \dfrac{z}{2}$

19. $\dfrac{A}{16} = \dfrac{2\frac{1}{2}}{10}$ **20.** $\dfrac{1.2}{2} = \dfrac{x}{2.4}$

21. $\dfrac{P}{100} = \dfrac{\frac{3}{4}}{15}$ **22.** $\dfrac{8}{\frac{1}{2}} = \dfrac{x}{4}$

23. $\dfrac{P}{100} = \dfrac{12\frac{1}{2}}{50}$ **24.** $\dfrac{9}{y} = \dfrac{3\frac{1}{3}}{\frac{1}{6}}$

25. $\dfrac{6\frac{1}{4}}{100} = \dfrac{1}{B}$ **26.** $\dfrac{3.1}{x} = \dfrac{10}{2.5}$

8.8D Solving Rational Equations That Simplify to Second-Degree Equations

In this section, the equation that remains after we remove denominators and grouping symbols from the rational equation will usually be a *second-degree equation* (a polynomial equation whose highest-degree term is second-degree), and it can be solved by factoring. The suggestions in the following box are based on using the addition, subtraction, multiplication, and division properties of equality.

Solving a rational equation that simplifies to a second-degree equation

1. Find the LCD, and find all excluded values.

2. Remove denominators by multiplying *both sides of the equation* (that is, by multiplying *every term*) by the LCD.

3. **a.** Remove all grouping symbols.
   b. Collect and combine like terms on each side of the equal sign.

4. If there are second-degree terms, solve by these quadratic methods:
   a. Move all nonzero terms to one side of the equal sign by adding the same expression to both sides. Only zero must remain on the other side. Then arrange the terms in descending powers.
   b. Factor the polynomial. (*If the polynomial cannot be factored, we cannot solve the equation at this time.* Recheck the work; a mistake has probably been made.)
   c. Set each factor equal to zero, and solve each resulting equation.

5. Reject any apparent solutions that are excluded values.

6. Check any other apparent solutions in the original equation.

 Note   If the equation from step 3 is a first-degree equation, solve it by using the methods discussed earlier.

EXAMPLE 12   Solve $\dfrac{2}{x} + \dfrac{3}{x^2} = 1$.

SOLUTION

Step 1.   The LCD is x^2. Zero is an excluded value, since $x = 0$ makes the denominators zero.

Step 2. $x^2 \left(\dfrac{2}{x} + \dfrac{3}{x^2} \right) = x^2 \,(1)$ Multiplying both sides by x^2

$\dfrac{x^2}{1} \left(\dfrac{2}{x} \right) + \dfrac{x^2}{1} \left(\dfrac{3}{x^2} \right) = x^2$ Using the distributive property on the left side

Step 3. $2x + 3 = x^2$ ⟵ A second-degree term

Step 4a. $2x + 3 - 2x - 3 = x^2 - 2x - 3$ Adding $-2x - 3$ to both sides

$0 = x^2 - 2x - 3$

Step 4b. $0 = (x - 3)(x + 1)$ Factoring the polynomial

Step 4c.

$x - 3 = 0$	$x + 1 = 0$	Setting each factor equal to zero
$x - 3 + 3 = 0 + 3$	$x + 1 - 1 = 0 - 1$	
$x = 3$	$x = -1$	

Step 5.   We do not reject either value, since the only excluded value is 0.

✓ **Step 6.**   ***Check for $x = 3$***   ***Check for $x = -1$***

$\dfrac{2}{x} + \dfrac{3}{x^2} = 1$	$\dfrac{2}{x} + \dfrac{3}{x^2} = 1$
$\dfrac{2}{3} + \dfrac{3}{3^2} \overset{?}{=} 1$	$\dfrac{2}{-1} + \dfrac{3}{(-1)^2} \overset{?}{=} 1$
$\dfrac{2}{3} + \dfrac{1}{3} \overset{?}{=} 1$	$-2 + 3 \overset{?}{=} 1$
$1 = 1$	$1 = 1$

The solutions are 3 and -1.

EXAMPLE 13 Solve $\dfrac{8}{x} = \dfrac{3}{x+1} + 3$.

SOLUTION

Step 1. The LCD is $x(x+1)$. Excluded values are 0 and -1.

Step 2.
$$x(x+1)\left(\frac{8}{x}\right) = x(x+1)\left(\frac{3}{x+1} + 3\right)$$

$$\left(\frac{x(x+1)}{1}\right)\left(\frac{8}{x}\right) = \left(\frac{x(x+1)}{1}\right)\left(\frac{3}{x+1}\right) + \left(\frac{x(x+1)}{1}\right)\left(\frac{3}{1}\right)$$

$$8(x+1) = 3x + 3x(x+1)$$

A second-degree term

Step 3a. $8x + 8 = 3x + 3x^2 + 3x$

Step 3b. $8x + 8 = 3x^2 + 6x$ Combining like terms

Step 4a. $8x + 8 - 8x - 8 = 3x^2 + 6x - 8x - 8$ Adding $-8x - 8$ to both sides

$0 = 3x^2 - 2x - 8$ Simplifying

Step 4b. $0 = (3x + 4)(x - 2)$ Factoring the polynomial

Setting each factor equal to 0

Step 4c.
$3x + 4 = 0$ | $x - 2 = 0$
$3x + 4 - 4 = 0 - 4$ | $x - 2 + 2 = 0 + 2$
$3x = -4$ | $x = 2$
$\dfrac{3x}{3} = \dfrac{-4}{3}$
$x = -\frac{4}{3}$

Step 5. We do not reject either value, since the only excluded values are 0 and -1.

Step 6. Checking will verify that $-\frac{4}{3}$ and 2 are the solutions.

✖ **A Word of Caution** When we multiply both sides of an *equation* by the LCD, the denominators are removed completely. When we add or subtract rational expressions, the denominators *cannot* be removed. A common error is to confuse an *addition* (or subtraction) *problem* such as $\dfrac{2}{x} + \dfrac{3}{x^2}$ with solving an *equation* such as $\dfrac{2}{x} + \dfrac{3}{x^2} = 1$.

The addition problem

Each rational expression is converted to an equivalent rational expression with the LCD as its denominator.

$$\frac{2}{x} + \frac{3}{x^2} \quad \text{LCD} = x^2$$

$$= \frac{2 \cdot x}{x \cdot x} + \frac{3}{x^2} = \frac{2x}{x^2} + \frac{3}{x^2} = \frac{2x+3}{x^2}$$

Here, the result is a rational expression that represents the sum of the given rational expressions.

The usual mistake is to multiply both terms of the *sum* by the LCD.

$$\frac{2}{x} + \frac{3}{x^2} = \frac{x^2}{1} \cdot \frac{2}{x} + \frac{x^2}{1} \cdot \frac{3}{x^2}$$

$$= 2x + 3 \neq \frac{2}{x} + \frac{3}{x^2}$$

The sum has been multiplied by x^2 and therefore no longer has its original value.

The equation

Both sides are multiplied by the LCD to *remove* the denominators.

$$\frac{2}{x} + \frac{3}{x^2} = 1 \quad \text{LCD} = x^2$$

$$\frac{x^2}{1} \cdot \frac{2}{x} + \frac{x^2}{1} \cdot \frac{3}{x^2} = x^2 \quad (1)$$

$$2x \quad + \quad 3 \quad = x^2$$

We solved this equation by factoring in Example 12. Here, the result is two numbers (-1 and 3) that are solutions of the given equation.

Exercises 8.8D
Set I

Solve each equation.

1. $z + \frac{1}{z} = \frac{17}{z}$

2. $y + \frac{3}{y} = \frac{12}{y}$

3. $\frac{2}{x} - \frac{2}{x^2} = \frac{1}{2}$

4. $\frac{3}{x} - \frac{4}{x^2} = \frac{1}{2}$

5. $\frac{x}{x+1} = \frac{4x}{3x+2}$

6. $\frac{x}{3x-4} = \frac{3x}{2x+2}$

7. $\frac{5}{x} - 1 = \frac{x+11}{x}$

8. $\frac{7}{x} - 1 = \frac{x+15}{x}$

9. $\frac{1}{x-1} + \frac{2}{x+1} = \frac{5}{3}$

10. $\frac{2}{3x+1} + \frac{1}{x-1} = \frac{7}{10}$

11. $\frac{3}{2x+5} + \frac{x}{4} = \frac{3}{4}$

12. $\frac{5}{2x-1} - \frac{x}{6} = \frac{4}{3}$

13. $\frac{4}{x+1} - \frac{3}{x} = \frac{1}{15}$

14. $\frac{1}{x+1} - \frac{3}{x} = \frac{1}{2}$

15. $\frac{6}{x^2-9} = \frac{1}{x-3} - \frac{1}{5}$

16. $\frac{6-x}{x^2-4} = \frac{x}{x+2} + 2$

Exercises 8.8D
Set II

Solve each equation.

1. $x + \frac{1}{x} = \frac{10}{x}$

2. $y + \frac{2}{y} = \frac{18}{y}$

3. $\frac{5}{x} - \frac{1}{x^2} = \frac{9}{4}$

4. $\frac{1}{2} - \frac{1}{2x} = \frac{6}{x^2}$

5. $\frac{2x}{3x+1} = \frac{4x}{5x+1}$

6. $\frac{x}{8} + \frac{x}{2} = \frac{10}{x}$

7. $\frac{8}{x} - 1 = \frac{x+18}{x}$

8. $\frac{5}{x-3} + \frac{1}{6} = \frac{7}{x-2}$

9. $\frac{4}{x+1} = \frac{3}{x} + \frac{1}{15}$

10. $\frac{6}{x+2} - 1 = \frac{x-4}{x+2}$

11. $\frac{1}{x-2} - \frac{4}{x+2} = \frac{1}{5}$

12. $\frac{1}{x-5} + \frac{3}{x+2} = \frac{5}{6}$

13. $\frac{4}{3x-1} - 1 = \frac{2}{x}$

14. $\frac{3}{1-2x} = \frac{2x+1}{x-2}$

15. $\frac{x+4}{x^2-16} + \frac{7}{x-4} = -1$

16. $\frac{x+25}{x^2-25} + \frac{12x}{x-5} = 1$

8.9   Solving Literal Equations

Literal equations are equations that contain more than one variable.

EXAMPLE 1   Examples of literal equations:

a. $3x + 4y = 12$ This is an equation in two variables. We might be asked to solve it for x or for y.

b. $\dfrac{4ab}{d} = 15$ This is an equation in three variables. We might be asked to solve it for a, for b, or for d.

c. $A = P(1 + rt)$ This is an equation in four variables. We might be asked to solve it for P, for r, or for t. (It has already been solved for A.)

Generally, when we solve a literal equation for one of its variables, the solution will contain the other variables as well as constants. We must isolate the variable we are solving for; that is, that variable must appear only once, all by itself, on one side of the equal sign. All other variables and all constants must be on the other side. The suggestions in the following box are based on the addition, subtraction, multiplication, and division properties of equality.

Solving a literal equation

1. Remove denominators (if there are any) by multiplying both sides of the equation by the LCD.

2. Remove any grouping symbols.

3. On each side of the equal sign, collect and combine any like terms.

4. Move all terms that contain the variable you are solving for to one side of the equal sign and all other terms to the other side.

5. Factor out the variable you are solving for (if it appears in more than one term).

6. Divide both sides of the equation by the coefficient of the variable you are solving for (or multiply both sides of the equation by the multiplicative inverse of that coefficient).

In Example 2a, we will compare solving a literal equation with solving a "regular" equation, so that you can see how similar the methods are.

EXAMPLE 2   Solve the given equation for the indicated variable.

a. Solve $3x + 32 = 12$ for x, and solve $3x + 4y = 12$ for x.

SOLUTION   In the final equation, x must appear only once, by itself, on one side of the equal sign. All other variables and all constants must be on the other side of the equal sign.

Exercises 8.9
Set I

Solve each equation for the
the semicolon.

1. $2x + y = 4; x$

3. $y - z = -8; z$

5. $2x - y = -4; y$

7. $2x - 3y = 6; x$

9. $2(x - 3y) = x + 4; x$

11. $PV = k; V$

13. $I = Prt; P$

15. $p = 2l + 2w; l$

Writing Pro

Express the answer in you
sentences.

1. Suppose that a student ha
equation for y and gives t
plain why this equation h

Exercises 8.9
Set II

Solve each equation for the
the semicolon.

1. $x + 2y = 5; x$

3. $x - y = -4; y$

5. $2x - y = -4; x$

7. $3x - 4y = 12; y$

9. $3(x + 2y) = x + 2; x$

11. $d = rt; t$

13. $I = Prt; r$

"Regular equation"		*Literal equation*
$3x + 32 = 12$		$3x + 4y = 12$

	Getting the x-term by itself on one side of the equal sign	
$3x + 32 - 32 = 12 - 32$		$3x + 4y - 4y = 12 - 4y$
$3x = 12 - 32$		$3x = 12 - 4y$

	Dividing both sides of the equation by 3	
$\dfrac{3x}{3} = \dfrac{12 - 32}{3}$		$\dfrac{3x}{3} = \dfrac{12 - 4y}{3}$
$x = \dfrac{12 - 32}{3}$		$x = \dfrac{12 - 4y}{3}$

$$x = -\frac{20}{3}$$ ← These equations have both been solved for x

Notice that in both cases we have *isolated* x.

b. Solve $3x + 4y = 12$ for y.

S O L U T I O N In the final equation, y must appear only once, by itself, on one side of the equal sign. All other variables and all constants must be on the other side of the equal sign.

$$3x + 4y = 12$$

$$-3x + 3x + 4y = -3x + 12 \qquad \text{Getting the } y\text{-term by itself on one side of the equal sign}$$

$$4y = -3x + 12$$

$$\frac{4y}{4} = \frac{-3x + 12}{4} \qquad \text{Dividing both sides of the equation by 4}$$

$$y = \frac{-3x + 12}{4} \qquad \text{This equation has been solved for } y$$

or $\qquad y = -\dfrac{3}{4}x + 3 \qquad$ We will see *this* form of an equation in x and y later, when we discuss writing equations of straight lines

c. Solve $\dfrac{4ab}{d} = 15$ for a.

S O L U T I O N In the final equation, a must appear only once, by itself, on one side of the equal sign. All other variables and all constants must be on the other side of the equal sign. The LCD is d.

$$\frac{4ab}{d} = 15$$

$$\overset{1}{d}\left(\frac{4ab}{\underset{1}{d}}\right) = (15)\, d \qquad \text{Multiplying both sides by } d$$

$$4ab = 15d \qquad \text{Simplifying}$$

$$\frac{4ab}{4b} = \frac{15d}{4b} \qquad \text{Dividing both sides by } 4b \text{ to get } a \text{ by itself on the left}$$

$$a = \frac{15d}{4b} \qquad \text{This equation has been solved for } a$$

8.10 Solving Applied Problems That Involve Rational Expressions

In this section, we discuss applied problems that involve rational expressions. All the kinds of applied problems discussed in previous chapters can lead to equations that contain rational expressions. The general method of solving such problems is the one given earlier, except that when we solve the equation, we will have to "clear fractions."

EXAMPLE 1   When the speed of the wind is 25 mph, a certain airplane flying *with* the wind can fly 875 mi in the same length of time that it takes that plane to fly 625 mi when it's flying *against* the wind. Find the speed of the plane in still air.

SOLUTION

Step 1. Let $\quad x =$ the speed of the plane in still air (in mph)
Then $x + 25 =$ the speed when it's flying *with* the wind (in mph)
and $\quad x - 25 =$ the speed when it's flying *against* the wind (in mph)

Think We use the formula $r \cdot t = d$. If we solve this formula for t (time), we get $t = \dfrac{d}{r}$. Therefore, the time during which the plane flies *with* the wind is $\dfrac{875}{x + 25}$, and the time during which the plane flies *against* the wind is $\dfrac{625}{x - 25}$.

Reread

The time flying with the wind	equals	the time flying against the wind

Step 2. $\qquad \dfrac{875}{x + 25} \qquad = \qquad \dfrac{625}{x - 25}$

Multiplying both sides by $(x + 25)(x - 25)$, the LCD

Step 3. $\overset{1}{(x + 25)}(x - 25) \left(\dfrac{875}{x + 25} \right) = (x + 25)\overset{1}{(x - 25)} \left(\dfrac{625}{x - 25} \right)$

$(x - 25)(875) = (x + 25)(625)$

$875x - 21{,}875 = 625x + 15{,}625$ This is a first-degree equation

$-625x + 875x - 21{,}875 = -625x + 625x + 15{,}625$ Adding $-625x$ to both sides

$250x - 21{,}875 = 15{,}625$

$250x - 21{,}875 + 21{,}875 = 15{,}625 + 21{,}875$ Adding 21,875 to both sides

$250x = 37{,}500$

$\dfrac{250x}{250} = \dfrac{37{,}500}{250}$ Dividing both sides by 250

Step 4. $\qquad\qquad\qquad\qquad x = 150$, the speed of the plane in still air

✓ **Step 5.** ***Check*** The speed flying *with* the wind is $(150 + 25)$ mph, or 175 mph; the *time* flying with the wind is $\frac{875}{175}$ mph, or 5 hr. The speed flying *against* the wind is $(150 - 25)$ mph, or 125 mph; the *time* flying against the wind is $\frac{625}{125}$ mph, or 5 hr. The times are equal.

Step 6. Therefore, the speed of the plane in still air is 150 mph.

Note In step 3 of Example 1, if we had divided both sides of the equation by 25, we would have had smaller numbers to work with, and a calculator would not have been needed.

EXAMPLE 2   The denominator of a fraction exceeds the numerator by 8. If 2 is added to the numerator and 4 is subtracted from the denominator, the value of the resulting fraction is $\frac{5}{6}$. What is the original fraction?

SOLUTION   In this example, we must let x equal the numerator *or* the denominator of the fraction—*not* the original fraction.

Step 1. Let x = the numerator of the original fraction
Then $x + 8$ = the denominator of the original fraction

Also, $\dfrac{x}{x + 8}$ = the original fraction

and $\dfrac{x + 2}{(x + 8) - 4}$ = the resulting (new) fraction

Reread   The value of the resulting fraction **is** $\frac{5}{6}$

Step 2. $\dfrac{x + 2}{(x + 8) - 4} = \dfrac{5}{6}$

Step 3. $\dfrac{x + 2}{x + 4} = \dfrac{5}{6}$ This is a proportion

$6(x + 2) = 5(x + 4)$ Cross-multiplying *or* multiplying both sides by the LCD

$6x + 12 = 5x + 20$ Using the distributive property

$6x + 12 - 5x - 12 = 5x + 20 - 5x - 12$ Adding $-5x - 12$ to both sides

$x = 8$ The numerator

Step 4. The numerator is 8, and the denominator is $x + 8 = 8 + 8 = 16$. Therefore, the *apparent* solution is that the original fraction is $\frac{8}{16}$.

✓ **Step 5. Check**   The denominator exceeds the numerator by 8. If we add 2 to the numerator and subtract 4 from the denominator of $\frac{8}{16}$, we have

$$\frac{8 + 2}{16 - 4} = \frac{10}{12} = \frac{5}{6}$$

Step 6. Therefore, the original fraction is $\frac{8}{16}$.

EXAMPLE 3   Yen drove for 270 mi at a constant rate. If her rate had been 9 mph faster, the trip would have taken 1 hr less. Find Yen's actual rate.

SOLUTION   We solve the formula $r \cdot t = d$ for t: $t = \dfrac{d}{r}$.

Step 1. Let r = Yen's actual rate (the slower rate) in miles per hour
Then $r + 9$ = the faster rate in miles per hour

Also, $\dfrac{270}{r}$ = Yen's time (in hours) at the slower rate $\left(t = \dfrac{d}{r} \right)$

and $\dfrac{270}{r + 9}$ = the time (in hours) if she goes faster $\left(t = \dfrac{d}{r} \right)$

Reread   Number of hours at slower rate **equals** number of hours at faster rate **plus** 1 hr

Step 2. $\dfrac{270}{r} = \dfrac{270}{r + 9} + 1$

Step 3.

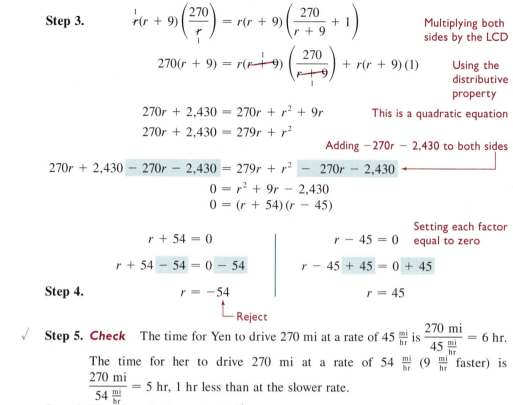

$$\overset{1}{r}(r + 9)\left(\frac{270}{r}\right) = r(r + 9)\left(\frac{270}{r + 9} + 1\right) \qquad \text{Multiplying both sides by the LCD}$$

$$270(r + 9) = r(\cancel{r + 9})\left(\frac{270}{\cancel{r + 9}}\right) + r(r + 9)(1) \qquad \text{Using the distributive property}$$

$$270r + 2{,}430 = 270r + r^2 + 9r \qquad \text{This is a quadratic equation}$$

$$270r + 2{,}430 = 279r + r^2$$

Adding $-270r - 2{,}430$ to both sides

$$270r + 2{,}430 \;\underline{-\; 270r \;-\; 2{,}430} = 279r + r^2 \;\underline{-\; 270r \;-\; 2{,}430}$$

$$0 = r^2 + 9r - 2{,}430$$

$$0 = (r + 54)(r - 45)$$

Setting each factor equal to zero

$r + 54 = 0$	$r - 45 = 0$
$r + 54 \underline{- 54} = 0 \underline{- 54}$	$r - 45 \underline{+ 45} = 0 \underline{+ 45}$

Step 4.

$r = -54$	$r = 45$

Reject

✓ **Step 5. Check**   The time for Yen to drive 270 mi at a rate of $45 \frac{\text{mi}}{\text{hr}}$ is $\dfrac{270 \text{ mi}}{45 \frac{\text{mi}}{\text{hr}}} = 6$ hr.

The time for her to drive 270 mi at a rate of $54 \frac{\text{mi}}{\text{hr}}$ $(9 \frac{\text{mi}}{\text{hr}}$ faster) is $\dfrac{270 \text{ mi}}{54 \frac{\text{mi}}{\text{hr}}} = 5$ hr, 1 hr less than at the slower rate.

Step 6. Therefore, Yen's rate is $45 \frac{\text{mi}}{\text{hr}}$, or 45 mph.

Solving Applied Problems That Lead to Proportions

Some applied problems are easily solved by setting up a proportion. The procedure for solving such problems is described in the following box.

Solving applied problems that lead to proportions

1. Represent the unknown quantity by x, and use the given conditions to form two ratios or two rates.

2. Form a proportion by setting the two ratios or rates equal to each other, being sure to put the *units* next to the numbers when you write the proportion. Be sure the units occupy corresponding positions in the two ratios (or rates) of the proportion. For example (if the problem involves miles and hours):

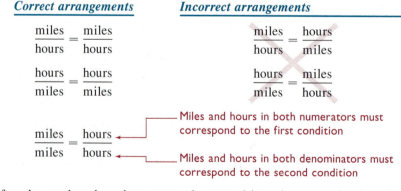

Correct arrangements   *Incorrect arrangements*

$$\frac{\text{miles}}{\text{hours}} = \frac{\text{miles}}{\text{hours}} \qquad\qquad \frac{\text{miles}}{\text{hours}} = \frac{\text{hours}}{\text{miles}}$$

$$\frac{\text{hours}}{\text{miles}} = \frac{\text{hours}}{\text{miles}} \qquad\qquad \frac{\text{hours}}{\text{miles}} = \frac{\text{miles}}{\text{hours}}$$

$$\frac{\text{miles}}{\text{miles}} = \frac{\text{hours}}{\text{hours}}$$

Miles and hours in both numerators must correspond to the first condition

Miles and hours in both denominators must correspond to the second condition

3. After the numbers have been correctly entered into the proportion by using the units as a guide, drop the units.

4. Solve the equation for x.

We will not show the checks or Step 1, Step 2, and so forth in Examples 4 and 5.

EXAMPLE 4

The scale on an architectural drawing is stated as "1 inch equals 8 feet." What are the dimensions of a room that measures $3\frac{1}{2}$ in. by 4 in. on the drawing?

SOLUTION There are really two unknowns. We must find the *width* of the room (it corresponds to $3\frac{1}{2}$ in.) and the *length* of the room (it corresponds to 4 in.).

Let x = the width of the room (in feet).

$$\frac{1 \text{ in.}}{8 \text{ ft}} = \frac{3\frac{1}{2} \text{ in.}}{x \text{ ft}} \qquad \text{Setting the two rates equal to each other}$$

$$\frac{1}{8} = \frac{3\frac{1}{2}}{x} \qquad \text{Dropping the units}$$

$$x = 8\left(3\frac{1}{2}\right) = 8\left(\frac{7}{2}\right) = 28 \qquad \text{Cross-multiplying}$$

Let y = the length of the room (in feet).

$$\frac{1 \text{ in.}}{8 \text{ ft}} = \frac{4 \text{ in.}}{y \text{ ft}}$$

$$\frac{1}{8} = \frac{4}{y} \qquad \text{Dropping the units}$$

$$y = 8(4) = 32 \qquad \text{Cross-multiplying}$$

Therefore, the room is 28 ft by 32 ft.

EXAMPLE 5

Jim knows that he can drive 406 mi on 14 gal of gasoline. At this rate, how far can he expect to drive on 35 gal of gasoline?

SOLUTION Let x = the number of miles Jim can drive on 35 gal of gasoline.

$$\frac{x \text{ mi}}{35 \text{ gal}} = \frac{406 \text{ mi}}{14 \text{ gal}} \qquad \text{Setting the two rates equal to each other}$$

$$\frac{x}{35} = \frac{406}{14} \qquad \text{Dropping the units}$$

$$14x = (406)(35) \qquad \text{Cross-multiplying}$$

$$x = \frac{(406)(35)}{14} = 1{,}015$$

Therefore, Jim can expect to drive 1,015 mi on 35 gal of gasoline.

Solving Work Problems

Work problems are similar to rate-time-distance problems. One basic relationship used to solve work problems is the following.

Rate $\times$ Time = Amount of work done

In symbols: $r \cdot t = w$

If we know the rate and the time, we can find the amount of work done. If we know the amount of work done in a certain length of time, we can find the rate. If we know the rate and the amount of work done, we can find the time.

EXAMPLE 6

If machine A produces brackets at the rate of thirty-five brackets per hour for 7 hr, how many brackets has it produced?

SOLUTION We use the formula $w = rt$.

$$w = 35 \ \frac{\text{brackets}}{\text{hr}} \times 7 \ \cancel{\text{hr}} = 245 \text{ brackets}$$

EXAMPLE 7

Machine B produces sprocket wheels at the rate of 175 sprocket wheels per hour. How long does it take the machine to produce 805 sprocket wheels?

SOLUTION Because $r \cdot t = w$, $t = \dfrac{w}{r}$. Therefore,

$$t = \frac{805 \text{ sprocket wheels}}{175 \ \dfrac{\text{spr. wh.}}{\text{hr}}} = 4.6 \text{ hr, or 4 hr 36 min}$$

The other basic relationship used to solve work problems is the following:

$$\begin{pmatrix} \text{Amount A} \\ \text{does} \\ \text{in time } x \end{pmatrix} + \begin{pmatrix} \text{Amount B} \\ \text{does} \\ \text{in time } x \end{pmatrix} = \begin{pmatrix} \text{Amount done} \\ \text{together} \\ \text{in time } x \end{pmatrix}$$

EXAMPLE 8

George can build a fence in 6 days. Brian can do that same job in 4 days. (a) What is George's rate? (b) What is Brian's rate? (c) How long would it take them to build the same fence if they worked together?

SOLUTION

a. Because rate × time = work $(r \cdot t = w)$, $r = \dfrac{w}{t}$. Therefore, George's rate is $\dfrac{1 \text{ fence}}{6 \text{ days}} = \dfrac{1}{6} \ \dfrac{\text{fence}}{\text{day}}$.

b. Brian's rate is $\dfrac{1 \text{ fence}}{4 \text{ days}} = \dfrac{1}{4} \ \dfrac{\text{fence}}{\text{day}}$.

c. Step 1. Let x = the number of days needed for Brian and George together to build the fence.

Think	George's rate	times	George's time	equals	portion of the fence that George builds
	$\dfrac{1 \text{ fence}}{6 \ \cancel{\text{day}}}$	$\cdot$	$x \ \cancel{\text{days}}$	$=$	$\dfrac{x}{6}$ fence

	Brian's rate	times	Brian's time	equals	portion of the fence that Brian builds
	$\dfrac{1 \text{ fence}}{4 \ \cancel{\text{day}}}$	$\cdot$	$x \ \cancel{\text{days}}$	$=$	$\dfrac{x}{4}$ fence

Amount George builds in x days	plus	amount Brian builds in x days	equals	amount they build together in x days

Step 2. $\dfrac{x}{6}$ $+$ $\dfrac{x}{4}$ $=$ 1 I fence is built

Step 3. $12\left(\dfrac{x}{6}+\dfrac{x}{4}\right)=12\,(1)$ Multiplying both sides by 12, the LCD

$$12\left(\dfrac{x}{6}\right)+12\left(\dfrac{x}{4}\right)=12$$ Using the distributive property

$$2x+3x=12$$

$$5x=12$$

$$\dfrac{5x}{5}=\dfrac{12}{5}$$

Step 4. $x=\dfrac{12}{5}$,   or $2\dfrac{2}{5}$

✓ **Step 5. Check**

George's work: $\dfrac{1}{6}\dfrac{\text{fence}}{\text{day}}\cdot\dfrac{12}{5}\text{ days}=\dfrac{2}{5}\text{ fence}$

Brian's work: $\dfrac{1}{4}\dfrac{\text{fence}}{\text{day}}\cdot\dfrac{12}{5}\text{ days}=\dfrac{3}{5}\text{ fence}$

Together: $\dfrac{2}{5}\text{ fence}+\dfrac{3}{5}\text{ fence}=1\text{ fence}$

Step 6. Therefore, it would take George and Brian $2\frac{2}{5}$ days to build the fence if they worked together.

EXAMPLE 9

In a film-processing lab, machine A can process 5,400 ft of film in 60 min. (a) What is the rate of machine A? (b) How long does it take machine B to process 4,400 ft of film if the two machines working together can process 14,240 ft in 80 min?

SOLUTION

a. The rate of machine A is $\dfrac{5{,}400\text{ ft}}{60\text{ min}}=90\,\dfrac{\text{ft}}{\text{min}}$.

b. Step 1. Let $x=$ the number of minutes for machine B to process 4,400 ft of film.

$$\dfrac{4{,}400\text{ ft}}{x\text{ min}}=\text{rate of machine B}$$

Reread

Number of feet A processes in 80 min	+	number of feet B processes in 80 min	=	number of feet they process together in 80 min

Rate · Time   Rate · Time

Step 2. $\left(90\,\dfrac{\text{ft}}{\text{min}}\right)(80\text{ min})+\left(\dfrac{4400}{x}\,\dfrac{\text{ft}}{\text{min}}\right)(80\text{ min})=14{,}240\text{ ft}$

$$7{,}200\quad+\quad\dfrac{352{,}000}{x}\quad=\quad14{,}240$$

Step 3. $x\left(7{,}200 + \dfrac{352{,}000}{x}\right) = x\,(14{,}240)$ Multiplying both sides by x, the LCD

$$x\,(7{,}200) + x\left(\dfrac{352{,}000}{x}\right) = 14{,}240x$$

$$7{,}200x + 352{,}000 = 14{,}240x$$

$$7{,}200x + 352{,}000 - 7{,}200x = 14{,}240x - 7{,}200x \qquad \text{\color{red}Adding } -7{,}200x \text{ to both sides}$$

$$352{,}000 = 7{,}040x$$

$$\dfrac{352{,}000}{7{,}040} = \dfrac{7{,}040x}{7{,}040}$$

Step 4. $x = 50$

✓ **Step 5. Check** If machine B processes 4,400 ft of film in 50 min, its rate is $\dfrac{4{,}400}{50}\dfrac{\text{ft}}{\text{min}}$, or $88\dfrac{\text{ft}}{\text{min}}$. The amount of film machine B processes in 80 min is $\left(88\dfrac{\text{ft}}{\text{m\cancel{in}}}\right)(80\ \cancel{\text{min}}) = 7{,}040$ ft. Machine A processes $\left(90\dfrac{\text{ft}}{\text{m\cancel{in}}}\right)(80\ \cancel{\text{min}}) = 7{,}200$ ft. The number of feet processed by machine A in 80 min plus the number of feet processed by machine B in 80 min (7,200 ft + 7,040 ft) is 14,240 ft.

Step 6. Therefore, it takes machine B 50 min to process 4,400 ft of film.

Exercises 8.10
Set I

Set up each problem algebraically, solve, and check. Be sure to state what your variables represent. (Note: These problems do not necessarily lead to equations involving fractions.)

1. The speed of the current of a river is 5 mph. If a boat can travel 999 mi with the current in the same time it could travel 729 mi against the current, what is the speed of the boat in still water?

2. The speed of the current in a river is 4 mph. If a boat can travel 288 mi with the current in the same time it could travel 224 mi against the current, what is the speed of the boat in still water?

3. The denominator of a fraction exceeds the numerator by 6. If 4 is added to the numerator and subtracted from the denominator, the resulting fraction equals $\frac{11}{10}$. What is the original fraction?

4. The denominator of a fraction exceeds the numerator by 4. If 6 is subtracted from the numerator and added to the denominator, the resulting fraction equals $\frac{5}{13}$. What is the original fraction?

5. The denominator of a fraction is twice the numerator. If 5 is added to the numerator and subtracted from the denominator, the value of the resulting fraction is $\frac{4}{5}$. What is the original fraction?

6. The denominator of a fraction is 3 times the numerator. If 5 is added to the numerator and subtracted from the denominator, the value of the resulting fraction is $\frac{1}{2}$. What is the original fraction?

7. The sum of a number and its reciprocal is $\frac{29}{10}$. Find the number.

8. The sum of a number and its reciprocal is $\frac{130}{33}$. Find the number.

For Exercises 9 and 10, use this information: The scale in an architectural drawing is 1 inch equals 8 feet.

9. Find the dimensions of a room that measures $2\frac{1}{2}$ in. by 3 in. on the drawing.

10. Find the dimensions of a room that measures $3\frac{1}{4}$ in. by $4\frac{1}{4}$ in. on the drawing.

11. The ratio of a person's weight on earth to that person's weight on the moon is 6:1. How much would a 150-lb person weigh on the moon?

12. The ratio of a person's weight on Mars to that person's weight on earth is 2:5. How much would a 196-lb person weigh on Mars?

13. The ratio of the weight of lead to the weight of an equal volume of aluminum is 21:5. If an aluminum bar weighs 150 lb, what would a lead bar of the same size weigh?

14. The ratio of the weight of platinum to the weight of an equal volume of copper is 12:5. If a platinum bar weighs 18 lb, what would a copper bar of the same size weigh?

15. On the first $6\frac{1}{2}$ hr of their shift, a fire crew built twenty-six chains of fire line. How much fire line can they build in the remainder of a 10-hr day if they continue to work at the same pace?

16. A city agency informed a local businesswoman that for every $5,000 worth of sales she made, her business license fee would be $15. If the woman's sales totaled $125,000 for the year, how much was her license fee?

17. A meat packer paid a cattle producer $3,420 for nine steers. Assuming that the rate remains unchanged, how much will the producer receive for sixteen steers of the same approximate weight?

18. A hog producer fed 1,320 hogs until they reached a marketing weight of 100 kg each. He received a check for their sale amounting to $59,400. How much did the producer receive per hog? What was the price paid for the hogs per kilogram of live weight?

19. Tricia can do a job in 8 hr. Mike can do the same job in 10 hr. How long will it take them to do the same job if they work together?

20. Joyce can paint a house in 5 days. Fred can paint the same house in 7 days. How long will it take them to paint the same house if they work together?

21. The sum of the numerator and denominator of a fraction is 42. If 2 is added to the numerator and 6 is added to the denominator, the resulting fraction is $\frac{2}{3}$. What is the original fraction?

22. The sum of the numerator and denominator of a fraction is 40. If 3 is added to the numerator and 2 is added to the denominator, the resulting fraction is $\frac{4}{5}$. What is the original fraction?

23. Ruth can proofread 220 pages of a deposition in 4 hr. How long does it take Jerry to proofread 210 pages if, when they work together, they can proofread 291 pages in 3 hr?

24. Barbara can type 95 pages of a manuscript in 3 hr. How long does it take Mike to type 110 pages if he and Barbara working together can type 322 pages in 6 hr?

25. Jim and Sandra live 63 mi apart. Both left their homes at 7 A.M. by bicycle, riding toward each other. They met at 10 A.M. If Jim's average speed was three-fourths of Sandra's, how fast did each cycle?

26. Rebecca and Jill live 75 mi apart. Both left their homes at 8 A.M. by bicycle, riding toward each other. They met at 10:30 A.M. If Jill's average speed was seven-eights of Rebecca's, how fast did each cycle?

27. Two numbers differ by 12. One-sixth the larger exceeds one-fifth the smaller by 2. Find the numbers.

28. Two numbers differ by 8. One-seventh the larger exceeds one-eighth the smaller by 2. Find the numbers.

29. Machine A, working at a constant rate, produces 24 widgets in a certain number of minutes. Machine B produces 2 more widgets per minute than machine A does. If it takes machine A 2 min more than machine B to produce 24 widgets, what is the rate of machine A?

30. A gardener's helper, working at a constant rate, can trim 30 ft of hedge in a certain number of minutes. The gardener trims hedges 1 ft per minute faster than her helper. If it takes the helper 1 min more than the gardener to trim 30 ft of hedge, what is the rate of the helper?

31. A cyclist traveled for 90 mi at a constant rate. If her rate had been 3 mph faster, the trip would have taken 1 hr less. Find the actual rate of the cyclist.

32. A skiboat traveled (in still water) for 120 mi at a constant rate. If the boat had gone 10 mph faster, the same trip would have taken 2 hr less. Find the actual rate of the boat.

Exercises 8.10
Set II

Set up each problem algebraically, solve, and check. Be sure to state what your variables represent. (Note: These problems do not necessarily lead to equations involving fractions.)

1. The speed of the current in a river is 5 mph. If a boat can travel 198 mi with the current in the same time it could travel 138 mi against the current, what is the speed of the boat in still water?

2. Eleanore paddled a kayak downstream for 3 hr. After having lunch, she paddled upstream for 5 hr. At that time she was still 6 mi short of getting back to her starting point. If the speed of the river was 2 mph, how fast did Eleanore's kayak move in still water? How far downstream did she travel?

3. The denominator of a fraction exceeds the numerator by 20. If 7 is added to the numerator and subtracted from the denominator, the resulting fraction equals $\frac{6}{7}$. Find the fraction.

4. Suppose the wind speed is 30 mph. If an airplane can fly 240 mi against the wind in the same time that it can fly 420 mi with the wind, what is the speed of the plane in still air?

5. The denominator of a fraction is twice the numerator. If 2 is added to the numerator and subtracted from the denominator, the value of the resulting fraction is $\frac{5}{7}$. What is the original fraction?

6. When a number is subtracted from its reciprocal, the difference is $\frac{21}{10}$. What is the number?

7. The sum of a number and its reciprocal is $\frac{34}{15}$. Find the number.

8. Mrs. Summers drove her motorboat upstream a certain distance while pulling her son Brian on a water ski. She returned to the starting point pulling her other son Derek. The round trip took 25 min of skiing time. On both legs of the trip, the speedometer read 30 mph. If the speed of the current is 6 mph, how far upstream did she travel?

For Exercises 9 and 10, use this information: The scale in an architectural drawing is 1 inch equals 8 feet.

9. Find the dimensions of a room that measures $2\frac{1}{4}$ in. by $3\frac{3}{4}$ in. on the drawing.

10. Find the dimensions of a room that measures $3\frac{1}{8}$ in. by $4\frac{3}{8}$ in. on the drawing.

11. An apartment house manager spent 22 hr painting three apartments. How much time can she expect to spend painting the remaining fifteen apartments?

12. Ralph drove 420 mi in $\frac{3}{4}$ of a day. At that same rate, about how far can he drive in $2\frac{1}{2}$ days?

13. The ratio of the width of a rectangle to its length is 4:9. If its area is 900 sq. m, find the width and the length.

14. A crew of ten men spent one week overhauling 25 trucks in a fleet of 100 trucks. How many men would need to work to complete the fleet overhaul in one additional week?

15. A car burns $2\frac{1}{2}$ qt of oil on a 1,800-mi trip. How many quarts of oil can the owner expect to use on a 12,000-mi trip?

16. Fifteen defective axles were found in 100,000 cars of a particular model. How many defective axles would you expect to find in the 2 million cars made of that same model?

17. Mr. Sanders has 300 Leghorn hens and needs 22.86 cm of roosting space per hen. How many meters of roosting space does he need?

18. The Forest Service must determine how many acres will be needed to make an addition of twenty-two campsites. If the existing 24-acre campground accommodates fifty-five campsites, how many additional acres will be needed?

19. Karla can do a job in 9 hr. Mark can do the same job in 11 hr. How long will it take them to do the same job if they work together?

20. Ben bought 120 stamps of three types—20¢, 15¢, and 3¢—at a total cost of $12.80. He bought twice as many 20¢ stamps as 15¢ stamps and 20 more 3¢ stamps than 20¢ stamps. How many of each kind did he buy?

21. The sum of the numerator and denominator of a fraction is 68. If 1 is added to the numerator and 12 is added to the denominator, the resulting fraction is $\frac{2}{7}$. What is the original fraction?

22. The sum of two numbers is 32. One-half the smaller exceeds one-third the larger by 1. Find the numbers.

23. David can type 65 pages of a manuscript in 2 hr. How long does it take Susan to type 450 pages if, when they work together, they can type 210 pages in 3 hr?

24. The sum of the first two of three consecutive odd integers added to the sum of the last two is 60. Find the integers.

25. Kim and Jerry live 80 mi apart. Both left their homes at 6 A.M. by bicycle, riding toward each other. They met at 10 A.M. If Kim's average speed was two-thirds of Jerry's, how fast did each cycle?

26. It takes Jim 3 times as long as Sherry to paint a certain house. Working together, Jim and Sherry could paint the same house in 3 days.
  a. How long would it take Sherry working alone to paint the house?
  b. How long would it take Jim working alone to paint the house?

27. Two numbers differ by 3. One-fifth the larger exceeds one-sixth the smaller by 1. Find the numbers.

28. Machine A takes 3 times as long as machine B to do a certain job. Both machines running together can do this same job in 4 hr. How long does it take each machine working alone to do the job?

29. Machine A, working at a constant rate, produces 36 widgets in a certain number of minutes. Machine B produces 3 more widgets per minute than machine A does. If it takes machine A 2 min more than machine B to produce 36 widgets, what is the rate of machine A?

30. A motorboat traveled (in still water) for 80 mi at a constant rate. If the boat had gone 4 mph faster, the same trip would have taken 1 hr less. Find the actual rate of the boat.

31. A jogger traveled for 48 mi at a constant rate. If she had been able to go 2 mph faster, she could have covered the same distance in 2 hr less. Find her actual rate.

32. A cyclist traveled for 80 mi at a constant rate. If his rate had been 6 mph slower, the same trip would have taken 3 hr longer. Find the actual rate of the cyclist.

8.11   Solving Variation Problems (Optional)

A **variation** is an equation that relates one variable to one or more other variables by means of multiplication, division, or both.

8.11A   Direct Variation

In a **direct variation**, one variable is some constant multiple of the other variable. For example, we can say "y varies directly as x" if there is some constant, k, such that $y = kx$. We call k the constant of variation, or, more commonly, the constant of proportionality.

The following statements *all* translate to the equation $y = kx$:

y varies directly as (or *with*) x

y varies as (or *with*) x

y is directly proportional to x

y is proportional to x

In a direct variation, as one variable *increases*, the other increases also, and as one variable decreases, the other decreases also.

For example, if someone works at a constant wage per hour (k), then the amount of money (A) that that person earns in a week is directly proportional to the number of hours (h) worked during the week. That is, $A = kh$, where k is the constant of proportionality. (In this case, k is the hourly rate of pay.) As the number of hours worked increases, the amount of money earned also increases.

If the number of miles per gallon of gasoline obtained by a certain car is a constant, then the number of miles driven in that car is directly proportional to the number of gallons of gasoline used. That is, if m is the number of miles driven and if g is the number of gallons of gasoline used, the equation of variation is $m = kg$.

EXAMPLE 1 y varies directly as x. If $y = 10$ when $x = 2$, (a) find k, the constant of proportionality, and (b) find y when $x = 3$.

SOLUTION   The variation equation is $y = kx$.

a. $10 = k(2)$ Substituting 10 for y and 2 for x

$$\frac{10}{2} = \frac{2k}{2}$$ Dividing both sides by 2

$5 = k$, or $k = 5$

b. $y = 5x$ Substituting 5 for k

$y = 5(3)$ Substituting 3 for x

$y = 15$

You might try some other values for x in Example 1 in order to see that as x increases, y increases also, and as x decreases, y decreases also.

EXAMPLE 2

The circumference, C, of a circle varies directly as its diameter, d. If $C \approx 12.56$ when $d = 4$, (a) find k, the constant of proportionality, and (b) find C when $d = 7$.

SOLUTION The equation of variation is $C = kd$.

a. $12.56 \approx k(4)$ Substituting 12.56 for C and 4 for d

$\dfrac{12.56}{4} \approx \dfrac{4k}{4}$ Dividing both sides by 4

$3.14 \approx k,$ or $k \approx 3.14$

b. $C \approx 3.14d$ Substituting 3.14 for k

     $C \approx 3.14(7)$ Substituting 7 for d

     $C \approx 21.98$

You might verify that as the diameter of a circle increases, the circumference increases also, and as the diameter of a circle decreases, the circumference decreases also.

Note There are several proportions associated with each direct variation problem, and the proportion enables us to find the missing variable *without* finding k. In Example 1, one possible proportion is $\dfrac{10}{2} = \dfrac{y}{3}$. If we solve this proportion for y, we get $y = 15$. In Example 2, one possible proportion is $\dfrac{12.56}{4} \approx \dfrac{C}{7}$. Solving for C, we get $C \approx 21.98$.

The relationship between x and y in a variation is not always linear. That is, y might be directly proportional to the *square* of x, or to the *cube* of x, and so forth.

EXAMPLE 3

y varies directly with the cube of x. If $y = 40$ when $x = 2$, (a) find k, the constant of proportionality, and (b) find y when $x = -3$.

SOLUTION The equation of variation is $y = kx^3$.

a. $40 = k(2^3)$ Substituting 40 for y and 2 for x

    $40 = 8k$

    $\dfrac{40}{8} = \dfrac{8k}{8}$ Dividing both sides by 8

     $5 = k,$ or $k = 5$

b. $y = 5x^3$ Substituting 5 for k

    $y = 5(-3)^3$ Substituting -3 for x

    $y = 5(-27) = -135$

EXAMPLE 4

Henry's gross salary for last week was $165, and he was paid for working 22 hr. His salary, S, for the week is directly proportional to h, the number of hours he worked during the week. (a) Write the variation equation. (b) Find k, the constant of proportionality. (c) Find how much Henry would have made if he had worked 27 hr.

SOLUTION

a. The variation equation is $S = kh$.

b. $165 = k(22)$ Substituting 165 for S and 22 for h

$\dfrac{165}{22} = \dfrac{22k}{22}$ Dividing both sides by 22

$7.5 = k$,   or $k = 7.5$ (Henry's hourly rate of pay is $7.50)

c. $S = 7.5h$ Substituting 7.5 for k

$S = 7.5(27)$ Substituting 27 for h

$S = 202.50$

Therefore, Henry would have earned $202.50 if he had worked 27 hr.

Exercises 8.11A
Set I

1. y varies directly as x. **(a)** Write the variation equation. If $y = 14$ when $x = 2$, **(b)** find k, the constant of proportionality, and **(c)** find y when $x = 5$.

2. z varies directly as w. **(a)** Write the variation equation. If $z = 28$ when $w = 4$, **(b)** find k, the constant of proportionality, and **(c)** find z when $w = 7$.

3. s varies directly as t. **(a)** Write the variation equation. If $s = -9$ when $t = -3$, **(b)** find k, the constant of proportionality, and **(c)** find s when $t = -12$.

4. u varies directly as v. **(a)** Write the variation equation. If $u = -6$ when $v = 3$, **(b)** find k, the constant of proportionality, and **(c)** find u when $v = -18$.

5. A varies directly as the square of s. **(a)** Write the variation equation. If $A = -20$ when $s = -2$, **(b)** find k, the constant of proportionality, and **(c)** find A when $s = 5$.

6. R varies directly as the square of x. **(a)** Write the variation equation. If $R = 90$ when $x = -3$, **(b)** find k, the constant of proportionality, and **(c)** find R when $x = 10$.

7. The *change* in the length l of a spring is directly proportional to the force F applied to the spring. **(a)** Write the variation equation. **(b)** If a 5-lb force stretches a spring 3 in., find the change in the length when a 2-lb force is applied.

8. If the number of miles per gallon (of gasoline) obtained by a certain car is a constant, then the number of miles driven m in that car is directly proportional to the number of gallons of gasoline g used. **(a)** Write the equation of variation. **(b)** If a car can go 252 mi on 14 gal of gasoline, find how many miles that car could go on 23 gal of gasoline.

9. The distance s through which an object falls (neglecting air resistance, etc.) varies directly as the square of the time t. **(a)** Write the equation of variation. **(b)** If an object falls 64 ft in 2 sec, how far will it fall in 3 sec?

10. The air resistance R on a car varies directly as the square of the car's velocity v. **(a)** Write the equation of variation. **(b)** If the air resistance is 400 lb at 60 mph, find the air resistance at 90 mph.

Exercises 8.11A
Set II

1. y varies directly as x. **(a)** Write the variation equation. If $y = 36$ when $x = 4$, **(b)** find k, the constant of proportionality, and **(c)** find y when $x = 3$.

2. u varies directly as v. **(a)** Write the variation equation. If $u = -28$ when $v = 2$, **(b)** find k, the constant of proportionality, and **(c)** find u when $v = -3$.

3. t varies directly as s. **(a)** Write the variation equation. If $t = -18$ when $s = -6$, **(b)** find k, the constant of proportionality, and **(c)** find t when $s = -4$.

4. w varies directly as t. **(a)** Write the variation equation. If $w = -6$ when $t = -\frac{1}{3}$, **(b)** find k, the constant of proportionality, and **(c)** find w when $t = 9$.

5. d varies directly as the square of s. **(a)** Write the variation equation. If $d = 12$ when $s = 2$, **(b)** find k, the constant of proportionality, and **(c)** find d when $s = 5$.

6. y varies directly as the cube of x. **(a)** Write the variation equation. If $y = 108$ when $x = 3$, **(b)** find k, the constant of proportionality, and **(c)** find y when $x = 2$.

7. In a business, if the price of an item is kept fixed, the revenue R varies directly as the number of items sold n. **(a)** Write the variation equation. **(b)** If the revenue is $14,000 when 700 items are sold, what is the revenue if 800 items are sold?

8. Stella's weekly salary S is directly proportional to the number of hours h she works during the week. **(a)** Write the variation equation. **(b)** If she made $116.40 for working 24 hr, how many hours would she have to work in order to earn $169.75?

9. The resistance R of a boat moving through water varies directly as the square of its speed s. **(a)** Write the variation equation. **(b)** If the resistance of a boat is 50 lb at a speed of 10 knots, what is the resistance at 20 knots?

10. In a business, if the price of an item is kept fixed, the revenue R varies directly as the number of items sold n. **(a)** Write the variation equation. **(b)** If the revenue is $12,000 when 800 items are sold, how many items must be sold for the revenue to be $15,000?

8.IIB Inverse Variation

In **inverse variation**, as one variable *increases*, the other *decreases*. We can say "y varies inversely as (or *with*) x" or "y is inversely proportional to x" if there is some constant, k, such that $y = \dfrac{k}{x}$. (Notice that x is in the *denominator* of the rational expression $\dfrac{k}{x}$.)

For example, suppose that for some reason, the area of a rectangle must be held constant, let's say at 72 sq. cm. Then the width w of the rectangle varies inversely as the length l. The variation equation would be $w = \dfrac{k}{l}$. That is, as the length of the rectangle increases (perhaps from 12 cm to 18 cm), the width must decrease (from 6 cm to 4 cm). (You might verify this.)

EXAMPLE 5 y varies inversely as x. If $y = 6$ when $x = 2$, (a) find k, the constant of proportionality, and (b) find y when $x = 3$.

SOLUTION   The variation equation is $y = \dfrac{k}{x}$.

a. $6 = \dfrac{k}{2}$ Substituting 6 for y and 2 for x

$2(6) = 2\left(\dfrac{k}{2}\right)$ Multiplying both sides by 2

$12 = k,$ or $k = 12$

b. $y = \dfrac{12}{x}$ Substituting 12 for k

$y = \dfrac{12}{3}$ Substituting 3 for x

$y = 4$

EXAMPLE 6 y is inversely proportional to the square of x. If $y = -3$ when $x = 4$, (a) find k, the constant of proportionality, and (b) find y when $x = -6$.

SOLUTION   The variation equation is $y = \dfrac{k}{x^2}$.

a. $-3 = \dfrac{k}{4^2}$ Substituting -3 for y and 4 for x

$16(-3) = 16\left(\dfrac{k}{16}\right)$ Multiplying both sides by 16

$-48 = k,$ or $k = -48$

b. $y = \dfrac{-48}{x^2}$ Substituting -48 for k

$y = \dfrac{-48}{(-6)^2}$ Substituting -6 for x

$y = \dfrac{-48}{36} = -\dfrac{4}{3}$

EXAMPLE 7   Under certain conditions, the pressure P of a gas varies inversely as the volume V.
(a) Write the variation equation. If $P = 30$ when $V = 500$, (b) find k, the constant of
proportionality, and (c) find P when $V = 200$.

SOLUTION

a. The variation equation is $P = \dfrac{k}{V}$.

b. $30 = \dfrac{k}{500}$ Substituting 30 for P and 500 for V

$500(30) = 500\left(\dfrac{k}{500}\right)$ Multiplying both sides by 500

$15{,}000 = k,$ or $k = 15{,}000$

c. $P = \dfrac{15{,}000}{V}$ Substituting 15,000 for k

$P = \dfrac{15{,}000}{200}$ Substituting 200 for V

$P = 75$

Exercises   8.11B
Set I

1. y varies inversely as x. **(a)** Write the variation equation.
If $y = 3$ when $x = 2$, **(b)** find k, the constant of proportionality, and **(c)** find y when $x = -10$.

2. z varies inversely as w. **(a)** Write the variation equation.
If $z = 2$ when $w = -3$, **(b)** find k, the constant of proportionality, and **(c)** find z when $w = \frac{1}{2}$.

3. s is inversely proportional to t. **(a)** Write the variation
equation. If $s = 3$ when $t = -2$, **(b)** find k, the constant
of proportionality, and **(c)** find s when $t = -5$.

4. u is inversely proportional to v. **(a)** Write the variation
equation. If $u = -\frac{1}{2}$ when $v = 16$, **(b)** find k, the constant of proportionality, and **(c)** find u when $v = 3$.

5. F varies inversely with the square of d. **(a)** Write the
variation equation. If $F = 3$ when $d = -4$, **(b)** find k,
the constant of proportionality, and **(c)** find F when
$d = 8$.

6. C varies inversely with the square of v. **(a)** Write the variation equation. If $C = 6$ when $v = -3$, **(b)** find k, the constant of proportionality, and **(c)** find C when $v = 6$.

7. Under certain conditions, the volume V of a gas varies inversely as its pressure P. **(a)** Write the variation equation. **(b)** If the volume is 1,600 cc at a pressure of 250 mm of mercury, what is the volume at a pressure of 400 mm of mercury?

8. The time t that it takes to get a certain job done varies inversely as the rate r at which the work is done. **(a)** Write the variation equation. **(b)** If it takes 14 hr to get a certain job done when the rate is $6 \dfrac{\text{jobs}}{\text{hr}}$, how long will it take to get that job done when the rate is $7 \dfrac{\text{jobs}}{\text{hr}}$?

9. Sound intensity I (loudness) is inversely proportional to the square of the distance d from the source of the sound. **(a)** Write the variation equation. **(b)** If a pneumatic drill has a sound intensity of 75 decibels at 50 ft, find its sound intensity at 150 ft.

10. The intensity I of light received from a light source varies inversely as the square of the distance d from the source. **(a)** Write the variation equation. **(b)** If the light intensity is 15 candela (or *candles*) at a distance of 10 ft from the light source, what is the light intensity at a distance of 15 ft?

Exercises 8.11B
Set II

1. y varies inversely as x. **(a)** Write the variation equation. If $y = -4$ when $x = -5$, **(b)** find k, the constant of proportionality, and **(c)** find y when $x = 10$.

2. P varies inversely as V. **(a)** Write the variation equation. If $P = 3$ when $V = 5$, **(b)** find k, the constant of proportionality, and **(c)** find P when $V = 12$.

3. R is inversely proportional to x. **(a)** Write the variation equation. If $R = 18$ when $x = 2$, **(b)** find k, the constant of proportionality, and **(c)** find R when $x = -6$.

4. T is inversely proportional to s. **(a)** Write the variation equation. If $T = -4$ when $s = 3$. **(b)** find k, the constant of proportionality, and **(c)** find T when $s = 5$.

5. L varies inversely with the square of r. **(a)** Write the variation equation. If $L = 16$ when $r = -3$, **(b)** find k, the constant of proportionality, and **(c)** find L when $r = 4$.

6. R varies inversely with the square of d. **(a)** Write the variation equation. If $R = 10$ when $d = 2$, **(b)** find k, the constant of proportionality, and **(c)** find R when $d = 15$.

7. Under certain conditions, the pressure P of a gas varies inversely as its volume V. **(a)** Write the variation equation. **(b)** If the pressure is 15 lb per square inch when the volume is 350 cu. in., find the pressure when the volume is 70 cu. in.

8. The rate r at which work is done varies inversely as the time t in which the work is done. **(a)** Write the variation equation. **(b)** If it takes 14 hr to get a certain job done when the rate is $6 \dfrac{\text{jobs}}{\text{hr}}$, what will the rate be when it takes 8 hr to get the same job done?

9. The gravitational attraction F between two bodies varies inversely as the square of the distance d separating them. **(a)** Write the variation equation. **(b)** If the attraction measures 36.5 units when the distance is 3.92 cm, find the attraction when the distance is 81.7 cm. (Round off the answer to three decimal places.)

10. The electrical resistance R of a wire varies inversely as the square of the diameter d of the wire. **(a)** Write the variation equation. **(b)** If the resistance of a wire that has a diameter of 0.0201 in. is 3.85 ohms, find the resistance of a wire of the same length and material that has a diameter of 0.0315 in. (Round off the answer to two decimal places.)

Sections 8.8–8.11

REVIEW

Excluded Values
8.8

Any value of the variable that would make any denominator zero must be *excluded*.

To Solve a Rational Equation
8.8

1. Find the LCD, and find all excluded values.

2. Remove denominators by multiplying *both sides of the equation* (that is, by multiplying *every term*) by the LCD.

3. a. Remove all grouping symbols.
   b. Collect and combine like terms on each side of the equal sign.

First-degree equations	*Second-degree equations*
4. a. Move all terms containing the variable to one side of the equal sign and all other terms to the other side. **b.** Divide both sides of the equation by the coefficient of the variable, or multiply both sides of the equation by the reciprocal of that coefficient.	**a.** Move *all* nonzero terms to one side of the equal sign. *Only zero must remain on the other side.* Then arrange the terms in descending powers. **b.** Factor the polynomial. **c.** Set each factor equal to zero, and solve each resulting equation.

5. Reject any apparent solutions that are excluded values.

6. Check any other apparent solutions in the original equation.

To Solve a Proportion
8.8

A **proportion** is a statement that two ratios or two rates are equal.

To solve a proportion, we can multiply both sides by the LCD, or we can "cross-multiply." That is, we can use this property: If $\dfrac{P}{Q} = \dfrac{S}{T}$, then $PT = QS$.

Literal Equations
8.9

Literal equations are equations that contain more than one variable.

To solve a literal equation, proceed in the same way used to solve an equation with a single variable. The variable we are solving for must appear only once, all by itself, on one side of the equal sign.

To Solve Applied Problems Using Proportions
8.10

1. Represent the unknown quantity by a variable, and use the given conditions to form two ratios or two rates.

2. Form a proportion by setting the two ratios or rates equal to each other; be sure the units occupy corresponding positions in the two ratios (or rates) of the proportion. Put the units next to the numbers when writing the proportion to be sure the proportion is set up correctly.

3. Drop the units.

4. Solve the equation for the variable.

Variation
8.11

A **variation** is an equation that relates one variable to one or more other variables by means of multiplication, division, or both.

In **direct variation**, one variable is some constant multiple of the other variable, and as one variable *increases*, the other increases also. The statements "*y* varies directly as (or *with*) *x*," "*y* varies as (or *with*) *x*," "*y* is directly proportional to *x*," and "*y* is proportional to *x*" are all statements of direct variation, and they *all* translate to the equation $y = kx$, where k is the constant of proportionality.

In **inverse variation**, as one variable *increases*, the other *decreases*. We can say "*y* varies inversely as (or *with*) *x*" or "*y* is inversely proportional to *x*" if there is some constant, k, such that $y = \dfrac{k}{x}$.

Sections 8.8–8.11   REVIEW EXERCISES   Set I

In Exercises 1 and 2, determine what value (or values) of the variables must be excluded.

1. $\dfrac{x-2}{x^2-9}$ **2.** $\dfrac{x+4}{x^3+x^2-2x}$

In Exercises 3–9, solve each equation.

3. $\dfrac{6}{m}=5$ **4.** $x-\dfrac{3x}{5}=2$

5. $\dfrac{z}{5}-\dfrac{z}{8}=3$ **6.** $\dfrac{3}{2}=\dfrac{3x+4}{5x-1}$

7. $\dfrac{4}{2z}+\dfrac{2}{z}=1$ **8.** $\dfrac{4}{x^2}-\dfrac{3}{x}=\dfrac{5}{2}$

9. $\dfrac{7}{2x-1}+\dfrac{1}{18}=\dfrac{x}{6}$

In Exercises 10–13, solve for the variable indicated after the semicolon.

10. $2x-7y=14;\ y$ **11.** $\dfrac{2m}{n}=P;\ n$

12. $V=\tfrac{1}{3}Bh;\ B$ **13.** $\dfrac{F-32}{C}=\dfrac{9}{5};\ C$

In Exercises 14–19, set up each problem algebraically, solve, and check. Be sure to state what your variables represent.

14. The numerator of a fraction exceeds the denominator by 4. If 3 is added to the denominator and subtracted from the numerator, the value of the resulting fraction is $\frac{7}{8}$. Find the original fraction.

15. It takes Mr. Maxwell 30 min to drive to work in the morning, but it takes him 45 min to return home over the same route during the evening rush hour. If his average morning speed is 10 mph faster than his average evening speed, how far is it from his home to his work?

16. The sum of a number and its reciprocal is $\frac{13}{6}$. Find the number.

17. Machine A can do a job in 6 hr. How long does it take machine B to do the same job if, when the two machines work together, they get the job done in 4 hr?

18. Justin drove 324 mi in three-fifths of a day. At that same rate, about how far can he drive in a day and a half?

19. Alan weaves $2\frac{1}{2}$ yd of fabric in five-eighths of a day. At that same rate, how many yards of fabric can he weave in half a day?

For those who studied Section 8.11, which was optional:

20. y varies directly as the cube of x. **(a)** Write the variation equation. If $y=54$ when $x=3$, **(b)** find k, the constant of proportionality, and **(c)** find y when $x=-4$.

21. y varies inversely as the square of x. **(a)** Write the variation equation. If $y=5$ when $x=3$, **(b)** find k, the constant of proportionality, and **(c)** find y when $x=5$.

Name _____

In Exercises 1 and 2, determine what value (or values) of the variables must be excluded.

1. $\dfrac{x+2}{x^2+2x-3}$

2. $\dfrac{x-2}{2x^2+x-3}$

In Exercises 3–9, solve each equation.

3. $\dfrac{3}{x}=4$

4. $\dfrac{x}{3}-\dfrac{x}{2}=2$

5. $\dfrac{x+2}{5}+\dfrac{2x}{3}=3$

6. $\dfrac{3x}{7}=\dfrac{x-1}{5}$

7. $\dfrac{x+2}{-2}=\dfrac{3}{x-3}$

8. $\dfrac{17}{6x}+\dfrac{5}{2x^2}=\dfrac{2}{3}$

9. $\dfrac{3}{x}-\dfrac{8}{x^2}=\dfrac{1}{4}$

In Exercises 10–13, solve for the variable indicated after the semicolon.

10. $\dfrac{P}{V}=C$; V

11. $V=lwh$; h

12. $V^2=2gS$; S

13. $5(x-2y)=14+3(2x-y)$; y

ANSWERS

1. _____

2. _____

3. _____

4. _____

5. _____

6. _____

7. _____

8. _____

9. _____

10. _____

11. _____

12. _____

13. _____

In Exercises 14–19, set up each problem algebraically, solve, and check. Be sure to state what your variables represent.

14. When a number is subtracted from its reciprocal, the difference is $\frac{40}{21}$. Find the number.

15. When the speed of the wind was 25 mph, a certain airplane could fly only 480 mi against the wind in the same time it could fly 780 mi with the wind. Find the speed of the plane in still air.

16. The denominator of a fraction exceeds the numerator by 7. If 10 is added to the numerator and 15 is subtracted from the denominator, the value of the resulting fraction is $\frac{11}{5}$. Find the original fraction.

17. Rebecca can paint a certain house in 5 days. How long would it take Karla alone to paint that house if the two women, working together, could paint the house in 3 days?

18. Jill drove 220 mi in two-fifths of a day. At that same rate, about how far can she drive in half a day?

19. Susan crochets $3\frac{1}{2}$ scarves in $2\frac{1}{3}$ days. At that same rate, how many scarves can she crochet in 6 days?

For those who studied Section 8.11, which was optional:

20. y varies directly as the square of x. **(a)** Write the variation equation. If $y = 75$ when $x = -5$, **(b)** find k, the constant of proportionality, and **(c)** find y when $x = -3$.

21. y varies inversely as the cube of x. **(a)** Write the variation equation. If $y = 4$ when $x = -2$, **(b)** find k, the constant of proportionality, and **(c)** find y when $x = 3$.

14. _____

15. _____

16. _____

17. _____

18. _____

19. _____

20a. _____

b. _____

c. _____

21a. _____

b. _____

c. _____

Chapter 8   DIAGNOSTIC TEST

The purpose of this test is to see how well you understand operations involving rational expressions and solving rational equations. We recommend that you work this diagnostic test *before* your instructor tests you on this chapter. Allow yourself about 50 minutes.

Complete solutions for all the problems on this test, together with section references, are given in the answer section in the back of this book. We suggest that you study the sections referred to for the problems you do incorrectly.

1. Determine whether $\dfrac{7x}{5y}$ and $\dfrac{7x-2}{5y-2}$ are equivalent rational expressions.

2. Reduce each rational expression to lowest terms.

   a. $\dfrac{x^2-9}{x^2-6x+9}$ b. $\dfrac{4x^2-23xy+15y^2}{20x^2y-15xy^2}$

3. Find each missing term.

   a. $-\dfrac{-8}{3}=\dfrac{-8}{?}$ b. $\dfrac{3}{x-5}=\dfrac{?}{5-x}$

In Problems 4–9, perform the indicated operations, reducing all answers to lowest terms.

4. $\dfrac{a}{a^2+5a+6}\cdot\dfrac{4a+8}{6a^3+18a^2}$ 5. $\dfrac{6x}{x-2}-\dfrac{3}{x-2}$

6. $\dfrac{2x}{x^2-9}\div\dfrac{4x^2}{x^2-6x+9}$ 7. $\dfrac{b}{b-1}-\dfrac{b+1}{b}$

8. $\dfrac{x+3}{x^2-10x+25}+\dfrac{x}{2x^2-7x-15}$

9. $\dfrac{x}{x+5}-\dfrac{x}{x-5}-\dfrac{50}{x^2-25}$

In Problems 10 and 11, simplify each complex fraction.

10. $\dfrac{\frac{6x^4}{11y^2}}{\frac{9x}{22y^4}}$ 11. $\dfrac{\frac{6}{x}+\frac{15}{x^2}}{2+\frac{5}{x}}$

12. What value(s) of the variable must be excluded, if any, in each of the following expressions?

   a. $\dfrac{3x}{x+4}$ b. $\dfrac{5x-4}{x^2+4x}$

In Problems 13–16, solve each equation.

13. $\dfrac{x+7}{3}=\dfrac{2x-1}{4}$ 14. $x-\dfrac{4}{x}=3$

15. $\dfrac{3}{4x+2}+\dfrac{x+1}{6}=1$ 16. $\dfrac{3x-5}{x}=\dfrac{x+1}{3}$

17. Solve for y: $5x+7y=18$.

In Problems 18–20, set up each problem algebraically, solve, and check. Be sure to state what your variables represent.

18. The denominator of a fraction is 6 more than the numerator. If 1 is subtracted from the numerator and 7 is added to the denominator, the value of the resulting fraction is $\frac{1}{3}$. What is the original fraction?

19. Mrs. Ames drove 21 mi in $\frac{2}{5}$ hr. At that same rate, how far can she drive in $2\frac{2}{3}$ hr?

20. David can type 128 pages of a manuscript in 4 hr. How long does it take Kenneth to type 111 pages if both men working together can type 345 pages in 5 hr?

Chapters 1–8   CUMULATIVE REVIEW EXERCISES

1. Evaluate $10-(3\sqrt{4}-5^2)$.

2. Using the formula $V=\dfrac{25}{8}\left(\dfrac{H}{D}-\dfrac{A}{R}\right)$, find V if $H=10$, $D=18$, $A=1$, and $R=5$.

3. Simplify $\left(\dfrac{24y^{-3}}{8y^{-1}}\right)^{-2}$.

In Exercises 4–9, perform the indicated operations and simplify.

4. $\dfrac{6x^2-2x}{2x}$ 5. $(15z^2+11z+4)\div(3z-2)$

6. $\dfrac{5x-5}{10}\cdot\dfrac{4x^3}{2x-2}$ 7. $\dfrac{2x}{x+1}-\dfrac{2x-1}{x+1}$

8. $5+\dfrac{3}{2x^3}$ 9. $\dfrac{2x-1}{x^2+9x+20}-\dfrac{x+5}{2x^2+7x-4}$

In Exercises 10 and 11, solve each equation.

10. $\dfrac{14}{3x}+\dfrac{42}{x}=1$

11. $\dfrac{3}{x}-\dfrac{8}{x^2}=\dfrac{1}{4}$

In Exercises 12–15, factor each expression completely, or write "not factorable."

12. x^2-x-42

13. $3w^2-48$

14. $20a^2-7ab-3b^2$

15. $3x^2-6x-2xy+4y$

In Exercises 16–20, set up each problem algebraically, solve, and check. Be sure to state what your variables represent.

16. The sum of two consecutive integers is 33. What are the integers?

17. The sum of two numbers is 5. Their product is -24. What are the numbers?

18. A dealer makes up a 15-lb mixture of oranges. One kind costs 78¢ per pound, and the other costs 99¢ per pound. How many pounds of each kind must be used in order for the mixture to cost 85¢ per pound?

19. Manny has twenty coins with a total value of $1.65. If the coins are all nickels and dimes, how many of each does he have?

20. Ricardo drove at a certain rate for 4 hr. If he had been able to drive 11 mph faster, the trip would have taken 3 hr. How fast did he drive? How far did he drive?

Critical Thinking and Problem-Solving Exercises

1. One day recently, Tad, Ted, and Adam were eating apples. Tad and Adam each ate the same number of apples, and they each ate at least one apple. Ted ate the most apples, and he ate fewer than 10. The *product* of the numbers of apples eaten by all three people was 12. How many apples did each person eat?

2. Julia, Ned, Sean, and Tiffany met for breakfast. Each ordered a different item for breakfast, and each was in a different type of business. The items ordered were pancakes, Belgian waffles, eggs Benedict, and oatmeal. Using the following clues, determine who sat in which chair, what business each person was in, and what each person ordered for lunch.

> The dentist was not sitting in seat 2 or 3.
>
> The machinist sat opposite the sales associate, who ordered eggs Benedict.
>
> Julia sat in seat 2.
>
> Tiffany sat in an odd-numbered seat, opposite the person who ordered pancakes.
>
> The machinist sat in seat 4 and ate oatmeal.
>
> Ned is a dentist.
>
> One person is an optician.

3. Three friends were comparing notes about reading. They discovered that in the last four months, the total number of books that Victor and Nina had read was 16, the total number of books that Nina and Ryan had read was 20, and the total number of books that Victor and Ryan had read was 22. How many books had each person (individually) read?

4. Consider the following algebraic expressions: $x + y$, $x - y$, $y - x$, $y + x$, $-y + x$, $-x + y$, $-x - y$, and $-y - x$.

  a. Name (if there are any) the *pairs* that are *equal* to each other.

  b. Name all the expressions that are the *additive inverses* of $x + y$.

  c. Name all the expressions that are the *additive inverses* of $x - y$.

5. The solution of each of the following problems contains an error. Find and describe the error, and solve each problem correctly.

  a. Simplify $\dfrac{4}{x} + \dfrac{2}{x + 2}$.

  LCD $= x(x + 2)$

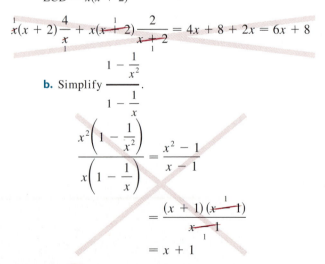

$$x(x + 2)\frac{4}{x} + x(x + 2)\frac{2}{x + 2} = 4x + 8 + 2x = 6x + 8$$

  b. Simplify $\dfrac{1 - \dfrac{1}{x^2}}{1 - \dfrac{1}{x}}$.

$$\frac{x^2\left(1 - \dfrac{1}{x^2}\right)}{x\left(1 - \dfrac{1}{x}\right)} = \frac{x^2 - 1}{x - 1}$$

$$= \frac{(x + 1)(x - 1)}{x - 1}$$

$$= x + 1$$

Radicals

CHAPTER 9

F inding roots of numbers and simplifying square roots of algebraic terms were introduced in previous chapters. In this chapter, we continue our discussion of simplifying square roots, and we discuss operations involving square roots. We also solve radical equations, and we solve some applied problems that lead to radical equations.

9.1   Simplifying Square Roots

Review of Square Roots

The *square root* of a number p is a number that, when squared, gives p. For example, $\sqrt{25} = 5$ *because* $5^2 = 25$. Every positive real number has both a positive and a negative square root, and the *positive* square root is called the *principal square root*. Recall that the symbol $\sqrt{p}$ represents the principal square root of p and that p is called the *radicand*:

$$\sqrt{p} \quad \text{Read "The square root of } p\text{"}$$

The radical sign ⎯⎤ ⎣⎯ The radicand

Just as the number 4 has two square roots, 2 and -2, the algebraic expression x^2 has two square roots, x and $-x$. Since x itself can be either a positive or a negative number, we don't know which is the principal square root. We do know, however, that $|x|$ must be positive (or zero); therefore, $|x|$ is the principal square root of x^2. In symbols, $\sqrt{x^2} = |x|$. However, in the remainder of this chapter, we assume that all variables represent positive numbers. For this reason, the absolute value symbol need not be used, and we will write $\sqrt{x^2} = x$.

Three important properties that involve square roots were given earlier; we repeat them here to refresh your memory:

Let a and b represent any nonnegative real numbers.

If $a = b$, then $\sqrt{a} = \sqrt{b}$.	The square root rule
$\sqrt{a^2} = a$	The rule for finding the square root of a perfect square
$\sqrt{ab} = \sqrt{a}\sqrt{b}$	The product rule for square roots

We will consider simplifying two kinds of radical expressions:

1. Square roots that do not involve division

2. Square roots that involve division

9.1A   Simplifying Square Roots That Do Not Involve Division

Squaring a Square Root   In Example 1, we square the square root of a number.

EXAMPLE 1

Examples of squaring the square root of a number:

— Replacing $\sqrt{25}$ with 5

a. $(\sqrt{25})^2 = (5)^2 = 25$; therefore, $(\sqrt{25})^2 = 25$.

— Replacing $\sqrt{16}$ with 4

b. $(\sqrt{16})^2 = (4)^2 = 16$; therefore, $(\sqrt{16})^2 = 16$.

— Replacing $\sqrt{121}$ with 11

c. $(\sqrt{121})^2 = (11)^2 = 121$; therefore, $(\sqrt{121})^2 = 121$.

The results obtained in Example 1 are generalized in the rule for squaring a square root, which is stated without proof.

The rule for squaring a square root

If a represents any nonnegative real number, then

$$(\sqrt{a})^2 = a$$

👉 **Note** The rule for finding the square root of a perfect square tells us that $\sqrt{a^2} = a$, and the rule for squaring a square root tells us that $(\sqrt{a})^2 = a$. Since $(\sqrt{a})^2$ and $\sqrt{a^2}$ both equal the same number (a), they must equal each other. That is, $(\sqrt{a})^2 = \sqrt{a^2} = a$.

EXAMPLE 2 Examples of simplifying expressions by using the rule for squaring a square root, $(\sqrt{a})^2 = a$:

a. $(\sqrt{7})^2 = 7$ **b.** $(\sqrt{13})^2 = 13$ **c.** $(\sqrt{x})^2 = x$
d. $(\sqrt{cd^4})^2 = cd^4$ **e.** $(\sqrt{x+3})^2 = x + 3$

The First Condition for Simplest Radical Form A number of conditions must be satisfied for a square root to be in *simplest radical form*. The *first* of these conditions is that no factor of the radicand can have an exponent greater than 1 when the radicand is expressed in prime-factored, exponential form.

When the radicand contains factors with *even* exponents, we can remove those factors from the radicand by dividing their exponents by 2. (Recall that $\sqrt{2^2x^4} = 2x^2$ because $(2x^2)^2 = 2^2x^4$, that $\sqrt{x^6y^8} = x^3y^4$ because $(x^3y^4)^2 = x^6y^8$, and so on.) See Example 3 for review.

EXAMPLE 3 Examples of simplifying radicals:

a. $\sqrt{5^2x^2y^4} = 5^{2/2}x^{2/2}y^{4/2} = 5xy^2$ **Check** $(5xy^2)^2 = 5^2x^2y^4$
b. $\sqrt{3^4z^6} = 3^{4/2}z^{6/2} = 3^2z^3$ **Check** $(3^2z^3)^2 = 3^4z^6$
c. $\sqrt{a^2} = a^{2/2} = a^1 = a$ **Check** $(a)^2 = a^2$

When any of the factors of the radicand contain *odd* exponents, we can use the following procedure:

Simplifying the square root of a product

1. Express the radicand in prime-factored, exponential form.
2. Find the square root of each factor of the radicand, as follows:
   a. If the exponent of a factor is an odd number greater than 1, write that factor as the product of two factors—one factor with an exponent of 1 and the other factor with an even exponent. Then proceed with steps b and c.
   b. If the exponent of a factor is an even number, remove that factor from the radical by dividing its exponent by 2 and dropping its radical sign.
   c. If the exponent of a factor is 1, that factor must remain under the radical sign.
3. The simplest radical form of the radical is the product of all the factors found in steps 2b and 2c.

We write any factors that do *not* contain a radical sign to the *left* of the radical sign.

EXAMPLE 4   Examples of simplifying square roots when the radicand contains odd exponents:

a. $\sqrt{x^7} = \sqrt{x^6 \cdot x^1} \leftarrow x^7 = x^6 \cdot x^1$

$\quad = \sqrt{x^6}\sqrt{x}$ ——— The factor x^6 is the highest power of x whose exponent (6) is exactly divisible by 2

$\quad = x^3\sqrt{x}$

b. $\sqrt{12x^4y^3} = \sqrt{2^2 \cdot 3 \cdot x^4 \cdot y^2 \cdot y}$

$\quad = \sqrt{2^2x^4y^2(\,3y\,)} \leftarrow$ Rearranging factors

$\quad = 2x^2y\sqrt{3y}$

$$\begin{array}{c|c} 2 & 12 \\ 2 & 6 \\ & 3 \quad 12 = 2^2 \cdot 3 \end{array}$$

c. $\sqrt{24a^5b^7} = \sqrt{2^2 \cdot 2 \cdot 3 \cdot a^4 \cdot a \cdot b^6 \cdot b}$

$\quad = \sqrt{2^2a^4b^6(\,2 \cdot 3ab\,)} \leftarrow$ Rearranging factors

$\quad = 2a^2b^3\sqrt{2 \cdot 3ab}$

$\quad = 2a^2b^3\sqrt{6ab}$

$$\begin{array}{c|c} 2 & 24 \\ 2 & 12 \\ 2 & 6 \\ & 3 \quad 24 = 2^3 \cdot 3 \end{array}$$

Exercises 9.1A
Set I

Simplify each expression.

1. $(\sqrt{9})^2$
2. $(\sqrt{100})^2$
3. $(\sqrt{15})^2$
4. $(\sqrt{29})^2$
5. $(\sqrt{3a^2})^2$
6. $(\sqrt{7y^2})^2$
7. $(\sqrt{x+y})^2$
8. $(\sqrt{u+v})^2$
9. $\sqrt{25x^2}$
10. $\sqrt{100y^2}$
11. $\sqrt{81s^4}$
12. $\sqrt{64t^6}$
13. $\sqrt{4x^2}$
14. $\sqrt{9y^2}$
15. $\sqrt{16z^4}$
16. $\sqrt{25b^6}$
17. $\sqrt{x^3}$
18. $\sqrt{y^5}$
19. $\sqrt{m^7}$
20. $\sqrt{n^9}$
21. $\sqrt{32x^2y^3}$

22. $\sqrt{98ab^3}$
23. $\sqrt{8s^4t^5}$
24. $\sqrt{44x^3y^6}$
25. $\sqrt{18a^2b^3}$
26. $\sqrt{27c^5d^2}$
27. $\sqrt{40x^6y^2}$
28. $\sqrt{135a^6b^8}$
29. $\sqrt{60h^5k^4}$
30. $\sqrt{90m^7n^6}$
31. $\sqrt{280y^5z^6}$
32. $\sqrt{270g^8h^9}$
33. $\sqrt{500x^7y^9}$
34. $\sqrt{216x^{11}y^{13}}$
35. $\sqrt{50u^5v^7w^2}$
36. $\sqrt{72a^3b^4c^5}$

Writing Problems

Express the answers in your own words and in complete sentences.

1. Explain why $2x\sqrt{12}$ is not in simplest radical form.

2. Find and describe the error in the following:
$$\sqrt{50x^2} = \sqrt{2 \cdot 5^2x^2} = 2x\sqrt{5}$$

Exercises 9.1A
Set II

Simplify each expression.

1. $(\sqrt{49})^2$
2. $(\sqrt{225})^2$
3. $(\sqrt{31})^2$
4. $(\sqrt{53})^2$
5. $(\sqrt{5r^2})^2$
6. $(\sqrt{2m^2})^2$
7. $(\sqrt{e+4})^2$
8. $(\sqrt{c+d})^2$
9. $\sqrt{36z^2}$
10. $\sqrt{120a^2}$
11. $\sqrt{25w^2}$
12. $\sqrt{64u^6}$
13. $\sqrt{16t^2}$
14. $\sqrt{12b^2}$
15. $\sqrt{25a^4}$
16. $\sqrt{84y^8}$
17. $\sqrt{a^5}$
18. $\sqrt{u^7}$

19. $\sqrt{z^5}$
20. $\sqrt{a^9}$
21. $\sqrt{12a^2b^5}$
22. $\sqrt{8x^5y^3}$
23. $\sqrt{18c^6d^7}$
24. $\sqrt{50s^3t^5}$
25. $\sqrt{84c^2d^4}$
26. $\sqrt{125xy^2}$
27. $\sqrt{28ab^5}$
28. $\sqrt{150x^3y}$
29. $\sqrt{80a^7b^4}$
30. $\sqrt{121x^4y^2}$
31. $\sqrt{88s^6t}$
32. $\sqrt{40p^3q^4}$
33. $\sqrt{54a^3b^4}$
34. $\sqrt{250x^5y^2z}$
35. $\sqrt{84a^4b^5c^6}$
36. $\sqrt{96u^6}$

9.IB Simplifying Square Roots That Involve Division

Earlier, we found square roots of fractions by inspection or by trial and error. For example, $\sqrt{\dfrac{9}{25}} = \dfrac{3}{5}$ because $\left(\dfrac{3}{5}\right)^2 = \dfrac{9}{25}$. Notice also that $\dfrac{\sqrt{9}}{\sqrt{25}} = \dfrac{3}{5}$. Since $\sqrt{\dfrac{9}{25}}$ and $\dfrac{\sqrt{9}}{\sqrt{25}}$ both equal $\dfrac{3}{5}$, they must equal each other. That is, $\sqrt{\dfrac{9}{25}} = \dfrac{\sqrt{9}}{\sqrt{25}}$. This fact is generalized in the quotient rule for square roots, which follows.

The quotient rule for square roots

If a represents any nonnegative real number and if b represents any positive real number, then

$$\sqrt{\dfrac{a}{b}} = \dfrac{\sqrt{a}}{\sqrt{b}}$$

The square root of a quotient	$=$	the quotient of the square roots

The Second Condition for Simplest Radical Form A *second* condition that must be satisfied for a square root to be in *simplest radical form* is that no radicand can contain a denominator; that is, the radicand cannot be a rational expression. The quotient rule for square roots can be used in simplifying radicals in which the radicand is a rational expression (see Example 5).

EXAMPLE 5

Examples of simplifying square roots in which the radicand is a rational expression (assume $y \neq 0$ and $k \neq 0$):

a. $\sqrt{\dfrac{4}{9}} = \dfrac{\sqrt{4}}{\sqrt{9}} = \dfrac{2}{3}$ Using the quotient rule for square roots

b. $\sqrt{\dfrac{25}{36}} = \dfrac{\sqrt{25}}{\sqrt{36}} = \dfrac{5}{6}$ Using the quotient rule for square roots

c. $\sqrt{\dfrac{x^4}{y^6}} = \dfrac{\sqrt{x^4}}{\sqrt{y^6}} = \dfrac{x^2}{y^3}$ Using the quotient rule for square roots

d. $\sqrt{\dfrac{50h^2}{2k^4}} = \sqrt{\dfrac{25h^2}{k^4}} = \dfrac{\sqrt{25h^2}}{\sqrt{k^4}} = \dfrac{5h}{k^2}$ We reduce $\frac{50}{2}$ to 25 before using the quotient rule for square roots

e. $\sqrt{\dfrac{3x^2}{4}} = \dfrac{\sqrt{3x^2}}{\sqrt{4}} = \dfrac{x\sqrt{3}}{2}$

In Example 5, all the denominators of the radicands were (or became, after the expression was simplified) perfect squares.

Rationalizing the Denominator

If a radicand contains a denominator that is *not* a perfect square, we must multiply the denominator by an expression that makes it a perfect square. The denominator then becomes *rational*; therefore, this procedure is called **rationalizing the denominator**.

Rationalizing a denominator that contains square roots and has one term

Multiply *both* numerator and denominator by an expression that makes the denominator a perfect square.

In Examples 6a and 6c, we show how to rationalize the denominator by extending the bar of the radical sign and multiplying the radicand by an expression that has a value of 1; that expression is placed *under* the radical sign. In Example 6b, we show how to rationalize the denominator by using the fundamental property of rational expressions $\left(\dfrac{a}{b} = \dfrac{a \cdot c}{b \cdot c} \right)$ on the radicand: We extend the fraction bar and the bar of the radical sign and multiply both numerator and denominator of the radicand by the same expression. Either method is acceptable.

EXAMPLE 6   Examples of rationalizing denominators:

Extending the bar of the radical sign to include $\frac{5}{5}$

a. $\sqrt{\dfrac{4}{5}} = \sqrt{\dfrac{4}{5} \cdot \dfrac{5}{5}} = \sqrt{\dfrac{4 \cdot 5}{5 \cdot 5}} = \dfrac{\sqrt{4 \cdot 5}}{\sqrt{5^2}} = \dfrac{\sqrt{4}\sqrt{5}}{5} = \dfrac{2\sqrt{5}}{5}$

Multiplying $\frac{4}{5}$ by $\frac{5}{5}$ in order to make the new denominator 5^2, a perfect square

b. $\sqrt{\dfrac{1}{7}} = \sqrt{\dfrac{1 \cdot 7}{7 \cdot 7}} = \dfrac{\sqrt{7}}{\sqrt{7^2}} = \dfrac{\sqrt{7}}{7}$

Multiplying both numerator and denominator of $\frac{1}{7}$ by 7 to make the denominator 7^2, a perfect square

c. $\sqrt{\dfrac{3}{20}} = \sqrt{\dfrac{3}{20} \cdot \dfrac{5}{5}} = \sqrt{\dfrac{3 \cdot 5}{20 \cdot 5}} = \dfrac{\sqrt{15}}{\sqrt{100}} = \dfrac{\sqrt{15}}{10}$

Multiplying $\frac{3}{20}$ by $\frac{5}{5}$ in order to make the new denominator 100, a perfect square

This problem could also have been done as follows:

$$\sqrt{\dfrac{3}{20}} = \sqrt{\dfrac{3}{20} \cdot \dfrac{20}{20}} = \sqrt{\dfrac{3 \cdot 2^2 \cdot 5}{20^2}} = \dfrac{\sqrt{2^2}\sqrt{3 \cdot 5}}{20} = \dfrac{\overset{1}{2}\sqrt{15}}{\underset{10}{20}} = \dfrac{\sqrt{15}}{10}$$

However, the arithmetic is easier the first way (that is, by multiplying numerator and denominator by the *smallest* number that will make the new denominator a perfect square).

The Third Condition for Simplest Radical Form   A *third* condition that must be satisfied for an algebraic expression to be in *simplest radical form* is that no denominator can contain a square root symbol.

In Example 7a, we show how to rationalize a denominator that contains a square root symbol by multiplying the given expression by an expression that equals 1. In Examples 7b–7d, we rationalize the denominator by using the fundamental property of rational expressions; that is, we multiply both numerator and denominator by the same expression. Either method is acceptable.

EXAMPLE 7   Examples of rationalizing denominators (assume $x > 0$):

a. $\dfrac{2}{\sqrt{5}} = \dfrac{2}{\sqrt{5}} \cdot \dfrac{\sqrt{5}}{\sqrt{5}} = \dfrac{2\sqrt{5}}{\sqrt{5}\sqrt{5}} = \dfrac{2\sqrt{5}}{(\sqrt{5})^2} = \dfrac{2\sqrt{5}}{5}$

The denominator is now a rational number

Multiplying by $\dfrac{\sqrt{5}}{\sqrt{5}}$ does not change the value of the expression, since $\dfrac{\sqrt{5}}{\sqrt{5}} = 1$; the product of the two denominators will be a rational number

The denominator is *not* a rational number

b. $\dfrac{5}{\sqrt{6}} = \dfrac{5 \cdot \sqrt{6}}{\sqrt{6} \cdot \boxed{\sqrt{6}}} = \dfrac{5\sqrt{6}}{(\sqrt{6})^2} = \dfrac{5\sqrt{6}}{6}$

— The denominator is now a rational number
— Multiplying both numerator and denominator by $\sqrt{6}$ does not change the value of the expression, and $\sqrt{6} \cdot \sqrt{6}$ will be a rational number

c. $\dfrac{6}{\sqrt{3}} = \dfrac{6 \cdot \sqrt{3}}{\sqrt{3} \cdot \boxed{\sqrt{3}}} = \dfrac{6\sqrt{3}}{(\sqrt{3})^2} = \dfrac{\overset{2}{6}\sqrt{3}}{\underset{1}{3}} = 2\sqrt{3}$

— The denominator is now a rational number
— Multiplying both numerator and denominator by $\sqrt{3}$ does not change the value of the expression, and $\sqrt{3} \cdot \sqrt{3}$ will be a rational number

d. $\dfrac{3xy}{\sqrt{x}} = \dfrac{3xy \cdot \sqrt{x}}{\sqrt{x} \cdot \boxed{\sqrt{x}}} = \dfrac{3xy\sqrt{x}}{(\sqrt{x})^2} = \dfrac{3xy\sqrt{x}}{x} = 3y\sqrt{x}$

— The denominator is now a rational number
— Multiplying both numerator and denominator by $\sqrt{x}$ does not change the value of the expression, and $\sqrt{x} \cdot \sqrt{x}$ will be a rational number

Note We could do the problems in Example 6 by using the quotient rule for square roots and then rationalizing the denominators by using the methods demonstrated in Example 7. For example, we could simplify $\sqrt{\frac{4}{5}}$ as follows:

$$\sqrt{\dfrac{4}{5}} = \dfrac{\sqrt{4}}{\sqrt{5}} = \dfrac{2}{\sqrt{5}} = \dfrac{2}{\sqrt{5}} \cdot \dfrac{\sqrt{5}}{\sqrt{5}} = \dfrac{2\sqrt{5}}{(\sqrt{5})^2} = \dfrac{2\sqrt{5}}{5}$$

Similarly, we can write

$$\sqrt{\dfrac{1}{7}} = \dfrac{\sqrt{1}}{\sqrt{7}} = \dfrac{1}{\sqrt{7}} = \dfrac{1}{\sqrt{7}} \cdot \dfrac{\sqrt{7}}{\sqrt{7}} = \dfrac{\sqrt{7}}{(\sqrt{7})^2} = \dfrac{\sqrt{7}}{7}$$

and so on.

If an algebraic *term* contains one square root, it will be in simplest radical form if it satisfies the conditions listed in the following box:

The simplified form of a term that contains square roots

1. When the radicand is in prime-factored, exponential form, no factor of the radicand has an exponent greater than 1.
2. No radicand contains a denominator.
3. No denominator contains a square root symbol.

Exercises 9.1B
Set I

Simplify each of the following expressions; assume that all variables in denominators are nonzero.

1. $\sqrt{\dfrac{9}{25}}$ 2. $\sqrt{\dfrac{36}{49}}$ 3. $\sqrt{\dfrac{16}{25}}$ 4. $\sqrt{\dfrac{81}{100}}$

5. $\sqrt{\dfrac{y^4}{x^2}}$ 6. $\sqrt{\dfrac{x^6}{v^8}}$ 7. $\sqrt{\dfrac{4x^2}{9}}$ 8. $\sqrt{\dfrac{16a^2}{49}}$

9. $\sqrt{\dfrac{2}{8k^2}}$ 10. $\sqrt{\dfrac{2m^2}{18}}$ 11. $\sqrt{\dfrac{4x^3y}{xy^3}}$ 12. $\sqrt{\dfrac{x^5y}{9xy^3}}$

13. $\sqrt{\dfrac{1}{5}}$ 14. $\sqrt{\dfrac{1}{3}}$ 15. $\sqrt{\dfrac{7}{13}}$ 16. $\sqrt{\dfrac{5}{17}}$ 25. $\sqrt{\dfrac{m^2}{3}}$ 26. $\sqrt{\dfrac{k^2}{5}}$ 27. $\sqrt{\dfrac{x^5z^4}{36y^2}}$ 28. $\sqrt{\dfrac{a^3c^6}{25b^2}}$

17. $\sqrt{\dfrac{1}{18}}$ 18. $\sqrt{\dfrac{1}{50}}$ 19. $\sqrt{\dfrac{7}{27}}$ 20. $\sqrt{\dfrac{5}{8}}$ 29. $\sqrt{\dfrac{3a^2b}{4b^3}}$ 30. $\sqrt{\dfrac{10uv^2}{8u}}$ 31. $\sqrt{\dfrac{b^2c^4}{16d^3}}$

21. $\dfrac{3}{\sqrt{7}}$ 22. $\dfrac{2}{\sqrt{3}}$ 23. $\dfrac{10}{\sqrt{5}}$ 24. $\dfrac{14}{\sqrt{2}}$ 32. $\sqrt{\dfrac{h^4k^8}{49p^5}}$ 33. $\sqrt{\dfrac{8m^2n}{2n^2}}$ 34. $\sqrt{\dfrac{18xy^2}{2x^2}}$

Writing Problems

Express the answers in your own words and in complete sentences.

1. Explain why $\sqrt{\dfrac{2x}{3}}$ is not in simplest radical form.

2. Explain why $\dfrac{8}{\sqrt{5}}$ is not in simplest radical form.

Exercises 9.1B
Set II

Simplify each of the following expressions; assume that all variables in denominators are nonzero.

1. $\sqrt{\dfrac{16}{49}}$ 2. $\sqrt{\dfrac{64}{81}}$ 3. $\sqrt{\dfrac{25}{64}}$ 4. $\sqrt{\dfrac{81}{121}}$

5. $\sqrt{\dfrac{a^2}{b^6}}$ 6. $\sqrt{\dfrac{c^8}{d^4}}$ 7. $\sqrt{\dfrac{36m^2}{25}}$ 8. $\sqrt{\dfrac{16a^4}{49}}$

9. $\sqrt{\dfrac{3}{27x^2}}$ 10. $\sqrt{\dfrac{18c^2}{2d^4}}$ 11. $\sqrt{\dfrac{h^3k^3}{16hk^5}}$ 12. $\sqrt{\dfrac{50a^4b}{2b^3}}$

13. $\sqrt{\dfrac{1}{11}}$ 14. $\sqrt{\dfrac{1}{2}}$ 15. $\sqrt{\dfrac{2}{19}}$ 16. $\sqrt{\dfrac{2}{18}}$

17. $\sqrt{\dfrac{1}{45}}$ 18. $\sqrt{\dfrac{1}{32}}$ 19. $\sqrt{\dfrac{2}{75}}$ 20. $\sqrt{\dfrac{3}{28}}$

21. $\dfrac{5}{\sqrt{10}}$ 22. $\dfrac{3}{\sqrt{75}}$ 23. $\dfrac{6}{\sqrt{15}}$ 24. $\dfrac{7}{\sqrt{28}}$

25. $\sqrt{\dfrac{e^2}{7}}$ 26. $\sqrt{\dfrac{f^2}{2}}$ 27. $\sqrt{\dfrac{r^3s^6}{9t^2}}$ 28. $\sqrt{\dfrac{50a}{2a^5b^4}}$

29. $\sqrt{\dfrac{15zw^4}{18z}}$ 30. $\sqrt{\dfrac{12ab^2}{5a^5b}}$ 31. $\sqrt{\dfrac{d^4e^6}{100f^3}}$

32. $\sqrt{\dfrac{125x^3y^4}{3xy^2}}$ 33. $\sqrt{\dfrac{45tu^2}{5t^2}}$ 34. $\sqrt{\dfrac{7cd^3}{5c^2d}}$

9.2   Multiplying and Dividing Square Roots

9.2A   Multiplying Square Roots

In earlier sections, we used the product rule for square roots to rewrite a single square root as a product of two or more square roots.

$$\sqrt{ab} = \sqrt{a}\,\sqrt{b}$$
$$\sqrt{4 \cdot 3} = \sqrt{4}\,\sqrt{3}$$

Using the product rule for square roots to simplify the square root of a product

$$= 2\sqrt{3}$$

In this section, we use that same rule to rewrite a product of two or more square roots as a single square root.

$$\sqrt{a}\sqrt{b} = \sqrt{ab}$$

$$\sqrt{2}\sqrt{8} = \sqrt{2 \cdot 8} \quad \text{Using the product rule for square roots}$$
$$\text{to find the product of square roots}$$

$$= \sqrt{16} = 4$$

In words: A product of square roots   equals   the square root of the product

In symbols: $\sqrt{a}\sqrt{b}$ $=$ $\sqrt{ab}$

A Word of Caution   Although a product of square roots equals the square root of the product, a *sum* of square roots does *not* equal the square root of the sum. A common error is shown below:

$$\sqrt{a} + \sqrt{b} = \sqrt{a + b}$$

This is incorrect. To see that it is incorrect, let's let $a = 9$ and $b = 16$. Then

$$\sqrt{a} + \sqrt{b} = \sqrt{9} + \sqrt{16} = 3 + 4 = 7$$
$$\sqrt{a + b} = \sqrt{9 + 16} = \sqrt{25} = 5$$

Since $7 \neq 5$, $\sqrt{9} + \sqrt{16} \neq \sqrt{9 + 16}$ and, in general, $\sqrt{a} + \sqrt{b} \neq \sqrt{a + b}$.

EXAMPLE 1   Examples of multiplying square roots:

A product of square roots   equals   the square root of the product

a. $\sqrt{5}\sqrt{20}$ $=$ $\sqrt{5 \cdot 20}$ $= \sqrt{100} = 10$

b. $\sqrt{2}\sqrt{8} = \sqrt{2 \cdot 8} = \sqrt{16} = 4$

c. $\sqrt{4x}\sqrt{x} = \sqrt{4x \cdot x} = \sqrt{4x^2} = 2x$

d. $\sqrt{3y}\sqrt{12y^3} = \sqrt{3y(12y^3)} = \sqrt{36y^4} = 6y^2$

e. $\sqrt{3}\sqrt{6}\sqrt{2} = \sqrt{3 \cdot 6 \cdot 2} = \sqrt{36} = 6$

f. $\sqrt{5z}\sqrt{10}\sqrt{2z^3} = \sqrt{5z(10)(2z^3)} = \sqrt{100z^4} = 10z^2$

Multiplying a Square Root by Itself

Let's see what happens if we multiply a square root by itself (see Example 2).

EXAMPLE 2   Examples of multiplying a square root by itself:

a. $\sqrt{16}\sqrt{16} = \sqrt{16 \cdot 16} = \sqrt{256} = 16$
  $\left.\begin{array}{l}\end{array}\right\}$ $\sqrt{16}\sqrt{16} = \sqrt{16 \cdot 16}$ by the
  or $\sqrt{16}\sqrt{16} = \sqrt{16 \cdot 16} = \sqrt{16^2} = 16$ product rule for square roots

  or $\sqrt{16}\sqrt{16} = 4 \cdot 4 = 16$

b. $\sqrt{36}\sqrt{36} = \sqrt{36 \cdot 36} = \sqrt{1{,}296} = 36$
  $\left.\begin{array}{l}\end{array}\right\}$ $\sqrt{36}\sqrt{36} = \sqrt{36 \cdot 36}$ by the
  or $\sqrt{36}\sqrt{36} = \sqrt{36 \cdot 36} = \sqrt{36^2} = 36$ product rule for square roots

  or $\sqrt{36}\sqrt{36} = 6 \cdot 6 = 36$

In this section, we use that same rule to rewrite a quotient of square roots as a single square root.

$$\frac{\sqrt{a}}{\sqrt{b}} = \sqrt{\frac{a}{b}} \quad (b \neq 0)$$

Using the quotient rule for square roots to rewrite a quotient of square roots as a single square root

$$\frac{\sqrt{50}}{\sqrt{2}} = \sqrt{\frac{50}{2}} = \sqrt{25} = 5$$

In words: A quotient of square roots   equals   the square root of the quotient

In symbols: $\dfrac{\sqrt{a}}{\sqrt{b}}$ $=$ $\sqrt{\dfrac{a}{b}}$ $(b \neq 0)$

EXAMPLE 6   Examples of dividing square roots (assume $a > 0$, $x > 0$, and $y > 0$):

| A quotient of square roots | equals | the square root of the quotient |

a. $\dfrac{\sqrt{8}}{\sqrt{2}}$ $=$ $\sqrt{\dfrac{8}{2}}$ $= \sqrt{4} = 2$

b. $\dfrac{\sqrt{a^5}}{\sqrt{a^3}} = \sqrt{\dfrac{a^5}{a^3}} = \sqrt{a^2} = a$

c. $\dfrac{\sqrt{27x}}{\sqrt{3x^3}} = \sqrt{\dfrac{27x}{3x^3}} = \sqrt{\dfrac{9}{x^2}} = \dfrac{3}{x}$

d. $\dfrac{\sqrt{28xy^3}}{\sqrt{7xy}} = \sqrt{\dfrac{28xy^3}{7xy}} = \sqrt{4y^2} = 2y$

In Example 7, we rationalize the denominator after the division has been performed.

EXAMPLE 7   Examples of finding and simplifying quotients of radicals (assume $x > 0$):

a. $\dfrac{\sqrt{5x}}{\sqrt{10x^2}} = \sqrt{\dfrac{5x}{10x^2}} = \sqrt{\dfrac{1}{2x}} = \dfrac{\sqrt{1}}{\sqrt{2x}} = \dfrac{1}{\sqrt{2x}} \cdot \dfrac{\sqrt{2x}}{\sqrt{2x}} = \dfrac{\sqrt{2x}}{2x}$ $\sqrt{2x}\sqrt{2x} = 2x$ by the rule for multiplying a square root by itself

b. $\dfrac{6\sqrt{25x^3}}{5\sqrt{3x}} = \dfrac{6}{5} \cdot \sqrt{\dfrac{25x^3}{3x}} = \dfrac{6}{5} \cdot \sqrt{\dfrac{25x^2}{3}} = \dfrac{6}{5} \cdot \sqrt{\dfrac{25x^2}{3} \cdot \dfrac{3}{3}} = \dfrac{6}{5} \cdot \sqrt{\dfrac{25x^2(3)}{3^2}}$

$$= \dfrac{\overset{2}{\cancel{6}}}{\underset{1}{\cancel{5}}} \cdot \dfrac{5x\sqrt{3}}{\underset{1}{\cancel{3}}} = 2x\sqrt{3}$$

Exercises 9.2B
Set I

Simplify each of the following expressions; assume that all variables in denominators are nonzero.

1. $\dfrac{\sqrt{20}}{\sqrt{5}}$

2. $\dfrac{\sqrt{7}}{\sqrt{28}}$

3. $\dfrac{\sqrt{32}}{\sqrt{2}}$

4. $\dfrac{\sqrt{98}}{\sqrt{2}}$

5. $\dfrac{\sqrt{4}}{\sqrt{5}}$

6. $\dfrac{\sqrt{9}}{\sqrt{7}}$

7. $\dfrac{\sqrt{15x}}{\sqrt{5x}}$

8. $\dfrac{\sqrt{18y}}{\sqrt{3y}}$

9. $\dfrac{\sqrt{75a^3}}{\sqrt{5a}}$

10. $\dfrac{\sqrt{24b^5}}{\sqrt{8b}}$

11. $\dfrac{\sqrt{35s^4}}{\sqrt{10s}}$

12. $\dfrac{\sqrt{42t^2}}{\sqrt{12t}}$

13. $\dfrac{\sqrt{8xy^3}}{\sqrt{12x^2y}}$

14. $\dfrac{\sqrt{10a^3b}}{\sqrt{15ab^2}}$

15. $\dfrac{\sqrt{72x^3y^2}}{\sqrt{2xy^2}}$

16. $\dfrac{\sqrt{27x^2y^3}}{\sqrt{3x^2y}}$

17. $\dfrac{\sqrt{x^4y}}{\sqrt{5y}}$

18. $\dfrac{\sqrt{m^6n}}{\sqrt{3n}}$

19. $\dfrac{4\sqrt{45m^3}}{3\sqrt{10m}}$

20. $\dfrac{6\sqrt{400x^4}}{5\sqrt{6x}}$

21. $\dfrac{5\sqrt{35a^5}}{7\sqrt{5a}}$

22. $\dfrac{8\sqrt{27x^7}}{3\sqrt{6x}}$

Exercises 9.2B
Set II

Simplify each of the following expressions; assume that all variables in denominators are nonzero.

1. $\dfrac{\sqrt{27}}{\sqrt{3}}$

2. $\dfrac{\sqrt{2}}{\sqrt{50}}$

3. $\dfrac{\sqrt{75}}{\sqrt{3}}$

4. $\dfrac{\sqrt{7}}{\sqrt{28}}$

5. $\dfrac{\sqrt{25}}{\sqrt{2}}$

6. $\dfrac{\sqrt{36}}{\sqrt{11}}$

7. $\dfrac{\sqrt{35x}}{\sqrt{5x}}$

8. $\dfrac{\sqrt{18m}}{\sqrt{6m}}$

9. $\dfrac{\sqrt{45x^3}}{\sqrt{5x}}$

10. $\dfrac{\sqrt{40y^4}}{\sqrt{5y}}$

11. $\dfrac{\sqrt{63a^4}}{\sqrt{14a}}$

12. $\dfrac{\sqrt{38b^3}}{\sqrt{57b}}$

13. $\dfrac{\sqrt{9ab^3}}{\sqrt{12a^2b}}$

14. $\dfrac{\sqrt{9x^3y}}{\sqrt{15xy^2}}$

15. $\dfrac{\sqrt{75a^5b^3}}{\sqrt{3a^3b^3}}$

16. $\dfrac{\sqrt{18x^3y}}{\sqrt{5xy^5}}$

17. $\dfrac{\sqrt{x^7y}}{\sqrt{3x}}$

18. $\dfrac{\sqrt{c^3d^4}}{\sqrt{7c^2d}}$

19. $\dfrac{9\sqrt{20a^6}}{2\sqrt{15a^3}}$

20. $\dfrac{12\sqrt{15b^3c}}{7\sqrt{3bc^5}}$

21. $\dfrac{9\sqrt{36x^5}}{4\sqrt{18x}}$

22. $\dfrac{6\sqrt{32y^3}}{5\sqrt{6y}}$

9.3   Adding and Subtracting Square Roots

Like square roots are square roots that have the same radicand.

EXAMPLE 1

Examples of like square roots:

a. $3\sqrt{5}$, $2\sqrt{5}$, $-7\sqrt{5}$ Only the coefficients can be different

b. $2\sqrt{x}$, $-9\sqrt{x}$, $11\sqrt{x}$

Unlike square roots are square roots that have different radicands.

EXAMPLE 2

Examples of unlike square roots:

a. $2\sqrt{15}$, $-6\sqrt{11}$, $8\sqrt{17}$

b. $5\sqrt{y}$, $3\sqrt{x}$, $-4\sqrt{13}$

Combining Like Square Roots

The addition of like square roots is very much like the addition of like terms; it is an application of the distributive property. For example,

$$3\sqrt{7} + 2\sqrt{7} = (3 + 2)\sqrt{7} = 5\sqrt{7}$$

Therefore, we add like square roots by adding their coefficients and then multiplying that sum by the square root.

EXAMPLE 3

Examples of combining like square roots:

a. $5\sqrt{2} + 3\sqrt{2} = (5 + 3)\sqrt{2} = 8\sqrt{2}$

This step need not be shown

b. $6\sqrt{3} - 4\sqrt{3} = (6 - 4)\sqrt{3} = 2\sqrt{3}$

c. $7\sqrt{x} - 3\sqrt{x} = (7 - 3)\sqrt{x} = 4\sqrt{x}$

This step need not be shown

d. $\dfrac{3}{2}\sqrt{5} + \dfrac{\sqrt{5}}{2} = \left(\dfrac{3}{2} + \dfrac{1}{2}\right)\sqrt{5} = 2\sqrt{5}$

$\dfrac{\sqrt{5}}{2} = \dfrac{1}{2}\sqrt{5}$, because $\dfrac{1}{2}\sqrt{5} = \dfrac{1}{2} \cdot \dfrac{\sqrt{5}}{1} = \dfrac{\sqrt{5}}{2}$

Combining Unlike Square Roots

When two or more *unlike* square roots are connected with *addition* or *subtraction* symbols, we usually cannot express the sum or difference with just one radical sign. That is, most unlike square roots cannot be combined. However, some *can* be combined. Sometimes terms that were not like square roots *become* like square roots after they have been expressed in simplest radical form; if this is the case, they can be combined (see Example 4).

Combining unlike square roots

1. Express each term in simplest radical form.
2. Combine any terms that have like square roots by adding their coefficients and multiplying that sum by the square root.

The Fifth Condition for Simplest Radical Form A *fifth* condition that must be satisfied for an algebraic expression involving square roots to be in *simplest radical form* is that all like square roots must be combined. For example, $2\sqrt{3} + 5\sqrt{3}$ is not in simplest radical form, because $2\sqrt{3}$ and $5\sqrt{3}$ *are* like square roots and have not been combined. The sum $2\sqrt{3} + 5\sqrt{3}$ must be rewritten as $7\sqrt{3}$, which *is* in simplest radical form.

EXAMPLE 4

Examples of simplifying and combining radicals:

a. $\sqrt{8} + \sqrt{18}$

$= \sqrt{4 \cdot 2} + \sqrt{9 \cdot 2} = 2\sqrt{2} + 3\sqrt{2} = 5\sqrt{2}$ $2\sqrt{2}$ and $3\sqrt{2}$ are like square roots

b. $\sqrt{12} - \sqrt{27} + 5\sqrt{3}$

$= \sqrt{4 \cdot 3} - \sqrt{9 \cdot 3} + 5\sqrt{3} = 2\sqrt{3} - 3\sqrt{3} + 5\sqrt{3} = 4\sqrt{3}$

c. $\sqrt{20} - 2\sqrt{45} - \sqrt{15}$

$= \sqrt{4 \cdot 5} - 2\sqrt{9 \cdot 5} - \sqrt{3 \cdot 5}$

$= 2\sqrt{5} - 2 \cdot 3\sqrt{5} - \sqrt{15}$ $2\sqrt{5}$ and $-2 \cdot 3\sqrt{5}$ are like square roots

$= 2\,\boxed{\sqrt{5}} - 6\,\boxed{\sqrt{5}} - \sqrt{15} = -4\sqrt{5} - \sqrt{15}$

d. $2\sqrt{\tfrac{1}{2}} - 6\sqrt{\tfrac{1}{8}} - 10\sqrt{\tfrac{4}{5}}$

$= 2\sqrt{\dfrac{1 \cdot 2}{2 \cdot 2}} - 6\sqrt{\dfrac{1 \cdot 2}{8 \cdot 2}} - 10\sqrt{\dfrac{4 \cdot 5}{5 \cdot 5}}$ Rationalizing the denominators

$= \dfrac{2}{1} \cdot \dfrac{\sqrt{2}}{\sqrt{4}} - \dfrac{6}{1} \cdot \dfrac{\sqrt{2}}{\sqrt{16}} - \dfrac{10}{1} \cdot \dfrac{\sqrt{4}\sqrt{5}}{\sqrt{25}}$

$= \dfrac{\overset{1}{\cancel{2}}}{1} \cdot \dfrac{\sqrt{2}}{\underset{1}{\cancel{2}}} - \dfrac{\overset{3}{\cancel{6}}\sqrt{2}}{\underset{2}{\cancel{4}}} - \dfrac{\overset{2}{\cancel{10}}}{1} \cdot \dfrac{2\sqrt{5}}{\underset{1}{\cancel{5}}}$

$= \sqrt{2} - \tfrac{3}{2}\sqrt{2} - 4\sqrt{5}$ $\sqrt{2}$ and $-\tfrac{3}{2}\sqrt{2}$ are like square roots

$= \left(1 - \tfrac{3}{2}\right)\sqrt{2} - 4\sqrt{5} = \left(\tfrac{2}{2} - \tfrac{3}{2}\right)\sqrt{2} - 4\sqrt{5} = -\tfrac{1}{2}\sqrt{2} - 4\sqrt{5}$

A Word of Caution It is important that you realize that in a sum such as $5 + 3\sqrt{2}$, the two terms 5 and $3\sqrt{2}$ are *not like radicals* and they cannot be combined. A common error is to add the 5 and the 3.

Correct method	Incorrect method
$5 + 3\sqrt{2}$ cannot be simplified	$5 + 3\sqrt{2} = 8\sqrt{2}$

(It *is* true that $5\sqrt{2} + 3\sqrt{2} = 8\sqrt{2}$, since $5\sqrt{2}$ and $3\sqrt{2}$ *are* like square roots.)

In a later chapter, you will need to be able to simplify expressions such as the ones in Examples 5 and 6.

EXAMPLE 5 Simplify $\dfrac{-4 + \sqrt{20}}{6}$.

SOLUTION We first simplify $\sqrt{20}$.

$\dfrac{-4 + \sqrt{20}}{6} = \dfrac{-4 + \sqrt{4 \cdot 5}}{6}$

$= \dfrac{-4 + 2\sqrt{5}}{6}$ Simplifying $\sqrt{20}$

$= \dfrac{\overset{1}{\cancel{2}}(-2 + \sqrt{5})}{\underset{3}{\cancel{6}}}$ Factoring 2 from the numerator and reducing the resulting expression

$= \dfrac{-2 + \sqrt{5}}{3}$

EXAMPLE 6   Simplify $\dfrac{8 - \sqrt{75}}{8}$.

SOLUTION   We first simplify $\sqrt{75}$.

$$\frac{8 - \sqrt{75}}{8} = \frac{8 - \sqrt{25 \cdot 3}}{8}$$

$$= \frac{8 - 5\sqrt{3}}{8} \qquad \text{This expression cannot be reduced}$$

A Word of Caution   A common error is to "cancel" the 8's in a problem like the one in Example 6.

Correct method

$\dfrac{8 - 5\sqrt{3}}{8}$ cannot be simplified

Incorrect method

$\dfrac{\overset{1}{8} - 5\sqrt{3}}{\underset{1}{8}} = 1 - 5\sqrt{3}$

8 is not a *factor* of the numerator.

Finding Decimal Approximations

Although we cannot combine the unlike square roots in an expression such as $-\frac{1}{2}\sqrt{2} - 4\sqrt{5}$, we can find a *decimal approximation* by using a calculator or Table I, found inside the back cover (see Example 7).

EXAMPLE 7   Approximate $-\frac{1}{2}\sqrt{2} - 4\sqrt{5}$, rounding off the answer to three decimal places.

SOLUTION

Using a Calculator   Because $\frac{1}{2} = 0.5$, or .5, the problem can be entered as follows:

$$\boxed{\,.\,}\ \boxed{5}\ \boxed{+/-}\ \boxed{\times}\ \boxed{2}\ \boxed{\sqrt{}}\ \boxed{-}\ \boxed{4}\ \boxed{\times}\ \boxed{5}\ \boxed{\sqrt{}}\ \boxed{=}$$

and the display is $\boxed{\quad-9.651378689}$. Rounding off this answer to three decimal places, we find that $-\frac{1}{2}\sqrt{2} - 4\sqrt{5} \approx -9.651$.

Using Table I $\sqrt{2} \approx 1.414$ and $\sqrt{5} \approx 2.236$. Therefore,

$$-\frac{1}{2}\sqrt{2} - 4\sqrt{5} \approx -\frac{1}{2}(1.414) - 4(2.236)$$

$$= -0.707 - 8.944$$

$$= -9.651 \qquad \text{No rounding off is necessary}$$

Notice that when the calculator answer is rounded off to the same number of decimal places as the answer obtained using Table I, we get the same number, -9.651. However, calculator answers sometimes differ slightly from table answers because of the rounding off process.

EXAMPLE 8   Find the approximate value of (a) $\sqrt{5} + \sqrt{7}$ and (b) $\sqrt{12}$. Determine whether $\sqrt{5} + \sqrt{7}$ equals $\sqrt{5 + 7}$, or $\sqrt{12}$.

SOLUTION

a. Using a calculator and rounding off each approximation to three decimal places, we have

$$\sqrt{5} + \sqrt{7} \approx 2.236 + 2.646 = 4.882$$

b. $\sqrt{12} \approx 3.464$

Because $4.882 \neq 3.464$, $\sqrt{5} + \sqrt{7} \neq \sqrt{12}$, or $\sqrt{5} + \sqrt{7} \neq \sqrt{5 + 7}$.

✖ **A Word of Caution** Remember that, in general,

$$\sqrt{a} + \sqrt{b} \neq \sqrt{a + b}$$

We saw in Example 8 that $\sqrt{5} + \sqrt{7} \neq \sqrt{5 + 7}$, and we saw earlier that $\sqrt{9} + \sqrt{16} \neq \sqrt{9 + 16}$.

A Summary of the Conditions for Simplest Radical Form

We summarize here *all* the conditions necessary for an expression with square roots to be in simplest radical form:

1. When the radicand is in prime-factored, exponential form, no factor of the radicand has an exponent greater than 1.

2. No radicand contains a denominator.

3. No denominator contains a square root symbol.

4. No term contains more than one radical sign.

5. All like square roots are combined.

Exercises 9.3
Set I

In Exercises 1–44, simplify each expression, if possible.

1. $2\sqrt{3} + 5\sqrt{3}$

2. $4\sqrt{2} + 3\sqrt{2}$

3. $3\sqrt{x} - \sqrt{x}$

4. $5\sqrt{a} - \sqrt{a}$

5. $\frac{3}{2}\sqrt{2} - \frac{\sqrt{2}}{2}$

6. $\frac{4}{3}\sqrt{3} - \frac{\sqrt{3}}{3}$

7. $5 \cdot 8\sqrt{5} + \sqrt{5}$

8. $3 \cdot 4\sqrt{7} + \sqrt{7}$

9. $\sqrt{25} + \sqrt{5}$

10. $\sqrt{16} + \sqrt{6}$

11. $\sqrt{50} + \sqrt{50}$

12. $\sqrt{32} + \sqrt{32}$

13. $\sqrt{68} - \sqrt{17}$

14. $\sqrt{52} - \sqrt{13}$

15. $\sqrt{45} - \sqrt{28}$

16. $\sqrt{90} - \sqrt{54}$

17. $2\sqrt{3} + \sqrt{12}$

18. $3\sqrt{2} + \sqrt{8}$

19. $2\sqrt{50} - \sqrt{32}$

20. $3\sqrt{24} - \sqrt{54}$

21. $3\sqrt{32} - \sqrt{8}$

22. $4\sqrt{27} - 3\sqrt{12}$

23. $\sqrt{\frac{1}{2}} + \sqrt{8}$

24. $\sqrt{\frac{1}{3}} + \sqrt{12}$

25. $\sqrt{24} - \sqrt{\frac{2}{3}}$

26. $\sqrt{45} - \sqrt{\frac{4}{5}}$

27. $10\sqrt{\frac{3}{5}} + \sqrt{60}$

28. $8\sqrt{\frac{3}{16}} + \sqrt{48}$

29. $\sqrt{\frac{25}{2} - \frac{3}{\sqrt{2}}}$

30. $5\sqrt{\frac{1}{5}} - \sqrt{\frac{9}{20}}$

31. $3\sqrt{\frac{1}{6}} + \sqrt{12} - 5\sqrt{\frac{3}{2}}$

32. $3\sqrt{\frac{5}{2}} + \sqrt{20} - 5\sqrt{\frac{1}{10}}$

33. $6 + 5\sqrt{3}$

34. $8 + 5\sqrt{2}$

35. $9 - 3\sqrt{50}$

36. $8 - 4\sqrt{44}$

37. $\dfrac{-4 + \sqrt{28}}{10}$

38. $\dfrac{-6 + \sqrt{48}}{8}$

39. $\dfrac{10 - \sqrt{96}}{6}$

40. $\dfrac{6 - \sqrt{80}}{4}$

41. $\dfrac{-12 - \sqrt{12}}{12}$

42. $\dfrac{-10 - \sqrt{32}}{10}$

43. $\dfrac{-3 + \sqrt{8}}{2}$

44. $\dfrac{-7 + \sqrt{18}}{2}$

In Exercises 45 and 46, find a decimal approximation for each expression, using a calculator or Table I. Round off the answers to three decimal places.

45. $\frac{1}{3}\sqrt{7} + 2\sqrt{3}$

46. $3\sqrt{6} + \frac{1}{4}\sqrt{5}$

Writing Problems

Express the answers in your own words and in complete sentences.

2. Explain why $\sqrt{36 + 64} \neq \sqrt{36} + \sqrt{64}$.

1. Explain why $\sqrt{52} - \sqrt{13}$ is not in simplest radical form.

Exercises 9.3
Set II

In Exercises 1–44, simplify each expression, if possible.

1. $3\sqrt{7} + 2\sqrt{7}$

2. $8\sqrt{14} + 7\sqrt{14}$

3. $4\sqrt{m} - \sqrt{m}$

4. $12\sqrt{c} - \sqrt{c}$

5. $\frac{2}{3}\sqrt{5} - \frac{\sqrt{5}}{3}$

6. $\frac{3}{2}\sqrt{6} - \frac{\sqrt{6}}{2}$

7. $3 \cdot 6\sqrt{3} + \sqrt{3}$

8. $5 \cdot 8\sqrt{11} + \sqrt{11}$

9. $\sqrt{9} + \sqrt{17}$

10. $\sqrt{25} + \sqrt{26}$

11. $\sqrt{18} + \sqrt{18}$

12. $\sqrt{48} + \sqrt{48}$

13. $\sqrt{44} - \sqrt{11}$

14. $\sqrt{117} - \sqrt{13}$

15. $\sqrt{52} - \sqrt{44}$

16. $\sqrt{160} - \sqrt{63}$

17. $5\sqrt{2} + \sqrt{18}$

18. $7\sqrt{7} + \sqrt{28}$

19. $\sqrt{45} - 3\sqrt{20}$

20. $\sqrt{48} - 5\sqrt{27}$

21. $4\sqrt{28} - \sqrt{63}$

22. $8\sqrt{45} - 3\sqrt{20}$

23. $\sqrt{20} + \sqrt{\frac{1}{5}}$

24. $\sqrt{18} + \sqrt{\frac{1}{2}}$

25. $\sqrt{75} - \sqrt{\frac{4}{3}}$

26. $\sqrt{\frac{1}{3}} - \sqrt{108}$

27. $6\sqrt{\frac{5}{9}} + \sqrt{45}$

28. $5\sqrt{8} + 3\sqrt{\frac{1}{2}}$

29. $\sqrt{\frac{16}{3} - \frac{6}{\sqrt{3}}}$

30. $\frac{4}{\sqrt{5}} - \sqrt{\frac{36}{5}}$

31. $10\sqrt{\frac{1}{15}} + \sqrt{60} - 8\sqrt{\frac{3}{5}}$

32. $7\sqrt{40} - 3\sqrt{\frac{5}{2}} + 8\sqrt{\frac{1}{10}}$

33. $8 + 3\sqrt{7}$

34. $9 + 4\sqrt{6}$

35. $7 - 2\sqrt{45}$

36. $10 - 3\sqrt{8}$

37. $\frac{-8 + \sqrt{52}}{10}$

38. $\frac{12 + \sqrt{288}}{12}$

39. $\frac{14 - \sqrt{72}}{4}$

40. $\frac{7 - \sqrt{128}}{4}$

41. $\frac{-8 - \sqrt{48}}{8}$

42. $\frac{-1 - \sqrt{50}}{2}$

43. $\frac{-5 + \sqrt{50}}{2}$

44. $\frac{-6 + \sqrt{98}}{2}$

In Exercises 45 and 46, find a decimal approximation for each expression, using a calculator or Table I. Round off the answers to three decimal places.

45. $\frac{1}{4}\sqrt{13} + 3\sqrt{6}$

46. $5\sqrt{5} + \frac{1}{5}\sqrt{7}$

9.4   Products and Quotients That Involve More Than One Term

As mentioned earlier, we find products and/or quotients in which some factors contain radicals and other factors do not by using methods similar to the methods we used in finding products and/or quotients that contain constants and variables. Therefore, when any of the factors contain more than one term, we must use the distributive property.

EXAMPLE 1   Simplify $\sqrt{2}(3\sqrt{2} - 5)$.

SOLUTION   Because the second factor contains two terms, we must use the distributive property.

$$\sqrt{2}(3\sqrt{2} - 5) = \sqrt{2} \cdot (3\sqrt{2}) - \sqrt{2}(5)$$
$$= 3\sqrt{2 \cdot 2} - 5\sqrt{2}$$
$$= 3 \cdot 2 - 5\sqrt{2}$$
$$= 6 - 5\sqrt{2}$$

The FOIL method can be used when we multiply two expressions that contain square roots if both contain two terms (see Example 2).

EXAMPLE 2   Simplify $(2\sqrt{3} - 5)(4\sqrt{3} - 6)$.

SOLUTION   We can use the FOIL method.

$$
\begin{array}{ccccccc}
& F & + & O & + & I & + & L
\end{array}
$$

$$
(2\sqrt{3} - 5)(4\sqrt{3} - 6) = [(2\sqrt{3})(4\sqrt{3})] + [(2\sqrt{3})(-6)] + [(-5)(4\sqrt{3})] + [(-5)(-6)]
$$

$$
= [2 \cdot 4(\sqrt{3})(\sqrt{3})] \quad -12\sqrt{3} \quad\quad -20\sqrt{3} \quad\quad +30
$$

$$
= \quad\quad 8 \cdot 3 \quad\quad\quad -32\sqrt{3} \quad\quad\quad +30
$$

$$
= \underline{24} - 32\sqrt{3} \underline{+ 30}
$$

$$
= 54 - 32\sqrt{3}
$$

Example 3 illustrates finding a product when the problem is in the form $(a + b)(a - b)$.

EXAMPLE 3   Simplify $(\sqrt{7} + 2)(\sqrt{7} - 2)$.

SOLUTION   We will use the formula $(a + b)(a - b) = a^2 - b^2$.

$$
(\sqrt{7} + 2)(\sqrt{7} - 2) = (\sqrt{7})^2 - (2)^2 = 7 - 4 = 3 \quad \text{Notice that the answer, 3, is \textit{rational}}
$$

(We would have obtained the same answer if we had used the FOIL method.)

Example 4 illustrates squaring an expression that contains square roots and has two terms.

EXAMPLE 4   Simplify $(3\sqrt{5} - 7x)^2$.

SOLUTION   Using the formula $(a - b)^2 = a^2 - 2ab + b^2$, we have

$$
(3\sqrt{5} - 7x)^2 = (3\sqrt{5})^2 - 2(3\sqrt{5})(7x) + (7x)^2 \quad \text{We put \textit{x} before } \sqrt{5} \text{ so it will be clear that \textit{x} is \textit{not} under the radical sign}
$$

$$
= 3^2(\sqrt{5})^2 - 42x\sqrt{5} + 7^2x^2
$$

$$
= 9 \cdot 5 - 42x\sqrt{5} + 49x^2
$$

$$
= 45 - 42x\sqrt{5} + 49x^2
$$

We obtain the same answer if we write $(3\sqrt{5} - 7x)^2$ as $(3\sqrt{5} - 7x)(3\sqrt{5} - 7x)$ and use the FOIL method.

In a division problem, when the dividend contains more than one term but the divisor contains only one term, we divide *each term* of the dividend by the divisor (see Example 5).

EXAMPLE 5 Simplify $\dfrac{3\sqrt{14} - \sqrt{8}}{\sqrt{2}}$.

SOLUTION

$$\frac{3\sqrt{14} - \sqrt{8}}{\sqrt{2}} = \frac{3\sqrt{14}}{\sqrt{2}} - \frac{\sqrt{8}}{\sqrt{2}} = 3\sqrt{\frac{14}{2}} - \sqrt{\frac{8}{2}} = 3\sqrt{7} - \sqrt{4} = 3\sqrt{7} - 2$$

Rationalizing a Denominator That Contains Two Terms

Recall that converting a denominator that is irrational to one that is rational is called *rationalizing the denominator*. In order to discuss division problems in which the divisor contains square roots and has two terms, we need a new definition.

Conjugate

> The **conjugate** of the algebraic expression $a + b$ is the algebraic expression $a - b$.

EXAMPLE 6 Examples of conjugates of algebraic expressions that contain square roots:

  a. The conjugate of $1 - \sqrt{2}$ is $1 + \sqrt{2}$.
  b. The conjugate of $2\sqrt{3} + 5$ is $2\sqrt{3} - 5$.
  c. The conjugate of $\sqrt{x} - \sqrt{y}$ is $\sqrt{x} + \sqrt{y}$.
  d. The conjugate of $-\sqrt{3} - 2$ is $-\sqrt{3} + 2$.   Notice that the sign of the *first* term does not change

In Examples 7 and 8, we use the fact that $(a + b)(a - b) = a^2 - b^2$.

EXAMPLE 7 Find the product of $1 - \sqrt{2}$ and its conjugate.

SOLUTION The conjugate of $1 - \sqrt{2}$ is $1 + \sqrt{2}$. Therefore, the product of $1 - \sqrt{2}$ and its conjugate is

$$(1 - \sqrt{2})(1 + \sqrt{2}) = 1^2 - (\sqrt{2})^2 = 1 - 2 = -1$$

Notice that the product of $(1 - \sqrt{2})$ and its conjugate is -1, a *rational number*.

When we multiply an algebraic expression that contains square roots and has two terms by its conjugate, we *always* get an expression that is rational. Because of this fact, the following procedure should be used whenever a denominator contains square roots and has two terms.

Rationalizing a denominator that contains square roots and has two terms

> Multiply both the numerator and the denominator by the conjugate of the denominator.

(Multiplying both numerator and denominator by the same number is equivalent to multiplying the given expression by an expression whose value is 1.)

EXAMPLE 8 Examples of rationalizing denominators that contain square roots and have two terms:

a. $\dfrac{2}{1 + \sqrt{3}} = \dfrac{2}{(1 + \sqrt{3})} \dfrac{(1 - \sqrt{3})}{(1 - \sqrt{3})} = \dfrac{2(1 - \sqrt{3})}{(1)^2 - (\sqrt{3})^2} = \dfrac{2(1 - \sqrt{3})}{1 - 3} = \dfrac{2(1 - \sqrt{3})}{-2}$

$= \dfrac{\overset{-1}{2}(1 - \sqrt{3})}{\underset{1}{-2}}$ Dividing both numerator and denominator by -2

$= -1 + \sqrt{3}$ or $\sqrt{3} - 1$

Multiplying both numerator and denominator by $1 - \sqrt{3}$ (the conjugate of the denominator $1 + \sqrt{3}$)

b. $\dfrac{6}{\sqrt{5} - \sqrt{3}} = \dfrac{6}{(\sqrt{5} - \sqrt{3})} \dfrac{(\sqrt{5} + \sqrt{3})}{(\sqrt{5} + \sqrt{3})} = \dfrac{6(\sqrt{5} + \sqrt{3})}{(\sqrt{5})^2 - (\sqrt{3})^2} = \dfrac{6(\sqrt{5} + \sqrt{3})}{5 - 3}$

$= \dfrac{\overset{3}{6}(\sqrt{5} + \sqrt{3})}{\underset{1}{2}} = 3\sqrt{5} + 3\sqrt{3}$

Multiplying both numerator and denominator by $\sqrt{5} + \sqrt{3}$ (the conjugate of the denominator $\sqrt{5} - \sqrt{3}$)

c. $\dfrac{z - 16}{\sqrt{z} - 4} = \dfrac{(z - 16)}{(\sqrt{z} - 4)} \dfrac{(\sqrt{z} + 4)}{(\sqrt{z} + 4)} = \dfrac{(z - 16)(\sqrt{z} + 4)}{(\sqrt{z})^2 - 4^2} = \dfrac{\overset{1}{(z - 16)}(\sqrt{z} + 4)}{\underset{1}{z - 16}}$

$= \sqrt{z} + 4$

Exercises 9.4

Set I

In Exercises 1–18, simplify each expression.

1. $\sqrt{2}(\sqrt{2} + 1)$
2. $\sqrt{3}(\sqrt{3} + 1)$
3. $\sqrt{3}(2\sqrt{3} + 1)$
4. $\sqrt{5}(3\sqrt{5} + 1)$
5. $\sqrt{x}(\sqrt{x} - 3)$
6. $\sqrt{y}(4 - \sqrt{y})$
7. $(\sqrt{7} + 2)(\sqrt{7} + 3)$
8. $(\sqrt{3} + 2)(\sqrt{3} + 4)$
9. $(\sqrt{8} - 3\sqrt{2})(\sqrt{8} + 2\sqrt{5})$
10. $(\sqrt{2} + 4\sqrt{3})(\sqrt{12} - 2\sqrt{3})$
11. $(\sqrt{2} + \sqrt{14})^2$
12. $(\sqrt{5} + \sqrt{11})^2$
13. $(\sqrt{2x} + 3)^2$
14. $(\sqrt{7x} + 4)^2$
15. $\dfrac{\sqrt{8} + \sqrt{18}}{\sqrt{2}}$
16. $\dfrac{\sqrt{12} + \sqrt{27}}{\sqrt{3}}$
17. $\dfrac{\sqrt{20} + 5\sqrt{10}}{\sqrt{5}}$
18. $\dfrac{2\sqrt{6} + \sqrt{14}}{\sqrt{6}}$

In Exercises 19 and 20, write the conjugate of each expression.

19. a. $2 + \sqrt{3}$ b. $2\sqrt{5} - 7$ c. $-6 - \sqrt{2}$

20. a. $3\sqrt{2} - 5$ b. $\sqrt{7} + 4$ c. $-\sqrt{5} - \sqrt{7}$

In Exercises 21–28, rationalize the denominators and simplify.

21. $\dfrac{3}{\sqrt{2} - 1}$
22. $\dfrac{5}{\sqrt{2} - 1}$
23. $\dfrac{6}{\sqrt{3} - \sqrt{2}}$
24. $\dfrac{8}{\sqrt{2} - \sqrt{3}}$
25. $\dfrac{6}{\sqrt{5} + \sqrt{2}}$
26. $\dfrac{4}{\sqrt{7} + \sqrt{5}}$
27. $\dfrac{x - 4}{\sqrt{x} + 2}$
28. $\dfrac{y - 9}{\sqrt{y} - 3}$ $(y \neq 9)$

In Exercises 29–32, approximate each expression, using a calculator or Table I. Round off the answers to two decimal places.

29. $\frac{1}{3}\sqrt{7} - 2\sqrt{3}$
30. $3\sqrt{6} - \frac{1}{4}\sqrt{5}$
31. $\dfrac{3 + 2\sqrt{11}}{6}$
32. $\dfrac{3 - 2\sqrt{11}}{6}$

Writing Problems

Express the answers in your own words and in complete sentences.

1. Explain why $(\sqrt{5} + \sqrt{2})^2 \neq 5 + 2$.

2. Explain why $5x + 3\sqrt{2}\sqrt{5}$ is not in simplest radical form.

3. Explain why $\dfrac{5\sqrt{6} - \sqrt{2}}{\sqrt{2}}$ is not in simplest radical form.

Explain how to rationalize the denominator if the denominator contains radicals and has one term.

4. Explain how to rationalize the denominator if the denominator contains radicals and has one term.

5. Explain how to rationalize the denominator if the denominator contains radicals and has two terms.

6. Explain how to rationalize the denominator if the radicand is a rational expression.

Exercises 9.4
Set II

In Exercises 1–18, simplify each expression.

1. $\sqrt{5}(\sqrt{5} + 1)$

2. $\sqrt{7}(\sqrt{7} - 5)$

3. $\sqrt{2}(3\sqrt{2} + 1)$

4. $\sqrt{5}(8\sqrt{5} - \sqrt{3})$

5. $\sqrt{z}(\sqrt{3} - \sqrt{z})$

6. $\sqrt{8}(\sqrt{5} + \sqrt{2})$

7. $(\sqrt{6} + 5)(\sqrt{6} + 2)$

8. $(\sqrt{11} - 2)(\sqrt{11} - 3)$

9. $(\sqrt{2} - 5\sqrt{8})(\sqrt{2} + 4\sqrt{8})$

10. $(\sqrt{7} + 3\sqrt{2})(3\sqrt{7} + 2\sqrt{2})$

11. $(\sqrt{7} + \sqrt{18})^2$

12. $(\sqrt{13} + \sqrt{12})^2$

13. $(3\sqrt{y} + 5)^2$

14. $(5\sqrt{z} + 2)^2$

15. $\dfrac{\sqrt{15} + \sqrt{10}}{\sqrt{5}}$

16. $\dfrac{\sqrt{21} - \sqrt{14}}{\sqrt{3}}$

17. $\dfrac{5\sqrt{32} + \sqrt{24}}{\sqrt{8}}$

18. $\dfrac{8\sqrt{48} - \sqrt{16}}{\sqrt{3}}$

In Exercises 19 and 20, write the conjugate of each expression.

19. **a.** $5 - \sqrt{7}$ **b.** $3\sqrt{11} - 2$ **c.** $-6 + \sqrt{13}$

20. **a.** $-2 + \sqrt{3}$ **b.** $-8 + \sqrt{5}$ **c.** $-\sqrt{14} - 1$

In Exercises 21–28, rationalize the denominators and simplify.

21. $\dfrac{10}{1 + \sqrt{2}}$

22. $\dfrac{8}{2 + \sqrt{12}}$

23. $\dfrac{13}{\sqrt{6} - \sqrt{5}}$

24. $\dfrac{21}{\sqrt{7} + \sqrt{3}}$

25. $\dfrac{15}{\sqrt{3} + \sqrt{8}}$

26. $\dfrac{12}{\sqrt{5} - 2}$

27. $\dfrac{z - 25}{\sqrt{z} + 5}$

28. $\dfrac{m^2 - 16}{2 - \sqrt{m}}$ $(m \neq 4)$

In Exercises 29–32, approximate each expression, using a calculator or Table I. Round off the answers to two decimal places.

29. $\frac{1}{4}\sqrt{8} - 3\sqrt{5}$

30. $\dfrac{\sqrt{2} + \sqrt{3}}{5}$

31. $\dfrac{5 + 6\sqrt{7}}{11}$

32. $\dfrac{-1 - 5\sqrt{3}}{3}$

9.5   Solving Radical Equations

A **radical equation** is an equation in which the variable appears in a radicand. In this text, we will consider only radical equations that contain square roots.

EXAMPLE 1   Examples of radical equations:

a. $\sqrt{x} = 7$ **b.** $\sqrt{x + 2} = 3$ **c.** $\sqrt{2x - 3} = \sqrt{x + 7} - 2$

The following property of real numbers is used in solving radical equations:

The squaring property

> If $a = b$, then $a^2 = b^2$.
>
> *In words:* If two numbers are equal, then their squares are equal.

We can remove square root signs from equations by using the squaring property; that is, we can square both sides of an equation. However, we sometimes introduce *extraneous roots* when we do so. This can occur because it is possible for the *squares* of two numbers to be equal when the numbers themselves were *not* equal. For example, $(-3)^2 = 3^2$ (the *squares* of -3 and 3 are equal), but $-3 \neq 3$ (the numbers themselves are *not* equal). Because extraneous roots can occur, *all apparent solutions must be checked in the original equation* whenever we square both sides of an equation.

Solving a radical equation

1. Arrange the terms so that one term with a radical is by itself on one side of the equation, if necessary.
2. Square both sides of the equation. That is, use the squaring property: If $a = b$, then $a^2 = b^2$.
3. Combine like terms on each side of the equal sign.
4. If a radical still remains, repeat steps 1, 2, and 3.
5. Solve the resulting equation.
6. Check apparent solutions in the original equation; there may be extraneous roots.

EXAMPLE 2   Solve $\sqrt{x} - 7 = 0$.

SOLUTION   We first get the radical by itself on one side of the equation.

$$\sqrt{x} - 7 = 0$$
$$\sqrt{x} - 7 + 7 = 0 + 7 \quad \text{Adding 7 to both sides}$$
$$\sqrt{x} = 7$$
$$(\sqrt{x})^2 = (7)^2 \quad \text{Squaring both sides}$$
$$x = 49$$

Check
$$\sqrt{x} - 7 = 0$$
$$\sqrt{49} - 7 \overset{?}{=} 0$$
$$7 - 7 \overset{?}{=} 0$$
$$0 = 0$$

The solution is 49.

EXAMPLE 3   Solve $\sqrt{3x - 7} = \sqrt{x + 5}$.

SOLUTION   One term with a radical is already by itself on one side of the equal sign.

$$\sqrt{3x - 7} = \sqrt{x + 5}$$
$$(\sqrt{3x - 7})^2 = (\sqrt{x + 5})^2 \quad \text{Squaring both sides}$$
$$3x - 7 = x + 5 \quad \text{Adding}$$
$$3x - 7 \; -x + 7 = x + 5 \; -x + 7 \quad -x + 7 \text{ to both sides}$$
$$2x = 12$$
$$\frac{2x}{2} = \frac{12}{2} \quad \text{Dividing both sides by 2}$$
$$x = 6$$

Check
$$\sqrt{3x - 7} = \sqrt{x + 5}$$
$$\sqrt{3(6) - 7} \overset{?}{=} \sqrt{(6) + 5}$$
$$\sqrt{18 - 7} \overset{?}{=} \sqrt{11}$$
$$\sqrt{11} = \sqrt{11}$$

The solution is 6.

EXAMPLE 4   Solve $\sqrt{x + 2} = 3$.

SOLUTION   One term with a radical is already by itself on one side of the equal sign.

$\sqrt{x + 2} = 3$	**Check**
$(\sqrt{x + 2})^2 = (3)^2$ Squaring both sides	$\sqrt{x + 2} = 3$
$x + 2 = 9$	$\sqrt{7 + 2} \overset{?}{=} 3$
$x + 2 \; -2 = 9 \; - 2$	$\sqrt{9} \overset{?}{=} 3$
$x = 7$	$3 = 3$

The solution is 7.

✗   **A Word of Caution**   If one side of the equation contains *more than one term*, squaring each *term* is not the same as squaring both sides of the equation. That is, a common error is to think that

$$\text{if} \quad a = b + c$$
$$\text{then} \quad a^2 = b^2 + c^2$$

However, if $a = b + c$, $a^2 \neq b^2 + c^2$. To verify this with numbers, suppose that $a = 7$, $b = 2$, and $c = 5$.

$$7 = 2 + 5 \quad \text{A true statement}$$

But

$$7^2 \neq 2^2 + 5^2$$

(Notice that $7^2 = 49$, but $2^2 + 5^2 = 4 + 25 = 29$; since $49 \neq 29$, $7^2 \neq 2^2 + 5^2$.)
It *is* true, however, that if $a = b + c$, then $a^2 = (b + c)^2$. Let's verify this with the same numbers:

$$7 = (2 + 5) \quad \text{A true statement}$$
$$7^2 = (2 + 5)^2$$
$$7^2 = (7)^2 \quad \text{True}$$

(In the last step, we substituted 7 for $2 + 5$.)

EXAMPLE 5   Solve $\sqrt{2x + 1} + 1 = x$.

SOLUTION   We must get the radical by itself on one side of the equal sign.

$\sqrt{2x + 1} + 1 = x$	
$\sqrt{2x + 1} + 1 \; -1 = x \; - 1$	Adding -1 to both sides to isolate the radical
$\sqrt{2x + 1} = x - 1$	
$(\sqrt{2x + 1})^2 = (x - 1)^2$	Squaring both sides
	When squaring $(x - 1)$, do not forget this middle term
$2x + 1 = x^2 \; - 2x \; + 1$	This is a quadratic equation
$2x + 1 \; - 2x - 1 = x^2 - 2x + 1 \; - 2x - 1$	Adding $-2x - 1$ to both sides
$0 = x^2 - 4x$	We now have 0 on one side
$0 = x(x - 4)$	Factoring the polynomial

$x = 0$	$x - 4 = 0$ Setting each factor equal to 0
	$x - 4 \; +4 \; = 0 \; +4$ Adding 4 to both sides
$x = 0$	$x = 4$

Check for x = 0

$$\sqrt{2x + 1} + 1 = x$$
$$\sqrt{2(0) + 1} + 1 \stackrel{?}{=} 0$$
$$\sqrt{0 + 1} + 1 \stackrel{?}{=} 0$$
$$\sqrt{1} + 1 \stackrel{?}{=} 0$$
$$1 + 1 \stackrel{?}{=} 0$$
$$2 \neq 0$$

Check for x = 4

$$\sqrt{2x + 1} + 1 = x$$
$$\sqrt{2(4) + 1} + 1 \stackrel{?}{=} 4$$
$$\sqrt{8 + 1} + 1 \stackrel{?}{=} 4$$
$$\sqrt{9} + 1 \stackrel{?}{=} 4$$
$$3 + 1 \stackrel{?}{=} 4$$
$$4 = 4$$

We see that 0 is *not* a solution, but 4 is a solution; therefore, 4 is the *only* solution of the equation.

EXAMPLE 6

Solve $\sqrt{2x - 3} = \sqrt{x + 7} - 2$.

SOLUTION We already have *one* term with a radical by itself on one side of the equal sign.

$$\sqrt{2x - 3} = \sqrt{x + 7} - 2$$

$$(\sqrt{2x - 3})^2 = (\sqrt{x + 7} - 2)^2 \quad \text{Squaring both sides}$$

When squaring ($\sqrt{x + 7} - 2$), do not forget this middle term

$$2x - 3 = x + 7 \; - 4\sqrt{x + 7} \; + 4$$

$$2x - 3 = x + 11 - 4\sqrt{x + 7} \quad \begin{array}{l}\text{We now need to get} \\ -4\sqrt{x + 7} \text{ by itself} \\ \text{on one side}\end{array}$$

$$2x - 3 \; -x - 11 \; = x + 11 - 4\sqrt{x + 7} \; -x - 11 \quad \begin{array}{l}\text{Adding } -x - 11 \\ \text{to both sides}\end{array}$$

$$x - 14 = -4\sqrt{x + 7} \quad \text{Simplifying}$$

$$(x - 14)^2 = (-4\sqrt{x + 7})^2 \quad \text{Squaring both sides again}$$

$$x^2 - 28x + 196 = 16(x + 7)$$

$$x^2 - 28x + 196 = 16x + 112 \quad \text{Simplifying}$$

$$x^2 - 28x + 196 \; -16x - 112 \; = 16x + 112 \; -16x - 112 \quad \begin{array}{l}\text{Adding } -16x - 112 \\ \text{to both sides}\end{array}$$

$$x^2 - 44x + 84 = 0 \quad \text{Simplifying}$$

$$(x - 2)(x - 42) = 0 \quad \text{Factoring the left side}$$

$x - 2 = 0$	$x - 42 = 0$
$x - 2 \; +2 \; = 0 \; +2$	$x - 42 \; +42 \; = 0 \; +42$
$x = 2$	$x = 42$

Check for $x = 2$

$$\sqrt{2x - 3} = \sqrt{x + 7} - 2$$

$$\sqrt{2(2) - 3} \overset{?}{=} \sqrt{(2) + 7} - 2$$

$$\sqrt{4 - 3} \overset{?}{=} \sqrt{9} - 2$$

$$\sqrt{1} \overset{?}{=} 3 - 2$$

$$1 = 1$$

Therefore, 2 is a solution because it does satisfy the equation.

Check for $x = 42$

$$\sqrt{2x - 3} = \sqrt{x + 7} - 2$$

$$\sqrt{2(42) - 3} \overset{?}{=} \sqrt{(42) + 7} - 2$$

$$\sqrt{84 - 3} \overset{?}{=} \sqrt{49} - 2$$

$$\sqrt{81} \overset{?}{=} 7 - 2$$

$$9 \neq 5$$

Therefore, 42 is *not* a solution because it *does not* satisfy the equation.

The only solution is 2.

Applied Problems That Involve Radical Equations

Many applications of algebra in business, engineering, the sciences, and so forth involve radical equations.

EXAMPLE 7   Find the amount of power, P, consumed if an appliance has a resistance of 16 ohms and draws 5 amps (amperes) of current. Use the following formula from electricity:

$$I = \sqrt{\frac{P}{R}}$$

where I, the current, is measured in amps (amperes); P, the power, is measured in watts; and R, the resistance, is measured in ohms.

SOLUTION   We must find the power, P, when $R = 16$ and $I = 5$.

$$I = \sqrt{\frac{P}{R}}$$

$$5 = \sqrt{\frac{P}{16}}$$

$$(5)^2 = \left(\sqrt{\frac{P}{16}}\right)^2 \quad \text{Squaring both sides}$$

$$25 = \frac{P}{16}$$

$$P = 400 \quad \text{Multiplying both sides by 16}$$

Check

$$I = \sqrt{\frac{P}{R}}$$

$$5 \overset{?}{=} \sqrt{\frac{400}{16}}$$

$$5 \overset{?}{=} \sqrt{25}$$

$$5 = 5 \quad \text{True}$$

Therefore, the amount of power consumed is 400 watts.

Exercises 9.5
Set I

In Exercises 1–26, solve each equation.

1. $\sqrt{x} = 5$
2. $\sqrt{x} = 4$
3. $\sqrt{2x} = 4$
4. $\sqrt{3x} = 6$
5. $\sqrt{x - 3} = 2$
6. $\sqrt{x + 4} = 6$
7. $\sqrt{2x + 1} = 9$
8. $\sqrt{5x - 4} = 4$
9. $\sqrt{3x + 1} = 5$
10. $\sqrt{7x + 8} = 6$
11. $\sqrt{x + 1} = \sqrt{2x - 7}$
12. $\sqrt{3x - 2} = \sqrt{x + 4}$
13. $\sqrt{3x - 2} = x$
14. $\sqrt{5x - 6} = x$
15. $\sqrt{4x - 1} = 2x$
16. $\sqrt{6x - 1} = 3x$
17. $\sqrt{x - 3} + 5 = x$
18. $\sqrt{x + 5} - 3 = x$
19. $\sqrt{3x + 7} + 5 = 3x$
20. $\sqrt{4x + 5} + 5 = 2x$
21. $\sqrt{5x + 14} - x = 4$

22. $\sqrt{3x + 16} - x = 6$

23. $\sqrt{2x + 2} = 1 + \sqrt{x + 2}$

24. $\sqrt{3x + 1} = 1 + \sqrt{x + 4}$

25. $\sqrt{3x - 5} = \sqrt{x + 6} - 1$

26. $\sqrt{5x - 1} = \sqrt{x + 14} - 1$

In Exercises 27–30, solve each problem for the indicated variable, using the formula $I = \sqrt{\dfrac{P}{R}}$, **as given in Example 7.**

27. Find the amount of power, P, consumed if an appliance has a resistance of 9 ohms and draws 7 amps of current.

28. Find the amount of power, P, consumed if an appliance has a resistance of 25 ohms and draws 2 amps of current.

29. Find R, the resistance in ohms, for an electrical system that consumes 500 watts and draws 5 amps of current.

30. Find R, the resistance in ohms, for an electrical system that consumes 600 watts and draws 5 amps of current.

In Exercises 31 and 32, use this formula from statistics: $\sigma = \sqrt{npq}$, **where** σ **(sigma) is the** *standard deviation*, n **is the number of trials,** p **is the probability of success, and** q **is the probability of failure.**

31. Find the probability of success, p, if the standard deviation is 3, the probability of failure is $\frac{3}{4}$, and the number of trials is 48.

32. Find the probability of success, p, if the standard deviation is $3\frac{1}{5}$, the probability of failure is $\frac{4}{5}$, and the number of trials is 64.

In Exercises 33 and 34, use this formula from geometry: $c = \sqrt{a^2 + b^2}$, **where** c **is the length of the longest side of a right triangle and** a **and** b **are the lengths of the other two sides.**

33. Find the length of one side of a right triangle if the length of the longest side is 5 cm and the length of one of the other sides is 3 cm.

34. Find the length of one side of a right triangle if the length of the longest side is 13 in. and the length of one of the other sides is 5 in.

Writing Problems

Express the answers in your own words and in complete sentences.

1. Explain why, when we square both sides of the equation $\sqrt{x - 2} = x + 2$, we *don't* get $x - 2 = x^2 + 2^2$.

2. Find and describe the error in the solution of the following:

Solve $x = \sqrt{x + 2}$.

Solution:
$$x = \sqrt{x + 2}$$
$$x^2 = x + 2$$
$$x^2 - x - 2 = 0$$
$$(x - 2)(x + 1) = 0$$
$$x - 2 = 0 \qquad x + 1 = 0$$
$$x = 2 \qquad x = -1$$

The solutions are 2 and -1.

Exercises 9.5
Set II

In Exercises 1–26, solve each equation.

1. $\sqrt{x} = 8$

2. $\sqrt{x} = 11$

3. $\sqrt{5x} = 10$

4. $\sqrt{3x} = 12$

5. $\sqrt{x - 6} = 3$

6. $\sqrt{x - 5} = 2$

7. $\sqrt{6x + 1} = 5$

8. $\sqrt{3x - 2} = 5$

9. $\sqrt{9x - 5} = 7$

10. $\sqrt{5x + 6} = 6$

11. $\sqrt{9 - 2x} = \sqrt{5x - 12}$

12. $\sqrt{6x - 1} = \sqrt{4x - 5}$

13. $\sqrt{3x + 10} = x$

14. $x = \sqrt{8x - 15}$

15. $\sqrt{3x + 2} = 3x$

16. $2x = \sqrt{40 - 12x}$

17. $\sqrt{x - 6} + 8 = x$

18. $1 + \sqrt{3x - 3} = x$

19. $\sqrt{4x + 13} + 1 = 2x$

20. $\sqrt{10 + 3x} - 4 - x = 0$

21. $\sqrt{6x - 2} - x = 1$

22. $\sqrt{6x + 15} - x = 4$

23. $\sqrt{3x + 1} = 2 + \sqrt{x + 1}$

24. $\sqrt{2x + 7} = 1 + \sqrt{x + 3}$

25. $\sqrt{2x - 1} = 1 + \sqrt{x - 1}$

26. $\sqrt{2x - 5} = \sqrt{x + 9} - 1$

In Exercises 27–30, solve each problem for the indicated variable, using the formula $I = \sqrt{\dfrac{P}{R}}$**, as given in Example 7.**

27. Find the amount of power, P, consumed if an appliance has a resistance of 12 ohms and draws 5 amps of current.

28. Find the amount of power, P, consumed if an appliance has a resistance of 20 ohms and draws 6 amps of current.

29. Find R, the resistance in ohms, for an electrical system that consumes 700 watts and draws 5 amps of current.

30. Find R, the resistance in ohms, for an electrical system that consumes 400 watts and draws 4 amps of current.

In Exercises 31 and 32, use this formula from statistics: $\sigma = \sqrt{npq}$**, where** σ **(sigma) is the** *standard deviation,* n **is the number of trials,** p **is the probability of success, and** q **is the probability of failure.**

31. Find the probability of success, p, if the standard deviation is $2\frac{2}{3}$, the probability of failure is $\frac{2}{3}$, and the number of trials is 32.

32. Find the probability of success, p, if the standard deviation is 5, the probability of failure is $\frac{5}{6}$, and the number of trials is 180.

In Exercises 33 and 34, use this formula from geometry: $c = \sqrt{a^2 + b^2}$**, where** c **is the length of the longest side of a right triangle and** a **and** b **are the lengths of the other two sides.**

33. Find the length of one side of a right triangle if the length of the longest side is 17 ft and the length of one of the other sides is 8 ft.

34. Find the length of one side of a right triangle if the length of the longest side is 10 m and the length of one of the other sides is 6 m.

Sections 9.1–9.5   REVIEW

Square Roots
9.1

The **square root** of a number p is a number that gives p when squared. A positive real number p has two square roots, a positive root called the **principal square root**, written $\sqrt{p}$, and a negative square root, written $-\sqrt{p}$.

Square Root Properties
9.1–9.2

In this text, it is understood that all variables in radicands represent nonnegative numbers.

$\sqrt{ab} = \sqrt{a}\,\sqrt{b}$ The product rule for square roots

$(\sqrt{a})^2 = \sqrt{a^2} = a$ The rule for squaring a square root

$\sqrt{\dfrac{a}{b}} = \dfrac{\sqrt{a}}{\sqrt{b}}$ $(b \ne 0)$ The quotient rule for square roots

$\sqrt{a}\,\sqrt{a} = a$ The rule for multiplying a square root by itself

To Simplify the Square Root of a Product
9.1

1. Express the radicand in prime-factored, exponential form.

2. Find the square root of each factor of the radicand, as follows:

     a. If the exponent of a factor is an odd number greater than 1, write that factor as the product of two factors—one factor with an exponent of 1 and the other factor with an even exponent. Then proceed with steps b and c.

     b. If the exponent of a factor is an even number, remove that factor from the radical by dividing its exponent by 2 and dropping its radical sign.

     c. If the exponent of a factor is 1, that factor must remain under the radical sign.

3. The simplest radical form of the radical is the product of all the factors found in steps 2b and 2c.

Simplest Radical Form for an Expression That Has Square Roots
9.1, 9.2, 9.3

1. When the radicand is in prime-factored, exponential form, no factor of the radicand has an exponent greater than 1.

2. No radicand contains a denominator.

3. No denominator contains a square root symbol.

4. No term contains more than one radical sign.

5. All like square roots are combined.

To Multiply Square Roots
9.2, 9.4

Use the product rule for square roots: $\sqrt{a}\sqrt{b} = \sqrt{ab}$. Simplify the result. If one of the factors has more than one term, use the distributive property. If both factors have two terms, the FOIL method can be used.

To Divide Square Roots
9.2, 9.4

Use the quotient rule for square roots: $\dfrac{\sqrt{a}}{\sqrt{b}} = \sqrt{\dfrac{a}{b}}$. You may, however, need to rationalize the denominator.

Conjugates
9.4

The **conjugate** of $a + b$ is $a - b$.

To Rationalize a Denominator
9.1, 9.4

A. With one term: Multiply both numerator and denominator by an expression that makes the denominator a perfect square.

B. With two terms: Multiply both numerator and denominator by the conjugate of the denominator.

To Combine Square Roots
9.3

A. Like square roots: Add the coefficients and multiply that sum by the square root.

B. Unlike square roots:

   1. Express each term in simplest radical form.

   2. Combine any terms that have like square roots by adding their coefficients and multiplying that sum by the square root.

To Solve a Radical Equation
9.5

1. Arrange the terms so that one term with a radical is by itself on one side of the equation.

2. Square both sides of the equation. (Use the squaring property: If $a = b$, then $a^2 = b^2$.)

3. Combine like terms on each side of the equal sign.

4. If a radical still remains, repeat steps 1, 2, and 3.

5. Solve the resulting equation.

6. Check apparent solutions in the original equation; there may be extraneous roots.

Sections 9.1–9.5   REVIEW EXERCISES   Set I

In Exercises 1–26, simplify each expression.

1. $(\sqrt{36})^2$

2. $(\sqrt{3})^2$

3. $(\sqrt{a+b})^2$

4. $\sqrt{64x^4y^6}$

5. $\sqrt{x^3}$

6. $\sqrt{36a^4b^2}$

7. $\sqrt{a^3b^3}$

8. $\sqrt{11}\sqrt{11}$

9. $\sqrt{2}\sqrt{32}$

10. $\sqrt{50}$

11. $\sqrt{180}$

12. $4\sqrt{2} - \sqrt{2}$

13. $\sqrt{18} - \sqrt{8}$

14. $\sqrt{27} - \sqrt{12}$

15. $\dfrac{1}{\sqrt{5}}$

16. $\sqrt{\dfrac{1}{7}}$

17. $\dfrac{6}{\sqrt{3}}$

18. $\sqrt{\dfrac{9}{2}}$

19. $\sqrt{8} - \sqrt{\dfrac{1}{2}}$ **20.** $\sqrt{54} - \sqrt{\dfrac{2}{3}}$

21. $\sqrt{2}(\sqrt{8} + \sqrt{18})$ **22.** $(5 + \sqrt{2})(5 - \sqrt{2})$

23. $(2\sqrt{3} + 1)^2$ **24.** $(3\sqrt{2} - 1)^2$

25. $\dfrac{8}{\sqrt{3} - 2}$ **26.** $\dfrac{6}{2 + \sqrt{5}}$

In Exercises 27–34, solve each equation.

27. $\sqrt{x} = 4$ **28.** $\sqrt{3a} = 6$

29. $\sqrt{2x - 1} = 5$ **30.** $\sqrt{5a - 4} = \sqrt{3a + 2}$

31. $\sqrt{7x - 6} = x$ **32.** $\sqrt{9x - 8} = x$

33. $\sqrt{2x + 7} = \sqrt{x} + 2$ **34.** $\sqrt{3x + 1} = \sqrt{x + 8} - 1$

In Exercise 35, use the formula $c = \sqrt{a^2 + b^2}$, where c is the length of the longest side of a right triangle.

35. Find the length of one side of a right triangle if the length of the longest side is 13 yd and the length of one of the other sides is 12 yd.

ANSWERS

Name

In Exercises 1–26, simplify each expression.

1. $(\sqrt{64})^2$ **2.** $\sqrt{121x^6y^2z^4}$ **3.** $(\sqrt{11})^2$

4. $(\sqrt{m+n})^2$ **5.** $\sqrt{w^7}$

6. $\sqrt{25t^4s^2}$ **7.** $\sqrt{u^3v^5}$

8. $\sqrt{6}\sqrt{6}$ **9.** $\sqrt{18}\sqrt{2}$

10. $\sqrt{68}$ **11.** $\sqrt{175}$

12. $5\sqrt{6} - \sqrt{6}$ **13.** $\sqrt{45} - \sqrt{20}$

14. $\sqrt{63} - \sqrt{28}$ **15.** $\dfrac{1}{\sqrt{13}}$

16. $\sqrt{\dfrac{1}{17}}$ **17.** $\dfrac{12}{\sqrt{6}}$

18. $\sqrt{\dfrac{16}{5}}$ **19.** $\sqrt{60} - \sqrt{\dfrac{3}{5}}$

1. _____

2. _____

3. _____

4. _____

5. _____

6. _____

7. _____

8. _____

9. _____

10. _____

11. _____

12. _____

13. _____

14. _____

15. _____

16. _____

17. _____

18. _____

19. _____

20. $\sqrt{24} - \sqrt{\dfrac{2}{3}}$

21. $\sqrt{5}(\sqrt{5} + \sqrt{20})$

22. $(\sqrt{6} - 2)(\sqrt{6} + 2)$

23. $(3 - 2\sqrt{7})^2$

24. $(\sqrt{2} - 5)^2$

25. $\dfrac{3}{1 - \sqrt{2}}$

26. $\dfrac{12}{\sqrt{5} + 3}$

In Exercises 27–34, solve each equation.

27. $\sqrt{z} = 5$

28. $\sqrt{5h} = 10$

29. $\sqrt{4x - 3} = 5$

30. $\sqrt{7 - 3m} = \sqrt{m + 3}$

31. $x = \sqrt{x + 20}$

32. $\sqrt{4x - 7} = \sqrt{x + 1}$

33. $\sqrt{5x - 4} = 2 + \sqrt{x}$

34. $\sqrt{2x + 5} = \sqrt{x + 14} - 1$

20. _____

21. _____

22. _____

23. _____

24. _____

25. _____

26. _____

27. _____

28. _____

29. _____

30. _____

31. _____

32. _____

33. _____

34. _____

35. _____

In Exercise 35, use the formula $c = \sqrt{a^2 + b^2}$, where c is the length of the longest side of a right triangle.

35. Find the length of one side of a right triangle if the length of the longest side is 17 in. and the length of one of the other sides is 15 in.

Chapter 9	D I A G N O S T I C T E S T

The purpose of this test is to see how well you understand operations involving radicals. We recommend that you work this diagnostic test *before* your instructor tests you on this chapter. Allow yourself about 50 minutes.

Complete solutions for all the problems on this test, together with section references, are given in the answer section in the back of the book. We suggest that you study the sections referred to for the problems you do incorrectly.

In Problems 1–21, simplify each expression; assume that all variables in denominators are nonzero.

1. $\sqrt{16z^2}$ **2.** $\sqrt{50}$ **3.** $\sqrt{x^6y^5}$

4. $\sqrt{\dfrac{3}{14}}$ **5.** $\sqrt{\dfrac{48}{3n^2}}$ **6.** $\sqrt{60}$

7. $\sqrt{13}\,\sqrt{13}$ **8.** $\sqrt{3}\,\sqrt{75y^2}$ **9.** $\sqrt{5}\,(2\sqrt{5}-3)$

10. $(\sqrt{13}+\sqrt{3})(\sqrt{13}-\sqrt{3})$

11. $(4\sqrt{x}+3)(3\sqrt{x}+4)$

12. $(8+\sqrt{10})^2$

13. $\dfrac{\sqrt{5}}{\sqrt{80}}$

14. $\dfrac{\sqrt{3}}{\sqrt{6}}$ **15.** $\dfrac{\sqrt{18}-\sqrt{36}}{\sqrt{3}}$

16. $\dfrac{\sqrt{x^4y}}{\sqrt{3y}}$ **17.** $\dfrac{6}{1+\sqrt{5}}$

18. $8\sqrt{t}-\sqrt{t}$ **19.** $7\sqrt{3}-\sqrt{27}$

20. $\sqrt{54}+4\sqrt{24}$ **21.** $\sqrt{45}+6\sqrt{\dfrac{4}{5}}$

In Problems 22–24, solve each equation.

22. $\sqrt{8x}=12$ **23.** $\sqrt{5x+11}=6$

24. $\sqrt{6x-5}=x$

Set up Problem 25 algebraically, solve, and check, using this formula from geometry: $c=\sqrt{a^2+b^2}$, where c is the length of the longest side of a right triangle and a and b are the lengths of the other two sides.

25. Find the length of one side of a right triangle if the length of the longest side is 17 cm and the length of one of the other sides is 8 cm.

Chapters 1–9	C U M U L A T I V E R E V I E W E X E R C I S E S

1. What is the additive inverse of $\frac{3}{5}$?

2. What is the multiplicative inverse of $-\frac{4}{7}$?

In Exercises 3–9, simplify each expression.

3. $\left(\dfrac{30x^4y^{-3}}{12y^{-1}}\right)^{-2}$ **4.** $(2h-6)^2-2[-4h(3-h)]$

5. $(\sqrt{2x})^2$ **6.** $(\sqrt{11}+\sqrt{14})^2$

7. $\sqrt{5}(3+\sqrt{80})$ **8.** $\sqrt{175}-\sqrt{28}$

9. $\dfrac{8}{\sqrt{6}-2}$

10. $-5+\sqrt{2}$ is called the _____ of $-5-\sqrt{2}$.

In Exercises 11–14, perform the indicated operations; express all answers in simplest form.

11. $\dfrac{x^2+x-2}{x^2-1}\div\dfrac{x^2-2x-8}{x^2-4x}$

12. $\dfrac{2x}{x+2}-\dfrac{5}{x-1}$

13. $\dfrac{4}{3x+1}-\dfrac{5}{3x^2+7x+2}+\dfrac{4}{3}$

14. $\dfrac{3x+2}{x^2+2x-15}+\dfrac{x+5}{3x^2+17x+10}$

In Exercises 15–22, solve for all x, or write either "identity" or "no solution."

15. $3(x-2)-5(3-x)=x+1$ **16.** $4x(x-1)=-1$

17. $\dfrac{2x-3}{2}=\dfrac{5x+4}{6}-\dfrac{5}{3}$

18. $6(2x+4)-5=8-3(7-4x)$

19. $4[3-5(x+1)]=12-20(x+1)$

20. $\sqrt{3x+2}=4$

21. $\dfrac{5}{x+2}-\dfrac{3}{x-5}=\dfrac{9}{4}$

22. $3(x-2)\le 6$

In Exercises 23–26, set up each problem algebraically, solve, and check. Be sure to state what your variables represent.

23. The area of a triangle is 16 sq. ft, and its altitude is 4 ft longer than its base. Find the lengths of the base and the altitude.

24. It takes Farah 20 min to drive to work in the morning, but it takes her 36 min to return home over the same route during the evening rush hour. If her average morning speed is 16 mph faster than her average evening speed, how far is it from her apartment to her office?

25. Sylvia has thirty-two coins with a total value of $2.45. If all the coins are nickels and dimes, how many of each does she have?

26. The sum of the first two of three consecutive odd integers added to the sum of the last two is 140. Find the integers.

Critical Thinking and Problem-Solving Exercises

1. What two numbers come next in this set of numbers:

$$4, 8, 9, 18, 14, 28, \underline{\hspace{1cm}}, \underline{\hspace{1cm}}$$

2. Ricardo's daughter is 5 years old and his son is 8 years old, so the *sum of the digits* of the children's ages is 13. How old will Ricardo's daughter be the next time that the *sum of the digits* of their ages is 13?

3. Annabel asked her brother Graham how much money he had. Graham, who likes to tease, replied, "The number of dollars I have is more than 270 but fewer than 360, and the number of dollars is divisible by 5 and also divisible by 9." How many dollars does Graham have?

4. Lavonne has six compact discs, one each of the music of her six favorite artists, and she has put these six CDs into two stacks. She put the Paula Abdul CD to the right of the Elton John CD, but not on top. She put the Kenny G CD on top of the Whitney Houston CD. She put the Elton John CD directly under the Whitney Houston CD. Finally, she put the Chuck Mangione CD between the Garth Brooks CD and the Paula Abdul CD. Figure out which CD is in which position.

5. The solution of each of the following problems contains an error. Find and describe the error, and solve the problem correctly.

a. Simplify $\sqrt{18} + 3\sqrt{32}$.

$$\sqrt{18} + 3\sqrt{32} = \sqrt{9 \cdot 2} + 3\sqrt{16 \cdot 2}$$
$$= 3\sqrt{2} + 7\sqrt{2}$$
$$= 10\sqrt{2}$$

b. Solve $\sqrt{3x - 2} = \sqrt{x} + 2$.

$$\sqrt{3x - 2} = \sqrt{x} + 2$$
$$(\sqrt{3x - 2})^2 = (\sqrt{x} + 2)^2$$
$$3x - 2 = x + 4$$
$$2x - 2 = 4$$
$$2x = 6$$
$$x = 3$$

6. In logic, a *statement* is a sentence that is either *true* or *false*, and we generally use a single *letter* to represent a statement. For example, p might be the statement "The moon is made of blue cheese," and q might be the statement "It is hot today." *Compound statements* in logic are made up by combining two statements with conjunctions such as "and" and "or." "The moon is made of blue cheese and it is hot today" is an example of a compound statement.

One particular compound statement is the statement "If p, then q." (For the statements given in the example in the paragraph above, this translates as "If the moon is made of blue cheese, then it is hot today.") The *converse* of the statement "If p, then q" is defined to be the statement "If q, then p." The converse of the statement "If the moon is made of blue cheese, then it is hot today" is "If it is hot today, then the moon is made of blue cheese."

When the statement "If p, then q" is true, its converse can be either true or false.

a. The statement "If $a = b$, then $a^2 = b^2$" is a true statement. (This was the statement of the squaring property.) What is the converse of "If $a = b$, then $a^2 = b^2$"?

b. Is the converse of "If $a = b$, then $a^2 = b^2$" always true? (Give an example that supports your answer.)

Quadratic Equations

CHAPTER 10

I n this chapter, we discuss several methods for solving quadratic equations. Earlier, we solved quadratic equations by factoring; in this chapter, we will review that method and introduce new methods, as well.

10.1 The General Form for Quadratic Equations

A **quadratic equation** is a polynomial equation whose highest-degree term is a second-degree term.

EXAMPLE 1

Examples of quadratic equations:

a. $8x^2 + 2x - 4 = 0$ **b.** $x^2 + 1 = 0$ **c.** $\frac{1}{3}x - 3x^2 = \frac{4}{5}$ **d.** $3x^2 = 2x$

The **general form** of a quadratic equation, as used in this text, is as follows:

The general form of a quadratic equation

$$ax^2 + bx + c = 0 \quad \text{or} \quad 0 = ax^2 + bx + c$$

where a, b, and c are integers and $a > 0$.

Notice that a is the coefficient of x^2, b is the coefficient of x, and c is the constant, and notice that 0 can be on the left side or on the right side of the equation.

We don't consider $\frac{2}{3}x^2 - x + 3 = 0$ to be in the general form, since $\frac{2}{3}$ (the coefficient of x^2) is not an integer. We don't consider $2x^2 + 7 = 4x$ to be in the general form, because we don't have zero on one side of the equal sign.

Converting a quadratic equation into the general form

1. Remove denominators, if there are any, by multiplying both sides of the equation by the LCD.
2. Remove grouping symbols, if there are any.
3. Collect and combine like terms on each side of the equal sign.
4. Arrange all nonzero terms in descending powers of the variable on one side of the equal sign, leaving only zero on the other side.

We had to convert quadratic equations into the general form (before we could solve them by factoring) in earlier sections, but we did not identify a, b, and c. We do identify a, b, and c in Example 2.

EXAMPLE 2

Convert each of the following quadratic equations into the general form, and identify a, b, and c.

a. $8x = -3 - 2x^2$

SOLUTION $8x = -3 - 2x^2$

$2x^2 + 3 + 8x = 2x^2 + 3 - 3 - 2x^2$ Adding $2x^2 + 3$ to both sides

$2x^2 + 8x + 3 = 0$ ⟵———————— The general form

————— The constant, c, is 3

————— The coefficient of x is 8; therefore, $b = 8$

————— The coefficient of x^2 is 2; therefore, $a = 2$

The general form is $2x^2 + 8x + 3 = 0$, and $a = 2$, $b = 8$, and $c = 3$.

b. $3x^2 = 5$

SOLUTION $3x^2 = 5$

$3x^2 \boxed{-5} = 5 \boxed{-5}$ Adding -5 to both sides

$3x^2 - 5 = 0$ There is no x-term, so b (the coefficient of x) is understood to be 0

or $\boxed{3}\, x^2 + \boxed{0}\, x + \boxed{(-5)} = 0$ Notice that c, the constant, is -5, not 5

The general form is $3x^2 - 5 = 0$, and $a = 3$, $b = 0$, and $c = -5$.

c. $8x = 7x^2$

SOLUTION We will add $-8x$ to both sides so that a will be positive.

$8x = 7x^2$

$8x \boxed{-8x} = 7x^2 \boxed{-8x}$ Adding $-8x$ to both sides

$0 = \boxed{7}\, x^2 + \boxed{(-8)}\, x + \boxed{0}$ The constant, c, is 0

The general form is $0 = 7x^2 - 8x$, and $a = 7$, $b = -8$, and $c = 0$.

d. $\frac{1}{2}x^2 - 3x = \frac{2}{3}$

SOLUTION We clear fractions first; the LCD is 6.

$\frac{1}{2}x^2 - 3x = \frac{2}{3}$

$\boxed{6}\left(\frac{1}{2}x^2 - 3x\right) = \boxed{6}\left(\frac{2}{3}\right)$ Multiplying both sides by 6

$6\left(\frac{1}{2}x^2\right) - 6(3x) = 4$ Using the distributive property on the left and performing the multiplication on the right

$3x^2 - 18x = 4$

$3x^2 - 18x \boxed{-4} = 4 \boxed{-4}$ Adding -4 to both sides

$\boxed{3}\, x^2 + \boxed{(-18)}\, x + \boxed{(-4)} = 0$

The general form is $3x^2 - 18x - 4 = 0$, and $a = 3$, $b = -18$, and $c = -4$.

e. $(x + 3)(x - 2) = 2x + 1$

SOLUTION We remove the parentheses first.

$(x + 3)(x - 2) = 2x + 1$

$x^2 + x - 6 = 2x + 1$ Using the FOIL method on the left

$x^2 + x - 6 \boxed{-2x - 1} = 2x + 1 \boxed{-2x - 1}$ Adding $-2x - 1$ to both sides

$\boxed{1}\, x^2 + \boxed{(-1)}\, x + \boxed{(-7)} = 0$

The general form is $x^2 - x - 7 = 0$, and $a = 1$, $b = -1$, and $c = -7$.

Exercises  10.1
Set I

Convert each of the following quadratic equations into the general form, and identify a, b, and c.

1. $2x^2 = 5x + 3$

2. $3x^2 = 4 - 2x$

3. $6x^2 = x$

4. $2x - 3x^2 = 0$

5. $\frac{3x}{2} + 5 = x^2$

6. $4 - x^2 = \frac{2x}{3}$

7. $x^2 - \frac{5x}{4} + \frac{2}{3} = 0$

8. $2x^2 + \frac{3x}{5} = \frac{1}{3}$

9. $x(x - 3) = 4$

10. $2x(x + 1) = 12$

11. $3x(x + 1) = (x + 1)(x + 2)$

12. $(x + 1)(x - 3) = 4x(x - 1)$

Writing Problems

Express the answer in your own words and in complete sentences.

1. Describe the first two steps you would use to get the equation $(x + 5)(x - 1) = 3x(x + 2)$ into the general form.

Exercises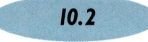
Set II

Convert each of the following quadratic equations into the general form, and identify a, b, and c.

1. $5x = 4x^2 + 1$

2. $3 = 5x - x^2$

3. $5x^2 = x$

4. $(x - 2)(2x + 4) = 0$

5. $\dfrac{4x}{3} + 2 = x^2$

6. $7x^2 = 3$

7. $3x^2 + \dfrac{x}{4} = \dfrac{1}{5}$

8. $\left(\dfrac{x}{2} - 3\right)\left(x + \dfrac{1}{4}\right) = 2$

9. $(x - 1)(3x) = 7$

10. $2 = 3x^2$

11. $3x(x + 1) = (x + 2)(x - 3)$

12. $(4x - 1)(x + 2) = x(x - 4)$

10.2   Solving Quadratic Equations by Factoring

Recall that the method of solving quadratic equations by factoring is as follows.

> **Solving a quadratic equation by factoring**
>
> **1.** Write the equation in the general form
> $$ax^2 + bx + c = 0 \quad \text{or} \quad 0 = ax^2 + bx + c$$
>
> **2.** Factor the polynomial.
>
> **3.** Use the zero-factor property; that is, set each factor equal to zero. Then solve each resulting equation.
>
> **4.** Check apparent solutions in the original equation.

EXAMPLE 1   Solve $5 - 2x^2 = 3x$.

SOLUTION   We first write the equation in the general form. We'll move all nonzero terms to the *right* so that a, the coefficient of x^2, will be positive.

$$5 - 2x^2 = 3x$$

$$5 - 2x^2 \boxed{+ 2x^2 - 5} = 3x \boxed{+ 2x^2 - 5} \qquad \text{Adding } 2x^2 - 5 \text{ to both sides}$$

$$0 = \boxed{2}x^2 + 3x - 5 \qquad \text{Arranging the polynomial in descending powers of } x; \text{ the coefficient of } x^2 \text{ is positive}$$

$$0 = (2x + 5)(x - 1) \qquad \text{Factoring the polynomial}$$

$$2x + 5 = 0 \qquad\qquad x - 1 = 0 \quad \text{Setting each factor equal to 0}$$

$$2x + 5 \;\boxed{-\;5} = 0 \;\boxed{-\;5} \qquad\qquad x - 1 \;\boxed{+\;1} = 0 \;\boxed{+\;1}$$

$$2x = -5 \qquad\qquad\qquad x = 1$$

$$\frac{2x}{2} = \frac{-5}{2}$$

$$x = -\tfrac{5}{2}$$

✓ **Check for $x = -\tfrac{5}{2}$**   **Check for $x = 1$**

$$5 - 2x^2 = 3x \qquad\qquad 5 - 2x^2 = 3x$$

$$5 - 2\left(-\tfrac{5}{2}\right)^2 \overset{?}{=} 3\left(-\tfrac{5}{2}\right) \qquad 5 - 2(1)^2 \overset{?}{=} 3(1)$$

$$5 - \overset{1}{2}\left(\tfrac{25}{4}\right) \overset{?}{=} -\tfrac{15}{2} \qquad\qquad 5 - 2(1) \overset{?}{=} 3$$

$$\tfrac{10}{2} - \tfrac{25}{2} \overset{?}{=} -\tfrac{15}{2} \qquad\qquad 5 - 2 \overset{?}{=} 3$$

$$-\tfrac{15}{2} = -\tfrac{15}{2} \quad \text{True} \qquad\qquad 3 = 3 \quad \text{True}$$

The solutions are $-\tfrac{5}{2}$ and 1.

EXAMPLE 2

Solve $x^2 = \dfrac{3 - 5x}{2}$.

SOLUTION   We can treat this equation as a proportion, or we can multiply both sides by 2 (the LCD). There are no excluded values.

$$\frac{x^2}{1} = \frac{3 - 5x}{2}$$

$$2x^2 = 3 - 5x \qquad\qquad \text{Multiplying both sides by 2 or cross-multiplying}$$

$$2x^2 \;\boxed{+\;5x\;-\;3} = 3 - 5x \;\boxed{+\;5x\;-\;3} \qquad \text{Adding } 5x - 3 \text{ to both sides}$$

$$2x^2 + 5x - 3 = 0 \qquad\qquad \text{The general form}$$

$$(2x - 1)(x + 3) = 0 \qquad\qquad \text{Factoring the polynomial}$$

$$2x - 1 = 0 \qquad\qquad x + 3 = 0 \quad \text{Setting each factor equal to 0}$$

$$2x - 1 \;\boxed{+\;1} = 0 \;\boxed{+\;1} \qquad\qquad x + 3 \;\boxed{-\;3} = 0 \;\boxed{-\;3}$$

$$2x = 1 \qquad\qquad\qquad x = -3$$

$$\frac{2x}{2} = \frac{1}{2}$$

$$x = \tfrac{1}{2}$$

You can verify that the solutions are $\tfrac{1}{2}$ and -3.

EXAMPLE 3

Solve $\dfrac{x-1}{x-3} = \dfrac{12}{x+1}$.

SOLUTION We can cross-multiply, since this equation is a proportion, or we can multiply both sides by the LCD, $(x-3)(x+1)$. Excluded values are 3 and -1.

$$\frac{x-1}{x-3} = \frac{12}{x+1}$$

$$(x-1)(x+1) = 12(x-3) \qquad \textcolor{red}{\text{Multiplying both sides by the LCD } or \text{ cross-multiplying}}$$

$$x^2 - 1 = 12x - 36$$

$$x^2 - 1 \;\boxed{-12x + 36} = 12x - 36 \;\boxed{-12x + 36} \qquad \textcolor{red}{\text{Adding } -12x + 36 \text{ to both sides}}$$

$$x^2 - 12x + 35 = 0 \qquad \textcolor{red}{\text{The general form}}$$

$$(x-5)(x-7) = 0 \qquad \textcolor{red}{\text{Factoring the polynomial}}$$

$$\textcolor{red}{\text{Setting each factor equal to 0}}$$

$x - 5 = 0$	$x - 7 = 0$
$x - 5 \;\boxed{+5} = 0 \;\boxed{+5}$	$x - 7 \;\boxed{+7} = 0 \;\boxed{+7}$
$x = 5$	$x = 7$

Checking will confirm that the solutions are 5 and 7.

EXAMPLE 4

Solve $(6x+2)(x-4) = 2 - 11x$.

SOLUTION

$$(6x+2)(x-4) = 2 - 11x$$

$$6x^2 - 22x - 8 = 2 - 11x$$

$$6x^2 - 22x - 8 \;\boxed{+11x - 2} = 2 - 11x \;\boxed{+11x - 2} \qquad \textcolor{red}{\text{Adding } 11x - 2 \text{ to both sides}}$$

$$6x^2 - 11x - 10 = 0 \qquad \textcolor{red}{\text{The general form}}$$

$$(3x+2)(2x-5) = 0 \qquad \textcolor{red}{\text{Factoring the polynomial}}$$

$$\textcolor{red}{\text{Setting each factor equal to 0}}$$

$3x + 2 = 0$	$2x - 5 = 0$
$3x + 2 \;\boxed{-2} = 0 \;\boxed{-2}$	$2x - 5 \;\boxed{+5} = 0 \;\boxed{+5}$
$3x = -2$	$2x = 5$
$\dfrac{3x}{3} = \dfrac{-2}{3}$	$\dfrac{2x}{2} = \dfrac{5}{2}$
$x = -\frac{2}{3}$	$x = \frac{5}{2}$

You can verify that the solutions are $-\frac{2}{3}$ and $\frac{5}{2}$.

EXAMPLE 5

Michelle bicycled from her home to the beach and back, a *total* distance of 120 mi. Her average speed returning was 3 mph slower than her average speed going to the beach. If her total bicycling time was 9 hr, what was her average speed going to the beach?

SOLUTION

Let $\quad x =$ the average speed (in mph) going to the beach

Then $x - 3 =$ the average speed (in mph) returning from the beach

The distance each way is 60 mi. We solve the formula $rt = d$ (the distance-rate-time formula) for t:

$$t = \frac{d}{r}$$

We then use this formula to find the time going to the beach and the time returning from the beach:

$$\frac{60 \text{ mi}}{x \, \frac{\text{mi}}{\text{hr}}} = \text{the time (in hours) going to the beach}$$

$$\frac{60 \text{ mi}}{(x - 3) \, \frac{\text{mi}}{\text{hr}}} = \text{the time (in hours) returning from the beach}$$

The sum of the two times equals 9 hr:

$$\frac{60}{x} + \frac{60}{x - 3} = 9$$

$$x(x - 3)\left(\frac{60}{x} + \frac{60}{x - 3}\right) = x(x - 3)\,(9) \qquad \text{Multiplying both sides by the LCD}$$

$$\overset{1}{x}(x - 3)\left(\frac{60}{\underset{1}{x}}\right) + x(\overset{1}{x - 3})\left(\frac{60}{\underset{1}{x - 3}}\right) = x(x - 3)\,(9) \qquad \text{Simplifying}$$

$$60(x - 3) + 60x = 9x(x - 3)$$

$$60x - 180 + 60x = 9x^2 - 27x$$

$$120x - 180 = 9x^2 - 27x \qquad \text{Combining like terms}$$

$$120x - 180 \; - 120x + 180 \; = 9x^2 - 27x \; - 120x + 180$$

$$0 = 9x^2 - 147x + 180 \qquad \text{The general form}$$

$$0 = 3(3x^2 - 49x + 60) \qquad \text{Factoring out 3}$$

$$0 = 3(3x - 4)\,(x - 15) \qquad \text{Factoring the polynomial}$$

Setting each factor equal to 0

$3 \neq 0$	$3x - 4 = 0$	$x - 15 = 0$
	$3x - 4 + 4 = 0 + 4$	$x - 15 + 15 = 0 + 15$
	$3x = 4$	$x = 15$
	$\dfrac{3x}{3} = \dfrac{4}{3}$	
	$x = \frac{4}{3}$	

✓ **Check for $x = \frac{4}{3}$**   If $\frac{4}{3}$ is the speed going to the beach, then the speed returning is

$$x - 3 = \tfrac{4}{3} - 3 = \tfrac{4}{3} - \tfrac{9}{3} = -\tfrac{5}{3}$$

But this is not possible, since a speed cannot be negative.

✓ **Check for $x = 15$**   If 15 is the speed going to the beach, then

$$x - 3 = 15 - 3 = 12 \qquad \text{The speed returning, in mph}$$

$$\frac{60}{x} = \frac{60}{15} = 4 \qquad \text{The time going to the beach, in hours}$$

$$\frac{60}{x - 3} = \frac{60}{12} = 5 \qquad \text{The time returning, in hours}$$

The sum of the two times is 9 hr. Therefore, Michelle's average speed going to the beach was 15 mph.

Exercises 10.2
Set I

In Exercises 1–22, solve each equation by factoring.

1. $x^2 + x - 6 = 0$

2. $x^2 - x - 6 = 0$

3. $x^2 + x = 12$

4. $x^2 + 2x = 15$

5. $2x^2 - x = 1$

6. $2x^2 + x = 1$

7. $\dfrac{x}{8} = \dfrac{2}{x}$

8. $\dfrac{x}{3} = \dfrac{12}{x}$

9. $\dfrac{x + 2}{3} = \dfrac{-1}{x - 2}$

10. $\dfrac{x + 3}{2} = \dfrac{-4}{x - 3}$

11. $x^2 + 9x + 8 = 0$

12. $x^2 + 7x + 6 = 0$

13. $2x^2 + 4x = 0$

14. $3x^2 - 9x = 0$

15. $x^2 = x + 2$

16. $x^2 = x + 6$

17. $\dfrac{x}{2} + \dfrac{2}{x} = \dfrac{5}{2}$

18. $\dfrac{x}{3} + \dfrac{2}{x} = \dfrac{7}{3}$

19. $\dfrac{x - 1}{4} + \dfrac{6}{x + 1} = 2$

20. $\dfrac{3x + 1}{5} + \dfrac{8}{3x - 1} = 3$

21. $2x^2 = \dfrac{2 - x}{3}$

22. $4x^2 = \dfrac{14x - 3}{2}$

In Exercises 23–32, set up each problem algebraically, solve, and check. Be sure to state what your variables represent.

23. The length of a rectangle is 5 m more than its width. If its area is 24 sq. m, find its dimensions.

24. The length of a rectangle is 4 yd more than its width. If its area is 77 sq. yd, find its dimensions.

25. If the product of two consecutive even integers is increased by 4, the result is 84. Find the integers.

26. If the product of two consecutive integers is decreased by 6, the result is 36. Find the integers.

27. Bruce drove from Los Angeles to the Mexican border and back to Los Angeles, a total distance of 240 mi. His average speed returning to Los Angeles was 20 mph faster than his average speed going to Mexico. If his total driving time was 5 hr, what was his average speed driving from Los Angeles to Mexico?

28. Nick drove from his home to San Diego and back, a total distance of 360 mi. His average speed returning to his home was 15 mph faster than his average speed going to San Diego. If his total driving time was 7 hr, what was his average speed driving from his home to San Diego?

29. The base of a triangle is 3 in. longer than its altitude. The area of the triangle is 20 sq. in. Find the lengths of the altitude and the base.

30. The base of a triangle is 3 cm longer than its altitude. The area of the triangle is 90 sq. cm. Find the lengths of the altitude and the base.

31. The sum of the base and the altitude of a triangle is 19 in. The area of the triangle is 42 sq. in. Find the lengths of the base and the altitude.

32. The sum of the base and the altitude of a triangle is 15 cm. The area of the triangle is 27 sq. cm. Find the lengths of the base and the altitude.

Exercises 10.2
Set II

In Exercises 1–22, solve each equation by factoring.

1. $x^2 + x - 12 = 0$

2. $x^2 + 2x - 8 = 0$

3. $x^2 + 4x = 5$

4. $x^2 + 2x = 63$

5. $3x^2 - 2x = 1$

6. $2x^2 = 3 - 5x$

7. $\dfrac{x}{27} = \dfrac{3}{x}$

8. $\dfrac{x - 2}{5} = \dfrac{1}{x + 2}$

9. $\dfrac{x + 1}{1} = \dfrac{3}{x - 1}$

10. $2x^2 + 7x + 3 = 0$

11. $x^2 + 6x + 5 = 0$

12. $x^2 + x = 0$

13. $2x^2 - 6x = 0$

14. $x^2 + 2x = 15$

15. $x^2 = x + 12$

16. $x + \dfrac{1}{4} = \dfrac{3}{4x}$

17. $\dfrac{x}{2} - \dfrac{3}{x} = \dfrac{5}{4}$

18. $\dfrac{x + 3}{4} + \dfrac{3}{x + 3} = \dfrac{61}{28}$

19. $\dfrac{2x + 1}{5} + \dfrac{10}{2x + 1} = 3$

20. $\dfrac{x}{2} = \dfrac{19}{10} + \dfrac{3}{x}$

21. $3x^2 = \dfrac{11x + 10}{2}$

22. $2x^2 = \dfrac{1 - 5x}{3}$

In Exercises 23–32, set up each problem algebraically, solve, and check. Be sure to state what your variables represent.

23. The length of a rectangle is 5 km more than its width. If its area is 36 sq. km, find its dimensions.

24. The length of a rectangle is 3 times its width. If the numerical sum of its area and perimeter is 80, find its dimensions.

25. If the product of two consecutive even integers is increased by 8, the result is 16. Find the integers.

26. If the product of two consecutive odd integers is decreased by 3, the result is 12. Find the integers.

27. Leon drove from his home to Lake Shasta and back, a total distance of 600 mi. His average speed returning to his home was 10 mph faster than his average speed going to Lake Shasta. If his total driving time was 11 hr, what was his average speed driving from his home to Lake Shasta?

28. Ruth drove from Creston to Des Moines, a distance of 90 mi. Then she continued on from Des Moines to Omaha, a distance of 120 mi. Her average speed was 10 mph faster on the second part of the journey than on the first part. If the total driving time was 6 hr, what was her average speed on the first leg of the journey?

29. The base of a triangle is 4 m longer than its altitude. The area of the triangle is 48 sq. m. Find the lengths of the altitude and the base.

30. The base of a triangle is 2 cm shorter than its altitude. The area of the triangle is 40 sq. cm. Find the lengths of the altitude and the base.

31. The sum of the base and the altitude of a triangle is 17 m. The area of the triangle is 36 sq. m. Find the lengths of the altitude and the base.

32. The difference of the base and the altitude of a triangle is 4 ft. (The altitude is greater than the base.) The area of the triangle is 16 sq. ft. Find the lengths of the altitude and the base.

10.3 Solving Incomplete Quadratic Equations

A quadratic equation is an equation that can be expressed in the form $ax^2 + bx + c = 0$, where $a \neq 0$. (If a were zero, the equation would not be a quadratic equation.) An **incomplete quadratic equation** is a quadratic equation in which $b = 0$, or $c = 0$, or both b and c are zero.

EXAMPLE 1   Examples of incomplete quadratic equations:

a. $12x^2 + 5 = 0$ $(b = 0)$ **b.** $7x^2 - 2x = 0$ $(c = 0)$ **c.** $3x^2 = 0$ $(b = 0$ and $c = 0)$

We don't need any special rules or methods for solving an incomplete quadratic equation when $c = 0$ (see Examples 2 and 3).

EXAMPLE 2   Solve $12x^2 = 3x$.

SOLUTION

$$12x^2 = 3x$$
$$12x^2 \boxed{-3x} = 3x \boxed{-3x} \qquad \text{Adding } -3x \text{ to both sides}$$
$$12x^2 - 3x = 0 \qquad \text{The general form}$$
$$3x(4x - 1) = 0 \qquad \text{Factoring the polynomial}$$

Setting each factor equal to 0

$3x = 0$	$4x - 1 = 0$
$\dfrac{3x}{3} = \dfrac{0}{3}$	$4x - 1 \boxed{+1} = 0 \boxed{+1}$
$x = 0$	$4x = 1$
	$\dfrac{4x}{4} = \dfrac{1}{4}$
	$x = \frac{1}{4}$

You can verify that the solutions are 0 and $\frac{1}{4}$.

A Word of Caution In solving the equation in Example 2, a common error is to divide both sides of the equation by $3x$:

$$12x^2 = 3x$$

$$\frac{12x^2}{3x} = \frac{3x}{3x}$$

$$4x = 1$$

$$x = \tfrac{1}{4}$$

In the incorrect method, we *lose* the solution 0 by dividing both sides of the equation by $3x$. Notice that when the equation $12x^2 = 3x$ is solved correctly in Example 2, there are *two* solutions, 0 and $\tfrac{1}{4}$.

Do not divide both sides of an equation by an expression containing the variable, because you may lose solutions if you do.

EXAMPLE 3   Solve $\tfrac{2}{5}x = 3x^2$.

SOLUTION   The LCD = 5.

$$\tfrac{2}{5}x = 3x^2$$

$$5\left(\tfrac{2}{5}x\right) = 5\,(3x^2) \qquad \text{Multiplying both sides by 5}$$

$$2x = 15x^2$$

$$2x - 2x = 15x^2 - 2x \qquad \text{Adding } -2x \text{ to both sides}$$

$$0 = 15x^2 - 2x \qquad \text{The general form}$$

$$0 = x(15x - 2) \qquad \text{Factoring the polynomial}$$

$$x = 0 \qquad\qquad 15x - 2 = 0 \quad \begin{array}{l}\text{Setting each factor}\\\text{equal to 0}\end{array}$$

$$15x - 2 + 2 = 0 + 2$$

$$15x = 2$$

$$\frac{15x}{15} = \frac{2}{15}$$

$$x = \tfrac{2}{15}$$

Checking will confirm that the solutions are 0 and $\tfrac{2}{15}$.

Some incomplete quadratic equations in which $b = 0$ can be solved by factoring. Some examples are $x^2 - 4 = 0$ (see Example 4, Method 2) and $4x^2 - 9 = 0$. Other incomplete quadratic equations in which $b = 0$ can be solved by using Property 1, which follows:

Property 1

If $x^2 = p$, then $\qquad\qquad x = \sqrt{p} \quad \text{or} \quad x = -\sqrt{p}$

We can combine the answers $\sqrt{p}$ and $-\sqrt{p}$ by using the $\pm$ symbol (read "plus or minus") introduced earlier.

To solve an incomplete quadratic equation in which $b = 0$ (that is, an equation that is of the form $ax^2 + c = 0$) by using Property 1, we use the following procedure:

Solving a quadratic equation of the form $ax^2 + c = 0$

1. Arrange the terms of the equation so that the second-degree term is on one side of the equal sign and the constant term is on the other side.
2. Divide both sides of the equation by a, the coefficient of x^2. (The equation should now be in the form $x^2 = p$.)
3. Use Property 1: If $x^2 = p$, then $x = \sqrt{p}$ or $x = -\sqrt{p}$.
4. Express the solutions in simplest form. When the radicand is *positive* or *zero*, the square roots are real numbers; when the radicand is *negative*, the square roots are *not* real numbers.
5. Check all apparent real solutions by substituting those values into the original equation.

In Example 4, we demonstrate the method given in the box above and also show that the equation can be solved by factoring.

EXAMPLE 4

Solve $x^2 - 4 = 0$.

SOLUTION $x^2 - 4 = 0$ is in the general form, and $b = 0$. Step 2 does not apply, because $a = 1$.

Method 1: Using Property 1

$$x^2 - 4 = 0$$

Step 1. $x^2 - 4 \boxed{+ 4} = 0 \boxed{+ 4}$

$$x^2 = 4$$

Step 3. $x = \pm\sqrt{4}$

Step 4. $x = \pm 2$

Method 2: Solving by factoring

$$x^2 - 4 = 0$$

$$(x + 2)(x - 2) = 0$$

$x + 2 = 0$ $x - 2 = 0$

$x + 2 \boxed{- 2} = 0 \boxed{- 2}$ $x - 2 \boxed{+ 2} = 0 \boxed{+ 2}$

$x = -2$ $x = 2$

Notice that we obtain both 2 and -2 as solutions when we use either method. You can verify that the solutions are 2 and -2.

In Example 5, we solve the equation $4x^2 - 9 = 0$ by using Property 1, although that equation *can* be solved by factoring over the integers.

EXAMPLE 5

Solve $4x^2 - 9 = 0$.

SOLUTION $4x^2 - 9 = 0$ is in the general form, and $b = 0$.

$$4x^2 - 9 = 0$$

Step 1. $4x^2 - 9 \boxed{+ 9} = 0 \boxed{+ 9}$ Adding 9 to both sides

$$4x^2 = 9$$

Step 2. $\dfrac{4x^2}{4} = \dfrac{9}{4}$ Dividing both sides by 4

$$x^2 = \tfrac{9}{4}$$

Step 3. $x = \pm\sqrt{\tfrac{9}{4}}$ Using Property 1

Step 4. $x = \pm\tfrac{3}{2}$

Checking will confirm that the solutions are $\tfrac{3}{2}$ and $-\tfrac{3}{2}$.

Examples 6 and 7 *cannot* be solved by factoring over the integers.

EXAMPLE 6

Solve $3x^2 - 5 = 0$. Express the answers in simplest radical form, and also find decimal approximations of the solutions, rounding off the approximations to two decimal places.

SOLUTION $3x^2 - 5 = 0$ is in the general form, and $b = 0$.

$$3x^2 - 5 = 0$$

Step 1. $3x^2 - 5 \boxed{+ 5} = 0 \boxed{+ 5}$ Adding 5 to both sides

$$3x^2 = 5$$

Step 2. $\dfrac{3x^2}{3} = \dfrac{5}{3}$ Dividing both sides by 3

$$x^2 = \tfrac{5}{3}$$ ——— Using Property 1

Step 3. $x = \pm\sqrt{\dfrac{5}{3}} = \pm\sqrt{\dfrac{5 \cdot 3}{3 \cdot 3}}$ Under the radical sign, we multiply numerator and denominator by 3 to rationalize the denominator

Step 4. $x = \pm\dfrac{\sqrt{15}}{3} \approx \pm\dfrac{3.873}{3} \approx \pm 1.29$ Using a calculator (or Table 1) to approximate $\sqrt{15}$

✓ **Step 5. Check for $x = +\dfrac{\sqrt{15}}{3}$**   **Check for $x = -\dfrac{\sqrt{15}}{3}$**

$$3x^2 - 5 = 0 \qquad\qquad\qquad 3x^2 - 5 = 0$$

$$3\left(\dfrac{\sqrt{15}}{3}\right)^2 - 5 \overset{?}{=} 0 \qquad\qquad 3\left(-\dfrac{\sqrt{15}}{3}\right)^2 - 5 \overset{?}{=} 0$$

$$3\left(\tfrac{15}{9}\right) - 5 \overset{?}{=} 0 \qquad\qquad 3\left(\tfrac{15}{9}\right) - 5 \overset{?}{=} 0$$

$$5 - 5 \overset{?}{=} 0 \qquad\qquad\qquad 5 - 5 \overset{?}{=} 0$$

$$0 = 0 \quad \text{True} \qquad\qquad\qquad 0 = 0 \quad \text{True}$$

The solutions are $\dfrac{\sqrt{15}}{3}$ and $-\dfrac{\sqrt{15}}{3}$, or approximately 1.29 and -1.29.

Note   The solutions $\pm\dfrac{\sqrt{15}}{3}$ are *exact* solutions, whereas ± 1.29 are approximations.

When we're solving equations (in an algebra book or in an algebra class), we're often interested in exact solutions. For *applied* problems, however, decimal approximations are often more useful than exact solutions.

In Example 7, the solutions *do exist*. However, they are not real numbers; they are *imaginary numbers*, and imaginary numbers are not discussed in this book.

EXAMPLE 7

Solve $x^2 + 25 = 0$, or write "not real."

SOLUTION

$$x^2 + 25 = 0$$

$$x^2 + 25 \boxed{- 25} = 0 \boxed{- 25} \quad \text{Adding } -25 \text{ to both sides}$$

Step 1. $x^2 = -25$

Step 3. $x = \pm\sqrt{-25}$ The radicand is negative

Since square roots of negative numbers are not real, the solutions are not real.

EXAMPLE 8   Solve $(x + 3)^2 = 7$ by using Property 1.

SOLUTION   This equation is in the *form* $x^2 = p$, so we start with step 3.

$$(x + 3)^2 = 7$$

This equation has not yet been solved for x

Step 3. $x + 3 = \pm\sqrt{7}$ Using Property I

$x + 3 - 3 = -3 \pm\sqrt{7}$ Adding -3 to both sides

Step 4. $x = -3 \pm\sqrt{7}$

✓ **Step 5. Check for $x = -3 + \sqrt{7}$**   **Check for $x = -3 - \sqrt{7}$**

$$(x + 3)^2 = 7 \qquad\qquad (x + 3)^2 = 7$$

$$(-3 + \sqrt{7} + 3)^2 \overset{?}{=} 7 \qquad (-3 - \sqrt{7} + 3)^2 \overset{?}{=} 7$$

$$(\sqrt{7})^2 \overset{?}{=} 7 \qquad\qquad (-\sqrt{7})^2 \overset{?}{=} 7$$

$$7 = 7 \quad\text{True} \qquad\qquad 7 = 7 \quad\text{True}$$

Therefore, the solutions are $-3 + \sqrt{7}$ and $-3 - \sqrt{7}$. Using a calculator (or Table I), we can find decimal approximations for the answers:

$$-3 + \sqrt{7} \approx -3 + 2.646 = -0.354 \quad\text{and}\quad -3 - \sqrt{7} \approx -3 - 2.646 = -5.646$$

Exercises   *10.3*
Set I

In Exercises 1–20, solve each equation, or write "not real." Check all real answers. Express any real, irrational answers in simplest radical form, and also give decimal approximations for those answers, rounding off the approximations to two decimal places.

1. $8x^2 = 4x$ **2.** $15x^2 = 5x$

3. $x^2 - 9 = 0$ **4.** $x^2 - 36 = 0$

5. $x^2 - 4x = 0$ **6.** $x^2 - 16x = 0$

7. $x^2 + 16 = 0$ **8.** $x^2 + 1 = 0$

9. $x^2 - 1 = 0$ **10.** $x^2 - 16 = 0$

11. $5x^2 = 4$ **12.** $3x^2 = 25$

13. $8 - 2x^2 = 0$ **14.** $27 - 3x^2 = 0$

15. $x(x + 3) = 3x - 4$ **16.** $x(x + 5) = 5x - 9$

17. $2(x + 3) = 6 + x(x + 2)$

18. $5(2x - 3) - x(2 - x) = 8(x - 1) - 7$

19. $2x(3x - 4) = 2(3 - 4x)$

20. $5x(2x - 3) = 3(4 - 5x)$

In Exercises 21–26, follow the method shown in Example 8 to solve the equations. Express the solutions in simplest radical form, and also give decimal approximations of those solutions, rounding off the approximations to two decimal places.

21. $(x - 4)^2 = 23$

22. $(x - 8)^2 = 5$

23. $(x + 3)^2 = 3$

24. $(x + 8)^2 = 11$

25. $(x + 4)^2 = 50$

26. $(x + 1)^2 = 12$

Writing Problems

Express the answers in your own words and in complete sentences.

1. Describe the first two steps you would use in solving the equation $7x^2 = 5x$.

2. Describe the first two steps you would use in solving the equation $6x^2 - 5 = 0$.

Exercises  10.3
Set II

In Exercises 1–20, solve each equation, or write "not real." Check all real answers. Express any real, irrational answers in simplest radical form, and also give decimal approximations for those answers, rounding off the approximations to two decimal places.

1. $8x^2 = 12x$

2. $3x = 6x^2$

3. $x^2 - 25 = 0$

4. $x^2 = 100$

5. $x^2 - 49x = 0$

6. $x^2 = 9x$

7. $x^2 + 81 = 0$

8. $x^2 + 4x = 0$

9. $x^2 - 121 = 0$

10. $5x^2 = 25$

11. $3x^2 = 7$

12. $x^2 - 25x = 0$

13. $64 - 4x^2 = 0$

14. $x(x + 7) = 7x - 1$

15. $x(x + 1) = x - 9$

16. $x^2 + 3(x - 1) = -3$

17. $3(x + 4) = 12 + x(x + 2)$

18. $x(x + 3) - 7 = 3(x - 1)$

19. $4x(5x - 2) = 2(40 - 4x)$

20. $5x(2x - 3) = 3(10 - 5x)$

In Exercises 21–26, follow the method shown in Example 8 to solve the equations. Express the solutions in simplest radical form, and also give decimal approximations of those solutions, rounding off the approximations to two decimal places.

21. $(x - 10)^2 = 13$

22. $(x + 3)^2 = 8$

23. $(x + 6)^2 = 6$

24. $(x - 5)^2 = 15$

25. $(x + 2)^2 = 10$

26. $(x + 11)^2 = 75$

10.4 Solving Quadratic Equations by Completing the Square

The methods we have shown in previous sections can be used to solve only *some* quadratic equations. The method we show in this section (completing the square) and the one we show in the next section can be used to solve *all* quadratic equations.

In completing the square to solve quadratic equations, we want to get the equation into the form $(x - k)^2 = p$. Then we will use Property 1 and the method shown in Example 8 of Section 10.3 to solve the equation for x.

Before we attempt to solve any equations, let's look at some factoring problems.

$$x^2 + 10x + 25 = (x + 5)^2$$

$$x^2 - 6x + 9 = (x - 3)^2$$

$$x^2 + 14x + 49 = (x + 7)^2 \quad \text{You can verify these facts}$$

What if the constants were missing? That is, what if we had

$$x^2 + 10x + ? = (x + ?)^2$$

$$x^2 - 6x + ? = (x - ?)^2$$

$$x^2 + 14x + ? = (x + ?)^2$$

and we needed to determine the numbers that should replace the question marks? What number needs to be inserted into the box in $x^2 + 10x + \boxed{}$ so that the expression will be the square of some binomial? A little experimenting and some thought will show that if we find $\frac{1}{2}$ of 10 (that is, half of the coefficient of x) and square that number, we'll have

$$x^2 + 10x + \boxed{25} = (x + \boxed{5})^2 \quad \tfrac{1}{2}(10) = 5, \text{ and } 5^2 = 25$$

When we find the constant that we can add to $x^2 + 10x$ (for example) to make the expression the *square* of a binomial, we say we are *completing the square*. The number we would have to add to $x^2 - 6x$ to complete the square would be 9 $\left[\frac{1}{2}\text{ of } -6 \text{ is } -3, \text{ and}\right.$ $\left.(-3)^2 = 9\right]$, and the number we would have to add to $x^2 + 14x$ to complete the square would be 49 $\left(\frac{1}{2}\text{ of } 14 \text{ is } 7, \text{ and } 7^2 = 49\right)$.

After we've decided what number must be added to one side of the equation to complete the square, we must use the addition property of equality and add that number to *both* sides of the equation.

The entire procedure for solving an equation by completing the square is given in the following box.

Solving a quadratic equation that is in the form $ax^2 + bx + c = 0$ by completing the square

1. If a, the coefficient of x^2, is not 1, divide both sides of the equation by a. (It is understood that $a \neq 0$.)

2. Get the x^2-term and the x-term on the left side of the equal sign and the constant on the right side.

3. Determine what number should be added to the left side to make the left side the square of a binomial, and add this number *to both sides of the equation*.

4. Factor the left side of the equation—the equation is now in the form $(x + k)^2 = p$.

5. Use Property 1 to solve the equation for $x + k$.

6. Add $-k$ to both sides of the equation to solve for x.

7. Simplify the right side, and if the two solutions are *rational*, separate and simplify them.

8. If the solutions are real, check them in the original equation.

EXAMPLE 1   Solve $x^2 - 6x + 1 = 0$ by completing the square.

SOLUTION   Because the coefficient of x^2 is 1, we start with step 2.

$$x^2 - 6x + 1 = 0$$

Step 2. $x^2 - 6x + 1 \;\boxed{-1} = 0 \;\boxed{-1}$ Adding -1 to both sides

One-half of -6 is -3, and $(-3)^2 = \boxed{9}$

$$x^2 \;\boxed{-6}\, x \qquad = -1$$

Step 3. $x^2 - 6x \;\boxed{+9} = -1 \;\boxed{+9}$ Adding $\boxed{9}$ to both sides to make the left side the square of a binomial

Step 4. $(x - 3)^2 = 8$ Factoring the left side

Step 5. $x - 3 = \pm\sqrt{8}$ Using Property 1

Step 6. $x - 3 \;\boxed{+3} = \boxed{3} \;\pm\sqrt{8}$ Adding 3 to both sides of the equation

Step 7. $x = 3 \pm 2\sqrt{2}$ $\sqrt{8} = \sqrt{4 \cdot 2} = \sqrt{4} \cdot \sqrt{2} = 2\sqrt{2}$

Step 8.

✓ **Check for $x = 3 + 2\sqrt{2}$**

$$x^2 - 6x + 1 = 0$$
$$(3 + 2\sqrt{2})^2 - 6(3 + 2\sqrt{2}) + 1 \overset{?}{=} 0$$
$$9 + 12\sqrt{2} + 8 - 18 - 12\sqrt{2} + 1 \overset{?}{=} 0$$
$$\text{True} \quad 0 = 0$$

Check for $x = 3 - 2\sqrt{2}$

$$x^2 - 6x + 1 = 0$$
$$(3 - 2\sqrt{2})^2 - 6(3 - 2\sqrt{2}) + 1 \overset{?}{=} 0$$
$$9 - 12\sqrt{2} + 8 - 18 + 12\sqrt{2} + 1 \overset{?}{=} 0$$
$$\text{True} \quad 0 = 0$$

Therefore, the solutions are $3 + 2\sqrt{2}$ and $3 - 2\sqrt{2}$.

EXAMPLE 2

Solve $x^2 - 5x + 6 = 0$ by completing the square.

SOLUTION   Because the coefficient of x^2 is 1, we begin with step 2.

$$x^2 - 5x + 6 = 0$$

Step 2. $x^2 - 5x + 6 \boxed{-6} = 0 \boxed{-6}$ Adding -6 to both sides

One-half of -5 is $-\frac{5}{2}$, and $\left(-\frac{5}{2}\right)^2 = \frac{25}{4}$

$$x^2 \boxed{-5}\, x \qquad = -6$$

Step 3. $x^2 - 5x \boxed{+\frac{25}{4}} = -\boxed{\frac{24}{4}} + \boxed{\frac{25}{4}}$ Adding $\frac{25}{4}$ to both sides to make the left side the square of a binomial

Remember that $-6 = -\frac{24}{4}$

Step 4. $\left(x - \frac{5}{2}\right)^2 = \frac{1}{4}$ Factoring the left side

Step 5. $x - \frac{5}{2} = \pm\sqrt{\frac{1}{4}}$ Using Property I

$$x - \frac{5}{2} = \pm\frac{1}{2} \qquad \pm\sqrt{\frac{1}{4}} = \pm\frac{1}{2}$$

Step 6. $x - \frac{5}{2} \boxed{+\frac{5}{2}} = \boxed{\frac{5}{2}} \pm \frac{1}{2}$ Adding $\frac{5}{2}$ to both sides of the equation

$$x = \frac{5}{2} \pm \frac{1}{2}$$

Separating the two answers

Step 7. $x = \frac{5}{2} + \frac{1}{2} = \frac{6}{2} = 3$ or $x = \frac{5}{2} - \frac{1}{2} = \frac{4}{2} = 2$

You can verify that the solutions are 3 and 2, and you might also verify that the equation *could* have been solved by factoring.

EXAMPLE 3

Solve $x^2 + 6x - 3 = 0$ by completing the square.

SOLUTION   Because the coefficient of x^2 is 1, we begin with step 2.

$$x^2 + 6x - 3 = 0$$

Step 2. $x^2 + 6x - 3 \boxed{+3} = 0 \boxed{+3}$ Adding 3 to both sides

$$x^2 + 6x \qquad = 3$$

Step 3. $x^2 + 6x \boxed{+9} = 3 \boxed{+9}$ Completing the square

Step 4. $(x + 3)^2 = 12$ Factoring the left side

Step 5. $x + 3 = \pm\sqrt{12}$ Using Property I

Step 6. $x + 3 \boxed{-3} = \boxed{-3} \pm \sqrt{12}$ Adding -3 to both sides

Step 7. $x = -3 \pm 2\sqrt{3}$ $\sqrt{12} = \sqrt{4 \cdot 3} = 2\sqrt{3}$

Checking will confirm that the solutions are $-3 + 2\sqrt{3}$ and $-3 - 2\sqrt{3}$.

EXAMPLE 4

Solve $4x^2 - 5x + 2 = 0$ by completing the square.

SOLUTION $4x^2 - 5x + 2 = 0$

Step 1. $\dfrac{4x^2 - 5x + 2}{4} = \dfrac{0}{4}$ Dividing both sides by 4, the coefficient of x^2

$$\tfrac{4}{4}x^2 - \tfrac{5}{4}x + \tfrac{2}{4} = 0$$

$$x^2 - \tfrac{5}{4}x + \tfrac{1}{2} = 0$$

Step 2. $x^2 - \frac{5}{4}x + \frac{1}{2} \boxed{- \frac{1}{2}} = 0 \boxed{- \frac{1}{2}}$ Adding $-\frac{1}{2}$ to both sides

$\qquad x^2 - \frac{5}{4}x \qquad = -\frac{1}{2}$ One-half of $-\frac{5}{4}$ is $-\frac{5}{8}$

Step 3. $\quad x^2 - \frac{5}{4}x \boxed{+ \frac{25}{64}} = -\frac{32}{64} \boxed{+ \frac{25}{64}}$ Completing the square and writing $-\frac{1}{2}$ as $-\frac{32}{64}$

Step 4. $\qquad \left(x - \frac{5}{8}\right)^2 = -\frac{7}{64}$ Factoring the left side

Step 5. $\qquad x - \frac{5}{8} = \pm\sqrt{-\frac{7}{64}}$ Using Property 1

Step 6. $\quad x - \dfrac{5}{8} \boxed{+ \dfrac{5}{8}} = \dfrac{5}{8} \pm \dfrac{\sqrt{-7}}{8}$ Adding $\frac{5}{8}$ to both sides

Step 7. $\qquad\qquad x = \dfrac{5}{8} \pm \dfrac{\sqrt{-7}}{8}$

The radicand is negative. Therefore, the solutions are not real.

In Example 5, we use the method of completing the square to solve the quadratic equation that arises from an applied problem. The equation *can* be solved by factoring, but many students find it easier to use a calculator and complete the square in problems with numbers such as the ones in this example.

EXAMPLE 5   The length of a rectangle is 20 cm more than its width. If the area of the rectangle is 4,125 sq. cm, find the width of the rectangle.

SOLUTION   Let $\qquad x = $ the width of the rectangle (in centimeters)
Then $\;x + 20 = $ the length of the rectangle (in centimeters)
and $x(x + 20) = $ the area of the rectangle (in square centimeters)

The area of the rectangle	is	4,125

$\qquad x(x + 20) \qquad = 4,125$

Step 2. $x^2 + 20x \qquad = 4,125$

Step 3. $x^2 + 20x \boxed{+ 100} = 4,125 \boxed{+ 100}$ Completing the square

Step 4. $\qquad (x + 10)^2 = 4,225$ Factoring the left side

Step 5. $\qquad x + 10 = \pm\sqrt{4,225}$ Using Property 1

Step 6. $\quad x + 10 \boxed{- 10} = \boxed{-10} \pm\sqrt{4,225}$ Adding -10 to both sides

Step 7. $\qquad\qquad x = -10 \pm 65$ Using a calculator: $\sqrt{4,225} = 65$

Separating the two answers

$x = -10 + 65 = 55$ $\qquad x = -10 - 65 = -75$ We reject this solution, since a length cannot be negative

$x + 20 = 55 + 20 = 75$

Step 8. If the width is 55 cm and the length is 75 cm, the area is

$$(55 \text{ cm})(75 \text{ cm}) = 4,125 \text{ cm}^2$$

Therefore, the width of the rectangle is 55 cm.

Exercises 10.4
Set I

In Exercises 1–16, solve each equation by completing the square. If the radicand is negative, write "not real." Express any real, irrational solutions in simplest radical form.

1. $x^2 - 4x - 12 = 0$ **2.** $x^2 - 2x - 48 = 0$

3. $x^2 + 8x - 9 = 0$ **4.** $x^2 + 2x - 8 = 0$

5. $x^2 - 4x - 2 = 0$ **6.** $x^2 - 2x - 4 = 0$

7. $x^2 + 8x - 2 = 0$ **8.** $x^2 + 10x - 8 = 0$

9. $3x^2 - x - 2 = 0$ **10.** $2x^2 + 3x - 2 = 0$

11. $4x^2 + 4x - 1 = 0$ **12.** $4x^2 + 3x - 2 = 0$

13. $3x^2 + 2x + 1 = 0$

14. $4x^2 + 3x + 2 = 0$

15. $x^2 + 24x - 2,881 = 0$

16. $x^2 + 28x - 2,613 = 0$

In Exercises 17–20, set up each problem algebraically, solve, and check. Be sure to state what your variables represent.

17. The length of a rectangle is 30 cm more than its width. If the area of the rectangle is 6,664 sq. cm, what is the width of the rectangle?

18. The width of a rectangle is 32 in. less than its length. If the area of the rectangle is 1,769 sq. in., what is the length of the rectangle?

19. Typing at a constant rate, Jaime types 825 words in a certain length of time. If he had typed 20 words per minute faster, he would have cut the time for typing the 825 words by 4 min. At what rate was he typing?

20. A solution is being pumped into a 667-gal tank at a constant rate, and it fills the tank in a certain length of time. If the rate of flow had been 6 gal per minute slower, it would have taken 6 min longer to fill the tank. At what rate is the pump working?

Writing Problems

Express the answers in your own words and in complete sentences.

1. Explain what your first two steps would be in solving the equation $x^2 + 6x + 800 = 0$ by completing the square.

2. Explain what your first two steps would be in solving the equation $5x^2 + 8x - 3 = 0$ by completing the square.

3. Explain why we include the statement "It is understood that $a \neq 0$" in the procedure for solving a quadratic equation by completing the square.

Exercises 10.4
Set II

In Exercises 1–16, solve each equation by completing the square. If the radicand is negative, write "not real." Express any real, irrational solutions in simplest radical form.

1. $x^2 - 2x - 35 = 0$ **2.** $x^2 - 4x - 60 = 0$

3. $x^2 + 6x - 7 = 0$ **4.** $x^2 + 8x - 180 = 0$

5. $x^2 - 4x + 2 = 0$ **6.** $x^2 - 3x + 3 = 0$

7. $x^2 + 2x - 6 = 0$ **8.** $x^2 + 7x - 8 = 0$

9. $3x^2 - 8x - 3 = 0$ **10.** $5x^2 + 3x + 1 = 0$

11. $9x^2 + 9x - 1 = 0$ **12.** $3x^2 - 2x + 4 = 0$

13. $4x^2 + 6x + 1 = 0$

14. $5x^2 + 2x + 3 = 0$

15. $x^2 + 38x - 3,608 = 0$

16. $x^2 - 70x + 741 = 0$

In Exercises 17–20, set up each problem algebraically, solve, and check. Be sure to state what your variables represent.

17. The length of a rectangle is 22 in. more than its width. If the area of the rectangle is 1,995 sq. in., what is the width of the rectangle?

18. The length of a rectangle is 26 cm more than its width. If the area of the rectangle is (numerically) 503 more than its perimeter, what is the width of the rectangle?

19. Painting at a constant rate, Cuong paints 900 sq. ft of a wall in a certain length of time. If he had painted 10 sq. ft per minute faster, he could have cut the time for painting the 900 sq. ft by 1 min. At what rate was he painting?

20. Typing at a constant rate, Leslie types 1,680 words in a certain length of time. If she had typed 14 words per minute faster, she could have cut the time for typing the 1,680 words by 10 min. At what rate was she typing?

10.5

Solving Quadratic Equations by Using the Quadratic Formula

The method of completing the square can be used to solve any quadratic equation. We now use it in solving the general quadratic equation $ax^2 + bx + c = 0$, and in this way we derive the **quadratic formula**.

Deriving the Quadratic Formula

$$ax^2 + bx + c = 0$$ The general form

Step 1. $$\frac{ax^2 + bx + c}{a} = \frac{0}{a}$$ Dividing both sides of the equation by a

$$\frac{ax^2}{a} + \frac{bx}{a} + \frac{c}{a} = 0$$ Remember: $\frac{0}{a} = 0$

$$x^2 + \frac{b}{a}x + \frac{c}{a} = 0$$

Step 2. $$x^2 + \frac{b}{a}x + \boxed{\frac{c}{a}} - \boxed{\frac{c}{a}} = 0 - \boxed{\frac{c}{a}}$$ Adding $-\frac{c}{a}$ to both sides

$$x^2 + \boxed{\frac{b}{a}}x = -\frac{c}{a}$$ One-half of $\frac{b}{a} = \frac{1}{2} \cdot \frac{b}{a} = \boxed{\frac{b}{2a}}$

Step 3. $$x^2 + \frac{b}{a}x + \boxed{\frac{b^2}{4a^2}} = \boxed{\frac{b^2}{4a^2}} - \frac{c}{a}$$ Adding $\left(\frac{b}{2a}\right)^2$, or $\boxed{\frac{b^2}{4a^2}}$, to both sides

Step 4. $$\left(x + \frac{b}{2a}\right)^2 = \frac{b^2}{4a^2} - \frac{4ac}{4a^2}$$ Factoring the left side; note also that $\frac{c}{a} = \frac{4a \cdot c}{4a \cdot a} = \frac{4ac}{4a^2}$

$$\left(x + \frac{b}{2a}\right)^2 = \frac{b^2 - 4ac}{4a^2}$$ Writing the right side as a single term

Step 5. $$x + \frac{b}{2a} = \pm\sqrt{\frac{b^2 - 4ac}{4a^2}}$$ Using Property 1

Step 6. $$x + \boxed{\frac{b}{2a}} - \boxed{\frac{b}{2a}} = -\boxed{\frac{b}{2a}} \pm \frac{\sqrt{b^2 - 4ac}}{2a}$$ Adding $-\frac{b}{2a}$ to both sides and simplifying the radical

Step 7. $$x = \frac{-b \pm \sqrt{b^2 - 4ac}}{2a}$$ Combining the terms on the right side

The quadratic formula	If $ax^2 + bx + c = 0$ and if $a \neq 0$, then $$x = \frac{-b \pm \sqrt{b^2 - 4ac}}{2a}$$

The procedure for using the quadratic formula can be summarized as follows:

Solving a quadratic equation by using the quadratic formula

1. Write the equation in the general form

$$ax^2 + bx + c = 0 \quad \text{or} \quad 0 = ax^2 + bx + c$$

2. Identify a, b, and c, and substitute these values into the *quadratic formula*:

$$x = \frac{-b \pm \sqrt{b^2 - 4ac}}{2a} \quad (a \neq 0)$$

3. Simplify the apparent solutions. (Simplify the *radical*, and reduce the resulting expression to lowest terms.)

4. If the solutions are *rational* (that is, if the radicand $b^2 - 4ac$ is a perfect square), write the two solutions separately.

5. If the apparent solutions are real numbers, check them in the original equation.

A Word of Caution Be sure that your fraction bar and the bar of your radical sign are long enough.

$$x = \frac{-b \pm \sqrt{b^2 - 4ac}}{2a} \quad \text{is incorrect,}$$

$$x = \frac{-b \pm \frac{\sqrt{b^2 - 4ac}}{2a}}{} \quad \text{is incorrect,}$$

and

$$x = \frac{-b \pm \sqrt{b^2} - 4ac}{2a} \quad \text{is incorrect.}$$

Using the Quadratic Formula

EXAMPLE 1 Solve $x^2 - 5x + 6 = 0$ by using the quadratic formula.

SOLUTION

Step 1. The equation is in the general form.

Step 2. $a = 1$, $b = -5$, and $c = 6$. We substitute these values into the formula

$$x = \frac{-b \pm \sqrt{b^2 - 4ac}}{2a}.$$

$$x = \frac{-(-5) \pm \sqrt{(-5)^2 - 4(1)(6)}}{2(1)}$$

Step 3. $x = \dfrac{5 \pm \sqrt{25 - 24}}{2} = \dfrac{5 \pm \sqrt{1}}{2} = \dfrac{5 \pm 1}{2}$ Next we must write the two solutions separately

Step 4. $x = \dfrac{5 + 1}{2} = \dfrac{6}{2} = 3$ or $x = \dfrac{5 - 1}{2} = \dfrac{4}{2} = 2$

✓ **Step 5.** **Check for $x = 3$**

$$x^2 - 5x + 6 = 0$$

$$(3)^2 - 5(3) + 6 \stackrel{?}{=} 0$$

$$9 - 15 + 6 \stackrel{?}{=} 0$$

$$0 = 0 \quad \text{True}$$

Check for $x = 2$

$$x^2 - 5x + 6 = 0$$

$$(2)^2 - 5(2) + 6 \stackrel{?}{=} 0$$

$$4 - 10 + 6 \stackrel{?}{=} 0$$

$$0 = 0 \quad \text{True}$$

Therefore, the solutions are 3 and 2.

You might verify that the equation in Example 1 *could* have been solved by factoring. We can usually solve a quadratic equation more quickly by factoring than by using the quadratic formula or by completing the square. Therefore, after you have mastered using the quadratic formula (and/or completing the square), we recommend that you always try to solve a quadratic equation first by factoring; if you can't easily factor the polynomial, then use one of the other methods. The equations in Examples 2–5 *cannot* be solved by factoring over the integers.

EXAMPLE 2   Solve $2x^2 - 6x - 3 = 0$.

SOLUTION

Step 1. The equation is in the general form.

Step 2. $a = 2$, $b = -6$, and $c = -3$. We substitute these values into the formula

$$x = \frac{-b \pm \sqrt{b^2 - 4ac}}{2a}.$$

$$x = \frac{-(-6) \pm \sqrt{(-6)^2 - 4(2)(-3)}}{2(2)}$$

Step 3. $x = \dfrac{6 \pm \sqrt{36 + 24}}{4} = \dfrac{6 \pm \sqrt{60}}{4} = \dfrac{6 \pm \sqrt{4 \cdot 15}}{4} = \dfrac{6 \pm 2\sqrt{15}}{4}$

$$= \frac{\overset{1}{2}(3 \pm \sqrt{15})}{\underset{2}{4}} \qquad \text{\color{red}Factoring the numerator and reducing the expression}$$

$$= \frac{3 \pm \sqrt{15}}{2}$$

Step 4. The solutions are not rational.

✓ **Step 5.** **Check for $x = \dfrac{3 + \sqrt{15}}{2}$**

$$2x^2 - 6x - 3 = 0$$

$$2\left(\frac{3 + \sqrt{15}}{2}\right)^2 - \overset{3}{6}\left(\frac{3 + \sqrt{15}}{\underset{1}{2}}\right) - 3 \stackrel{?}{=} 0$$

$$2\left(\frac{9 + 6\sqrt{15} + 15}{4}\right) - 3(3 + \sqrt{15}) - 3 \stackrel{?}{=} 0$$

$$\overset{1}{2}\left(\frac{24 + 6\sqrt{15}}{\underset{2}{4}}\right) - 9 - 3\sqrt{15} - 3 \stackrel{?}{=} 0$$

$$12 + 3\sqrt{15} - 12 - 3\sqrt{15} \stackrel{?}{=} 0$$

$$0 = 0 \quad \text{True}$$

We recommend that you check the solution $\dfrac{3 - \sqrt{15}}{2}$.

The solutions are $\dfrac{3 + \sqrt{15}}{2}$ and $\dfrac{3 - \sqrt{15}}{2}$.

EXAMPLE 3

Solve $\frac{1}{4}x^2 = 1 - x$.

SOLUTION

Step 1. We will multiply both sides by 4 to clear fractions:

$$\frac{1}{4}x^2 = 1 - x$$

$$4\left(\frac{1}{4}x^2\right) = 4(1 - x) \qquad \text{Multiplying both sides by 4}$$

$$x^2 = 4 - 4x$$

$$x^2 \;+\; 4x \;-\; 4 = 4 - 4x \;+\; 4x \;-\; 4 \qquad \text{Adding } 4x - 4 \text{ to both sides}$$

$$x^2 + 4x - 4 = 0 \qquad \text{The general form}$$

Step 2. $a = 1$, $b = 4$, and $c = -4$. We substitute these values into the formula $x = \dfrac{-b \pm \sqrt{b^2 - 4ac}}{2a}$.

$$x = \frac{-(4) \pm \sqrt{(4)^2 - 4(1)(-4)}}{2(1)}$$

Step 3. $x = \dfrac{-4 \pm \sqrt{16 + 16}}{2} = \dfrac{-4 \pm \sqrt{32}}{2} = \dfrac{-4 \pm \sqrt{16 \cdot 2}}{2} = \dfrac{-4 \pm 4\sqrt{2}}{2}$

$$= \frac{\overset{2}{4}(-1 \pm \sqrt{2})}{\underset{1}{2}} \qquad \text{Factoring the numerator and reducing the expression}$$

$$= 2(-1 \pm \sqrt{2}) \quad \text{or} \quad -2 \pm 2\sqrt{2}$$

Step 4. The solutions are not rational.

Step 5.

✓

Check for $x = -2 + 2\sqrt{2}$	**Check for $x = -2 - 2\sqrt{2}$**
$\frac{1}{4}x^2 = 1 - x$	$\frac{1}{4}x^2 = 1 - x$
$\frac{1}{4}(-2 + 2\sqrt{2})^2 \overset{?}{=} 1 - (-2 + 2\sqrt{2})$	$\frac{1}{4}(-2 - 2\sqrt{2})^2 \overset{?}{=} 1 - (-2 - 2\sqrt{2})$
$\frac{1}{4}(4 - 8\sqrt{2} + 8) \overset{?}{=} 1 + 2 - 2\sqrt{2}$	$\frac{1}{4}(4 + 8\sqrt{2} + 8) \overset{?}{=} 1 + 2 + 2\sqrt{2}$
$\frac{1}{4}(12 - 8\sqrt{2}) \overset{?}{=} 3 - 2\sqrt{2}$	$\frac{1}{4}(12 + 8\sqrt{2}) \overset{?}{=} 3 + 2\sqrt{2}$
$3 - 2\sqrt{2} = 3 - 2\sqrt{2}$ True	$3 + 2\sqrt{2} = 3 + 2\sqrt{2}$ True

The solutions are $-2 + 2\sqrt{2}$ and $-2 - 2\sqrt{2}$.

EXAMPLE 4

Solve $4x^2 - 5x + 2 = 0$.

SOLUTION

Step 1. The equation is in the general form.

Step 2. $a = 4$, $b = -5$, and $c = 2$. We substitute these values into the formula $x = \dfrac{-b \pm \sqrt{b^2 - 4ac}}{2a}$.

$$x = \frac{-(-5) \pm \sqrt{(-5)^2 - 4(4)(2)}}{2(4)}$$

Step 3. $x = \dfrac{5 \pm \sqrt{25 - 32}}{8} = \dfrac{5 \pm \sqrt{-7}}{8}$

The solutions $\left(\dfrac{5 + \sqrt{-7}}{8} \text{ and } \dfrac{5 - \sqrt{-7}}{8}\right)$ are not real numbers, because the radicand (-7) is negative.

It is important that you realize that lengths of lines, objects, and so forth cannot be negative, but *they can be irrational*. When lengths are irrational, we sometimes want decimal approximations for the solutions, and we can use a calculator (or Table I) to find the approximations. For example, the length of a line might be $2 + \sqrt{15}$, with a decimal approximation of $2 + 3.873$, or about 5.873. However, the length of a line *cannot* be $2 - \sqrt{15}$, since $2 - \sqrt{15}$ is negative (the decimal approximation is $2 - 3.873$, or about -1.873).

EXAMPLE 5

The width of a rectangle is 5 m less than its length. If the area of the rectangle is 7 sq. m, find the width of the rectangle. Give an exact answer and also a decimal approximation, rounded off to two decimal places.

SOLUTION   Let $\quad x =$ the width of the rectangle (in meters)
Then $\quad x + 5 =$ the length of the rectangle (in meters)
and $x(x + 5) =$ the area of the rectangle (in square meters)

The area of the rectangle	is	7
$x(x + 5)$	$=$	7

Step 1. $x^2 + 5x \;\boxed{-\ 7} = \boxed{7}\ -\ 7$ Adding -7 to both sides

$x^2 + 5x - 7 = 0$

Step 2. $a = 1$, $b = 5$, and $c = -7$. We substitute these values into the formula
$$x = \frac{-b \pm \sqrt{b^2 - 4ac}}{2a}.$$

$$x = \frac{-(5) \pm \sqrt{(5)^2 - 4(1)(-7)}}{2(1)}$$

Step 3. $x = \dfrac{-5 \pm \sqrt{25 + 28}}{2} = \dfrac{-5 \pm \sqrt{53}}{2}$

Separating the two answers

$$x = \frac{-5 + \sqrt{53}}{2} \approx \frac{-5 + 7.280}{2} = \frac{2.280}{2} = 1.14 \qquad x = \frac{-5 - \sqrt{53}}{2}$$

We reject this solution, since a length can't be negative

$$x + 5 = \frac{-5 + \sqrt{53}}{2} + \frac{10}{2} = \frac{5 + \sqrt{53}}{2} \approx \frac{5 + 7.280}{2}$$

$$= \frac{12.280}{2} = 6.14$$

Step 5. If the width is $\dfrac{-5 + \sqrt{53}}{2}$ m and the length is $\dfrac{5 + \sqrt{53}}{2}$ m, the area is

$$\left(\frac{-5 + \sqrt{53}}{2}\ \text{m}\right)\left(\frac{5 + \sqrt{53}}{2}\ \text{m}\right) = \frac{(-5 + \sqrt{53})(5 + \sqrt{53})}{4}\ \text{sq. m}$$

$$= \frac{-25 + 53}{4}\ \text{sq. m} = \frac{28}{4}\ \text{sq. m} = 7\ \text{sq. m}$$

Also, if the width is about 1.14 m and the length is close to 6.14 m, then the area is approximately (1.14 m)(6.14 m), or 6.9996 sq. m, which is close to 7 sq. m.

Therefore, the width of the rectangle is $\dfrac{-5 + \sqrt{53}}{2}$ m, or about 1.14 m.

Note To evaluate $\dfrac{-5 + \sqrt{53}}{2}$ with a calculator, it's probably easiest to (mentally) use the commutative property of addition in the numerator, thinking of the expression as $\dfrac{\sqrt{53} - 5}{2}$. The keystrokes are as follows:

On most calculators, the first equal sign is *essential*; if it is omitted, the calculator interprets the problem as $\sqrt{53} - \frac{5}{2}$ and, following the correct order of operations for that problem, performs the division before the subtraction.

To evaluate $\dfrac{5 + \sqrt{53}}{2}$, the following keystrokes can be used:

										Display
5	+	5	3	√	=	÷	2	=		6.140054945

As we mentioned earlier, *exact* irrational answers are often desired when we are simply solving equations, but decimal approximations are often more useful when we're solving applied problems.

Exercises 10.5
Set I

In Exercises 1–16, use the quadratic formula to solve each equation. If the radicand is negative, write "not real." Express any real, irrational solutions in simplest radical form.

1. $3x^2 - x - 2 = 0$

2. $2x^2 + 3x - 2 = 0$

3. $x^2 - 4x + 1 = 0$

4. $x^2 - 4x - 1 = 0$

5. $\dfrac{x}{2} + \dfrac{2}{x} = \dfrac{5}{2}$

6. $\dfrac{x}{3} + \dfrac{2}{x} = \dfrac{7}{3}$

7. $2x^2 = 8x - 5$

8. $3x^2 = 6x - 2$

9. $\dfrac{1}{x} + \dfrac{x}{x - 1} = 3$

10. $\dfrac{2}{x} - \dfrac{x}{x + 1} = 4$

11. $x^2 + x + 1 = 0$

12. $x^2 + x + 2 = 0$

13. $4x^2 + 4x = 1$

14. $9x^2 - 6x = 2$

15. $3x^2 + 2x + 1 = 0$

16. $4x^2 + 3x + 2 = 0$

In Exercises 17 and 18, set up each problem algebraically and solve. Express all answers in simplest radical form.

17. A number less its reciprocal is $\frac{2}{3}$. What is the number?

18. If a number is subtracted from twice its reciprocal, the result is $\frac{5}{4}$. What is the number?

In Exercises 19–24, find the solutions and express them in simplest radical form. Also, use a calculator or Table I to approximate the solutions, rounding off the approximations to two decimal places. (In Exercises 21–24, set up each problem algebraically and solve.)

19. $4x^2 = 1 - 4x$

20. $9x^2 = 2 + 6x$

21. The length of a rectangle is 2 ft more than its width. If its area is 2 sq. ft, find its dimensions.

22. The length of a rectangle is 4 cm more than its width. If its area is 6 sq. cm, find its dimensions.

23. The perimeter of a square is numerically 6 less than its area. Find the length of a side of the square.

24. The area of a square is numerically 2 more than its perimeter. Find the length of a side of the square.

✏️ Writing Problems

Express the answer in your own words and in complete sentences.

1. Suppose that we solve a quadratic equation by using the quadratic formula and find that the solutions are $\dfrac{-2 \pm \sqrt{5}}{2}$. Explain what is wrong with saying that these solutions simplify to $-1 \pm \sqrt{5}$.

Exercises 10.5
Set II

In Exercises 1–16, use the quadratic formula to solve each equation. If the radicand is negative, write "not real." Express any real, irrational solutions in simplest radical form.

1. $3x^2 - 8x - 3 = 0$

2. $x^2 - 3x + 3 = 0$

3. $x^2 - 4x + 2 = 0$

4. $2x^2 = 2x - 5$

5. $\dfrac{x}{3} + \dfrac{1}{x} = \dfrac{7}{6}$

6. $\dfrac{x}{2} + \dfrac{1}{x} = \dfrac{33}{8}$

7. $3x^2 + 4 = 2x$

8. $5x^2 + 3x + 1 = 0$

9. $\dfrac{2}{x} = \dfrac{3x}{x+2} - 1$

10. $\dfrac{3}{x} = \dfrac{x}{x+2} + 1$

11. $x^2 + 3x + 5 = 0$

12. $x^2 - 3x + 4 = 0$

13. $9x^2 + 9x = 1$

14. $5x(x+5) = 1$

15. $5x^2 + 7 = 2x$

16. $x^2 = \dfrac{3 - 5x}{2}$

In Exercises 17 and 18, set up each problem algebraically and solve. Express all answers in simplest radical form.

17. A number less its reciprocal is $\frac{1}{4}$. What is the number?

18. A number is equal to the sum of its reciprocal and $\frac{2}{5}$. What is the number?

In Exercises 19–24, find the solutions and express them in simplest radical form. Also, use a calculator or Table I to approximate the solutions, rounding off the approximations to two decimal places. (In Exercises 21–24, set up each problem algebraically and solve.)

19. $5x^2 + 4x = 3$

20. $3x^2 = 5 + 6x$

21. The length of a rectangle is 2 m more than its width. If its area is 6 sq. m, find its dimensions.

22. The length of a rectangle is 5 in. more than its width. If its area is 3 sq. in., find its dimensions.

23. The perimeter of a square is numerically 3 more than its area. Find the length of a side of the square.

24. The area of a square is numerically 3 more than its perimeter. Find the length of a side of the square.

10.6   The Pythagorean Theorem

Right Triangles   A triangle that has a right angle (a 90° angle) is called a **right triangle**. The side opposite the right angle of a right triangle is called the **hypotenuse**; it is always the longest side of the triangle. The other two sides are often called the **legs** of the right triangle.

The diagonal of a rectangle divides the rectangle into two right triangles. The parts of a right triangle are shown below, along with a rectangle and one diagonal of the rectangle.

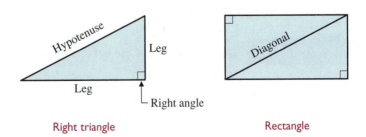

Right triangle   Rectangle

Using the Pythagorean Theorem

The Pythagorean theorem, which follows, applies only to right triangles.

The Pythagorean theorem

The square of the length of the hypotenuse of a right triangle is equal to the sum of the squares of the lengths of the two legs:

$$c^2 = a^2 + b^2$$

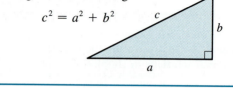

We can use the Pythagorean theorem to determine whether or not a given triangle is a right triangle. To do this, we use the rule as follows: If the square of the longest side of a triangle equals the sum of the squares of the two shorter sides, the triangle is a right triangle.

EXAMPLE 1   Determine whether each triangle is a right triangle.

SOLUTION

a.

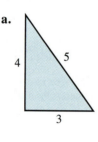

$$5^2 \overset{?}{=} 3^2 + 4^2$$

$$25 \overset{?}{=} 9 + 16$$

$$25 = 25$$

Therefore, the given triangle *is* a right triangle.

b.

$$(\sqrt{29})^2 \overset{?}{=} 5^2 + 2^2$$

$$29 \overset{?}{=} 25 + 4$$

$$29 = 29$$

Therefore, the given triangle *is* a right triangle.

c.

$$13^2 \overset{?}{=} 12^2 + 6^2$$

$$169 \overset{?}{=} 144 + 36$$

$$169 \neq 180$$

Therefore, the given triangle *is not* a right triangle.

The Pythagorean theorem can also be used to find one side of a right triangle when the other two sides are known. In such problems, we can consider *either* leg to be *a*, but the hypotenuse *must* be *c*.

EXAMPLE 2   Find the length of the hypotenuse of a right triangle whose legs are 8 cm and 6 cm.

SOLUTION   Let x = the length of the hypotenuse (in centimeters).

$$c^2 = a^2 + b^2 \quad \text{The Pythagorean theorem}$$

$$x^2 = 8^2 + 6^2$$

$$x^2 = 64 + 36 = 100$$

$$x = \pm\sqrt{100}$$

$$x = \pm 10$$

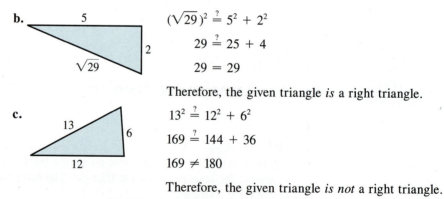

Note Checking will show that both 10 and -10 are solutions of the *equation*. However, -10 cannot be a solution of the *geometric problem*, because we consider lengths in geometric figures to be positive numbers. For this reason, in the problems of this section we will consider only the positive (*principal*) square root.

Therefore, the length of the hypotenuse is 10 cm.

As we mentioned earlier, lengths of geometric figures can be irrational, although they cannot be negative. In fact, in a square whose sides measure 1 in., the *exact* length of the diagonal is $\sqrt{2}$ in. The decimal approximation of this length is 1.41 in. (see Example 3).

EXAMPLE 3 The length of each side of a square is 1 in. Find the length of the diagonal of the square.

SOLUTION Let $x =$ the length of the diagonal of the square (in inches).

$$x^2 = 1^2 + 1^2$$

$$x^2 = 1 + 1$$

$$x^2 = 2$$

$$x = \sqrt{2} \approx 1.414$$

√ **Check** $(\sqrt{2})^2 \stackrel{?}{=} 1^2 + 1^2$

$$2 = 1 + 1 \quad \text{True}$$

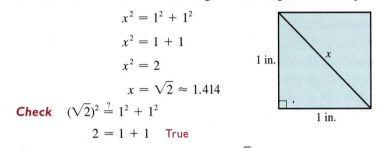

Therefore, the length of the diagonal is $\sqrt{2}$ in., or about 1.41 in.

EXAMPLE 4 Use the Pythagorean theorem to find x in the given right triangle.

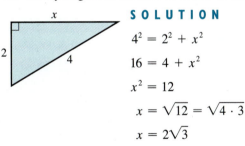

SOLUTION

$$4^2 = 2^2 + x^2$$

$$16 = 4 + x^2$$

$$x^2 = 12$$

$$x = \sqrt{12} = \sqrt{4 \cdot 3}$$

$$x = 2\sqrt{3}$$

Checking will confirm that the *exact* length of the side is $2\sqrt{3}$; the *approximate* length is 3.46.

EXAMPLE 5 Find the exact length of the diagonal of a rectangle that has a length of 6 ft and a width of 4 ft.

SOLUTION Let $x =$ the length of the diagonal (in feet).

$$x^2 = 6^2 + 4^2$$

$$x^2 = 36 + 16 = 52$$

$$x = \sqrt{52} = \sqrt{4 \cdot 13}$$

$$x = 2\sqrt{13}$$

Checking will show that the length of the diagonal is $2\sqrt{13}$ ft.

EXAMPLE 6

The length of a rectangle is 2 m more than its width. If the length of its diagonal is 10 m, find the dimensions of the rectangle.

SOLUTION Let x = the width of the rectangle (in meters)
Then $x + 2$ = the length of the rectangle (in meters)

$$(10)^2 = (x + 2)^2 + (x)^2 \qquad \text{Using the Pythagorean theorem}$$

$$100 = x^2 + 4x + 4 + x^2$$

$$100 = 2x^2 + 4x + 4$$

$$100 - 100 = 2x^2 + 4x + 4 - 100 \qquad \text{Adding } -100 \text{ to both sides}$$

$$0 = 2x^2 + 4x - 96 \qquad \text{Simplifying}$$

$$0 = 2(x^2 + 2x - 48) \qquad \text{Factoring out 2, the GCF}$$

$$0 = 2(x + 8)(x - 6) \qquad \text{Factoring the polynomial}$$

$2 \neq 0$ | $x + 8 = 0$ | $x - 6 = 0$

$x + 8 - 8 = 0 - 8$ | $x - 6 + 6 = 0 + 6$

$x = -8$ | $x = 6$ The width

We reject this solution | $x + 2 = 6 + 2 = 8$ The length

✓ **Check** $(6 \text{ m})^2 + (8 \text{ m})^2 \overset{?}{=} (10 \text{ m})^2$

$36 \text{ m}^2 + 64 \text{ m}^2 = 100 \text{ m}^2$ True

Therefore, the rectangle is 6 m wide and 8 m long.

Exercises 10.6
Set I

In Exercises 1–6, the lengths of the three sides of a triangle are given. Determine whether the triangle is a right triangle.

1. $a = 6$ cm, $b = 8$ cm, $c = 11$ cm

2. $a = 1$ in., $b = 2$ in., $c = 3$ in.

3. $a = 9$ m, $b = 12$ m, $c = 15$ m

4. $a = 15$ yd, $b = 8$ yd, $c = 17$ yd

5. $a = \sqrt{7}$ ft, $b = \sqrt{18}$ ft, $c = 5$ ft

6. $a = \sqrt{6}$ in., $b = \sqrt{30}$ in., $c = 6$ in.

In Exercises 7–20, use the Pythagorean theorem to find x in the given right triangle. If a solution is irrational, express it in simplest radical form *and* find a decimal approximation of the solution (rounded off to two decimal places).

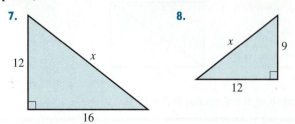

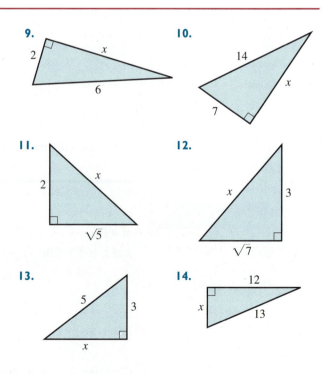

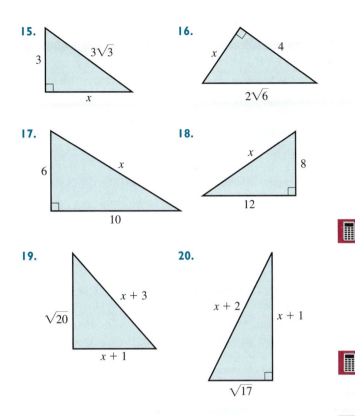

15.

16.

17.

18.

19.

20.

In Exercises 21–32, set up each problem algebraically and solve. If a solution is irrational, express it in simplest radical form unless other instructions are given.

21. The length of a rectangle is 24 cm, and the length of its diagonal is 25 cm. Find the width of the rectangle.

22. The length of a rectangle is 40 in., and the length of its diagonal is 41 in. Find the width of the rectangle.

23. The length of one leg of a right triangle is 4 m less than twice the length of the other leg. If the length of the hypotenuse is 10 m, how long are the two legs?

24. The length of a rectangle is 3 yd more than its width. If the length of the diagonal is 15 yd, find the dimensions of the rectangle.

25. The length of each side of a square is $\sqrt{6}$ in. Find the length of the diagonal of the square.

26. The length of each side of a square is $\sqrt{14}$ m. Find the length of the diagonal of the square.

27. The diagonal of a square is $4\sqrt{2}$ cm long. Find the length of one side of the square.

28. The diagonal of a square is $5\sqrt{2}$ cm long. Find the length of one side of the square.

29. A maintenance specialist needs to run a wire from the top of a pole that is 50 ft high to a point that is 30 ft from the base of the pole. What is the length of the wire that the maintenance specialist needs? Round off the answer to the nearest hundredth of a foot.

30. Television screens are measured diagonally. What would be the size of a TV screen that is 16 in. wide and 12 in. high?

31. Anne and Leticia left an intersection at the same time. Anne bicycled due north at a rate of 8 mph, and Leticia bicycled due east at a rate of 6 mph. How far apart were the women one-half hour after leaving the intersection?

32. Two small airplanes left Boston at the same time. One plane flew west at a rate of 120 mph, and the other flew south at a rate of 160 mph. How far apart were the planes 2 hr after they left Boston?

In Exercises 33 and 34, set up each problem algebraically and solve. Give decimal approximations for the answers, using a calculator or Table I, and round off the answers to two decimal places.

33. The length of each side of a square is 41.6 cm. Find the length of the diagonal of the square.

34. The length of a rectangle is 3.67 m, and the length of its diagonal is 4.53 m. Find the width of the rectangle.

Exercises   10.6
Set II

In Exercises 1–6, the lengths of the three sides of a triangle are given. Determine whether the triangle is a right triangle.

1. $a = 5$ m, $b = 12$ m, $c = 14$ m

2. $a = \sqrt{7}$ ft, $b = \sqrt{11}$ ft, $c = 3\sqrt{2}$ ft

3. $a = 12$ yd, $b = 5$ yd, $c = 13$ yd

4. $a = \sqrt{11}$ m, $b = \sqrt{24}$ m, $c = 6$ m

5. $a = \sqrt{7}$ ft, $b = \sqrt{29}$ ft, $c = 6$ ft

6. $a = 2\sqrt{2}$ cm, $b = 2\sqrt{2}$ cm, $c = 4$ cm

In Exercises 7–20, use the Pythagorean theorem to find x in the given right triangle. If a solution is irrational, express it in simplest radical form *and* find a decimal approximation of the solution (rounded off to two decimal places).

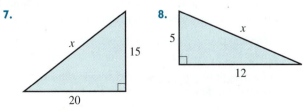

7.

8.

9.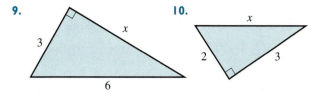

10.

11.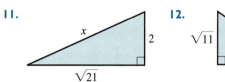

12.

13.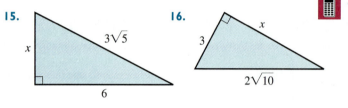

14.

15.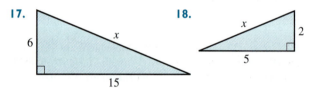

16.

17.

18.

19.

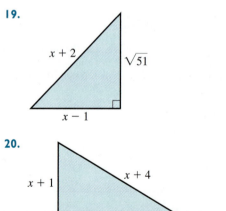

20.

In Exercises 21–32, set up each problem algebraically and solve. If a solution is irrational, express it in simplest radical form unless other instructions are given.

21. The length of a rectangle is 15 cm, and the length of its diagonal is 17 cm. Find the width of the rectangle.

22. The length of a rectangle is 15 ft, and the length of its diagonal is $10\sqrt{3}$ ft. Find the width of the rectangle.

23. The length of one leg of a right triangle is 2 cm less than twice the length of the other leg. If the length of the hypotenuse is 5 cm, how long are the two legs?

24. The width of a rectangle is 4 cm less than its length. If the length of the diagonal is $4\sqrt{5}$ cm, find the dimensions of the rectangle.

25. The length of each side of a square is $\sqrt{3}$ cm. Find the length of the diagonal of the square.

26. The length of each side of a square is $\sqrt{15}$ ft. Find the length of the diagonal of the square.

27. The diagonal of a square is $7\sqrt{2}$ cm long. Find the length of one side of the square.

28. The diagonal of a square is 24 in. long. Find the length of one side of the square.

29. Dixie needs to run a pipe diagonally across a wall of a storage cupboard in her widget factory. If the wall is 5 ft wide and 12 ft tall, how long must the pipe be?

30. A baseball diamond is *really* a square. If the distance between second base and home plate is about 127.3 ft, what is the distance between first base and home plate? (Round off the answer to one decimal place.)

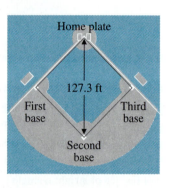

31. Two airplanes left Salt Lake City at the same time. One flew due west at a rate of 150 mph, and the other flew due north at a rate of 200 mph. How far apart were the planes 2 hr after they left Salt Lake City?

32. Two airplanes left Denver at the same time, one heading east and the other south. One flew 140 mph faster than the other, and 2 hr after leaving Denver, they were 680 mi apart. What was the speed of the slower plane?

In Exercises 33 and 34, set up each problem algebraically and solve. Give decimal approximations for the answers, using a calculator or Table I, and round off the answers to two decimal places.

33. The length of each side of a square is 7.13 in. Find the length of the diagonal of the square.

34. The diagonal of a rectangle is 7.81 cm long, and the length of the rectangle is 2.68 cm. Find the width of the rectangle.

**Sections
10.1–10.6**

REVIEW

**Quadratic
Equations
10.1**

A **quadratic equation** is a polynomial equation whose highest-degree term is a second-degree term. The **general form** of a quadratic equation as used in this text is

$$ax^2 + bx + c = 0 \quad \text{or} \quad 0 = ax^2 + bx + c$$

where a, b, and c are integers and $a > 0$.

**Methods of
Solving Quadratic
Equations
10.2**

Solving quadratic equations by factoring:

1. Write the equation in the general form.

2. Factor the polynomial.

3. Using the zero-factor property, set each factor equal to zero; then solve for the variable.

4. Check apparent solutions in the original equation.

10.3

Solving incomplete quadratic equations:

a. *When $c = 0$:*
Find the greatest common factor (GCF); then solve by factoring.

b. *When $b = 0$:*

1. Get the second-degree term on one side of the equal sign and the constant on the other side.

2. Divide both sides of the equation by the coefficient of x^2.

3. Use Property 1: If $x^2 = p$, then $x = \pm\sqrt{p}$.

4. Express the solutions in simplest form.

5. Check all apparent real solutions in the original equation.

10.4

Solving quadratic equations of the form $ax^2 + bx + c = 0$ by completing the square:

1. If a, the coefficient of x^2, is not 1, divide both sides of the equation by a.

2. Get the x^2-term and the x-term on the left side of the equal sign and the constant on the right side.

3. Determine what number should be added to the left side to make the left side the square of a binomial, and add this number *to both sides of the equation.*

4. Factor the left side of the equation—the equation is now in the form $(x + k)^2 = p$.

5. Use Property 1 to solve the equation for $x + k$.

6. Add $-k$ to both sides of the equation to solve for x.

7. Simplify the right side, and if the two solutions are *rational*, separate and simplify them.

8. If the solutions are real, check them in the original equation.

10.5

Solving quadratic equations by using the quadratic formula:

1. Write the equation in the general form.

2. Identify a, b, and c, and substitute these values into the *quadratic formula*:

$$x = \frac{-b \pm \sqrt{b^2 - 4ac}}{2a} \quad (a \neq 0)$$

3. Simplify the apparent solutions.

4. If the solutions are *rational* (that is, if the radicand $b^2 - 4ac$ is a perfect square), write the two solutions separately.

5. If the apparent solutions are real numbers, check them in the original equation.

**The Pythagorean
Theorem
10.6**

The square of the length of the hypotenuse of a right triangle is equal to the sum of the squares of the lengths of the two legs:

$$c^2 = a^2 + b^2$$

Sections 10.1–10.6   REVIEW EXERCISES   Set I

In Exercises 1–14, solve each equation by any convenient method, or write "not real." Express any real, irrational answers in simplest radical form.

1. $x^2 + x = 6$

2. $x^2 - 49x = 0$

3. $x^2 - 2x - 4 = 0$

4. $x^2 - 4x + 1 = 0$

5. $x^2 = 5x$

6. $\dfrac{3x}{5} = \dfrac{5}{12x}$

7. $\dfrac{x + 2}{3} = \dfrac{1}{x - 2} + \dfrac{2}{3}$

8. $x^2 + 144 = 0$

9. $5(x + 2) = x(x + 5)$

10. $3(x + 4) = x(x + 3)$

11. $\dfrac{2}{x} + \dfrac{x}{x + 1} = 5$

12. $\dfrac{3}{x} - \dfrac{x}{x + 2} = 2$

13. $3x^2 + 2x + 1 = 0$

14. $(2x - 1)(3x + 5) = x(x + 7) + 4$

15. Use the Pythagorean theorem to solve for x in the figure.

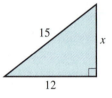

In Exercises 16–20, set up each problem algebraically and solve. Express any real, irrational answers in simplest radical form.

16. One leg of a right triangle is 7 in. longer than the other leg. The length of the hypotenuse is 13 in. Find the length of the longer leg.

17. The length of a side of a square is $\sqrt{10}$ m. Find the length of the diagonal of the square.

18. A solution is being pumped into an 880-gal tank at a constant rate, and it fills the tank in a certain length of time. If the rate of flow had been 30 gal per minute faster, it would have taken 3 min less to fill the tank. At what rate is the pump working?

19. The base of a triangle is 3 in. more than its altitude. The area of the triangle is 35 sq. in. Find the lengths of the base and the altitude.

20. A 10-ft ladder is leaning against a building. If the foot of the ladder is 6 ft away from the building, how high is the top of the ladder?

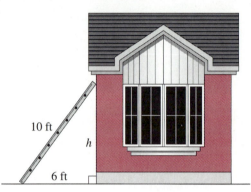

In Exercises 21 and 22, set up each problem algebraically and solve. Express each answer in simplest radical form, and also use a calculator (or Table I) to find decimal approximations for the answers, rounding them off to two decimal places.

21. A rectangle is 8 in. long and 4 in. wide. Find the length of the diagonal.

22. The width of a rectangle is 6 ft less than its length, and the area of the rectangle is 6 sq. ft. Find the dimensions of the rectangle.

Name _____

In Exercises 1–14, solve each equation by any convenient method, or write "not real."
Express any real, irrational answers in simplest radical form.

1. $x^2 + x = 12$　　　　　**2.** $x^2 - 36 = 0$

3. $x^2 - 2x - 2 = 0$　　　　**4.** $x^2 = 11x$

5. $x^2 + 1 = 0$　　　　　　**6.** $x^2 + x + 1 = 0$

7. $\dfrac{2x}{3} = \dfrac{3}{8x}$　　　　　**8.** $\dfrac{x + 2}{4} = \dfrac{1}{x - 2} + 1$

9. $2(x + 1) = x(x - 4)$　　**10.** $\dfrac{3}{x} + \dfrac{x}{x - 1} = 2$

11. $3x^2 - x = 0$　　　　　**12.** $4x^2 = 2x - 1$

13. $x^2 + 2x + 5 = 0$　　　**14.** $x^2 + 2x - 5 = 0$

15. Use the Pythagorean theorem to solve for x in the figure.

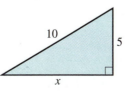

1. _____

2. _____

3. _____

4. _____

5. _____

6. _____

7. _____

8. _____

9. _____

10. _____

11. _____

12. _____

13. _____

14. _____

15. _____

In Exercises 16–20, set up each problem algebraically and solve. Express any answer that is not rational in simplest radical form.

16. The length of the diagonal of a square is $\sqrt{32}$ yd. What is the length of a side of the square?

17. Find the width of a rectangle that has a length of 12 in. and whose diagonal measures 13 in.

18. The base of a triangle is 5 cm more than its altitude. The area of the triangle is 7 sq. cm. Find the length of the altitude.

19. A solution is being pumped into a 2,508-gal tank at a constant rate, and it fills the tank in a certain length of time. If the rate of flow had been 18 gal per minute faster, it would have taken 3 min less to fill the tank. At what rate is the pump working?

20. Gale's rectangular living room is 1 yd longer than it is wide. The cost of carpeting it with carpet that costs $35 per square yard is $1,050. Find the dimensions of the room.

In Exercises 21 and 22, set up each problem algebraically and solve. Express each answer in simplest radical form, and also use a calculator (or Table I) to find decimal approximations for the answers, rounding them off to two decimal places.

21. A rectangle is 8 cm wide and 10 cm long. Find the length of its diagonal.

22. One leg of a right triangle is 2 m longer than the other leg. The length of the hypotenuse is $2\sqrt{2}$ m. Find the length of each leg.

16. _____

17. _____

18. _____

19. _____

20. _____

21. _____

22. _____

_____

Chapter 10   DIAGNOSTIC TEST

The purpose of this test is to see how well you understand solving quadratic equations. We recommend that you work this diagnostic test *before* your instructor tests you on this chapter. Allow yourself about 50 minutes.

Complete solutions for all the problems on this test, together with section references, are given in the answer section in the back of the book. We suggest that you study the sections referred to for the problems you do incorrectly.

In Problem 1, write each quadratic equation in the general form, and identify a, b, and c in each equation.

1. a. $5 + 3x^2 = 7x$ **b.** $3(x + 5) = 3(4 - x^2)$

In Problems 2–7, solve each equation, or write "not real." Use any convenient method. If necessary, use the quadratic formula, which is as follows:

If $ax^2 + bx + c = 0$, then $x = \dfrac{-b \pm \sqrt{b^2 - 4ac}}{2a}$

2. $x^2 = 8x + 15$ **3.** $5x^2 = 17x$

4. $x^2 + 2x + 6 = 0$ **5.** $6x^2 = 54$

6. $3(x + 1) = x^2$ **7.** $x(x + 3) = x$

8. Use the Pythagorean theorem to solve for x in the right triangle in the figure.

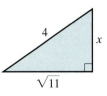

In Problems 9 and 10, set up each problem algebraically, solve, and check. Express any real, irrational answers in simplest radical form.

9. The length of each side of a square is 7 m. Find the length of a diagonal of the square.

10. The length of a rectangle is 6 ft more than its width. Its area is 78 sq. ft. Find the width and the length of the rectangle.

Chapters 1–10   CUMULATIVE REVIEW EXERCISES

1. Is division associative?

2. Is subtraction commutative?

3. What is the additive identity?

4. Is 0 a real number?

In Exercises 5–14, perform the indicated operation, or write "not defined." Express all answers in simplest form.

5. $73 \div 0$

6. $(2\sqrt{3} - 3)(4\sqrt{3} + 5)$

7. $(4x^2 + 3x - 2)(5x - 3)$

8. $(\sqrt{21} - \sqrt{5})^2$

9. $\dfrac{x^2 + 3x}{2x^2 + 7x + 5} \div \dfrac{x^2 - 9}{x^2 - 2x - 3}$

10. $(x^2 - 14x - 5) \div (x - 3)$ (Use long division)

11. $(x^2 + 2x - 8) \div \dfrac{x^2 - 4}{x^2 + 2x}$

12. $\dfrac{x + 3}{x - 3} - \dfrac{x^2 + 9}{x^2 - 9}$

13. $\dfrac{x^2 + 3x - 2}{x^2 + 5x + 6} - \dfrac{x - 1}{2x^2 + 7x + 3}$

14. $5\sqrt{18} - \sqrt{\dfrac{9}{2}}$

In Exercises 15–23, solve for x, or write either "identity" or "no solution."

15. $7 - (2x + 3) = 3x - 2 - (4x - 3)$

16. $\sqrt{6x + 13} = 3$

17. $\sqrt{5x + 3} = 10x$

18. $\dfrac{3x - 5}{4} + 2 = \dfrac{3 + 4x}{3}$

19. $12x - 3 \le 7x - 13$

20. $x^2 + 5x = 14$

21. $3 + 4x^2 - 12x = 0$

22. $x(3x - 1) = 2x + 6$

23. $7x^2 - 1 = 0$

24. Rationalize the denominator and simplify: $\dfrac{14}{3 - \sqrt{2}}$.

In Exercises 25–28, factor completely, or write "not factorable."

25. $3x^2 - 12x - 15$

26. $5x^2 - 45$

27. $6x^2 - 12x + 5$

28. $7x^2 - 13x - 2$

In Exercises 29–32, set up each problem algebraically, solve, and check. Be sure to state what your variables represent.

29. On a scale drawing, $\frac{1}{4}$ in. represents 1 ft. What will be the size of the drawing of a room that is 12 ft by 18 ft?

30. Two airplanes left Tulsa at the same time, one heading west and the other north. One flew 100 mph faster than the other, and 1 hr after leaving Tulsa, they were 500 mi apart. What was the speed of the slower plane?

31. Mr. McAllister invested part of $27,300 at 7% and the remainder at 10.5%. His total yearly income from these investments was $2,324. How much was invested at each rate?

32. The manager of a hardware store bought twelve saws. If she had, instead, bought eight saws of a higher quality, she would have paid $2.75 more per saw for the same total cost. Find the cost (per saw) of the less expensive saws.

Critical Thinking and Problem-Solving Exercises

1. Suppose that you need to memorize the following area code and phone number: (909) 369-8172. What, if any, patterns do you see that might help you remember the number?

2. Felicia is thinking of a number. Can you guess what that number is by using the following clues?

The number is a prime number.

The number is between 16 and 100.

The sum of the digits of the number is 14.

3. Tracy runs a nursery school, and she has 20 students in her school. When she allowed the children to help themselves to cookies last Wednesday, one-fifth of the children took five cookies each; one-fourth of the *remaining* children took four cookies each; one-sixth of the *remaining* children took three cookies each; one-half of the *remaining* students took two cookies each; one child took exactly one cookie. How many, if any, children did not take *any* cookies? How many cookies were taken altogether?

4. Tom was looking at his musical instruments the other day. He faced the row of instruments and made the following observations:

The clarinet was to the left of the saxophone and the piccolo.

The bassoon was not next to the piccolo.

The oboe and the piccolo were in the two center positions.

The saxophone and the flute were on the two ends.

List (in order from left to right) the order in which the instruments were arranged.

5. Amy has three jackets that she is particularly fond of (one is powder blue, one is red, and one is black), and three pairs of shoes that she likes really well (a black pair, a red pair, and a white pair). If she can wear any one of the three jackets with any of the pairs of shoes, how many different combinations are possible?

6. Discuss the *kinds* of numbers the solutions of a quadratic equation are if

a. $b^2 - 4ac$ is positive and is a perfect square

b. $b^2 - 4ac$ is positive but is *not* a perfect square

c. $b^2 - 4ac$ is zero

d. $b^2 - 4ac$ is negative

7. The "solution" of the following equation contains two errors. Find and describe the errors, and correctly solve the equation.

Solve $2x^2 - 4x - 1 = 0$.

Solution: $a = 2$, $b = -4$, $c = -1$. Using the quadratic formula, we have

$$x = \frac{-(-4) \pm \sqrt{(-4)^2 - 4(2)(-1)}}{2(2)}$$

$$= \frac{4 \pm \sqrt{16 - 8}}{4}$$

$$= \frac{4 \pm \sqrt{8}}{4} = \frac{\overset{1}{4} \pm 2\sqrt{2}}{\underset{1}{4}}$$

$$= 1 \pm 2\sqrt{2}$$

Formula	Description	Page
$A = lw$	The area of a rectangle, where A is the area, l is the length, and w is the width	182
$P = 2l + 2w$	The perimeter of a rectangle, where P is the perimeter, l is the length, and w is the width	182
$A = s^2$	The area of a square, where A is the area and s is the length of a side	182
$P = 4s$	The perimeter of a square, where P is the perimeter and s is the length of a side	182
$A = \frac{1}{2}bh$	The area of a triangle, where A is the area, b is the base, and h is the altitude	115
$P = a + b + c$	The perimeter of a triangle, where P is the perimeter and a, b, and c are the lengths of the sides	182
$A = bh$	The area of a parallelogram, where A is the area, b is the base, and h is the altitude	182
$A = \pi r^2$	The area of a circle, where A is the area, r is the radius, and $\pi \approx 3.14$	117
$C = 2\pi r$	The circumference of a circle, where C is the circumference, r is the radius, and $\pi \approx 3.14$	182
$V = lwh$	The volume of a rectangular solid, where V is the volume, l is the length, w is the width, and h is the height	182
$S = 2(lw + lh + wh)$	The surface area of a rectangular solid, where S is the surface area, l is the length, w is the width, and h is the height	182
$V = e^3$	The volume of a cube, where V is the volume and e is the length of an edge	183
$S = 6e^2$	The surface area of a cube, where S is the surface area and e is the length of an edge	183
$V = \frac{4}{3}\pi r^3$	The volume of a sphere, where V is the volume, r is the radius, and $\pi \approx 3.14$	117
$S = 4\pi r^2$	The surface area of a sphere, where S is the surface area, r is the radius, and $\pi \approx 3.14$	183
$V = \pi r^2 h$	The volume of a right circular cylinder, where V is the volume, r is the radius, h is the altitude (or height), and $\pi \approx 3.14$	183
$V = \frac{1}{3}\pi r^2 h$	The volume of a right circular cone, where V is the volume, r is the radius of the base, h is the altitude, and $\pi \approx 3.14$	125
$I = Prt$	Simple interest, where I is the interest, P is the principal, r is the rate, and t is the time	117
$A = P(1 + rt)$	The amount of money in a savings account, where the interest is simple interest, A is the amount, P is the amount originally invested, r is the annual interest rate, and t is the number of years	115
$A = P(1 + i)^n$	Compound interest, where A is the compounded amount, P is the principal, i is the interest rate per period, and n is the number of periods	117
$V = C - Crt$	The value of a depreciated item, where V is the present value, C is the original cost, r is the rate of depreciation, and t is the time	117
$d = rt$	The distance traveled, where d is the distance, r is the rate (or speed), and t is the time	213
$s = \frac{1}{2}gt^2$	The distance a freely falling object falls, where s is the distance, g is the force due to gravity, and t is the time	115
$T = \pi\sqrt{\dfrac{L}{g}}$	The time for a single swing of a pendulum, where T is the time, L is the length, g is the force due to gravity, and $\pi \approx 3.14$	116
$C = \frac{5}{9}(F - 32)$	Degrees Celsius when degrees Fahrenheit are given	115
$F = \frac{9}{5}C + 32$	Degrees Fahrenheit when degrees Celsius are given	117
$S = \dfrac{a(1 - r^n)}{1 - r}$	The sum of a geometric series, where S is the sum, a is the first term, n is the number of terms, and r is the common ratio	116
$I = \dfrac{E}{R}$	The current, where I is the current, E is the electromotive force, and R is the resistance	116
$\sigma = \sqrt{npq}$	The standard deviation of a binomial distribution, where σ is the standard deviation, n is the number of trials, p is the probability of success, and q is the probability of failure	117
$C = \dfrac{a}{a + 12} \cdot A$	A child's dosage of medicine, where C is the child's dosage, a is the age of the child, and A is the adult dosage	117

Systems of Equations

CHAPTER 12

O
n previous chapters, we showed how to solve a single equation for a single variable. In this chapter, we show how to solve systems of two linear equations in two variables. Systems that have *more* than two equations and variables are not discussed in this book.

12.1   Basic Definitions

A **system of two equations** in (the same) two variables is a set of two equations considered together, or simultaneously. The **solution**, if one exists, of a system of two equations in two variables is the set of ordered pairs that satisfies *both* equations.

EXAMPLE 1   Examples of systems of two equations in two variables and their solutions:

a. $\begin{cases} x - y = 6 \\ x + y = 2 \end{cases}$ is a system of two equations in x and y. $(4, -2)$ is a solution of the system, because when we substitute 4 for x and -2 for y into the first equation, we have

$$4 - (-2) = 6 \quad \text{True}$$

and when we substitute 4 for x and -2 for y in the second equation, we have

$$4 + (-2) = 2 \quad \text{True}$$

This can be stated as "$(4, -2)$ satisfies the system $\begin{cases} x - y = 6 \\ x + y = 2 \end{cases}$."

b. $\begin{cases} x + y = 4 \\ 2x - y = 2 \end{cases}$ is a system of two equations in x and y. $(2, 2)$ is a solution of the system, because when we substitute 2 for x and 2 for y in the first equation, we have

$$2 + 2 = 4 \quad \text{True}$$

and substituting 2 for x and 2 for y in the second equation gives

$$2(2) - 2 = 2 \quad \text{True}$$

EXAMPLE 2   Determine whether $(0, 3)$ is a solution of the system $\begin{cases} x + y = 3 \\ 3x - y = 5 \end{cases}$.

SOLUTION

$$0 + 3 = 3 \quad \text{Substituting 0 for } x \text{ and 3 for } y \text{ in the first equation}$$
$$3 = 3 \quad \text{True}$$
$$3(0) - 3 = 5 \quad \text{Substituting 0 for } x \text{ and 3 for } y \text{ in the second equation}$$
$$-3 = 5 \quad \text{False}$$

Because a false statement results when we substitute the coordinates of the ordered pair into the second equation, $(0, 3)$ does not satisfy *both* equations. Therefore, $(0, 3)$ is not a solution of the given system of equations.

Exercises  12.1
Set I

Determine whether each ordered pair is a solution of the given system of equations.

1. $\begin{cases} 3x + y = 8 \\ 2x - y = 2 \end{cases}$ **a.** $(0, 8)$ **b.** $(1, 0)$ **c.** $(2, 2)$

2. $\begin{cases} 4x - y = 3 \\ 3x + y = 11 \end{cases}$ **a.** $(0, -3)$ **b.** $(2, 5)$ **c.** $(1, 1)$

3. $\begin{cases} 2x + 3y = 6 \\ 4x + 6y = 12 \end{cases}$ **a.** $(3, 2)$ **b.** $(0, 2)$ **c.** $(3, 0)$

4. $\begin{cases} 3x - 2y = 6 \\ 6x - 4y = 12 \end{cases}$ **a.** $(2, 0)$ **b.** $(0, -3)$ **c.** $(-2, -6)$

5. $\begin{cases} x + y = 2 \\ x + y = 4 \end{cases}$ **a.** $(0, 4)$ **b.** $(0, 2)$ **c.** $(4, 0)$

6. $\begin{cases} 2x - y = 4 \\ 2x - y = 2 \end{cases}$ **a.** $(0, -4)$ **b.** $(2, 0)$ **c.** $(1, 0)$

Writing Problems

Express the answers in your own words and in complete sentences.

1. Explain why $(2, -1)$ is a solution for the system $\begin{cases} x - 3y = 5 \\ 4x - 2y = 10 \end{cases}$.

2. Explain why $(4, 4)$ is not a solution for the system $\begin{cases} x - 3y = -8 \\ 3x - y = -6 \end{cases}$.

Exercises 12.1
Set II

Determine whether each ordered pair is a solution of the given system of equations.

1. $\begin{cases} 2x + y = 6 \\ 3x - y = -1 \end{cases}$ **a.** $(0, 6)$ **b.** $(1, 4)$ **c.** $(3, 0)$

2. $\begin{cases} 4x - 2y = 8 \\ 2x - y = 4 \end{cases}$ **a.** $(0, -4)$ **b.** $(2, 0)$ **c.** $(6, 8)$

3. $\begin{cases} x + 3y = 6 \\ 2x + 6y = 12 \end{cases}$ **a.** $(6, 0)$ **b.** $(0, 2)$ **c.** $(1, 3)$

4. $\begin{cases} 3x - y = 3 \\ 6x - 2y = 3 \end{cases}$ **a.** $(1, 0)$ **b.** $(0, -3)$ **c.** $\left(\frac{1}{2}, 0\right)$

5. $\begin{cases} 4x + y = 4 \\ 8x + 2y = 4 \end{cases}$ **a.** $(0, 4)$ **b.** $(1, 0)$ **c.** $\left(\frac{1}{2}, 0\right)$

6. $\begin{cases} 2x + y = 6 \\ 2x - y = 2 \end{cases}$ **a.** $(0, 6)$ **b.** $(2, 2)$ **c.** $(1, 0)$

 12.2   # The Graphical Method of Solving a System of Linear Equations

The **graphical method** of solving a system of equations is not an exact method of solution, but it can sometimes be used successfully in solving such systems. Recall that the graph of a linear equation in two variables is a straight line.

Solving a system of linear equations by the graphical method

1. Graph the solution set of each equation on the same set of axes.

2. Three outcomes are possible:

    a. The lines intersect in exactly one point (see Figure 1 in Example 1). The solution is the ordered pair that represents the point of intersection of the two lines.

    b. The lines are parallel; that is, the lines will never intersect, even if they are extended indefinitely (see Figure 2 in Example 2). There is *no solution* for the system of equations.

    c. The lines coincide; that is, both equations are of the same line (see Figure 3 in Example 3). There are *infinitely many solutions*; any ordered pair that represents a point on the line is one of the solutions.

3. If the lines graphed in step 1 intersect in exactly one point, check the coordinates of that point in *both* equations.

EXAMPLE 1

Solve the system $\begin{cases} x + 2y = 8 \\ x - y = 2 \end{cases}$ graphically.

SOLUTION   We graph the solution set of each equation on the same set of axes (Figure 1). (Recall that we can choose *any* values for the checkpoints except those that are already in the table. For line 1, we arbitrarily let x be 2 for the checkpoint, and for line 2, we let x be -2.)

Line 1 $x + 2y = 8$

y-intercept: If $x = 0$, then $y = 4$.

x-intercept: If $y = 0$, then $x = 8$.

Checkpoint: If $x = 2$, then $y = 3$.

x	y
0	4
8	0
2	3

Line 2 $x - y = 2$

y-intercept: If $x = 0$, then $y = -2$.

x-intercept: If $y = 0$, then $x = 2$.

Checkpoint: If $x = -2$, then $y = -4$.

x	y
0	-2
2	0
-2	-4

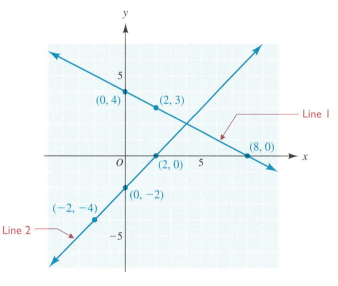

Figure 1

The lines *appear* to intersect at the point $(4, 2)$. We check that solution by substituting 4 for x and 2 for y in *both* equations.

$$(1) \quad x + 2y = 8 \qquad\qquad (2)\ x - y = 2$$
$$4 + 2(2) \overset{?}{=} 8 \qquad\qquad 4 - 2 \overset{?}{=} 2$$
$$8 = 8 \quad \text{True} \qquad\qquad 2 = 2 \quad \text{True}$$

Therefore, the solution of the system is $(4, 2)$.

The coordinates of any point on line 1 satisfy the equation of line 1. The coordinates of any point on line 2 satisfy the equation of line 2. The only point that lies on *both* lines is $(4, 2)$. Therefore, it is the only point whose coordinates satisfy *both* equations.

When a system of equations has a solution, it is called a **consistent system**. When each equation in the system has a different graph, the system is called an **independent system**. The system in Example 1 is a consistent, independent system.

EXAMPLE 2   Solve the system $\left\{\begin{array}{l} 3x + 2y = 6 \\ 6x + 4y = 24 \end{array}\right\}$ graphically.

SOLUTION   We graph the solution set of each equation on the same set of axes (Figure 2).

Line 1 $3x + 2y = 6$

y-intercept: If $x = 0$, then $y = 3$.

x-intercept: If $y = 0$, then $x = 2$.

Checkpoint: If $x = -2$, then $y = 6$.

x	y
0	3
2	0
-2	6

Line 2 $6x + 4y = 24$

y-intercept: If $x = 0$, then $y = 6$.

x-intercept: If $y = 0$, then $x = 4$.

Checkpoint: If $x = 2$, then $y = 3$.

x	y
0	6
4	0
2	3

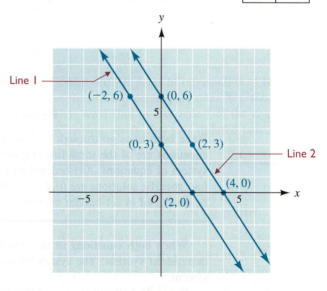

Figure 2

The lines will never meet; they are parallel. (You might verify that their slopes are equal.) There is *no solution* for the system of equations.

EXAMPLE 5   Solve the system $\begin{Bmatrix} x - y = 6 \\ y + x = 2 \end{Bmatrix}$ by the elimination method.

SOLUTION   In this example, the coefficients of the y's are already the additive inverses of each other. Therefore, we can omit step 2; when we add the equations, the y's will be eliminated. We must be sure to write both equations in standard form as we begin.

$$\begin{array}{ll} \text{Equation 1:} & x - y = 6 \\ \text{Equation 2:} & \underline{x + y = 2} \\ & 2x = 8 \end{array}$$ Writing both equations in standard form

$2x = 8$ Adding the equations

$\dfrac{2x}{2} = \dfrac{8}{2}$ Dividing both sides by 2

$x = 4$

Substituting 4 for x in Equation 1, we have

$$\begin{array}{ll} x - y = 6 & \text{Equation 1} \\ 4 - y = 6 & \text{Substituting 4 for } x \\ -4 + 4 - y = 6 - 4 & \text{Adding } -4 \text{ to both sides} \\ -y = 2 \\ y = -2 & \text{Changing all the signs of the equation} \end{array}$$

The system $\begin{Bmatrix} x = 4 \\ y = -2 \end{Bmatrix}$ is equivalent to the system $\begin{Bmatrix} x - y = 6 \\ y + x = 2 \end{Bmatrix}$, because both systems have the same solution set.

Checking will confirm that the solution is $(4, -2)$.

EXAMPLE 6   Solve the system $\begin{Bmatrix} 3y + 2x = 4 \\ 5x + 6y = 11 \end{Bmatrix}$.

SOLUTION   We first write both equations in standard form.

$$\begin{array}{llll} 5 \,]\ 2x + 3y = 4 & \Rightarrow & 10x + 15y = 20 \\ -2 \,]\ 5x + 6y = 11 & \Rightarrow & \underline{-10x - 12y = -22} \\ & & 3y = -2 & \text{Adding the equations} \\ & & y = -\tfrac{2}{3} & \text{Dividing both sides by 3} \end{array}$$

Substituting $-\tfrac{2}{3}$ for y in $2x + 3y = 4$, we have

$$2x + 3y = 4$$

$$2x + \frac{\overset{1}{3}}{1}\left(\frac{-2}{\underset{1}{3}}\right) = 4$$

$$2x - 2 = 4$$

$$2x - 2 + 2 = 4 + 2 \qquad \text{Adding 2 to both sides}$$

$$2x = 6$$

$$\frac{2x}{2} = \frac{6}{2} \qquad \text{Dividing both sides by 2}$$

$$x = 3$$

Checking will confirm that the solution of the system is $\left(3, -\tfrac{2}{3}\right)$.

In all the examples so far, when we attempted to solve the system by the elimination method, we had one variable left after we added the equations. Therefore, there was exactly one solution for each system. Each system was consistent and independent, and the lines representing the equations intersected at exactly one point.

Solving Systems with No Solution by the Elimination Method

Sometimes when we attempt to solve a system of equations by the elimination method, both variables drop out and we're left with a *false statement*. In this case, there is no solution for the system. The system is inconsistent, and the graphs of the equations are parallel lines.

EXAMPLE 7   Solve the system $\left\{ \begin{array}{l} 4x - 2y = 5 \\ 2x - y = 2 \end{array} \right\}$ by the elimination method or write "inconsistent."

SOLUTION   Both equations are in standard form.

Equation 1: $\boxed{-2}$] -1] $4x - 2y = 5 \Rightarrow \boxed{-4} \; x \; + \boxed{2} \; y = -5$

Equation 2: $\boxed{4}$] 2] $2x - \; y = 2 \Rightarrow \underline{\boxed{4} \; x \; - \boxed{2} \; y = \quad 4}$

$$0 = -1 \quad \text{False}$$

This is the pair we'll use

Because both variables were eliminated and we were left with a *false* statement, there is no solution for the system. The system is independent and *inconsistent*.

A Word of Caution   The answer for Example 7 is "inconsistent," *not* "$0 = -1$."

Solving Systems with More Than One Solution by the Elimination Method

Sometimes when we attempt to solve a system of equations by the elimination method, both variables drop out and we're left with a *true statement*. In this case, the system is *dependent*. There are many solutions for the system, and the graphs of the equations are lines that coincide.

EXAMPLE 8   Solve the system $\left\{ \begin{array}{l} 4x - 2y = 4 \\ 2x - y = 2 \end{array} \right\}$ by the elimination method or write "dependent."

SOLUTION   Both equations are in standard form.

Equation 1: $\boxed{-2}$] -1] $4x - 2y = 4 \Rightarrow \boxed{-4} \; x \; + \boxed{2} \; y = -4$

Equation 2: $\boxed{4}$] 2] $2x - \; y = 2 \Rightarrow \underline{\boxed{4} \; x \; - \boxed{2} \; y = \quad 4}$

$$0 = \quad 0 \quad \text{True}$$

This is the pair we'll use

Because both variables were eliminated and we were left with a *true* statement, there are many solutions for the system. Any ordered pair that satisfies Equation 1 or Equation 2 is a solution of the system. The system of equations is consistent and *dependent*.

A Word of Caution   The answer for Example 8 is "dependent," *not* "$0 = 0$."

EXAMPLE 9

Solve the system $\begin{cases} \dfrac{x-3}{y+4} = \dfrac{1}{2} \\ \dfrac{x}{y} = \dfrac{3}{4} \end{cases}$.

SOLUTION   Before we can solve this system of equations, we must write both equations in standard form by clearing fractions, simplifying, and moving the variables to one side of each equation and the constant to the other side. (We note that -4 and 0 are excluded values for y.)

Equation 1: $\dfrac{x-3}{y+4} = \dfrac{1}{2} \Rightarrow 2(x-3) = 1(y+4) \Rightarrow 2x-6 = y+4 \Rightarrow 2x-y = 10$

Equation 2: $\dfrac{x}{y} = \dfrac{3}{4} \Rightarrow 4x = 3y \Rightarrow 4x - 3y = 0$

Equation 1: $-3 \,]\, 2x - y = 10 \Rightarrow -6x + 3y = -30$

Equation 2: $4x - 3y = 0 \Rightarrow \underline{ 4x - 3y = 0}$

$ -2x = -30$ Adding the equations

$ \dfrac{-2x}{-2} = \dfrac{-30}{-2}$ Dividing both sides by -2

$ x = 15$

Substituting 15 for x in $2x - y = 10$, we have

$2x - y = 10$

$2(\,15\,) - y = 10$ Substituting 15 for x

$30 - y = 10$

$-30 + 30 - y = 10 - 30$ Adding -30 to both sides

$-y = -20$

$y = 20$ Changing all the signs *or* multiplying both sides by -1

You can verify that the solution is $(15, 20)$.

Exercises  12.3
Set 1

Find the solution of each system of equations by the elimination method, and check the solution. Write "inconsistent" if no solution exists. Write "dependent" if many solutions exist.

1. $\begin{cases} 2x - y = -4 \\ x + y = -2 \end{cases}$

2. $\begin{cases} 2x + y = 6 \\ x - y = 0 \end{cases}$

3. $\begin{cases} x - 2y = 10 \\ x + y = 4 \end{cases}$

4. $\begin{cases} x + 4y = 4 \\ x - 2y = -2 \end{cases}$

5. $\begin{cases} x - 3y = 6 \\ 4x + 3y = 9 \end{cases}$

6. $\begin{cases} 2x + 5y = 2 \\ 3x - 5y = 3 \end{cases}$

7. $\begin{cases} x + y = 2 \\ 3x - 2y = -9 \end{cases}$

8. $\begin{cases} x + 2y = -4 \\ 2x - y = -3 \end{cases}$

9. $\begin{cases} x + 2y = 0 \\ y - 2x = 0 \end{cases}$

10. $\begin{cases} x - 3y = 0 \\ 3y - x = 0 \end{cases}$

11. $\begin{cases} 4x + 3y = 2 \\ 3x + 5y = -4 \end{cases}$

12. $\begin{cases} 5x + 7y = 1 \\ 3x + 4y = 1 \end{cases}$

13. $\begin{cases} 6x - 10y = 6 \\ 9x - 15y = -4 \end{cases}$

14. $\begin{cases} 7x - 2y = 7 \\ 21x - 6y = 6 \end{cases}$

15. $\begin{cases} 3x - 5y = -2 \\ 10y - 6x = 4 \end{cases}$

16. $\begin{cases} 15x - 9y = -3 \\ 6y - 10x = 2 \end{cases}$

17. $\begin{cases} \dfrac{x}{y} = \dfrac{5}{8} \\ \dfrac{x+3}{y+4} = \dfrac{9}{14} \end{cases}$

18. $\begin{cases} \dfrac{x}{y} = \dfrac{2}{3} \\ \dfrac{x+4}{y+3} = \dfrac{3}{4} \end{cases}$

19. $\begin{cases} \dfrac{x+3}{y+7} = \dfrac{4}{3} \\ \dfrac{x-1}{5-y} = \dfrac{2}{3} \end{cases}$

20. $\begin{cases} \dfrac{x+12}{y+11} = \dfrac{5}{7} \\ \dfrac{x+8}{y+9} = \dfrac{1}{2} \end{cases}$

Writing Problems

Express the answers in your own words and in complete sentences.

1. Explain what *equivalent systems of equations* are.

Explain why $\begin{cases} x = 2 \\ y = 4 \end{cases}$ and $\begin{cases} x - 2y = -6 \\ 4x + 3y = 20 \end{cases}$ are equivalent systems of equations.

Exercises 12.3
Set II

Find the solution of each system of equations by the elimination method, and check the solution. Write "inconsistent" if no solution exists. Write "dependent" if many solutions exist.

1. $\begin{cases} x - y = -5 \\ 3x + y = -3 \end{cases}$

2. $\begin{cases} x - 2y = 4 \\ -2x + 4y = 3 \end{cases}$

3. $\begin{cases} x + y = 5 \\ x - 3y = 3 \end{cases}$

4. $\begin{cases} x - 2y = 3 \\ 3x + 7y = -4 \end{cases}$

5. $\begin{cases} x - 2y = 6 \\ 3x - 2y = 12 \end{cases}$

6. $\begin{cases} 3x + 5y = 4 \\ 2x + 3y = 4 \end{cases}$

7. $\begin{cases} 2x + 6y = 2 \\ 3x + 9y = 3 \end{cases}$

8. $\begin{cases} 4x - 3y = 7 \\ 2x + 7y = 12 \end{cases}$

9. $\begin{cases} y - 5x = 0 \\ x - 5y = 0 \end{cases}$

10. $\begin{cases} 5x + 2y = 7 \\ 10x - 4y = 14 \end{cases}$

11. $\begin{cases} 7x + 2y = 14 \\ 2x - 3y = 4 \end{cases}$

12. $\begin{cases} 6x + 3y = 4 \\ y + 2x = 2 \end{cases}$

13. $\begin{cases} 2x - 5y = 4 \\ 6x - 15y = 8 \end{cases}$

14. $\begin{cases} 5x + 3y = -12 \\ 3y - 6x = 10 \end{cases}$

15. $\begin{cases} 6x + 4y = 2 \\ 3x - 2y = 3 \end{cases}$

16. $\begin{cases} 9x + 3y = 0 \\ 2x - 5y = 0 \end{cases}$

17. $\begin{cases} \dfrac{x}{y} = \dfrac{4}{5} \\ \dfrac{x + 3}{y + 3} = \dfrac{5}{6} \end{cases}$

18. $\begin{cases} \dfrac{x}{y} = \dfrac{4}{7} \\ \dfrac{x + 2}{y} = \dfrac{2}{3} \end{cases}$

19. $\begin{cases} \dfrac{x + 10}{y + 7} = \dfrac{2}{3} \\ \dfrac{3 - x}{y + 12} = \dfrac{1}{2} \end{cases}$

20. $\begin{cases} \dfrac{x + 4}{2y + 6} = \dfrac{1}{2} \\ \dfrac{x - 1}{y + 6} = \dfrac{1}{2} \end{cases}$

12.4 The Substitution Method of Solving a System of Linear Equations

All linear systems of two equations in two variables can be solved by the elimination method shown in the previous section. However, it is important that you also become familiar with the substitution method of solution; some problems lend themselves very nicely to the substitution method, and some higher-degree systems *can't* be solved with the elimination method.

Solving a linear system of two equations by the substitution method

1. Solve one equation for one of the variables in terms of the other by using the method given earlier for solving literal equations.

2. Substitute the expression obtained in step 1 into the *other* equation. Simplify both sides of the resulting equation.

3. Three outcomes are possible:

   a. *One variable remains.* Solve the equation resulting from step 2 for its variable, and substitute that value into either equation to find the value of the other variable. There is one solution for the system. The system is consistent and independent.

   b. *Both variables are eliminated, and a false statement results.* There is no solution; the system is inconsistent and independent.

   c. *Both variables are eliminated, and a true statement results.* There are many solutions; the system is consistent and dependent.

4. If one solution was found in step 3, check it in both equations.

If you have decided to use (or have been told to use) the substitution method, how do you know which equation to start with and which variable to solve for? Sometimes one of the equations is already solved for a variable (this is the kind of equation that lends itself *very* well to the substitution method); if so, use that equation and that variable (see Examples 1 and 4). Sometimes in one of the equations, the coefficient of one of the variables is 1; if so, solve that equation for the variable with a coefficient of 1 (see Examples 2 and 5). Otherwise, solve for the variable with the smallest coefficient if the substitution method *must* be used (see Examples 3 and 6).

EXAMPLE 1　Examples of systems with one equation already solved for one variable:

a. $\begin{cases} x + y = 4 \\ y = x + 2 \end{cases}$

└─ Already solved for y

┌─ Already solved for x

b. $\begin{cases} x = -3 - y \\ 2x + 5y = 0 \end{cases}$

Note　It is very easy to use the substitution method in problems like the ones in Example 1, because one equation has already been solved for one variable.

EXAMPLE 2　Examples of systems in which one variable has a coefficient of 1:

┌─ y has a coefficient of 1

a. $\begin{cases} 2x + y = 5 \\ 3x - 2y = 18 \end{cases} \Rightarrow y = 5 - 2x$

b. $\begin{cases} 2x + 6y = 3 \\ x - 4y = 2 \end{cases} \Rightarrow x = 4y + 2$

↑

x has a coefficient of 1

EXAMPLE 3　Examples of choosing the variable with the smallest coefficient:

┌─ The smallest of the four coefficients

a. $\begin{cases} 6x + 3y = -1 \\ 4x + 9y = 4 \end{cases} \Rightarrow y = \dfrac{-1 - 6x}{3}$　　We solved for y by using the method for solving literal equations

└─ The smallest possible denominator

┌─ The smallest of the four coefficients

b. $\begin{cases} 3x + 9y = 7 \\ 12x - 8y = 15 \end{cases} \Rightarrow x = \dfrac{7 - 9y}{3}$　　We solved for x by using the method for solving literal equations

└─ The smallest possible denominator

Note　We believe that it is better to solve problems such as the ones in Examples 3 and 6 by using the elimination method rather than the substitution method. We demonstrate in this section, however, that it is *possible* to use the substitution method on such problems.

Solving Systems with Exactly One Solution by the Substitution Method

EXAMPLE 4 Solve the system $\left\{\begin{array}{l} x + y = 4 \\ y = x + 2 \end{array}\right\}$ by the substitution method.

SOLUTION

$$\text{Equation 1: } x + y = 4$$

$$\text{Equation 2: } y = x + 2$$

Step 1. It will be easier to substitute *into* Equation 1, since Equation 2 has already been solved for y: $y = \boxed{x + 2}$.

Step 2. We substitute $\boxed{x + 2}$ for y in Equation 1:

$$x + \quad y \quad = 4 \qquad \text{Equation I}$$

$$x + (\, \boxed{x + 2} \,) = 4 \qquad \text{Substituting } x + 2 \text{ for } y$$

Step 3.

$$2x + 2 = 4 \qquad\qquad\qquad \text{Simplifying the left side of the equation}$$

$$2x + 2 - 2 = 4 - 2 \qquad\qquad \text{Adding } -2 \text{ to both sides}$$

$$2x = 2$$

$$\frac{2x}{2} = \frac{2}{2} \qquad\qquad\qquad \text{Dividing both sides by 2}$$

$$x = 1$$

To find y, we substitute 1 for x in Equation 2:

$$y = x + 2$$

$$y = \boxed{1} + 2$$

$$y = 3$$

✓ **Step 4. Check** (1) $x + y = 4$ (2) $y = x + 2$

$\qquad\qquad\qquad\qquad\quad 1 + 3 \overset{?}{=} 4 \qquad\qquad\qquad\quad 3 \overset{?}{=} 1 + 2$

$\qquad\qquad\qquad\qquad\qquad 4 = 4 \quad \text{True} \qquad\qquad\quad 3 = 3 \qquad\qquad \text{True}$

Therefore, $(1, 3)$ is the solution for this system.

EXAMPLE 5 Solve the system $\left\{\begin{array}{l} 2x + y = 5 \\ 3x - 2y = 18 \end{array}\right\}$ by the substitution method.

SOLUTION

$$\text{Equation 1: } 2x + y = 5$$

$$\text{Equation 2: } 3x - 2y = 18$$

Step 1. In Equation 1, the coefficient of y is 1; therefore, we solve Equation 1 for y:

$$2x + \boxed{y} = 5 \qquad\qquad \text{Equation I}$$

$$-2x + 2x + y = 5 - 2x \qquad \text{Adding } -2x \text{ to both sides}$$

$$y = \boxed{5 - 2x} \qquad \text{This equation has been solved for } y$$

Step 2. We substitute $5 - 2x$ for y in Equation 2:

$$3x - \quad 2 \quad y \quad = 18 \qquad \text{Equation 2}$$

$$3x - 2(\; 5 - 2x \;) = 18 \qquad \text{Substituting } 5 - 2x \text{ for } y$$

$$3x - 10 + 4x = 18 \qquad \text{Using the distributive property}$$

$$7x - 10 = 18 \qquad \text{Combining like terms}$$

Step 3. $\qquad\qquad 7x - 10 + 10 = 18 + 10 \qquad \text{Adding 10 to both sides}$

$$7x = 28$$

$$\frac{7x}{7} = \frac{28}{7} \qquad \text{Dividing both sides by 7}$$

$$x = 4$$

To find y, we substitute 4 for x in the equation $y = 5 - 2x$.

$$y = 5 - 2x$$

$$y = 5 - 2(\; 4 \;)$$

$$y = 5 - 8$$

$$y = -3$$

You can verify that $(4, -3)$ is the solution for this system.

EXAMPLE 6   Solve the system $\left\{\begin{array}{l} 6x + 3y = -1 \\ 4x + 9y = 4 \end{array}\right\}$ by the substitution method.

SOLUTION

The smallest of the four coefficients

$$\text{Equation 1: } 6x + 3y = -1$$

$$\text{Equation 2: } 4x + 9y = 4$$

Step 1. No variable has a coefficient of 1. We solve Equation 1 for y because it has the smallest coefficient.

$$6x + 3y = -1$$

$$-6x + 6x + 3y = -1 - 6x \qquad \text{Adding } -6x \text{ to both sides}$$

$$3y = -1 - 6x$$

$$\frac{3y}{3} = \frac{-1 - 6x}{3} \qquad \text{Dividing both sides by 3}$$

$$y = \frac{-1 - 6x}{3} \qquad \text{This equation has been solved for } y$$

Step 2. We substitute $\dfrac{-1 - 6x}{3}$ for y in Equation 2:

$$4x + 9 \qquad y \qquad = 4 \qquad \text{Equation 2}$$

$$4x + \overset{3}{9}\left(\frac{-1 - 6x}{3} \right) = 4 \qquad \text{Substituting } \frac{-1 - 6x}{3} \text{ for } y$$

$$4x + 3(-1 - 6x) = 4$$

$$4x - 3 - 18x = 4 \qquad \text{Using the distributive property}$$

$$-3 - 14x = 4$$

$$3 - 3 - 14x = 4 + 3 \qquad \text{Adding 3 to both sides}$$

$$-14x = 7$$

$$\frac{-14x}{-14} = \frac{7}{-14} \qquad \text{Dividing both sides by } -14$$

$$x = -\tfrac{1}{2}$$

Step 3. We substitute $-\frac{1}{2}$ for x in $y = \dfrac{-1 - 6x}{3}$.

$$y = \frac{-1 - \overset{2}{6}\left(-\dfrac{1}{\underset{1}{2}}\right)}{3} = \frac{-1 - 3(-1)}{3} = \frac{-1 + 3}{3} = \frac{2}{3}$$

You can verify that $\left(-\frac{1}{2}, \frac{2}{3}\right)$ is the solution for this system.

Solving Systems with No Solution by the Substitution Method

Example 7 shows how to identify systems that have no solution when you are using the substitution method.

EXAMPLE 7 Solve the system $\begin{Bmatrix} x + 3y = 4 \\ 2x + 6y = 4 \end{Bmatrix}$ using the substitution method or write "inconsistent."

SOLUTION

The coefficient is 1

Equation 1: $x + 3y = 4$

Equation 2: $2x + 6y = 4$

Step 1. We solve Equation 1 for x:

$$x + 3y = 4$$

$$x + 3y - 3y = 4 - 3y \qquad \text{Adding } -3y \text{ to both sides}$$

$$x = 4 - 3y \qquad \text{This equation has been solved for } x$$

Step 2. We substitute $4 - 3y$ for x in Equation 2:

$$2\,x + 6y = 4 \qquad \text{Equation 2}$$

$$2(4 - 3y) + 6y = 4$$

$$8 - 6y + 6y = 4$$

Step 3. $8 = 4 \qquad \text{A false statement}$

Because both variables were eliminated and we were left with a *false* statement, there is *no solution* for this system of equations. The system is *inconsistent*. (The graphs of the equations are parallel lines.)

Solving Systems with More Than One Solution by the Substitution Method

Example 8 shows how to identify systems that have more than one solution when you are using the substitution method.

EXAMPLE 8 Solve the system $\begin{Bmatrix} 6x + 3y = 6 \\ 2x + y = 2 \end{Bmatrix}$ using the substitution method or write "dependent."

SOLUTION

Equation 1: $6x + 3y = 6$

Equation 2: $2x + y = 2$

The coefficient is 1

Step 1. We solve Equation 2 for y:

$$2x + y = 2$$

$$-2x + 2x + y = 2 - 2x \quad \text{Adding } -2x \text{ to both sides}$$

$$y = \boxed{2 - 2x} \quad \text{This equation has been solved for } y$$

Step 2. We substitute $2 - 2x$ for y in Equation 1:

$$6x + 3\,y = 6 \quad \text{Equation 1}$$

$$6x + 3(\,2 - 2x\,) = 6$$

$$6x + 6 - 6x = 6$$

Step 3.

$$6 = 6 \quad \text{A true statement}$$

Because both variables were eliminated and we were left with a *true* statement, this system of equations has *many* solutions. Any ordered pair that satisfies Equation 1 also satisfies Equation 2. The system is *dependent*. (The graphs of the equations are lines that coincide.)

Exercises 12.4
Set 1

Find the solution of each system of equations. Write "inconsistent" if no solution exists. Write "dependent" if many solutions exist. In Exercises 1–12, use the substitution method.

1. $\begin{cases} 2x - 3y = 1 \\ x = y + 2 \end{cases}$

2. $\begin{cases} y = 2x + 3 \\ 3x + 2y = 20 \end{cases}$

3. $\begin{cases} 3x + 4y = 2 \\ y = x - 3 \end{cases}$

4. $\begin{cases} 2x + 3y = 11 \\ x = y - 2 \end{cases}$

5. $\begin{cases} y = 2 - 4x \\ 7x + 3y = 1 \end{cases}$

6. $\begin{cases} 5x + 7y = 1 \\ x = -4y - 5 \end{cases}$

7. $\begin{cases} 4x - y = 3 \\ 8x - 2y = 6 \end{cases}$

8. $\begin{cases} x - 3y = 2 \\ 3x - 9y = 6 \end{cases}$

9. $\begin{cases} x + 3 = 0 \\ 3x - 2y = 6 \end{cases}$

10. $\begin{cases} y - 4 = 0 \\ 3y - 5x = 15 \end{cases}$

11. $\begin{cases} x = 3y - 4 \\ 5y - 2x = 2 \end{cases}$

12. $\begin{cases} 3y - 4x = 11 \\ y = 2x - 1 \end{cases}$

In Exercises 13–20, use any convenient method.

13. $\begin{cases} 8x + 4y = 7 \\ x = 2 - 2y \end{cases}$

14. $\begin{cases} x = 1 - y \\ 4x + 8y = 5 \end{cases}$

15. $\begin{cases} 3x - 2y = 8 \\ 2y - 3x = 4 \end{cases}$

16. $\begin{cases} 4x - 5y = 15 \\ 5y - 4x = 10 \end{cases}$

17. $\begin{cases} 5x - 4y = 2 \\ y = 1 + 2x \end{cases}$

18. $\begin{cases} 4x - 9y = 7 \\ x = y + 3 \end{cases}$

19. $\begin{cases} 4x + 4y = 3 \\ 6x + 12y = -6 \end{cases}$

20. $\begin{cases} 4x + 9y = -11 \\ 10x + 6y = 11 \end{cases}$

Writing Problems

Express the answers in your own words and in complete sentences.

1. Explain why it would be convenient to solve the system $\begin{cases} 3x - 7y = 12 \\ y = 2x - 6 \end{cases}$ by substitution.

2. Which method would you use in solving the system $\begin{cases} 5x - 4y = 3 \\ 3x + 5y = 1 \end{cases}$? Explain why you would use that method.

Exercises 12.4
Set II

Find the solution of each system of equations. Write "inconsistent" if no solution exists. Write "dependent" if many solutions exist. In Exercises 1–12, use the substitution method.

1. $\begin{cases} x - y = 1 \\ y = 2x - 3 \end{cases}$

2. $\begin{cases} x + 3y = 4 \\ x = 2 - y \end{cases}$

3. $\begin{cases} x + 2y = -1 \\ x = 5 + y \end{cases}$

4. $\begin{cases} x + y = 1 \\ y = x + 7 \end{cases}$

5. $\begin{cases} x = 9 + 4y \\ 3x + 8y = 7 \end{cases}$

6. $\begin{cases} 3x + 4y = 18 \\ y = 5x - 7 \end{cases}$

7. $\begin{cases} 5x + y = 6 \\ 10x + 2y = 12 \end{cases}$

8. $\begin{cases} 2x + 5y = -5 \\ 3y + x = -2 \end{cases}$

9. $\begin{cases} 5x + y = 0 \\ 3x + 2y = 7 \end{cases}$

10. $\begin{cases} x + y = 4 \\ 2x - y = 2 \end{cases}$

11. $\begin{cases} y = 5 - x \\ 3x + 5y = 19 \end{cases}$

12. $\begin{cases} 5y - 2x = 7 \\ x = 2y - 4 \end{cases}$

In Exercises 13–20, use any convenient method.

13. $\begin{cases} 7x + 5y = -4 \\ x = 2y - 6 \end{cases}$

14. $\begin{cases} 4x + 7y = 9 \\ 6x + 5y = -3 \end{cases}$

15. $\begin{cases} 2x - y = 3 \\ y - 2x = 1 \end{cases}$

16. $\begin{cases} x + 2y = 3 \\ 4x + 8y = 12 \end{cases}$

17. $\begin{cases} 8x + 3y = 4 \\ x = 3y + 5 \end{cases}$

18. $\begin{cases} x + 3y = 0 \\ x = 2y + 10 \end{cases}$

19. $\begin{cases} 3x + 2y = 7 \\ 6x - 4y = 7 \end{cases}$

20. $\begin{cases} 7x - 2y = 5 \\ 5x + 3y = 8 \end{cases}$

12.5 Using Systems of Equations to Solve Applied Problems

In solving applied problems involving more than one unknown, it is sometimes difficult to represent each unknown in terms of a single variable. In this section, we eliminate that difficulty by using a different variable for each unknown. When we use two variables, we must find two equations to represent the given facts; we then use a system of equations to solve the problem.

Solving an applied problem by using a system of equations

Read First read the problem very carefully.

Think Determine what *type* of problem it is, if possible. Determine what is unknown.

Sketch Draw a sketch *with labels*, if it would be helpful.

Step 1. Represent each unknown number by a *different* variable.

Reread Reread the problem, breaking it up into key words and phrases.

Step 2. Translate each key word or phrase into an algebraic expression; fit these expressions together into *two different equations* that represent the conditions given in the problem.

Step 3. Solve the *system* of equations for *both* variables, using the elimination method, the substitution method, or the graphical method.

Step 4. Be sure to answer *all* the questions asked.

Step 5. Check the solutions *in the word statements.*

Step 6. State the results clearly.

In Example 1, the problem is solved on the left by using a single variable and a single equation. On the right, it is solved by using two variables and a system of equations. (We will not show Step 1, Step 2, and so on, in this example.)

EXAMPLE 1

The sum of two numbers is 20. Their difference is 6. What are the numbers?

SOLUTION

Using one variable

Let $x =$ the larger number

Then $20 - x =$ the smaller number

The difference   is   6

$$x - (20 - x) = 6$$

$$x - 20 + x = 6$$

$$2x - 20 + 20 = 6 + 20$$

$$2x = 26$$

The larger number $x = 13$

The smaller $20 - x = 7$
number

The difficulty in using the one-variable method to solve this problem is that some people cannot decide whether to represent the second unknown number by $x - 20$ or by $20 - x$.

Using two variables

Let $x =$ the larger number

Let $y =$ the smaller number

The sum of two numbers is 20

(1) $x + y$ $= 20$

The difference   is   6

(2) $x - y$ $= 6$

Using the elimination method:

$$\begin{matrix} (1) \\ (2) \end{matrix} \left\{ \begin{matrix} x + y = 20 \\ x - y = 6 \end{matrix} \right\}$$

$2x \quad = 26$ Adding Equations 1 and 2

$x = 13$ The larger number

Substituting 13 for x in Equation 1, we have

$$x + y = 20$$

$$13 + y = 20$$

$$y = 7 \quad \text{The smaller number}$$

$\checkmark$ **Check** $13 + 7 = 20$ The sum of the numbers is 20

$13 - 7 = 6$ The difference of the numbers is 6

Therefore, the numbers are 13 and 7.

EXAMPLE 2

Kelly invested $9,000, part of it at 5% simple annual interest and part of it at 3% simple annual interest. If the total income for one year was $420, how much was invested at each rate?

SOLUTION

Step 1. Let $x =$ the amount invested at 5% interest

Let $y =$ the amount invested at 3% interest

Think   We use the formula $I = Prt$, where I is the amount of interest earned, P is the principal, r is the annual interest rate (changed to decimal form), and t is the number of years. The amount of interest earned at 5% is $x(0.05)(1)$, or $0.05x$, and the amount of interest earned at 3% is $y(0.03)(1)$, or $0.03y$.

Reread	Total amount invested		is	$9,000
Step 2.	x	$+$ y	$=$	9,000

Reread	Amount of interest earned at 5% interest	plus	amount of interest earned at 3% interest	equals	$420
Step 2.	$0.05x$	$+$	$0.03y$	$=$	420

Step 3. Using the elimination method, we have

$$(1) \begin{cases} x + y = 9{,}000 \\ (2) 0.05x + 0.03y = 420 \end{cases}$$

$$\begin{array}{ll} -3x - 3y = -27{,}000 & \text{Equation I, multiplied by } -3 \\ \underline{5x + 3y = 42{,}000} & \text{Equation 2, multiplied by I00} \\ 2x = 15{,}000 & \text{Adding the equations} \end{array}$$

$$\frac{2x}{2} = \frac{15{,}000}{2} \qquad \text{Dividing both sides by 2}$$

$$x = 7{,}500 \qquad \text{The number of dollars invested at 5\%}$$

Substituting 7,500 for x in Equation 1, we have

$$\begin{array}{ll} x + y = 9{,}000 & \text{Equation I} \\ 7{,}500 + y = 9{,}000 & \text{Substituting 7,500 for } x \text{ in} \\ & \text{Equation I} \\ -7{,}500 + 7{,}500 + y = -7{,}500 + 9{,}000 & \text{Adding } -7{,}500 \text{ to both sides} \end{array}$$

Step 4. $\qquad\qquad\qquad\qquad y = 1{,}500$ — The number of dollars invested at 3%

✓ **Step 5. Check** $\$7{,}500 + \$1{,}500 = \$9{,}000$ There was \$9,000 invested altogether

$$(\$7{,}500)\,(0.05)\,(1) + (\$1{,}500)\,(0.03)\,(1) = \$375 + \$45$$
$$= \$420 \qquad \text{The income was \$420}$$

Step 6. Therefore, Kelly invested \$7,500 at 5% interest and \$1,500 at 3% interest.

EXAMPLE 3

A fraction has the value $\frac{2}{3}$. If 3 is added to the numerator and 2 is added to the denominator, the value of the resulting fraction is $\frac{3}{4}$. Find the original fraction.

SOLUTION

Step 1. Let $x =$ the numerator of the original fraction

Let $y =$ the denominator of the original fraction

Step 2. $\qquad (1) \begin{cases} \dfrac{x}{y} = \dfrac{2}{3} & \text{The value of the original fraction is } \frac{2}{3} \\[2mm] (2)\ \dfrac{x+3}{y+2} = \dfrac{3}{4} & \text{The value of the new fraction is } \frac{3}{4} \end{cases}$

Before we can solve the system of equations, we must clear fractions and line up the like terms. (We note that 0 and -2 are excluded values for y.)

Step 3. (1) $\qquad 3x = 2y \qquad\qquad$ Multiplying both sides by the LCD or cross-multiplying

(2) $4(x + 3) = 3(y + 2)$ Multiplying both sides by the LCD or cross-multiplying

(1) $\qquad 3x = 2y \qquad\qquad \Rightarrow 3x - 2y = 0$

(2) $4(x + 3) = 3(y + 2) \Rightarrow 4x + 12 = 3y + 6 \Rightarrow 4x - 3y = -6$

Therefore, the system is

$$(1) \begin{cases} 3x - 2y = 0 \\ (2) 4x - 3y = -6 \end{cases}$$

We now solve the system:

$$\begin{array}{lll} (1) & 3]\ 3x - 2y = 0 & \Rightarrow 9x - 6y = 0 \\ (2) & -2]\ 4x - 3y = -6 & \Rightarrow \underline{-8x + 6y = 12} \\ & & x = 12 \end{array}$$

Substituting 12 for x in $3x = 2y$, we have

$$3x = 2y$$

$$3(\boxed{12}) = 2y \qquad \text{Substituting 12 for } x$$

$$36 = 2y$$

$$\frac{36}{2} = \frac{2y}{2} \qquad \text{Dividing both sides by 2}$$

$$y = 18$$

Step 4. The *original fraction* is apparently $\frac{12}{18}$.

✓ **Step 5. Check** $\dfrac{12}{18} = \dfrac{2}{3}$ The value of the original fraction is $\frac{2}{3}$

$$\frac{12 + 3}{18 + 2} = \frac{15}{20} = \frac{3}{4} \quad \begin{array}{l}\text{If 3 is added to the numerator and 2 is added to the}\\ \text{denominator, the value of the resulting fraction is } \frac{3}{4}\end{array}$$

Step 6. Therefore, the original fraction *is* $\frac{12}{18}$.

Some applied problems should be solved by using one variable, while others are best solved by using two variables and a system of equations.

How to choose which method to use for solving an applied problem	**1.** Read the problem completely and determine how many unknown numbers there are. **2.** If there is only one unknown number, use the one-variable method. **3.** If there are two unknown numbers, try to represent both of them in terms of one variable. If this is too difficult, represent each unknown number by a different variable and then solve, using a system of equations.

Systems that have more than two equations or more than two variables are not discussed in this book.

Exercises  12.5
Set I

Set up each of the following problems algebraically *using two variables*, solve, and check. Be sure to state what your variables represent.

1. The sum of two numbers is 80. Their difference is 12. What are the numbers?

2. The sum of two numbers is 90. Their difference is 4. What are the numbers?

3. The sum of two angles is 90°. Their difference is 60°. What are the angles?

4. The sum of two angles is 180°. Their difference is 32°. What are the angles?

5. Dixie invested $8,000, part of it at 4% simple annual interest and part of it at 3% simple annual interest. If the total income for one year was $255, how much was invested at each rate?

6. Marty invested $6,000, part of it at 5% simple annual interest and part of it at 4% simple annual interest. If the total income for one year was $252, how much was invested at each rate?

7. Find two numbers such that twice the smaller plus 3 times the larger is 34, and 5 times the smaller minus twice the larger is 9.

8. Find two numbers such that 5 times the larger plus 3 times the smaller is 47, and 4 times the larger minus twice the smaller is 20.

9. Jason paid $16.88 for a 6-lb mixture of granola and dried apple chunks. If the granola cost $2.10 per pound and the dried apple chunks cost $4.24 per pound, how many pounds of each did he buy?

10. A 100-lb mixture of two different grades of coffee costs $351.50. If grade A costs $3.80 per pound and grade B costs $3.30 per pound, how many pounds of each grade were used?

11. Don spent $4.48 for twenty-two stamps. If he bought only 22¢ and 18¢ stamps, how many of each kind did he buy?

12. Sue spent $12.38 for fifty stamps. If she bought only 22¢ and 45¢ stamps, how many of each type did she buy?

13. The length of a rectangle is 1 ft 6 in. longer than its width. Its perimeter is 19 ft. Find its dimensions.

14. The length of a rectangle is 2 ft 6 in. longer than its width. Its perimeter is 25 ft. Find its dimensions.

15. A fraction has the value $\frac{2}{3}$. If 4 is added to the numerator and if the denominator is decreased by 2, the resulting fraction has the value $\frac{6}{7}$. Find the original fraction.

16. A fraction has the value $\frac{3}{4}$. If 4 is added to its numerator and 8 is subtracted from its denominator, the value of the resulting fraction is 1. Find the original fraction.

17. It took a boat 7 hr to travel 252 mi with the current, and it took the same boat 9 hr to travel the same distance against the same current. Find the average speed of the boat in still water and the average speed of the current.

18. It took a plane 5 hr to fly 450 mi against the wind and only 3 hr to return with the same wind. Find the average speed of the plane in still air and the average speed of the wind.

19. A tie and a pin cost $1.10. The tie cost $1 more than the pin. What was the cost of each?

20. A number of birds are resting on two limbs of a tree. One limb is above the other. A bird on the lower limb says to the birds on the upper limb, "If one of you will come down here, we will have an equal number on each limb." A bird from above replies, "If one of you will come up here, we will have twice as many up here as you will have down there." How many birds are sitting on each limb?

Exercises  *12.5*
Set II

Set up each of the following problems algebraically *using two variables*, solve, and check. Be sure to state what your variables represent.

1. The sum of two numbers is 42. Their difference is 12. What are the numbers?

2. Half the sum of two numbers is 15. Half their difference is 8. Find the numbers.

3. The sum of two angles is 90°. Their difference is 16°. Find the angles.

4. A 20-lb mixture of product A and product B costs $19.75. If product A costs 85¢ per pound and product B costs $1.40 per pound, find the number of pounds of each.

5. Edmund invested $5,000, part of it at 4% simple annual interest and part of it at 2% simple annual interest. If the total income for one year was $144, how much was invested at each rate?

6. Kate put twice as much money into a savings account that paid 4% simple annual interest as she put into another account that paid 3% simple annual interest. If the total income for one year was $462, how much was invested at each rate?

7. Find two numbers such that twice the smaller plus 4 times the larger is 66, and 6 times the smaller minus 3 times the larger is 3.

8. Beatrice has seventeen coins with a total value of $3.05. If these coins are all nickels and quarters, how many of each kind does she have?

9. A 20-lb mixture of almonds and hazelnuts costs $53.00. If almonds cost $2.50 per pound and hazelnuts cost $3.00 per pound, find the number of pounds of each.

10. A fraction has the value $\frac{1}{2}$. If 2 is added to the numerator and 1 is subtracted from the denominator, the value of the resulting fraction is $\frac{4}{7}$. Find the original fraction.

11. Rachelle spent $4.55 for fifteen stamps. If she bought only 25¢ and 35¢ stamps, how many of each kind did she buy?

12. A class received $233 for selling 200 tickets to a school play. If tickets for students cost $1 each and tickets for nonstudents cost $2 each, how many tickets were sold to nonstudents?

13. The length of a rectangle is 3 ft 6 in. longer than its width. The perimeter of the rectangle is 13 ft. Find the dimensions of the rectangle.

14. Jill spent $59.50 for fourteen tickets to a movie. Some were for children and some for adults. If children's tickets cost $3.50 and adults' tickets cost $5.25, how many of each kind did she buy?

15. A fraction has the value $\frac{1}{2}$. If 4 is added to the numerator and 2 is subtracted from the denominator, the resulting fraction has the value $\frac{2}{3}$. Find the original fraction.

16. Several families went to a school play together. They spent $10.60 for eight tickets. If adults' tickets cost $1.95 and children's tickets cost 95¢, how many of each kind of ticket were bought?

17. It took Wing 6 hr to ride his bicycle 36 mi against the wind but only 2 hr to return to his starting point with the same wind. Find Wing's average riding speed in still air and the average speed of the wind.

18. A 100-lb mixture of product A and product B costs $22. If product A costs 20¢ per pound and product B costs 25¢ per pound, how many pounds of each kind are there in the mixture?

19. A bracelet and a pin cost $11. One cost $10 more than the other. What was the cost of each?

20. It took a plane 7 hr to fly 875 mi against the wind but only 5 hr to return to its starting point with the same wind. Find the average speed of the plane in still air and the average speed of the wind.

Sections 12.1–12.5

REVIEW

Solution of a System of Equations 12.1

A **solution** of a system of two equations in two variables is an ordered pair that satisfies both equations.

Solving a System of Equations 12.2, 12.3, 12.4

Graphical method: Graph the solution set of each equation on the same set of axes. If the lines intersect in exactly one point, the ordered pair representing that point is the solution.

Elimination method:

1. Express both equations in standard form: $Ax + By = C$.

2. If necessary, multiply one or both equations by numbers that make the coefficients of one of the variables become the additive inverses of each other; write the equations one under the other *with like terms lined up*.

3. Add the equations from step 2 together; *if one variable remains*, there is exactly one solution for the system. To find it, solve the resulting equation for its variable, and substitute that value into either equation to find the value of the other variable.

Substitution method:

1. Solve one equation for one of its variables in terms of the other.

2. Substitute the expression obtained in step 1 into the *other* equation. Simplify both sides of the resulting equation.

3. *If one variable remains*, there is exactly one solution for the system. To find it, solve the resulting equation for its variable, and substitute that value into either equation to find the value of the other variable.

Other possible outcomes:

There may be no solution. For the *graphical* method, the lines are parallel. For the *algebraic* methods, both variables drop out and a false statement results.

There may be many solutions. For the *graphical* method, the lines coincide. For the *algebraic* methods, both variables drop out and a true statement results.

To Solve an Applied Problem Using a System of Equations
12.5

Read   First read the problem very carefully.

Think   Determine what type of problem it is, if possible. Determine what is unknown.

Sketch   Draw a sketch *with labels*, if it would be helpful.

Step 1. Represent each unknown number by a *different* variable.

Reread   Reread the problem, breaking it up into key words and phrases.

Step 2. Translate each key word or phrase into an algebraic expression; fit these expressions together into *two different equations*.

Step 3. Solve the *system* of equations for *both* variables, using the elimination method, the substitution method, or the graphical method.

Step 4. Be sure to answer *all* the questions asked.

Step 5. Check the solutions *in the word statements*.

Step 6. State the results clearly.

Sections 12.1–12.5   REVIEW EXERCISES   Set 1

In Exercises 1–3, find the solution of each system graphically. Write "inconsistent" if no solution exists. Write "dependent" if many solutions exist.

1. $\begin{cases} x + y = 6 \\ x - y = 4 \end{cases}$

2. $\begin{cases} x + y = 5 \\ x - y = -3 \end{cases}$

3. $\begin{cases} 3x - y = 6 \\ 6x - 2y = 12 \end{cases}$

In Exercises 4–6, find the solution of each system, using the elimination method. Write "inconsistent" if no solution exists. Write "dependent" if many solutions exist.

4. $\begin{cases} 4x + 5y = 22 \\ 3x + y = 11 \end{cases}$

5. $\begin{cases} 4x - 8y = 4 \\ 3x - 6y = 3 \end{cases}$

6. $\begin{cases} 4x - 7y = 28 \\ 7y - 4x = 20 \end{cases}$

In Exercises 7–9, find the solution of each system, using the substitution method. Write "inconsistent" if no solution exists. Write "dependent" if many solutions exist.

7. $\begin{cases} x = y + 2 \\ 4x - 5y = 3 \end{cases}$

8. $\begin{cases} 7x - 3y = 1 \\ y = x + 5 \end{cases}$

9. $\begin{cases} x + y = 4 \\ 2x - y = 2 \end{cases}$

In Exercise 10, solve the system by any convenient method. Write "inconsistent" if no solution exists. Write "dependent" if many solutions exist.

10. $\begin{cases} 5x - 4y = -7 \\ -6x + 8y = 2 \end{cases}$

In Exercises 11–14, set up each problem algebraically *using two variables*, solve, and check. Be sure to state what your variables represent.

11. The sum of two numbers is 84. Their difference is 22. What are the numbers?

12. A fraction has the value $\frac{4}{5}$. If 3 is subtracted from the numerator and 5 is added to the denominator, the value of the resulting fraction is $\frac{3}{5}$. Find the original fraction.

13. An office manager paid $107.75 for twenty boxes of paper for the copy machine and the laser printer. If paper for the laser printer costs $7.50 per box and paper for the copy machine costs $4.25 per box, how many boxes of each kind were purchased?

14. It took a boat 7 hr to travel 231 mi with the current but 11 hr to travel the same distance against the same current. Find the average speed of the boat in still water and the average speed of the current.

_____

Name

ANSWERS

In Exercises 1 and 2, find the solution of each system graphically. Write "inconsistent" if no solution exists. Write "dependent" if many solutions exist.

1. $\begin{cases} 3x + y = -9 \\ 3x - 2y = 0 \end{cases}$

2. $\begin{cases} 2x - 3y = 3 \\ 3y - 2x = 6 \end{cases}$

1. _____

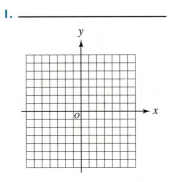

2. _____

In Exercises 3–5, find the solution of each system, using the elimination method. Write "inconsistent" if no solution exists. Write "dependent" if many solutions exist.

3. $\begin{cases} 3x - 2y = 10 \\ 5x + 4y = 24 \end{cases}$

4. $\begin{cases} 6x + 4y = -1 \\ 4x + 6y = -9 \end{cases}$

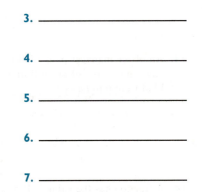

3. _____

5. $\begin{cases} 6x + 4y = 13 \\ 8x + 10y = 1 \end{cases}$

4. _____

5. _____

6. _____

7. _____

In Exercises 6 and 7, solve each system, using the substitution method. Write "inconsistent" if no solution exists. Write "dependent" if many solutions exist.

6. $\begin{cases} x - 3y = 15 \\ 5x + 7y = -13 \end{cases}$

7. $\begin{cases} 4x - 6y = 2 \\ 6x - 9y = 3 \end{cases}$

In Exercises 27–30, set up each problem algebraically, solve, and check. Be sure to state what your variables represent.

27. A 5-lb mixture of beef back ribs and beef stew meat costs $7.35. If the stew meat is $2.10 per pound and the back ribs are $1.20 per pound, how many pounds of each kind are there?

28. Find two consecutive integers whose product is 11 more than their sum.

29. Elaine left home at 8 A.M., driving toward the mountains. Her sister Heather left their home one hour later, driving on the same road. By driving 6 mph faster, Heather overtook Elaine at 4 P.M. (the same day).

a. How fast was Elaine driving?

b. How fast was Heather driving?

c. How far had they gone when Heather overtook Elaine?

In Exercise 30, solve, using *two* variables.

30. The width of a rectangle is 5 m less than its length. Its perimeter is 58 m. Find the dimensions of the rectangle.

Critical Thinking and Problem-Solving Exercises

1. Five people are attending a meeting. If each person shakes hands with each other person exactly once, how many handshakes will occur? How many handshakes will occur if ten people attend the meeting?

2. The sum of the measures of the angles of a three-sided polygon is 180°; the sum of the measures of the angles of a four-sided polygon is 360°; the sum of the measures of the angles of a five-sided polygon is 540°. What is the sum of the measures of the angles of a seven-sided polygon?

3. Suppose you need to construct a rectangle whose perimeter *has* to be 24 cm, and suppose the length and the width have to be in whole centimeters, not in fractional parts of centimeters. Make an organized list of possible lengths and widths for the rectangle, and include the *area* for each rectangle. Which dimensions give the largest area?

4. Brooke, Eric, and Trevor are business associates who share an office. Trevor is in the office every ninth day, Eric is in the office every fourth day, and Brooke is in the office every sixth day.

a. How often are Brooke and Eric in the office on the same day?

b. How often are Brooke and Trevor in the office on the same day?

c. How often are Eric and Trevor in the office on the same day?

d. How often are all three people in the office on the same day?

5. Lucinda stores her accessories in seven numbered, colored boxes on a shelf in her room. The accessories are scarves, pins, belts, rings, necklaces, gloves, and hats. Using the sketch and the following clues, determine the color of each box and the contents of each box:

Box 3 is peach.

The pins are at the top of the stack.

The pink box is immediately below the belts.

The box on the bottom right holds the necklaces; the other bottom box is white and does not hold the rings.

The orchid-colored box is immediately above the hats on the left side of the stack.

The white box of gloves is on the left side of the shelf; the blue box is on the right and one level higher than the gloves.

The aqua box, which has a number less than 6, is to the immediate left of the belts.

The blue box is directly below the scarves.

The box that contains the belts has a number that is 4 more than that of the yellow box.

6. Make up an applied problem that can be solved by using a system of equations. (If you're working in groups, let other students in your group solve *your* problem, and *you* solve *their* problems.)

Sets

APPENDIX A

3. Given that $A = \{$Bob, John$\}$, $B = \{$Charles, Tom, Bob$\}$, $C = \{$Tom, John, Dick$\}$, and $D = \{$Ray, Bob$\}$, find any two sets that are disjoint.

4. Given the Venn diagram below, write each of the following sets in roster notation. (In the figure, it is understood that $a \in U$, $a \notin P$, $a \notin Q$, $c \in P$, $j \in Q$, and so forth.)

  a. P

  b. Q

  c. $P \cup Q$

  d. $P \cap Q$

  e. U

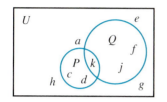

5. Given $X = \{2, 5, 6, 11\}$ and $Y = \{7, 5, 11, 13\}$:

  a. Find $X \cap Y$.

  b. Find $Y \cap X$.

  c. Does $X \cap Y = Y \cap X$?

6. Given $K = \{a, 4, 7, b\}$, $L = \{m, 4, 6, b\}$, and $M = \{n, 4, 7, t\}$:

  a. Write $K \cap L$ in roster notation.

  b. Find $n(K \cap L)$.

  c. Write $L \cup M$ in roster notation.

  d. Find $n(L \cup M)$.

7. Shade $A \cup B$ in the Venn diagram given here.

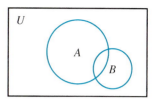

8. Shade $P \cap Q$.

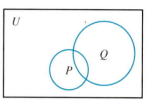

9. Write the name of the set represented by the shaded area.

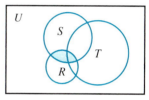

10. Write the name of the set represented by the shaded area.

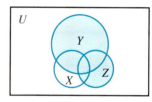

Brief Review of Arithmetic

APPENDIX
B

Order of Operations

1. If there are any grouping symbols in the expression, that part of the expression within a set of grouping symbols is evaluated first.

2. The evaluation then proceeds in this order:
First: Powers and roots are done.
Next: Multiplication and division are done in order from left to right.
Last: Addition and subtraction are done in order from left to right.

Following are two examples showing the correct order of operations:

$$12 \div 3 \times 4 \qquad\qquad 4 \times 3^2 \div 6 - 2$$
$$4 \quad \times 4 = 16 \qquad\qquad 4 \times 9 \div 6 - 2$$
$$36 \quad \div 6 - 2$$
$$6 \quad - 2 = 4$$

Fractions

An expression such as $\frac{2}{3}$ is called a **fraction**, and $\frac{2}{3}$ can be interpreted as the division problem $2 \div 3$. We call 2 the *numerator* and 3 the *denominator*; the denominator cannot be zero.

The fundamental property of fractions permits us to divide both the numerator and the denominator of a fraction by the same nonzero number (in symbols: $\frac{a}{b} = \frac{a \div d}{b \div d}$, if $d \neq 0$), and it permits us to multiply both the numerator and the denominator of a fraction by the same nonzero number (in symbols: $\frac{a}{b} = \frac{a \cdot c}{b \cdot c}$, if $c \neq 0$).

Reducing a Fraction to Lower Terms   To reduce a fraction to lower terms, we must divide both numerator and denominator by a number that is a divisor of both. The fundamental property of fractions permits us to do this. For example,

$$\frac{10}{15} = \frac{10 \div 5}{15 \div 5} = \frac{2}{3}$$

When we use the fundamental property of fractions, the new fraction is **equivalent** to the original one; therefore, $\frac{10}{15}$ and $\frac{2}{3}$ are equivalent fractions.

We sometimes show the work of reducing a fraction as follows:

$$\frac{\overset{\overset{1}{3}}{\cancel{15}}}{\underset{\underset{5}{15}}{\cancel{75}}} = \frac{1}{5} \qquad \text{\color{red}{We divided numerator and denominator first by 5 and then by 3}}$$

A fraction is in *lowest terms* when the only natural number that is a divisor of both the numerator and the denominator is 1; therefore, $\frac{1}{5}$ is in lowest terms.

Raising a Fraction to Higher Terms   When we add or subtract fractions, it is often necessary to raise a fraction to higher terms. The fundamental property of fractions permits us to do this. For example,

$$\frac{2}{3} = \frac{2 \cdot 5}{3 \cdot 5} = \frac{10}{15}$$

Improper Fractions and Mixed Numbers   The numerator of an **improper fraction** is greater than or equal to the denominator; for example, $\frac{8}{5}$ and $\frac{7}{7}$ are improper fractions. A **mixed number** consists of a whole number followed by a fraction; there is an understood plus sign between the two parts. Thus, the mixed number $5\frac{1}{2}$ *means* $5 + \frac{1}{2}$.

Writing an Improper Fraction as a Whole or Mixed Number If the denominator of an improper fraction divides exactly into the numerator, that improper fraction can be written as a whole number. For example, $\frac{18}{6} = 18 \div 6 = 3$. If the denominator of an improper fraction does *not* divide exactly into the numerator, that improper fraction can be written as a mixed number, as follows:

1. Divide the numerator by the denominator.

2. The *quotient* is the whole number part of the mixed number.

3. The fractional part of the mixed number is written as $\dfrac{\text{remainder}}{\text{denominator}}$.

For example, $\frac{17}{3} = 5\frac{2}{3}$ because

The same denominator

$$\frac{17}{3} = 3{\overline{)17}}^{\,5} \quad \begin{array}{l} \text{2 is the numerator of the fractional part} \\ \text{of the answer, and 3 is the denominator} \end{array}$$
$$\phantom{\frac{17}{3} = } \underline{15}$$
$$\phantom{\frac{17}{3} = 3} 2$$

Writing a Mixed Number as an Improper Fraction We can write a mixed number as an improper fraction as follows:

1. Multiply the whole number by the denominator of the fraction.

2. Add the numerator of the fraction to the product found in step 1.

3. Write the sum found in step 2 as the numerator of an improper fraction; the denominator of that improper fraction is the same as the denominator of the fractional part of the mixed number.

$$\frac{\text{whole number} \times \text{denominator} + \text{numerator}}{\text{denominator}}$$

$$2\frac{3}{5} = \frac{2 \times 5 + 3}{5} = \frac{10 + 3}{5} = \frac{13}{5}$$

Adding and Subtracting Fractions To add and subtract fractions, we use the rules $\frac{a}{c} + \frac{b}{c} = \frac{a+b}{c}$ and $\frac{a}{c} - \frac{b}{c} = \frac{a-b}{c}$. Therefore, we can add and subtract fractions only when their denominators are the same.

$$\frac{2}{7} + \frac{3}{7} = \frac{2+3}{7} = \frac{5}{7} \qquad \text{and} \qquad \frac{3}{8} - \frac{1}{8} = \frac{3-1}{8} = \frac{2}{8} = \frac{1}{4}$$

When the denominators are not the same, we find the *least common denominator* (*LCD*) (the smallest number that all the denominators divide into), and then we raise the fractions to higher terms, making all the denominators equal the LCD.
We can find the LCD as follows:

Finding the least common denominator

1. Factor each denominator, expressing any repeated factors in exponential form.

2. Write down each different base that appears in any denominator.

3. Raise each base to the *highest* power to which it occurs in *any* of the denominators.

4. The LCD is the product of all the factors found in step 3.

For example, to find the LCD for $\frac{1}{96} + \frac{5}{72} + \frac{7}{48}$, we can proceed as follows:

Step 1. $96 = 2^5 \cdot 3$; $72 = 2^3 \cdot 3^2$; $48 = 2^4 \cdot 3$ The denominators in factored form
Step 2. 2, 3   Writing all the different bases
Step 3. 2^5, 3^2 Raising each base to the *highest* power to which it occurs in any of the factorizations
Step 4. LCD $= 2^5 \cdot 3^2 = 32 \cdot 9 = 288$ The *product* of all the expressions from step 3

(This means that 288 is the *smallest* number that 96, 72, and 48 *all* divide into exactly.)

To add $\frac{1}{96} + \frac{5}{72} + \frac{7}{48}$, we first determine that 288 is the LCD; then we proceed as follows:

$$\frac{1}{96} = \frac{1 \times 3}{96 \times 3} = \frac{3}{288} \qquad 288 \div 96 = 3$$

$$\frac{5}{72} = \frac{5 \times 4}{72 \times 4} = \frac{20}{288} \qquad 288 \div 72 = 4$$

$$+\frac{7}{48} = +\frac{7 \times 6}{48 \times 6} = +\frac{42}{288} \qquad 288 \div 48 = 6$$

$$\frac{65}{288}$$

Multiplying Fractions   The numerator of the product is the product of the numerators, and the denominator of the product is the product of the denominators. In symbols: $\frac{a}{b} \times \frac{c}{d} = \frac{a \times c}{b \times d}$. It may then be possible to reduce the answer to lower terms.

$$\frac{1}{5} \times \frac{3}{7} = \frac{1 \times 3}{5 \times 7} = \frac{3}{35} \qquad \Bigg| \qquad \frac{6}{7} \times \frac{14}{15} = \frac{\overset{2}{6} \times \overset{2}{14}}{\underset{1}{7} \times \underset{5}{15}} = \frac{4}{5}$$

The Reciprocal of a Number   The **reciprocal** of a number is 1 divided by that number. For example, the reciprocal of 5 is $\frac{1}{5}$. To find the reciprocal of a fraction, we can simply invert the fraction; thus, the reciprocal of $\frac{2}{3}$ is $\frac{3}{2}$.

Dividing Fractions   To divide one fraction by another, we multiply the dividend (the number we're dividing *into*) by the *reciprocal* of the divisor (the number we're dividing *by*). In symbols: $\frac{a}{b} \div \frac{c}{d} = \frac{a}{b} \times \frac{d}{c} = \frac{a \times d}{b \times c}$.

$$\frac{3}{4} \div \frac{7}{5} = \frac{3}{4} \times \frac{5}{7} = \frac{3 \times 5}{4 \times 7} = \frac{15}{28} \qquad \Bigg| \qquad \frac{3}{5} \div \frac{6}{11} = \frac{3}{5} \times \frac{11}{6} = \frac{\overset{1}{3} \times 11}{5 \times \underset{2}{6}} = \frac{11}{10}, \text{ or } 1\frac{1}{10}$$

Adding and Subtracting Mixed Numbers   To add mixed numbers, we add the whole number parts and add the fractional parts. (If the sum of the fractional parts is itself an improper fraction, that fraction must be changed to a mixed number and its whole number part must then be added to the other whole number.)

The LCD of 8 and 12 is 24

$$\begin{array}{c} 1\dfrac{3}{5} \\[2mm] +\,3\dfrac{1}{5} \\ \hline 4\dfrac{4}{5} \end{array} \qquad \Bigg| \qquad \begin{array}{c} 4\dfrac{7}{8} = 4\dfrac{7 \times 3}{8 \times 3} = 4\dfrac{21}{24} \\[2mm] +\,5\dfrac{7}{12} = +\,5\dfrac{7 \times 2}{12 \times 2} = +\,5\dfrac{14}{24} \\ \hline 9\dfrac{35}{24} \end{array}$$

$$9\frac{35}{24} = 9 + \frac{35}{24}$$

$$9\frac{35}{24} = 9 + 1\frac{11}{24} = 10\frac{11}{24}$$

To subtract mixed numbers, we "borrow" if the fractional part of the subtrahend (the number we're subtracting) is larger than the fractional part of the minuend (the number we're subtracting *from*); then we subtract the whole number parts and subtract

the fractional parts. In the example below, we must borrow because $\frac{3}{6}$ is smaller than $\frac{4}{6}$. To borrow, we write 13 as $12 + 1$, and then we rewrite the 1 as $\frac{6}{6}$ so that we can add it to $\frac{3}{6}$.

The LCD of 2 and 3 is 6

$$13\frac{1}{2} = 13\frac{1 \times 3}{2 \times 3} = 13\frac{3}{6} = 12 + 1 + \frac{3}{6} = 12 + \frac{6}{6} + \frac{3}{6} = 12\frac{9}{6}$$

$$-9\frac{2}{3} = -9\frac{2 \times 2}{3 \times 2} = -9\frac{4}{6} \qquad\qquad\qquad = -9\frac{4}{6}$$

$$\overline{\qquad\qquad\qquad\qquad\qquad\qquad\qquad\qquad\qquad\qquad 3\frac{5}{6}}$$

Multiplying and Dividing Mixed Numbers Mixed numbers must be changed to improper fractions before they can be multiplied or divided.

$$\left(2\frac{3}{5}\right) \div \left(1\frac{3}{8}\right) = \frac{13}{5} \div \frac{11}{8} = \frac{13}{5} \times \frac{8}{11} = \frac{104}{55}, \text{ or } 1\frac{49}{55}$$

 Note Mixed numbers *can* be changed to improper fractions before they are added or subtracted, also. If subtraction problems are done this way, it is never necessary to "borrow."

Simplifying a Complex Fraction Divide the numerator of the complex fraction by the denominator.

$$\frac{\dfrac{5}{12}}{\dfrac{15}{16}} = \frac{5}{12} \div \frac{15}{16} = \frac{\overset{1}{\cancel{5}}}{\underset{3}{\cancel{12}}} \times \frac{\overset{4}{\cancel{16}}}{\underset{3}{\cancel{15}}} = \frac{4}{9}$$

$$\frac{3 - \dfrac{1}{5}}{\dfrac{1}{3} + \dfrac{1}{5}} = \frac{\dfrac{15}{5} - \dfrac{1}{5}}{\dfrac{5}{15} + \dfrac{3}{15}} = \frac{\dfrac{14}{5}}{\dfrac{8}{15}} = \frac{14}{5} \div \frac{8}{15} = \frac{\overset{7}{\cancel{14}}}{\underset{1}{\cancel{5}}} \times \frac{\overset{3}{\cancel{15}}}{\underset{4}{\cancel{8}}} = \frac{21}{4} = 5\frac{1}{4}$$

Converting a Fraction to a Decimal Divide the numerator by the denominator.

$$\frac{5}{12} = \begin{array}{r} 0.416 \approx 0.42 \\ 12\overline{)5.000} \\ \underline{4\ 8} \\ 20 \\ \underline{12} \\ 80 \\ \underline{72} \\ 8 \end{array}$$

Rounded to two decimal places
When the answer is to be rounded off to two decimal places, we need three decimal places in the dividend

Converting a Decimal to a Fraction If the last digit of the decimal number is in the tenth's place, the denominator of the fraction will be 10; if the last digit is in the hundredth's place, the denominator of the fraction will be 100; and so on. Therefore,

$$0.7 = \tfrac{7}{10}, \quad 0.48 = \tfrac{48}{100} = \tfrac{12}{25}, \quad \text{and} \quad 0.003 = \tfrac{3}{1,000}$$

Percents

The Meaning of a Percent *Percent* means "per hundred." For example, 5% of a quantity is $\frac{5}{100}$, or $\frac{1}{20}$, of that quantity.

Converting a Decimal to a Percent   To convert a decimal to a percent, we move the decimal point two places to the *right* and attach the percent symbol to the right of the number.

$$0.13 = 13\% \qquad 0.003 = 0.3\% \qquad 2.5 = 2.50 = 250\%$$

Converting a Percent to a Decimal   To convert a percent to a decimal, we move the decimal point two places to the *left* and drop the percent symbol.

$$83\% = 0.83 \qquad 3\% = 03\% = 0.03 \qquad 115\% = 1.15$$

Converting a Fraction to a Percent   To convert a fraction to a percent, we first change the fraction to its decimal form (rounding off, if necessary); then we move the decimal point two places to the right and attach the percent symbol to the right of the number. In the second example below, we round off the decimal form to four decimal places, which means that the percent form is rounded off to two decimal places.

$$\tfrac{1}{4} = 0.25 = 25\% \qquad \tfrac{5}{12} \approx 0.4167 = 41.67\%$$

Converting a Percent to a Fraction   To convert a percent to a fraction, we write the number with a denominator of 100 and drop the percent symbol. We then reduce the resulting fraction, if possible.

$$32\% = \tfrac{32}{100} = \tfrac{8}{25}, \quad 105\% = \tfrac{105}{100} = \tfrac{21}{20}, \quad \text{and} \quad 0.07\% = \tfrac{0.07}{100} = \tfrac{7}{10,000}$$

Exercises   *B.1*

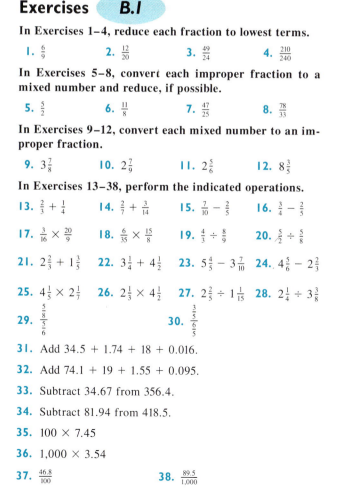

In Exercises 1–4, reduce each fraction to lowest terms.

1. $\frac{6}{9}$ **2.** $\frac{12}{20}$ **3.** $\frac{49}{24}$ **4.** $\frac{210}{240}$

In Exercises 5–8, convert each improper fraction to a mixed number and reduce, if possible.

5. $\frac{5}{2}$ **6.** $\frac{11}{8}$ **7.** $\frac{47}{25}$ **8.** $\frac{78}{33}$

In Exercises 9–12, convert each mixed number to an improper fraction.

9. $3\frac{7}{8}$ **10.** $2\frac{7}{9}$ **11.** $2\frac{5}{6}$ **12.** $8\frac{3}{5}$

In Exercises 13–38, perform the indicated operations.

13. $\frac{2}{3} + \frac{1}{4}$ **14.** $\frac{2}{7} + \frac{3}{14}$ **15.** $\frac{7}{10} - \frac{2}{5}$ **16.** $\frac{3}{4} - \frac{2}{5}$

17. $\frac{3}{16} \times \frac{20}{9}$ **18.** $\frac{6}{35} \times \frac{15}{8}$ **19.** $\frac{4}{3} \div \frac{8}{9}$ **20.** $\frac{5}{2} \div \frac{5}{8}$

21. $2\frac{2}{3} + 1\frac{3}{5}$ **22.** $3\frac{1}{4} + 4\frac{1}{2}$ **23.** $5\frac{4}{5} - 3\frac{7}{10}$ **24.** $4\frac{5}{6} - 2\frac{2}{3}$

25. $4\frac{1}{5} \times 2\frac{1}{7}$ **26.** $2\frac{1}{3} \times 4\frac{1}{2}$ **27.** $2\frac{2}{5} \div 1\frac{1}{15}$ **28.** $2\frac{1}{4} \div 3\frac{3}{8}$

29. $\dfrac{\frac{5}{8}}{\frac{5}{6}}$ **30.** $\dfrac{\frac{3}{5}}{\frac{6}{5}}$

31. Add $34.5 + 1.74 + 18 + 0.016$.

32. Add $74.1 + 19 + 1.55 + 0.095$.

33. Subtract 34.67 from 356.4.

34. Subtract 81.94 from 418.5.

35. 100×7.45

36. $1,000 \times 3.54$

37. $\frac{46.8}{100}$ **38.** $\frac{89.5}{1,000}$

In Exercises 39–42, perform the indicated operations; then round off the answer to the indicated place.

39. 70.9×94.78 (one decimal place)

40. 40.8×68.59 (one decimal place)

41. $6.007 \div 7.25$ (two decimal places)

42. $8.009 \div 4.67$ (two decimal places)

43. Convert $5\frac{3}{4}$ to a decimal.

44. Convert $4\frac{3}{5}$ to a decimal.

45. Convert 0.65 to a fraction.

46. Convert 0.25 to a fraction.

47. Convert 5.9 to a mixed number.

48. Convert 4.3 to a mixed number.

In Exercises 49–52, convert each number to a percent.

49. 4.7   **50.** 0.001   **51.** 0.258   **52.** 12

In Exercises 53–56, convert each percent to a decimal.

53. 3.5%   **54.** 0.02%   **55.** 157%   **56.** 17.8%

In Exercises 57–60, convert each fraction to a percent. In Exercises 59 and 60, round off to two decimal places.

57. $\frac{1}{8}$ **58.** $\frac{3}{5}$ **59.** $\frac{7}{12}$ **60.** $\frac{1}{7}$

In Exercises 61–64, convert each percent to a fraction reduced to lowest terms.

61. 35%   **62.** 80%   **63.** 12%   **64.** 18%

Functions

APPENDIX

C

C.I   Definitions

Functions   We live in a world of functions. If we are paid at a fixed hourly rate, then our weekly salary is a function of the number of hours we work each week. At any given time, the cost of sending a letter first class is fixed; therefore, the cost of mailing several letters at once (assuming none of the letters is overweight) is a function of the number of letters being mailed.

A function can be described in any of several ways: by a rule (that is, by a written statement), by a table of values, by a graph, by a set of ordered pairs of numbers, or by an equation. The formal definition of a function is as follows:

> A function is a set of ordered pairs in which no two of the ordered pairs have the *same first coordinate* but *different second coordinates.*

EXAMPLE 1   Examples of sets of ordered pairs:

a. The set $\{(3,7),(-4,2),(6,7),(-1,-1)\}$ is a set of ordered pairs, and it *is* a function, because no two of the ordered pairs have the same *first* coordinate.

b. The set $\{(6,2),(-3,4),(6,7)\}$ is a set of ordered pairs; however, it is *not* a function, because the ordered pairs $(6,2)$ and $(6,7)$ have the same first coordinate but different second coordinates.

A function can be thought of as being a rule that assigns one and only one value to y for each value of x. Because the value of y *depends* on the value of x, we often call y the **dependent variable** and x the **independent variable**. In the equation $y = 3x + 7$, for example, we usually assign various values to x (the independent variable) and each corresponding value of y *depends* on what value we chose for x; that is, if $x = 1$, then $y = 3(1) + 7 = 10$.

The Domain of a Function   For the functions discussed in this book, the **domain of a function** is the set of all the values that x can have so that y is a real number; it is the set of all the *first coordinates*, or x-values, of the ordered pairs. For example, the domain of the function $\{(-4,2),(-1,-1),(3,7),(6,7)\}$ is $\{-4,-1,3,6\}$. (Recall that we enclose the elements of a set within braces; because the domain of a function is a *set*, we enclose its elements within braces.)

The Range of a Function   The **range of a function** is the set of all the values that y can have; it is the set of all the *second coordinates*, or y-values, of the ordered pairs. For example, the range of the function $\{(-1,-1),(-4,2),(3,7),(6,7)\}$ is $\{-1,2,7\}$. (Because the range is a *set*, we enclose its elements within braces.)

EXAMPLE 2   Examples of functions and their domains:

a. $\{(3,7),(-2,7),(0,-5)\}$

This is a set of three ordered pairs; they all have different first coordinates. Therefore, it is a function. The domain of $\{(3,7),(-2,7),(0,-5)\}$ is the set $\{3,-2,0\}$.

b. $y = x + 1$

For any value of x we choose, there will be one and only one corresponding value for y. Therefore, this is a function. Because we can choose *any* value for x, the domain of this function is the set of all real numbers.

c. $y = \dfrac{3}{x - 1}$

Recall that for the rational expression $\dfrac{3}{x - 1}$, 1 is an excluded value. Therefore, the domain is the set of all real numbers except 1. For any value of x that we choose from the domain, there will be one and only one corresponding value for y. Therefore, $y = \dfrac{3}{x - 1}$ is a function.

d. $y = \sqrt{x + 4}$

We learned earlier that square roots of negative numbers are not real numbers. Therefore, if y is to be a real number, the radicand $x + 4$ must be greater than or equal to zero. We must, then, solve the inequality $x + 4 \geq 0$:

$$x + 4 \geq 0$$

$$x + 4 - 4 \geq 0 - 4 \quad \text{Adding } -4 \text{ to both sides}$$

$$x \geq -4$$

Therefore, the domain is the set of all real numbers greater than or equal to -4. If we let x be any number greater than or equal to -4, we get one and only one corresponding value for y. Therefore, $y = \sqrt{x + 4}$ is a function.

The procedure for finding the domain of a function when an *equation* for the function has been given is as follows:

> The domain of the function will be the set of all real numbers unless:
>
> **1.** The domain is restricted by some statement accompanying the equation.
>
> **2.** There are variables in a denominator or variables with negative exponents.
>
> **3.** Variables occur under a radical sign when the index of the radical is an even number.

The Graph of a Function   The graph of a function is simply the graph of all the ordered pairs of the function.

EXAMPLE 3   Find the domain of each of the following functions, and graph each function.

a. $\{(2, 4), (-1, 3), (0, 3), (3, 1)\}$

SOLUTION   The domain of $\{(2, 4), (-1, 3), (0, 3), (3, 1)\}$ is $\{2, -1, 0, 3\}$.

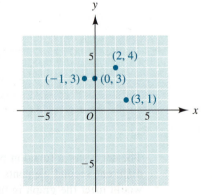

b. $y = x + 1, x \le 2$

> **SOLUTION** In this example, there *are* restrictions on the domain—namely, $x \le 2$. Therefore, the domain is the set of all real numbers less than or equal to 2. For the graph, we make a table of values, taking care to choose x-values only from the domain.

x	y
-2	-1
0	1
2	3

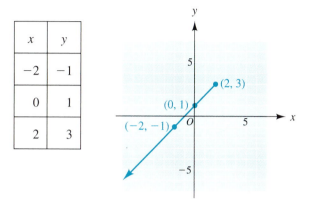

> We then plot those points and connect them with *part* of a straight line; the straight line does *not* extend to the right of the point where $x = 2$.

 Note It is possible to graph *any* set of ordered pairs, whether or not it is a function.

The following statement is true because of the way in which a function is defined:

> No vertical line can meet the graph of a function in more than one point.

 **EXAMPLE 4** Examples of graphs of sets of ordered pairs (these sets of ordered pairs are *not* functions):

a. $\{(3, -2), (4, 1), (3, 3)\}$

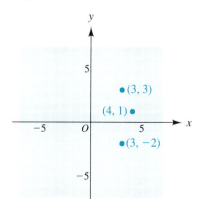

The set is *not* a function because the ordered pairs $(3, -2)$ and $(3, 3)$ have two different second elements but the same first element (3). The vertical line $x = 3$ would meet the graph in more than one point.

b. $x = 1$

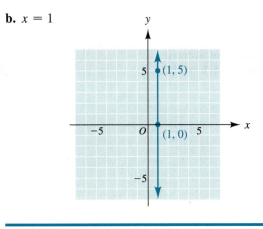

$x = 1$ is *not* a function, because there are ordered pairs satisfying the equation $x = 1$ that have the same first element but different second elements. Two such pairs are $(1, 5)$ and $(1, 0)$.

EXAMPLE 5

Determine from the graph whether each of the following is a function.

SOLUTION

a.

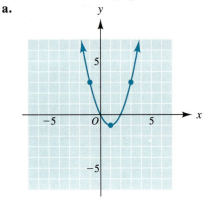

This *is* the graph of a function, because no vertical line can meet the graph in more than one point.

b.

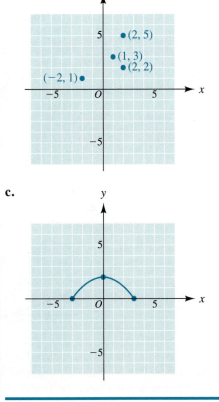

This is *not* the graph of a function, because a vertical line through $(2, 5)$ would also pass through $(2, 2)$.

c.

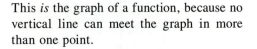

This *is* the graph of a function, because no vertical line can meet the graph in more than one point.

Exercises C.1

In Exercises 1–6, determine whether each set of ordered pairs is a function, and find the domain of each of the functions.

1. $\{(4, -1), (2, 5), (3, 0), (0, 4)\}$

2. $\{(6, 3), (-2, 5), (4, -1), (0, 0)\}$

3. $\{(2, -5), (3, 4), (2, 0), (7, -4), (-1, 3)\}$

4. $\{(3, 7), (-1, 4), (5, -2), (3, 0), (-8, 2)\}$

5. $\{(-8, -2), (3, -4), (6, -2), (9, -4)\}$

6. $\{(-3, -7), (-1, -7), (0, -2), (3, 0)\}$

In Exercises 7–14, find the domain of each function.

7. $y = \dfrac{7}{x - 12}$

8. $y = \dfrac{5}{x - 2}$

9. $y = 3x + 4$

10. $y = 2x + 5$

11. $y = \sqrt{x - 5}$

12. $y = \sqrt{x - 10}$

13. $y = x + 3, \ x \geq 0$

14. $y = x + 7, \ x \geq -3$

15. Are the following graphs of functions?

a.

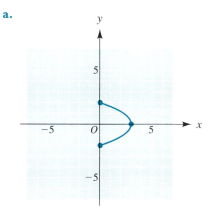

b.

c.

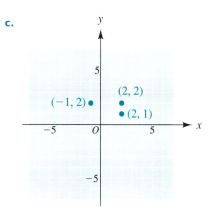

16. Are the following graphs of functions?

a.

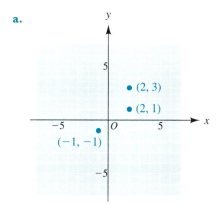

b.

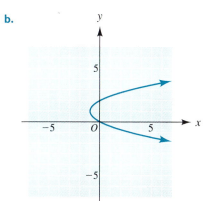

c.

C.2

Functional Notation

Functional Notation When we have a rule that assigns one and only one value to y for each value of x, we say that y is a function of x. This can be written $y = f(x)$, which is read "y equals f of x."

> ✕ **A Word of Caution** $f(x)$ does *not* mean f times x. It is simply an alternative way of writing y when y is a function of x.

The equations $y = x + 1$ and $f(x) = x + 1$ mean exactly the same thing, and we often combine the equations in the following manner:

$$y = f(x) = x + 1$$

Evaluating a Function To *evaluate* a function means to determine what value $f(x)$ or y has for a particular x-value. The notation $f(a)$ (read "f of a") means that the function $f(x)$ is to be evaluated at $x = a$.

EXAMPLE 1 If $f(x) = 3x + 2$, find $f(4)$.

SOLUTION Since $f(x) = 3x + 2$,

$$f(4) = 3(4) + 2 \quad \text{Substituting 4 for } x$$
$$f(4) = 12 + 2$$
$$f(4) = 14$$

The statement $f(4) = 14$ is read "f of 4 equals 14," and it means that $y = 14$ when $x = 4$.

EXAMPLE 2 If $f(x) = \dfrac{3x^2 + 1}{x - 2}$, find $f(1)$, $f(-3)$, and $f(0)$.

SOLUTION $f(x) = \dfrac{3x^2 + 1}{x - 2}$

$$f(1) = \frac{3(1)^2 + 1}{1 - 2} = \frac{3 + 1}{-1} = \frac{4}{-1} = -4$$

$$f(-3) = \frac{3(-3)^2 + 1}{-3 - 2} = \frac{3(9) + 1}{-5} = \frac{27 + 1}{-5} = -\frac{28}{5}$$

$$f(0) = \frac{3(0)^2 + 1}{0 - 2} = \frac{0 + 1}{-2} = \frac{1}{-2} = -\frac{1}{2}$$

The statement $f(1) = -4$ means that $y = -4$ when $x = 1$; the statement $f(-3) = -\frac{28}{5}$ means that $y = -\frac{28}{5}$ when $x = -3$; and the statement $f(0) = -\frac{1}{2}$ means that $y = -\frac{1}{2}$ when $x = 0$.

Note that letters other than f can be used to name functions, variables other than x can be used for the independent variable, and variables other than y can be used for the dependent variable.

EXAMPLE 3   Examples of the use of letters other than f, x, and y to represent functions and variables:

a. $A = F(r) = \pi r^2$ A is a function of r.
b. $s = g(t) = 4t - 1$ s is a function of t.
c. $V = h(s) = s^3$ V is a function of s.

Exercises   C.2

Evaluate the functions.

1. If $f(x) = (x + 2)^2$, find the following:

   a. $f(0)$ **b.** $f(2)$ **c.** $f(-3)$ **d.** $f(1)$

2. If $f(x) = (2x + 3)^2$, find the following:

   a. $f(0)$ **b.** $f(1)$ **c.** $f(-2)$ **d.** $f(2)$

3. If $g(x) = x^2 + 4x + 4$, find the following:

   a. $g(0)$ **b.** $g(2)$ **c.** $g(-3)$ **d.** $g(1)$

4. If $h(x) = 4x^2 + 12x + 9$, find the following:

   a. $h(0)$ **b.** $h(1)$ **c.** $h(-2)$ **d.** $h(2)$

5. If $F(x) = x^2 + 4$, find the following:

   a. $F(0)$ **b.** $F(2)$ **c.** $F(-3)$ **d.** $F(1)$

6. If $G(x) = 4x^2 + 9$, find the following:

   a. $G(0)$ **b.** $G(1)$ **c.** $G(-2)$ **d.** $G(2)$

7. If $f(x) = 3x^2 + x - 1$, find the following:

   a. $f(0)$ **b.** $f(-2)$ **c.** $f(5)$ **d.** $f(1)$

8. If $f(x) = 2x^2 + 3x - 5$, find the following:

   a. $f(0)$ **b.** $f(-3)$ **c.** $f(4)$ **d.** $f(1)$

9. If $f(x) = x^3 - 1$, find the following:

   a. $f(0)$ **b.** $f(-1)$ **c.** $f(2)$ **d.** $f(1)$

10. If $f(x) = x^3 + 1$, find the following:

   a. $f(0)$ **b.** $f(-2)$ **c.** $f(1)$ **d.** $f(-1)$

11. If $g(t) = \dfrac{4t^2 + 5t - 1}{t + 2}$, find the following:

   a. $g(0)$ **b.** $g(-1)$ **c.** $g(2)$ **d.** $g(4)$

12. If $h(t) = \dfrac{-2t^2 - 3t + 1}{t + 3}$, find the following:

   a. $h(0)$ **b.** $h(-2)$ **c.** $h(1)$ **d.** $h(-1)$

A N S W E R S

Answers to Set I Exercises (including Solutions to Odd-Numbered Exercises), Diagnostic Tests, and Cumulative Review Exercises

Exercises 1.1   (page 6)

I. 5   **2.** 5   **3.** 1   **4.** 0   **5.** 9   **6.** 99   **7.** 10
8. 100   **9.** >   **10.** <   **II.** >   **12.** <   **13.** Yes
14. Yes   **15.** Yes   **16.** Yes   **17.** No   **18.** No
19. No   **20.** Yes   **21.** Yes   **22.** No   **23.** Three
24. Four

Exercises 1.2   (page 10)

I. Negative seventy-five (or minus seventy-five)
2. Negative forty-nine (or minus forty-nine)
3. -54 **4.** -14
5. -2, because -2 is to the right of -4 on the number line
6. 0
7. -5, because -5 is to the right of -10 on the number line
8. -15 **9.** -1 **10.** Yes   **II.** Yes   **12.** Yes
13. Yes   **14.** Yes   **15.** Yes   **16.** There is none
17. 1, 2, 3, 4   **18.** 7, 8, 9   **19.** 0, 2, 4   **20.** 1, 3, 5, 7
21. >, because 0 is to the right of -3 on the number line
22. >
23. <, because -5 is to the left of 2 on the number line
24. <
25. >, because -2 is to the right of -10 on the number line
26. <   **27.** -62 **28.** $-45°F$

Exercises 1.3   (page 18)

I. $|-73| = 73$; the absolute value of a number is never negative.
2. 55   **3.** $|-48| = 48$; therefore, $-|-48| = -(48) = -48$
4. -26
5. Since signs are same, add absolute values: $4 + 5 = 9$. Sum has sign of both numbers: $+$; so $4 + 5 = 9$.
6. 8

7. Since signs are same, add absolute values: $3 + 4 = 7$. Sum has sign of both numbers: $-$; so $-3 + (-4) = -7$.
8. -8
9. Since signs are different, subtract absolute values: $6 - 5 = 1$. Sum has sign of number with larger absolute value: $-$. Therefore, $-6 + 5 = -1$.
10. -5
II. Since signs are different, subtract absolute values: $7 - 3 = 4$. Sum has sign of number with larger absolute value: $+$. Therefore, $7 + (-3) = 4$.
12. 5
13. Since signs are same, add absolute values: $8 + 4 = 12$. Sum has sign of both numbers: $-$; so $-8 + (-4) = -12$.
14. -11
15. Since signs are different, subtract absolute values: $9 - 3 = 6$. Sum has sign of number with larger absolute value: $-$. Therefore, $3 + (-9) = -6$.
16. -4 **17.** $-5 + 0 = -5$, because 0 is the additive identity.
18. -17
19. Since signs are different, subtract absolute values: $9 - 7 = 2$. Sum has sign of number with larger absolute value: $+$. Therefore, $-7 + 9 = 2$.
20. 3
21. Since signs are same, add absolute values: $2 + 11 = 13$. Sum has sign of both numbers: $-$; so $-2 + (-11) = -13$.
22. -9
23. Since signs are different, subtract absolute values: $15 - 5 = 10$. Sum has sign of number with larger absolute value: $-$. Therefore, $5 + (-15) = -10$.
24. -8
25. Since signs are different, subtract absolute values: $9 - 8 = 1$. Sum has sign of number with larger absolute value: $+$. Therefore, $-8 + 9 = 1$.
26. 6
27. Since signs are different, subtract absolute values: $4 - 4 = 0$. Zero is neither positive nor negative. Sum $-4 + 4 = 0$.
28. 0
29. Since signs are same, add absolute values: $27 + 13 = 40$. Sum has sign of both numbers: $-$; so $-27 + (-13) = -40$.
30. -54
31. Since signs are different, subtract absolute values: $121 - 80 = 41$. Sum has sign of number with larger absolute value: $+$. Therefore, $-80 + 121 = 41$.
32. 65

33. Since signs are different, subtract absolute values: $105 - 73 = 32$. Sum has sign of number with larger absolute value: $+$. Therefore, $105 + (-73) = 32$.

34. 105

35. Since signs are same, add absolute values. Sum has sign of both numbers: $-$.

$\dfrac{3}{2} = \dfrac{15}{10}$
$\dfrac{17}{5} = \dfrac{34}{10}$ $\Big\rangle$ Add

$\dfrac{49}{10}$ Therefore, $-\dfrac{3}{2} + \left(-\dfrac{17}{5}\right) = -\dfrac{49}{10}$, or $-4\dfrac{9}{10}$.

36. $-\dfrac{31}{4}$, or $-7\dfrac{3}{4}$

37. Since signs are different, subtract absolute values. Sum has sign of number with larger absolute value: $-$.

$\dfrac{29}{6} = \dfrac{29}{6}$
$-\dfrac{4}{3} = -\dfrac{8}{6}$ $\Big\rangle$ Subtract

$\dfrac{21}{6}$, or $\dfrac{7}{2}$, or $3\dfrac{1}{2}$ Therefore, $-\dfrac{29}{6} + \dfrac{4}{3} = -\dfrac{7}{2}$, or $-3\dfrac{1}{2}$.

38. $-\dfrac{37}{8}$, or $-4\dfrac{5}{8}$

39. Since signs are different, subtract absolute values. Sum has sign of number with larger absolute value: $+$.

$5\dfrac{1}{4} = 5\dfrac{3}{12} = 4\dfrac{15}{12}$
$-2\dfrac{1}{3} = -2\dfrac{4}{12} = -2\dfrac{4}{12}$ $\Big\rangle$ Subtract

$2\dfrac{11}{12}$ Therefore, $5\dfrac{1}{4} + \left(-2\dfrac{1}{3}\right) = 2\dfrac{11}{12}$.

40. $4\dfrac{7}{10}$

41. Since signs are different, subtract absolute values. Sum has sign of number with larger absolute value: $-$.

$8 = 7\dfrac{8}{8}$
$-2\dfrac{5}{8} = -2\dfrac{5}{8}$ $\Big\rangle$ Subtract

$5\dfrac{3}{8}$ Therefore, $2\dfrac{5}{8} + (-8) = -5\dfrac{3}{8}$.

42. $-3\dfrac{2}{5}$

43. Since signs are different, subtract absolute values: $6.075 - 3.146 = 2.929$. Sum has sign of number with larger absolute value: $+$. Therefore, $6.075 + (-3.146) = 2.929$.

44. 88.373

45. Since signs are different, subtract absolute values: $5.200 - 2.345 = 2.855$. Sum has sign of number with larger absolute value: $-$. Therefore, $-5.2 + 2.345 = -2.855$.

46. -0.775

47. $|-6| = 6$. Therefore, $|-6| + (-2) = 6 + (-2)$. Since signs are different, subtract absolute values: $6 - 2 = 4$. Sum has sign of number with larger absolute value: $+$. Therefore, $|-6| + (-2) = 6 + (-2) = 4$.

48. 3

49. $|-17| = 17$. Therefore, $-8 + |-17| = -8 + 17$. Since signs are different, subtract absolute values: $17 - 8 = 9$. Sum has sign of number with larger absolute value: $+$. Therefore, $-8 + |-17| = -8 + 17 = 9$.

50. 21

51. $|23| = 23$. Therefore, $-9 + |23| = -9 + 23$. Since signs are different, subtract absolute values: $23 - 9 = 14$. Sum has sign of number with larger absolute value: $+$. Therefore, $-9 + |23| = -9 + 23 = 14$.

52. 31

53. We must add $53°$ to $-35°$. Since signs are different, subtract absolute values: $53 - 35 = 18$. Sum has sign of number with larger absolute value: $+$. Therefore, sum is $18°F$.

54. $17°F$ **55.** 6   **56.** $-\dfrac{2}{3}$

Exercises 1.4   (page 23)

1. $6 - (+14) = 6 + (-14) = -8$ **2.** -7

3. $-3 - (-2) = -3 + (+2) = -1$ **4.** -1

5. $-6 - (+2) = -6 + (-2) = -8$ **6.** -13

7. $9 - (-5) = 9 + (+5) = 14$ **8.** 10

9. $2 - (-7) = 2 + (+7) = 9$ **10.** 8

11. $-5 - (-9) = -5 + (+9) = 4$ **12.** 6

13. $-4 - (3) = -4 + (-3) = -7$ **14.** -15

15. $6 - (+11) = 6 + (-11) = -5$ **16.** -4

17. $-9 - (-4) = -9 + (+4) = -5$ **18.** -3

19. $4 - (-7) = 4 + (+7) = 11$ **20.** 17

21. $-15 - (+11) = -15 + (-11) = -26$ **22.** -40

23. $0 - 7 = -7$ **24.** -15

25. $0 - (-9) = 0 + (+9) = 9$ **26.** 25

27. $16 - 0 = 16$ **28.** 10

29. $156 - (-97) = 156 + (+97) = 253$ **30.** 373

31. $-26.3 - (-3.84) = -26.30 + (+3.84) = -22.46$

32. 29.46   **33.** $2.009 - (+7) = 2.009 + (-7.000) = -4.991$

34. -8.11 **35.** $-\dfrac{13}{2} - \left(-\dfrac{11}{3}\right) = -\dfrac{39}{6} + \left(+\dfrac{22}{6}\right) = -\dfrac{17}{6}$, or $-2\dfrac{5}{6}$

36. $-\dfrac{11}{2}$, or $-5\dfrac{1}{2}$

37. $6\dfrac{1}{3} - \left(+8\dfrac{1}{4}\right) = 6\dfrac{1}{3} + \left(-8\dfrac{1}{4}\right) = 6\dfrac{4}{12} + \left(-8\dfrac{3}{12}\right) = 6\dfrac{4}{12} + \left(-7\dfrac{15}{12}\right)$
$= -1\dfrac{11}{12}$

38. $2\dfrac{7}{10}$ **39.** $(+5) - (-2) = (+5) + (+2) = 7$ **40.** -5

41. $\left(-5\dfrac{1}{4}\right) - \left(+2\dfrac{1}{2}\right) = \left(-5\dfrac{1}{4}\right) + \left(-2\dfrac{1}{2}\right) = \left(-5\dfrac{1}{4}\right) + \left(-2\dfrac{2}{4}\right) = -7\dfrac{3}{4}$

42. $-10\dfrac{1}{2}$ **43.** $-8 - (+5) = -8 + (-5) = -13$ **44.** -9

45. $7 - 12 = 7 + (-12) = -5$ **46.** -10

47. $0 - 4 = 0 + (-4) = -4$ **48.** -8

49. $\$473.29 - \$238.43 = \$473.29 + (-\$238.43) = \$234.86$

50. $\$578.89$

51. $42 - (-7) = 42 + (+7) = 49$. Therefore, the rise in temperature was $49°F$.

52. $43.1°F$ **53.** $-141 - 68 = -141 + (-68) = -209$ ft

54. 9,800 ft

55. $29,029 \text{ ft} - (-35,840 \text{ ft}) = 29,029 \text{ ft} + (+35,840 \text{ ft})$
$= 64,869 \text{ ft}$

56. 4,010 ft

Exercises 1.5   (page 29)

1. $3(-2) = -(3 \times 2) = -6$ **2.** -24

3. $-5(2) = -(5 \times 2) = -10$ **4.** -35

5. $-8(-2) = +(8 \times 2) = 16$ **6.** 42

7. $8(-4) = -(8 \times 4) = -32$ **8.** -45

9. $-7(9) = -(7 \times 9) = -63$ **10.** -48

11. $-10(-10) = +(10 \times 10) = 100$ **12.** 81

13. $8(-7) = -(8 \times 7) = -56$ **14.** -72

15. $-26(10) = -(26 \times 10) = -260$ **16.** -132

17. $-20(-10) = +(20 \times 10) = 200$ **18.** 600

19. $75(-15) = -(75 \times 15) = -1,125$ **20.** $-1,118$

21. $-30(+5) = -(30 \times 5) = -150$ **22.** -300

23. $-7(-20) = +(7 \times 20) = 140$ **24.** 360

25. $\left(-5\frac{1}{2}\right)(0) = 0$ **26.** 0 **27.** $\left(\frac{27}{4}\right)\left(-\frac{33}{4}\right) = -\frac{891}{16}$, or $-55\frac{11}{16}$

28. $-\frac{1,826}{27}$, or $-67\frac{17}{27}$

29. $\left(-3\frac{3}{5}\right)\left(-8\frac{1}{5}\right) = \left(-\frac{18}{5}\right)\left(-\frac{41}{5}\right) = \frac{738}{25}$, or $29\frac{13}{25}$

30. $\frac{684}{35}$, or $19\frac{19}{35}$ **31.** $-3.5(-1.4) = +(3.5 \times 1.4) = 4.90$

32. 7.52 **33.** $2.74(-100) = -(2.74 \times 100) = -274$

34. -304 **35.** $\left(2\frac{1}{3}\right)\left(-3\frac{1}{2}\right) = \left(\frac{7}{3}\right)\left(-\frac{7}{2}\right) = -\frac{49}{6} = -8\frac{1}{6}$

36. $13\frac{13}{20}$ **37.** $0\left(-\frac{2}{3}\right) = 0$ **38.** 0 **39.** $-\frac{1}{12}$ **40.** $\frac{7}{5}$

41. 12 **42.** $-\frac{5}{7}$ **43.** $-\frac{1}{3}$ **44.** $-\frac{1}{2}$

45. No; their product is -1, not 1.

46. No; their product is -1, not 1.

Exercises 1.6 (page 33)

1. 2. Positive because numbers have same sign. **2.** 3

3. -4. Negative because numbers have different signs.

4. -2 **5.** -5 **6.** -2 **7.** 2 **8.** 5 **9.** -5

10. -6 **11.** -4 **12.** -5 **13.** 3 **14.** 3

15. -3 **16.** -4 **17.** 9 **18.** 7 **19.** -15

20. -2.5 **21.** -3 **22.** -7 **23.** -3 **24.** -3

25. Not defined **26.** Not defined **27.** Not defined

28. Not defined **29.** 0 **30.** 0 **31.** Not defined

32. Not defined **33.** $\frac{-15}{6} = \frac{-5}{2} = -2\frac{1}{2}$ **34.** $-2\frac{1}{4}$

35. $\frac{7.5}{-0.5} = \frac{75}{-5} = \frac{15}{-1} = -15$ **36.** -5

37. $\frac{-6.3}{-0.9} = \frac{-63}{-9} = 7$ **38.** 8 **39.** $\frac{-367}{100} = -3.67$

40. -4.86 **41.** $-\frac{3}{8} \div \frac{2}{5} = -\frac{3}{8} \cdot \frac{5}{2} = -\frac{15}{16}$ **42.** 4

43. -5.6 **44.** 6.1

Sections 1.1–1.6 Review Exercises (page 36)

1. 8, 9 **2.** 10 **3.** -9 **4.** -1

5. a. $<$, because -3 is to the left of 8 on the number line.

  b. $>$, because 5 is to the right of -2 on the number line.

6. a. $>$ **b.** $>$ **7.** 1 **8.** -4

9. Since signs are different, subtract absolute values: $3 - 2 = 1$. Sum has same sign as number with larger absolute value: $+$. Therefore, sum is 1.

10. -1 **11.** 3. Positive because numbers have same sign.

12. 2 **13.** $-5 - (-3) = -5 + (+3) = -2$ **14.** -5

15. $+5 - |-2| = +5 - (+2) = +5 + (-2) = 3$ **16.** 5

17. 12. Positive because numbers have same sign. **18.** 20

19. $-7 - 3 = -7 + (-3) = -10$ **20.** -4.37 **21.** -8

22. -2

23. -120. Negative because numbers have different signs.

24. -90

25. -8. Negative because numbers have different signs.

26. $-2\frac{1}{3}$ **27.** $9 - (-4) = 9 + (+4) = 13$ **28.** 11

29. $\left(-2\frac{2}{3}\right)\left(2\frac{1}{2}\right) = \left(-\frac{8}{3}\right)\left(\frac{5}{2}\right) = -\left(\frac{\overset{4}{\cancel{8}}}{3} \times \frac{5}{\underset{1}{\cancel{2}}}\right) = -\frac{20}{3}$

$= -6\frac{2}{3}$

30. $4\frac{3}{10}$ **31.** -3 **32.** -3 **33.** -12 **34.** -16.05

35. 5. Positive because numbers have same sign. **36.** 8

37. 0. Zero divided by any number other than zero is zero.

38. 0 **39.** $-10 - (-6) = -10 + (+6) = -4$ **40.** 3

Sections 1.1–1.6 Diagnostic Test (page 39)

Following each problem number is the number (in parentheses) of the textbook section where that kind of problem is discussed.

1. (1.1) False; zero is not a natural number. **2.** (1.2) True

3. (1.3) False **4.** (1.2) False **5.** (1.3) False

6. (1.2) True **7.** (1.1) False **8.** (1.2) False

9. (1.2) True **10.** (1.5) False **11.** (1.2) $<$

12. (1.2) $>$ **13.** (1.5) Factors

14. (1.3) $-|-56| = -(+56) = -56$ (Notice that $|-56| = +56$.)

15. (1.4) $3 - (+17) = 3 + (-17) = -14$

16. (1.4) $-12 - (-5) = -12 + (+5) = -7$

17. (1.4) $6 - (-2) = 6 + (+2) = 8$ **18.** (1.3) $-\frac{5}{8}$

19. (1.3) 6 **20.** (1.5) $-\frac{1}{6}$

21. (1.3) $8 + (-26) = -18$ (Signs are different; we subtract absolute values.)

22. (1.3) $-13 + (-5) = -18$ (Signs are same; we add absolute values.)

23. (1.3) $-21 + (-5) = -26$ (Signs are same; we add absolute values.)

24. (1.3) $-\frac{2}{5} + \frac{3}{10} = -\frac{4}{10} + \frac{3}{10} = -\frac{1}{10}$ (Signs are different; we subtract absolute values.)

25. (1.3) $6.16 + (-8.3) = -2.14$ (Signs are different; we subtract absolute values.)

26. (1.3) $-\frac{7}{2} + \frac{17}{8} = -\frac{28}{8} + \frac{17}{8} = -\frac{11}{8}$, or $-1\frac{3}{8}$ (Signs are different; we subtract absolute values.)

27. (1.4) $-8 - (-3) = -8 + (+3) = -5$

28. (1.4) $6 - (-12) = 6 + (+12) = 18$

29. (1.4) $-4 - 1 = -4 + (-1) = -5$

30. (1.4) $8 - 37 = 8 + (-37) = -29$

31. (1.4) $-5\frac{2}{3} - \left(-2\frac{8}{9}\right) = -5\frac{6}{9} + \left(+2\frac{8}{9}\right) = -4\frac{15}{9} + 2\frac{8}{9} = -2\frac{7}{9}$

32. (1.4) $3\frac{1}{3} - 8\frac{1}{6} = 3\frac{2}{6} + \left(-8\frac{1}{6}\right) = 3\frac{2}{6} + \left(-7\frac{7}{6}\right) = -4\frac{5}{6}$

33. (1.4) $-2.325 - (-6.3) = -2.325 + (+6.3) = 3.975$

34. (1.4) $0 - 12 = -12$ **35.** (1.5) $0(-12) = 0$

36. (1.5) $-4(-1) = 4$ **37.** (1.5) $12(-6) = -72$

38. (1.5) $-16(0) = 0$ **39.** (1.5) $-\frac{5}{6}\left(-\frac{3}{10}\right) = \frac{1}{4}$

40. (1.5) $-\frac{4}{3}\left(\frac{7}{4}\right) = -\frac{7}{3}$, or $-2\frac{1}{3}$

41. (1.5) $(6.32)(-0.1) = -0.632$

42. (1.6) $8 \div (-2) = -4$ **43.** (1.6) $\frac{17}{0}$ is not defined.

44. (1.6) $\frac{-24}{10} = -2.4$ **45.** (1.6) $-36 \div (-12) = 3$

46. (1.6) $\frac{18}{-24} = -\frac{3}{4}$ **47.** (1.6) $0 \div (-2) = 0$

48. (1.6) $\frac{0}{0}$ is not defined. **49.** (1.6) $-54 \div 9 = -6$

50. (1.6) $\frac{-4.9}{-7} = 0.7$

Exercises 1.7 (page 45)

1. True; commutative property of addition

2. True; commutative property of addition

3. True; associative property of addition

4. True; associative property of addition **5.** False

6. False **7.** True; associative property of multiplication

8. True; associative property of multiplication **9.** False

10. False **11.** True; commutative property of multiplication

12. True; commutative property of multiplication

13. True; commutative property of addition

14. True; commutative property of addition

15. True; commutative property of addition

16. True; commutative property of addition   **17.** False

18. True; distributive property

19. True; commutative property of addition

20. True; commutative property of addition

21. True; 1 is the multiplicative identity

22. True; 0 is the additive identity

23. True; additive inverse property

24. True; additive inverse property

25. True; distributive property

26. False   **27.** False   **28.** False

29. True; associative and commutative properties of addition

30. True; commutative and associative properties of multiplication

31. False   **32.** False

33. True; associative property of multiplication

34. True; associative property of multiplication

35. True; commutative property of addition

36. True; commutative property of addition   **37.** $3 + (-7)$

38. $-5 + 12$ **39.** $[4(-3)](6)$ **40.** $(-8 \cdot 4)(-2)$

41. $(3 \cdot 5) - (3 \cdot 8)$ **42.** $(6 \cdot 15) + (6 \cdot 2)$ **43.** $3(-4)$

44. $-3(6)$ **45.** $[4 + (-3)] + 6$ **46.** $-3 + [4 + (-2)]$

47. $(5 + 4) + ([-2] + [-8] + [-5]) = 9 + [-15] = -6$

48. -11 **49.** $8 + (-11) = -3$ **50.** -12

51. $(-5)(-4)(-2) = -(5 \times 4 \times 2) = -40$ **52.** -48

53. $(2)(-5)(-9) = +(2 \times 5 \times 9) = 90$ **54.** 24

55. $(-2)(-3)(-5)(-4) = +(2 \times 3 \times 5 \times 4) = 120$ **56.** 56

Exercises 1.8   (page 50)

1. Base: 3; exponent: 3; $3^3 = 3 \cdot 3 \cdot 3 = 27$

2. Base: 2; exponent: 4; 16

3. Base: -5; exponent: 2; $(-5)^2 = (-5)(-5) = 25$

4. Base: -6; exponent: 3; -216

5. Base: 7; exponent: 2; $7^2 = 7 \cdot 7 = 49$

6. Base: 3; exponent: 4; 81

7. Base: 0; exponent: 3; $0^3 = 0 \cdot 0 \cdot 0 = 0$

8. Base: 0; exponent: 4; 0   **9.** -10 **10.** 100

11. $10^3 = 10 \cdot 10 \cdot 10 = 1,000$ **12.** 10,000

13. $(-10)^5 = (-10)(-10)(-10)(-10)(-10) = -100,000$

14. 1,000,000   **15.** 2   **16.** 32

17. $(-2)^6 = (-2)(-2)(-2)(-2)(-2)(-2) = 64$ **18.** -128

19. $2^8 = 2 \cdot 2 \cdot 2 \cdot 2 \cdot 2 \cdot 2 \cdot 2 \cdot 2 = 256$ **20.** 625

21. $40^3 = 40 \cdot 40 \cdot 40 = 64,000$ **22.** 0

23. $(-12)^3 = (-12)(-12)(-12) = -1,728$ **24.** 225

25. $(-1)^5 = (-1)(-1)(-1)(-1)(-1) = -1$ **26.** -1

27. $-2^2 = -(2 \cdot 2) = -4$ **28.** -9 **29.** $(-1)^{99} = -1$

30. 1   **31.** $-[(-1)^5] = -(-1) = 1$ **32.** -1

33. $-[(-9)^2] = -(81) = -81$ **34.** -25 **35.** $0^8 = 0$

36. 0   **37.** 161.29   **38.** 237.16   **39.** 0.024336

40. 0.007569   **41.** $\left(-\frac{5}{8}\right)^2 = \left(-\frac{5}{8}\right)\left(-\frac{5}{8}\right) = \frac{25}{64}$ **42.** $\frac{16}{81}$

43. $\left(-\frac{1}{4}\right)^3 = \left(-\frac{1}{4}\right)\left(-\frac{1}{4}\right)\left(-\frac{1}{4}\right) = -\frac{1}{64}$ **44.** $-\frac{1}{243}$ **45.** $\frac{3}{16}$

46. $-\frac{3}{5}$ **47.** -2.5 **48.** -0.7

Exercises 1.9   (page 55)

1. P; $\{1, 5\}$ **2.** C; $\{1, 2, 4, 8\}$ **3.** P; $\{1, 13\}$

4. C; $\{1, 3, 5, 15\}$ **5.** C; $\{1, 2, 3, 4, 6, 12\}$

6. P; $\{1, 11\}$ **7.** C; $\{1, 3, 7, 21\}$ **8.** P; $\{1, 23\}$

9. C; $\{1, 5, 11, 55\}$ **10.** P; $\{1, 41\}$ **11.** C; $\{1, 7, 49\}$

12. P; $\{1, 31\}$ **13.** C; $\{1, 3, 17, 51\}$

14. C; $\{1, 2, 3, 6, 7, 14, 21, 42\}$ **15.** C; $\{1, 3, 37, 111\}$

16. P; $\{1, 101\}$ **17.** $\pm 1, \pm 2, \pm 4$ **18.** $\pm 1, \pm 3, \pm 9$

19. $\pm 1, \pm 2, \pm 5, \pm 10$ **20.** $\pm 1, \pm 2, \pm 7, \pm 14$

21. $\pm 1, \pm 3, \pm 5, \pm 15$ **22.** $\pm 1, \pm 2, \pm 4, \pm 8, \pm 16$

23. $\pm 1, \pm 2, \pm 3, \pm 6, \pm 9, \pm 18$ **24.** $\pm 1, \pm 2, \pm 4, \pm 5, \pm 10, \pm 20$

25. $\pm 1, \pm 3, \pm 7, \pm 21$ **26.** $\pm 1, \pm 2, \pm 11, \pm 22$

27. $2 \lfloor 14$
$ 7$
$14 = 2 \cdot 7$ **28.** $3 \cdot 5$ **29.** $3 \lfloor 21$
$ 7$
$21 = 3 \cdot 7$

30. $2 \cdot 11$ **31.** $2 \lfloor 26$
$ 13$
$26 = 2 \cdot 13$ **32.** 3^3 **33.** 29   **34.** 31

35. $2 \lfloor 32$
$2 \lfloor 16$
$2 \lfloor 8$
$2 \lfloor 4$
$ 2$
$32 = 2^5$ **36.** $3 \cdot 11$ **37.** $2 \lfloor 34$
$ 17$
$34 = 2 \cdot 17$ **38.** $5 \cdot 7$

39. $2 \lfloor 84$
$2 \lfloor 42$
$3 \lfloor 21$
$ 7$
$84 = 2^2 \cdot 3 \cdot 7$ **40.** $3 \cdot 5^2$ **41.** $2 \lfloor 144$
$2 \lfloor 72$
$2 \lfloor 36$
$2 \lfloor 18$
$3 \lfloor 9$
$ 3$
$144 = 2^4 \cdot 3^2$

42. $2^2 \cdot 3^2 \cdot 5$

43. The prime numbers greater than 17 and less than 37 are 19, 23, 29, and 31. The only one of these numbers that yields a remainder of 1 when divided by 5 is 31.

44. 37

Exercises 1.10A   (page 59)

1. Radicand: 16; $\sqrt{16} = 4$, because $4^2 = 16$.

2. Radicand: 25; 5

3. Radicand: 4; $\sqrt{4} = 2$. Therefore, $-\sqrt{4} = -2$.

4. Radicand: 9; -3

5. Radicand: 81; $\sqrt{81} = 9$, because $9^2 = 81$.

6. Radicand: 36; 6   **7.** $\sqrt{100} = 10$, because $10^2 = 100$.

8. 12   **9.** $\sqrt{81} = 9$. Therefore, $-\sqrt{81} = -9$.

10. -11 **11.** Radicand is negative; not a real number

12. Not a real number

13. Radicand is negative; not a real number

14. Not a real number

15. $\sqrt{0.25} = 0.5$, because $(0.5)^2 = 0.25$.   **16.** 0.8

17. $\sqrt{1.44} = 1.2$, because $(1.2)^2 = 1.44$.   **18.** 1.1

19. $\sqrt{\frac{1}{16}} = \frac{1}{4}$, because $\left(\frac{1}{4}\right)^2 = \frac{1}{16}$.   **20.** $\frac{1}{11}$

21. $\sqrt{\frac{9}{25}} = \frac{3}{5}$, because $\left(\frac{3}{5}\right)^2 = \frac{9}{25}$.   **22.** $\frac{4}{9}$

Exercises 1.10B   (page 63)

1. $\sqrt{2^8} = 2^{8/2} = 2^4$, or 16   **2.** 7^3, or 343

3. $\sqrt{3^6 \cdot 11^2} = \sqrt{3^6} \cdot \sqrt{11^2} = 3^{6/2} \cdot 11^{2/2} = 3^3 \cdot 11^1 = 27 \cdot 11$
$= 297$

4. $2^4 \cdot 5^2$, or 400

5. $\sqrt{529} = \sqrt{23^2} = 23$. If trial and error is used, try 20: $20^2 = 400$, and $400 < 529$; so 20 is too small. Try 23: $23^2 = 529$; thus, $\sqrt{529} = 23$.

6. 19

7. $\sqrt{441} = \sqrt{3^2 \cdot 7^2} = \sqrt{3^2} \cdot \sqrt{7^2} = 3^{2/2} \cdot 7^{2/2} = 3^1 \cdot 7^1 = 21$. If trial and error is used, try 20: $20^2 = 400$, and $400 < 441$; so 20 is too small. Try 21: $21^2 = 441$; thus, $\sqrt{441} = 21$.

8. 25

9. $\sqrt{289} = \sqrt{17^2} = 17$. If trial and error is used, try 10: $10^2 = 100$, and $100 < 289$; so 10 is too small. Try 13: $13^2 = 169$, and $169 < 289$; so 13 is too small. Try 17: $17^2 = 289$; thus, $\sqrt{289} = 17$.

10. 18

11. $\sqrt{729} = \sqrt{3^6} = 3^{6/2} = 3^3$, or 27. If trial and error is used, try 30: $30^2 = 900$, and $900 > 729$; so 30 is too large. Try 27: $27^2 = 729$; thus, $\sqrt{729} = 27$.

12. 36

Exercises 1.10C   (page 65)

1. $\sqrt{2^9} = \sqrt{2^8 \cdot 2^1} = \sqrt{2^8}\sqrt{2} = 2^4\sqrt{2} = 16\sqrt{2}$

2. $5^3\sqrt{5}$, or $125\sqrt{5}$

3. $\sqrt{3^3 \cdot 5^6} = \sqrt{3^2 \cdot 3^1 \cdot 5^6} = \sqrt{3^2} \cdot \sqrt{3} \cdot \sqrt{5^6} = 3(\sqrt{3})(5^3)$ $= 3 \cdot 125\sqrt{3} = 375\sqrt{3}$

4. $196\sqrt{7}$ **5.** $\sqrt{98} = \sqrt{49 \cdot 2} = \sqrt{49}\sqrt{2} = 7\sqrt{2}$

6. $2\sqrt{5}$ **7.** $\sqrt{18} = \sqrt{9 \cdot 2} = \sqrt{9}\sqrt{2} = 3\sqrt{2}$

8. $3\sqrt{5}$ **9.** $\sqrt{8} = \sqrt{4 \cdot 2} = \sqrt{4}\sqrt{2} = 2\sqrt{2}$ **10.** $4\sqrt{2}$

11. $\sqrt{44} = \sqrt{4 \cdot 11} = \sqrt{4}\sqrt{11} = 2\sqrt{11}$ **12.** $3\sqrt{15}$

13. $\sqrt{450} = \sqrt{2 \cdot 3^2 \cdot 5^2} = \sqrt{2}\sqrt{3^2}\sqrt{5^2} = 3 \cdot 5\sqrt{2} = 15\sqrt{2}$

14. $8\sqrt{3}$

15. $\sqrt{108} = \sqrt{2^2 \cdot 3^3} = \sqrt{2^2 \cdot 3^2 \cdot 3^1} = \sqrt{2^2} \cdot \sqrt{3^2} \cdot \sqrt{3}$ $= 2 \cdot 3\sqrt{3} = 6\sqrt{3}$

16. $21\sqrt{2}$

Exercises 1.10D   (page 67)

1. 3.606   **2.** 4.243   **3.** 6.083   **4.** 7.071   **5.** 8.888

6. 7.746   **7.** 9.274   **8.** 9.592   **9.** 683   **10.** 821

11. 522   **12.** 299

Exercises 1.10E   (page 73)

1. Using trial and error: $2^3 = 8$, so $\sqrt[3]{64} \neq 2$; $3^3 = 27$, so $\sqrt[3]{64} \neq 3$. $4^3 = 64$; therefore, $\sqrt[3]{64} = 4$.

2. 3   **3.** Since $3^3 = 27$, $\sqrt[3]{27} = 3$.   **4.** 5

5. Since $\sqrt[3]{27} = 3$, $-(\sqrt[3]{27}) = -(3) = -3$.   **6.** -5

7. $-\sqrt[4]{1} = -(\sqrt[4]{1}) = -(1) = -1$ **8.** 2

9. -5, because $(-5)^3 = -125$.   **10.** -2

11. Not a real number   **12.** Not a real number

13. -10, because $(-10)^3 = -1{,}000$.   **14.** -4 **15.** -1

16. -2 **17.** 3, because $3^6 = 729$.   **18.** 6

19. Not a real number   **20.** Not a real number

21. a. No natural numbers   **b.** Integer: -12
   c. Rational numbers: $\frac{1}{2}$, -12, $0.\overline{26}$
   d. Irrational numbers: $\sqrt[3]{13}$, $0.196732468\ldots$
   e. Real numbers: $\sqrt[3]{13}$, $\frac{1}{2}$, -12, $0.\overline{26}$, $0.196732468\ldots$
   f. Number that is not real: $\sqrt{-15}$

22. a. Natural number: 18   **b.** Integer: 18
   c. Rational numbers: 18, $\frac{11}{32}$, $0.\overline{37}$
   d. Irrational numbers: $0.67249713\ldots$, $\sqrt[3]{12}$
   e. Real numbers: $0.67249713\ldots$, 18, $\frac{11}{32}$, $0.\overline{37}$, $\sqrt[3]{12}$
   f. Number that is not real: $\sqrt{-27}$

23. $\sqrt{3} \approx 1.7$, $-\frac{9}{4} = -2\frac{1}{4}$, and $\sqrt[3]{-1} = -1$

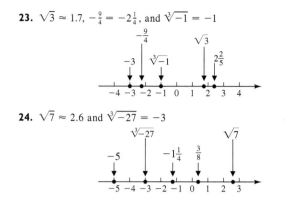

24. $\sqrt{7} \approx 2.6$ and $\sqrt[3]{-27} = -3$

Exercises 1.11   (page 81)

1. $12 - 8 - 6 = 4 - 6 = -2$ **2.** 2

3. $17 - 11 + 13 - 9 = 6 + 13 - 9 = 19 - 9 = 10$ **4.** 12

5. $5 - 8 - 12 - 3 - 14 + 6 - 21 = 11 + (-58) = -47$

6. -12 **7.** $7 + 2 \cdot 4 = 7 + 8 = 15$ **8.** 28

9. $9 - 3 \cdot 2 = 9 - 6 = 3$ **10.** -10

11. $10 \div 2 \cdot 5 = 5 \cdot 5 = 25$ **12.** 60

13. $12 \div 6 \div 2 = 2 \div 2 = 1$ **14.** $\frac{1}{3}$

15. $(-12) \div 2 \cdot (-3) = (-6) \cdot (-3) = 18$ **16.** -36

17. $8 \cdot 5^2 = 8 \cdot 25 = 200$ **18.** 96   **19.** 0   **20.** 0

21. $12 \cdot 4 + 16 \div 8 = 48 + 16 \div 8 = 48 + 2 = 50$ **22.** 15

23. $28 \div 4 \cdot 2(6) = 7 \cdot 2(6) = 14(6) = 84$ **24.** -48

25. $(-2)^2 + (-4)(5) - (-3)^2 = 4 + (-4)(5) - 9$ $= 4 + (-20) - (9) = -16 + (-9) = -25$

26. -3

27. $2 \cdot 3 + 3^2 - 4 \cdot 2 = 2 \cdot 3 + 9 - 4 \cdot 2 = 6 + 9 - 8$ $= 15 + (-8) = 7$

28. 624

29. $(10^2)\sqrt{16} + 5(4) - 80 = (100)(4) + 5(4) - 80$ $= 400 + 20 - 80 = 420 - 80 = 340$

30. 39

31. $2 \cdot (-6) \div 3 \cdot (8 - 4) = 2 \cdot (-6) \div 3 \cdot (4) = -12 \div 3 \cdot 4$ $= -4 \cdot 4 = -16$

32. -50 **33.** $24 - [(-6) + 18] = 24 - [12] = 12$

34. 11   **35.** $[12 - (-19)] - 16 = [31] - 16 = 15$ **36.** 6

37. $[11 - (5 + 8)] - 24 = [11 - (13)] - 24 = [-2] - 24$ $= -26$

38. -25

39. $20 - [5 - (7 - 10)] = 20 - [5 - (-3)] = 20 - [8] = 12$

40. 3   **41.** $\dfrac{7 + (-12)}{8 - 3} = \dfrac{-5}{5} = -1$ **42.** -4

43. $15 - \{4 - [2 - 3(6 - 4)]\} = 15 - \{4 - [2 - 3(2)]\}$ $= 15 - \{4 - [2 - 6]\} = 15 - \{4 - [-4]\} = 15 - \{8\} = 7$

44. 30

45. $32 \div (-2)^3 - 5\left\{7 - \dfrac{6 - 2}{5}\right\} = 32 \div (-2)^3 - 5\left\{7 - \dfrac{4}{5}\right\}$

$= 32 \div (-2)^3 - \overset{1}{5}\left\{\dfrac{31}{\underset{1}{5}}\right\} = 32 \div (-8) - 31 = (-4) - 31$

$= -35$

46. -16 **47.** $\sqrt{3^2 + 4^2} = \sqrt{9 + 16} = \sqrt{25} = 5$ **48.** 12

33. Let x = smallest integer
Then $x + 1$ = next integer
and $x + 2$ = largest integer
The sum is $x + (x + 1) + (x + 2)$.

34. $x + (x + 1)$, where x = the smaller integer and $x + 1$ = the larger integer

35. Let x = first odd integer
Then $x + 2$ = second odd integer
and $x + 4$ = third odd integer
and $x + 6$ = fourth odd integer
The sum is $x + (x + 2) + (x + 4) + (x + 6)$.

36. $x + (x + 2) + (x + 4)$, where x = first even integer, $x + 2$ = second even integer, and $x + 4$ = third even integer

37. Let h = height (in cm) or Let r = radius (in cm)
Then $h - 4$ = radius (in cm)   Then $r + 4$ = height (in cm)
Volume = $\pi(h - 4)^2 h$ Volume = $\pi r^2(r + 4)$

38. Area = $\frac{1}{2}(h - 5)h$,   or Area = $\frac{1}{2}b(b + 5)$,
where h = altitude (in m)   where b = base (in m)
and $h - 5$ = base (in m)   and $b + 5$ = altitude (in m)

39. Let r = radius   or Let h = height

Then $4r$ = height   Then $\dfrac{h}{4}$ = radius

Volume = $\frac{1}{3}\pi r^2(4r)$ Volume = $\frac{1}{3}\pi\left(\dfrac{h}{4}\right)^2 h$

40. Volume = $\pi(6h)^2 h$, or Volume = $\pi r^2\left(\dfrac{r}{6}\right)$,

where h = height   where radius = r

and $6h$ = radius   and height = $\dfrac{r}{6}$

Exercises 4.2   (page 187)

1. Let x = unknown number
$13 + 2x = 25$

2. Let x = unknown number
$25 + 3x = 34$

3. Let x = unknown number
$5x - 8 = 22$

4. Let x = unknown number
$4x - 5 = 15$

5. Let x = unknown number
$7 - x = x + 1$

6. Let x = unknown number
$6 + x = 12 - x$

7. Let x = unknown number
$\frac{1}{5}x = 4$

8. Let x = unknown number
$\dfrac{x}{12} = 6$

9. Let x = unknown number
$\frac{1}{2}x - 4 = 6$

10. Let x = unknown number
$\frac{1}{3}x - 5 = 4$

11. Let x = unknown number
$2(5 + x) = 26$

12. Let x = unknown number
$4(9 + x) = 18$

13. Let x = unknown number
$(x + x)(3) = 24$

14. Let x = unknown number
$5(x + x) = 40$

15. Let x = length of shorter piece of rope (in m)
Then $x + 13$ = length of longer piece (in m)
$x + (x + 13) = 75$

16. Let x = length of shorter piece of wire (in m)
Then $x + 8$ = length of longer piece (in m)
$x + (x + 8) = 46$

17. Let x = length of shorter piece of wire (in cm)
Then $2x$ = length of longer piece of wire (in cm)
$x + 2x = 72$

18. Let x = length of shorter piece of wire (in cm)
Then $3x$ = length of longer piece of wire (in cm)
$x + 3x = 72$

19. Let x = number of cans of pears
Then $x + 8$ = number of cans of peaches
$x + (x + 8) = 42$

20. Let x = number of cans of corn
Then $x + 11$ = number of cans of peas
$x + (x + 11) = 49$

21. Let x = first integer
Then $x + 1$ = second integer
and $x + 2$ = third integer
and $x + 3$ = fourth integer
$x + (x + 1) + (x + 2) + (x + 3) = 106$

22. Let x = smallest integer
Then $x + 1$ = next integer
and $x + 2$ = third integer
$x + (x + 1) + (x + 2) = -72$

23. Let w = width (in cm) or Let l = length (in cm)
Then $w + 4$ = length (in cm)   Then $l - 4$ = width (in cm)
$2w + 2(w + 4) = 36$ $2l + 2(l - 4) = 36$

24. Let w = width (in ft) or Let l = length (in ft)
Then $w + 6$ = length (in ft)   Then $l - 6$ = width (in ft)
$2w + 2(w + 6) = 64$ $2l + 2(l - 6) = 64$

Exercises 4.3   (page 193)

(The checks will not be shown.)

1. Let x = unknown number
$$13 + 2x = 25$$
$$-13 + 13 + 2x = -13 + 25$$
$$2x = 12$$
$$\frac{2x}{2} = \frac{12}{2}$$
$$x = 6 \quad \text{The number is 6.}$$

2. 3

3. Let x = unknown number
$$5x - 8 = 22$$
$$5x - 8 + 8 = 22 + 8$$
$$5x = 30$$
$$\frac{5x}{5} = \frac{30}{5}$$
$$x = 6 \quad \text{The number is 6.}$$

4. 5

5. Let x = unknown number
$$7 - x = x + 1$$
$$7 - x - x = x + 1 - x$$
$$7 - 2x = 1$$
$$-7 + 7 - 2x = -7 + 1$$
$$-2x = -6$$
$$\frac{-2x}{-2} = \frac{-6}{-2}$$
$$x = 3 \quad \text{The number is 3.}$$

6. 3

7. Let x = unknown number
$$\tfrac{1}{2}x - 4 = 6$$
$$\tfrac{1}{2}x - 4 + 4 = 6 + 4$$
$$\tfrac{1}{2}x = 10$$
$$2\left(\tfrac{1}{2}x\right) = 2(10)$$
$$x = 20 \quad \text{The number is 20.}$$

8. 27

9. Let x = unknown number
$$2(5 + x) = 26$$
$$10 + 2x = 26$$
$$-10 + 10 + 2x = -10 + 26$$
$$2x = 16$$
$$\frac{2x}{2} = \frac{16}{2}$$
$$x = 8 \quad \text{The number is 8.}$$

10. -4

11. Let x = length of shorter piece of wire (in cm)
Then $x + 10$ = length of longer piece of wire (in cm)
$$x + (x + 10) = 36$$
$$x + x + 10 = 36$$
$$2x + 10 = 36$$
$$2x + 10 - 10 = 36 - 10$$
$$2x = 26$$
$$\frac{2x}{2} = \frac{26}{2}$$
$$x = 13 \qquad \text{The shorter piece is 13 cm long.}$$
$$x + 10 = 13 + 10 = 23 \quad \text{The longer piece is 23 cm long.}$$

12. 4 yd and 6 yd

13. Let x = number of packages of apricots
Then $x + 7$ = number of packages of apples
$$x + (x + 7) = 15$$
$$x + x + 7 = 15$$
$$2x + 7 = 15$$
$$2x + 7 - 7 = 15 - 7$$
$$2x = 8$$
$$\frac{2x}{2} = \frac{8}{2}$$
$$x = 4$$
$$x + 7 = 4 + 7 = 11$$
Rebecca buys 4 packages of apricots and 11 packages of apples.

14. 5 skeins of green yarn; 8 skeins of blue yarn

15. $A = \frac{1}{2}bh$
$A = \frac{1}{\cancel{2}} \, (\overset{13}{\cancel{26}} \text{ cm}) \, (13 \text{ cm}) = 169 \text{ sq. cm}$

16. $V = 30$ cu. m; $S = 62$ sq. m

17. $V = \frac{4}{3}\pi r^3 \qquad S = 4\pi r^2$
$V \approx \frac{4}{3}(3.14)(6 \text{ yd})^3 \quad S \approx 4(3.14)(6 \text{ yd})^2$
$V \approx 904.32$ cu. yd $\quad S \approx 452.16$ sq. yd

18. $A \approx 50.24$ sq. ft
$C \approx 25.12$ ft

19. Let x = first odd integer
Then $x + 2$ = second odd integer
and $x + 4$ = third odd integer
$$x + (x + 2) + (x + 4) = 177$$
$$x + x + 2 + x + 4 = 177$$
$$3x + 6 = 177$$
$$3x + 6 - 6 = 177 - 6$$
$$3x = 171$$
$$\frac{3x}{3} = \frac{171}{3}$$
$$x = 57$$
$$x + 2 = 57 + 2 = 59$$
$$x + 4 = 57 + 4 = 61$$
The odd integers are 57, 59, and 61.

20. $-50, -48,$ and -46

21. Let x = first integer
Then $x + 1$ = second integer
and $x + 2$ = third integer
$$4[x + (x + 2)] = 140 + (x + 1)$$
$$4[x + x + 2] = 140 + x + 1$$
$$4[2x + 2] = 141 + x$$
$$8x + 8 = 141 + x$$
$$-x + 8x + 8 = -x + 141 + x$$
$$7x + 8 = 141$$
$$7x + 8 - 8 = 141 - 8$$
$$7x = 133$$
$$\frac{7x}{7} = \frac{133}{7}$$
$$x = 19$$
$$x + 1 = 19 + 1 = 20$$
$$x + 2 = 19 + 2 = 21$$
The integers are 19, 20, and 21.

22. 14, 15, and 16

23. Let x = number of degrees in third angle
$$62 + 47 + x = 180$$
$$109 + x = 180$$
$$-109 + 109 + x = -109 + 180$$
$$x = 71$$
The third angle measures 71°.

24. 125°

25. Let w = width of rectangle (in ft)
Then $2w$ = length (in ft)
$$2w + 2(2w) = 102$$
$$2w + 4w = 102$$
$$6w = 102$$
$$\frac{\overset{1}{\cancel{6}}w}{\cancel{6}} = \frac{102}{6}$$
$$w = 17$$
$$2w = 2(17) = 34$$
The rectangle is 34 ft long and 17 ft wide.

26. Length = 64 cm; width = 16 cm

27. Let x = length of rectangle (in ft)
$$2x + 2(5) = 44$$
$$2x + 10 = 44$$
$$2x + 10 - 10 = 44 - 10$$
$$2x = 34$$
$$\frac{2x}{2} = \frac{34}{2}$$
$$x = 17 \quad \text{The rectangle is 17 ft long.}$$

28. 12 yd

29. Let x = unknown number
$$2(4 + x) + x = x + 10$$
$$8 + 2x + x = x + 10$$
$$8 + 3x = x + 10$$
$$8 + 3x - x = x + 10 - x$$
$$8 + 2x = 10$$
$$-8 + 8 + 2x = -8 + 10$$
$$2x = 2$$
$$\frac{2x}{2} = \frac{2}{2}$$
$$x = 1 \quad \text{The number is 1.}$$

30. 5

31. Let x = unknown number
$$3(8 + 2x) = 4(3x + 8)$$
$$24 + 6x = 12x + 32$$
$$24 + 6x - 12x = 12x + 32 - 12x$$
$$24 - 6x = 32$$
$$-24 + 24 - 6x = -24 + 32$$
$$-6x = 8$$
$$\frac{-6x}{-6} = \frac{8}{-6}$$
$$x = -\frac{4}{3} \quad \text{The number is } -\frac{4}{3}.$$

32. $\frac{10}{11}$

33. Let x = unknown number
$$10x - 3(4 + x) = 5(9 + 2x)$$
$$10x - 12 - 3x = 45 + 10x$$
$$7x - 12 = 45 + 10x$$
$$-10x + 7x - 12 = -10x + 45 + 10x$$
$$-3x - 12 = 45$$
$$-3x - 12 + 12 = 45 + 12$$
$$-3x = 57$$
$$\frac{-3x}{-3} = \frac{57}{-3}$$
$$x = -19 \quad \text{The number is } -19.$$

34. $-2\frac{1}{2}$

Vertical addition is shown for Exercise 35.

35. Let x = unknown number
$$8.66x - 5.75(6.94 + x) = 4.69(8.55 + 3.48x)$$
$$8.66x - 39.905 - 5.75x = 40.0995 + 16.3212x$$
$$2.91x - 39.905 = 40.0995 + 16.3212x$$
$$\underline{-2.91x - 40.0995 = -40.0995 - 2.91 \quad x}$$
$$-80.0045 = 13.4112x$$
$$\frac{-80.0045}{13.4112} = \frac{\overset{1}{\cancel{13.4112}}x}{\cancel{13.4112}}$$
$$x \approx -5.97$$
The number is approx. -5.97.

4

36. ≈ -32.58

37. Let x = unknown number
$$x + 18 \geq 5$$
$$x + 18 - 18 \geq 5 - 18$$
$$x \geq -13 \quad \text{The unknown number can be any number}$$
greater than or equal to -13.

38. The unknown number can be any number greater than or equal to 35.

39. Let x = unknown length (in m)
$$37 + x < 80$$
$$-37 + 37 + x < -37 + 80$$
$$x < 43 \quad \text{The second piece must be less than}$$
43 m (and greater than 0 m).

40. Any length greater than 17 m

Exercises 4.4A   (page 197)

1. $7(5 \text{ cents}) = 35$ cents   **2.** 275 cents

3. $9(50 \text{ cents}) = 450$ cents   **4.** 120 cents

5. $x(25 \text{ cents}) = 25x$ cents   **6.** $5y$ cents

7. $7(\$1.50) + 5(\$0.75) = \$10.50 + \$3.75 = \$14.25$

8. $21.25   **9.** $x(\$3.50) + y(\$1.90) = (3.50x + 1.90y)$ dollars

10. $(2.75x + 1.50y)$ dollars

11. $x(25 \text{ cents}) + y(6 \text{ cents}) = (25x + 6y)$ cents

12. $(4x + 2y)$ cents

Exercises 4.4B   (page 200)

(The checks will not be shown.)

1. Let x = number of $10 bills   **2.** 5 $5 bills,
Then $13 - x$ = number of $5 bills   6 $10 bills
$$10x + 5(13 - x) = 95$$
$$10x + 65 - 5x = 95$$
$$5x + 65 = 95$$
$$5x + 65 - 65 = 95 - 65$$
$$5x = 30$$
$$\frac{5x}{5} = \frac{30}{5}$$
$$x = 6$$
$$13 - x = 13 - 6 = 7$$
Bill has 6 $10 bills and 7 $5 bills.

3. Let x = number of nickels   **4.** 10 nickels,
Then $12 - x$ = number of quarters   8 quarters
$$5x + 25(12 - x) = 220$$
$$5x + 300 - 25x = 220$$
$$-20x + 300 = 220$$
$$-20x + 300 - 300 = 220 - 300$$
$$-20x = -80$$
$$\frac{-20x}{-20} = \frac{-80}{-20}$$
$$x = 4$$
$$12 - x = 12 - 4 = 8$$
Jennifer has 4 nickels and 8 quarters.

5. Let x = number of nickels
Then $x + 4$ = number of quarters
and $3x$ = number of dimes
$$5x + 25(x + 4) + 10(3x) = 400$$
$$5x + 25x + 100 + 30x = 400$$
$$60x + 100 = 400$$
$$60x + 100 - 100 = 400 - 100$$
$$60x = 300$$
$$\frac{60x}{60} = \frac{300}{60}$$
$$x = 5$$
$$x + 4 = 5 + 4 = 9$$
$$3x = 3(5) = 15$$
Derek has 5 nickels, 9 quarters, and 15 dimes.

6. 2 nickels, 9 dimes, and 18 quarters

7. Let x = number of quarters
Then $x - 3$ = number of dimes
and $2x - 3$ = number of nickels
$$25x + 10(x - 3) + 5(2x - 3) = 225$$
$$25x + 10x - 30 + 10x - 15 = 225$$
$$45x - 45 = 225$$
$$45x - 45 + 45 = 225 + 45$$
$$45x = 270$$
$$\frac{45x}{45} = \frac{270}{45}$$
$$x = 6$$
$$x - 3 = 6 - 3 = 3$$
$$2x - 3 = 2(6) - 3 = 9$$
Michael has 6 quarters, 3 dimes, and 9 nickels.

8. 7 $1 bills, 12 $5 bills, 19 $10 bills

9. Let x = number of box seats sold
Then $5x$ = number of balcony seats sold
This leaves $1,080 - (x + 5x)$, or $1,080 - 6x$, for the number of orchestra seats.
$$30x + 12(5x) + 21(1,080 - 6x) = 19,800$$
$$30x + 60x + 22,680 - 126x = 19,800$$
$$-36x + 22,680 = 19,800$$
$$-36x + 22,680 - 22,680 = 19,800 - 22,680$$
$$-36x = -2,880$$
$$\frac{-36x}{-36} = \frac{-2,880}{-36}$$
$$x = 80$$
$$5x = 5(80) = 400$$
$$1,080 - 6x = 1,080 - 6(80) = 600$$
There were 80 box seat tickets, 400 balcony seat tickets, and 600 orchestra seat tickets sold.

10. 4,400 box seat tickets, 17,600 reserved seat tickets, 35,200 general admission seat tickets

11. Let x = number of 12¢ stamps
Then $2x$ = number of 10¢ stamps
This leaves $60 - (x + 2x)$, or $60 - 3x$, for the number of 2¢ stamps.
$$12x + 10(2x) + 2(60 - 3x) = 380$$
$$12x + 20x + 120 - 6x = 380$$
$$26x + 120 = 380$$
$$26x + 120 - 120 = 380 - 120$$
$$26x = 260$$
$$\frac{26x}{26} = \frac{260}{26}$$
$$x = 10$$
$$2x = 2(10) = 20$$
$$60 - 3x = 60 - 3(10) = 30$$
Christy bought 10 12¢ stamps, 20 10¢ stamps, and 30 2¢ stamps.

12. 10 8¢ stamps, 30 6¢ stamps, and 60 12¢ stamps

Exercises 4.5   (page 206)

(The checks will not be shown.)

1. Let $4x =$ larger number
Let $3x =$ smaller number
$$3x + 4x = 35$$
$$7x = 35$$
$$\frac{\overset{1}{\cancel{7}}x}{\cancel{7}} = \frac{35}{7}$$
$$x = 5$$
$$3x = 3(5) = 15 \quad \text{Smaller number}$$
$$4x = 4(5) = 20 \quad \text{Larger number}$$

2. 91, 39

3. Let $4x =$ width
Let $9x =$ length
$$2(9x) + 2(4x) = 78$$
$$18x + 8x = 78$$
$$26x = 78$$
$$\frac{\overset{1}{\cancel{26}}x}{\cancel{26}} = \frac{78}{26}$$
$$x = 3$$
$$9x = 9(3) = 27 \quad \text{Length}$$
$$4x = 4(3) = 12 \quad \text{Width}$$

4. 28, 8

5. Let $4x =$ amount spent for food
Let $5x =$ amount spent for rent
Let $x =$ amount spent for clothing
$$4x + 5x + x = 850$$
$$10x = 850$$
$$\frac{\overset{1}{\cancel{10}}x}{\cancel{10}} = \frac{850}{10}$$
$$x = \$85 \qquad \text{For clothing}$$
$$4x = 4(\$85) = \$340 \quad \text{For food}$$
$$5x = 5(\$85) = \$425 \quad \text{For rent}$$

6. $5,400 for tuition, $5,400 for housing, $1,350 for food

7. Let $6x =$ one part
Let $5x =$ other part
$$6x + 5x = 88$$
$$11x = 88$$
$$\frac{\overset{1}{\cancel{11}}x}{\cancel{11}} = \frac{88}{11}$$
$$x = 8$$
$$6x = 6(8) = 48$$
$$5x = 5(8) = 40$$
The numbers are 48 and 40.

8. 22, 77

9. Let $4x =$ length of shortest side (in ft)
Let $5x =$ length of next side (in ft)
Let $6x =$ length of longest side (in ft)
$$4x + 5x + 6x = 90$$
$$15x = 90$$
$$\frac{\overset{1}{\cancel{15}}x}{\cancel{15}} = \frac{90}{15}$$
$$x = 6$$
$$4x = 4(6) = 24$$
$$5x = 5(6) = 30$$
$$6x = 6(6) = 36$$
The lengths of the sides are 24 ft, 30 ft, and 36 ft.

10. 20 yd, 24 yd, 28 yd

11. Let $2x =$ amount received by first niece
Let $3x =$ amount received by second niece
Let $4x =$ amount received by third niece

$$2x + 3x + 4x = \$27{,}000$$
$$9x = \$27{,}000$$
$$\frac{\overset{1}{\cancel{9}}x}{\cancel{9}} = \frac{\$27{,}000}{9}$$
$$x = \$3{,}000$$
$$2x = 2(\$3{,}000) = \$6{,}000$$
$$3x = 3(\$3{,}000) = \$9{,}000$$
$$4x = 4(\$3{,}000) = \$12{,}000$$
The first niece received $6,000, the second $9,000, and the third $12,000.

12. $75,000; $15,000; $60,000

13. Let $4x =$ number of cu. yd of gravel
Let $2\frac{1}{2}x =$ number of cu. yd of sand
Let $1x =$ number of cu. yd of cement
$$4x = 5 \qquad\qquad \text{Mrs. Mora used 5 cu. yd of gravel}$$
$$\frac{\overset{1}{\cancel{4}}x}{\cancel{4}} = \frac{5}{4}$$
$$x = \frac{5}{4} = 1\frac{1}{4} \qquad \text{Amount of cement}$$
$$2\frac{1}{2}x = \frac{5}{2}\left(\frac{5}{4}\right) = \frac{25}{8} = 3\frac{1}{8} \quad \text{Amount of sand}$$
Mrs. Mora used $3\frac{1}{8}$ cu. yd of sand and $1\frac{1}{4}$ cu. yd of cement.

14. 519 seafood dinners

15. Let $7x =$ length
Let $4x =$ width
$$2(7x) + 2(4x) < 88$$
$$14x + 8x < 88$$
$$22x < 88$$
$$\frac{\overset{1}{\cancel{22}}x}{\cancel{22}} < \frac{88}{22}$$
$$x < 4$$
$$4x < 16$$
The width must be less than 16 (and greater than 0).

16. The length must be greater than 24.

17. Let $2x =$ length of shortest side
Let $4x =$ length of next side
Let $7x =$ length of longest side
$$2x + 4x + 7x > 117$$
$$13x > 117$$
$$\frac{\overset{1}{\cancel{13}}x}{\cancel{13}} > \frac{117}{13}$$
$$x > 9$$
$$7x > 63$$
The longest side must be greater than 63.

18. The length of the shortest side must be less than 24 (and greater than 0).

19. $\dfrac{360 \text{ miles}}{7 \text{ hours}}$

20. $\dfrac{35 \text{ miles}}{4 \text{ hours}}$

21. $\dfrac{3 \text{ afghans}}{25 \text{ days}}$

22. $\dfrac{500 \text{ words}}{11 \text{ minutes}}$

23. Let $x =$ number of square yards
$$x \; \cancel{\text{sq yd}}\left(23 \frac{\text{dollars}}{\cancel{\text{sq yd}}}\right) = 805 \text{ dollars}$$
$$23x = 805$$
$$\frac{\overset{1}{\cancel{23}}x}{\cancel{23}} = \frac{805}{23}$$
$$x = 35 \text{ sq. yd}$$

24. 36 hours

25. Let $x =$ number of gallons of gasoline
$$x \; \cancel{\text{gallons}}\left(23 \frac{\text{miles}}{\cancel{\text{gallon}}}\right) = 368 \text{ miles}$$
$$23x = 368$$
$$\frac{\overset{1}{\cancel{23}}x}{\cancel{23}} = \frac{368}{23}$$
$$x = 16$$
Jianula will use 16 gal of gasoline.

26. 12 rolls

4

Exercises 4.6   (page 212)

1. Let x = number

$0.30x = 15$

$$x = \frac{15}{0.30} = 50$$

2. 80

3. Let x = percent

$250x = 115$

$$x = \frac{115}{250} = 0.46 = 46\%$$

4. $\approx 146.67\%$

5. Let x = unknown number

$x = 0.25(40) = 10$

6. 29.25

7. Let x = number

$0.15x = 127.5$

$$x = \frac{127.5}{0.15} = 850$$

8. 800

9. Let x = percent

$8x = 17$

$$x = \frac{17}{8} = 2.125 = 212.5\%$$

10. 200%

11. Let x = unknown number

$x = 0.63(48) = 30.24$

12. 42.63

13. Let x = number

$1.25x = 750$

$$x = \frac{750}{1.25} = 600$$

14. 250

15. Let x = percent

$16x = 23$

$$x = \frac{23}{16} = 1.4375 = 143.75\%$$

16. $\approx 247.83\%$

17. Let x = unknown number

$x = 2.00(12) = 24$

18. 27

19. Let x = number

$0.15x = 37.5$

$x = 250$

20. \$36.45

21. Let x = number

$0.66\frac{2}{3}x = 42$ $\left(0.66\frac{2}{3} = \frac{66\frac{2}{3}}{100} = \frac{\frac{200}{3}}{100} = \frac{2}{3}\right)$

$\frac{2}{3}x = 42$

$\frac{3}{2}\left(\frac{2}{3}\right)x = \frac{3}{2}(42)$

$x = 63$

22. 216

23. Sixty-eight is 80% of what number?

Let x = number

$0.80x = 68$

$$x = \frac{68}{0.80} = 85$$

The team has played 85 games.

24. 32,200

25. Seven is what percent of 42?

Let x = percent

$42x = 7$

$$x = \frac{7}{42} \approx 0.1667 = 16.67\%$$

26. \$57.50

27. Fifty-four is what percent of 210?

Let x = percent

$210x = 54$

$$x = \frac{54}{210} \approx 0.2571 = 25.71\%$$

28. 11%

29. The markup is 35% of \$125.

Let x = number

$x = 0.35(\$125) = \$43.75$ Markup

Selling price = Cost + Markup

Selling price = \$125 + \$43.75 = \$168.75

30. \$86.40

31. Total weight of steers = $15 \times 1{,}027$ lb = 15,405 lb

Total weight of heifers = 18×956 lb = 17,208 lb

If 3% of the weight is lost in shipping, then 97% of the weight remains at time of sale.

Let x = amount of money from sale of steers

$x = 0.97 \times 15{,}405 \times 0.84 = 12{,}551.9940$

Let y = amount of money from sale of heifers

$y = 0.97 \times 17{,}208 \times 0.78 = 13{,}019.5728$

$x + y = 25{,}571.5668 \approx 25{,}571.57$

Amount of check = \$25,571.57

32. $\approx 38.60\%$

Exercises 4.7   (page 218)

(The checks will not be shown.)

1. Let x = number of hours

$\left(51\,\frac{mi}{hr}\right)(x\ hr) = 408$ mi

$51x = 408$

$$\frac{\overset{1}{\cancel{51}}x}{\underset{1}{\cancel{51}}} = \frac{408}{51}$$

$x = 8$ Hours for Robbie to drive 408 mi

2. 9 hours

3. $47.72\ \cancel{sec}\left(\dfrac{1\ min}{60\ \cancel{sec}}\right) = \dfrac{47.72}{60}$ min ≈ 0.7953 min; therefore,

27 min 47.72 sec ≈ 27.7953 min.

Let x = the rate in meters per minute

$$x = \frac{10{,}000\ m}{27.7953\ min} \approx 359.773\ \frac{m}{min}$$

$27\ \cancel{min}\left(\dfrac{60\ sec}{1\ \cancel{min}}\right) = (27)(60)$ sec $= 1{,}620$ sec; therefore,

27 min 47.72 sec = 1,620 sec + 47.72 sec = 1,667.72 sec.

Let y = rate in meters per second

$$y = \frac{10{,}000\ m}{1{,}667.72\ sec} \approx 5.996\ \frac{m}{sec}$$

4. 2.7 steps per sec; 162.2 steps per min

5. Let x = number of hours until the planes will be 1,400 mi apart

$210x + 190x = 1{,}400$

$400x = 1{,}400$

$$\frac{400x}{400} = \frac{1{,}400}{400}$$

$x = \frac{7}{2}$, or $3\frac{1}{2}$

The planes will be 1,400 mi apart $3\frac{1}{2}$ hr after leaving Denver.

6. $2\frac{1}{4}$ hr

7. Let x = Jennie's speed (in mph) with no traffic

Then $x - 10$ = Jennie's speed (in mph) with heavy traffic

$4x = 5(x - 10)$

$4x = 5x - 50$

$-5x + 4x = -5x + 5x - 50$

$-x = -50$

$x = 50$

a. Jennie's average speed is 50 mph when there is no traffic.

b. The distance is $\left(50\,\dfrac{mi}{hr}\right)(4\ hr) = 200$ mi.

8. a. 60 mph   **b.** 300 mi

9. Let x = Mr. Robinson's speed (in mph)

Then $x + 9$ = Mr. Reid's speed (in mph)

$5x = 4(x + 9)$

$5x = 4x + 36$

$-4x + 5x = -4x + 4x + 36$

$x = 36$

$x + 9 = 36 + 9 = 45$

a. Mr. Robinson's speed is 36 mph.

b. Mr. Reid's speed is 45 mph.

c. The distance is $\left(36\,\dfrac{mi}{hr}\right)(5\ hr) = 180$ mi.

10. a. 45 mph   **b.** 54 mph   **c.** 270 mi

11. Let $\quad x = $ speed of boat (in mph) in still water
Then $x + 3 = $ speed downstream (in mph)
and $\quad x - 3 = $ speed upstream (in mph)
$$3(x + 3) = 4(x - 3)$$
$$3x + 9 = 4x - 12$$
$$-3x + 3x + 9 = -3x + 4x - 12$$
$$9 = x - 12$$
$$9 + 12 = x - 12 + 12$$
$$21 = x$$
$$x = 21$$

The speed of the boat in still water was $21\,\dfrac{\text{mi}}{\text{hr}}$.

12. 6 ft per sec

13. Let $\quad x = $ speed of boat (in mph) in still water
Then $x + 4 = $ speed downstream (in mph)
and $\quad x - 4 = $ speed upstream (in mph)
$$3(x + 4) = 4(x - 4) + 6$$
$$3x + 12 = 4x - 16 + 6$$
$$3x + 12 = 4x - 10$$
$$-3x + 3x + 12 = -3x + 4x - 10$$
$$12 = x - 10$$
$$12 + 10 = x - 10 + 10$$
$$22 = x$$
$$x = 22$$

The speed of the boat in still water was $22\,\dfrac{\text{mi}}{\text{hr}}$.

14. 24 mph

15. Let $\quad t = $ number of hours driven on Saturday
Then $\quad 17 - t = $ number of hours driven the next day
and $\quad 54t = $ distance driven on Saturday
and $48(17 - t) = $ distance driven the next day
$$54t = 48(17 - t) \quad \text{The distances are equal.}$$
$$54t = 816 - 48t$$
$$54t + 48t = 816 - 48t + 48t$$
$$102t = 816$$
$$\frac{102t}{102} = \frac{816}{102}$$
$$t = 8$$

a. It took Matthew 8 hr to get to his friend's house.

b. The distance is $\left(54\,\dfrac{\text{mi}}{\text{hr}}\right)(8\ \text{hr}) = 432$ mi.

16. a. 3 hr   **b.** 48 mi

Exercises 4.8   (page 224)

1. Let $\quad x = $ number of pounds of apple chunks
Then $10 - x = $ number of pounds of granola
$$4.2x + 2.2(10 - x) = 3(10)$$
$$42x + 22(10 - x) = 30(10)$$
$$42x + 220 - 22x = 300$$
$$20x + 220 = 300$$
$$20x + 220 - 220 = 300 - 220$$
$$20x = 80$$
$$\frac{20x}{20} = \frac{80}{20}$$
$$x = 4$$
$$10 - x = 10 - 4 = 6$$

The merchant should use 4 lb of apple chunks and 6 lb of granola.

2. 23 lb of macadamia nuts and 27 lb of peanuts

3. Let $\quad x = $ number of pounds of apples at 95¢ per pound
Then $30 - x = $ number of pounds of apples at 65¢ per pound
$$95x + 65(30 - x) = 78(30)$$
$$95x + 1{,}950 - 65x = 2{,}340$$
$$30x + 1{,}950 = 2{,}340$$
$$30x + 1{,}950 - 1{,}950 = 2{,}340 - 1{,}950$$
$$30x = 390$$
$$\frac{30x}{30} = \frac{390}{30}$$
$$x = 13$$
$$30 - x = 30 - 13 = 17$$

Alice should use 13 lb of the more expensive apples and 17 lb of the cheaper ones.

4. 35 lb of English toffee and 25 lb of peanut brittle

5. Let $\quad x = $ number of pounds of gumdrops
Then $27 - x = $ number of pounds of caramels
$$2.8x + 2.2(27 - x) = 66$$
$$28x + 22(27 - x) = 660$$
$$28x + 594 - 22x = 660$$
$$6x + 594 = 660$$
$$6x + 594 - 594 = 660 - 594$$
$$6x = 66$$
$$\frac{6x}{6} = \frac{66}{6}$$
$$x = 11$$
$$27 - x = 27 - 11 = 16$$

Margie should use 11 lb of gumdrops and 16 lb of caramels.

6. 23 lb of Brand C and 17 lb of Brand D

7. Let $x = $ number of pounds of candy
(There will be $x + 50$ pounds of the mixture.)
$$3x + 2.4(50) = 2.5(x + 50)$$
$$30x + 24(50) = 25(x + 50)$$
$$30x + 1{,}200 = 25x + 1{,}250$$
$$-25x + 30x + 1{,}200 = -25x + 25x + 1{,}250$$
$$5x + 1{,}200 = 1{,}250$$
$$5x + 1{,}200 - 1{,}200 = 1{,}250 - 1{,}200$$
$$5x = 50$$
$$\frac{5x}{5} = \frac{50}{5}$$
$$x = 10$$

Dorothy should use 10 lb of candy.

8. 20 lb of cashews

Exercises 4.9   (page 229)

(The checks will not be shown.)

1. What is 20% of 16 mL?   **2.** 0.4 L
Let $x = $ unknown number
$x = (0.20)(16\ \text{mL}) = 3.2$ mL
There is 3.2 mL of pure alcohol in the solution.

3. 43 is what percent of 500?   **4.** 5.4%
Let $x = $ percent
$$500x = 43$$
$$x = \tfrac{43}{500} = 0.086 = 8.6\%$$
The solution is an 8.6% solution.

5. 24 is 40% of what number?   **6.** 50 mL
Let $x = $ unknown number
$$0.40x = 24$$
$$x = \tfrac{24}{0.40} = 60$$
There is 60 mL of solution.

7. Let x = number of cc of 20% solution   **8.** 20 pt
(There will be [100 + x] cc altogether.)
$$0.50(100) + 0.20(x) = 0.25(100 + x)$$
$$50(100) + 20(x) = 25(100 + x)$$
$$5,000 + 20x = 2,500 + 25x$$
$$5,000 + 20x - 25x = 2,500 + 25x - 25x$$
$$5,000 - 5x = 2,500$$
$$-5,000 + 5,000 - 5x = -5,000 + 2,500$$
$$-5x = -2,500$$
$$\frac{-5x}{-5} = \frac{-2,500}{-5}$$
$$x = 500$$
500 cc of the 20% solution should be added.

9. Let x = number of mL of water   **10.** 120 mL
(There will be [x + 500] mL altogether.)
$$0.40(500) = 0.25(x + 500)$$
$$40(500) = 25(x + 500)$$
$$20,000 = 25x + 12,500$$
$$20,000 - 12,500 = 25x + 12,500 - 12,500$$
$$7,500 = 25x$$
$$\frac{7,500}{25} = \frac{25x}{25}$$
$$x = 300$$
300 mL of water should be added.

11. Let x = number of L of pure alcohol   **12.** 5 L
(There will be [x + 10] L altogether.)
$$1x + 0.2(10) = 0.5(x + 10)$$
$$10x + 2(10) = 5(x + 10)$$
$$10x + 20 = 5x + 50$$
$$-5x + 10x + 20 = -5x + 5x + 50$$
$$5x + 20 = 50$$
$$5x + 20 - 20 = 50 - 20$$
$$5x = 30$$
$$\frac{5x}{5} = \frac{30}{5}$$
$$x = 6$$
6 L of pure alcohol should be added.

Exercises 4.10   (page 230)

(The checks will not be shown.)

1. Use the formula $I = Prt$, with $I = 756$, $r = 0.07$, and $t = 3$.
Let P = amount invested originally
$$I = Prt$$
$$756 = P(0.07)(3)$$
$$756 = 0.21P$$
$$\frac{756}{0.21} = \frac{0.21P}{0.21}$$
$$P = 3,600$$
Roy had invested $3,600 originally.

2. 200 members

3. Use the formula $V = \pi r^2 h$, with $V = 100\pi$ and $r = 5$.
Let h = height (in in.)
$$V = \pi r^2 h$$
$$100\pi = \pi(5^2)h$$
$$\frac{100\pi}{25\pi} = \frac{25\pi h}{25\pi}$$
$$4 = h$$
$$h = 4$$
The height of the cylinder is 4 in.

4. 720 cu. in.

5. Let x = amount invested at 5% interest
Then $5,000 - x$ = amount invested at 6% interest
$$0.05x + 0.06(5,000 - x) = 272$$
$$5x + 6(5,000 - x) = 27,200$$
$$5x + 30,000 - 6x = 27,200$$
$$-x + 30,000 = 27,200$$
$$-x + 30,000 - 30,000 = 27,200 - 30,000$$
$$-x = -2,800$$
$$x = 2,800$$
$2,800 was invested at 5%.

6. 120 mi

7. Let x = measure of the supplement of angle A
Then $4x$ = measure of angle A
$$x + 4x = 180$$
$$5x = 180$$
$$\frac{5x}{5} = \frac{180}{5}$$
$$x = 36$$
$$4x = 4(36) = 144$$
The measure of angle A is 144°.

8. $7.20 per hour.

9. Let x = larger number
Then $x - 16$ = smaller number
$$x + (x - 16) = 84$$
$$2x - 16 = 84$$
$$2x - 16 + 16 = 84 + 16$$
$$2x = 100$$
$$\frac{2x}{2} = \frac{100}{2}$$
$$x = 50 \qquad \text{The larger number}$$
$$x - 16 = 50 - 16 = 34 \quad \text{The smaller number}$$

Sections 4.1–4.10 Review Exercises   (page 232)

(The checks will not be shown.)

1. Let $6x$ = one number   **2.** $1,102.50
Let $7x$ = other number
$$6x + 7x = 52$$
$$13x = 52$$
$$\frac{\cancel{13}x}{\cancel{13}} = \frac{52}{13}$$
$$x = 4$$
$$6x = 6(4) = 24 \quad \text{One number}$$
$$7x = 7(4) = 28 \quad \text{Other number}$$

3. Let x = number   **4.** 37.5%
$$0.31x = 77.5$$
$$x = \tfrac{77.5}{0.31} = 250$$

5. The amount of increase is 5% of $380.
Let x = increase
$$x = 0.05(\$380) = \$19$$
New rent = $380 + $19 = $399

6. 9 oz lead and 15 oz tin

7. Let $3x$ = length of shortest side
Let $4x$ = length of next side
Let $5x$ = length of longest side
$$3x + 4x + 5x = 108$$
$$12x = 108$$
$$\frac{12x}{12} = \frac{108}{12}$$
$$x = 9$$
$$3x = 3(9) = 27 \quad \text{The shortest side}$$
$$4x = 4(9) = 36 \quad \text{The next side}$$
$$5x = 5(9) = 45 \quad \text{The longest side}$$

8. 15 $10 bills; 7 $50 bills

9. Let $\quad x$ = number of 10¢ stamps
Then $\quad 2x$ = number of 12¢ stamps
and $2(2x)$ = number of 2¢ stamps
$10x + 12(2x) + 2(4x) = 210$
$\qquad 10x + 24x + 8x = 210$
$\qquad\qquad\quad 42x = 210$
$$\frac{\overset{1}{\cancel{42}}x}{\underset{1}{\cancel{42}}} = \frac{210}{42}$$
$\qquad\qquad x = 5 \qquad$ 10¢ stamps
$\qquad\quad 2x = 2(5) = 10 \quad$ 12¢ stamps
$\qquad\quad 4x = 4(5) = 20 \quad$ 2¢ stamps

10. 1,000 box seats, 5,500 reserved seats, 22,000 general admission seats

11. We first find 24% of 1,800.
Let x = that number
$x = 0.24(1,800) = 432$
The number of minority students needed is $432 - 256 = 176$.

12. 5 mph

13. Let $\quad x$ = Mrs. Koontz's speed (in mph)
Then $x + 5$ = Mrs. Fowler's speed (in mph)
$\qquad\quad 10x = 9(x + 5)$
$\qquad\quad 10x = 9x + 45$
$\quad 10x - 9x = 9x + 45 - 9x$
$\qquad\qquad x = 45$
$\qquad x + 5 = 45 + 5 = 50$

   a. Mrs. Koontz's speed is 45 mph.

   b. Mrs. Fowler's speed is 50 mph.

   c. The distance is $\left(45\ \dfrac{mi}{hr}\right)(10\ hr) = 450\ mi$.

14. 12 lb at 85¢ per pound and 18 lb at 95¢ per pound

15. Let x = number of cc of water added
$\qquad 0.25(500) + 0 = 0.05(500 + x)$
$\qquad\quad 25(500) = 5(500 + x)$
$\qquad\quad 12,500 = 2,500 + 5x$
$-2,500 + 12,500 = -2,500 + 2,500 + 5x$
$\qquad\quad 10,000 = 5x$
$$\frac{10,000}{5} = \frac{5x}{5}$$
$\qquad\qquad x = 2,000$
2,000 cc of water must be added.

Chapter 4 Diagnostic Test   (page 235)

Following each problem number is the number (in parentheses) of the textbook section where that kind of problem is discussed.

   1. (4.6) 28 is 40% of what number?
   Let x = unknown number
   $0.40x = 28$
   $\quad x = \frac{28}{0.40} = 70\ mL$
   Check: 40% of 70 mL = $(0.40)(70\ mL)$
   $\qquad\qquad\qquad\quad = 28\ mL$

   2. (4.6) 17 is what percent of 20?
   Let x = percent $\qquad$ *Check*: 85% of 20 = $(0.85)(20)$
   $20x = 17 \qquad\qquad\qquad\qquad\qquad = 17$
   $\quad x = \frac{17}{20} = 0.85 = 85\%$

3. (4.3) Let x = unknown number
$\qquad\qquad 16 + 3x = 37$
$-16 + 16 + 3x = -16 + 37 \quad$ *Check*: $\quad 16 + 3x \overset{?}{=} 37$
$\qquad\qquad\quad 3x = 21 \qquad\qquad\qquad\qquad 16 + 3(7) \overset{?}{=} 37$
$\qquad\qquad\quad \frac{3x}{3} = \frac{21}{3} \qquad\qquad\qquad\qquad 16 + 21 \overset{?}{=} 37$
$\qquad\qquad\qquad x = 7 \qquad\qquad\qquad\qquad\qquad\quad 37 = 37$
The number is 7.

4. (4.4) Let $\qquad x$ = number of $10 bills
$\qquad\quad$ Then $25 - x$ = number of $5 bills
$\qquad\quad$ Therefore, $10x$ = value of the $10 bills
$\qquad\quad$ and $5(25 - x)$ = value of the $5 bills
$\qquad 10x + 5(25 - x) = 165$
$\qquad 10x + 125 - 5x = 165$
$\qquad\qquad 5x + 125 = 165$
$\quad 5x + 125 - 125 = 165 - 125$
$\qquad\qquad\qquad 5x = 40$
$$\qquad\qquad\qquad \frac{5x}{5} = \frac{40}{5}$$
$\qquad\qquad\qquad\quad x = 8$
$\qquad\quad 25 - x = 25 - 8 = 17$
Check: 8($10) = $\quad$ $80
$\qquad\quad$ 17($5) = $\quad$ $85
$\qquad\qquad\qquad\qquad$ $165
Therefore, Ellen has 8 $10 bills and 17 $5 bills.

5. (4.8) Let $\qquad x$ = number of lb of apricots
$\qquad\quad$ Then $50 - x$ = number of lb of granola
$\qquad\quad 2.70x + 2.20(50 - x) = 2.34(50)$
$\qquad\quad 270x + 220(50 - x) = 234(50)$
$\qquad 270x + 11,000 - 220x = 11,700$
$\qquad\qquad 50x + 11,000 = 11,700$
$\quad 50x + 11,000 - 11,000 = 11,700 - 11,000$
$\qquad\qquad\qquad 50x = 700$
$$\qquad\qquad\qquad \frac{50x}{50} = \frac{700}{50}$$
$\qquad\qquad\qquad\quad x = 14$
$\qquad\qquad 50 - x = 50 - 14 = 36$
Check: $2.70(14) = \quad $37.80 \qquad\qquad $2.34(50) = $117.00
$\qquad\quad $2.20(36) = \quad $79.20
$\qquad\qquad\qquad\qquad\qquad $117.00
There should be 14 lb of apricots and 36 lb of granola.

6. (4.7) Let $\qquad x$ = Kevin's average speed (in mph)
$\qquad\quad$ Then $x + 9$ = Jason's average speed (in mph)
$\qquad\quad$ Therefore, $6x$ = Kevin's distance
$\qquad\quad$ and $5(x + 9)$ = Jason's distance
$\qquad\qquad 6x = 5(x + 9) \quad$ The distances are equal.
$\qquad\qquad 6x = 5x + 45$
$\quad 6x - 5x = 5x + 45 - 5x$
$\qquad\qquad x = 45$
$\qquad x + 9 = 45 + 9 = 54$

Check: $\left(45\ \dfrac{mi}{hr}\right)(6\ hr) = 270\ mi$

$\qquad\qquad \left(54\ \dfrac{mi}{hr}\right)(5\ hr) = 270\ mi$

   a. Kevin's speed is 45 mph.

   b. Jason's speed is 54 mph.

   c. The distance is $\left(45\ \dfrac{mi}{hr}\right)(6\ hr) = 270\ mi$.

4

7. (4.8) Let x = number of lb of nuts to be added
(There will be $(x + 30)$ lb in the mixture.)
$$3.50x + 8.00(30) = 4.85(x + 30)$$
$$350x + 800(30) = 485(x + 30)$$
$$350x + 24,000 = 485x + 14,550$$
$$-485x + 350x + 24,000 = -485x + 485x + 14,550$$
$$-135x + 24,000 = 14,550$$
$$-135x + 24,000 - 24,000 = 14,550 - 24,000$$
$$-135x = -9,450$$
$$\frac{-135x}{-135} = \frac{-9,450}{-135}$$
$$x = 70 \text{ lb of nuts to be added}$$
(The mixture will contain 70 lb + 30 lb, or 100 lb.)
$$Check: 3.50 \frac{\text{dollars}}{\text{lb}} (70 \text{ lb}) + 8.00 \frac{\text{dollars}}{\text{lb}} (30 \text{ lb})$$
$$\overset{?}{=} 4.85 \frac{\text{dollars}}{\text{lb}} (100 \text{ lb})$$
$$\$245 + \$240 \overset{?}{=} \$485$$
$$\$485 = \$485$$

8. (4.9) Let x = number of mL of 80% solution to be added
(There will be $[x + 140]$ mL altogether.)
$$0.80x + 0.30(140) = 0.45(x + 140)$$
$$80x + 30(140) = 45(x + 140)$$
$$80x + 4,200 = 45x + 6,300$$
$$-45x + 80x + 4,200 = -45x + 45x + 6,300$$
$$35x + 4,200 - 4,200 = 6,300 - 4,200$$
$$35x = 2,100$$
$$\frac{35x}{35} = \frac{2,100}{35}$$
$$x = 60$$
$$Check: 0.80(60 \text{ mL}) + 0.30(140 \text{ mL}) = 48 \text{ mL} + 42 \text{ mL}$$
$$= 90 \text{ mL}$$
$$0.45(60 \text{ mL} + 140 \text{ mL}) = 0.45(200 \text{ mL})$$
$$= 90 \text{ mL}$$
60 mL of the 80% solution should be added.

9. (4.5) Let $7x$ = length of shortest side (in m)
Let $9x$ = length of next side (in m)
Let $11x$ = length of longest side (in m)
$$7x + 9x + 11x = 135$$
$$27x = 135$$
$$\frac{\overset{1}{\cancel{27}}x}{\underset{1}{\cancel{27}}} = \frac{135}{27}$$
$$x = 5$$
$$7x = 7(5) = 35 \text{ m}$$
$$9x = 9(5) = 45 \text{ m}$$
$$11x = 11(5) = 55 \text{ m}$$
$$Check: 35 \text{ m} + 45 \text{ m} + 55 \text{ m} \overset{?}{=} 135 \text{ m}$$
$$135 \text{ m} = 135 \text{ m}$$
The lengths are 35 m, 45 m, and 55 m.

10. (4.6) We must first find the markup, which is 30% of $240.
Let x = markup
$x = 0.30(\$240) = \$72$
The selling price is $240 + $72 = $312.

Chapters 1–4 Cumulative Review Exercises (page 235)

1. Not defined   **2.** 98

3. $46 - 2\{4 - [3(5 - 8) - 10]\} = 46 - 2\{4 - [3(-3) - 10]\}$
$= 46 - 2\{4 - [-9 - 10]\} = 46 - 2\{4 - [-19]\}$
$= 46 - 2\{23\} = 46 - 46 = 0$

4. 45   **5.** $S = 4\pi r^2$ **6.** 11
$$S \approx 4(3.14)(5^2)$$
$$S \approx 4(3.14)(25)$$
$$S \approx 314$$

7. $8 + 12\{3 - 2(x + 4)\} = 8 + 12\{3 - 2x - 8\}$
$= 8 + 12\{-5 - 2x\} = 8 - 60 - 24x = -52 - 24x$

8. $-13x + 10$

9. $\quad 32 - 4x = 15$ **10.** -33
$$-32 + 32 - 4x = -32 + 15$$
$$-4x = -17$$
$$\frac{-4x}{-4} = \frac{-17}{-4}$$
$$x = \tfrac{17}{4}, \text{ or } 4\tfrac{1}{4}$$
The solution is $4\tfrac{1}{4}$.

11. $\quad 12w - 12 = -8w + 3$ **12.** -3
$$8w + 12w - 12 = 8w - 8w + 3$$
$$20w - 12 = 3$$
$$20w - 12 + 12 = 3 + 12$$
$$20w = 15$$
$$\frac{20w}{20} = \frac{15}{20}$$
$$w = \tfrac{3}{4}$$
The solution is $\tfrac{3}{4}$.

13. $\quad 6z - 22 = 2[8 - 20z + 4]$ **14.** 0
$$6z - 22 = 2[12 - 20z]$$
$$6z - 22 = 24 - 40z$$
$$40z + 6z - 22 = 40z + 24 - 40z$$
$$46z - 22 = 24$$
$$46z - 22 + 22 = 24 + 22$$
$$46z = 46$$
$$\frac{46z}{46} = \frac{46}{46}$$
$$z = 1$$
The solution is 1.

15. 0   **16.** Yes   **17.** Yes   **18.** Yes   **19.** 1

20. 88 and 72   **21.** Let $\quad x$ = number of quarters
Then $15 - x$ = number of nickels
$$25x + 5(15 - x) = 175$$
$$25x + 75 - 5x = 175$$
$$20x + 75 = 175$$
$$20x + 75 - 75 = 175 - 75$$
$$20x = 100$$
$$\frac{20x}{20} = \frac{100}{20}$$
$$x = 5$$
$$15 - x = 15 - 5 = 10$$
Patty has 5 quarters and 10 nickels.

22. 60 mph

23. Let $\quad x$ = number of lb of candy at $1.80 per lb
Then $10 - x$ = number of lb of candy at $1.20 per lb
$$1.80x + 1.20(10 - x) = 15.90$$
$$180x + 120(10 - x) = 1,590$$
$$180x + 1,200 - 120x = 1,590$$
$$60x + 1,200 = 1,590$$
$$60x + 1,200 - 1,200 = 1,590 - 1,200$$
$$60x = 390$$
$$\frac{60x}{60} = \frac{390}{60}$$
$$x = 6.5$$
$$10 - x = 10 - 6.5 = 3.5$$
There should be 6.5 lb of the $1.80 per lb candy and 3.5 lb of the $1.20 per lb candy.

24. 20 L

25. Let $\quad x =$ first even integer

Then $x + 2 =$ second even integer

and $\quad x + 4 =$ third even integer

and $\quad x + 6 =$ fourth even integer

$$x + x + 2 + x + 4 + x + 6 = 108$$
$$4x + 12 = 108$$
$$4x + 12 - 12 = 108 - 12$$
$$4x = 96$$
$$\frac{4x}{4} = \frac{96}{4}$$
$$x = 24$$
$$x + 2 = 24 + 2 = 26$$
$$x + 4 = 24 + 4 = 28$$
$$x + 6 = 24 + 6 = 30$$

The integers are 24, 26, 28, and 30.

Exercises 5.1   (page 239)

1. $x^{3+4} = x^7$ **2.** x^{11} **3.** $y^{1+3} = y^4$ **4.** z^5

5. $m^{2+1} = m^3$ **6.** a^4 **7.** $10^{2+3} = 10^5$ **8.** 10^7

9. $5^{1+4+5} = 5^{10}$ **10.** 3^6 **11.** $x^{1+3+4} = x^8$ **12.** y^9

13. $x^2 y^5$ **14.** $a^3 b^2$ **15.** $3^2 \cdot 5^3$ or 1,125

16. $2^3 \cdot 3^2$ or 72   **17.** $a^4 + a^2$ (cannot be rewritten)

18. $x^3 + x^4$ **19.** a^{x+w} **20.** x^{a+b}

21. $x^y y^x$ (cannot be rewritten)   **22.** $a^b b^a$ **23.** $x^{2+5} y^3 = x^7 y^3$

24. $w^2 z^7$ **25.** $a^{2+5} b^3 = a^7 b^3$ **26.** $x^{12} y$

27. $s^7 + s^4$ (cannot be rewritten)   **28.** $t^3 + t^8$

29. $7^{3+5} = 7^8$ **30.** 6^{19} **31.** $2x^3$ (terms are like terms)

32. $2y^6$ **33.** $2(2^{14}) = 2^1(2^{14}) = 2^{1+14} = 2^{15}$ **34.** $2(3^{18})$

35. $2z^7$ (terms are like terms)   **36.** $2w^2$

Exercises 5.2   (page 241)

1. $-(2 \cdot 4)(a \cdot a^2) = -8a^3$ **2.** $-15x^4$

3. $+(5 \cdot 6)(h^2 h^3) = 30h^5$ **4.** $48k^4$

5. $(-5x^3)(-5x^3) = +(5 \cdot 5)(x^3 x^3) = 25x^6$ **6.** $49y^8$

7. $+(2 \cdot 4 \cdot 3)(a^3 a a^4) = 24a^8$ **8.** $-48b^6$

9. $+(9 \cdot 2)(mm^5 m^2) = 18m^8$ **10.** $28n^7$

11. $-(5 \cdot 7)x^2 y = -35x^2 y$ **12.** $-12x^3 y$

13. $+(6 \cdot 4)(m^3 m)(n^2 n^2) = 24m^4 n^4$ **14.** $-40h^6 k^4$

15. $-(2 \cdot 3)(x^{10} \cdot x^{12})(y^2 \cdot y^7) = -6x^{22} y^9$ **16.** $10a^{12} b^{15}$

17. $(3xy^2)(3xy^2) = (3 \cdot 3)(x \cdot x)(y^2 \cdot y^2) = 9x^2 y^4$ **18.** $16x^4 y^6$

19. $-(5)(x^4)(y^5 \cdot y^4)(z \cdot z^7) = -5x^4 y^9 z^8$ **20.** $-21E^2 F^{11} G^{18}$

21. $-(2^3 \cdot 2^2)(R \cdot R^5)S^2 T^4 = -32R^6 S^2 T^4$ **22.** $-243x^9 yz^5$

23. $+(5 \cdot 4)(c^2 \cdot c^5)(d \cdot d)(e^3 \cdot e^2) = 20c^7 d^2 e^5$ **24.** $24m^4 n^7 r^5$

25. $-(2 \cdot 3 \cdot 7)(x^2 \cdot x)(y^2 \cdot y)(z \cdot z) = -42x^3 y^3 z^2$

26. $420x^4 y^3 z^5$ **27.** $-(3 \cdot 5)(x \cdot x^2 \cdot x^3)(y \cdot y^2 \cdot y^3) = -15x^6 y^6$

28. $-6x^2 y^2 z^2$ **29.** $-(2 \cdot 5)(a \cdot a)(b \cdot b)(c \cdot c) = -10a^2 b^2 c^2$

30. $6x^3 y^3 z^3$ **31.** $(5 \cdot 2)(x^2 \cdot x)(y \cdot y)(z^3 \cdot z) = 10x^3 y^2 z^4$

32. $6a^5 b^7$

33. $-(5 \cdot 7)(x^2 \cdot x^5 \cdot x)(y^2 \cdot y^3 \cdot y)(z \cdot z \cdot z^5) = -35x^8 y^6 z^7$

34. $-32R^9 S^7 T^{13}$

35. $-(3 \cdot 7)(h^2 \cdot h^4)(k^1 \cdot k^5)(m^3 \cdot m^1) = -21h^6 k^6 m^4$

36. $-48k^3 m^3 n^3$

Exercises 5.3   (page 244)

1. $(-3x^2)(2x^2) + (-3x^2)(-4x) + (-3x^2)(5)$
$\quad = -6x^4 + 12x^3 - 15x^2$

2. $-15x^5 + 10x^4 + 35x^3$

3. $(4x)(3x^2) + (4x)(-6) = 12x^3 - 24x$ **4.** $15x^3 - 30x$

5. $(-2x)(5x^2) + (-2x)(3x) + (-2x)(-4) = -10x^3 - 6x^2 + 8x$

6. $-8x^3 + 20x^2 - 12x$

7. $(y^2)(7y^3) + (-4y)(7y^3) + (3)(7y^3) = 7y^5 - 28y^4 + 21y^3$

8. $-45z^2 + 5z^3 - 10z^4$

9. $(2x^2)(4x) + (-3x)(4x) + (5)(4x) = 8x^3 - 12x^2 + 20x$

10. $15w^3 + 10w^2 - 40w$ **11.** $(x)(xy) + (x)(-3) = x^2 y - 3x$

12. $a^2 b - 4a$ **13.** $(3a)(ab) + (3a)(-2a^2) = 3a^2 b - 6a^3$

14. $12x^2 - 8xy^2$

15. $(-2x)(-3y) + (4x^2 y)(-3y) = 6xy - 12x^2 y^2$

16. $6az - 4a^2 z^2$

17. $(-2xy)(x^2 y) + (-2xy)(-y^2 x) + (-2xy)(-y) + (-2xy)(-5)$
$\quad = -2x^3 y^2 + 2x^2 y^3 + 2xy^2 + 10xy$

18. $-24ab + 3a^3 b + 3ab^3 - 3a^2 b^2$

19. $(3x^3)(-2xy) + (-2x^2 y)(-2xy) + (y^3)(-2xy)$
$\quad = -6x^4 y + 4x^3 y^2 - 2xy^4$

20. $-8yz^4 + 2y^2 z^3 + 2y^4 z$

21. $(2xy^2 z)(-5xz^3) + (-7x^2 z^2)(-5xz^3) = -10x^2 y^2 z^4 + 35x^3 z^5$

22. $-9a^3 bc^4 + 12a^2 b^3 c^3$

23. $(5x^2 y^3 z)(-4xz^2) + (-2xz^3)(-4xz^2) + (y^4)(-4xz^2)$
$\quad = -20x^3 y^3 z^3 + 8x^2 z^5 - 4xy^4 z^2$

24. $-12x^3 y^3 z^2 + 6xy^3 z^3 + 8x^2 y^2 z^4$

25. $6z^2 + (-3z^2)(4) + (-3z^2)(-3z) + (-3z^2)(-2z^3)$
$\quad = 6z^2 - 12z^2 + 9z^3 + 6z^5 = -6z^2 + 9z^3 + 6z^5$

26. $9x^3 - 12x^4 + 8x^5$

27. $5x - 2x(3x^2) + (-2x)(7x) + (-2x)(-3)$
$\quad = \underline{5x} - 6x^3 - 14x^2 + \underline{6x} = 11x - 14x^2 - 6x^3$

28. $35y - 18y^2$ **29.** $3x^2 \underline{- 4x} + 2 \underline{- 5x} = 3x^2 - 9x + 2$

30. $14y^2 + 5y - 4$

31. $(3x^2)(-5x) - 4x(-5x) + 2(-5x) = -15x^3 + 20x^2 - 10x$

32. $42y^3 + 6y^2 - 12y$ **33.** $(3)(2)x^3 xy = 6x^4 y$ **34.** $30x^3 z^2$

35. $4x^2(2x^2) + (4x^2)(-3x) + (4x^2)(7) + (-4x)(x^3) + (-4x)(2x)$
$\quad + (-4x)(-5)$
$\quad = \underline{8x^4} - 12x^3 + \underline{\underline{28x^2}} - 4x^4 - \underline{\underline{8x^2}} + 20x$
$\quad = 4x^4 - 12x^3 + 20x^2 + 20x$

36. $-15y^4 + 3y^3 + 9y^2 + 3y$

37. $8x^2(7x^2) + 8x^2(-3x) + 8x^2(1) + (-5x)(6x^2) + (-5x)(-8x)$
$\quad + (-5x)(9)$
$\quad = 56x^4 \underline{- 24x^3} + \underline{\underline{8x^2}} - 30x^3 + \underline{\underline{40x^2}} - 45x$
$\quad = 56x^4 - 54x^3 + 48x^2 - 45x$

38. $42a^5 + 18a^4 + 33a^3 - 15a^2$

39. $3x - 5x[8 + (-2)(4x) + (-2)(-y)]$
$\quad = 3x - 5x[8 + (-8x) + 2y]$
$\quad = 3x + (-5x)(8) + (-5x)(-8x) + (-5x)(2y)$
$\quad = 3x - 40x + 40x^2 - 10xy = -37x + 40x^2 - 10xy$

40. $-4x + 6x^2 - 24xy$

41. $3 - 2[7x^2 + (-3x)(8x) + (-3x)(2)]$
$\quad = 3 - 2(7x^2 - 24x^2 - 6x)$
$\quad = 3 - 2(-17x^2 - 6x) = 3 + (-2)(-17x^2) + (-2)(-6x)$
$\quad = 34x^2 + 12x + 3$

42. $-4y^2 - 32y + 7$

43. $8x - 2x[6 - 3(2x) + (-3)(-1)] = 8x - 2x[6 - 6x + 3]$
$\quad = 8x - 2x[9 - 6x] = 8x + (-2x)(9) + (-2x)(-6x)$
$\quad = \underline{8x} - 18x + 12x^2 = -10x + 12x^2$

44. $7y + 12y^2$

45. $7x + 4x(3y) + 4x(-5) + 4x(8x) = \underline{7x} + 12xy \underline{- 20x} + 32x^2$
$\quad = 32x^2 + 12xy - 13x$

46. $65y + 32yz - 72y^2$

Exercises 5.4A (page 246)

1. $y^{2\cdot5} = y^{10}$ 2. N^{12} 3. $x^{8\cdot2} = x^{16}$ 4. z^{28}

5. $2^{4\cdot2} = 2^8$ 6. 3^8 7. x^5y^5 8. a^4b^4 9. 2^6c^6

10. 3^4x^4 11. $x^{4\cdot7} = x^{28}$ 12. v^{24} 13. $10^{2\cdot3} = 10^6$

14. 10^{14} 15. $(-6)^2 = 36$ 16. 144 17. $2(-27) = -54$

18. -24 19. $-4 \cdot 25 = -100$ 20. -48

Exercises 5.4B (page 250)

1. $x^{7-2} = x^5$ 2. y^2 3. $x^4 - x^2$ (cannot be rewritten)

4. $s^8 - s^3$ 5. $a^{5-1} = a^4$ 6. b^6 7. $10^{11-1} = 10^{10}$

8. 5^5 9. $\dfrac{x^8}{y^4}$ (cannot be rewritten) 10. $\dfrac{a^4}{b^3}$

11. $\dfrac{a^9}{a^4} = a^{9-4} = a^5$ 12. b^8

13. $\dfrac{a^3}{b^2}$ (cannot be simplified because the bases are different)

14. $\dfrac{x^5}{y^3}$ 15. $\dfrac{s^5}{s^4} = s^{5-4} = s^1 = s$ 16. t

17. $\dfrac{x^{5a}}{x^{3a}} = x^{5a-3a} = x^{2a}$ 18. M^{4x} 19. $\dfrac{y^{6c}}{y^{4c}} = y^{6c-4c} = y^{2c}$

20. z^{2t} 21. $\dfrac{s^7}{t^7}$ 22. $\dfrac{x^9}{y^9}$ 23. $\dfrac{2^4}{x^4}$ 24. $\dfrac{3^2}{z^2}$ 25. $\dfrac{x^6}{2^6}$

26. $\dfrac{c^3}{5^3}$ 27. $a^{2\cdot2}b^{3\cdot2} = a^4b^6$ 28. $x^{12}y^{15}$

29. $2^{1\cdot2}z^{3\cdot2} = 2^2z^6$ 30. 3^3w^6 31. $\dfrac{x^{1\cdot2}y^{4\cdot2}}{z^{2\cdot2}} = \dfrac{x^2y^8}{z^4}$

32. $\dfrac{a^9b^3}{c^6}$ 33. $\dfrac{5^{1\cdot4}y^{3\cdot4}}{2^{1\cdot4}x^{2\cdot4}} = \dfrac{5^4y^{12}}{2^4x^8}$ 34. $\dfrac{6^3b^{12}}{7^3c^6}$

35. $\left(\dfrac{2^2x^3}{y^5z}\right)^3 = \dfrac{2^{2\cdot3}x^{3\cdot3}}{y^{5\cdot3}z^{1\cdot3}} = \dfrac{2^6x^9}{y^{15}z^3}$ 36. $\dfrac{3^4a^8}{b^6c^2}$

37. $\dfrac{(-4)^2}{-4^2} = \dfrac{16}{-16} = -1$ 38. -1 39. $\dfrac{(-3)^3}{-3^3} = \dfrac{-27}{-27} = 1$

40. 1

Exercises 5.5A (page 256)

1. 1 2. 1 3. $10^{-4} = \dfrac{1}{10^4}$, or $\dfrac{1}{10{,}000}$ 4. $\dfrac{1}{12^2}$, or $\dfrac{1}{144}$

5. $5^{-2} = \dfrac{1}{5^2}$, or $\dfrac{1}{25}$ 6. $\dfrac{1}{2^5}$, or $\dfrac{1}{32}$ 7. $4^{-3} = \dfrac{1}{4^3}$, or $\dfrac{1}{64}$

8. $\dfrac{1}{13^2}$, or $\dfrac{1}{169}$ 9. $-10^{-5} = -\dfrac{1}{10^5}$, or $-\dfrac{1}{100{,}000}$

10. $-\dfrac{1}{4^3}$, or $-\dfrac{1}{64}$ 11. $-12^{-2} = -\dfrac{1}{12^2}$, or $-\dfrac{1}{144}$

12. $-\dfrac{1}{13^2}$, or $-\dfrac{1}{169}$ 13. $(-12)^{-2} = \dfrac{1}{(-12)^2} = \dfrac{1}{144}$

14. $\dfrac{1}{169}$ 15. $(-5)^{-3} = \dfrac{1}{(-5)^3} = \dfrac{1}{-125} = -\dfrac{1}{125}$ 16. $-\dfrac{1}{64}$

17. 1 18. 1 19. 1 20. 1 21. $6r^0 = 6 \cdot 1 = 6$

22. 9 23. $\dfrac{1}{x^4}$ 24. $\dfrac{1}{y^7}$ 25. $(-8)t^{-2} = -8\left(\dfrac{1}{t^2}\right) = -\dfrac{8}{t^2}$

26. $-\dfrac{5}{y^2}$ 27. $(-2z)^{-3} = \dfrac{1}{(-2z)^3} = \dfrac{1}{-8z^3} = -\dfrac{1}{8z^3}$

28. $-\dfrac{1}{32a^5}$ 29. $(-8t)^{-2} = \dfrac{1}{(-8t)^2} = \dfrac{1}{64t^2}$ 30. $\dfrac{1}{25y^2}$

31. $-2z^{-3} = -2\left(\dfrac{1}{z^3}\right) = -\dfrac{2}{z^3}$ 32. $-\dfrac{2}{a^5}$

33. $5^0x^{-3} = 1\left(\dfrac{1}{x^3}\right) = \dfrac{1}{x^3}$ 34. $\dfrac{1}{y^4}$ 35. $\dfrac{1}{a^{-4}} = a^4$

36. b^5 37. $\left(\dfrac{x}{y}\right)^{-3} = \left(\dfrac{y}{x}\right)^3 = \dfrac{y^3}{x^3}$ 38. $\dfrac{t^2}{s^2}$

39. $\left(\dfrac{5}{x}\right)^{-2} = \left(\dfrac{x}{5}\right)^2 = \dfrac{x^2}{25}$ 40. $\dfrac{81}{t^4}$

41. $\dfrac{1}{(4x)^{-3}} = (4x)^3 = 4^3x^3$, or $64x^3$ 42. 6^2z^2, or $36z^2$

43. $(xy)^{-2} = \dfrac{1}{(xy)^2} = \dfrac{1}{x^2y^2}$ 44. $\dfrac{1}{a^4b^4}$

45. $r^{-4}st^{-2} = \dfrac{1}{r^4} \cdot \dfrac{s}{1} \cdot \dfrac{1}{t^2} = \dfrac{s}{r^4t^2}$ 46. $\dfrac{t}{r^5s^3}$

47. $ab^{-2}c^0 = \dfrac{a}{1} \cdot \dfrac{1}{b^2} \cdot 1 = \dfrac{a}{b^2}$ 48. $\dfrac{z}{x^3}$

49. $\dfrac{h^2}{k^{-4}} = \dfrac{h^2}{1} \cdot \dfrac{1}{k^{-4}} = \dfrac{h^2}{1} \cdot \dfrac{k^4}{1} = \dfrac{h^2k^4}{1} = h^2k^4$ 50. m^3n^2

51. $\dfrac{x^{-4}}{y} = \dfrac{x^{-4}}{1} \cdot \dfrac{1}{y} = \dfrac{1}{x^4} \cdot \dfrac{1}{y} = \dfrac{1 \cdot 1}{x^4y} = \dfrac{1}{x^4y}$ 52. $\dfrac{1}{a^5b}$

53. $\dfrac{a^3b^0}{c^{-2}} = \dfrac{a^3}{1} \cdot \dfrac{b^0}{1} \cdot \dfrac{1}{c^{-2}} = \dfrac{a^3}{1} \cdot \dfrac{1}{1} \cdot \dfrac{c^2}{1} = \dfrac{a^3 \cdot c^2}{1} = a^3c^2$

54. e^2f^3 55. $\dfrac{p^4r^{-1}}{t^{-2}} = \dfrac{p^4}{1} \cdot \dfrac{r^{-1}}{1} \cdot \dfrac{1}{t^{-2}} = \dfrac{p^4}{1} \cdot \dfrac{1}{r^1} \cdot \dfrac{t^2}{1} = \dfrac{p^4t^2}{r}$

56. $\dfrac{u^5w^3}{v^2}$

57. $\dfrac{-3x^{-2}y^{-1}}{z^{-3}} = \dfrac{-3}{1} \cdot \dfrac{x^{-2}}{1} \cdot \dfrac{y^{-1}}{1} \cdot \dfrac{1}{z^{-3}} = \dfrac{-3}{1} \cdot \dfrac{1}{x^2} \cdot \dfrac{1}{y^1} \cdot \dfrac{z^3}{1} = \dfrac{-3z^3}{x^2y}$,

or $-\dfrac{3z^3}{x^2y}$

58. $-\dfrac{5u^5}{s^3t^2}$

59. $\dfrac{-2uv^{-7}}{w^{-4}} = \dfrac{-2}{1} \cdot \dfrac{u}{1} \cdot \dfrac{v^{-7}}{1} \cdot \dfrac{1}{w^{-4}} = \dfrac{-2}{1} \cdot \dfrac{u}{1} \cdot \dfrac{1}{v^7} \cdot \dfrac{w^4}{1} = \dfrac{-2uw^4}{v^7}$,

or $-\dfrac{2uw^4}{v^7}$

60. $-\dfrac{4x^3z^2}{y^4}$ 61. $\dfrac{1}{x^2} = x^{-2}$ 62. y^{-3}

63. $\dfrac{h}{k} = \dfrac{h}{1} \cdot \dfrac{1}{k} = \dfrac{h}{1} \cdot \dfrac{k^{-1}}{1} = hk^{-1}$ 64. mn^{-1}

65. $\dfrac{x^2}{yz^5} = \dfrac{x^2}{1} \cdot \dfrac{1}{y} \cdot \dfrac{1}{z^5} = \dfrac{x^2}{1} \cdot \dfrac{y^{-1}}{1} \cdot \dfrac{z^{-5}}{1} = x^2y^{-1}z^{-5}$

66. $a^3b^{-2}c^{-1}$

Exercises 5.5B (page 260)

1. $x^{-3+4} = x^1 = x$ 2. y^4 3. $x^{2(-4)} = x^{-8} = \dfrac{1}{x^8}$

4. $\dfrac{1}{z^6}$ 5. $a^{-2(3)} = a^{-6} = \dfrac{1}{a^6}$ 6. $\dfrac{1}{b^{10}}$ 7. $y^{(-3)(-4)} = y^{12}$

8. s^{14} 9. $m^{-2(4)}n^{1(4)} = m^{-8}n^4 = \dfrac{1}{m^8} \cdot \dfrac{n^4}{1} = \dfrac{n^4}{m^8}$ 10. $\dfrac{r^5}{p^{15}}$

11. $x^{(-2)(-4)}y^{(3)(-4)} = x^8y^{-12} = \dfrac{x^8}{1} \cdot \dfrac{1}{y^{12}} = \dfrac{x^8}{y^{12}}$ 12. $\dfrac{w^6}{z^8}$

13. $x^{6-9} = x^{-3} = \dfrac{1}{x^3}$ 14. $\dfrac{1}{t^4}$ 15. $y^{-2-5} = y^{-7} = \dfrac{1}{y^7}$

16. $\frac{1}{z^4}$ 17. $d^{4-(-5)} = d^{4+(+5)} = d^9$ 18. c^7

19. $\left(\frac{1}{M^2N^3}\right)^4 = \frac{1^4}{M^{2\cdot4}N^{3\cdot4}} = \frac{1}{M^8N^{12}}$ 20. $R^{15}S^{12}$

21. $x^{3m+(-m)} = x^{2m}$ 22. y^{3n} 23. $x^{3b(-2)} = x^{-6b} = \frac{1}{x^{6b}}$

24. $\frac{1}{y^{6a}}$ 25. $x^{2a-(-5a)} = x^{2a+(+5a)} = x^{7a}$ 26. a^{8x}

27. $b^{-2n-3n} = b^{-5n} = \frac{1}{b^{5n}}$ 28. $\frac{1}{y^{5m}}$ 29. $10^{3+(-2)} = 10^1 = 10$

30. $\frac{1}{2}$ 31. $10^{4+(-2)} = 10^2$, or 100   32. 3

33. $10^{2(-1)} = 10^{-2} = \frac{1}{10^2}$, or $\frac{1}{100}$ 34. $\frac{1}{2^6}$, or $\frac{1}{64}$

35. $1 \cdot 49 = 49$ 36. 4^3, or 64   37. $10^{2-(-5)} = 10^{2+(+5)} = 10^7$

38. 2^5, or 32   39. $\frac{1}{10^2}$, or $\frac{1}{100}$ 40. 5^2, or 25

41. $10^{-3+2-5} = 10^{-6} = \frac{1}{10^6}$, or $\frac{1}{1,000,000}$ 42. $\frac{1}{2^3}$, or $\frac{1}{8}$

43. $3^{4-2-(-8)} = 3^{4-2+(+8)} = 3^{10}$ 44. 7^4

Exercises 5.6   (page 266)

1. 8,060   2. 31,400   3. 0.00132   4. 0.00082
5. 5.26   6. 9.11   7. 3.53×10^4 8. 8.25×10^5
9. 3.12×10^{-3} 10. 1.45×10^{-4} 11. 8.97×10^0
12. 2.497×10^0 13. 8.15×10^{-1} 14. 2.74×10^{-1}
15. 2×10^{-4} 16. 6×10^{-3} 17. 4.5×10^1
18. 1.2×10^1

19. $\frac{(5\times10^6)\times(3\times10^{-6})}{(6\times10^{-4})\times(2\times10^4)} = \frac{(5\times3)\times(10^6\times10^{-6})}{(6\times2)\times(10^{-4}\times10^4)} = \frac{\overset{5}{\cancel{15}}\times10^0}{\underset{4}{\cancel{12}}\times10^0}$
$= \frac{5}{4} = 1.25 \times 10^0$

20. 0.175, or 1.75×10^{-1} 21. 5.418×10^{11}
22. 3×10^{-10} 23. 9×10^{-4} 24. 1.5×10^{-3}

25. $\left(6.02 \times 10^{23} \frac{\text{molecules}}{\cancel{\text{mole}}}\right)(700 \; \cancel{\text{moles}})$

$= 4,214 \times 10^{23}$ molecules
$= (4.214 \times 10^3) \times 10^{23}$ molecules $= 4.214 \times 10^{26}$ molecules
26. \$9,120,000,000

27. $\frac{4,400,000,000 \text{ mi}}{12 \; \cancel{\text{yr}}} \times \frac{1 \; \cancel{\text{yr}}}{365 \; \cancel{\text{days}}} \times \frac{1 \; \cancel{\text{day}}}{24 \text{ hr}} \approx 41,857 \, \frac{\text{mi}}{\text{hr}}$
28. 1.60704×10^{10} mi, or 16,070,400,000 mi

Sections 5.1–5.6 Review Exercises   (page 268)

1. $m^{2+3} = m^5$ 2. y^8 3. $2^{1+2} = 2^3$ 4. x^{7y}
5. 10^{1+y} 6. a^2 7. $5x(x^2) + 5x(7) = 5x^3 + 35x$
8. $-10x^3y - 6x^2 + 2x$
9. $(-5)(-1)(-10)(-2)(ee^4e)(f^2f^7f)(g^3g^2) = 100e^6f^{10}g^5$
10. $27x^3$ 11. $x^{2\cdot4}y^{3\cdot4} = x^8y^{12}$ 12. n^{15}
13. $p^{-3\cdot5} = p^{-15} = \frac{1}{p^{15}}$ 14. $\frac{1}{16c^4}$ 15. 1   16. $\frac{t^3}{s^{12}}$
17. $(-2)^2x^{4\cdot2} = 4x^8$ 18. $-\frac{125}{a^{12}}$ 19. $\frac{1}{(-10)^3} = -\frac{1}{1,000}$
20. r^2 21. $\frac{1}{x^5x^4} = \frac{1}{x^9}$ 22. $\frac{c^4}{d^4}$ 23. $1\left(\frac{1}{m^{-3}}\right) = m^3$

24. $x^4 + x^2$ (cannot be simplified)

25. $n^{-6}\left(\frac{1}{n^0}\right) = \left(\frac{1}{n^6}\right)(1) = \frac{1}{n^6}$ 26. $\frac{x^{10}y^{15}}{32z^{20}}$

27. $\left(\frac{b^3c^0}{a^{-4}}\right)^5 = (b^3a^4)^5 = a^{20}b^{15}$ 28. $\frac{1}{x^8}$ 29. $\left(\frac{t^3}{s^5r^6}\right)^4 = \frac{t^{12}}{r^{24}s^{20}}$

30. 1   31. $5^{-2}a^{-6}b^8 = \frac{b^8}{5^2a^6}$, or $\frac{b^8}{25a^6}$ 32. $-63x^{12}y^8$

33. $2x(3x^2 - x) - 1(3x^2 - 4) = 6x^3 - 2x^2 - 3x^2 + 4$
$= 6x^3 - 5x^2 + 4$

34. $21m^3n^3 - 14m^2n^2$ 35. $\frac{1}{4^2}$, or $\frac{1}{16}$ 36. $\frac{1}{10^4}$, or $\frac{1}{10,000}$

37. $2^{0-(-3)} = 2^{0+(+3)} = 2^3$, or 8   38. 9

39. $\frac{(-8)^2}{-8^2} = \frac{64}{-64} = -1$ 40. a^3b^{-2}

41. $\frac{m^2}{1} \cdot \frac{1}{n^{-3}} = \frac{m^2}{1} \cdot \frac{n^3}{1} = \frac{m^2n^3}{1} = m^2n^3$ 42. $10^{-2}u^{-4}v^3w^5$

43. 4.53×10^4 44. 3.156×10^{-2}

Chapter 5 Diagnostic Test   (page 271)

Following each problem number is the number (in parentheses) of the textbook section where that kind of problem is discussed.

1. (5.2) $(-3xy)(5x^3y)(-2xy^4) = (-3)(5)(-2)(xx^3x)(yyy^4)$
$= 30x^5y^6$
(Because an even number of factors are negative, the answer is positive.)

2. (5.3) $2xy^2(x^2 - 3y - 4) = 2xy^2(x^2) + 2xy^2(-3y) + 2xy^2(-4)$
$= 2x^3y^2 - 6xy^3 - 8xy^2$

3. (5.1) $x^3 \cdot x^4 = x^{3+4} = x^7$ 4. (5.4) $(x^2)^3 = x^{2\cdot3} = x^6$

5. (5.4) $\frac{x^5}{x^2} = x^{5-2} = x^3$ 6. (5.5) $x^{-4} = \frac{1}{x^4}$

7. (5.5) $x^2y^{-3} = \frac{x^2}{1} \cdot \frac{1}{y^3} = \frac{x^2}{y^3}$

8. (5.5) $\frac{a^{-3}}{b} = \frac{a^{-3}}{1} \cdot \frac{1}{b} = \frac{1}{a^3} \cdot \frac{1}{b} = \frac{1}{a^3b}$

9. (5.4) $\frac{x^{5a}}{x^{3a}} = x^{5a-3a} = x^{2a}$ 10. (5.5) $(4^{3x})^0 = 1$

11. (5.4) $(x^2y^4)^3 = x^{2\cdot3}y^{4\cdot3} = x^6y^{12}$

12. (5.5) $(a^{-3}b)^2 = a^{-3\cdot2}b^{1\cdot2} = a^{-6}b^2 = \frac{1}{a^6} \cdot \frac{b^2}{1} = \frac{b^2}{a^6}$

13. (5.4) $\left(\frac{p^3}{q^2}\right)^2 = \frac{p^{3\cdot2}}{q^{2\cdot2}} = \frac{p^6}{q^4}$

14. (5.5) $\left(\frac{x}{y^2}\right)^{-3} = \frac{x^{1(-3)}}{y^{2(-3)}} = \frac{x^{-3}}{y^{-6}} = \frac{1}{x^3} \cdot \frac{y^6}{1} = \frac{y^6}{x^3}$

or $\left(\frac{x}{y^2}\right)^{-3} = \left(\frac{y^2}{x}\right)^3 = \frac{y^{2\cdot3}}{x^{1\cdot3}} = \frac{y^6}{x^3}$

15. (5.5) $\left(\frac{3x^{-2}}{y^{-3}}\right)^{-1} = \left(\frac{y^{-3}}{3x^{-2}}\right)^1 = \frac{y^{-3}}{1} \cdot \frac{1}{3} \cdot \frac{1}{x^{-2}} = \frac{1}{y^3} \cdot \frac{1}{3} \cdot \frac{x^2}{1}$

$= \frac{x^2}{3y^3}$ or $\left(\frac{3x^{-2}}{y^{-3}}\right)^{-1} = \left(\frac{3}{1} \cdot \frac{x^{-2}}{1} \cdot \frac{1}{y^{-3}}\right)^{-1} = \left(\frac{3}{1} \cdot \frac{1}{x^2} \cdot \frac{y^3}{1}\right)^{-1}$

$= \left(\frac{3y^3}{x^2}\right)^{-1} = \frac{x^2}{3y^3}$ or $\left(\frac{3x^{-2}}{y^{-3}}\right)^{-1} = \frac{3^{-1}x^{(-2)(-1)}}{y^{(-3)(-1)}} = \frac{3^{-1}x^2}{y^3}$

$= \frac{3^{-1}}{1} \cdot \frac{x^2}{1} \cdot \frac{1}{y^3} = \frac{1}{3} \cdot \frac{x^2}{1} \cdot \frac{1}{y^3} = \frac{x^2}{3y^3}$

5

16. (5.3)
$$3h(2k^2 - 5h) - h(2h - 3k^2) = 6hk^2 - 15h^2 - 2h^2 + 3hk^2$$
$$= 9hk^2 - 17h^2$$

17. (5.3) $x(x^2 + 2x + 4) - 2(x^2 + 2x + 4)$
$$= x^3 + 2x^2 + 4x - 2x^2 - 4x - 8 = x^3 - 8$$

18. (5.1) $2^3 \cdot 2^2 = 2^{3+2} = 2^5 = 32$

19. (5.5) $10^{-4} \cdot 10^2 = 10^{-4+2} = 10^{-2} = \dfrac{1}{10^2} = \dfrac{1}{100}$

20. (5.5) $5^{-2} = \dfrac{1}{5^2} = \dfrac{1}{25}$

21. (5.5) $(2^{-3})^2 = 2^{(-3)(2)} = 2^{-6} = \dfrac{1}{2^6} = \dfrac{1}{64}$

22. (5.5) $\dfrac{10^{-3}}{10^{-4}} = 10^{-3-(-4)} = 10^{-3+(+4)} = 10^1 = 10$

23. (5.5) $(5^0)^2 = 1^2 = 1$

24. (5.5) $\dfrac{a^3}{b} = \dfrac{a^3}{1} \cdot \dfrac{1}{b} = \dfrac{a^3}{1} \cdot \dfrac{b^{-1}}{1} = \dfrac{a^3 b^{-1}}{1} = a^3 b^{-1}$

25. (5.6) **a.** 1.326×10^3 **b.** 5.27×10^{-1}

Chapters 1–5 Cumulative Review Exercises   (page 271)

1. False   **2.** True   **3.** False   **4.** False   **5.** False

6. False   **7.** $\dfrac{-9}{3} = -3$ **8.** 12

9. $6(-3) - 4(6) = -18 - 24 = -42$ **10.** -28 **11.** 0

12. -25 **13.** $A = P(1 + rt)$
$$A = 1{,}200[1 + 0.15(4)]$$
$$A = 1{,}200[1.6]$$
$$A = 1{,}920$$

14. $x = -4$ **15.** $(-5)(4)(pp^3) = -20p^4$ **16.** $200h^9 j^{10} k^5$

17. $7 - 12xy + 12 = 19 - 12xy$ **18.** $-26x^2 y^2 + 9x^2 y^3$

19. Let $x = $ number of adults' tickets   **20.** -2
Then $9 - x = $ number of children's tickets
$$2.50x + 1.25(9 - x) = 16.25$$
$$250x + 125(9 - x) = 1{,}625$$
$$250x + 1{,}125 - 125x = 1{,}625$$
$$125x + 1{,}125 = 1{,}625$$
$$125x + 1{,}125 - 1{,}125 = 1{,}625 - 1{,}125$$
$$125x = 500$$
$$\dfrac{125x}{125} = \dfrac{500}{125}$$
$$x = 4$$
$$9 - x = 9 - 4 = 5$$
There were 4 adults' tickets and 5 children's tickets purchased.

21. Let $x = $ width of the rectangle (in cm)
Then $x + 7 = $ length of the rectangle (in cm)
and $2x + 2(x + 7) = $ perimeter of the rectangle (in cm)
$$2x + 2(x + 7) = 66$$
$$2x + 2x + 14 = 66$$
$$4x + 14 = 66$$
$$4x + 14 - 14 = 66 - 14$$
$$4x = 52$$
$$\dfrac{4x}{4} = \dfrac{52}{4}$$
$$x = 13$$
$$x + 7 = 13 + 7 = 20$$
The width is 13 cm, and the length is 20 cm.

22. 24 mph

Exercises 6.1   (page 277)

1. a. First degree   **2. a.** First degree
  b. Second degree   **b.** Third degree
3. Not a polynomial   **4.** Not a polynomial
5. a. Third degree   **6. a.** Third degree
  b. Sixth degree   **b.** Third degree
7. a. Second degree   **8. a.** Second degree
  b. Fourth degree   **b.** Third degree
9. Not a polynomial   **10.** Not a polynomial
11. Not a polynomial   **12.** Not a polynomial
13. a. Second degree   **14. a.** Third degree
  b. Second degree   **b.** Third degree
15. Not a polynomial   **16.** Not a polynomial
17. $8x^5 + 7x^3 - 4x - 5$; leading coefficient is 8.
18. $-3y^5 - 2y^3 + 4y^2 + 10$; leading coefficient is -3.
19. $xy^3 + 8xy^2 - 4x^2 y$; leading coefficient is 1.
20. $x^4 y^2 + 3x^3 y - 3xy^3$; leading coefficient is 1.

Exercises 6.2   (page 281)

1. $2m^2 - m + 4 + 3m^2 + m - 5 = 5m^2 - 1$

2. $11n^2 + 2n + 3$

3. $2x^3 - 4 + 4x^2 + 8x - 9x + 7 = 2x^3 + 4x^2 - x + 3$

4. $9z^2 + 9$

5. $(3x^2 + 4x - 10) + (-5x^2 + 3x - 7)$
$$= 3x^2 + 4x - 10 - 5x^2 + 3x - 7 = -2x^2 + 7x - 17$$

6. $-a^2 - 7a + 14$

7. $(8b^2 + 2b - 14) - (-5b^2 + 4b + 8)$
$$= (8b^2 + 2b - 14) + (+5b^2 - 4b - 8)$$
$$= 8b^2 + 2b - 14 + 5b^2 - 4b - 8 = 13b^2 - 2b - 22$$
or
$$\begin{array}{r} 8b^2 + 2b - 14 \\ -(-5b^2 + 4b + 8) \end{array} \rightarrow \begin{array}{r} 8b^2 + 2b - 14 \\ +(+5b^2 - 4b - 8) \\ \hline 13b^2 - 2b - 22 \end{array}$$

8. $19c^2 + 5c + 1$

9. $6a - 5a^2 + 6 + 4a^2 + 6 - 3a = -a^2 + 3a + 12$

10. $11b^2 + 3$

11. $(4a^2 + 6 - 3a) - (5a + 3a^2 - 4)$
$$= (4a^2 + 6 - 3a) + (-5a - 3a^2 + 4)$$
$$= 4a^2 + 6 - 3a - 5a - 3a^2 + 4 = a^2 - 8a + 10$$
or
$$\begin{array}{r} 4a^2 - 3a + 6 \\ -(3a^2 + 5a - 4) \end{array} \rightarrow \begin{array}{r} 4a^2 - 3a + 6 \\ +(-3a^2 - 5a + 4) \\ \hline a^2 - 8a + 10 \end{array}$$

12. $2b^2 - 9b + 15$ **13.** $17a^3 + 8a^2 - 2a$

14. $-20b^4 + b^3 + 5b^2 - 1$ **15.** $12x^2 y^3 - 9xy^2$

16. $9a^2 b + 2ab^2 - 10ab$

17. $\begin{array}{r} 15x^3 - 4x^2 \quad\quad + 12 \\ -(8x^3 \quad\quad + 9x - 5) \end{array} \rightarrow \begin{array}{r} 15x^3 - 4x^2 \quad\quad + 12 \\ +(- 8x^3 \quad\quad - 9x + 5) \\ \hline 7x^3 - 4x^2 - 9x + 17 \end{array}$

18. $-7y^3 - 28y^2 + 19y - 24$

19. $\begin{array}{r} 10a^2 b - 6ab + 5ab^2 \\ -(3a^2 b + 6ab - 7ab^2) \end{array} \rightarrow \begin{array}{r} 10a^2 b - 6ab + 5ab^2 \\ +(- 3a^2 b - 6ab + 7ab^2) \\ \hline 7a^2 b - 12ab + 12ab^2 \end{array}$

20. $22m^3 n^2 - 4m^2 n^2 - 9mn$

21. $7m^8 - 4m^4 + 4m^4 + m^5 + 8m^8 - m^5 = 15m^8$

22. $-h^6 + 17h$

23. $\underline{6r^3t} + \underline{\underline{14r^2t}} \dashuline{- 11} + \dashuline{19} - \underline{\underline{8r^2t}} + \underline{r^3t} + \dashuline{8} - \underline{\underline{6r^2t}} = 7r^3t + 16$

24. $4m^2n^2 + 11$

25. $(7x^2y^2 - 3x^2y + xy + 7) + (-3x^2y^2 + 5xy - 4 - 7x^2y)$
$= \underline{7x^2y^2} - \underline{\underline{3x^2y}} + \dashuline{xy} + \doubleline{7} - \underline{3x^2y^2} + \dashuline{5xy} - \doubleline{4} - \underline{\underline{7x^2y}}$
$= 4x^2y^2 - 10x^2y + 6xy + 3$

26. $9x^2y^2 - 2x^2y - 4xy - 13$

27. $(x^2 + 4) - [(x^2 - 5) + (-3x^2 - 1)]$
$= (x^2 + 4) - [x^2 - 5 - 3x^2 - 1]$
$= (x^2 + 4) + [-x^2 + 5 + 3x^2 + 1]$
$= \underline{x^2} + \underline{\underline{4}} - \underline{x^2} + \underline{\underline{5}} + \underline{3x^2} + \underline{\underline{1}} = 3x^2 + 10$

28. $6x^2 - 7$

29. $[(5x^2 - 2x + 1) + (-4x^2 + 6x - 8)] - (2x^2 - 4x + 3)$
$= [5x^2 - 2x + 1 - 4x^2 + 6x - 8] + (-2x^2 + 4x - 3)$
$= \underline{5x^2} - \underline{\underline{2x}} + \dashuline{1} - \underline{4x^2} + \underline{\underline{6x}} - \dashuline{8} - \underline{2x^2} + \underline{\underline{4x}} - \dashuline{3}$
$= -x^2 + 8x - 10$

30. $-4y$

31. $[(5 + xy^2 + x^3y) + (-6 - 3xy^2 + 4x^3y)] - [(x^3y + 3xy^2 - 4) + (2x^3y - xy^2 + 5)]$
$= [5 + xy^2 + x^3y - 6 - 3xy^2 + 4x^3y]$
$\quad - [x^3y + 3xy^2 - 4 + 2x^3y - xy^2 + 5]$
$= [5x^3y - 2xy^2 - 1] - [3x^3y + 2xy^2 + 1]$
$= [5x^3y - 2xy^2 - 1] + [-3x^3y - 2xy^2 - 1]$
$= \underline{5x^3y} - \underline{\underline{2xy^2}} - \dashuline{1} - \underline{3x^3y} - \underline{\underline{2xy^2}} - \dashuline{1} = 2x^3y - 4xy^2 - 2$

32. $6m^2n + 6$

33. $\underline{7.239x^2} - \underline{\underline{4.028x}} + \dashuline{6.205} - \underline{2.846x^2} + \underline{\underline{8.096x}} + \dashuline{5.307}$
$= 4.393x^2 + 4.068x + 11.512$

34. $37.528x^2 + 6.15x - 52.64$

Exercises 6.3A (page 287)

1. $15xy + 20x + 6y + 8$ **2.** $14xy + 7x + 12y + 6$

3. $40xy - 24x - 35y + 21$ **4.** $8y^2 - 35y + 12$

5. $15x^3 + 5x^2y - 6xy - 2y^2$ **6.** $2s^3 - 8st + 3s^2t - 12t^2$

7. $x^2 - 2x + 3x - 6 = x^2 + x - 6$ **8.** $a^2 - a - 12$

9. $y^2 - 9y + 8y - 72 = y^2 - y - 72$ **10.** $z^2 + 7z - 30$

11. $-56x^4y^3$ **12.** $-24a^5b^5$

13. $(2x)(3x^2) + (2x)(7) = 6x^3 + 14x$ **14.** $12y^4 - 6y$

15. $x^2 + 4x + 1x + 4 = x^2 + 5x + 4$ **16.** $x^2 + 4x + 3$

17. $a^2 + 2a + 5a + 10 = a^2 + 7a + 10$ **18.** $a^2 + 8a + 7$

19. $(5a^2)(2a) + (-b^2)(2a) = 10a^3 - 2ab^2$ **20.** $21c^3 - 3cd^2$

21. $x^2 - 8x + 8x - 64 = x^2 - 64$ **22.** $z^2 - 100$

23. $m^2 - 2m - 8$ **24.** $n^2 + 4n - 21$ **25.** $y^2 - 3y - 108$

26. $z^2 - 16z + 55$ **27.** $-abx - aby$ **28.** $abx - aby$

29. $ax - bx + ay - by$ **30.** $ax + bx - ay - by$ **31.** $16x^2$

32. $9x^2$ **33.** $(4 + x)(4 + x) = 16 + 8x + x^2$

34. $9 + 6x + x^2$ **35.** $(b - 4)(b - 4) = b^2 - 8b + 16$

36. $b^2 - 12b + 36$ **37.** $3x^2 + 7x + 2$ **38.** $2x^2 + 7x + 6$

39. $8x^2 + 10x - 12$ **40.** $6x^2 + 7x - 5$

41. $20x^2 - 38x + 12$ **42.** $15x^2 - 32x + 16$

43. $(2x + 5)(2x + 5) = 4x^2 + 20x + 25$

44. $9x^2 + 24x + 16$ **45.** $6x^3 + 3x^2 - 4x - 2$

46. $24a^3 - 16a^2 + 3a - 2$ **47.** $8x^2 + 26xy - 7y^2$

48. $12x^2 + 7xy - 10y^2$ **49.** $49x^2 - 140xy + 100y^2$

50. $16u^2 - 72uv + 81v^2$

51. $(3a + 2b)(3a + 2b) = 9a^2 + 12ab + 4b^2$

52. $4x^2 - 24xy + 36y^2$ **53.** $16c^2 - 9d^2$ **54.** $25e^2 - 4f^2$

Exercises 6.3B (page 292)

1.
$$\begin{array}{r} 2x^2 + x - 1 \\ x - 3 \\ \hline -6x^2 - 3x + 3 \\ 2x^3 + x^2 - x \\ \hline 2x^3 - 5x^2 - 4x + 3 \end{array}$$

2. $3x^3 - 5x^2 - 3x + 2$

3.
$$\begin{array}{r} x^2 + x + 1 \\ x^2 + x + 1 \\ \hline x^2 + x + 1 \\ x^3 + x^2 + x \\ x^4 + x^3 + x^2 \\ \hline x^4 + 2x^3 + 3x^2 + 2x + 1 \end{array}$$

4. $x^4 - 2x^3 - x^2 + 2x + 1$ **5.** $4z^3 - 16z^2 + 64z$

6. $-5a^3 - 25a^2 - 125a$

7.
$$\begin{array}{r} z^2 - 4z + 16 \\ z + 4 \\ \hline 4z^2 - 16z + 64 \\ z^3 - 4z^2 + 16z \\ \hline z^3 + 64 \end{array}$$

8. $a^3 - 125$

9.
$$\begin{array}{r} -3z^3 + z^2 - 5z + 4 \\ -z + 4 \\ \hline -12z^3 + 4z^2 - 20z + 16 \\ 3z^4 - z^3 + 5z^2 - 4z \\ \hline 3z^4 - 13z^3 + 9z^2 - 24z + 16 \end{array}$$

10. $v^4 - 4v^3 + 5v + 6$

11. $6x^3y^2 - 2x^2y^3 + 8xy^4$ **12.** $12x^2y^3 + 3x^3y^2 - 9x^4y$

13.
$$\begin{array}{r} 2x^2 - x + 4 \\ x^2 - 5x - 3 \\ \hline -6x^2 + 3x - 12 \\ -10x^3 + 5x^2 - 20x \\ 2x^4 - x^3 + 4x^2 \\ \hline 2x^4 - 11x^3 + 3x^2 - 17x - 12 \end{array}$$

14. $12a^4 - 14a^3 + 18a^2 + 19a - 15$

15.
$$\begin{array}{r} 3y^2 - 2y + 7 \\ 4y^2 + 8y - 3 \\ \hline -9y^2 + 6y - 21 \\ 24y^3 - 16y^2 + 56y \\ 12y^4 - 8y^3 + 28y^2 \\ \hline 12y^4 + 16y^3 + 3y^2 + 62y - 21 \end{array}$$

16. $6x^4 - 17x^3 + 47x^2 + 21x - 36$

17.
$$\begin{array}{r} 5x - 2 \\ 5x - 2 \\ \hline -10x + 4 \\ 25x^2 - 10x \\ \hline 25x^2 - 20x + 4 \end{array}$$

18. $4x^2 - 20x + 25$

19. $(x + y)^2 = (x + y)(x + y) = x^2 + 2xy + y^2$ and
$(x - y)^2 = (x - y)(x - y) = x^2 - 2xy + y^2$
Then $(x + y)^2(x - y)^2 = (x^2 + 2xy + y^2)(x^2 - 2xy + y^2)$
$$\begin{array}{r} x^2 + 2xy + y^2 \\ x^2 - 2xy + y^2 \\ \hline x^2y^2 + 2xy^3 + y^4 \\ -2x^3y - 4x^2y^2 - 2xy^3 \\ x^4 + 2x^3y + x^2y^2 \\ \hline x^4 - 2x^2y^2 + y^4 \end{array}$$

20. $x^4 - 8x^2 + 16$

21. $(x+2)^3 = \underbrace{(x+2)(x+2)}(x+2)$

First find $(x+2)^2$

$$
\begin{array}{r}
x+2 \\
x+2 \\
\hline
2x+4 \\
x^2+2x \\
\hline
x^2+4x+4
\end{array}
\qquad
\begin{array}{r}
x^2+4x+4 \\
x+2 \\
\hline
2x^2+8x+8 \\
x^3+4x^2+4x \\
\hline
x^3+6x^2+12x+8
\end{array}
$$

22. $x^3 + 9x^2 + 27x + 27$

23. $(x^2+2x-3)^2 = (x^2+2x-3)(x^2+2x-3)$

$$
\begin{array}{r}
x^2+2x-3 \\
x^2+2x-3 \\
\hline
-3x^2-6x+9 \\
+2x^3+4x^2-6x \\
x^4+2x^3-3x^2 \\
\hline
x^4+4x^3-2x^2-12x+9
\end{array}
$$

24. $y^4 - 8y^3 + 6y^2 + 40y + 25$

25. $(x+3)^2 = (x+3)(x+3) = x^2 + 6x + 9$

Then $(x+3)^4 = (x+3)^2(x+3)^2$

$= (x^2 + 6x + 9)(x^2 + 6x + 9)$

$$
\begin{array}{r}
x^2+6x+9 \\
x^2+6x+9 \\
\hline
9x^2+54x+81 \\
6x^3+36x^2+54x \\
x^4+6x^3+9x^2 \\
\hline
x^4+12x^3+54x^2+108x+81
\end{array}
$$

26. $x^4 + 8x^3 + 24x^2 + 32x + 16$

Exercises 6.4A   (page 294)

1. $(x)^2 - (3)^2 = x^2 - 9$ **2.** $z^2 - 16$

3. $(w)^2 - (6)^2 = w^2 - 36$ **4.** $y^2 - 25$

5. $(5a)^2 - (4)^2 = 25a^2 - 16$ **6.** $36a^2 - 25$

7. $(2u)^2 - (5v)^2 = 4u^2 - 25v^2$ **8.** $9m^2 - 49n^2$

9. $(4b)^2 - (9c)^2 = 16b^2 - 81c^2$ **10.** $49a^2 - 64b^2$

11. $(2x^2)^2 - (9)^2 = 4x^4 - 81$ **12.** $100y^4 - 9$

13. $(1)^2 - (8z^3)^2 = 1 - 64z^6$ **14.** $81v^8 - 1$

15. $(5xy)^2 - (z)^2 = 25x^2y^2 - z^2$ **16.** $100a^2b^2 - c^2$

17. $(7mn)^2 - (2rs)^2 = 49m^2n^2 - 4r^2s^2$ **18.** $64h^2k^2 - 25e^2f^2$

Exercises 6.4B   (page 296)

1. $(x)^2 - 2(x)(1) + (1)^2 = x^2 - 2x + 1$ **2.** $x^2 - 10x + 25$

3. $(x)^2 + 2(x)(3) + (3)^2 = x^2 + 6x + 9$ **4.** $x^2 + 8x + 16$

5. $(4x)^2 - 2(4x)(1) + (1)^2 = 16x^2 - 8x + 1$

6. $49x^2 - 14x + 1$

7. $(12x)^2 + 2(12x)(1) + 1^2 = 144x^2 + 24x + 1$

8. $121x^2 + 22x + 1$

9. $(2s)^2 + 2(2s)(4t) + (4t)^2 = 4s^2 + 16st + 16t^2$

10. $9u^2 + 42uv + 49v^2$

11. $(5x)^2 - 2(5x)(3y) + (3y)^2 = 25x^2 - 30xy + 9y^2$

12. $16x^2 - 56xy + 49y^2$

13. $(3x)^2 + 2(3x)(2z) + (2z)^2 = 9x^2 + 12xz + 4z^2$

14. $4x^2 + 28xs + 49s^2$

15. $(7x)^2 - 2(7x)(8y) + (8y)^2 = 49x^2 - 112xy + 64y^2$

16. $81x^2 - 126xy + 49y^2$

Exercises 6.5A   (page 299)

1. $\dfrac{6}{3} \cdot \dfrac{x^5}{x^2} = 2x^{5-2} = 2x^3$ **2.** $4y^4$

3. $\dfrac{4}{12} \cdot \dfrac{a^2}{a} \cdot \dfrac{b}{1} \cdot \dfrac{c^2}{c} = \dfrac{4 \cdot 1}{4 \cdot 3} \cdot a^{2-1} \cdot b \cdot c^{2-1} = \dfrac{1}{3}a^1bc^1 = \dfrac{1}{3}abc$

4. $\dfrac{1}{3}x^2yz$

5. $\dfrac{-5}{-10} \cdot \dfrac{x^3}{x^2} \cdot \dfrac{y^2}{y} = \dfrac{-5 \cdot 1}{-5 \cdot 2} \cdot x^{3-2} \cdot y^{2-1} = \dfrac{1}{2}x^1y^1 = \dfrac{1}{2}xy$

6. $\dfrac{1}{3}st^2$

7. $-\dfrac{8}{12} \cdot \dfrac{u^5}{u} \cdot \dfrac{v^2}{v^3} \cdot \dfrac{w}{w^4} = -\dfrac{4 \cdot 2}{4 \cdot 3} \cdot u^{5-1} \cdot v^{2-3} \cdot w^{1-4}$

$= -\dfrac{2}{3} \cdot u^4 \cdot v^{-1} \cdot w^{-3} = -\dfrac{2}{3} \cdot \dfrac{u^4}{1} \cdot \dfrac{1}{v^1} \cdot \dfrac{1}{w^3} = -\dfrac{2u^4}{3vw^3}$

8. $-\dfrac{2c^2d^2}{3e^3}$

9. $\dfrac{3x}{3} + \dfrac{6}{3} = x + 2$ **10.** $2x + 3$ **11.** $\dfrac{4}{4} + \dfrac{8x}{4} = 1 + 2x$

12. $1 - 2x$ **13.** $\dfrac{6x}{2} + \dfrac{-8y}{2} = 3x - 4y$ **14.** $x - 2y$

15. $\dfrac{2x^2}{x} + \dfrac{3x}{x} = 2x + 3$ **16.** $4y - 3$

17. $\dfrac{15x^3}{5x^2} + \dfrac{-5x^2}{5x^2} = 3x - 1$ **18.** $2y^2 - 1$

19. $\dfrac{3a^2b}{ab} + \dfrac{-ab}{ab} = 3a - 1$ **20.** $5n - 1$

21. $\dfrac{8x^7}{4x^2} + \dfrac{4x^5}{4x^2} + \dfrac{-12x^3}{4x^2} = 2x^5 + x^3 - 3x$ **22.** $y^4 + 3y^2 - 2y$

23. $\dfrac{5x^5}{-5x^2} + \dfrac{-4x^3}{-5x^2} + \dfrac{10x^2}{-5x^2} = -x^3 + \frac{4}{5}x - 2$

24. $-y^2 + \frac{5}{7}y - 2$ **25.** $\dfrac{-15x^2y^2z^2}{-5xyz} + \dfrac{-30xyz}{-5xyz} = 3xyz + 6$

26. $3abc + 2$ **27.** $\dfrac{13x^4y^2}{13x^2y^2} + \dfrac{-26x^2y^3}{13x^2y^2} + \dfrac{39x^2y^2}{13x^2y^2} = x^2 - 2y + 3$

28. $3n^3 - 5m - 2$

29. $\dfrac{5x^3}{-5x^2} + \dfrac{-8x^2}{-5x^2} + \dfrac{3}{-5x^2} = -x + \dfrac{8}{5} - \dfrac{3}{5x^2}$

30. $-y + \dfrac{9}{7} - \dfrac{8}{7y^2}$

31. $\dfrac{15x^3y^2}{15x^2y^2} + \dfrac{-30xy^3}{15x^2y^2} + \dfrac{20xy}{15x^2y^2} = x - \dfrac{2y}{x} + \dfrac{4}{3xy}$

32. $2n - \dfrac{5m}{2} - \dfrac{1}{mn}$

Exercises 6.5B   (page 307)

Note: Sign changes for the subtractions are not shown here.

1.
$$
\begin{array}{r}
x+3 \\
x+2\overline{)x^2+5x+6} \\
\underline{x^2+2x} \\
3x+6 \\
\underline{3x+6} \\
0
\end{array}
$$
   2. $x + 2$

3.
$$
\begin{array}{r}
x+3 \\
x-4\overline{)x^2-x-12} \\
\underline{x^2-4x} \\
3x-12 \\
\underline{3x-12} \\
0
\end{array}
$$

4. $x - 4$

5.
$$3x - 2 \overline{)6x^2 + 5x - 6}$$
$$\underline{6x^2 - 4x}$$
$$9x - 6$$
$$\underline{9x - 6}$$
$$0$$
with quotient $2x + 3$

6. $4x + 5$

7.
$$5v - 7 \overline{)15v^2 + 19v + 10}$$
$$\underline{15v^2 - 21v}$$
$$40v + 10$$
$$\underline{40v - 56}$$
$$66$$
with quotient $3v + 8$ R 66

8. $5v - 7$ R 52

9.
$$2x - 3 \overline{)6x^2 + x - 15}$$
$$\underline{6x^2 - 9x}$$
$$10x - 15$$
$$\underline{10x - 15}$$
$$0$$
with quotient $3x + 5$

10. $4x - 2$

11.
$$-x + 2 \overline{)-4x^3 + 0x^2 + 8x + 10}$$
$$\underline{-4x^3 + 8x^2}$$
$$-8x^2 + 8x$$
$$\underline{-8x^2 + 16x}$$
$$-8x + 10$$
$$\underline{-8x + 16}$$
$$-6$$
with quotient $4x^2 + 8x + 8$ R -6

12. $x^2 + 3x - 3$ R -6

13.
$$2a + 3b \overline{)6a^2 + 5ab + b^2}$$
$$\underline{6a^2 + 9ab}$$
$$-4ab + b^2$$
$$\underline{-4ab - 6b^2}$$
$$7b^2$$
with quotient $3a - 2b$ R $7b^2$

14. $2a + 3b$ R $5b^2$

15.
$$a - 2 \overline{)a^3 + 0a^2 + 0a - 8}$$
$$\underline{a^3 - 2a^2}$$
$$2a^2 + 0a$$
$$\underline{2a^2 - 4a}$$
$$4a - 8$$
$$\underline{4a - 8}$$
$$0$$
with quotient $a^2 + 2a + 4$

16. $c^2 + 3c + 9$

17.
$$x - 4 \overline{)x^3 + 0x^2 - 8x - 15}$$
$$\underline{x^3 - 4x^2}$$
$$4x^2 - 8x$$
$$\underline{4x^2 - 16x}$$
$$8x - 15$$
$$\underline{8x - 32}$$
$$17$$
with quotient $x^2 + 4x + 8$ R 17

18. $2x^2 - 3x + 6$ R -3

19.
$$x^2 + x - 1 \overline{)x^4 + 2x^3 - x^2 - 2x + 4}$$
$$\underline{x^4 + x^3 - x^2}$$
$$x^3 + 0x^2 - 2x$$
$$\underline{x^3 + x^2 - x}$$
$$-x^2 - x + 4$$
$$\underline{-x^2 - x + 1}$$
$$3$$
with quotient $x^2 + x - 1$ R 3

20. $x^2 - x + 1$ R 6

21.
$$x^2 + 2x + 3 \overline{)x^4 + 3x^3 + 6x^2 + 5x - 5}$$
$$\underline{x^4 + 2x^3 + 3x^2}$$
$$x^3 + 3x^2 + 5x$$
$$\underline{x^3 + 2x^2 + 3x}$$
$$x^2 + 2x - 5$$
$$\underline{x^2 + 2x + 3}$$
$$-8$$
with quotient $x^2 + x + 1$ R -8

22. $x^2 + 2x + 3$ R 2

Exercises 6.6 (page 308)

1. If x is the smaller integer, the larger one is $x + 2$.
$x(x + 2) = x^2 + 2x$

2. $3x + 6$

3. The length is $4x$, and the height is $(3 + x)$. The volume is
$4x(x)(3 + x) = 4x^2(3 + x) = 12x^2 + 4x^3$

4. $(3x - 4)(4 + x)(x) = 3x^3 + 8x^2 - 16x$

5. The length of the photograph is $(x + 2)$, so the area of the photograph is $x(x + 2)$. The width of the mat is $x + 2 + 2$ (or $x + 4$), and the length of the mat is $(x + 2) + 3 + 3$ (or $x + 8$). The *required* area is
$(x + 4)(x + 8) - x(x + 2) = x^2 + 12x + 32 - x^2 - 2x$
$= 10x + 32$

6. $8x + 28$

7. The height will be x, and the width and length will both be $(10 - 2x)$. The volume is
$x(10 - 2x)(10 - 2x) = x(100 - 40x + 4x^2)$
$= 100x - 40x^2 + 4x^3$

8. $168x - 52x^2 + 4x^3$

Sections 6.1–6.6 Review Exercises   (page 310)

1. a. Third degree   **b.** Third degree

2. Not a polynomial   **3.** Not a polynomial

4. a. Fourth degree   **b.** First degree

5. a. $3x^4 + 7x^2 + x - 6$ **b.** 3

6. a. $y^4 + 3xy^2 + 3x^2y + x^3$ **b.** 1

7. $5x^2y + 3xy^2 - 4y^3 + 2xy^2 + 4y^3 + 3x^2y = 8x^2y + 5xy^2$

8. $-18x^4y^4$ **9.** $9xy^2(x^2y) + 9xy^2(-2xy) = 9x^3y^3 - 18x^2y^3$

10. $5a^2b + 9ab^2 - 2b^2 + 4a - 8$

11. $2a + (-4)(a) + (-4)(-b) = 2a - 4a + 4b = -2a + 4b$

12. $c + 2d$

13. $x^2(x^2) + x^2(y^2) + (-y^2)(x^2) + (-y^2)(y^2)$
$= x^4 + x^2y^2 - x^2y^2 - y^4 = x^4 - y^4$

14. $x^3 + 27$

15. $-30x^2y^2$ **16.** $-30abc^2$ **17.** $6x^2y - 15xy^2$

18. $10abc - 6ac^2$ **19.** $-15xy^2 - 10xy$ **20.** $-6ac^2 - 15bc$

21. $3xy + 2x - 5y$ **22.** $2ac + 5b - 3c$

23. $3x^3y^3 - x^2y^2 - 8x^2y^2 + 2x^3y^3 = 5x^3y^3 - 9x^2y^2$

24. $-4a^2b^2 + 2a^2b^3$ **25.** $\dfrac{8x}{2} + \dfrac{2}{2} = 4x + 1$

26. $-20x^3y^3 + 12xy^3$ **27.** $x^2 + 2x - 35$ **28.** $x^3 - 8$

29.
$$2x - 3 \overline{)6x^2 - 9x + 10}$$
$$\underline{6x^2 - 9x}$$
$$+ 10$$
with quotient $3x + 0$ R 10

30. $x^2 - 16x + 64$

31. $(x^2 + 3 - 5x) - (4x^3 - x + 4x^2 - 1)$
$= (x^2 + 3 - 5x) + (-4x^3 + x - 4x^2 + 1)$
$= \underline{x^2} + \underline{3} \underline{\underline{- 5x}} - 4x^3 \underline{\underline{+ x}} - \underline{4x^2} + \underline{1} =$
$= -4x^3 - 3x^2 - 4x + 4$

32. $a^2 - 1$

33. $(2m^2 - 5) - [(7 - m^2) + (-4m^2 + 3)]$
$= (2m^2 - 5) - [\underline{7} - m^2 - 4m^2 + \underline{3}]$
$= (2m^2 - 5) - [10 - 5m^2]$
$= (2m^2 - 5) + [-10 + 5m^2] = \underline{2m^2} - \underline{5} \underline{\underline{- 10}} + \underline{5m^2}$
$= 7m^2 - 15$

34. $-14x^2 + 22x - 2$ **35.** $12x^3y^4 - 15x^2y^3 - 30xy^2$

36. $x^3 - 6x^2 + 12x - 8$ **37.** $9a^2 - 25$

38. $9 + 6x + 6y + x^2 + 2xy + y^2$

39.
$$\begin{array}{r} -3y + 2 \\ 3y + 2)\overline{-9y^2 + 0y + 4} \\ \underline{-9y^2 - 6y} \\ 6y + 4 \\ \underline{6y + 4} \end{array}$$
    40. $3ab^2 - \frac{4}{5}b + 2$

41. $-3xyz - z^2$ **42.** $3xy$

43. $4^2 + 2(4)(x) + x^2 = x^2 + 8x + 16$ **44.** $16x^2$

45. $(2^2x^2y^4)(3y) = 2^2 \cdot 3x^2y^5 = 12x^2y^5$ **46.** $3c - 16d$

47.
$$\begin{array}{r} 7a^2 - 3ab + 5 - b^2 \\ \underline{-(9a^2 + 2ab - 8 + b^2)} \end{array} \rightarrow \begin{array}{r} 7a^2 - 3ab + 5 - b^2 \\ \underline{+(-9a^2 - 2ab + 8 - b^2)} \\ -2a^2 - 5ab + 13 - 2b^2 \end{array}$$

48. $6x^3 - 7x^2 + x - 4$

49. If x is the smallest integer, the second integer is $x + 2$ and the third is $x + 4$.
$x(x + 2)(x + 4) = x(x^2 + 6x + 8) = x^3 + 6x^2 + 8x$

50. $9\pi x^3$

Chapter 6 Diagnostic Test   (page 315)

Following each problem number is the number (in parentheses) of the textbook section where that kind of problem is discussed.

1. (6.1) **a.** Second degree   **b.** Third degree
    c. Third degree

2. (6.1) **a.** -4 **b.** 8

3. (6.2)
$$\begin{array}{r} -7x^3 + 4x^2 \qquad + 3 \\ 3x^3 \qquad + 6x - 5 \\ 7x^2 - 4x + 8 \\ \hline -4x^3 + 11x^2 + 2x + 6 \end{array}$$

4. (6.2) $(6xy^2 - 5xy) + (17xy - 7x^2y) + (3xy^2 - y^3)$
$= \underline{6xy^2} \underline{\underline{- 5xy}} \underline{\underline{+ 17xy}} - 7x^2y + \underline{3xy^2} - y^3$
$= 9xy^2 + 12xy - 7x^2y - y^3$

5. (6.2) $(-3x^2 - 6x + 9) - (8 - 2x + 5x^2)$
$= (-3x^2 - 6x + 9) + (-8 + 2x - 5x^2)$
$= \underline{-3x^2} \underline{\underline{- 6x}} \underline{+ 9} \underline{- 8} \underline{\underline{+ 2x}} - 5x^2 = -8x^2 - 4x + 1$

6. (6.2)
$$\begin{array}{r} -6a^3 + 5a^2 \qquad + 4 \\ \underline{-(4a^3 \qquad + 6a - 7)} \end{array} \rightarrow \begin{array}{r} -6a^3 + 5a^2 \qquad + 4 \\ \underline{+(-4a^3 \qquad - 6a + 7)} \\ -10a^3 + 5a^2 - 6a + 11 \end{array}$$

7. (6.3) $(8x - 3) - (10x - 5) + (9 - 5x)$
$= (8x - 3) + (-10x + 5) + (9 - 5x)$
$= \underline{8x} \underline{\underline{- 3}} \underline{- 10x} \underline{\underline{+ 5}} \underline{\underline{+ 9}} - 5x = -7x + 11$

8. (6.3) $(9x - 7)(8x + 9) = 72x^2 + 81x - 56x - 63$
    $= 72x^2 + 25x - 63$

9. (6.3) $-6xy(3x^2 - 5xy^2 + 8y)$
    $= -6xy(3x^2) + (-6xy)(-5xy^2) + (-6xy)(8y)$
    $= -18x^3y + 30x^2y^3 - 48xy^2$

10. (6.4) $(6x + 8)(6x - 8) = (6x)^2 - (8)^2 = 36x^2 - 64$

11. (6.3) $5x - 7x(3x^2 + 8x - 6)$
    $= 5x + (-7x)(3x^2) + (-7x)(8x) + (-7x)(-6)$
    $= 5x - 21x^3 - 56x^2 + 42x$
    $= -21x^3 - 56x^2 + 47x$

12. (6.3) $(4abc^2)(3b)(-2a^2c) = (4)(3)(-2)(aa^2)(bb)(c^2c)$
    $= -24a^3b^2c^3$

13. (6.3) $(4abc^2 + 3b)(-2a^2c) = (4abc^2)(-2a^2c) + (3b)(-2a^2c)$
    $= -8a^3bc^3 - 6a^2bc$

14. (6.4) $(3x - 8)^2 = (3x)^2 - 2(3x)(8) + (8)^2 = 9x^2 - 48x + 64$

15. (6.3)
$$\begin{array}{r} 2x^2 - \quad x - 4 \\ x^2 + 3x - 5 \\ \hline -10x^2 + 5x + 20 \\ 6x^3 - 3x^2 - 12x \\ 2x^4 - x^3 - 4x^2 \\ \hline 2x^4 + 5x^3 - 17x^2 - 7x + 20 \end{array}$$

16. (6.4) $(3xy^2 - 5z^3)(3xy^2 + 5z^3) = (3xy^2)^2 - (5z^3)^2$
    $= 9x^2y^4 - 25z^6$

17. (6.5) $\dfrac{8x^4 - 4x^3 + 12x^2}{4x^2} = \dfrac{8x^4}{4x^2} + \dfrac{-4x^3}{4x^2} + \dfrac{12x^2}{4x^2}$
    $= 2x^2 - x + 3$

18. (6.5)
$$\begin{array}{r} 5x + 2 \quad \text{R } 3 \\ 3x - 1)\overline{15x^2 + x + 1} \\ \underline{15x^2 - 5x} \\ 6x + 1 \\ \underline{6x - 2} \\ 3 \end{array}$$

19. (6.5)
$$\begin{array}{r} x^2 - 2x - 8 \quad \text{R} -20 \\ x - 4)\overline{x^3 - 6x^2 + 0x + 12} \\ \underline{x^3 - 4x^2} \\ -2x^2 + 0x \\ \underline{-2x^2 + 8x} \\ -8x + 12 \\ \underline{-8x + 32} \\ -20 \end{array}$$

20. (6.6) Let x = first even integer
    Then $x + 2$ = second even integer
    and $x + 4$ = third even integer
    and $x + 6$ = fourth even integer
    The product of the four integers is
    $x(x + 2)(x + 4)(x + 6)$
    $= [x(x + 2)][(x + 4)(x + 6)]$
    $= [x^2 + 2x][x^2 + 10x + 24]$
    $= [x^2 + 2x](x^2) + [x^2 + 2x](10x) + [x^2 + 2x](24)$
    $= x^4 + 2x^3 + 10x^3 + 20x^2 + 24x^2 + 48x$
    $= x^4 + 12x^3 + 44x^2 + 48x$

Chapters 1–6 Cumulative Review Exercises   (page 315)

1. $F = \frac{9}{5}C + 32$ **2.** A term of   **3.** $294 = 2 \cdot 3 \cdot 7^2$
$F = \frac{9}{5}(-20) + 32$
$F = \frac{-180}{5} + 32$
$F = -36 + 32$
$F = -4$

4. -13

5. $5[11 - 2(-3)] - 4(8) = 5[11 + 6] - 32 = 5[17] - 32$
$= 85 - 32 = 53$

6. $12\frac{1}{5}$

7. $-25 \cdot 4 - 15 \div 3(5) = -25 \cdot 4 - 5(5) = -100 - 25$
$= -125$

8. $\frac{1}{11}$ **9.** $\quad 4(4x + 2) = 8[5 - 2(2 - x)]$ **10.** 3
$16x + 8 = 8[5 - 4 + 2x]$
$16x + 8 = 8[1 + 2x]$
$16x + 8 = 8 + 16x$
$-16x + 16x + 8 = -16x + 8 + 16x$
$8 = 8$ *True*; identity

11. $5(x - 6) + 2(3 - 4x) = 6 - (2x + 8)$ **12.** No solution
$5x - 30 + 6 - 8x = 6 - 2x - 8$
$-3x - 24 = -2x - 2$
$2x - 3x - 24 = 2x - 2x - 2$
$-x - 24 = -2$
$-x - 24 + 24 = -2 + 24$
$-x = 22$
$x = -22$

13. $3(7x - 10) - 6(1 + 3x) = 4(3 - 3x)$
$21x - 30 - 6 - 18x = 12 - 12x$
$-36 + 3x = 12 - 12x$
$-36 + 3x + 12x = 12 - 12x + 12x$
$-36 + 15x = 12$
$36 - 36 + 15x = 36 + 12$
$15x = 48$
$\dfrac{15x}{15} = \dfrac{48}{15} = \dfrac{3 \cdot 16}{3 \cdot 5}$
$x = \frac{16}{5}$, or $3\frac{1}{5}$

14. 10^5, or 100,000 **15.** $x^{3c-c} = x^{2c}$ **16.** $\dfrac{8a^6}{b^3}$

17. $\left(\dfrac{6y^{-1}}{3y^3}\right)^2 = (2y^{-1-3})^2 = (2y^{-4})^2 = 2^2y^{-8} = \dfrac{4}{y^8}$ **18.** $\dfrac{x^4}{9}$

19. $2x(3x^2) + (2x)(6x) + (2x)(8) + (-4)(3x^2) + (-4)(5x)$
$+ (-4)(-6)$
$= 6x^3 + 12x^2 + 16x - 12x^2 - 20x + 24 = 6x^3 - 4x + 24$

20. a^6b^5 **21.** x^8y^8 **22.** y^{16} **23.** $2y^8$

24. $6x^3 - 19x^2 + 16x - 15$ **25.** $(9x)^2 - (2)^2 = 81x^2 - 4$

26. $25x^2 - 30x + 9$ **27.** $\begin{array}{r} 5x^2 + 3x + 7 \quad \text{R } 6 \\ x - 1\overline{)5x^3 - 2x^2 + 4x - 1} \\ \underline{5x^3 - 5x^2} \\ 3x^2 + 4x \\ \underline{3x^2 - 3x} \\ 7x - 1 \\ \underline{7x - 7} \\ 6 \end{array}$

28. $6x^2 - 11x - 35$ **29.** $x^3 - 3x^2 + 5x - 15$

30. $4x^2 + 28x + 49$ **31.** $6x - 5 < 13$ **32.** $x \geq 4$
$6x - 5 + 5 < 13 + 5$
$6x < 18$
$\dfrac{6x}{6} < \dfrac{18}{6}$
$x < 3$

33. Twenty-two is what percent of 25? **34.** $175
Let x = percent
$25x = 22$
$\dfrac{25x}{25} = \dfrac{22}{25}$
$x = \frac{22}{25} = 0.88 = 88\%$
Susan's score is 88%.

35. Let $\quad x$ = number of pounds of walnuts
Then $10 - x$ = number of pounds of almonds
$4.50x + 3.10(10 - x) = 3.52(10)$
$450x + 310(10 - x) = 352(10)$
$450x + 3,100 - 310x = 3,520$
$140x + 3,100 = 3,520$
$140x + 3,100 - 3,100 = 3,520 - 3,100$
$140x = 420$
$\dfrac{140x}{140} = \dfrac{420}{140}$
$x = 3$
$10 - x = 10 - 3 = 7$
Three pounds of walnuts and 7 lb of almonds should be used.

36. The lengths of the sides are 15, 25, and 35.

Exercises 7.1A (page 325)

1. $12 = 2^2 \cdot 3$; $24 = 2^3 \cdot 3$; GCF $= 2^2 \cdot 3$, or 12 **2.** 8
3. $21 = 3 \cdot 7$; $10 = 2 \cdot 5$; GCF $= 1$ **4.** 1
5. $32 = 2^5$; $24 = 2^3 \cdot 3$; $40 = 2^3 \cdot 5$; GCF $= 2^3$, or 8 **6.** 6
7. $18 = 2 \cdot 3^2$; $12 = 2^2 \cdot 3$; $48 = 2^4 \cdot 3$; GCF $= 2 \cdot 3$, or 6
8. 4 **9.** $4(3x + 2)$ **10.** $3(2x + 3)$ **11.** Not factorable
12. Not factorable **13.** $2(x + 4)$ **14.** $3(x + 3)$
15. $5(a - 2)$ **16.** $7(b - 2)$ **17.** $3(2y - 1)$
18. $5(3z - 1)$ **19.** $3x(3x + 1)$ **20.** $4y(2y - 1)$
21. $5a^2(2a - 5)$ **22.** $9b^2(3 - 2b^2)$ **23.** Not factorable
24. Not factorable **25.** $2ab(a + 2b)$ **26.** $3mn^2(1 + 2m)$
27. $6c^2d^2(2c - 3d)$ **28.** $15ab^3(1 - 3ab)$
29. $4x(x^2 - 3 - 6x)$ **30.** $6y(3 - y - 5y^2)$
31. Not factorable **32.** Not factorable
33. $2x(4x^2 - 3x + 1)$ **34.** $3y^2(3y^2 + 2y - 1)$
35. $8(3a^4 + a^2 - 5)$ **36.** $15(3b^3 - b^4 - 2)$
37. $14xy^3(-x^7y^6 + 3x^4y - 2)$ **38.** $7uv^5(-3u^6v^3 - 9 + 5u)$
39. Not factorable **40.** Not factorable
41. $11a^{10}b^4(-4a^4b^3 - 3b + 2a)$
42. $13e^8f^5(-2f + e^2f^3 - 3e^4)$ **43.** $2(9u^{10}v^5 + 12 - 7u^{10}v^6)$
44. $15(2a^3b^4 - 1 + 3a^8b^7)$ **45.** $6y^2(3x^3y^2 - 2z^3 - 8x^4y)$
46. $8m^3(4m^2n^7 - 3m^5p^9 - 5n^6)$

Exercises 7.1B (page 328)

1. GCF is $(s + t)$; $(s + t)(c + b)$ **2.** $(b + c)(a + d)$
3. GCF is $(a - b)$; $(a - b)(x + 5)$ **4.** $(s - t)(y + 7)$
5. GCF is $(u + v)$; $(u + v)(x - 3)$ **6.** $(t + u)(s - 2)$
7. GCF is $(x - y)$; $(x - y)(8 - a)$ **8.** $(a - b)(7 - c)$
9. GCF is $(s - t)$; $(s - t)(4 + u - v)$
10. $(x - y)(a + 5 - b)$
11. GCF is $3xy^2(a + b)$; $3xy^2(a + b)(xy + 3)$
12. $5ab(s + t^2)(2b^2 + 3a)$
13. GCF is $2u^3v^4(x - y)$; $2u^3v^4(x - y)(2v + 3u)$
14. $5x^2y^3(3a - 2b)(2y^2 + 3x^3)$

Exercises 7.2 (page 331)

1. $(\sqrt{m^2} + \sqrt{n^2})(\sqrt{m^2} - \sqrt{n^2}) = (m + n)(m - n)$
2. $(u + v)(u - v)$
3. $(\sqrt{x^2} + \sqrt{9})(\sqrt{x^2} - \sqrt{9}) = (x + 3)(x - 3)$
4. $(x + 5)(x - 5)$
5. $(\sqrt{a^2} + \sqrt{1})(\sqrt{a^2} - \sqrt{1}) = (a + 1)(a - 1)$

7

6. $(1 + b)(1 - b)$

7. $(\sqrt{4c^2} + \sqrt{1})(\sqrt{4c^2} - \sqrt{1}) = (2c + 1)(2c - 1)$

8. $(4d + 1)(4d - 1)$

9. $(\sqrt{16x^2} + \sqrt{9y^2})(\sqrt{16x^2} - \sqrt{9y^2}) = (4x + 3y)(4x - 3y)$

10. $(5a + 2b)(5a - 2b)$ **11.** Not factorable

12. Not factorable   **13.** GCF is $2x$; $2x(2x^3 - 1)$

14. $3a(3a^3 - 1)$ **15.** Not factorable   **16.** Not factorable

17. $(\sqrt{49u^4} + \sqrt{36v^4})(\sqrt{49u^4} - \sqrt{36v^4})$
$= (7u^2 + 6v^2)(7u^2 - 6v^2)$

18. $(9m^3 + 10n^2)(9m^3 - 10n^2)$

19. $x^6 - a^4 = (\sqrt{x^6} + \sqrt{a^4})(\sqrt{x^6} - \sqrt{a^4}) = (x^3 + a^2)(x^3 - a^2)$

20. $(b + y^3)(b - y^3)$

21. $2x^2 - 18 = 2(x^2 - 9) = 2(x + 3)(x - 3)$

22. $3(x + 2)(x - 2)$ **23.** Not factorable

24. Not factorable

25. $(\sqrt{a^2b^2} + \sqrt{c^2d^2})(\sqrt{a^2b^2} - \sqrt{c^2d^2}) = (ab + cd)(ab - cd)$

26. $(mn + rs)(mn - rs)$

27. $(\sqrt{49} + \sqrt{25w^2z^2})(\sqrt{49} - \sqrt{25w^2z^2})$
$= (7 + 5wz)(7 - 5wz)$

28. $(6 + 5uv)(6 - 5uv)$

29. $(\sqrt{4h^4k^4} + \sqrt{1})(\sqrt{4h^4k^4} - \sqrt{1}) = (2h^2k^2 + 1)(2h^2k^2 - 1)$

30. $(3x^2y^2 + 1)(3x^2y^2 - 1)$

31. $(\sqrt{81a^4b^6} + \sqrt{16m^2n^8})(\sqrt{81a^4b^6} - \sqrt{16m^2n^8})$
$= (9a^2b^3 + 4mn^4)(9a^2b^3 - 4mn^4)$

32. $(7c^4d^2 + 10e^3f)(7c^4d^2 - 10e^3f)$ **33.** $7x^2(7x^2y^2 - 1)$

34. $5y^2(5x^2y^2 - 1)$ **35.** $5x(x^2 - 9y^2) = 5x(x + 3y)(x - 3y)$

36. $11r(r + 2s)(r - 2s)$

37. $225x^2s^2 - 100x^2t^2 = 25x^2(9s^2 - 4t^2)$
$= 25x^2(3s + 2t)(3s - 2t)$

38. $16y^2(2u + 3v)(2u - 3v)$

Exercises 7.3A   (page 339)

1. $(x + 2)(x + 4)$ **2.** $(x + 1)(x + 8)$ **3.** $(x + 1)(x + 4)$

4. $(x + 2)(x + 2)$ or $(x + 2)^2$ **5.** $(k + 1)(k + 6)$

6. $(k + 2)(k + 3)$ **7.** $u^2 + 7u + 10 = (u + 2)(u + 5)$

8. $(u + 1)(u + 10)$ **9.** Not factorable   **10.** Not factorable

11. $(b - 2)(b - 7)$ **12.** $(b - 1)(b - 14)$

13. $(z - 4)(z - 5)$ **14.** $(z - 2)(z - 10)$

15. $x^2 - 11x + 18 = (x - 2)(x - 9)$ **16.** $(x - 3)(x - 6)$

17. $(x + 10)(x - 1)$ **18.** $(y - 5)(y + 2)$

19. $(z - 3)(z + 2)$ **20.** $(m + 6)(m - 1)$

21. GCF is $5x$; $5x(x + 2)$ **22.** $4y^2(2y + 1)$

23. $(x + 5)(x - 1)$ **24.** $(y + 7)(y - 1)$

25. Not factorable   **26.** Not factorable

27. $z^3(z^2 + 9z - 10) = z^3(z + 10)(z - 1)$

28. $x^2(x + 8)(x - 1)$ **29.** Not factorable

30. Not factorable

31. $(u^2 - 4)(u^2 + 16) = (u + 2)(u - 2)(u^2 + 16)$

32. $(v^2 + 2)(v^2 - 32)$

33. $(4 - v)(4 - v)$ or $(4 - v)^2$ or $(v - 4)^2$

34. $(2 - v)(8 - v)$ or $(v - 2)(v - 8)$ **35.** $(b + 4d)(b - 15d)$

36. $(c + 20x)(c - 3x)$ **37.** $(r + 3s)(r - 16s)$

38. $(s + 24t)(s - 2t)$

39. $x^2(x^2 + 2x - 35) = x^2(x + 7)(x - 5)$

40. $x^2(x + 8)(x - 6)$

41. $x(x^2 + 14x - 15) = x(x + 15)(x - 1)$

42. $x^2(x + 9)(x - 1)$

43. $3(x^2 + 2x - 8) = 3(x + 4)(x - 2)$

44. $5(x + 5)(x - 2)$

45. $4(x^2 - 4x + 3) = 4(x - 3)(x - 1)$

46. $2(x - 3)(x - 4)$ **47.** $x^2(x^2 + 6x + 1)$

48. $y^2(y^2 + 5y + 1)$ **49.** $(x^2 + 3)(x^2 - 3)$

50. $(a^2 + 5)(a^2 - 5)$

Exercises 7.3B   (page 345)

1. $(x + 2)(3x + 1)$ **2.** $(x + 1)(3x + 2)$

3. $(x + 1)(5x + 2)$ **4.** $(x + 2)(5x + 1)$

5. $4x^2 + 7x + 3 = (x + 1)(4x + 3)$ **6.** $(x + 3)(4x + 1)$

7. Not factorable   **8.** Not factorable

9. $(a - 3)(5a - 1)$ **10.** $(m - 1)(5m - 3)$

11. $(b - 7)(3b - 1)$ **12.** $(u - 1)(3u - 7)$

13. $(z - 7)(5z - 1)$ **14.** $(z - 1)(5z - 7)$

15. $3n^2 + 14n - 5 = (n + 5)(3n - 1)$

16. $(k - 1)(5k + 7)$ **17.** $(3x + 7)(3x - 7)$

18. $(4y + 1)(4y - 1)$ **19.** $(x + 3y)(7x + 2y)$

20. $(a + 6b)(7a + b)$ **21.** $(h - k)(7h - 4k)$

22. $(h - 2k)(7h - 2k)$ **23.** Not factorable

24. Not factorable   **25.** $(7x - 3)(7x - 3)$ or $(7x - 3)^2$

26. $(5x - 2)(5x - 2)$ or $(5x - 2)^2$

27. $3u(6u^2 + 13u - 5) = 3u(2u + 5)(3u - 1)$

28. $2y(2y - 1)(3y + 5)$ **29.** $4(x^2 + x + 1)$

30. $6(x^2 + x + 3)$ **31.** $7(x^2 - 7)$ **32.** $5(x^2 - 5)$

33. $(3 - v)(2 - 5v)$ **34.** $(1 - v)(6 - 5v)$

35. $3x^2 + 20x + 12 = (3x + 2)(x + 6)$ **36.** $(3x + 4)(x + 3)$

37. $5(9x^2 - 24x + 16) = 5(3x - 4)(3x - 4)$, or $5(3x - 4)^2$

38. $4(4x - 3)(4x - 3)$, or $4(4x - 3)^2$ **39.** $(2x - 3)(4x + 5)$

40. $(8x + 5)(x - 3)$ **41.** $(2y - 5)(3y - 2)$

42. $(2y - 1)(3y - 10)$ **43.** $(3a - 2)(3a + 10)$

44. $(3b + 5)(3b - 4)$ **45.** $(2e^2 - 5)(3e^2 + 4)$

46. $(5f^2 + 3)(2f^2 - 7)$ **47.** Not factorable

48. Not factorable

Sections 7.1–7.3 Review Exercises   (page 346)

1. $4(2x - 1)$ **2.** Not factorable   **3.** $(m + 2)(m - 2)$

4. Not factorable   **5.** $(x + 3)(x + 7)$ **6.** Not factorable

7. $2u(u + 2)$ **8.** $3b(1 - 2b + 4b^2)$ **9.** $(z + 2)(z - 9)$

10. Not factorable   **11.** $(4x - 1)(x - 6)$

12. $(2x + 5)(2x + 5)$ or $(2x + 5)^2$

13. $9(k^2 - 16) = 9(k + 4)(k - 4)$ **14.** Not factorable

15. $2(4 - a^2) = 2(2 + a)(2 - a)$ **16.** $2(5c - 2)(c - 4)$

17. $3uv(5u - 1)$ **18.** $5ab(b - 2a - 1)$

19. $(2x + 3y)(5x - 8y)$ **20.** $4x^2y^2(2x^3 - 3y^2 - 4)$

21. $3(y^2 - 100) = 3(y + 10)(y - 10)$ **22.** $(t + 8)(t - 9)$

23. $(5u + 7t)(u - t)$ **24.** Not factorable

25. $(7x + 1)(x + 1)$ **26.** $6(x + 3)(x + 5)$

27. Not factorable   **28.** $(s - 7)(s - 9)$

29. $(u + v)(3 - t)$ **30.** Not factorable

31. $(4m - 1)(m + 6)$ **32.** $(w - 5)(v - 8)$

Exercises 7.4 (page 352)

1. $am + bm + an + bn = m(a + b) + n(a + b)$
$= (a + b)(m + n)$

2. $(u + v)(c + d)$

3. $st + 4t + 3su + 12u = t(s + 4) + 3u(s + 4)$
$= (s + 4)(t + 3u)$

4. $(c + 1)(5z + 7t)$

5. $3xr - 6yr + 4x - 8y = 3r(x - 2y) + 4(x - 2y)$
$= (x - 2y)(3r + 4)$

6. $(2s - 3t)(2m + 5n)$

7. $mx - nx - my + ny = x(m - n) - y(m - n)$
$= (m - n)(x - y)$

8. $(h - k)(a - b)$

9. $xy + x - y - 1 = x(y + 1) - 1(y + 1) = (y + 1)(x - 1)$

10. $(a - 1)(d + 1)$

11. $3a^2 - 6ab + 2a - 4b = 3a(a - 2b) + 2(a - 2b)$
$= (a - 2b)(3a + 2)$

12. $(h - 3k)(2h + 5)$

13. $6e^2 - 2ef - 9e + 3f = 2e(3e - f) - 3(3e - f)$
$= (3e - f)(2e - 3)$

14. $(2m - n)(4m - 3)$

15. $h^2 - k^2 + 2h + 2k = (h + k)(h - k) + 2(h + k)$
$= (h + k)(h - k + 2)$

16. $(x + y)(x - y + 4)$

17. $x^3 + 3x^2 - 4x + 12 = x^2(x + 3) - 4(x - 3)$; not factorable

18. Not factorable

19. $a^3 - 2a^2 - 4a + 8 = a^2(a - 2) - 4(a - 2)$
$= (a - 2)(a^2 - 4) = (a - 2)(a + 2)(a - 2)$
$= (a - 2)^2(a + 2)$

20. $(x - 3)(x + 3)(x - 3) = (x - 3)^2(x + 3)$

21. $10xy - 15y + 8x - 12 = 5y(2x - 3) + 4(2x - 3)$
$= (2x - 3)(5y + 4)$

22. $(7 + 3n)(5 - 6m)$

23. $a^2 - 4 + ab - 2b = (a + 2)(a - 2) + b(a - 2)$
$= (a - 2)(a + 2 + b)$

24. $(x - 5)(x + 5 - y)$

Exercises 7.5 (page 354)

1. $MP = 3 \cdot 2 = 6$
$6 = (-1)(-6) = \boxed{(1)(6)}$ and $(1) + (6) = 7$
$= (-2)(-3) = (2)(3)$
$\underbrace{3x^2 + 1x}_{} + \underbrace{6x + 2}_{}$
$= x(3x + 1) + 2(3x + 1)$
$= (3x + 1)(x + 2)$

2. $(x + 1)(3x + 2)$

3. $MP = 5 \cdot 2 = 10$
$10 = (-1)(-10) = (1)(10)$
$= (-2)(-5) = \boxed{(2)(5)}$ and $(2) + (5) = 7$
$\underbrace{5x^2 + 2x}_{} + \underbrace{5x + 2}_{}$
$= x(5x + 2) + 1(5x + 2)$
$= (5x + 2)(x + 1)$

4. $(x + 2)(5x + 1)$

5. $4x^2 + 7x + 3$
$MP = 4 \cdot 3 = 12$

$12 = (-1)(-12) = (1)(12)$
$= (-2)(-6) = (2)(6)$
$= (-3)(-4) = \boxed{(3)(4)}$ and $(3) + (4) = 7$
$\underbrace{4x^2 + 3x}_{} + \underbrace{4x + 3}_{}$
$= x(4x + 3) + 1(4x + 3)$
$= (4x + 3)(x + 1)$

6. $(x + 3)(4x + 1)$

7. $MP = 5 \cdot 4 = 20$
$20 = (1)(20) = (-1)(-20)$ None of the
$= (2)(10) = (-2)(-10)$ sums of these
$= (4)(5) = (-4)(-5)$ pairs is $+20$.
Not factorable

8. Not factorable

9. $MP = 5 \cdot 3 = 15$
$15 = (1)(15) = \boxed{(-1)(-15)}$ and $(-1) + (-15) = -16$
$= (3)(5) = (-3)(-5)$
$\underbrace{5a^2 - 1a}_{} - \underbrace{15a + 3}_{}$
$= a(5a - 1) - 3(5a - 1)$
$= (5a - 1)(a - 3)$

10. $(m - 1)(5m - 3)$

11. $MP = 3 \cdot 7 = 21$
$21 = (1)(21) = \boxed{(-1)(-21)}$ and $(-1) + (-21) = -22$
$= (3)(7) = (-3)(-7)$
$\underbrace{3b^2 - 1b}_{} - \underbrace{21b + 7}_{}$
$= b(3b - 1) - 7(3b - 1)$
$= (3b - 1)(b - 7)$

12. $(u - 1)(3u - 7)$

13. $MP = 5 \cdot 7 = 35$
$35 = (1)(35) = \boxed{(-1)(-35)}$ and $(-1) + (-35) = -36$
$= (5)(7) = (-5)(-7)$
$\underbrace{5z^2 - 1z}_{} - \underbrace{35z + 7}_{}$
$= z(5z - 1) - 7(5z - 1)$
$= (5z - 1)(z - 7)$

14. $(z - 1)(5z - 7)$

15. $3n^2 + 14n - 5$
$MP = 3 \cdot (-5) = -15$
$-15 = (1)(-15) = \boxed{(-1)(15)}$ and $(-1) + (15) = 14$
$= (3)(-5) = (-3)(5)$
$\underbrace{3n^2 - 1n}_{} + \underbrace{15n - 5}_{}$
$= n(3n - 1) + 5(3n - 1)$
$= (3n - 1)(n + 5)$

16. $(k - 1)(5k + 7)$

17. (MP method need not be used.) $(3x + 7)(3x - 7)$

18. $(4y + 1)(4y - 1)$

19. $MP = 7 \cdot 6 = 42$
$42 = (-1)(-42) = (1)(42)$
$= (-2)(-21) = \boxed{(2)(21)}$ and $(2) + (21) = 23$
$= (-3)(-14) = (3)(14)$
$= (-6)(-7) = (6)(7)$
$\underbrace{7x^2 + 2xy}_{} + \underbrace{21xy + 6y^2}_{}$
$= x(7x + 2y) + 3y(7x + 2y)$
$= (7x + 2y)(x + 3y)$

20. $(a + 6b)(7a + b)$

21. $MP = 7 \cdot 4 = 28$
$28 = (1)(28) = (-1)(-28)$
$= (2)(14) = (-2)(-14)$
$= (4)(7) = \boxed{(-4)(-7)}$ and $(-4) + (-7) = -11$
$\underbrace{7h^2 - 4hk}_{} - \underbrace{7hk + 4k^2}_{}$
$= h(7h - 4k) - k(7h - 4k)$
$= (7h - 4k)(h - k)$

22. $(h - 2k)(7h - 2k)$

23. $\text{MP} = 3(-6) = -18$
$$-18 = (1)(-18) = (-1)(18)$$
$$ = (2)(-9) = (-2)(9)$$
$$ = (3)(-6) = (-3)(6)$$
None of the sums of these pairs is -18.
Not factorable

24. Not factorable

25. $\text{MP} = 49 \cdot 9 = 441$
$$441 = (1)(441) = (-1)(-441)$$
$$ = (3)(147) = (-3)(-147)$$
$$ = (7)(63) = (-7)(-63)$$
$$ = (9)(49) = (-9)(-49)$$
$$ = (21)(21) = \boxed{(-21)(-21)} \text{ and } (-21) + (-21) = -42$$
$$\underbrace{49x^2 - 21x}\; \underbrace{- 21x + 9}$$
$$= 7x(7x - 3) - 3(7x - 3)$$
$$= (7x - 3)(7x - 3), \text{ or } (7x - 3)^2$$

26. $(5x - 2)(5x - 2)$ or $(5x - 2)^2$

27. $3u(6u^2 + 13u - 5)$
To factor $6u^2 + 13u - 5$:
$\text{MP} = 6 \cdot (-5) = -30$
$$-30 = (1)(-30) = (-1)(30)$$
$$ = (2)(-15) = \boxed{(-2)(15)} \text{ and } (-2) + (15) = 13$$
$$ = (3)(-10) = (-3)(10)$$
$$ = (5)(-6) = (-5)(6)$$
$$\underbrace{6u^2 - 2u}\; \underbrace{+ 15u - 5}$$
$$= 2u(3u - 1) + 5(3u - 1)$$
$$= (3u - 1)(2u + 5)$$
The final answer is $3u(2u + 5)(3u - 1)$.

28. $2y(2y - 1)(3y + 5)$

29. $4(x^2 + x + 1)$
To factor $x^2 + x + 1$:
$\text{MP} = 1 \cdot 1 = 1$
$$1 = (1)(1) = (-1)(-1)$$
The sum of neither pair is 1. Therefore, $x^2 + x + 1$ is not factorable. The final answer is $4(x^2 + x + 1)$.

30. $6(x^2 + x + 3)$

Exercises 7.6   (page 356)

1. $2(x^2 - 4y^2) = 2(\sqrt{x^2} + \sqrt{4y^2})(\sqrt{x^2} - \sqrt{4y^2})$
$= 2(x + 2y)(x - 2y)$

2. $3(x + 3y)(x - 3y)$

3. $5(a^4 - 4b^2) = 5(\sqrt{a^4} + \sqrt{4b^2})(\sqrt{a^4} - \sqrt{4b^2})$
$= 5(a^2 + 2b)(a^2 - 2b)$

4. $6(m + 3n^2)(m - 3n^2)$

5. $(\sqrt{x^4} + \sqrt{y^4})(\sqrt{x^4} - \sqrt{y^4}) = (x^2 + y^2)(x^2 - y^2)$
$= (x^2 + y^2)(x + y)(x - y)$

6. $(a^2 + 4)(a + 2)(a - 2)$

7. $2(2v^2 + 7v - 4) = 2(v + 4)(2v - 1)$ **8.** $3(v - 5)(2v + 1)$

9. $4(2z^2 - 3z - 2) = 4(z - 2)(2z + 1)$

10. $3(2z - 3)(3z + 1)$

11. $4(x^2 - 25) = 4(\sqrt{x^2} + \sqrt{25})(\sqrt{x^2} - \sqrt{25})$
$= 4(x + 5)(x - 5)$

12. $9(x + 2)(x - 2)$

13. $2(6x^2 + 5x - 4) = 2(2x - 1)(3x + 4)$

14. $3(3x + 2)(5x - 4)$

15. $a(b^2 - 2b + 1) = a(b - 1)(b - 1) = a(b - 1)^2$

16. $a(u - 1)^2$

17. $(\sqrt{x^4} + \sqrt{81})(\sqrt{x^4} - \sqrt{81}) = (x^2 + 9)(x^2 - 9)$
$= (x^2 + 9)(x + 3)(x - 3)$

18. $(4y^4 + z^2)(2y^2 + z)(2y^2 - z)$ **19.** $16(x^2 + 1)$

20. $25(b^2 + 4)$ **21.** $2u(u^2 + uv - 6v^2) = 2u(u + 3v)(u - 2v)$

22. $3m(m + 3n)(m - 4n)$

23. $4h(2h^2 - 5hk + 3k^2) = 4h(2h - 3k)(h - k)$

24. $5k(h - 2k)(3h - k)$

25. $a^3b^2(a^2 - 4b^2) = a^3b^2(a + 2b)(a - 2b)$

26. $x^2y^2(y + 10x)(y - 10x)$

27. $2a(x^2 - 4a^2y^2) = 2a(\sqrt{x^2} + \sqrt{4a^2y^2})(\sqrt{x^2} - \sqrt{4a^2y^2})$
$= 2a(x + 2ay)(x - 2ay)$

28. $3b^2(x^2 + 2y)(x^2 - 2y)$

29. $4(3 + x) - x^2(3 + x) = (3 + x)(4 - x^2)$
$= (3 + x)(2 + x)(2 - x)$

30. $(5 - z)(3 + z)(3 - z)$

31. $6my - \underline{4nz} + 15mz - \underline{5zn} = 6my - 9nz + 15mz$
$= 3(2my - 3nz + 5mz)$

32. $4(xy + mn)$

33. $(x^2 - 4)(x^2 - 4) = (x + 2)(x - 2)(x + 2)(x - 2)$
$= (x + 2)^2(x - 2)^2$

34. $(y + 3)^2(y - 3)^2$

35. $(x^4 + 1)(x^4 - 1) = (x^4 + 1)(x^2 + 1)(x^2 - 1)$
$= (x^4 + 1)(x^2 + 1)(x + 1)(x - 1)$

36. $(a^4 + b^4)(a^2 + b^2)(a + b)(a - b)$

37. $(x + a)(x - a) - 4(x - a) = (x - a)(x + a - 4)$

38. $(m - 5)(m + 5 + n)$

39. $6[ac + bc - ad - bd] = 6[c(a + b) - d(a + b)]$
$= 6(a + b)(c - d)$

40. $(2c + d)(5y - 3z)$

Exercises 7.7   (page 362)

(The checks will not be shown.)

1. $(x - 5)(x + 4) = 0$ **2.** $-7, 2$
$$x - 5 = 0 \quad | \quad x + 4 = 0$$
$$x = 5 \quad | \quad x = -4$$
The solutions are 5 and -4

3. $3x(x - 4) = 0$ **4.** $0, -6$
$$3x = 0 \quad | \quad x - 4 = 0$$
$$x = 0 \quad | \quad x = 4$$
The solutions are 0 and 4.

5. $(x + 10)(2x - 3) = 0$ **6.** $8, -\frac{2}{3}$
$$x + 10 = 0 \quad | \quad 2x - 3 = 0$$
$$x = -10 \quad | \quad 2x = 3$$
$$\quad | \quad x = \frac{3}{2} = 1\frac{1}{2}$$
The solutions are -10 and $1\frac{1}{2}$.

7. $(x + 1)(x + 8) = 0$ **8.** $-2, -4$
$$x + 1 = 0 \quad | \quad x + 8 = 0$$
$$x = -1 \quad | \quad x = -8$$
The solutions are -1 and -8.

9. $(x - 4)(x + 3) = 0$ **10.** $-4, 3$
$$x - 4 = 0 \quad | \quad x + 3 = 0$$
$$x = 4 \quad | \quad x = -3$$
The solutions are 4 and -3.

11. $x^2 - 64 = 0$ **12.** $12, -12$
$(x + 8)(x - 8) = 0$
$$x + 8 = 0 \quad | \quad x - 8 = 0$$
$$x = -8 \quad | \quad x = 8$$
The solutions are -8 and 8.

13. $2x(3x - 5) = 0$ **14.** $0, \frac{7}{2}$
$$2x = 0 \quad | \quad 3x - 5 = 0$$
$$x = 0 \quad | \quad 3x = 5$$
$$\quad | \quad x = \frac{5}{3}$$
The solutions are 0 and $\frac{5}{3}$.

15. $0 = 4w^2 - 24w$
$0 = 4w(w - 6)$

$4w = 0$	$w - 6 = 0$
$w = 0$	$w = 6$

The solutions are 0 and 6.

16. $0, 4$

17. $5a^2 - 16a + 3 = 0$
$(a - 3)(5a - 1) = 0$

$a - 3 = 0$	$5a - 1 = 0$
$a = 3$	$5a = 1$
	$a = \frac{1}{5}$

The solutions are 3 and $\frac{1}{5}$.

18. $7, \frac{1}{3}$

19. $3u^2 - 2u - 5 = 0$
$(u + 1)(3u - 5) = 0$

$u + 1 = 0$	$3u - 5 = 0$
$u = -1$	$3u = 5$
	$u = \frac{5}{3} = 1\frac{2}{3}$

The solutions are -1 and $1\frac{2}{3}$.

20. $7, -\frac{1}{5}$

21. $x^2 - 5x + 6 = 2$
$x^2 - 5x + 4 = 0$
$(x - 1)(x - 4) = 0$

$x - 1 = 0$	$x - 4 = 0$
$x = 1$	$x = 4$

The solutions are 1 and 4.

22. $2, 6$

23. $x^2 - 4x = 12$
$x^2 - 4x - 12 = 0$
$(x - 6)(x + 2) = 0$

$x - 6 = 0$	$x + 2 = 0$
$x = 6$	$x = -2$

The solutions are 6 and -2.

24. $-3, 5$

25. $4x(2x - 1)(3x + 7) = 0$

$4x = 0$	$2x - 1 = 0$	$3x + 7 = 0$
$x = 0$	$2x = 1$	$3x = -7$
	$x = \frac{1}{2}$	$x = -\frac{7}{3} = -2\frac{1}{3}$

The solutions are $0, \frac{1}{2}$, and $-2\frac{1}{3}$.

26. $0, \frac{3}{4}, \frac{6}{7}$

27. $2x^3 + x^2 - 3x = 0$
$x(2x^2 + x - 3) = 0$
$x(x - 1)(2x + 3) = 0$

$x = 0$	$x - 1 = 0$	$2x + 3 = 0$
	$x = 1$	$2x = -3$
		$x = -\frac{3}{2} = -1\frac{1}{2}$

The solutions are $0, 1$, and $-1\frac{1}{2}$.

28. $0, -5, \frac{1}{2}$

29. $2a^2(a - 5) = 0$
$2aa(a - 5) = 0$

$2 \neq 0$	$a = 0$	$a = 0$	$a - 5 = 0$
		$\uparrow$	$a = 5$

$a = 0$ need not be listed twice.
The solutions are 0 and 5.

30. $0, 6$

Exercises 7.8 (page 369)

1. Let $x = $ smaller number
Then $x + 5 = $ larger number
$x(x + 5) = 14$
$x^2 + 5x = 14$
$x^2 + 5x - 14 = 0$
$(x + 7)(x - 2) = 0$

$x + 7 = 0$	$x - 2 = 0$
$x = -7$	$x = 2$
$x + 5 = -2$	$x + 5 = 7$

First answer: The numbers are 2 and 7. Second answer: The numbers are -7 and -2.

2. First answer: The numbers are 3 and 9. Second answer: The numbers are -9 and -3.

3. Let $x = $ one number
Then $12 - x = $ other number
$x(12 - x) = 35$
$12x - x^2 = 35$
$0 = x^2 - 12x + 35$
$0 = (x - 5)(x - 7)$

$x - 5 = 0$	$x - 7 = 0$
$x = 5$	$x = 7$
$12 - x = 7$	$12 - x = 5$

The numbers are 5 and 7.

4. 2 and -6

5. Let $x = $ first integer
Then $x + 1 = $ second integer
and $x + 2 = $ third integer
$x(x + 1) + (x + 1)(x + 2) = 8$
$x^2 + x + x^2 + 3x + 2 = 8$
$2x^2 + 4x - 6 = 0$
$2(x^2 + 2x - 3) = 0$
$2(x - 1)(x + 3) = 0$

$2 \neq 0$	$x - 1 = 0$	$x + 3 = 0$
	$x = 1$	$x = -3$
	$x + 1 = 2$	$x + 1 = -2$
	$x + 2 = 3$	$x + 2 = -1$

First answer: The integers are 1, 2, and 3.
Second answer: The integers are $-3, -2$, and -1.

6. The integers are 2, 3, and 4.

7. Let $x = $ first even integer
Then $x + 2 = $ second even integer
and $x + 4 = $ third even integer
$2x(x + 2) = 16 + (x + 2)(x + 4)$
$2x^2 + 4x = 16 + x^2 + 6x + 8$
$x^2 - 2x - 24 = 0$
$(x + 4)(x - 6) = 0$

$x + 4 = 0$	$x - 6 = 0$
$x = -4$	$x = 6$
$x + 2 = -2$	$x + 2 = 8$
$x + 4 = 0$	$x + 4 = 10$

First answer: The integers are $-4, -2$, and 0. Second answer: The integers are 6, 8, and 10.

8. First answer: The integers are 5, 7, and 9. Second answer: The integers are $-15, -13$, and -11.

9. Let $w = $ width (in ft)
Then $w + 5 = $ length (in ft)
Area $= (w + 5)w$
$(w + 5)w = 84$
$w^2 + 5w = 84$
$w^2 + 5w - 84 = 0$
$(w + 12)(w - 7) = 0$

$w + 12 = 0$	$w - 7 = 0$
$w = -12$ Reject	$w = 7$
	$w + 5 = 12$

The width is 7 ft, and the length is 12 ft.

10. Length $= 7$ ft, width $= 4$ ft

11. Let $x = $ length of side of larger square (in cm)
Then $x - 3 = $ length of side of smaller square (in cm)
$x^2 = 4(x - 3)^2$
$x^2 = 4(x^2 - 6x + 9)$
$x^2 = 4x^2 - 24x + 36$
$0 = 3x^2 - 24x + 36$
$0 = 3(x^2 - 8x + 12)$
$0 = 3(x - 2)(x - 6)$

7

The checks will not be shown for Exercises 17–25.

17. $5x^2 + 9x - 2 = 0$ **18.** $-\frac{11}{17}$
$(5x - 1)(x + 2) = 0$
$5x - 1 = 0 \;\mid\; x + 2 = 0$
$5x = 1 \;\mid\; x = -2$
$x = \frac{1}{5}$
The solutions are $\frac{1}{5}$ and -2.

19. $2x^2 + 12 = 11x$ **20.** $\frac{12}{5}$
$2x^2 - 11x + 12 = 0$
$(2x - 3)(x - 4) = 0$
$2x - 3 = 0 \;\mid\; x - 4 = 0$
$2x = 3 \;\mid\; x = 4$
$x = \frac{3}{2}$
The solutions are $\frac{3}{2}$ and 4.

21. Let $\quad x$ = smallest odd integer
Then $x + 2$ = next odd integer
$x + (x + 2) = -22$
$2x + 2 = -22$
$2x = -24$
$x = -12$
$x + 2 = -10$
The integers are -12 and -10.

22. The width is 10 m, and the length is 15 m.

23. Let x = number of pounds of cashews
[There will be $(20 + x)$ lb in the mixture.]
$7.50x + 3.50(20) = 5.90(20 + x)$
$75x + 35(20) = 59(20 + x)$
$75x + 700 = 1{,}180 + 59x$
$16x = 480$
$x = 30$
Therefore, 30 lb of cashews should be used.

24. 45 L

25. Let $\quad x$ = number of quarters
Then $x + 1$ = number of dimes
and $\quad 3x$ = number of nickels
$25x + 10(x + 1) + 5(3x) = 260$
$25x + 10x + 10 + 15x = 260$
$50x + 10 = 260$
$50x = 250$
$x = 5$
$x + 1 = 6$
$3x = 15$
Margaret has 5 quarters, 6 dimes, and 15 nickels.

Exercises 8.1   (page 380)

1. Yes, because if we multiply both x and $2y$ by 5, we get $5x$ and $10y$.

2. Yes

3. No; we can't get the second rational expression from the first by multiplying both x and $2y$ by the same number.

4. No

5. Yes, because if we multiply both $(x + 1)$ and $2(3x - 2)$ by 6, we get $6(x + 1)$ and $12(3x - 2)$.

6. Yes

7. Yes, because if we multiply both $s - t$ and $t - s$ by -1, we get $t - s$ and $s - t$.

8. Yes

Exercises 8.2   (page 385)

1. $\dfrac{\overset{1}{2x}}{\underset{3x^2}{6x^3}} = \dfrac{1}{3x^2}$ **2.** $\dfrac{1}{3y}$ **3.** $\dfrac{\overset{2b}{6ab^2}}{\underset{1}{3ab}} = 2b$ **4.** $2m$

5. $\dfrac{\overset{3}{\overset{}{9x^4}}}{\underset{2}{\underset{}{6x^6}}} = \dfrac{3}{2x^2}$ **6.** $\dfrac{3}{2y^4}$ **7.** $\dfrac{\overset{2}{10x^4}}{\underset{3}{15x^3}} = \dfrac{2x}{3}$ **8.** $\dfrac{5y^3}{3}$

9. $\dfrac{\overset{3\;\;h^2\;k^2}{12h^4k^3}}{\underset{2\;\;1\;\;1}{8h^2k}} = \dfrac{3h^2k^2}{2}$ **10.** $\dfrac{4a^4b}{3}$ **11.** $\dfrac{\overset{5\;\;x^2\;1}{15x^4yz^4}}{\underset{6\;\;1\;\;y^2}{18xy^3}} = \dfrac{5x^2z^4}{6y^2}$

12. $\dfrac{3a^2b^2}{7c^2}$ **13.** $\dfrac{\overset{8\;\;1\;\;t^3\;1}{24st^4u^2}}{\underset{5\;\;s\;\;1\;\;u}{15s^2tu^3}} = \dfrac{8t^3}{5su}$ **14.** $\dfrac{9u^2v^4}{4w}$

15. $\dfrac{\overset{3\;\;\;x\;y^4}{21x^3y^5}}{\underset{2\;\;\;1}{14yz^3}} = \dfrac{3x^3y^4}{2z^3}$ **16.** $\dfrac{5a}{7b}$ **17.** $\dfrac{5(\overset{1}{x-2})}{\underset{1}{x-2}} = 5$

18. 3   **19.** $\dfrac{7(\overset{1}{x-3})}{\underset{1}{15x(x-3)}} = \dfrac{7}{15x}$ **20.** $\dfrac{3}{y^2}$

21. $\dfrac{\overset{x^2}{6x^3y}}{\underset{1}{6xy(5x^2 - 3y)}} = \dfrac{x}{5x^2 - 3y}$ **22.** $\dfrac{1}{2a^2 - 4b}$

23. $\dfrac{\overset{2}{4xy(2x+3y)}}{\underset{3x}{6x^2y(2x+3y)}} = \dfrac{2}{3x}$ **24.** $\dfrac{3s}{4t}$ **25.** $-\dfrac{5x - 6}{6 - 5x} = -(-1) = 1$

26. -1 **27.** $\dfrac{\overset{1}{5x}(x + 6)}{\underset{2}{10x}(x - 4)} = \dfrac{x + 6}{2(x - 4)}$ **28.** $\dfrac{x}{3}$

29. $\frac{6}{4} = \frac{3}{2}$; incorrect to "cancel" the 4's   **30.** 4

31. Cannot be reduced   **32.** Cannot be reduced

33. $\dfrac{\overset{1}{(x+1)}(x - 1)}{\underset{1}{(x+1)}} = x - 1$ **34.** $x + 2$

35. $\dfrac{(3x - 2)(\overset{1}{2x+1})}{(5x - 1)(\underset{1}{2x+1})} = \dfrac{3x - 2}{5x - 1}$ **36.** $\dfrac{2x - 3}{3x + 2}$

37. $\dfrac{(\overset{1}{x+y})(x - y)}{(\underset{1}{x+y})(x + y)} = \dfrac{x - y}{x + y}$ **38.** $\dfrac{a + 3b}{a - 3b}$

39. $\dfrac{(2y - 3x)(y + 2x)}{(3x - 2y)(x + y)} = -1\left(\dfrac{y + 2x}{x + y}\right) = -\dfrac{2x + y}{x + y}$

40. $-\dfrac{3x + 2y}{2x + 3y}$

41. $\dfrac{2(4x^2 - y^2)}{a(2x - y) + b(2x - y)} = \dfrac{2(2x + y)(\overset{1}{2x-y})}{(\underset{1}{2x-y})(a + b)} = \dfrac{2(2x + y)}{a + b}$

42. $\dfrac{3(x - 2y)}{a + b}$ **43.** $\dfrac{\overset{1}{8-z}}{\underset{1}{8-z}} = 1$ **44.** 1

45. $\dfrac{(a - 2b)(\overset{1}{a-b})}{(2a + b)(\underset{1}{a-b})} = \dfrac{a - 2b}{2a + b}$ **46.** $\dfrac{8}{3n + m}$

47. $\dfrac{(3 + 4x)(3 - 4x)}{(4x - 3)(4x - 3)} = \left(\dfrac{3 + 4x}{4x - 3}\right)(-1) = -\dfrac{3 + 4x}{4x - 3}$

48. $-\dfrac{5 + 3x}{3x - 5}$

49. $\dfrac{(5 + 3x)(2 - x)}{(2x + 5)(x - 2)} = \left(\dfrac{5 + 3x}{2x + 5}\right)(-1) = -\dfrac{5 + 3x}{2x + 5}$

50. $-\dfrac{4 - 5x}{3x + 4}$

51. $\dfrac{3(6-x-x^2)}{6(x^2+x-6)} = \dfrac{(3+x)(2-x)}{2(x+3)(x-2)} = \dfrac{1}{2}(1)(-1) = -\dfrac{1}{2}$

52. $-\dfrac{1}{4}$ **53.** $\dfrac{8x^{-3}}{12x} = \dfrac{\overset{2}{\cancel{8}}}{\underset{3}{\cancel{12}}}\cdot\dfrac{x^{-3}}{x} = \dfrac{2}{3}x^{-3-1} = \dfrac{2}{3}x^{-4} = \dfrac{2}{3}\cdot\dfrac{1}{x^4} = \dfrac{2}{3x^4}$

54. $\dfrac{3}{2y^3}$

55. $\dfrac{24u^4v^{-3}w^0}{27u^{-2}v^3w^6} = \dfrac{\overset{8}{\cancel{24}}}{\underset{9}{\cancel{27}}}\cdot\dfrac{u^4}{u^{-2}}\cdot\dfrac{v^{-3}}{v^3}\cdot\dfrac{w^0}{w^6} = \dfrac{8}{9}\cdot u^{4-(-2)}\cdot v^{-3-3}\cdot\dfrac{1}{w^6}$

$= \dfrac{8}{9}u^6v^{-6}\cdot\dfrac{1}{w^6} = \dfrac{8u^6}{9v^6w^6}$

56. $\dfrac{5b^6}{6a^3c^{11}}$

57. $\dfrac{8xy^3z^{-2}}{22x^0y^{-2}} = \dfrac{\overset{4}{\cancel{8}}}{\underset{11}{\cancel{22}}}\cdot\dfrac{x}{x^0}\cdot\dfrac{y^3}{y^{-2}}\cdot\dfrac{z^{-2}}{1} = \dfrac{4}{11}\cdot\dfrac{x}{1}\cdot y^{3-(-2)}\cdot\dfrac{1}{z^2}$

$= \dfrac{4}{11}\cdot\dfrac{x}{1}\cdot\dfrac{y^5}{1}\cdot\dfrac{1}{z^2} = \dfrac{4xy^5}{11z^2}$

58. $\dfrac{3t^5}{7s^4u^3}$

59. $\left(\dfrac{a^2b^{-4}}{b^{-5}}\right)^2 = (a^2b^{-4-(-5)})^2 = (a^2b^1)^2 = a^{2\cdot2}b^{1\cdot2} = a^4b^2$

60. x^3y^6

61. $\left(\dfrac{mn^{-1}}{m^3}\right)^{-2} = \left(\dfrac{m^3}{mn^{-1}}\right)^2 = \left(\dfrac{m^3}{m}\cdot\dfrac{1}{n^{-1}}\right)^2 = (m^{3-1}n^1)^2 = (m^2n^1)^2$

$= m^{2\cdot2}n^{1\cdot2} = m^4n^2$

62. a^3b^6

63. $\left(\dfrac{x^4}{x^{-1}y^{-2}}\right)^{-1} = \left(\dfrac{x^{-1}y^{-2}}{x^4}\right)^1 = \dfrac{x^{-1}}{x^4}\cdot\dfrac{y^{-2}}{1} = x^{-1-4}\cdot\dfrac{1}{y^2} = x^{-5}\cdot\dfrac{1}{y^2}$

$= \dfrac{1}{x^5}\cdot\dfrac{1}{y^2} = \dfrac{1}{x^5y^2}$

64. $\dfrac{1}{x^5y^4}$

65. $\left(\dfrac{\overset{1}{\cancel{3}}m^{-1}}{\underset{3}{9}n^3}\right)^4 = \left(\dfrac{1}{3}\cdot m^{-1}\cdot\dfrac{1}{n^3}\right)^4 = \left(\dfrac{1}{3}\cdot\dfrac{1}{m}\cdot\dfrac{1}{n^3}\right)^4 = \left(\dfrac{1}{3mn^3}\right)^4$

$= \dfrac{1^4}{3^4m^4n^{3\cdot4}} = \dfrac{1}{81m^4n^{12}}$

66. $\dfrac{1}{16x^2y^4}$

67. $\left(\dfrac{6x^{-3}}{9y^{-2}}\right)^{-3} = \left(\dfrac{\overset{3}{\cancel{9}}y^{-2}}{\underset{2}{6}x^{-3}}\right)^3 = \left(\dfrac{3}{2}\cdot y^{-2}\cdot\dfrac{1}{x^{-3}}\right)^3 = \left(\dfrac{3}{2}\cdot\dfrac{1}{y^2}\cdot\dfrac{x^3}{1}\right)^3$

$= \left(\dfrac{3x^3}{2y^2}\right)^3 = \dfrac{3^3x^{3\cdot3}}{2^3y^{2\cdot3}} = \dfrac{27x^9}{8y^6}$

68. $\dfrac{64a^6}{27b^9}$

69. $\left(\dfrac{8x^2y^{-3}}{14x^3y^{-2}}\right)^{-2} = \left(\dfrac{\overset{7}{\cancel{14}}x^3y^{-2}}{\underset{4}{\cancel{8}}x^2y^{-3}}\right)^2 = \left(\dfrac{7}{4}\cdot\dfrac{x^3}{x^2}\cdot\dfrac{y^{-2}}{y^{-3}}\right)^2$

$= \left(\dfrac{7}{4}\cdot x^{3-2}\cdot y^{-2-(-3)}\right)^2 = \left(\dfrac{7}{4}\cdot x^1\cdot y^1\right)^2 = \left(\dfrac{7}{4}\cdot\dfrac{x}{1}\cdot\dfrac{y}{1}\right)^2$

$= \left(\dfrac{7xy}{4}\right)^2 = \dfrac{7^2x^2y^2}{4^2}$, or $\dfrac{49x^2y^2}{16}$

70. $\dfrac{27a^3b^6}{8}$

71. $\left(\dfrac{x^4y^7}{x^3y^3}\right)^3 = (x^{4-3}y^{7-3})^3 = (x^1y^4)^3 = x^{1\cdot3}y^{4\cdot3} = x^3y^{12}$

72. $27a^{24}b^6$

73. $\left(\dfrac{-3x^6y^8}{xy^3}\right)^3 = \left(\dfrac{-3}{1}\cdot\dfrac{x^6}{x}\cdot\dfrac{y^8}{y^3}\right)^3 = (-3x^5y^5)^3 = (-3)^3x^{5\cdot3}y^{5\cdot3}$

$= -27x^{15}y^{15}$

74. $16m^{20}n^{32}$

75. $\left(\dfrac{-2x^4y^6}{xy^2}\right)^3 = \left(\dfrac{-2}{1}\cdot\dfrac{x^4}{x}\cdot\dfrac{y^6}{y^2}\right)^3 = (-2x^3y^4)^3 = (-2)^3x^{3\cdot3}y^{4\cdot3}$

$= -8x^9y^{12}$

76. $-8s^6t^9$

Exercises 8.3   (page 394)

1. $\dfrac{5}{6}\div\dfrac{5}{3} = \dfrac{5}{6}\cdot\dfrac{3}{5} = \dfrac{1}{2}$ **2.** $\dfrac{3}{14}$ **3.** $\dfrac{\overset{1}{\cancel{4}a^3}}{\underset{1}{\cancel{5b^2}}}\cdot\dfrac{\overset{2}{\cancel{10}b}}{\underset{2}{\cancel{8a^2}}} = \dfrac{a}{b}$

4. $\dfrac{c}{d}$ **5.** $\dfrac{3x^2}{16}\div\dfrac{x}{8} = \dfrac{3x^2}{\underset{2}{\cancel{16}}}\cdot\dfrac{\overset{1}{\cancel{8}}}{x} = \dfrac{3x}{2}$ **6.** $\dfrac{3y}{2}$

7. $\dfrac{\overset{1}{\cancel{3}}x^4y^2z}{\underset{6}{\underset{2}{\cancel{18}}}xy}\cdot\dfrac{\overset{5}{\cancel{15}}z}{x^3yz^2} = \dfrac{5x^4y^2z^2}{2x^4y^2z^2} = \dfrac{5}{2}$ **8.** $\dfrac{9}{2}$

9. $\dfrac{x}{x+2}\cdot\dfrac{5(\overset{1}{\cancel{x+2}})}{x^3} = \dfrac{5}{x^2}$ **10.** $\dfrac{3}{b^3}$

11. $\dfrac{\overset{1}{\cancel{y-2}}}{y}\cdot\dfrac{6}{3(\underset{1}{\cancel{y-2}})} = \dfrac{2}{y}$ **12.** $\dfrac{2}{a}$

13. $\dfrac{b^3}{a+3}\div\dfrac{4b^2}{2a+6} = \dfrac{b^3}{\underset{1}{\cancel{a+3}}}\cdot\dfrac{\overset{1}{2(\cancel{a+3})}}{\underset{2}{\cancel{4b^2}}} = \dfrac{b}{2}$ **14.** $\dfrac{m}{2}$

15. $\dfrac{5s-15}{30s}\div\dfrac{s-3}{45s^2} = \dfrac{\overset{1}{5(\cancel{s-3})}}{\underset{6}{\underset{2}{\cancel{30s}}}}\cdot\dfrac{\overset{15}{\cancel{45s^2}}}{\underset{1}{\cancel{s-3}}} = \dfrac{15s}{2}$ **16.** $4n$

17. $\dfrac{a+4}{a-4}\div\dfrac{a^2+8a+16}{a^2-16} = \dfrac{\overset{1}{\cancel{a+4}}}{\underset{1}{\cancel{a-4}}}\cdot\dfrac{\overset{1}{(\cancel{a+4})}\overset{1}{(\cancel{a-4})}}{\underset{1}{(\cancel{a+4})(\cancel{a+4})}} = 1$

18. 1

19. $\dfrac{5}{\underset{1}{\cancel{z+4}}}\cdot\dfrac{\overset{1}{(\cancel{z+4})}\overset{1}{(\cancel{z-4})}}{(z-4)(\underset{1}{\cancel{z-4}})} = \dfrac{5}{z-4}$ **20.** $\dfrac{7}{x-3}$

21. $\dfrac{3(\overset{1}{\cancel{a-b}})}{4(\underset{2}{\cancel{c+d}})}\cdot\dfrac{\overset{1}{2(\cancel{c+d})}}{\underset{-1}{\cancel{b-a}}} = -\dfrac{3}{2}$, or $-1\dfrac{1}{2}$ **22.** $-\dfrac{7}{2}$, or $-3\dfrac{1}{2}$

23. $\dfrac{\overset{1}{4(\cancel{x-2})}}{\underset{1}{\cancel{4}}}\cdot\dfrac{x+2}{(\cancel{x+2})(\underset{1}{\cancel{x-2}})} = 1$ **24.** 1

25. $\dfrac{4a+4b}{ab^2}\div\dfrac{3a+3b}{a^2b} = \dfrac{4(\overset{1}{\cancel{a+b}})}{ab^2}\cdot\dfrac{a^2b}{3(\underset{1}{\cancel{a+b}})} = \dfrac{4a}{3b}$ **26.** $\dfrac{5y}{4x}$

27. $\dfrac{a^2-9b^2}{a^2-6ab+9b^2}\div\dfrac{a+3b}{a-3b} = \dfrac{\overset{1}{(\cancel{a+3b})}\overset{1}{(\cancel{a-3b})}}{(\underset{1}{\cancel{a-3b}})(\underset{1}{\cancel{a-3b}})}\cdot\dfrac{\overset{1}{\cancel{a-3b}}}{\underset{1}{\cancel{a+3b}}} = 1$

28. 1

25. LCD $= (x - 1)(x + 1)$

$$\frac{(x + 1)(x + 1)}{(x - 1)(x + 1)} + \frac{-(x - 1)(x - 1)}{(x + 1)(x - 1)}$$

$$= \frac{x^2 + 2x + 1}{(x - 1)(x + 1)} + \frac{-x^2 + 2x - 1}{(x - 1)(x + 1)} = \frac{4x}{x^2 - 1}$$

26. $\dfrac{16x}{x^2 - 16}$

27. LCD $= xy(x - y)$

$$\frac{y}{x(x - y)} \cdot \frac{y}{y} + \frac{-x}{y(x - y)} \cdot \frac{x}{x} = \frac{y^2 - x^2}{xy(x - y)}$$

$$= -\frac{x^2 - y^2}{xy(x - y)} = -\frac{(x + y)(x - y)}{xy(x - y)} = -\frac{x + y}{xy}$$

28. $\dfrac{a + b}{ab}$

29. LCD $= (x - 3)(x + 3)$

$$\frac{2x(x + 3)}{(x - 3)(x + 3)} + \frac{-2x(x - 3)}{(x + 3)(x - 3)} + \frac{36}{(x + 3)(x - 3)}$$

$$= \frac{2x^2 + 6x - 2x^2 + 6x + 36}{(x - 3)(x + 3)} = \frac{12x + 36}{(x - 3)(x + 3)}$$

$$= \frac{12(x + 3)}{(x - 3)(x + 3)} = \frac{12}{x - 3}$$

30. $-\dfrac{8}{x - 4}$, or $\dfrac{8}{4 - x}$

31. LCD $= (x + 2)^2$

$$\frac{x}{x^2 + 4x + 4} + \frac{1}{(x + 2)} \frac{(x + 2)}{(x + 2)} = \frac{x + x + 2}{(x + 2)^2} = \frac{2x + 2}{(x + 2)^2}$$

32. $\dfrac{5 - 3x}{(x - 1)^2}$

33. LCD $= (x + 5)(x - 1)(2x + 3)$

$$\frac{(2x + 3)(2x + 3)}{(x + 5)(x - 1)(2x + 3)} - \frac{(x + 5)(x + 5)}{(2x + 3)(x - 1)(x + 5)}$$

$$= \frac{(2x + 3)(2x + 3) - (x + 5)(x + 5)}{(x + 5)(x - 1)(2x + 3)}$$

$$= \frac{4x^2 + 12x + 9 - 1(x^2 + 10x + 25)}{(x + 5)(x - 1)(2x + 3)}$$

$$= \frac{4x^2 + 12x + 9 - x^2 - 10x - 25}{(x + 5)(x - 1)(2x + 3)}$$

$$= \frac{3x^2 + 2x - 16}{(x + 5)(x - 1)(2x + 3)}$$

34. $\dfrac{8x^2 - 2x - 45}{(x - 1)(x + 7)(3x + 2)}$

35. LCD $= (x + 4)^2(x - 4)$

$$\frac{x^2(x - 4)}{(x + 4)^2(x - 4)} - \frac{(x + 1)(x + 4)}{(x + 4)(x - 4)(x + 4)}$$

$$= \frac{x^3 - 4x^2 - 1(x^2 + 5x + 4)}{(x + 4)^2(x - 4)} = \frac{x^3 - 4x^2 - x^2 - 5x - 4}{(x + 4)^2(x - 4)}$$

$$= \frac{x^3 - 5x^2 - 5x - 4}{(x + 4)^2(x - 4)}$$

36. $\dfrac{x^3 - 4x^2 - 5x - 6}{(x + 3)^2(x - 3)}$

37. LCD $= x(2x + 5)(x - 1)(x + 1)$

$$\frac{(x + 1)(x + 1)}{x(2x + 5)(x - 1)(x + 1)} - \frac{(x - 1)(x - 1)}{x(2x + 5)(x + 1)(x - 1)}$$

$$= \frac{x^2 + 2x + 1 - 1(x^2 - 2x + 1)}{x(2x + 5)(x - 1)(x + 1)}$$

$$= \frac{x^2 + 2x + 1 - x^2 + 2x - 1}{x(2x + 5)(x - 1)(x + 1)} = \frac{4\overset{1}{x}}{\underset{1}{x}(2x + 5)(x - 1)(x + 1)}$$

$$= \frac{4}{(2x + 5)(x - 1)(x + 1)}$$

38. $\dfrac{4}{(3x + 4)(x - 1)(x + 1)}$

39. LCD $= 8x(4x + 3)$

$$\frac{5(4x)}{2(4x + 3)(4x)} - \frac{5(4x + 3)}{8x(4x + 3)} + \frac{4(8)}{x(4x + 3)(8)}$$

$$= \frac{20x}{8x(4x + 3)} - \frac{20x + 15}{8x(4x + 3)} + \frac{32}{8x(4x + 3)}$$

$$= \frac{20x - (20x + 15) + 32}{8x(4x + 3)} = \frac{20x - 20x - 15 + 32}{8x(4x + 3)}$$

$$= \frac{17}{8x(4x + 3)}$$

40. $\dfrac{7}{9x(3x + 5)}$

41. LCD $= 36x^2(2x - 5)$

$$\frac{8(4x^2)}{9(2x - 5)(4x^2)} - \frac{1(3x)}{12x(2x - 5)(3x)} - \frac{7[3(2x - 5)]}{12x^2[3(2x - 5)]}$$

$$= \frac{32x^2}{36x^2(2x - 5)} + \frac{-3x}{36x^2(2x - 5)} + \frac{-21(2x - 5)}{36x^2(2x - 5)}$$

$$= \frac{32x^2 - 3x - 42x + 105}{36x^2(2x - 5)} = \frac{32x^2 - 45x + 105}{36x^2(2x - 5)}$$

42. $\dfrac{28x^2 - 235x + 45}{40x^2(5x - 1)}$

43. $\dfrac{3}{8x^2} - \dfrac{x + 3}{4x^2 - 8x} \div \dfrac{x^2 + 2x - 3}{x^2 - 2x}$

$$= \frac{3}{8x^2} - \frac{x + 3}{4x(x - 2)} \cdot \frac{x(x - 2)}{(x + 3)(x - 1)}$$

$$= \frac{3}{8x^2} - \frac{1}{4(x - 1)} \quad \text{(the LCD is } 8x^2(x - 1)\text{)}$$

$$= \frac{3(x - 1)}{8x^2(x - 1)} - \frac{1(2x^2)}{4(x - 1)(2x^2)} = \frac{3x - 3 - 2x^2}{8x^2(x - 1)}, \text{ or}$$

$$\frac{-2x^2 + 3x - 3}{8x^2(x - 1)}$$

44. $\dfrac{-3x^2 + 8x + 24}{9x^2(x + 3)}$

45. $\dfrac{x + 3}{x - 5} - \dfrac{x - 2}{x^2 - x - 12} \div \dfrac{x + 7}{x^2 + 3x - 28}$

$$= \frac{x + 3}{x - 5} - \frac{x - 2}{(x - 4)(x + 3)} \cdot \frac{(x + 7)(x - 4)}{x + 7}$$

$$= \frac{x + 3}{x - 5} - \frac{x - 2}{x + 3} \quad \text{(the LCD is } (x - 5)(x + 3)\text{)}$$

$$= \frac{(x + 3)(x + 3)}{(x - 5)(x + 3)} - \frac{(x - 2)(x - 5)}{(x + 3)(x - 5)}$$

$$= \frac{x^2 + 6x + 9 - (x^2 - 7x + 10)}{(x - 5)(x + 3)}$$

$$= \frac{x^2 + 6x + 9 - x^2 + 7x - 10}{(x - 5)(x + 3)} = \frac{13x - 1}{(x - 5)(x + 3)}$$

46. $\dfrac{16x + 4}{(x - 2)(x + 4)}$

8

Exercises 8.7   (page 414)

1. $\dfrac{\frac{3}{4}}{\frac{5}{6}} = \dfrac{3}{4} \div \dfrac{5}{6} = \dfrac{3}{4} \cdot \dfrac{\overset{3}{6}}{\underset{2}{5}} = \dfrac{9}{10}$ 2. $\frac{4}{5}$

3. $\dfrac{\frac{2}{3}}{\frac{4}{9}} = \dfrac{2}{3} \div \dfrac{4}{9} = \dfrac{\overset{1}{2}}{\underset{1}{3}} \cdot \dfrac{\overset{3}{9}}{\underset{2}{4}} = \dfrac{3}{2}$, or $1\frac{1}{2}$ 4. $1\frac{1}{2}$

5. LCD of secondary denominators is 8.   6. $\frac{9}{7} = 1\frac{2}{7}$

$\dfrac{8}{8} \cdot \dfrac{\frac{3}{4} - \frac{1}{2}}{\frac{5}{8} + \frac{1}{4}} = \dfrac{\frac{8}{1}\left(\frac{3}{4}\right) + \frac{8}{1}\left(-\frac{1}{2}\right)}{\frac{8}{1}\left(\frac{5}{8}\right) + \frac{8}{1}\left(\frac{1}{4}\right)} = \dfrac{6 - 4}{5 + 2} = \dfrac{2}{7}$

7. LCD of secondary denominators is 20.

$\dfrac{20}{20} \cdot \dfrac{\frac{3}{5} + \frac{2}{1}}{\frac{2}{1} - \frac{3}{4}} = \dfrac{\overset{4}{20}\left(\frac{3}{5}\right) + 20\left(\frac{2}{1}\right)}{20\left(\frac{2}{1}\right) + \overset{5}{20}\left(-\frac{3}{4}\right)} = \dfrac{12 + 40}{40 - 15} = \dfrac{52}{25} = 2\frac{2}{25}$

8. $1\frac{1}{82}$ 9. $\dfrac{\frac{5x^3}{3y^4}}{\frac{10x}{9y}} = \dfrac{5x^3}{3y^4} \div \dfrac{10x}{9y} = \dfrac{\overset{1}{5x^3}}{\underset{1}{3y^4}} \cdot \dfrac{\overset{3}{9y}}{\underset{2}{10x}} = \dfrac{3x^2}{2y^3}$ 10. $6ab$

11. $\dfrac{\frac{18cd^2}{5a^3b}}{\frac{12cd^2}{15ab^2}} = \dfrac{18cd^2}{5a^3b} \div \dfrac{12cd^2}{15ab^2} = \dfrac{\overset{3}{18cd^2}}{\underset{a^2}{5a^3b}} \cdot \dfrac{\overset{3b}{15ab^2}}{\underset{1}{12cd^2}} = \dfrac{9b}{2a^2}$ 12. $\dfrac{2xz^2}{y}$

13. $\dfrac{\frac{x+3}{5}}{\frac{2x+6}{10}} = \dfrac{x+3}{5} \div \dfrac{2x+6}{10} = \dfrac{x+3}{5} \cdot \dfrac{\overset{2}{10}}{2(x+3)} = 1$

14. $1\frac{1}{2}$

15. $\dfrac{\frac{x+2}{2x}}{\frac{x+1}{4x^2}} = \dfrac{x+2}{2x} \div \dfrac{x+1}{4x^2} = \dfrac{x+2}{2x} \cdot \dfrac{\overset{2x}{4x^2}}{x+1} = \dfrac{2x^2+4x}{x+1}$

16. $\dfrac{3x-9}{x^2-9x}$ 17. LCD = b

$\dfrac{b}{b} \cdot \dfrac{\frac{a}{b}+1}{\frac{a}{b}-1} = \dfrac{b\left(\frac{a}{b}\right)+b(1)}{b\left(\frac{a}{b}\right)-b(1)} = \dfrac{a+b}{a-b}$

18. $\dfrac{2y+x}{2y-x}$ 19. LCD = x

$\dfrac{x}{x} \cdot \dfrac{\frac{1}{x}+x}{\frac{1}{x}-x} = \dfrac{x\left(\frac{1}{x}\right)+x(x)}{x\left(\frac{1}{x}\right)-x(x)} = \dfrac{1+x^2}{1-x^2}$

20. $\dfrac{a^2-4}{a^2+4}$ 21. LCD = d^2

$\dfrac{d^2}{d^2} \cdot \dfrac{\frac{c}{d}+2}{\frac{c^2}{d^2}-4} = \dfrac{d^2\left(\frac{c}{d}\right)+d^2(2)}{d^2\left(\frac{c^2}{d^2}\right)-d^2(4)} = \dfrac{cd+2d^2}{c^2-4d^2}$

$= \dfrac{d(c+2d)}{(c-2d)(c+2d)} = \dfrac{d}{c-2d}$

22. $\dfrac{x+y}{y}$

23. LCD = y

$\dfrac{y}{y} \cdot \dfrac{x+\frac{x}{y}}{1+\frac{1}{y}} = \dfrac{y(x)+\overset{1}{y}\left(\frac{x}{y}\right)}{y(1)+\overset{1}{y}\left(\frac{1}{y}\right)} = \dfrac{yx+x}{y+1} = \dfrac{x(y+1)}{y+1} = x$

24. $\frac{1}{3}$ 25. LCD = x^2y^2

$\dfrac{x^2y^2}{x^2y^2} \cdot \dfrac{\frac{1}{x^2}-\frac{1}{y^2}}{\frac{1}{x}+\frac{1}{y}} = \dfrac{x^2y^2\left(\frac{1}{x^2}\right)-x^2y^2\left(\frac{1}{y^2}\right)}{x^2y^2\left(\frac{1}{x}\right)+x^2y^2\left(\frac{1}{y}\right)}$

$= \dfrac{y^2-x^2}{xy^2+x^2y} = \dfrac{(y+x)(y-x)}{xy(y+x)} = \dfrac{y-x}{xy}$

26. $\dfrac{a+2}{2a}$

27. LCD = x^2

$\dfrac{x^2}{x^2} \cdot \dfrac{\frac{2}{x}-\frac{4}{x^2}}{\frac{1}{x}-\frac{2}{x^2}} = \dfrac{x^2\left(\frac{2}{x}\right)-x^2\left(\frac{4}{x^2}\right)}{x^2\left(\frac{1}{x}\right)-x^2\left(\frac{2}{x^2}\right)} = \dfrac{2x-4}{x-2} = \dfrac{2(x-2)}{x-2}$

$= 2$

28. $\frac{1}{4}$ 29. LCD = $3x(x+1)$

$\dfrac{3x(x+1)}{3x(x+1)} \cdot \dfrac{\frac{x}{x+1}+\frac{4}{3x}}{\frac{x}{x+1}-\frac{3}{x}}$

$= \dfrac{\left(\frac{3x(x+1)}{1}\right)\left(\frac{x}{x+1}\right) + \left(\frac{3x(x+1)}{1}\right)\left(\frac{4}{3x}\right)}{\left(\frac{3x(x+1)}{1}\right)\left(\frac{x}{x+1}\right) - \left(\frac{3x(x+1)}{1}\right)\left(\frac{3}{x}\right)}$

$= \dfrac{3x^2+4(x+1)}{3x^2-9(x+1)} = \dfrac{3x^2+4x+4}{3x^2-9x-9}$

30. $\dfrac{8x^2+4x+1}{20x+4}$

31. LCD = $4x(x-6)$

$\dfrac{4x(x-6)}{4x(x-6)} \cdot \dfrac{\frac{1}{4x}+\frac{x}{x-6}}{\frac{x-1}{x}-\frac{3}{x-6}}$

$= \dfrac{\left(\frac{4x(x-6)}{1}\right)\left(\frac{1}{4x}\right) + \left(\frac{4x(x-6)}{1}\right)\left(\frac{x}{x-6}\right)}{\left(\frac{4x(x-6)}{1}\right)\left(\frac{x-1}{x}\right) - \left(\frac{4x(x-6)}{1}\right)\left(\frac{3}{x-6}\right)}$

$= \dfrac{x-6+4x^2}{4(x^2-7x+6)-12x} = \dfrac{4x^2+x-6}{4x^2-28x+24-12x}$

$= \dfrac{4x^2+x-6}{4x^2-40x+24}$

32. $\dfrac{2x^2-2+2y}{2y+x-1}$

Sections 8.1–8.7 Review Exercises   (page 416)

1. Yes, because if we multiply both 1 and 2 by $(x + 1)$, we get $x + 1$ and $2x + 2$.

2. No

3. $x - 6$: The signs of the rational expressions are different, and the signs of the denominators are different; therefore, the signs of the numerators must be *the same*.

4. 2

5. -4: The signs of the rational expressions are the same, and the signs of the numerators are different; therefore, the signs of the denominators must be different.

6. $\dfrac{2b}{a^2}$ **7.** $\dfrac{\overset{1}{\cancel{2}}(1 + 2m)}{\underset{1}{\cancel{2}}} = 1 + 2m$ **8.** $1 - 2n$

9. $\dfrac{(\overset{1}{\cancel{a + 2}})(a - 2)}{\underset{1}{\cancel{a + 2}}} = a - 2$ **10.** $\dfrac{1}{x + 2}$

11. $\dfrac{a - b}{a(x + y) - b(x + y)} = \dfrac{\overset{1}{\cancel{a - b}}}{(x + y)(\underset{1}{\cancel{a - b}})} = \dfrac{1}{x + y}$

12. $\dfrac{6y^3}{5x^4}$

13. $\left(\dfrac{6a^4 b^{-2}}{8a^{-1}b^3}\right)^{-2} = \left(\dfrac{\overset{4}{\cancel{8}}a^{-1}b^3}{\underset{3}{\cancel{6}}a^4 b^{-2}}\right)^2 = \left(\dfrac{4}{3} \cdot a^{-1-4}b^{3-(-2)}\right)^2$

$= \left(\dfrac{4}{3}a^{-5}b^5\right)^2 = \left(\dfrac{4}{3} \cdot \dfrac{1}{a^5} \cdot \dfrac{b^5}{1}\right)^2$

$= \left(\dfrac{4b^5}{3a^5}\right)^2 = \dfrac{4^2 b^{5 \cdot 2}}{3^2 a^{5 \cdot 2}} = \dfrac{16b^{10}}{9a^{10}}$

14. $\dfrac{10}{k}$

15. LCD $= 2x$

$\dfrac{5(2x)}{1(2x)} - \dfrac{3}{2x} = \dfrac{10x - 3}{2x}$

16. 1   **17.** $\dfrac{\overset{a}{\cancel{-5a^2}}}{\underset{1}{\cancel{3b}}} \cdot \dfrac{\overset{3b}{\cancel{9b^2}}}{\underset{2}{\cancel{10a}}} = -\dfrac{3ab}{2}$

18. $-6x^2 y^2$ **19.** $\dfrac{\overset{1}{\cancel{3}}(\overset{1}{\cancel{x + 2}})}{\underset{2}{\cancel{6}}} \cdot \dfrac{2x^2}{4(\underset{1}{\cancel{x + 2}})} = \dfrac{x^2}{4}$ **20.** $\dfrac{y - 17}{6}$

21. LCD $= (a - 1)(a + 2)$

$\dfrac{(a - 2)(a + 2)}{(a - 1)(a + 2)} + \dfrac{(a + 1)(a - 1)}{(a + 2)(a - 1)} = \dfrac{a^2 - 4 + a^2 - 1}{(a + 2)(a - 1)}$

$= \dfrac{2a^2 - 5}{(a + 2)(a - 1)}$

22. $-\dfrac{5}{(k - 2)(k - 3)}$

23. $\dfrac{2x^2 - 6x}{x + 2} \div \dfrac{x}{4x + 8} = \dfrac{2\overset{1}{\cancel{x}}(x - 3)}{\underset{1}{\cancel{x + 2}}} \cdot \dfrac{4(\overset{1}{\cancel{x + 2}})}{\underset{1}{\cancel{x}}} = 8(x - 3)$

24. $8z$

25. LCD $= (2x + 3)(x + 2)(3x + 4)$

$\dfrac{(3x + 4)(3x + 4)}{(2x + 3)(x + 2)(3x + 4)} - \dfrac{(x + 2)(2x + 3)}{(3x + 4)(x + 2)(2x + 3)}$

$= \dfrac{9x^2 + 24x + 16 - 1(2x^2 + 7x + 6)}{(2x + 3)(x + 2)(3x + 4)}$

$= \dfrac{9x^2 + 24x + 16 - 2x^2 - 7x - 6}{(2x + 3)(x + 2)(3x + 4)}$

$= \dfrac{7x^2 + 17x + 10}{(2x + 3)(x + 2)(3x + 4)}$

26. $\dfrac{-3x^3 + x^2 - 248x + 400}{15x(x + 4)^2(x - 4)}$

27. LCD $= (2x + 1)(x - 6)(x + 6)$

$\dfrac{x + 6}{(2x + 1)(x - 6)} + \dfrac{x - 6}{(2x + 1)(x + 6)}$

$= \dfrac{(x + 6)(x + 6)}{(2x + 1)(x - 6)(x + 6)} + \dfrac{(x - 6)(x - 6)}{(2x + 1)(x + 6)(x - 6)}$

$= \dfrac{x^2 + 12x + 36 + x^2 - 12x + 36}{(2x + 1)(x + 6)(x - 6)}$

$= \dfrac{2x^2 + 72}{(2x + 1)(x + 6)(x - 6)}$

28. $\dfrac{-11x^2 + 14x + 1}{(3x - 5)(2x - 1)(x + 1)}$

29. LCD of secondary denominators is $9m^2$.

$\dfrac{\dfrac{5k^2}{3m^2}}{\dfrac{10k}{9m}} = \dfrac{5k^2}{3m^2} \div \dfrac{10k}{9m} = \dfrac{5\overset{k}{\cancel{k^2}}}{3\underset{m}{\cancel{m^2}}} \cdot \dfrac{\overset{3}{\cancel{9m}}}{\underset{2}{\cancel{10k}}} = \dfrac{3k}{2m}$

30. $\dfrac{3b - a}{2b + a}$

31. LCD $= b^2$

$\dfrac{b^2}{b^2} \cdot \dfrac{\left(\dfrac{a^2}{b^2} - 1\right)}{\left(\dfrac{a}{b} - 1\right)} = \dfrac{\overset{1}{\cancel{b^2}}\left(\dfrac{a^2}{\cancel{b^2}}\right) - b^2(1)}{\overset{b}{\cancel{b^2}}\left(\dfrac{a}{\cancel{b}}\right) - b^2(1)} = \dfrac{a^2 - b^2}{ab - b^2}$

$= \dfrac{(a + b)(\overset{1}{\cancel{a - b}})}{b(\underset{1}{\cancel{a - b}})} = \dfrac{a + b}{b}$

32. $\dfrac{x + 4}{x}$

Sections 8.1–8.7 Diagnostic Test (page 419)

Following each problem number is the number (in parentheses) of the textbook section where that kind of problem is discussed.

1. (8.1) No; we can't get the second rational expression from the first by multiplying or dividing both numerator and denominator by the same expression.

2. (8.2) $\dfrac{12a^6 b^3}{8a^2 b^5} = \dfrac{\overset{3}{\cancel{12}}}{\underset{2}{\cancel{8}}} \cdot \dfrac{a^6}{a^2} \cdot \dfrac{b^3}{b^5} = \dfrac{3}{2}a^{6-2}b^{3-5} = \dfrac{3}{2}a^4 b^{-2}$

$= \dfrac{3}{2} \cdot \dfrac{a^4}{1} \cdot \dfrac{1}{b^2} = \dfrac{3a^4}{2b^2}$

3. (8.2) $\dfrac{5x + 1}{25x^2 + 10x + 1} = \dfrac{\overset{1}{\cancel{5x + 1}}}{(5x + 1)(\underset{1}{\cancel{5x + 1}})} = \dfrac{1}{5x + 1}$

4. (8.2) $\dfrac{x^2 - 4y^2}{2x^2 + 3xy - 2y^2} = \dfrac{(\overset{1}{\cancel{x + 2y}})(x - 2y)}{(2x - y)(\underset{1}{\cancel{x + 2y}})} = \dfrac{x - 2y}{2x - y}$

5. (8.3) **a.** $3x$: The signs of the numerators are different, and the signs of the rational expressions are the same; therefore, the signs of the denominators must be different.

     b. -17: The signs of the denominators are different, and the signs of the rational expressions are the same; therefore, the signs of the numerators must be different.

6. (8.5) (1) $2 \cdot 3 \cdot x^3 \cdot y,\ 3 \cdot 5 \cdot x^2 \cdot y^2$ are the factored polynomials.
     (2) $2, 3, 5, x, y$ are all the different bases.
     (3) $2^1, 3^1, 5^1, x^3, y^2$ are the highest powers of each base.
     (4) LCM $= 2^1 \cdot 3^1 \cdot 5^1 \cdot x^3 \cdot y^2 = 30x^3 y^2$

7. (8.3) $\dfrac{x}{3y^2} \cdot \dfrac{\overset{2}{\cancel{6}}x\overset{1}{\cancel{y}}}{5y^3} = \dfrac{2x^2}{5y^4}$

8. (8.3) $\dfrac{x}{x^2 + 6x + 8} \cdot \dfrac{3x + 12}{12x^2 + 24x} = \dfrac{\overset{1}{\cancel{x}}}{(x + 2)\,(\cancel{x + 4})} \cdot \dfrac{\overset{1}{3}(\cancel{x + 4})}{\underset{4}{\cancel{12}}\,\underset{1}{x}(x + 2)}$

$$= \dfrac{1}{4(x + 2)^2}$$

9. (8.3) $\dfrac{4}{x - 5} \div \dfrac{8}{x^2 - 5x} = \dfrac{4}{x - 5} \cdot \dfrac{x^2 - 5x}{8}$

$$= \dfrac{\overset{1}{\cancel{4}}}{\cancel{x - 5}} \cdot \dfrac{x(\cancel{x - 5})}{\underset{2}{\cancel{8}}} = \dfrac{x}{2}$$

10. (8.3) $\dfrac{7y}{7y + y^2} \div \dfrac{1 - y^2}{7 + 8y + y^2} = \dfrac{7y}{7y + y^2} \cdot \dfrac{7 + 8y + y^2}{1 - y^2}$

$$= \dfrac{\overset{1}{7}\cancel{y}}{\underset{1}{\cancel{y}}(\cancel{7 + y})} \cdot \dfrac{(\cancel{7 + y})\,\overset{1}{(\cancel{1 + y})}}{(1 - y)(\cancel{1 + y})} = \dfrac{7}{1 - y}$$

11. (8.4) $\dfrac{8}{3y^2} - \dfrac{2}{3y^2} = \dfrac{8 - 2}{3y^2} = \dfrac{\overset{2}{\cancel{6}}}{\underset{1}{\cancel{3}}y^2} = \dfrac{2}{y^2}$

12. (8.4) $\dfrac{8x}{x + 7} + \dfrac{3}{x + 7} = \dfrac{8x + 3}{x + 7}$

13. (8.4) $\dfrac{8x}{4x - 1} - \dfrac{2}{4x - 1} = \dfrac{8x - 2}{4x - 1} = \dfrac{2(\overset{1}{\cancel{4x - 1}})}{\underset{1}{\cancel{4x - 1}}} = \dfrac{2}{1} = 2$

14. (8.6) LCD $= (x - 1)(x + 3)$

$$\dfrac{3x\,(x + 3)}{(x - 1)\,(x + 3)} + \dfrac{4\,(x - 1)}{(x + 3)\,(x - 1)}$$

$$= \dfrac{3x^2 + 9x + 4x - 4}{(x + 3)(x - 1)} = \dfrac{3x^2 + 13x - 4}{(x + 3)(x - 1)}$$

15. (8.6) LCD $= (x + 5)(x - 3)$

$$\dfrac{(x - 3)\,(x - 3)}{(x + 5)\,(x - 3)} + \dfrac{(x + 5)\,(x + 5)}{(x + 5)\,(x - 3)}$$

$$= \dfrac{x^2 - 6x + 9 + x^2 + 10x + 25}{(x + 5)(x - 3)} = \dfrac{2x^2 + 4x + 34}{(x + 5)(x - 3)}$$

16. (8.6) $x^2 + 7x + 12 = (x + 3)(x + 4);$

$$x^2 + 5x + 6 = (x + 2)(x + 3)$$

$$\text{LCD} = (x + 3)(x + 4)(x + 2)$$

$$\dfrac{(x + 14)\,(x + 2)}{(x + 3)\,(x + 4)\,(x + 2)} - \dfrac{(x + 5)\,(x + 4)}{(x + 2)\,(x + 3)\,(x + 4)}$$

$$= \dfrac{(x^2 + 16x + 28) - 1(x^2 + 9x + 20)}{(x + 3)(x + 4)(x + 2)}$$

$$= \dfrac{x^2 + 16x + 28 - x^2 - 9x - 20}{(x + 3)(x + 4)(x + 2)}$$

$$= \dfrac{7x + 8}{(x + 3)(x + 4)(x + 2)}$$

17. (8.6) LCD $= (x - 4)(x + 4)$, since $x^2 - 16 = (x - 4)(x + 4)$

$$\dfrac{x\,(x + 4)}{(x - 4)\,(x + 4)} - \dfrac{x\,(x - 4)}{(x + 4)\,(x - 4)} - \dfrac{32}{(x + 4)\,(x - 4)}$$

$$= \dfrac{(x^2 + 4x) - 1(x^2 - 4x) - (32)}{(x - 4)(x + 4)}$$

$$= \dfrac{x^2 + 4x - x^2 + 4x - 32}{(x - 4)(x + 4)}$$

$$= \dfrac{8x - 32}{(x - 4)(x + 4)} = \dfrac{8(\overset{1}{\cancel{x - 4}})}{(\cancel{x - 4})(x + 4)} = \dfrac{8}{x + 4}$$

18. (8.6) LCD $= (3x - 1)(x - 7)(x + 7)$, since $3x^2 - 22x + 7$

$$= (3x - 1)(x - 7) \text{ and } 3x^2 + 20x - 7 = (3x - 1)(x + 7)$$

$$\dfrac{(x + 7)(x + 7)}{(3x - 1)(x - 7)(x + 7)} - \dfrac{(x - 7)(x - 7)}{(3x - 1)(x + 7)(x - 7)}$$

$$= \dfrac{x^2 + 14x + 49 - 1(x^2 - 14x + 49)}{(3x - 1)(x + 7)(x - 7)}$$

$$= \dfrac{x^2 + 14x + 49 - x^2 + 14x - 49}{(3x - 1)(x + 7)(x - 7)}$$

$$= \dfrac{28x}{(3x - 1)(x + 7)(x - 7)}$$

19. (8.7) $\dfrac{\dfrac{4x^3}{9y^4}}{\dfrac{2x}{27y^2}} = \dfrac{4x^3}{9y^4} \div \dfrac{2x}{27y^2} = \dfrac{\overset{2x^2}{\cancel{4x^3}}}{9\cancel{y^4}\underset{y^2}{\,}} \cdot \dfrac{\overset{3}{\cancel{27y^2}}}{\underset{1}{\cancel{2x}}} = \dfrac{6x^2}{y^2}$

20. (8.7) LCD $= x^2$

$$\dfrac{\dfrac{6}{x} - \dfrac{4}{x^2}}{5 + \dfrac{2}{x}} = \dfrac{\dfrac{6}{x} - \dfrac{4}{x^2}}{5 + \dfrac{2}{x}} \cdot \dfrac{x^2}{x^2} = \dfrac{\left(\dfrac{\overset{x}{\cancel{x^2}}}{1}\right)\left(\dfrac{6}{\cancel{x}}\right)_{\!1} - \left(\dfrac{\overset{1}{\cancel{x^2}}}{1}\right)\left(\dfrac{4}{\cancel{x^2}}\right)_{\!1}}{\left(\dfrac{x^2}{1}\right)\left(\dfrac{5}{1}\right) + \left(\dfrac{\overset{x}{\cancel{x^2}}}{1}\right)\left(\dfrac{2}{\cancel{x}}\right)_{\!1}}$$

$$= \dfrac{6x - 4}{5x^2 + 2x}$$

Exercises 8.8A   (page 420)

1. If $x - 2 = 0$, $x = 2$, so 2 would make the denominator zero and is an excluded value.

2. -3

3. None. (There are no variables in the denominator.)

4. None

5. If $3x^2 - 6x = 0$, $x = 0$ or $x = 2$, so 0 and 2 would make the denominator zero and are excluded values.

6. 0, 3

7. If $x^2 - x - 2 = 0$, $x = -1$ or $x = 2$, so -1 and 2 would make the denominator zero and are excluded values.

8. 3, -4

Exercises 8.8B   (page 424)

(The checks usually will not be shown.)

1. LCD $= 12$. Multiply both sides of the equation by 12.

$$(12)\left(\dfrac{x}{3}\right) + (12)\left(\dfrac{x}{4}\right) = (12)(7)$$

$$4x + 3x = 84$$

$$7x = 84$$

$$x = 12$$

The solution is 12.

2. 15

3. LCD $= c$. Multiply both sides of the equation by c.   **4.** $1\frac{1}{2}$

$$\left(\dfrac{10}{c}\right)(c) = (2)(c)$$

$$10 = 2c$$

$$5 = c, \text{ or } c = 5$$

The solution is 5.

5. LCD $= 10$ **6.** 63

$$(10)\dfrac{a}{2} + (10)\dfrac{-a}{5} = (10)(6)$$

$$5a - 2a = 60$$

$$3a = 60$$

$$a = 20$$

The solution is 20.

8

8

7. LCD $= 2x$

$$\left(\frac{9}{2x}\right)(2x) = (3)(2x)$$

$$9 = 6x$$
$$\frac{9}{6} = x$$
$$x = \frac{3}{2} = 1\frac{1}{2}$$

The solution is $1\frac{1}{2}$.

8. $\frac{2}{3}$

9. LCD $= 15$

$$(15)\left(\frac{M-2}{5}\right) + (15)\left(\frac{M}{3}\right) = (15)\left(\frac{1}{5}\right)$$

$$3(M-2) + 5M = 3$$
$$3M - 6 + 5M = 3$$
$$8M = 9$$
$$M = \frac{9}{8} = 1\frac{1}{8}$$

The solution is $1\frac{1}{8}$.

10. $-\frac{5}{9}$

11. LCD $= x(x+4)$

$$\left(\frac{7}{x+4}\right)(x)(x+4) = \left(\frac{3}{x}\right)(x)(x+4)$$

$$7x = 3(x+4)$$
$$7x = 3x + 12$$
$$4x = 12$$
$$x = 3$$

The solution is 3.

12. 4

13. LCD $= x - 2$

$$\left(\frac{3x}{x-2}\right)(x-2) = 5(x-2)$$

$$3x = 5(x-2)$$
$$3x = 5x - 10$$
$$-2x = -10$$
$$x = 5$$

The solution is 5.

14. 3

15. LCD $= x - 2$

$$\left(\frac{3x-1}{x-2}\right)(x-2) = 3(x-2) + \left(\frac{2x+1}{x-2}\right)(x-2)$$

$$3x - 1 = 3(x-2) + (2x+1)$$
$$3x - 1 = 3x - 6 + 2x + 1$$
$$3x - 1 = 5x - 5$$
$$4 = 2x$$
$$x = 2 \quad 2 \text{ is an excluded value.}$$

No solution. (If you try to check the apparent solution, you will have zeros in denominators.)

16. No solution

17. LCD $= (x^2 + 1)(1 + 2x)$

$$\left(\frac{x}{x^2+1}\right)(x^2+1)(1+2x) = \left(\frac{2}{1+2x}\right)(x^2+1)(1+2x)$$

$$x(1+2x) = 2(x^2+1)$$
$$x + 2x^2 = 2x^2 + 2$$
$$x = 2$$

The solution is 2.

18. $\frac{2}{3}$

19. LCD $= 12$

$$(\overset{4}{\cancel{12}})\left(\frac{2x-1}{\underset{1}{\cancel{3}}}\right) + (\overset{3}{\cancel{12}})\left(\frac{3x}{\underset{1}{\cancel{4}}}\right) = (\overset{2}{\cancel{12}})\left(\frac{5}{\underset{1}{\cancel{6}}}\right)$$

$$4(2x-1) + 3(3x) = 10$$
$$8x - 4 + 9x = 10$$
$$17x = 14$$
$$x = \frac{14}{17}$$

The solution is $\frac{14}{17}$.

20. $1\frac{1}{9}$

21. LCD $= 10$

$$(\overset{2}{\cancel{10}})\left(\frac{2(m-3)}{\underset{1}{\cancel{5}}}\right) + (\overset{5}{\cancel{10}})\left(\frac{-3(m+2)}{\underset{1}{\cancel{2}}}\right) = (10)\left(\frac{7}{10}\right)$$

$$4(m-3) - 15(m+2) = 7$$
$$4m - 12 - 15m - 30 = 7$$
$$-11m = 49$$
$$m = \frac{49}{-11} = -4\frac{5}{11}$$

The solution is $-4\frac{5}{11}$.

22. $7\frac{4}{11}$

23. LCD $= x - 2$

$$\left(\frac{x-2}{1}\right)\left(\frac{x}{x-2}\right) = \left(\frac{x-2}{1}\right)\left(\frac{2}{x-2}\right) + \left(\frac{x-2}{1}\right)(5)$$

$$x = 2 + 5x - 10$$
$$8 = 4x$$
$$x = 2 \quad 2 \text{ is an excluded value.}$$

No solution. (If you try to check the apparent solution, you will have zeros in denominators.)

24. No solution

We will not show multiplication by the LCD in Exercises 25 and 27.

25. LCD $= 6(x+3)(x-2)(x+2)$

$$\frac{4}{(x+3)(x-2)} + \frac{3}{(x+3)(x+2)} = \frac{23}{6(x+2)(x-2)}$$

$$4(6)(x+2) + 3(6)(x-2) = 23(x+3)$$
$$24x + 48 + 18x - 36 = 23x + 69$$
$$42x + 12 = 23x + 69$$
$$19x = 57$$
$$x = 3$$

26. 2

27. LCD $= 63(x+3)(x-3)(x+1)$

$$\frac{7}{9(x+3)(x-3)} - \frac{5}{(x+3)(x+1)} = \frac{-10}{63(x+1)(x-3)}$$

$$7(7)(x+1) - 5(63)(x-3) = -10(x+3)$$
$$49x + 49 - 315x + 945 = -10x - 30$$
$$-266x + 994 = -10x - 30$$
$$1,024 = 256x$$
$$x = 4$$

28. 3

Exercises 8.8C   (page 430)

1. $\dfrac{x}{14} = \dfrac{-3}{7}$

$$7x = -3(14)$$
$$x = \frac{-3(14)}{7} = -6$$

The solution is -6.

2. -10

3. $\dfrac{x}{4} = \dfrac{2}{3}$

$$3x = 2(4) = 8$$
$$x = \frac{8}{3} = 2\frac{2}{3}$$

The solution is $2\frac{2}{3}$.

4. $7\frac{1}{2}$

5. $\dfrac{8}{x} = \dfrac{4}{5}$

$$4x = 8(5) = 40$$
$$x = 10$$

The solution is 10.

6. $2\frac{2}{3}$

7. $\dfrac{4}{7} = \dfrac{x}{21}$

$$7x = 4(21)$$
$$x = \frac{4(21)}{7} = 12$$

The solution is 12.

8. $11\frac{1}{4}$

9. $\dfrac{100}{x} = \dfrac{40}{30} = \dfrac{4}{3}$

$$4x = 300$$
$$x = 75$$

The solution is 75.

10. 24   **11.** $\dfrac{x+1}{x-1} = \dfrac{3}{2}$ **12.** 4

$$2(x+1) = 3(x-1)$$
$$2x + 2 = 3x - 3$$
$$5 = x; \; x = 5$$

The solution is 5.

13. $\dfrac{2x+7}{9} = \dfrac{2x+3}{5}$ **14.** -2

$$5(2x+7) = 9(2x+3)$$
$$10x + 35 = 18x + 27$$
$$8 = 8x$$
$$x = 1$$

The solution is 1.

15. $\dfrac{5x-10}{10} = \dfrac{3x-5}{7}$ **16.** $2\frac{2}{7}$

$$7(5x-10) = 10(3x-5)$$
$$35x - 70 = 30x - 50$$
$$5x = 20$$
$$x = 4$$

The solution is 4.

17. $\dfrac{\frac{3}{4}}{6} = \dfrac{P}{16}$ **18.** $2\frac{1}{2}$

$$6P = \dfrac{3}{\overset{}{4}} \cdot \dfrac{\overset{4}{\cancel{16}}}{1} = 12$$

$$\dfrac{\overset{1}{\cancel{6}}P}{\cancel{6}} = \dfrac{12}{6} = 2$$

The solution is 2.

19. $\dfrac{A}{9} = \dfrac{3\frac{1}{3}}{5}$ **20.** 1

$$5A = 9\left(3\tfrac{1}{3}\right)$$

$$5A = \dfrac{\overset{3}{\cancel{9}}}{1} \cdot \dfrac{10}{\underset{1}{\cancel{3}}} = 30$$

$$\dfrac{\overset{1}{\cancel{5}}A}{\cancel{5}} = \dfrac{30}{5} = 6$$

The solution is 6.

21. $\dfrac{7.7}{B} = \dfrac{\overset{0.7}{\cancel{3.5}}}{\underset{1}{\cancel{5}}}$ **22.** 22.96   **23.** $\dfrac{P}{100} = \dfrac{\frac{3}{2}}{15}$

$$0.7B = 7.7$$

$$\dfrac{\overset{1}{\cancel{0.7}}B}{\cancel{0.7}} = \dfrac{7.7}{0.7} = 11$$

The solution is 11.

$$15P = \dfrac{\overset{50}{\cancel{100}}}{1} \cdot \dfrac{3}{\underset{1}{\cancel{2}}}$$
$$15P = 150$$
$$P = 10$$

The solution is 10.

24. 4   **25.** $\dfrac{12\frac{1}{2}}{100} = \dfrac{A}{48}$ **26.** 54

$$100A = \left(12\tfrac{1}{2}\right)(48)$$

$$100A = \dfrac{25}{\underset{1}{\cancel{2}}} \cdot \dfrac{\overset{24}{\cancel{48}}}{1}$$

$$\dfrac{\overset{1}{\cancel{100}}A}{\cancel{100}} = \dfrac{\overset{6}{\cancel{600}}}{\cancel{100}} = 6$$

The solution is 6.

Exercises 8.8D   (page 433)

(Checks will usually not be shown.)

1. LCD $= z$ **2.** $-3, 3$

$$(z)z + (\overset{1}{\cancel{z}})\left(\dfrac{1}{\underset{1}{\cancel{z}}}\right) = (\overset{1}{\cancel{z}})\left(\dfrac{17}{\underset{1}{\cancel{z}}}\right)$$

$$z^2 + 1 = 17$$
$$z^2 + 1 - 17 = 0$$
$$z^2 - 16 = 0$$
$$(z+4)(z-4) = 0$$

$$z + 4 = 0 \quad | \quad z - 4 = 0$$
$$z = -4 \quad | \quad z = 4$$

The solutions are -4 and 4.

3. LCD $= 2x^2$ **4.** 2, 4

$$\left(\dfrac{2x^2}{1}\right)\left(\dfrac{2}{x}\right) + \left(\dfrac{2x^2}{1}\right)\left(-\dfrac{2}{x^2}\right) = \left(\dfrac{2x^2}{1}\right)\left(\dfrac{1}{2}\right)$$

$$4x - 4 = x^2$$
$$0 = x^2 - 4x + 4$$
$$0 = (x-2)(x-2)$$

$$x - 2 = 0 \quad | \quad x - 2 = 0$$
$$x = 2 \quad | \quad x = 2$$

The solution is 2.

5. $\dfrac{x}{x+1} = \dfrac{4x}{3x+2}$ **6.** 0, 2

$$4x(x+1) = x(3x+2)$$
$$4x^2 + 4x = 3x^2 + 2x$$
$$x^2 + 2x = 0$$
$$x(x+2) = 0$$

$$x = 0 \quad | \quad x + 2 = 0$$
$$\quad \quad | \quad x = -2$$

The solutions are 0 and -2.

7. LCD $= x$ **8.** -4

$$\left(\dfrac{5}{x}\right)(x) - 1(x) = \left(\dfrac{x+11}{x}\right)(x)$$

$$5 - x = x + 11$$
$$-6 = 2x$$
$$x = -3$$

The solution is -3.

9. LCD $= 3(x+1)(x-1)$

$$3(x+1)(\overset{1}{\cancel{x-1}})\left(\dfrac{1}{\underset{1}{\cancel{x-1}}}\right) + 3(\overset{1}{\cancel{x+1}})(x-1)\left(\dfrac{2}{\underset{1}{\cancel{x+1}}}\right)$$

$$= \overset{1}{\cancel{3}}(x+1)(x-1)\left(\dfrac{5}{\underset{1}{\cancel{3}}}\right)$$

$$3(x+1) + 6(x-1) = 5(x^2-1)$$
$$3x + 3 + 6x - 6 = 5x^2 - 5$$
$$0 = 5x^2 - 9x - 2$$
$$0 = (5x+1)(x-2)$$

$$5x + 1 = 0 \quad | \quad x - 2 = 0$$
$$x = -\tfrac{1}{5} \quad | \quad x = 2$$

The solutions are $-\tfrac{1}{5}$ and 2.

10. $3, \frac{1}{21}$

8

11. LCD = $4(2x + 5)$

$$\frac{\overset{1}{4(2x+5)}}{1}\cdot\frac{3}{2x+5} + \frac{\overset{1}{4(2x+5)}}{1}\cdot\frac{x}{4} = \frac{\overset{1}{4(2x+5)}}{1}\cdot\frac{3}{4}$$

$$4(3) + x(2x + 5) = 3(2x + 5)$$
$$12 + 2x^2 + 5x = 6x + 15$$
$$2x^2 - x - 3 = 0$$
$$(x + 1)(2x - 3) = 0$$

$x + 1 = 0$ $\quad\big|\quad$ $2x - 3 = 0$
$\quad x = -1$ $\qquad$ $2x = 3$
$\qquad\qquad\qquad\qquad$ $x = \frac{3}{2}$

The solutions are -1 and $\frac{3}{2}$.

12. $2, -\frac{19}{2}$

13. LCD = $15x(x + 1)$

$$\frac{\overset{1}{15x(x+1)}}{1}\cdot\frac{4}{x+1} - \frac{\overset{1}{15x}(x+1)}{1}\cdot\frac{3}{x} = \frac{\overset{1}{15x(x+1)}}{1}\cdot\frac{1}{15}$$

$$60x - 45(x + 1) = x^2 + x$$
$$60x - 45x - 45 = x^2 + x$$
$$15x - 45 = x^2 + x$$
$$0 = x^2 - 14x + 45$$
$$0 = (x - 5)(x - 9)$$

$x - 5 = 0$ $\quad\big|\quad$ $x - 9 = 0$
$\quad x = 5$ $\qquad\quad$ $x = 9$

The solutions are 5 and 9.

14. $-2, -3$

15. $x^2 - 9 = (x + 3)(x - 3)$; LCD = $5(x + 3)(x - 3)$

$$\frac{\overset{1}{5(x+3)(x-3)}}{1}\cdot\frac{6}{(x+3)(x-3)}$$

$$= \frac{5(x+3)\overset{1}{(x-3)}}{1}\cdot\frac{1}{x-3} - \frac{\overset{1}{5}(x+3)(x-3)}{1}\cdot\frac{1}{5}$$

$$30 = 5x + 15 - 1(x^2 - 9)$$
$$30 = 5x + 15 - x^2 + 9$$
$$30 = 5x + 24 - x^2$$
$$x^2 - 5x + 6 = 0$$
$$(x - 2)(x - 3) = 0$$

$x - 2 = 0$ $\quad\big|\quad$ $x - 3 = 0$
$\quad x = 2$ $\qquad\quad$ $x = 3$, but 3 cannot be a solution, since it
$\qquad\qquad\qquad\qquad\qquad$ is an excluded value.

The only solution is 2.

16. $\frac{7}{3}$

Exercises 8.9   (page 437)

1. $2x + y = 4$
$\quad 2x = 4 - y$
$\quad\; x = \frac{4 - y}{2}$

2. $y = \frac{6 - x}{3}$

3. $y - z = -8$
$\quad -z = -8 - y$
$\qquad z = 8 + y$

4. $n = m + 5$

5. $2x - y = -4$
$\quad -y = -2x - 4$
$\qquad y = 2x + 4$

6. $z = 3y + 5$

7. $2x - 3y = 6$
$\quad 2x = 3y + 6$
$\quad\; x = \frac{3y + 6}{2}$

8. $x = \frac{2y + 6}{3}$

9. $2(x - 3y) = x + 4$
$\quad 2x - 6y = x + 4$
$\qquad\quad x = 6y + 4$

10. $x = \frac{2y + 14}{7}$

11. $PV = k$

$$\frac{\overset{1}{PV}}{P} = \frac{k}{P}$$

$$V = \frac{k}{P}$$

12. $R = \frac{E}{I}$

13. $I = Prt$

$$\frac{I}{rt} = \frac{P\overset{1}{rt}}{rt}$$

$$P = \frac{I}{rt}$$

14. $l = \frac{V}{wh}$

15. $\quad p = 2l + 2w$
$p - 2w = 2l$

$$\frac{p - 2w}{2} = \frac{\overset{1}{2l}}{2}$$

$$l = \frac{p - 2w}{2}$$

16. $w = \frac{P - 2l}{2}$

17. $\quad y = mx + b$
$y - b = mx$

$$\frac{y - b}{m} = \frac{\overset{1}{mx}}{m}$$

$$x = \frac{y - b}{m}$$

18. $t = \frac{V - k}{g}$

19. $\quad\dfrac{S}{1} = \dfrac{a}{1 - r}$
$S(1 - r) = a$
$S - Sr = a$
$\quad -Sr = a - S$

$$\frac{-Sr}{-S} = \frac{a - S}{-S}$$

$$r = \frac{S - a}{S}$$

20. $R = \frac{E - Ir}{I}$

21. LCD = 9

$$9(C) = \frac{\overset{1}{9}}{1}\left(\frac{5(F - 32)}{9}\right)$$

$$9C = 5F - 160$$
$$9C + 160 = 5F$$
$$F = \frac{9C + 160}{5}$$

22. $B = \frac{2A - hb}{h}$

23. $\quad L = a + (n - 1)d$
$\qquad L = a + nd - d$
$L - a + d = nd$

$$\frac{L - a + d}{d} = \frac{n\overset{1}{d}}{d}$$

$$n = \frac{L - a + d}{d}$$

24. $h = \frac{A - 2\pi r^2}{2\pi r}$

25.
$$\frac{z}{1} = \frac{Rr}{R + r}$$
$$z(R + r) = 1(Rr)$$
$$zR + zr = Rr$$
$$zr = Rr - zR$$
$$zr = R(r - z)$$
$$\frac{zr}{r - z} = \frac{R(\overset{1}{\cancel{r - z}})}{\cancel{r - z}_1}$$
$$R = \frac{zr}{r - z}$$

26. $b = \dfrac{ca}{a - c}$

27. LCD = Fuv
$$\left(\frac{Fuv}{1}\right)\left(\frac{1}{F}\right) = \left(\frac{Fuv}{1}\right)\left(\frac{1}{u}\right) + \left(\frac{Fuv}{1}\right)\left(\frac{1}{v}\right)$$
$$uv = Fv + Fu$$
$$uv - Fu = Fv$$
$$u(v - F) = Fv$$
$$\frac{u(v - F)}{v - F} = \frac{Fv}{v - F}$$
$$u = \frac{Fv}{v - F}$$

28. $a = \dfrac{bc}{b - c}$

Exercises 8.10   (page 444)

(The checks will not be shown.)

1. Let x = speed of the boat in still water (in mph)
Then $x + 5$ = speed with current (in mph)
and $x - 5$ = speed against current (in mph)
$$\frac{999}{x + 5} = \text{time going with current} \quad \left(\text{Since } rt = d, t = \frac{d}{r}.\right)$$
$$\frac{729}{x - 5} = \text{time going against current}$$
$$\frac{999}{x + 5} = \frac{729}{x - 5} \quad \text{The times are equal.}$$
$$999(x - 5) = 729(x + 5)$$
$$999x - 4{,}995 = 729x + 3{,}645$$
$$270x = 8{,}640$$
$$x = 32$$
The speed of the boat in still water is 32 mph.

2. 32 mph

3. Let n = numerator
Then $n + 6$ = denominator
$$\frac{n + 4}{n + 6 - 4} = \frac{11}{10}$$
$$10(n + 4) = 11(n + 2)$$
$$10n + 40 = 11n + 22$$
$$18 = n$$
The original fraction is $\dfrac{n}{n + 6} = \dfrac{18}{18 + 6} = \dfrac{18}{24}$.

4. $\frac{16}{20}$

5. Let x = numerator
Then $2x$ = denominator
$$\frac{x + 5}{2x - 5} = \frac{4}{5}$$
$$(x + 5)(5) = 4(2x - 5)$$
$$5x + 25 = 8x - 20$$
$$45 = 3x$$
$$x = 15 \quad \text{Numerator}$$
$$2x = 30 \quad \text{Denominator}$$
The fraction is $\frac{15}{30}$.

6. $\frac{15}{45}$

7. Let x = the fraction
Then $\dfrac{1}{x}$ = its reciprocal
$$x + \frac{1}{x} = \frac{29}{10} \quad \text{LCD is } 10x.$$
$$x(10x) + \left(\frac{1}{x}\right)(10x) = \left(\frac{29}{10}\right)(10x)$$
$$10x^2 + 10 = 29x$$
$$10x^2 - 29x + 10 = 0$$
$$(2x - 5)(5x - 2) = 0$$

$2x - 5 = 0$	$5x - 2 = 0$
$2x = 5$	$5x = 2$
$x = \frac{5}{2}$	$x = \frac{2}{5}$
$\frac{1}{x} = \frac{2}{5}$	$\frac{1}{x} = \frac{5}{2}$

First answer: The number is $\frac{2}{5}$ and its reciprocal is $\frac{5}{2}$. Second answer: The number is $\frac{5}{2}$ and its reciprocal is $\frac{2}{5}$.

8. $\frac{3}{11}$, or $\frac{11}{3}$

9. Let w = width of room (in ft)   Let l = length of room (in ft)
$$\frac{1 \text{ in.}}{8 \text{ ft}} = \frac{2\frac{1}{2} \text{ in.}}{w \text{ ft}} \qquad \frac{1 \text{ in.}}{8 \text{ ft}} = \frac{3 \text{ in.}}{l \text{ ft}}$$
$$\frac{1}{8} = \frac{\frac{5}{2}}{w} \qquad \frac{1}{8} = \frac{3}{l}$$
$$w = 8\left(\frac{5}{2}\right) = 20 \qquad l = 8(3) = 24$$
Therefore, the room is 20 ft by 24 ft.

10. 26 ft by 34 ft

11. Let x = person's weight on moon (in lb)
$$\frac{6 \text{ earth}}{1 \text{ moon}} = \frac{150 \text{ earth}}{x \text{ moon}}$$
$$\frac{6}{1} = \frac{150}{x}$$
$$6x = 150$$
$$x = \frac{150}{6} = 25$$
The person would weigh 25 lb on the moon.

12. 78.4 lb

13. Let x = weight of lead bar (in lb)
$$\frac{21 \text{ lead}}{5 \text{ aluminum}} = \frac{x \text{ lead}}{150 \text{ aluminum}}$$
$$\frac{21}{5} = \frac{x}{150}$$
$$5x = 21(150)$$
$$x = \frac{21(150)}{5} = 630$$
The lead bar would weigh 630 lb.

14. 7.5 lb

15. Let x = number of chains of fire line
$10 \text{ hr} - 6\frac{1}{2} \text{ hr} = 3\frac{1}{2} \text{ hr}$
$$\frac{26 \text{ chains}}{6\frac{1}{2} \text{ hr}} = \frac{x \text{ chains}}{3\frac{1}{2} \text{ hr}}$$
$$6\frac{1}{2}x = 26\left(3\frac{1}{2}\right)$$
$$\frac{13x}{2} = \frac{91}{1}$$
$$13x = 182$$
$$\frac{13x}{13} = \frac{182}{13}$$
$$x = 14$$
The crew could build 14 chains of fire line.

16. $375

17. Let x = amount received for 16 steers
$$\frac{3{,}420 \text{ dollars}}{9 \text{ steers}} = \frac{x \text{ dollars}}{16 \text{ steers}}$$
$$9x = 16(3{,}420)$$
$$\frac{9x}{9} = \frac{54{,}720}{9}$$
$$x = 6{,}080$$
The producer will receive $6,080.

8

18. \$45 per hog; 45¢ per kg

19. Tricia's rate is $\dfrac{1 \text{ job}}{8 \text{ hr}}$, or $\dfrac{1}{8} \dfrac{\text{job}}{\text{hr}}$. Mike's rate is $\dfrac{1 \text{ job}}{10 \text{ hr}}$, or $\dfrac{1}{10} \dfrac{\text{job}}{\text{hr}}$.

Let x = number of hr for Tricia and Mike together to do the job

Tricia and Mike work for x hr.

$\dfrac{1}{8}x + \dfrac{1}{10}x = 1$ LCD is 40.

$\dfrac{1}{8}x(40) + \dfrac{1}{10}x(40) = 1(40)$

$5x + 4x = 40$

$9x = 40$

$x = \dfrac{40}{9}$, or $4\dfrac{4}{9}$

The job will take $4\dfrac{4}{9}$ hr.

20. $2\dfrac{11}{12}$ days

21. Let x = numerator   **22.** $\dfrac{17}{23}$

Then $42 - x$ = denominator

$\dfrac{x + 2}{(42 - x) + 6} = \dfrac{2}{3}$

$\dfrac{x + 2}{48 - x} = \dfrac{2}{3}$

$3x + 6 = 96 - 2x$

$5x = 90$

$x = 18$ Numerator

$42 - x = 24$ Denominator

The fraction is $\dfrac{18}{24}$.

23. Ruth's rate is $\dfrac{220 \text{ pages}}{4 \text{ hour}}$, or $55 \dfrac{\text{pages}}{\text{hr}}$.

Let t = number of hr for Jerry to proofread 210 pages

Then Jerry's rate is $\dfrac{210 \text{ pg}}{t \text{ hr}}$.

Ruth and Jerry read for 3 hr.

$\left(\dfrac{210}{t}\right)(3) + 55(3) = 291$ Number of pages read

$\left(\dfrac{630}{t}\right)(t) + 165(t) = 291(t)$

$630 + 165t = 291t$

$630 = 126t$

$t = 5$

It takes Jerry 5 hr to proofread 210 pages.

24. 5 hr

25. Let x = Sandra's speed (in mph)

Then $\dfrac{3}{4}x$ = Jim's speed (in mph)

They each ride 3 hr.

$x(3) + \dfrac{3}{4}x(3) = 63$ LCD is 4

$3x(4) + \dfrac{9}{4}x(4) = 63(4)$

$12x + 9x = 252$

$21x = 252$

$x = 12$

$\dfrac{3}{4}x = 9$

Sandra's speed was 12 mph, and Jim's was 9 mph.

26. Rebecca's speed was 16 mph, and Jill's was 14 mph.

27. Let x = larger number   **28.** 48 and 56

Then $x - 12$ = smaller number

$\dfrac{1}{6}x - \dfrac{1}{5}(x - 12) = 2$ LCD is 30

$\dfrac{1}{6}x(30) - \dfrac{1}{5}(x - 12)(30) = 2(30)$

$5x - 6(x - 12) = 60$

$5x - 6x + 72 = 60$

$-x = -12$

$x = 12$

$x - 12 = 0$

The numbers are 12 and 0.

29. Let x = rate of machine A (in widgets per minute)

Then $x + 2$ = rate of machine B (in widgets per minute)

Solving the formula $rt = w$ for t, we have $t = \dfrac{w}{r}$.

Then $\dfrac{24}{x}$ = time for machine A to produce 24 widgets

and $\dfrac{24}{x + 2}$ = time for machine B to produce 24 widgets

Time for machine A = time for machine B plus 2 minutes

$\dfrac{24}{x} = \dfrac{24}{x + 2} + 2$

$\left(\dfrac{24}{\cancel{x}}\right)^{1}(\cancel{x})(x + 2) = \left(\dfrac{24}{\cancel{x+2}}\right)(x)(\cancel{x + 2})^{1} + 2(x)(x + 2)$

$24x + 48 = 24x + 2x^2 + 4x$

$0 = 2x^2 + 4x - 48$

$0 = 2(x^2 + 2x - 24)$

$0 = 2(x + 6)(x - 4)$

$2 \neq 0$ | $x + 6 = 0$ | $x - 4 = 0$

            $x = -6$ Reject | $x = 4$

Machine A works at the rate of 4 widgets per minute.

30. 5 ft per minute

31. Let x = cyclist's actual rate (in mph)

Then $x + 3$ = cyclist's faster rate (in mph)

Solving the formula $rt = d$ for t, we have $t = \dfrac{d}{r}$.

Then $\dfrac{90}{x}$ = actual time for 90 mi

and $\dfrac{90}{x + 3}$ = time if cyclist went faster

Actual time = faster time plus 1 hr

$\dfrac{90}{x} = \dfrac{90}{x + 3} + 1$

$\left(\dfrac{90}{\cancel{x}}\right)^{1}(\cancel{x})(x + 3) = \left(\dfrac{90}{\cancel{x+3}}\right)(x)(\cancel{x + 3})^{1} + 1(x)(x + 3)$

$90x + 270 = 90x + x^2 + 3x$

$0 = x^2 + 3x - 270$

$0 = (x + 18)(x - 15)$

$x + 18 = 0$ | $x - 15 = 0$

$x = -18$ Reject | $x = 15$

The cyclist's actual rate was 15 mph.

32. 20 mph

Exercises 8.11A   (page 449)

1. a. $y = kx$ **b.** $14 = k(2)$ **c.** $y = (7)(5)$
                     $k = 7$ $= 35$

2. a. $z = kw$ **b.** $k = 7$ **c.** 49

3. a. $s = kt$ **b.** $-9 = k(-3)$ **c.** $s = (3)(-12)$
                      $k = 3$ $= -36$

4. a. $u = kv$ **b.** $k = -2$ **c.** 36

5. a. $A = ks^2$ **b.** $-20 = k(-2)^2$ **c.** $A = -5(5)^2$
                      $-20 = 4k$ $= -5(25)$
                        $k = -5$ $= -125$

6. a. $R = kx^2$ **b.** $k = 10$ **c.** 1,000

7. a. $l = kF$
   b. Since $3 = k(5)$, $k = \dfrac{3}{5}$. Then $l = \dfrac{3}{5}(2) = \dfrac{6}{5} = 1.2$. The
      length changes 1.2 in. when a 2-lb force is applied.

8. a. $m = kg$ **b.** 414 mi

9. a. $s = kt^2$
   b. Since $64 = k(2)^2$, $k = 16$. Then $s = 16(3)^2 = 144$. The
      object will fall 144 ft in 3 sec.

10. a. $R = kv^2$ **b.** 900 lb

Exercises 8.11B   (page 451)

1. a. $y = \dfrac{k}{x}$ **b.** $3 = \dfrac{k}{2}$ **c.** $y = \dfrac{6}{-10}$
    $k = 6$ $= -\dfrac{3}{5}$

2. a. $z = \dfrac{k}{w}$ **b.** $k = -6$ **c.** -12

3. a. $s = \dfrac{k}{t}$ **b.** $3 = \dfrac{k}{-2}$ **c.** $s = \dfrac{-6}{-5}$
    $k = -6$ $= \dfrac{6}{5}$, or 1.2

4. a. $u = \dfrac{k}{v}$ **b.** $k = -8$ **c.** $-\dfrac{8}{3}$

5. a. $F = \dfrac{k}{d^2}$ **b.** $3 = \dfrac{k}{(-4)^2}$ **c.** $F = \dfrac{48}{(8)^2}$
    $k = 3(16) = 48$ $= \dfrac{48}{64} = \dfrac{3}{4}$

6. a. $C = \dfrac{k}{v^2}$ **b.** $k = 54$ **c.** $\dfrac{3}{2}$, or 1.5

7. a. $V = \dfrac{k}{P}$

   b. Since $1{,}600 = \dfrac{k}{250}$, $k = 400{,}000$. Then

   $V = \dfrac{400{,}000}{400} = 1{,}000$. The volume is $1{,}000$ cc at a

   pressure of 400 mm of mercury.

8. a. $t = \dfrac{k}{r}$ **b.** 12 hr

9. a. $I = \dfrac{k}{d^2}$

   b. Since $75 = \dfrac{k}{50^2}$, $k = 187{,}500$. Then

   $I = \dfrac{187{,}500}{150^2} = \dfrac{187{,}500}{22{,}500} = \dfrac{25}{3}$, or $8\dfrac{1}{3}$.

10. a. $I = \dfrac{k}{d^2}$ **b.** $\dfrac{20}{3}$, or $6\dfrac{2}{3}$

Sections 8.8–8.11 Review Exercises (page 454)

1. Setting the denominator equal to zero, we have
    $x^2 - 9 = 0$
    $(x + 3)(x - 3) = 0$
    $x + 3 = 0 \quad | \quad x - 3 = 0$
    $\quad x = -3 \quad | \quad \quad x = 3$
    Therefore, -3 and 3 must be excluded.

(The checks will not be shown.)

2. $0, 1, -2$

3. LCD $= m$ **4.** 5
    $(m)\left(\dfrac{6}{m}\right) = (m)(5)$
    $6 = 5m$
    $\dfrac{6}{5} = m$
    $m = 1\dfrac{1}{5}$
    The solution is $1\dfrac{1}{5}$.

5. LCD $= 40$ **6.** $\dfrac{11}{9}$
    $(\overset{8}{\cancel{40}})\left(\dfrac{z}{\cancel{5}}\right) + (\overset{5}{\cancel{40}})\left(\dfrac{-z}{\cancel{8}}\right) = (40)(3)$
    $8z - 5z = 120$
    $3z = 120$
    $z = 40$
    The solution is 40.

7. LCD $= 2z$
    $(2z)\left(\dfrac{4}{2z}\right) + (2z)\left(\dfrac{2}{z}\right) = (2z)(1)$
    $4 + 4 = 2z$
    $8 = 2z$
    $z = 4$
    The solution is 4.

8. $-2, \dfrac{4}{5}$

9. LCD $= 18(2x - 1)$
    $\left(\dfrac{7}{2x - 1}\right)(18)(2x - 1) + \left(\dfrac{1}{18}\right)(18)(2x - 1) = \left(\dfrac{x}{6}\right)(18)(2x - 1)$
    $126 + 2x - 1 = 3x(2x - 1)$
    $125 + 2x = 6x^2 - 3x$
    $0 = 6x^2 - 5x - 125$
    $0 = (6x + 25)(x - 5)$

    $6x + 25 = 0 \quad | \quad x - 5 = 0$
    $\quad 6x = -25 \quad | \quad \quad x = 5$
    $\quad \; x = -\dfrac{25}{6} \quad |$
    The solutions are $-\dfrac{25}{6}$ and 5.

10. $y = \dfrac{2x - 14}{7}$ **11.** LCD $= n$
    $(n)\dfrac{2m}{n} = (n)P$
    $2m = nP$
    $\dfrac{2m}{P} = n$
    $n = \dfrac{2m}{P}$

12. $B = \dfrac{3V}{h}$

13. $\dfrac{F - 32}{C} = \dfrac{9}{5}$ **14.** $\dfrac{17}{13}$
    $9C = 5(F - 32)$
    $C = \dfrac{5}{9}(F - 32)$

15. Let $x =$ speed returning from work (in mph)
    $\dfrac{1}{2}(x + 10) = \dfrac{3}{4}(x)$
    $\dfrac{1}{2}x + 5 = \dfrac{3}{4}x$
    $5 = \dfrac{1}{4}x$
    $x = 20$

    Distance $=$ (rate) $\times$ (time) $= \left(20 \, \dfrac{\text{mi}}{\text{hr}}\right)\left(\dfrac{3}{4} \, \text{hr}\right) = 15$ mi
    The distance is 15 mi.

16. First answer: The number is $\dfrac{2}{3}$ and its reciprocal is $\dfrac{3}{2}$. Second answer: The number is $\dfrac{3}{2}$ and its reciprocal is $\dfrac{2}{3}$.

17. Let $x =$ number of hours for machine B to do the job
    Machine A's rate is $\dfrac{1 \text{ job}}{6 \text{ hr}}$. Machine B's rate is $\dfrac{1 \text{ job}}{x \text{ hr}}$.
    Both machines worked 4 hr.
    $\left(\dfrac{1 \text{ job}}{6 \text{ hr}}\right)(4 \text{ hr}) + \left(\dfrac{1 \text{ job}}{x \text{ hr}}\right)(4 \text{ hr}) = 1 \text{ job}$ One job was completed.
    $\left(\dfrac{1}{6}\right)(4)(6x) + \left(\dfrac{1}{x}\right)(4)(6x) = 1(6x)$
    $4x + 24 = 6x$
    $24 = 2x$
    $x = 12$
    It would take machine B 12 hr to do the job.

18. 810 mi

8

19. Let x = number of yards in $\frac{1}{2}$ day

$$\frac{2\frac{1}{2}\text{ yd}}{\frac{5}{8}\text{ day}} = \frac{x\text{ yd}}{\frac{1}{2}\text{ day}}$$

$$\frac{\frac{5}{2}}{\frac{5}{8}} = \frac{x}{\frac{1}{2}}$$

$$\frac{5}{8}x = \left(\frac{5}{2}\right)\left(\frac{1}{2}\right)$$

$$\frac{5}{8}x = \frac{5}{4}$$

$$\frac{5}{8}x(8) = \frac{5}{4}(8)$$

$$5x = 10$$

$$x = 2$$

Alan can weave 2 yd in half a day.

20. a. $y = kx^3$ **b.** $k = 2$ **c.** -128

21. a. $y = \dfrac{k}{x^2}$ **b.** $5 = \dfrac{k}{3^2}$, so $k = 5(9) = 45$.

  c. $y = \dfrac{45}{5^2} = \dfrac{45}{25} = \dfrac{9}{5}$, or $1\frac{4}{5}$, or 1.8

Chapter 8 Diagnostic Test (page 457)

Following each problem number is the number (in parentheses) of the textbook section where that kind of problem is discussed.

1. (8.1) No; we can't get the second rational expression from the first by multiplying or dividing both numerator and denominator by the same expression.

2. (8.2) **a.** $\dfrac{x^2-9}{x^2-6x+9} = \dfrac{(x+3)\overset{1}{(\cancel{x-3})}}{(x-3)\underset{1}{(\cancel{x-3})}} = \dfrac{x+3}{x-3}$

  b. $\dfrac{4x^2-23xy+15y^2}{20x^2y-15xy^2} = \dfrac{\overset{1}{(\cancel{4x-3y})}(x-5y)}{5xy\underset{1}{(\cancel{4x-3y})}}$

   $= \dfrac{x-5y}{5xy}$

3. (8.3) **a.** The signs of the numerators are the same, and the signs of the rational expressions are different; the signs of the denominators must be different. The missing term must be -3.
  b. -3: The signs of the denominators are different, and the signs of the rational expressions are the same; therefore, the signs of the numerators must be different.

4. (8.3) $\dfrac{a}{a^2+5a+6} \cdot \dfrac{4a+8}{6a^3+18a^2}$

  $= \dfrac{\overset{1}{\cancel{a}}}{(\cancel{a+2})(a+3)} \cdot \dfrac{\overset{2}{4}(\overset{1}{\cancel{a+2}})}{\underset{3a}{6a^2}(a+3)} = \dfrac{2}{3a(a+3)^2}$

5. (8.4) $\dfrac{6x}{x-2} - \dfrac{3}{x-2} = \dfrac{6x-3}{x-2}$, or $\dfrac{3(2x-1)}{x-2}$

6. (8.3) $\dfrac{2x}{x^2-9} \div \dfrac{4x^2}{x^2-6x+9} = \dfrac{2x}{(x+3)(x-3)} \cdot \dfrac{(x-3)(x-3)}{4x^2}$

  $= \dfrac{x-3}{2x(x+3)}$

7. (8.6) LCD = $b(b-1)$

  $\dfrac{b \cdot b}{(b-1) \cdot b} - \dfrac{(b+1)(b-1)}{b(b-1)} = \dfrac{b^2-(b^2-1)}{b(b-1)}$

  $= \dfrac{b^2-b^2+1}{b(b-1)} = \dfrac{1}{b(b-1)}$

8. (8.6) LCD = $(x-5)^2(2x+3)$

  $\dfrac{(x+3)(2x+3)}{(x-5)^2(2x+3)} + \dfrac{x(x-5)}{(x-5)(2x+3)(x-5)}$

  $= \dfrac{2x^2+9x+9+x^2-5x}{(x-5)^2(2x+3)} = \dfrac{3x^2+4x+9}{(x-5)^2(2x+3)}$

9. (8.6) LCD = $(x+5)(x-5)$

  $\dfrac{x(x-5)}{(x+5)(x-5)} - \dfrac{x(x+5)}{(x-5)(x+5)} - \dfrac{50}{(x+5)(x-5)}$

  $= \dfrac{x^2-5x-1(x^2+5x)-(50)}{(x+5)(x-5)}$

  $= \dfrac{x^2-5x-x^2-5x-50}{(x+5)(x-5)}$

  $= \dfrac{-10x-50}{(x+5)(x-5)} = \dfrac{-10(x+5)}{(x+5)(x-5)} = \dfrac{-10}{x-5}$

10. (8.7) $\dfrac{\frac{6x^4}{11y^2}}{\frac{9x}{22y^4}} = \dfrac{6x^4}{11y^2} \div \dfrac{9x}{22y^4} = \dfrac{6x^4}{11y^2} \cdot \dfrac{22y^4}{9x} = \dfrac{4x^3y^2}{3}$

11. (8.7) LCD = x^2

  $\dfrac{x^2\left(\frac{6}{x}+\frac{15}{x^2}\right)}{x^2\left(2+\frac{5}{x}\right)} = \dfrac{6x+15}{2x^2+5x} = \dfrac{3(2x+5)}{x(2x+5)} = \dfrac{3}{x}$

12. (8.8) **a.** If $x+4=0$, $x=-4$, so -4 must be excluded (it makes the denominator zero).
  b. If $x^2+4x=0$, $x=0$ or $x=-4$, so 0 and -4 must be excluded (they make the denominator zero).

The checks will not be shown for Problems 13–20.

13. (8.8) $\dfrac{x+7}{3} = \dfrac{2x-1}{4}$

  $4(x+7) = 3(2x-1)$

  $4x+28 = 6x-3$

  $31 = 2x$

  $x = \frac{31}{2}$

  The solution is $\frac{31}{2}$.

14. (8.8) LCD = x

  $x - \dfrac{4}{x} = 3$

  $x(x) - \left(\dfrac{4}{x}\right)(x) = 3(x)$

  $x^2-3x-4 = 0$

  $(x-4)(x+1) = 0$

  $x-4=0 \mid x+1=0$

  $x=4 \mid x=-1$

  The solutions are 4 and -1.

15. (8.8) Since $4x+2 = 2(2x+1)$, the LCD is $6(2x+1)$.

  $6(2x+1)\left(\dfrac{3}{2(2x+1)}\right) + 6(2x+1)\left(\dfrac{x+1}{6}\right) = 1(6)(2x+1)$

  $9 + 2x^2+3x+1 = 12x+6$

  $2x^2+3x+10 = 12x+6$

  $2x^2-9x+4 = 0$

  $(x-4)(2x-1) = 0$

  $x-4=0 \mid 2x-1=0$

  $x=4 \mid 2x=1$

   $\mid x=\frac{1}{2}$

  The solutions are 4 and $\frac{1}{2}$.

16. (8.8) $\dfrac{3x-5}{x} = \dfrac{x+1}{3}$

  $3(3x-5) = x(x+1)$

  $9x-15 = x^2+x$

  $0 = x^2-8x+15$

  $0 = (x-3)(x-5)$

  $x-3=0 \mid x-5=0$

  $x=3 \mid x=5$

  The solutions are 3 and 5.

17. (8.9) Solve for y:

$$5x + 7y = 18$$
$$7y = 18 - 5x$$
$$y = \frac{18 - 5x}{7}$$

18. (8.10) Let $\quad x = $ numerator

Then $x + 6 = $ denominator

$$\frac{x - 1}{(x + 6) + 7} = \frac{1}{3}$$
$$\frac{x - 1}{x + 13} = \frac{1}{3}$$
$$3(x - 1) = 1(x + 13)$$
$$3x - 3 = x + 13$$
$$2x = 16$$
$$x = 8 \quad \text{Numerator}$$
$$x + 6 = 14 \quad \text{Denominator}$$

The fraction is $\frac{8}{14}$.

19. (8.10) Let $x = $ number of mi covered in $2\frac{2}{3}$ hr $\left(\text{or } \frac{8}{3} \text{ hr}\right)$

$$\frac{21 \text{ mi}}{\frac{2}{5} \text{ hr}} = \frac{x \text{ mi}}{\frac{8}{3} \text{ hr}}$$
$$\frac{2}{5}x = 21\left(\frac{8}{3}\right)$$
$$\frac{2}{5}x = 56$$
$$(5)\left(\frac{2}{5}x\right) = 56(5)$$
$$2x = 280$$
$$x = 140$$

Mr. Ames can drive 140 mi in $2\frac{2}{3}$ hr.

20. (8.10) Let $t = $ number of hours for Kenneth to type 111 pages

Then Kenneth's rate is $\dfrac{111 \text{ pages}}{t \text{ hr}}$.

David's rate is $\dfrac{128 \text{ pages}}{4 \text{ hr}}$, or $32 \dfrac{\text{pages}}{\text{hr}}$.

Kenneth and David type for 5 hr.

$$\left(\frac{111}{t}\right)(5) + 32(5) = 345 \quad \text{Number of pages typed}$$
$$\left(\frac{555}{t}\right)(t) + 160(t) = 345(t)$$
$$555 + 160t = 345t$$
$$555 = 185t$$
$$t = 3$$

It takes Kenneth 3 hr to type 111 pages.

Chapters 1–8 Cumulative Review Exercises   (page 457)

1. $10 - (3 \cdot 2 - 25) = 10 - (6 - 25) = 10 - (-19) = 29$

2. $V = \frac{10}{9}$

3. $\left(\dfrac{24y^{-3}}{8y^{-1}}\right)^{-2} = (3y^{-2})^{-2} = 3^{-2}y^4 = \dfrac{y^4}{3^2} = \dfrac{y^4}{9}$

4. $3x - 1$

5. $3z - 2 \overline{)15z^2 + 11z + 4} \quad \dfrac{5z + 7}{} \text{ R } 18$

$$\underline{15z^2 - 10z}$$
$$21z + 4$$
$$\underline{21z - 14}$$
$$18$$

6. x^3

7. $\dfrac{2x - (2x - 1)}{x + 1} = \dfrac{2x - 2x + 1}{x + 1} = \dfrac{1}{x + 1}$

8. $\dfrac{10x^3 + 3}{2x^3}$

9. LCD $= (x + 4)(x + 5)(2x - 1)$

$$\frac{(2x - 1)(2x - 1)}{(x + 4)(x + 5)(2x - 1)} - \frac{(x + 5)(x + 5)}{(2x - 1)(x + 4)(x + 5)}$$
$$= \frac{(4x^2 - 4x + 1) - (x^2 + 10x + 25)}{(x + 4)(x + 5)(2x - 1)}$$
$$= \frac{4x^2 - 4x + 1 - x^2 - 10x - 25}{(x + 4)(x + 5)(2x - 1)}$$
$$= \frac{3x^2 - 14x - 24}{(x + 4)(x + 5)(2x - 1)}$$

10. $\frac{140}{3}$

11. LCD $= 4x^2$

$$(4x^2)\left(\frac{3}{x}\right) - (4x^2)\left(\frac{8}{x^2}\right) = \left(\frac{1}{4}\right)(4x^2)$$
$$12x - 32 = x^2$$
$$0 = x^2 - 12x + 32$$
$$0 = (x - 4)(x - 8)$$

$$\begin{array}{c|c} x - 4 = 0 & x - 8 = 0 \\ x = 4 & x = 8 \end{array}$$

(The check will not be shown.) The solutions are 4 and 8.

12. $(x + 6)(x - 7)$ **13.** $3(w^2 - 16) = 3(w + 4)(w - 4)$

14. $(5a - 3b)(4a + b)$

15. $3x^2 - 6x - 2xy + 4y = 3x(x - 2) - 2y(x - 2)$
$\quad = (x - 2)(3x - 2y)$

16. 16 and 17

17. Let $\quad x = $ one number

Then $5 - x = $ other number

$$x(5 - x) = -24$$
$$5x - x^2 = -24$$
$$0 = x^2 - 5x - 24$$
$$0 = (x - 8)(x + 3)$$

$$\begin{array}{c|c} x - 8 = 0 & x + 3 = 0 \\ x = 8 & x = -3 \\ 5 - x = -3 & 5 - x = 8 \end{array}$$

(The check will not be shown.) The numbers are 8 and -3.

18. 5 lb of oranges at 99¢ per pound and 10 lb of oranges at 78¢ per pound

19. Let $\quad n = $ number of nickels

Then $20 - n = $ number of dimes

$$5n + 10(20 - n) = 165$$
$$5n + 200 - 10n = 165$$
$$-5n = -35$$
$$n = 7$$
$$20 - n = 13$$

Manny has 7 nickels and 13 dimes.

20. 33 mph; 132 mi

Exercises 9.1A   (page 462)

1. 9   **2.** 100   **3.** 15   **4.** 29   **5.** $3a^2$

6. $7y^2$ **7.** $x + y$ **8.** $u + v$ **9.** $\sqrt{25}\sqrt{x^2} = 5x$

10. $10y$ **11.** $\sqrt{81}\sqrt{s^4} = 9s^2$ **12.** $8t^3$

13. $\sqrt{4}\sqrt{x^2} = 2x$ **14.** $3y$ **15.** $\sqrt{16}\sqrt{z^4} = 4z^2$

16. $5b^3$ **17.** $\sqrt{x^2 \cdot x} = \sqrt{x^2}\sqrt{x} = x\sqrt{x}$ **18.** $y^2\sqrt{y}$

19. $\sqrt{m^6 \cdot m} = \sqrt{m^6}\sqrt{m} = m^3\sqrt{m}$ **20.** $n^4\sqrt{n}$

21. $\sqrt{16 \cdot 2x^2y^2y} = \sqrt{16}\sqrt{2}\sqrt{x^2}\sqrt{y^2}\sqrt{y} = 4xy\sqrt{2y}$

22. $7b\sqrt{ab}$

23. $\sqrt{4 \cdot 2s^4t^4t} = \sqrt{4}\sqrt{2}\sqrt{s^4}\sqrt{t^4}\sqrt{t} = 2s^2t^2\sqrt{2t}$

24. $2xy^3\sqrt{11x}$

25. $\sqrt{9 \cdot 2a^2b^2b} = \sqrt{9}\sqrt{2}\sqrt{a^2}\sqrt{b^2}\sqrt{b} = 3ab\sqrt{2b}$

26. $3c^2d\sqrt{3c}$

9

27. $\sqrt{4 \cdot 10x^6y^2} = \sqrt{4}\,\sqrt{10}\,\sqrt{x^6}\,\sqrt{y^2} = 2x^3y\sqrt{10}$

28. $3a^3b^4\sqrt{15}$

29. $\sqrt{2^2 \cdot 3 \cdot 5h^4hk^4} = \sqrt{2^2}\,\sqrt{15}\,\sqrt{h^4}\,\sqrt{h}\,\sqrt{k^4} = 2h^2k^2\sqrt{15h}$

30. $3m^3n^3\sqrt{10m}$

31. $\sqrt{2^2 \cdot 2 \cdot 5 \cdot 7y^4yz^6} = \sqrt{2^2}\,\sqrt{70}\,\sqrt{y^4}\,\sqrt{y}\,\sqrt{z^6} = 2y^2z^3\sqrt{70y}$

32. $3g^4h^4\sqrt{30h}$

33. $\sqrt{100 \cdot 5x^6y^8y} = \sqrt{100}\,\sqrt{5}\,\sqrt{x^6}\,\sqrt{x}\,\sqrt{y^8}\,\sqrt{y} = 10x^3y^4\sqrt{5xy}$

34. $6x^5y^6\sqrt{6xy}$

35. $\sqrt{25 \cdot 2u^4uv^6vw^2} = \sqrt{25}\,\sqrt{2}\,\sqrt{u^4}\,\sqrt{u}\,\sqrt{v^6}\,\sqrt{v}\,\sqrt{w^2}$
$= 5u^2v^3w\sqrt{2uv}$

36. $6ab^2c^2\sqrt{2ac}$

Exercises 9.1B   (page 465)

1. $\dfrac{\sqrt{9}}{\sqrt{25}} = \dfrac{3}{5}$ **2.** $\dfrac{6}{7}$ **3.** $\dfrac{\sqrt{16}}{\sqrt{25}} = \dfrac{4}{5}$ **4.** $\dfrac{9}{10}$

5. $\dfrac{\sqrt{y^4}}{\sqrt{x^2}} = \dfrac{y^2}{x}$ **6.** $\dfrac{x^3}{v^4}$ **7.** $\dfrac{\sqrt{4x^2}}{\sqrt{9}} = \dfrac{2x}{3}$ **8.** $\dfrac{4a}{7}$

9. $\sqrt{\dfrac{1}{4k^2}} = \dfrac{\sqrt{1}}{\sqrt{4k^2}} = \dfrac{1}{2k}$ **10.** $\dfrac{m}{3}$

11. $\sqrt{\dfrac{4x^2}{y^2}} = \dfrac{\sqrt{4x^2}}{\sqrt{y^2}} = \dfrac{2x}{y}$ **12.** $\dfrac{x^2}{3y}$

13. $\sqrt{\dfrac{1 \cdot 5}{5 \cdot 5}} = \dfrac{\sqrt{5}}{\sqrt{5^2}} = \dfrac{\sqrt{5}}{5}$ **14.** $\dfrac{\sqrt{3}}{3}$

15. $\sqrt{\dfrac{7 \cdot 13}{13 \cdot 13}} = \dfrac{\sqrt{91}}{\sqrt{13^2}} = \dfrac{\sqrt{91}}{13}$ **16.** $\dfrac{\sqrt{85}}{17}$

17. $\sqrt{\dfrac{1 \cdot 2}{18 \cdot 2}} = \dfrac{\sqrt{2}}{\sqrt{36}} = \dfrac{\sqrt{2}}{6}$ **18.** $\dfrac{\sqrt{2}}{10}$

19. $\sqrt{\dfrac{7 \cdot 3}{27 \cdot 3}} = \dfrac{\sqrt{21}}{\sqrt{81}} = \dfrac{\sqrt{21}}{9}$ **20.** $\dfrac{\sqrt{10}}{4}$

21. $\dfrac{3}{\sqrt{7}} \cdot \dfrac{\sqrt{7}}{\sqrt{7}} = \dfrac{3\sqrt{7}}{7}$ **22.** $\dfrac{2\sqrt{3}}{3}$

23. $\dfrac{10}{\sqrt{5}} \cdot \dfrac{\sqrt{5}}{\sqrt{5}} = \dfrac{\overset{2}{\cancel{10}}\sqrt{5}}{\underset{1}{\cancel{5}}} = 2\sqrt{5}$ **24.** $7\sqrt{2}$

25. $\sqrt{\dfrac{m^2(3)}{3(3)}} = \dfrac{m\sqrt{3}}{3}$ **26.** $\dfrac{k\sqrt{5}}{5}$ **27.** $\dfrac{\sqrt{x^4xz^4}}{\sqrt{36y^2}} = \dfrac{x^2z^2\sqrt{x}}{6y}$

28. $\dfrac{ac^3\sqrt{a}}{5b}$ **29.** $\sqrt{\dfrac{3a^2}{4b^2}} = \dfrac{a\sqrt{3}}{2b}$ **30.** $\dfrac{v\sqrt{5}}{2}$

31. $\sqrt{\dfrac{b^2c^4d}{16d^3d}} = \dfrac{bc^2\sqrt{d}}{4d^2}$ **32.** $\dfrac{h^2k^4\sqrt{p}}{7p^3}$

33. $\sqrt{\dfrac{4m^2n}{n^2}} = \dfrac{2m\sqrt{n}}{n}$ **34.** $\dfrac{3y\sqrt{x}}{x}$

Exercises 9.2A   (page 469)

1. 3   **2.** 7   **3.** 4   **4.** 9   **5.** $\sqrt{36} = 6$

6. 8   **7.** $\sqrt{9x^2} = 3x$ **8.** $5y$

9. $\sqrt{5 \cdot 10 \cdot 2} = \sqrt{100} = 10$ **10.** 12

11. $\sqrt{4 \cdot 2 \cdot 2 \cdot 9 \cdot 4 \cdot 5} = \sqrt{4}\,\sqrt{2 \cdot 2}\,\sqrt{9}\,\sqrt{4}\,\sqrt{5}$
$= 2 \cdot 2 \cdot 3 \cdot 2\sqrt{5} = 24\sqrt{5}$

12. $60\sqrt{7}$ **13.** $\sqrt{5 \cdot 5 \cdot 2 \cdot x \cdot x^2} = 5 \cdot x\sqrt{2 \cdot x} = 5x\sqrt{2x}$

14. $11y\sqrt{3y}$

15. $\sqrt{3 \cdot 2 \cdot 6 \cdot 5 \cdot x \cdot x^2 \cdot y^2} = 6 \cdot x \cdot y\sqrt{5x} = 6xy\sqrt{5x}$

16. $14ab\sqrt{5ab}$ **17.** $\sqrt{100a^2b^2b} = 10ab\sqrt{b}$ **18.** $9xy\sqrt{x}$

19. $\sqrt{2a \cdot 6 \cdot 3a} = \sqrt{36a^2} = 6a$ **20.** $4h^2$

21. $(5 \cdot 2)\sqrt{2x \cdot 8x^3 \cdot 3x^5} = 10\sqrt{48x^9} = 10\sqrt{3 \cdot 16 \cdot x^8 \cdot x}$
$= 10 \cdot 4 \cdot x^4\sqrt{3x} = 40x^4\sqrt{3x}$

22. $72M^3\sqrt{2M}$

23. $(8 \cdot 3)\sqrt{5x(15x^3)(6x)} = 24\sqrt{450x^5} = 24\sqrt{225 \cdot 2 \cdot x^4 \cdot x}$
$= 24 \cdot 15 \cdot x^2 \cdot \sqrt{2x} = 360x^2\sqrt{2x}$

24. $450a^3\sqrt{3}$

Exercises 9.2B   (page 471)

1. $\sqrt{\dfrac{20}{5}} = \sqrt{4} = 2$ **2.** $\dfrac{1}{2}$ **3.** $\sqrt{\dfrac{32}{2}} = \sqrt{16} = 4$ **4.** 7

5. $\dfrac{\sqrt{4}}{\sqrt{5}} = \dfrac{2}{\sqrt{5}} \cdot \dfrac{\sqrt{5}}{\sqrt{5}} = \dfrac{2\sqrt{5}}{5}$ **6.** $\dfrac{3\sqrt{7}}{7}$ **7.** $\sqrt{\dfrac{15x}{5x}} = \sqrt{3}$

8. $\sqrt{6}$ **9.** $\sqrt{\dfrac{75a^3}{5a}} = \sqrt{15a^2} = a\sqrt{15}$ **10.** $b^2\sqrt{3}$

11. $\sqrt{\dfrac{35s^4}{10s}} = \sqrt{\dfrac{7s^3}{2}} = \dfrac{\sqrt{7s^3}}{\sqrt{2}} \cdot \dfrac{\sqrt{2}}{\sqrt{2}} = \dfrac{s\sqrt{14s}}{2}$ **12.** $\dfrac{\sqrt{14t}}{2}$

13. $\sqrt{\dfrac{8xy^3}{12x^2y}} = \sqrt{\dfrac{2y^2}{3x}} = \dfrac{\sqrt{2y^2}}{\sqrt{3x}} = \dfrac{\sqrt{2y^2}}{\sqrt{3x}} \cdot \dfrac{\sqrt{3x}}{\sqrt{3x}} = \dfrac{y\sqrt{6x}}{3x}$

14. $\dfrac{a\sqrt{6b}}{3b}$ **15.** $\sqrt{\dfrac{72x^3y^2}{2xy^2}} = \sqrt{36x^2} = 6x$ **16.** $3y$

17. $\sqrt{\dfrac{x^4y}{5y}} = \dfrac{\sqrt{x^4}}{\sqrt{5}} \cdot \dfrac{\sqrt{5}}{\sqrt{5}} = \dfrac{x^2\sqrt{5}}{5}$ **18.** $\dfrac{m^3\sqrt{3}}{3}$

19. $\dfrac{4}{3}\sqrt{\dfrac{45m^3}{10m}} = \dfrac{4}{3}\sqrt{\dfrac{9m^2}{2}} = \dfrac{4}{3} \cdot \dfrac{\sqrt{9m^2}}{\sqrt{2}} = \dfrac{4 \cdot \overset{1}{\cancel{3}}m}{\underset{1}{\cancel{3}}\sqrt{2}} \cdot \dfrac{\sqrt{2}}{\sqrt{2}} = \dfrac{\overset{2}{\cancel{4}}m\sqrt{2}}{\underset{1}{\cancel{2}}}$
$= 2m\sqrt{2}$

20. $4x\sqrt{6x}$ **21.** $\dfrac{5}{7}\sqrt{\dfrac{35a^5}{5a}} = \dfrac{5}{7}\sqrt{7a^4} = \dfrac{5}{7}a^2\sqrt{7}$

22. $4x^3\sqrt{2}$

Exercises 9.3   (page 475)

1. $(2 + 5)\sqrt{3} = 7\sqrt{3}$ **2.** $7\sqrt{2}$ **3.** $(3 - 1)\sqrt{x} = 2\sqrt{x}$

4. $4\sqrt{a}$ **5.** $\left(\dfrac{3}{2} - \dfrac{1}{2}\right)\sqrt{2} = \dfrac{2}{2}\sqrt{2} = \sqrt{2}$ **6.** $\sqrt{3}$

7. $40\sqrt{5} + \sqrt{5} = (40 + 1)\sqrt{5} = 41\sqrt{5}$ **8.** $13\sqrt{7}$

9. $5 + \sqrt{5}$ **10.** $4 + \sqrt{6}$

11. $5\sqrt{2} + 5\sqrt{2} = (5 + 5)\sqrt{2} = 10\sqrt{2}$ **12.** $8\sqrt{2}$

13. $2\sqrt{17} - \sqrt{17} = (2 - 1)\sqrt{17} = \sqrt{17}$ **14.** $\sqrt{13}$

15. $3\sqrt{5} - 2\sqrt{7}$ **16.** $3\sqrt{10} - 3\sqrt{6}$

17. $2\sqrt{3} + \sqrt{4 \cdot 3} = 2\sqrt{3} + 2\sqrt{3} = 4\sqrt{3}$ **18.** $5\sqrt{2}$

19. $2\sqrt{25 \cdot 2} - \sqrt{16 \cdot 2} = 2 \cdot 5\sqrt{2} - 4\sqrt{2} = 10\sqrt{2} - 4\sqrt{2}$
$= 6\sqrt{2}$

20. $3\sqrt{6}$

21. $3\sqrt{16 \cdot 2} - \sqrt{4 \cdot 2} = 3 \cdot 4\sqrt{2} - 2\sqrt{2} = 12\sqrt{2} - 2\sqrt{2}$
$= 10\sqrt{2}$

22. $6\sqrt{3}$

23. $\sqrt{\dfrac{1 \cdot 2}{2 \cdot 2}} + \sqrt{4 \cdot 2} = \dfrac{1}{2}\sqrt{2} + 2\sqrt{2} = \left(\dfrac{1}{2} + 2\right)\sqrt{2} = \dfrac{5\sqrt{2}}{2}$

24. $\dfrac{7\sqrt{3}}{3}$

25. $\sqrt{4 \cdot 6} - \sqrt{\dfrac{2 \cdot 3}{3 \cdot 3}} = 2\sqrt{6} - \dfrac{\sqrt{6}}{3} = \left(2 - \dfrac{1}{3}\right)\sqrt{6} = \dfrac{5\sqrt{6}}{3}$

26. $\dfrac{13\sqrt{5}}{5}$

27. $10\sqrt{\dfrac{3 \cdot 5}{5 \cdot 5}} + \sqrt{4 \cdot 15} = \dfrac{10\sqrt{15}}{5} + 2\sqrt{15} = 2\sqrt{15} + 2\sqrt{15}$
   $= 4\sqrt{15}$

28. $6\sqrt{3}$ **29.** $\dfrac{\sqrt{25}}{\sqrt{2}} - \dfrac{3}{\sqrt{2}} = \dfrac{5-3}{\sqrt{2}} = \dfrac{2}{\sqrt{2}} \cdot \dfrac{\sqrt{2}}{\sqrt{2}} = \dfrac{2\sqrt{2}}{2}$
   $= \sqrt{2}$

30. $\dfrac{7\sqrt{5}}{10}$

31. $3\sqrt{\dfrac{1 \cdot 6}{6 \cdot 6}} + \sqrt{4 \cdot 3} - 5\sqrt{\dfrac{3 \cdot 2}{2 \cdot 2}}$

$= \dfrac{\overset{1}{\cancel{3}}}{1} \cdot \dfrac{\sqrt{6}}{\underset{2}{\cancel{6}}} + \sqrt{4}\sqrt{3} - \dfrac{5}{1} \cdot \dfrac{\sqrt{6}}{2} = \dfrac{1}{2}\sqrt{6} + 2\sqrt{3} - \dfrac{5}{2}\sqrt{6}$

$= \left(\dfrac{1}{2} - \dfrac{5}{2}\right)\sqrt{6} + 2\sqrt{3} = -2\sqrt{6} + 2\sqrt{3}$

32. $\sqrt{10} + 2\sqrt{5}$

33. $6 + 5\sqrt{3}$ is simplified.   **34.** $8 + 5\sqrt{2}$ is simplified.

35. $9 - 3\sqrt{25 \cdot 2} = 9 - 3 \cdot 5\sqrt{2} = 9 - 15\sqrt{2}$

36. $8 - 8\sqrt{11}$

37. $\dfrac{-4 + \sqrt{4 \cdot 7}}{10} = \dfrac{-4 + 2\sqrt{7}}{10} = \dfrac{\overset{1}{\cancel{2}}(-2 + \sqrt{7})}{\underset{5}{\cancel{10}}} = \dfrac{-2 + \sqrt{7}}{5}$

38. $\dfrac{-3 + 2\sqrt{3}}{4}$

39. $\dfrac{10 - \sqrt{16 \cdot 6}}{6} = \dfrac{10 - 4\sqrt{6}}{6} = \dfrac{\overset{1}{\cancel{2}}(5 - 2\sqrt{6})}{\underset{3}{\cancel{6}}} = \dfrac{5 - 2\sqrt{6}}{3}$

40. $\dfrac{3 - 2\sqrt{5}}{2}$

41. $\dfrac{-12 - \sqrt{4 \cdot 3}}{12} = \dfrac{-12 - 2\sqrt{3}}{12} = \dfrac{\overset{1}{\cancel{2}}(-6 - \sqrt{3})}{\underset{6}{\cancel{12}}} = \dfrac{-6 - \sqrt{3}}{6}$

42. $\dfrac{-5 - 2\sqrt{2}}{5}$ **43.** $\dfrac{-3 + \sqrt{4 \cdot 2}}{2} = \dfrac{-3 + 2\sqrt{2}}{2}$

44. $\dfrac{-7 + 3\sqrt{2}}{2}$

45. $\approx \dfrac{1}{3}(2.646) + 2(1.732) = 0.882 + 3.464 = 4.346$

46. 7.907 if a calculator is used; 7.906 if tables are used

Exercises 9.4   (page 479)

1. $\sqrt{2}(\sqrt{2}) + \sqrt{2}(1) = 2 + \sqrt{2}$ **2.** $3 + \sqrt{3}$

3. $\sqrt{3}(2\sqrt{3}) + \sqrt{3}(1) = 6 + \sqrt{3}$ **4.** $15 + \sqrt{5}$

5. $\sqrt{x}(\sqrt{x}) + \sqrt{x}(-3) = x - 3\sqrt{x}$ **6.** $4\sqrt{y} - y$

7. $\sqrt{7}\sqrt{7} + 5\sqrt{7} + 2(3) = 7 + 5\sqrt{7} + 6 = 13 + 5\sqrt{7}$

8. $11 + 6\sqrt{3}$

9. $\sqrt{8}\sqrt{8} - 3\sqrt{2}\sqrt{8} + 2\sqrt{8}\sqrt{5} - 3\sqrt{2}\ (2\sqrt{5})$
   $= 8 - 3\sqrt{16} + 2\sqrt{40} - 6\sqrt{10}$
   $= 8 - 3(4) + 2\sqrt{4 \cdot 10} - 6\sqrt{10}$
   $= 8 - 12 + 2 \cdot 2\sqrt{10} - 6\sqrt{10} = -4 + 4\sqrt{10} - 6\sqrt{10}$
   $= -4 - 2\sqrt{10}$

10. 0

11. $(\sqrt{2} + \sqrt{14})^2 = (\sqrt{2})^2 + 2\sqrt{2}\sqrt{14} + (\sqrt{14})^2$
   $= 2 + 2\sqrt{2^2 \cdot 7} + 14 = 16 + 2 \cdot 2\sqrt{7}$
   $= 16 + 4\sqrt{7}$

12. $16 + 2\sqrt{55}$

13. $(\sqrt{2x} + 3)^2 = (\sqrt{2x})^2 + 2\sqrt{2x}(3) + (3)^2$
   $= 2x + 6\sqrt{2x} + 9$

14. $7x + 8\sqrt{7x} + 16$

15. $\dfrac{\sqrt{8}}{\sqrt{2}} + \dfrac{\sqrt{18}}{\sqrt{2}} = \sqrt{\dfrac{8}{2}} + \sqrt{\dfrac{18}{2}} = \sqrt{4} + \sqrt{9} = 2 + 3 = 5$

16. 5

17. $\dfrac{\sqrt{20}}{\sqrt{5}} + \dfrac{5\sqrt{10}}{\sqrt{5}} = \sqrt{\dfrac{20}{5}} + 5\sqrt{\dfrac{10}{5}} = \sqrt{4} + 5\sqrt{2} = 2 + 5\sqrt{2}$

18. $2 + \dfrac{\sqrt{21}}{3}$ **19. a.** $2 - \sqrt{3}$ **b.** $2\sqrt{5} + 7$ **c.** $-6 + \sqrt{2}$

20. a. $3\sqrt{2} + 5$ **b.** $\sqrt{7} - 4$ **c.** $-\sqrt{5} + \sqrt{7}$

21. $\dfrac{3(\sqrt{2} + 1)}{(\sqrt{2} - 1)(\sqrt{2} + 1)} = \dfrac{3(\sqrt{2} + 1)}{(\sqrt{2})^2 - 1^2} = \dfrac{3(\sqrt{2} + 1)}{2 - 1}$
   $= 3(\sqrt{2} + 1) = 3\sqrt{2} + 3$

22. $5\sqrt{2} + 5$

23. $\dfrac{6(\sqrt{3} + \sqrt{2})}{(\sqrt{3} - \sqrt{2})(\sqrt{3} + \sqrt{2})} = \dfrac{6(\sqrt{3} + \sqrt{2})}{(\sqrt{3})^2 - (\sqrt{2})^2} = \dfrac{6(\sqrt{3} + \sqrt{2})}{3 - 2}$
   $= 6(\sqrt{3} + \sqrt{2}) = 6\sqrt{3} + 6\sqrt{2}$

24. $-8\sqrt{2} - 8\sqrt{3}$

25. $\dfrac{6(\sqrt{5} - \sqrt{2})}{(\sqrt{5} + \sqrt{2})(\sqrt{5} - \sqrt{2})} = \dfrac{6(\sqrt{5} - \sqrt{2})}{5 - 2} = \dfrac{\overset{2}{\cancel{6}}(\sqrt{5} - \sqrt{2})}{\underset{1}{\cancel{3}}}$
   $= 2\sqrt{5} - 2\sqrt{2}$

26. $2\sqrt{7} - 2\sqrt{5}$

27. $\dfrac{(x - 4)(\sqrt{x} - 2)}{(\sqrt{x} + 2)(\sqrt{x} - 2)} = \dfrac{\overset{1}{\cancel{(x - 4)}}(\sqrt{x} - 2)}{\underset{1}{\cancel{(x - 4)}}} = \sqrt{x} - 2$

28. $\sqrt{y} + 3$

29. $\approx \dfrac{1}{3}(2.646) - 2(1.732) = 0.882 - 3.464 = -2.582 \approx -2.58$

30. ≈ 6.79 **31.** $\approx \dfrac{3 + 2(3.317)}{6} = \dfrac{3 + 6.634}{6} = \dfrac{9.634}{6} \approx 1.61$

32. ≈ -0.61

Exercises 9.5   (page 484)

(The checks will usually not be shown.)

1. $\sqrt{x} = 5$ **2.** 16
   $(\sqrt{x})^2 = 5^2$
   $x = 25$
   The solution is 25.

3. $\sqrt{2x} = 4$ **4.** 12
   $(\sqrt{2x})^2 = 4^2$
   $2x = 16$
   $x = 8$
   The solution is 8.

5. $\sqrt{x - 3} = 2$ **6.** 32
   $(\sqrt{x - 3})^2 = 2^2$
   $x - 3 = 4$
   $x = 7$
   The solution is 7.

7. $(\sqrt{2x + 1})^2 = 9^2$ **8.** 4
   $2x + 1 = 81$
   $2x = 80$
   $x = 40$
   The solution is 40.

9. $(\sqrt{3x + 1})^2 = 5^2$ **10.** 4
   $3x + 1 = 25$
   $3x = 24$
   $x = 8$
   The solution is 8.

11. $(\sqrt{x + 1})^2 = (\sqrt{2x - 7})^2$ **12.** 3
   $x + 1 = 2x - 7$
   $8 = x$
   $x = 8$
   The solution is 8.

9

9

13. $(\sqrt{3x-2})^2 = (x)^2$ **14.** 2, 3
$$3x - 2 = x^2$$
$$0 = x^2 - 3x + 2$$
$$0 = (x-2)(x-1)$$
$$x - 2 = 0 \quad | \quad x - 1 = 0$$
$$x = 2 \quad | \quad x = 1$$
The solutions are 2 and 1.

15. $(\sqrt{4x-1})^2 = (2x)^2$ **16.** $\frac{1}{3}$
$$4x - 1 = 4x^2$$
$$0 = 4x^2 - 4x + 1$$
$$0 = (2x-1)(2x-1)$$
$$2x - 1 = 0$$
$$2x = 1$$
$$x = \frac{1}{2}$$
The solution is $\frac{1}{2}$.

17. $\sqrt{x-3} = x - 5$ **18.** −1
$$(\sqrt{x-3})^2 = (x-5)^2$$
$$x - 3 = x^2 - 10x + 25$$
$$0 = x^2 - 11x + 28$$
$$0 = (x-4)(x-7)$$
$$x - 4 = 0 \quad | \quad x - 7 = 0$$
$$x = 4 \quad | \quad x = 7$$
Check for $x = 4$:   Check for $x = 7$:
$$\sqrt{x-3}+5 = x \qquad \sqrt{x-3}+5 = x$$
$$\sqrt{4-3}+5 \overset{?}{=} 4 \qquad \sqrt{7-3}+5 \overset{?}{=} 7$$
$$\sqrt{1}+5 \overset{?}{=} 4 \qquad \sqrt{4}+5 \overset{?}{=} 7$$
$$1+5 \overset{?}{=} 4 \qquad 2+5 \overset{?}{=} 7$$
$$6 \neq 4 \qquad\qquad 7 = 7$$
4 does not check. The only solution is 7.

19. $\sqrt{3x+7} = 3x - 5$ **20.** 5
$(\sqrt{3x+7})^2 = (3x-5)^2$ Squaring both sides
$$3x + 7 = 9x^2 - 30x + 25$$
$$0 = 9x^2 - 33x + 18$$
$$0 = 3(3x^2 - 11x + 6)$$
$$0 = 3(3x-2)(x-3)$$
$$3 \neq 0 \quad | \quad 3x - 2 = 0 \quad | \quad x - 3 = 0$$
$$\qquad\qquad 3x = 2 \qquad\quad x = 3$$
$$\qquad\qquad x = \frac{2}{3}$$
Check for $x = \frac{2}{3}$:   Check for $x = 3$:
$$\sqrt{3x+7}+5 = 3x \qquad \sqrt{3x+7}+5 = 3x$$
$$\sqrt{3\left(\frac{2}{3}\right)+7}+5 \overset{?}{=} 3\left(\frac{2}{3}\right) \qquad \sqrt{3(3)+7}+5 \overset{?}{=} 3(3)$$
$$\sqrt{2+7}+5 \overset{?}{=} 2 \qquad \sqrt{9+7}+5 \overset{?}{=} 9$$
$$\sqrt{9}+5 \overset{?}{=} 2 \qquad \sqrt{16}+5 \overset{?}{=} 9$$
$$3+5 \overset{?}{=} 2 \qquad 4+5 \overset{?}{=} 9$$
$$8 \neq 2 \qquad\qquad 9 = 9$$
$\frac{2}{3}$ does not check. The only solution is 3.

21. $\sqrt{5x+14} = x + 4$ **22.** −4, −5
$(\sqrt{5x+14})^2 = (x+4)^2$ Squaring both sides
$$5x + 14 = x^2 + 8x + 16$$
$$0 = x^2 + 3x + 2$$
$$0 = (x+1)(x+2)$$
$$x + 1 = 0 \quad | \quad x + 2 = 0$$
$$x = -1 \quad | \quad x = -2$$
Both solutions check; the solutions are −1 and −2.

23. $(\sqrt{2x+2})^2 = (1 + \sqrt{x+2})^2$ Squaring both sides
$$2x + 2 = 1 + 2\sqrt{x+2} + (x+2)$$
$$2x + 2 = 3 + 2\sqrt{x+2} + x$$
$$x - 1 = 2\sqrt{x+2}$$
$$(x-1)^2 = (2\sqrt{x+2})^2$$ Squaring both sides again
$$x^2 - 2x + 1 = 4(x+2)$$
$$x^2 - 2x + 1 = 4x + 8$$
$$x^2 - 6x - 7 = 0$$
$$(x-7)(x+1) = 0$$

$$x - 7 = 0 \quad | \quad x + 1 = 0$$
$$x = 7 \quad | \quad x = -1$$
Check for $x = 7$:   Check for $x = -1$:
$$\sqrt{2x+2} = 1 + \sqrt{x+2} \qquad \sqrt{2x+2} = 1 + \sqrt{x+2}$$
$$\sqrt{2(7)+2} \overset{?}{=} 1 + \sqrt{(7)+2} \qquad \sqrt{2(-1)+2} \overset{?}{=} 1 + \sqrt{(-1)+2}$$
$$\sqrt{16} \overset{?}{=} 1 + \sqrt{9} \qquad\qquad \sqrt{0} \overset{?}{=} 1 + \sqrt{1}$$
$$4 \overset{?}{=} 1 + 3 \qquad\qquad 0 \overset{?}{=} 1 + 1$$
$$4 = 4 \qquad\qquad\qquad 0 \neq 2$$
−1 does not check. The only solution is 7.

24. 5

25. $(\sqrt{3x-5})^2 = (\sqrt{x+6}-1)^2$ Squaring both sides
$$3x - 5 = (x+6) - 2\sqrt{x+6} + 1$$
$$3x - 5 = x + 7 - 2\sqrt{x+6}$$
$$2x - 12 = -2\sqrt{x+6}$$
$$(2x-12)^2 = (-2\sqrt{x+6})^2$$ Squaring both sides again
$$4x^2 - 48x + 144 = 4(x+6)$$
$$4x^2 - 48x + 144 = 4x + 24$$
$$4x^2 - 52x + 120 = 0$$
$$4(x^2 - 13x + 30) = 0$$
$$4(x-3)(x-10) = 0$$
$$4 \neq 0 \quad | \quad x - 3 = 0 \quad | \quad x - 10 = 0$$
$$\qquad\qquad x = 3 \qquad\quad x = 10$$
Check for $x = 3$:   Check for $x = 10$:
$$\sqrt{3x-5} = \sqrt{x+6}-1 \qquad \sqrt{3x-5} = \sqrt{x+6}-1$$
$$\sqrt{3(3)-5} \overset{?}{=} \sqrt{(3)+6}-1 \qquad \sqrt{3(10)-5} \overset{?}{=} \sqrt{(10)+6}-1$$
$$\sqrt{4} \overset{?}{=} \sqrt{9}-1 \qquad\qquad \sqrt{25} \overset{?}{=} \sqrt{16}-1$$
$$2 \overset{?}{=} 3-1 \qquad\qquad 5 \overset{?}{=} 4-1$$
$$2 = 2 \qquad\qquad\qquad 5 \neq 3$$
10 does not check. The only solution is 3.

26. 2

27. $I = \sqrt{\dfrac{P}{R}}$; $R = 9, I = 7$ **28.** 100 watts

$$7 = \sqrt{\dfrac{P}{9}}$$

$$(7)^2 = \left(\sqrt{\dfrac{P}{9}}\right)^2$$

$$49 = \dfrac{P}{9}$$

$$441 = P$$
The amount of power consumed is 441 watts.

29. $I = \sqrt{\dfrac{P}{R}}$; $I = 5, P = 500$ **30.** 24 ohms

$$5 = \sqrt{\dfrac{500}{R}}$$

$$(5)^2 = \left(\sqrt{\dfrac{500}{R}}\right)^2$$

$$25 = \dfrac{500}{R}$$

$$25R = 500$$

$$R = 20$$
The resistance is 20 ohms.

31. $\sigma = \sqrt{npq}$; $\sigma = 3, q = \frac{3}{4}, n = 48$ **32.** $\frac{1}{5}$
$$3 = \sqrt{48\left(\frac{3}{4}\right)p}$$
$$3 = \sqrt{36p}$$
$$(3)^2 = (\sqrt{36p})^2$$ Squaring both sides
$$9 = 36p$$
$$p = \frac{1}{4}$$
The probability of success is $\frac{1}{4}$.

33. Let a = length of one side (in cm)
$$c = \sqrt{a^2 + b^2}; \quad c = 5, b = 3$$
$$5 = \sqrt{a^2 + 3^2}$$
$$(5)^2 = (\sqrt{a^2 + 9})^2$$
$$25 = a^2 + 9$$
$$0 = a^2 - 16$$
$$0 = (a - 4)(a + 4)$$

$a - 4 = 0$	$a + 4 = 0$
$a = 4$	$a = -4$ Reject
	A length cannot be negative.

The length of one side is 4 cm.

34. 12 in.

Sections 9.1–9.5 Review Exercises (page 487)

1. 36 **2.** 3 **3.** $a + b$ **4.** $8x^2y^3$

5. $\sqrt{x^2 \cdot x} = \sqrt{x^2}\sqrt{x} = x\sqrt{x}$ **6.** $6a^2b$

7. $\sqrt{a^2 \cdot a \cdot b^2 \cdot b} = \sqrt{a^2}\sqrt{a}\sqrt{b^2}\sqrt{b} = ab\sqrt{ab}$ **8.** 11

9. $\sqrt{64} = 8$ **10.** $5\sqrt{2}$

11. $\sqrt{2^2 \cdot 3^2 \cdot 5} = \sqrt{2^2}\sqrt{3^2}\sqrt{5} = 2 \cdot 3\sqrt{5} = 6\sqrt{5}$

12. $3\sqrt{2}$ **13.** $3\sqrt{2} - 2\sqrt{2} = (3 - 2)\sqrt{2} = \sqrt{2}$

14. $\sqrt{3}$ **15.** $\dfrac{1 \cdot \sqrt{5}}{\sqrt{5} \cdot \sqrt{5}} = \dfrac{\sqrt{5}}{5}$ **16.** $\dfrac{\sqrt{7}}{7}$

17. $\dfrac{6 \cdot \sqrt{3}}{\sqrt{3} \cdot \sqrt{3}} = \dfrac{\overset{2}{6}\sqrt{3}}{\underset{1}{3}} = 2\sqrt{3}$ **18.** $\dfrac{3\sqrt{2}}{2}$

19. $\sqrt{4 \cdot 2} - \sqrt{\dfrac{1 \cdot 2}{2 \cdot 2}} = 2\sqrt{2} - \dfrac{\sqrt{2}}{2} = 2\sqrt{2} - \dfrac{1}{2}\sqrt{2} = \dfrac{3\sqrt{2}}{2}$

20. $\dfrac{8\sqrt{6}}{3}$

21. $\sqrt{2}(\sqrt{8}) + \sqrt{2}(\sqrt{18}) = \sqrt{16} + \sqrt{36} = 4 + 6 = 10$

22. 23

23. $(2\sqrt{3})^2 + 2(2\sqrt{3})(1) + (1)^2 = 12 + 4\sqrt{3} + 1 = 13 + 4\sqrt{3}$

24. $19 - 6\sqrt{2}$

25. $\dfrac{8 \cdot (\sqrt{3} + 2)}{(\sqrt{3} - 2)(\sqrt{3} + 2)} = \dfrac{8(\sqrt{3} + 2)}{3 - 4} = -8(\sqrt{3} + 2)$
$$= -8\sqrt{3} - 16$$

26. $6\sqrt{5} - 12$

27. $\sqrt{x} = 4$
$$(\sqrt{x})^2 = 4^2$$
$$x = 16$$
The solution is 16.

28. 12

29. $(\sqrt{2x - 1})^2 = 5^2$
$$2x - 1 = 25$$
$$2x = 26$$
$$x = 13$$
The solution is 13.

30. 3

31. $(\sqrt{7x - 6})^2 = (x)^2$
$$7x - 6 = x^2$$
$$0 = x^2 - 7x + 6$$
$$0 = (x - 6)(x - 1)$$

$x - 6 = 0$	$x - 1 = 0$
$x = 6$	$x = 1$

The solutions are 6 and 1.

32. 1, 8

33. $(\sqrt{2x + 7})^2 = (\sqrt{x} + 2)^2$
$$2x + 7 = x + 4\sqrt{x} + 4$$
$$x + 3 = 4\sqrt{x}$$
$$(x + 3)^2 = (4\sqrt{x})^2$$
$$x^2 + 6x + 9 = 16x$$
$$x^2 - 10x + 9 = 0$$
$$(x - 9)(x - 1) = 0$$

$x - 9 = 0$	$x - 1 = 0$
$x = 9$	$x = 1$

The solutions are 9 and 1.

34. 1

35. Let a = length of one side (in yd)
$$c = \sqrt{a^2 + b^2}; \quad c = 13, b = 12$$
$$13 = \sqrt{a^2 + 12^2}$$
$$(13)^2 = (\sqrt{a^2 + 144})^2$$
$$169 = a^2 + 144$$
$$0 = a^2 - 25$$
$$0 = (a - 5)(a + 5)$$

$a - 5 = 0$	$a + 5 = 0$
$a = 5$	$a = -5$ Reject
	A length cannot be negative.

The length of one side is 5 yd.

Chapter 9 Diagnostic Test (page 491)

Following each problem number is the number (in parentheses) of the textbook section where that kind of problem is discussed.

1. (9.1) $\sqrt{16z^2} = \sqrt{16}\sqrt{z^2} = 4z$

2. (9.1) $\sqrt{50} = \sqrt{25 \cdot 2} = \sqrt{25}\sqrt{2} = 5\sqrt{2}$

3. (9.1) $\sqrt{x^6y^5} = \sqrt{x^6 \cdot y^4 \cdot y} = \sqrt{x^6}\sqrt{y^4}\sqrt{y} = x^3y^2\sqrt{y}$

4. (9.1) $\sqrt{\dfrac{3}{14}} = \sqrt{\dfrac{3 \cdot 14}{14 \cdot 14}} = \dfrac{\sqrt{42}}{\sqrt{14^2}} = \dfrac{\sqrt{42}}{14}$

5. (9.1) $\sqrt{\dfrac{48}{3n^2}} = \sqrt{\dfrac{16}{n^2}} = \dfrac{\sqrt{16}}{\sqrt{n^2}} = \dfrac{4}{n}$

6. (9.1) $\sqrt{60} = \sqrt{4 \cdot 15} = \sqrt{4}\sqrt{15} = 2\sqrt{15}$

7. (9.2) $\sqrt{13}\sqrt{13} = 13$

8. (9.2) $\sqrt{3}\sqrt{75y^2} = \sqrt{225y^2} = \sqrt{225}\sqrt{y^2} = 15y$

9. (9.5) $\sqrt{5}(2\sqrt{5} - 3) = \sqrt{5}(2\sqrt{5}) - \sqrt{5}(3)$
$$= 2 \cdot \sqrt{5 \cdot 5} - 3\sqrt{5}$$
$$= 2 \cdot 5 - 3\sqrt{5} = 10 - 3\sqrt{5}$$

10. (9.4) $(\sqrt{13} + \sqrt{3})(\sqrt{13} - \sqrt{3}) = (\sqrt{13})^2 - (\sqrt{3})^2$
$$= 13 - 3 = 10$$

11. (9.4) $(4\sqrt{x} + 3)(3\sqrt{x} + 4)$
$$= (4\sqrt{x})(3\sqrt{x}) + 4(4\sqrt{x}) + 3(3\sqrt{x}) + (3)(4)$$
$$= 12\sqrt{x \cdot x} + 16\sqrt{x} + 9\sqrt{x} + 12$$
$$= 12x + 25\sqrt{x} + 12$$

12. (9.4) $(8 + \sqrt{10})^2$
$$= (8)^2 + 2(8)(\sqrt{10}) + (\sqrt{10})^2$$
$$= 64 + 16\sqrt{10} + 10$$
$$= 74 + 16\sqrt{10}$$

13. (9.2) $\dfrac{\sqrt{5}}{\sqrt{80}} = \sqrt{\dfrac{5}{80}} = \sqrt{\dfrac{1}{16}} = \dfrac{1}{4}$

14. (9.2) $\dfrac{\sqrt{3}}{\sqrt{6}} = \sqrt{\dfrac{3}{6}} = \sqrt{\dfrac{1}{2}} = \sqrt{\dfrac{1 \cdot 2}{2 \cdot 2}} = \dfrac{\sqrt{2}}{2}$

15. (9.4) $\dfrac{\sqrt{18} - \sqrt{36}}{\sqrt{3}} = \dfrac{\sqrt{18}}{\sqrt{3}} - \dfrac{\sqrt{36}}{\sqrt{3}} = \sqrt{\dfrac{18}{3}} - \sqrt{\dfrac{36}{3}}$
$$= \sqrt{6} - \sqrt{12} = \sqrt{6} - 2\sqrt{3}$$

16. (9.2) $\dfrac{\sqrt{x^4y}}{\sqrt{3y}} = \sqrt{\dfrac{x^4y}{3y}} = \sqrt{\dfrac{x^4(3)}{3(3)}}$
$$= \dfrac{\sqrt{x^4}\sqrt{3}}{\sqrt{9}} = \dfrac{x^2\sqrt{3}}{3} \text{ or } \dfrac{x^2}{3}\sqrt{3}$$

17. (9.2) $\dfrac{6}{1 + \sqrt{5}} = \dfrac{6(1 - \sqrt{5})}{(1 + \sqrt{5})(1 - \sqrt{5})} = \dfrac{6(1 - \sqrt{5})}{1^2 - (\sqrt{5})^2}$
$$= \dfrac{6(1 - \sqrt{5})}{1 - 5} = \dfrac{\overset{3}{6}(1 - \sqrt{5})}{\underset{2}{-4}} = -\dfrac{3}{2} + \dfrac{3\sqrt{5}}{2}$$

9

18. (9.3) $8\sqrt{t} - \sqrt{t} = (8 - 1)\sqrt{t} = 7\sqrt{t}$

19. (9.3) $7\sqrt{3} - \sqrt{27} = 7\sqrt{3} - 3\sqrt{3} = (7 - 3)\sqrt{3} = 4\sqrt{3}$

20. (9.3) $\sqrt{54} + 4\sqrt{24} = 3\sqrt{6} + 4 \cdot 2\sqrt{6}$
$$= 3\sqrt{6} + 8\sqrt{6} = 11\sqrt{6}$$

21. (9.3) $\sqrt{45} + 6\sqrt{\dfrac{4}{5}} = 3\sqrt{5} + 6\sqrt{\dfrac{4 \cdot 5}{5 \cdot 5}} = 3\sqrt{5} + 6\left(\dfrac{2}{5}\right)\sqrt{5}$
$$= \left(3 + \dfrac{12}{5}\right)\sqrt{5} = \left(\dfrac{15}{5} + \dfrac{12}{5}\right)\sqrt{5}$$
$$= \left(\dfrac{27}{5}\right)\sqrt{5}, \text{ or } \dfrac{27\sqrt{5}}{5}$$

22. (9.5) $\sqrt{8x} = 12$ *Check:* $\sqrt{8x} = 12$
$(\sqrt{8x})^2 = (12)^2$ $\sqrt{8(18)} \overset{?}{=} 12$
$8x = 144$ $\sqrt{144} \overset{?}{=} 12$
$x = 18$ $12 = 12$
The solution is 18.

23. (9.5) $\sqrt{5x + 11} = 6$ *Check:* $\sqrt{5x + 11} = 6$
$(\sqrt{5x + 11})^2 = 6^2$ $\sqrt{5(5) + 11} \overset{?}{=} 6$
$5x + 11 = 36$ $\sqrt{36} \overset{?}{=} 6$
$5x = 25$ $6 = 6$
$x = 5$
The solution is 5.

24. (9.5) $\sqrt{6x - 5} = x$
$(\sqrt{6x - 5})^2 = x^2$
$6x - 5 = x^2$
$0 = x^2 - 6x + 5$
$0 = (x - 5)(x - 1)$
$x - 5 = 0 \quad | \quad x - 1 = 0$
$x = 5 \quad | x = 1$
Check for x = 5:   *Check for x = 1:*
$\sqrt{6x - 5} = x$ $\sqrt{6x - 5} = x$
$\sqrt{6(5) - 5} \overset{?}{=} 5$ $\sqrt{6(1) - 5} \overset{?}{=} 1$
$\sqrt{25} \overset{?}{=} 5$ $\sqrt{1} \overset{?}{=} 1$
$5 = 5$ $1 = 1$
The solutions are 5 and 1.

25. (9.5) Let x = unknown length (in cm)
$c = \sqrt{a^2 + b^2}$
$17 = \sqrt{x^2 + 8^2}$
$(17)^2 = (\sqrt{x^2 + 8^2})^2$
$289 = x^2 + 64$ *Check:*
$0 = x^2 - 225$ $c = \sqrt{a^2 + b^2}$
$0 = (x - 15)(x + 15)$ $17 \overset{?}{=} \sqrt{(15)^2 + 8^2}$
$x - 15 = 0 \quad | \quad x + 15 = 0$ $17 \overset{?}{=} \sqrt{225 + 64}$
$x = 15 \quad | x = -15$ Reject $17 \overset{?}{=} \sqrt{289}$
$ |$ A length cannot be negative. $17 = 17$
Therefore, the length of one side of the triangle is 15 cm.

Chapters 1–9 Cumulative Review Exercises   (page 491)

1. $-\dfrac{3}{5}$ **2.** $-\dfrac{7}{4}$

3. $\left(\dfrac{\overset{5}{\cancel{30}}x^4 y^{-3}}{\underset{2}{\cancel{12}} y^{-1}}\right)^{-2} = \left(\dfrac{5x^4 y^{-2}}{2}\right)^{-2} = \dfrac{5^{-2} x^{-8} y^4}{2^{-2}} = \dfrac{2^2 y^4}{5^2 x^8} = \dfrac{4y^4}{25x^8}$

4. $-4h^2 + 36$ **5.** $2x$ **6.** $25 + 2\sqrt{154}$

7. $\sqrt{5}(3) + \sqrt{5}(\sqrt{80}) = 3\sqrt{5} + \sqrt{400} = 3\sqrt{5} + 20$

8. $3\sqrt{7}$

9. $\dfrac{8(\sqrt{6} + 2)}{(\sqrt{6} - 2)(\sqrt{6} + 2)} = \dfrac{8(\sqrt{6} + 2)}{(\sqrt{6})^2 - (2)^2} = \dfrac{8(\sqrt{6} + 2)}{6 - 4}$
$$= \dfrac{\overset{4}{\cancel{8}}(\sqrt{6} + 2)}{\underset{1}{\cancel{2}}} = 4\sqrt{6} + 8$$

10. Conjugate   **11.** $\dfrac{(x - 1)(x + 2)}{(x - 1)(x + 1)} \cdot \dfrac{x(x - 4)}{(x + 2)(x - 4)} = \dfrac{x}{x + 1}$

12. $\dfrac{2x^2 - 7x - 10}{(x + 2)(x - 1)}$

13. LCD $= 3(3x + 1)(x + 2)$
$$\dfrac{4}{3x + 1} - \dfrac{5}{(3x + 1)(x + 2)} + \dfrac{4}{3}$$
$$= \dfrac{4(3)(x + 2)}{(3x + 1)(3)(x + 2)} - \dfrac{5(3)}{(3x + 1)(x + 2)(3)} + \dfrac{4(3x + 1)(x + 2)}{3(3x + 1)(x + 2)}$$
$$= \dfrac{12x + 24 - 15 + 12x^2 + 28x + 8}{3(3x + 1)(x + 2)} = \dfrac{12x^2 + 40x + 17}{3(3x + 1)(x + 2)}$$

14. $\dfrac{10x^2 + 14x - 11}{(x + 5)(x - 3)(3x + 2)}$ **15.** $3x - 6 - 15 + 5x = x + 1$
$8x - 21 = x + 1$
$7x = 22$
$x = \dfrac{22}{7}$

16. $\dfrac{1}{2}$ **17.** LCD $= 6$
$$\dfrac{\overset{3}{\cancel{6}}}{1} \cdot \dfrac{2x - 3}{\underset{1}{\cancel{2}}} = \dfrac{\overset{1}{\cancel{6}}}{1} \cdot \dfrac{5x + 4}{\underset{1}{\cancel{6}}} - \dfrac{\overset{2}{\cancel{6}}}{1} \cdot \dfrac{5}{\underset{1}{\cancel{3}}}$$
$$3(2x - 3) = 1(5x + 4) - 2(5)$$
$$6x - 9 = 5x + 4 - 10$$
$$x = 3$$
The solution is 3.

18. No solution

19. $4[3 - 5x - 5] = 12 - 20x - 20$ **20.** $\dfrac{14}{3}$
$4[-5x - 2] = -8 - 20x$
$-20x - 8 = -8 - 20x$
$-8 = -8$ *True*; identity

21. LCD $= 4(x + 2)(x - 5)$
$$4(\cancel{x + 2})(x - 5)\left(\dfrac{5}{\cancel{x + 2}}\right) - 4(x + 2)(\cancel{x - 5})\left(\dfrac{3}{\cancel{x - 5}}\right)$$
$$= \overset{1}{\cancel{4}}(x + 2)(x - 5)\left(\dfrac{9}{\underset{1}{\cancel{4}}}\right)$$
$$4(x - 5)(5) - 4(x + 2)(3) = 9(x + 2)(x - 5)$$
$$20x - 100 - 12x - 24 = 9(x^2 - 3x - 10)$$
$$-124 + 8x = 9x^2 - 27x - 90$$
$$0 = 9x^2 - 35x + 34$$
$$0 = (x - 2)(9x - 17)$$
$x - 2 = 0 \quad | \quad 9x - 17 = 0$
$x = 2 \quad | 9x = 17$
$ | x = \dfrac{17}{9}$
Checking verifies that the solutions are 2 and $\dfrac{17}{9}$.

22. $x \le 4$

In Exercises 23–26, the checks will not be shown.

23. Let $ x$ = base of the triangle (in ft)   **24.** 12 mi
Then $ x + 4$ = altitude of the triangle (in ft)
and $\frac{1}{2}x(x + 4)$ = area of the triangle
$$\tfrac{1}{2}x(x + 4) = 16$$
$$\overset{1}{\cancel{2}}\left\{\dfrac{1}{\underset{1}{\cancel{2}}}x(x + 4)\right\} = 2(16)$$
$$x^2 + 4x - 32 = 0$$
$$(x - 4)(x + 8) = 0$$
$x - 4 = 0 \quad | \quad x + 8 = 0$
$ x = 4 \quad | x = -8$ Reject
$x + 4 = 8 \quad |$
The base is 4 ft, and the altitude is 8 ft.

25. Let $\quad x = $ number of dimes
Then $32 - x = $ number of nickels
$$10x + 5(32 - x) = 245$$
$$10x + 160 - 5x = 245$$
$$5x + 160 = 245$$
$$5x = 85$$
$$x = 17$$
$$32 - x = 15$$
Sylvia has 17 dimes and 15 nickels.

26. 33, 35, and 37

Exercises 10.1 (page 495)

1. $2x^2 - 5x - 3 = 0; a = 2, b = -5, c = -3$

2. $3x^2 + 2x - 4 = 0; a = 3, b = 2, c = -4$

3. $6x^2 - x + 0 = 0; a = 6, b = -1, c = 0$

4. $3x^2 - 2x + 0 = 0; a = 3, b = -2, c = 0$

5. LCD $= 2$
$$\overset{1}{\underset{1}{\cancel{2}}} \cdot \frac{3x}{\underset{1}{\cancel{2}}} + 2 \cdot 5 = 2 \cdot x^2$$
$$3x + 10 = 2x^2$$
$$0 = 2x^2 - 3x - 10; a = 2, b = -3, c = -10$$

6. $3x^2 + 2x - 12 = 0; a = 3, b = 2, c = -12$

7. LCD $= 12$
$$12(x^2) + \frac{\overset{3}{\cancel{12}}}{1}\left(\frac{-5x}{\underset{1}{\cancel{4}}}\right) + \frac{\overset{4}{\cancel{12}}}{1}\left(\frac{2}{\underset{1}{\cancel{3}}}\right) = 0$$
$$12x^2 - 15x + 8 = 0; a = 12, b = -15, c = 8$$

8. $30x^2 + 9x - 5 = 0; a = 30, b = 9, c = -5$

9. $x^2 - 3x = 4$
$x^2 - 3x - 4 = 0; a = 1, b = -3, c = -4$

10. $2x^2 + 2x - 12 = 0; a = 2, b = 2, c = -12$

11. $\quad 3x^2 + 3x = x^2 + 3x + 2$
$2x^2 + 0x - 2 = 0; a = 2, b = 0, c = -2$

12. $3x^2 - 2x + 3 = 0; a = 3, b = -2, c = 3$

Exercises 10.2 (page 500)

(The checks will not be shown.)

1. $(x + 3)(x - 2) = 0$
$$x + 3 = 0 \quad | \quad x - 2 = 0$$
$$x = -3 \quad | \quad x = 2$$
The solutions are -3 and 2.

2. $3, -2$

3. $\quad x^2 + x - 12 = 0$
$(x + 4)(x - 3) = 0$
$$x + 4 = 0 \quad | \quad x - 3 = 0$$
$$x = -4 \quad | \quad x = 3$$
The solutions are -4 and 3.

4. $3, -5$

5. $\quad 2x^2 - x - 1 = 0$
$(2x + 1)(x - 1) = 0$
$$2x + 1 = 0 \quad | \quad x - 1 = 0$$
$$x = -\frac{1}{2} \quad | \quad x = 1$$
The solutions are $-\frac{1}{2}$ and 1.

6. $\frac{1}{2}, -1$

7. $\frac{x}{8} = \frac{2}{x}$ A proportion
$$x^2 = 16 \quad \text{Cross-multiplying}$$
$$x^2 - 16 = 0$$
$(x - 4)(x + 4) = 0$
$$x - 4 = 0 \quad | \quad x + 4 = 0$$
$$x = 4 \quad | \quad x = -4$$
The solutions are 4 and -4.

8. $6, -6$

9. $\frac{x + 2}{3} = \frac{-1}{x - 2}$ A proportion
$$(x + 2)(x - 2) = 3(-1) \quad \text{Cross-multiplying}$$
$$x^2 - 4 = -3$$
$$x^2 - 1 = 0$$
$(x + 1)(x - 1) = 0$
$$x + 1 = 0 \quad | \quad x - 1 = 0$$
$$x = -1 \quad | \quad x = 1$$
The solutions are -1 and 1.

10. $1, -1$

11. $(x + 1)(x + 8) = 0$
$$x + 1 = 0 \quad | \quad x + 8 = 0$$
$$x = -1 \quad | \quad x = -8$$
The solutions are -1 and -8.

12. $-1, -6$

13. $2x(x + 2) = 0$
$$2x = 0 \quad | \quad x + 2 = 0$$
$$x = 0 \quad | \quad x = -2$$
The solutions are 0 and -2.

14. $0, 3$

15. $\quad x^2 - x - 2 = 0$
$(x - 2)(x + 1) = 0$
$$x - 2 = 0 \quad | \quad x + 1 = 0$$
$$x = 2 \quad | \quad x = -1$$
The solutions are 2 and -1.

16. $3, -2$

17. LCD $= 2x$
$$\frac{\overset{1}{\cancel{2x}}}{1} \cdot \frac{x}{\underset{1}{\cancel{2}}} + \frac{\overset{1}{\cancel{2x}}}{1} \cdot \frac{2}{\underset{1}{\cancel{x}}} = \frac{\overset{1}{\cancel{2x}}}{1} \cdot \frac{5}{\underset{1}{\cancel{2}}}$$
$$x^2 + 4 = 5x$$
$$x^2 - 5x + 4 = 0$$
$(x - 1)(x - 4) = 0$
$$x - 1 = 0 \quad | \quad x - 4 = 0$$
$$x = 1 \quad | \quad x = 4$$
The solutions are 1 and 4.

18. $1, 6$

19. LCD $= 4(x + 1)$
$$\frac{\overset{1}{\cancel{4(x + 1)}}}{1} \cdot \frac{x - 1}{\underset{1}{\cancel{4}}} + \frac{\overset{1}{\cancel{4(x + 1)}}}{1} \cdot \frac{6}{\cancel{(x + 1)}} = 4(x + 1)(2)$$
$$(x + 1)(x - 1) + 24 = 8(x + 1)$$
$$x^2 - 1 + 24 = 8x + 8$$
$$x^2 - 8x + 15 = 0$$
$(x - 3)(x - 5) = 0$
$$x - 3 = 0 \quad | \quad x - 5 = 0$$
$$x = 3 \quad | \quad x = 5$$
The solutions are 3 and 5.

20. $2, 3$

21. $\frac{2x^2}{1} = \frac{2 - x}{3}$ A proportion
$$3(2x^2) = 1(2 - x) \quad \text{Cross-multiplying}$$
$$6x^2 = 2 - x$$
$$6x^2 + x - 2 = 0$$
$(2x - 1)(3x + 2) = 0$
$$2x - 1 = 0 \quad | \quad 3x + 2 = 0$$
$$2x = 1 \quad | \quad 3x = -2$$
$$x = \frac{1}{2} \quad | \quad x = -\frac{2}{3}$$
The solutions are $\frac{1}{2}$ and $-\frac{2}{3}$.

22. $1\frac{1}{2}, \frac{1}{4}$

23. Let $\quad w = $ width (in m)
Then $w + 5 = $ length (in m)
Area $= w(w + 5)$
$$w(w + 5) = 24$$
$$w^2 + 5w = 24$$
$$w^2 + 5w - 24 = 0$$
$(w + 8)(w - 3) = 0$
$$w + 8 = 0 \quad | \quad w - 3 = 0$$
$$w = -8 \quad \text{Reject} \quad | \quad w = 3$$
A length cannot be negative. $\quad | \quad w + 5 = 3 + 5 = 8$
The width is 3 m, and the length is 8 m.

24. Width $= 7$ yd; length $= 11$ yd

25. Let $\quad x =$ first even integer
Then $x + 2 =$ second even integer
$$x(x + 2) + 4 = 84$$
$$x^2 + 2x + 4 = 84$$
$$x^2 + 2x - 80 = 0$$
$$(x + 10)(x - 8) = 0$$

$x + 10 = 0 \quad | \quad x - 8 = 0$
$\qquad x = -10 \quad | \qquad x = 8$
$x + 2 = -8 \quad | \quad x + 2 = 10$

First answer: The integers are -10 and -8. Second answer: The integers are 8 and 10.

26. First answer: The integers are 6 and 7. Second answer: The integers are -7 and -6.

27. Let $\quad x =$ rate from LA to Mexico (in mph)
Then $x + 20 =$ rate from Mexico to LA (in mph)

$r \cdot t = d$. Therefore, $t = \dfrac{d}{r}$.

Time from LA to Mexico $= \dfrac{120}{x}$; time from Mexico to

LA $= \dfrac{120}{x + 20}$

Total time is 5 hr.

$$\frac{120}{x} + \frac{120}{x + 20} = 5 \quad \text{LCD is } x(x + 20).$$

$$\overset{1}{x}(x + 20)\frac{120}{\overset{}{x}} + x(\overset{1}{x + 20})\frac{120}{x + 20} = x(x + 20)(5)$$

$$120x + 2{,}400 + 120x = 5x^2 + 100x$$
$$240x + 2{,}400 = 5x^2 + 100x$$
$$0 = 5x^2 - 140x - 2{,}400$$
$$0 = 5(x^2 - 28x - 480)$$
$$0 = 5(x + 12)(x - 40)$$

$5 \neq 0 \quad | \quad x + 12 = 0 \quad | \quad x - 40 = 0$
$\qquad\qquad | \qquad x = -12 \quad \text{Reject} \quad | \qquad x = 40$

Therefore, average speed to Mexico is 40 mph.

28. 45 mph

29. Let $\quad x =$ altitude (in in.)
Then $\quad x + 3 =$ base (in in.)
and $\frac{1}{2}x(x + 3) =$ area
$$\frac{1}{2}x(x + 3) = 20$$

$$\overset{1}{2}\left\{\frac{1}{\overset{}{2}}x(x + 3)\right\} = 2(20)$$

$$x^2 + 3x - 40 = 0$$
$$(x - 5)(x + 8) = 0$$

$x - 5 = 0 \quad | \quad x + 8 = 0$
$\qquad x = 5 \quad | \qquad x = -8 \quad \text{Reject}$
$x + 3 = 8 \quad | \quad$ A length cannot be negative.

The altitude is 5 in. and the base is 8 in.

30. The altitude is 12 cm, and the base is 15 cm.

31. Let $\quad x =$ base (in in.)
Then $\quad 19 - x =$ altitude (in in.)
and $\frac{1}{2}x(19 - x) =$ area
$$\frac{1}{2}x(19 - x) = 42$$

$$\overset{1}{2}\left\{\frac{1}{\overset{}{2}}x(19 - x)\right\} = 2(42)$$

$$19x - x^2 = 84$$
$$0 = x^2 - 19x + 84$$
$$0 = (x - 12)(x - 7)$$

$x - 12 = 0 \quad | \quad x - 7 = 0$
$\qquad x = 12 \quad | \qquad x = 7$
$19 - x = 7 \quad | \quad 19 - x = 12$

First solution: The base is 12 in. and the altitude is 7 in.
Second solution: The base is 7 in. and the altitude is 12 in.

32. First solution: The base is 6 cm and the altitude is 9 cm.
Second solution: The base is 9 cm and the altitude is 6 cm.

Exercises 10.3 (page 505)

1. $8x^2 - 4x = 0$
$4x(2x - 1) = 0$

$4x = 0 \quad | \quad 2x - 1 = 0$
$x = 0 \quad | \qquad 2x = 1$
$\qquad\quad | \qquad\quad x = \frac{1}{2}$

The solutions are 0 and $\frac{1}{2}$.

2. $0, \frac{1}{3}$

3. $x^2 = 9$
$x = \pm\sqrt{9}$
$x = \pm 3$
The solutions are 3 and -3.

4. $6, -6$

5. $x(x - 4) = 0$

$x = 0 \quad | \quad x - 4 = 0$
$\qquad\quad | \qquad x = 4$

The solutions are 0 and 4.

6. $0, 16$

7. $x^2 = -16$
$x = \pm\sqrt{-16}$
Not real

8. Not real

9. $x^2 = 1$
$x = \pm\sqrt{1}$
$x = \pm 1$
The solutions are 1 and -1.

10. ± 4

11. $x^2 = \frac{4}{5}$

$x = \pm\sqrt{\frac{4}{5}}$

$x = \pm\dfrac{2}{\sqrt{5}} = \pm\dfrac{2}{\sqrt{5}} \cdot \dfrac{\sqrt{5}}{\sqrt{5}}$

$x = \pm\dfrac{2\sqrt{5}}{5}$

The solutions are $\dfrac{2\sqrt{5}}{5}$ and $-\dfrac{2\sqrt{5}}{5}$.

12. $\pm\dfrac{5\sqrt{3}}{3}$

13. $2x^2 = 8$
$x^2 = 4$
$x = \pm\sqrt{4}$
$x = \pm 2$
The solutions are 2 and -2.

14. ± 3

15. $x^2 + 3x = 3x - 4$
$\quad\quad x^2 = -4$
$\quad\quad x = \pm\sqrt{-4}$
Not real

16. Not real

17. $2x + 6 = 6 + x^2 + 2x$
$\quad\quad x^2 = 0$
$\quad\quad x = \pm\sqrt{0}$
$\quad\quad x = 0$
The solution is 0.

18. 0

19. $6x^2 - 8x = 6 - 8x$
$\quad\quad 6x^2 = 6$
$\quad\quad x^2 = 1$
$\quad\quad x = \pm\sqrt{1}$
$\quad\quad x = \pm 1$
The solutions are 1 and -1.

20. $\pm\frac{1}{5}\sqrt{30}$

21. $(x - 4)^2 = 23$
$\quad x - 4 = \pm\sqrt{23}$
$\qquad\quad x = 4 \pm \sqrt{23}$
$\qquad\quad x = 4 + \sqrt{23} \approx 4 + 4.796 \approx 8.80$
or $\qquad x = 4 - \sqrt{23} \approx 4 - 4.796 \approx -0.80$
The solutions are $4 + \sqrt{23}$ and $4 - \sqrt{23}$, or approximately 8.80 and -0.80.

22. $8 \pm \sqrt{5}$, or approximately 10.24 and 5.76

23. $(x + 3)^2 = 3$
$x + 3 = \pm\sqrt{3}$
$x = -3 \pm \sqrt{3}$
$x = -3 + \sqrt{3} \approx -3 + 1.732 \approx -1.27$
or $x = -3 - \sqrt{3} \approx -3 - 1.732 \approx -4.73$
The solutions are $-3 + \sqrt{3}$ and $-3 - \sqrt{3}$, or approximately -1.27 and -4.73.

24. $-8 \pm \sqrt{11}$, or approximately -4.68 and -11.32

25. $(x + 4)^2 = 50$
$x + 4 = \pm\sqrt{50}$
$x = -4 \pm 5\sqrt{2}$
$x = -4 + 5\sqrt{2} \approx -4 + 7.071 \approx 3.07$
or $x = -4 - 5\sqrt{2} \approx -4 - 7.071 \approx -11.07$
The solutions are $-4 + 5\sqrt{2}$ and $-4 - 5\sqrt{2}$, or approximately 3.07 and -11.07.

26. $-1 \pm 2\sqrt{3}$, or approximately 2.46 and -4.46

Exercises 10.4   (page 510)

1. $x^2 - 4x = 12$
$x^2 - 4x + 4 = 12 + 4$
$(x - 2)^2 = 16$
$x - 2 = \pm\sqrt{16}$
$x = 2 \pm 4$
$x = 2 + 4 = 6 \mid x = 2 - 4 = -2$
The solutions are 6 and -2.

2. $-6, 8$

3. $x^2 + 8x = 9$
$x^2 + 8x + 16 = 9 + 16$
$(x + 4)^2 = 25$
$x + 4 = \pm\sqrt{25}$
$x = -4 \pm 5$
$x = -4 + 5 = 1 \mid x = -4 - 5 = -9$
The solutions are 1 and -9.

4. $-4, 2$

5. $x^2 - 4x = 2$
$x^2 - 4x + 4 = 2 + 4$
$(x - 2)^2 = 6$
$x - 2 = \pm\sqrt{6}$
$x = 2 \pm\sqrt{6}$
The solutions are $2 + \sqrt{6}$ and $2 - \sqrt{6}$.

6. $1 \pm \sqrt{5}$

7. $x^2 + 8x = 2$
$x^2 + 8x + 16 = 2 + 16$
$(x + 4)^2 = 18$
$x + 4 = \pm\sqrt{18}$
$x = -4 \pm 3\sqrt{2}$
The solutions are $-4 + 3\sqrt{2}$ and $-4 - 3\sqrt{2}$.

8. $-5 \pm \sqrt{33}$

9. $\dfrac{3x^2 - x - 2}{3} = \dfrac{0}{3}$
$x^2 - \frac{1}{3}x - \frac{2}{3} = 0$
$x^2 - \frac{1}{3}x = \frac{2}{3}$
$x^2 - \frac{1}{3}x + \frac{1}{36} = \frac{2}{3} + \frac{1}{36}$
$\left(x - \frac{1}{6}\right)^2 = \frac{25}{36}$
$x - \frac{1}{6} = \pm\sqrt{\frac{25}{36}}$
$x = \frac{1}{6} \pm \frac{5}{6}$
$x = \frac{1}{6} + \frac{5}{6} = \frac{6}{6} = 1 \mid x = \frac{1}{6} - \frac{5}{6} = -\frac{4}{6} = -\frac{2}{3}$
The solutions are 1 and $-\frac{2}{3}$.

10. $-2, \frac{1}{2}$

11. $\dfrac{4x^2 + 4x - 1}{4} = \dfrac{0}{4}$
$x^2 + x - \frac{1}{4} = 0$
$x^2 + x = \frac{1}{4}$
$x^2 + x + \frac{1}{4} = \frac{1}{4} + \frac{1}{4}$
$\left(x + \frac{1}{2}\right)^2 = \frac{2}{4} = \frac{1}{2}$
$x + \frac{1}{2} = \pm\sqrt{\frac{1}{2}} = \pm\sqrt{\frac{1 \cdot 2}{2 \cdot 2}}$
$x = -\frac{1}{2} \pm \frac{\sqrt{2}}{2}$
The solutions are $-\frac{1}{2} + \frac{\sqrt{2}}{2}$ and $-\frac{1}{2} - \frac{\sqrt{2}}{2}$.

12. $-\dfrac{3}{8} \pm \dfrac{\sqrt{41}}{8}$

13. $\dfrac{3x^2 + 2x + 1}{3} = \dfrac{0}{3}$
$x^2 + \frac{2}{3}x + \frac{1}{3} = 0$
$x^2 + \frac{2}{3}x = -\frac{1}{3}$
$x^2 + \frac{2}{3}x + \frac{1}{9} = -\frac{1}{3} + \frac{1}{9}$
$\left(x + \frac{1}{3}\right)^2 = -\frac{2}{9}$
$x + \frac{1}{3} = \pm\sqrt{-\frac{2}{9}}$
$x = -\frac{1}{3} \pm \sqrt{-\frac{2}{9}}$
Not real

14. Not real

15. $x^2 + 24x \qquad = 2,881$
$x^2 + 24x + 144 = 2,881 + 144$
$(x + 12)^2 = 3,025$
$x + 12 = \pm\sqrt{3,025}$
$x = -12 \pm 55$
$x = -12 + 55 = 43 \mid x = -12 - 55 = -67$
The solutions are 43 and -67.

16. $39, -67$

17. Let $\quad x$ = width (in cm)
Then $x + 30$ = length (in cm)
and $x(x + 30)$ = area
$x^2 + 30x = 6,664$
$x^2 + 30x + 225 = 6,664 + 225$
$(x + 15)^2 = 6,889$
$x + 15 = \pm\sqrt{6,889}$
$x = -15 \pm 83$
$x = -15 + 83 = 68 \mid x = -15 - 83$ Reject
The width of the rectangle is 68 cm.

18. 61 in.

19. Let $\quad x$ = Jaime's actual rate (in words per minute)
Then $x + 20$ = rate if he had typed faster

We solve the formula $rt = w$ for t, getting $t = \dfrac{w}{r}$.

So $\quad \dfrac{825}{x}$ = time at the slower rate

and $\dfrac{825}{x + 20}$ = time at the faster rate

The number of minutes at the slower rate equals the number of minutes at the faster rate plus 4 min.

$\dfrac{825}{x} = \dfrac{825}{x + 20} + 4 \quad$ LCD is $x(x + 20)$.

$x(x + 20)\left(\dfrac{825}{x}\right) = x(x + 20)\left(\dfrac{825}{x + 20} + 4\right)$

$\overset{1}{\cancel{x}}(x + 20)\left(\dfrac{825}{\cancel{x}}\right) = x(\overset{1}{\cancel{x + 20}})\left(\dfrac{825}{\cancel{x + 20}}\right) + x(x + 20)(4)$

$825x + 16,500 = 825x + 4x^2 + 80x$
$-4x^2 - 80x + 16,500 = 0$
$\dfrac{-4x^2 - 80x + 16,500}{-4} = \dfrac{0}{-4}$
$x^2 + 20x - 4,125 = 0$
$x^2 + 20x \qquad = 4,125$
$x^2 + 20x + 100 \quad = 4,125 + 100$
$(x + 10)^2 = 4,225$
$x + 10 = \pm\sqrt{4,225}$
$x = -10 \pm 65$
$x = -10 + 65 = 55 \mid x = -10 - 65 = -75$ Reject
Jaime typed at the rate of 55 words per minute.

20. 29 gal per min

10

Exercises 10.5 (page 516)

(Checks will not be shown.)

1. $3x^2 - 1x - 2 = 0$; $a = 3$, $b = -1$, $c = -2$

$$x = \frac{-(-1) \pm \sqrt{(-1)^2 - 4(3)(-2)}}{2(3)}$$

$$= \frac{1 \pm \sqrt{1 + 24}}{6} = \frac{1 \pm \sqrt{25}}{6}$$

$$x = \frac{1 \pm 5}{6} = \begin{cases} \dfrac{1 + 5}{6} = \dfrac{6}{6} = 1 \\ \dfrac{1 - 5}{6} = \dfrac{-4}{6} = -\dfrac{2}{3} \end{cases}$$

The solutions are 1 and $-\frac{2}{3}$.

2. $\frac{1}{2}$, -2

3. $1x^2 - 4x + 1 = 0$; $a = 1$, $b = -4$, $c = 1$

$$x = \frac{-(-4) \pm \sqrt{(-4)^2 - 4(1)(1)}}{2(1)} = \frac{4 \pm \sqrt{16 - 4}}{2(1)} = \frac{4 \pm \sqrt{12}}{2}$$

$$= \frac{4 \pm 2\sqrt{3}}{2} = \frac{2(2 \pm \sqrt{3})}{2} = 2 \pm \sqrt{3}$$

The solutions are $2 + \sqrt{3}$ and $2 - \sqrt{3}$.

4. $2 \pm \sqrt{5}$

5. LCD $= 2x$

$$\frac{\overset{1}{\cancel{2x}}}{1} \cdot \frac{x}{\underset{1}{\cancel{2}}} + \frac{\overset{1}{\cancel{2x}}}{1} \cdot \frac{2}{\underset{1}{\cancel{x}}} = \frac{\overset{1}{\cancel{2x}}}{1} \cdot \frac{5}{\underset{1}{\cancel{2}}}$$

$$x^2 + 4 = 5x$$

$$1x^2 - 5x + 4 = 0; \ a = 1, \ b = -5, \ c = 4$$

$$x = \frac{-(-5) \pm \sqrt{(-5)^2 - 4(1)(4)}}{2(1)} = \frac{5 \pm \sqrt{25 - 16}}{2} = \frac{5 \pm \sqrt{9}}{2}$$

$$x = \frac{5 \pm 3}{2} = \begin{cases} \dfrac{5 + 3}{2} = \dfrac{8}{2} = 4 \\ \dfrac{5 - 3}{2} = \dfrac{2}{2} = 1 \end{cases}$$

The solutions are 4 and 1.

6. 1, 6

7. $2x^2 - 8x + 5 = 0$; $a = 2$, $b = -8$, $c = 5$

$$x = \frac{-(-8) \pm \sqrt{(-8)^2 - 4(2)(5)}}{2(2)}$$

$$= \frac{8 \pm \sqrt{64 - 40}}{4} = \frac{8 \pm \sqrt{24}}{4} = \frac{8 \pm 2\sqrt{6}}{4}$$

$$= \frac{\overset{1}{\cancel{2}}(4 \pm \sqrt{6})}{\underset{2}{\cancel{4}}} = \frac{4 \pm \sqrt{6}}{2}$$

The solutions are $\dfrac{4 + \sqrt{6}}{2}$ and $\dfrac{4 - \sqrt{6}}{2}$.

8. $\dfrac{3 + \sqrt{3}}{3}$, $\dfrac{3 - \sqrt{3}}{3}$

9. LCD $= x(x - 1)$

$$\frac{\overset{1}{\cancel{x}}(x - 1)}{1} \cdot \frac{1}{\underset{1}{\cancel{x}}} + \frac{x\overset{1}{\cancel{(x - 1)}}}{1} \cdot \frac{x}{\underset{1}{\cancel{x - 1}}} = \frac{x(x - 1)}{1} \cdot \frac{3}{1}$$

$$(x - 1) + x^2 = 3x(x - 1)$$

$$x - 1 + x^2 = 3x^2 - 3x$$

$$0 = 2x^2 - 4x + 1$$

$$a = 2, \ b = -4, \ c = 1$$

$$x = \frac{-(-4) \pm \sqrt{(-4)^2 - 4(2)(1)}}{2(2)}$$

$$= \frac{4 \pm \sqrt{16 - 8}}{4} = \frac{4 \pm \sqrt{8}}{4}$$

$$= \frac{4 \pm 2\sqrt{2}}{4} = \frac{2(2 \pm \sqrt{2})}{4} = \frac{2 \pm \sqrt{2}}{2}$$

The solutions are $\dfrac{2 + \sqrt{2}}{2}$ and $\dfrac{2 - \sqrt{2}}{2}$.

10. $\dfrac{-1 + \sqrt{11}}{5}$, $\dfrac{-1 - \sqrt{11}}{5}$

11. $x^2 + x + 1 = 0$; $a = 1$, $b = 1$, $c = 1$

$$x = \frac{-1 \pm \sqrt{(1)^2 - 4(1)(1)}}{2(1)} = \frac{-1 \pm \sqrt{1 - 4}}{2} = \frac{-1 \pm \sqrt{-3}}{2}$$

Not real

12. Not real

13. $4x^2 + 4x - 1 = 0$; $a = 4$, $b = 4$, $c = -1$

$$x = \frac{-(4) \pm \sqrt{(4)^2 - 4(4)(-1)}}{2(4)} = \frac{-4 \pm \sqrt{16 + 16}}{8}$$

$$= \frac{-4 \pm \sqrt{32}}{8} = \frac{-4 \pm 4\sqrt{2}}{8} = \frac{4(-1 \pm \sqrt{2})}{8} = \frac{-1 \pm \sqrt{2}}{2}$$

The solutions are $\dfrac{-1 + \sqrt{2}}{2}$ and $\dfrac{-1 - \sqrt{2}}{2}$.

14. $\dfrac{1 + \sqrt{3}}{3}$, $\dfrac{1 - \sqrt{3}}{3}$

15. $3x^2 + 2x + 1 = 0$; $a = 3$, $b = 2$, $c = 1$

$$x = \frac{-(2) \pm \sqrt{(2)^2 - 4(3)(1)}}{2(3)}$$

$$= \frac{-2 \pm \sqrt{4 - 12}}{6} = \frac{-2 \pm \sqrt{-8}}{6}$$

Not real

16. Not real

17. Let $x = $ number

Then $\dfrac{1}{x} = $ reciprocal of number

$$x - \frac{1}{x} = \frac{2}{3} \quad \text{LCD is } 3x$$

$$(3x)x - (3x)\left(\frac{1}{x}\right) = \left(\frac{2}{3}\right)(3x)$$

$$3x^2 - 3 = 2x$$

$$3x^2 - 2x - 3 = 0; \ a = 3, \ b = -2, \ c = -3$$

$$x = \frac{-(-2) \pm \sqrt{(-2)^2 - 4(3)(-3)}}{2(3)}$$

$$= \frac{2 \pm \sqrt{4 + 36}}{6}$$

$$= \frac{2 \pm \sqrt{40}}{6} = \frac{2 \pm 2\sqrt{10}}{6} = \frac{2(1 \pm \sqrt{10})}{6} = \frac{1 \pm \sqrt{10}}{3}$$

First answer: The number is $\dfrac{1 + \sqrt{10}}{3}$. Second answer: The number is $\dfrac{1 - \sqrt{10}}{3}$.

18. First answer: The number is $\dfrac{-5 + 3\sqrt{17}}{8}$. Second answer: The number is $\dfrac{-5 - 3\sqrt{17}}{8}$.

19. $4x^2 + 4x - 1 = 0$; $a = 4$, $b = 4$, $c = -1$

$$x = \frac{-(4) \pm \sqrt{(4)^2 - 4(4)(-1)}}{2(4)} = \frac{-4 \pm \sqrt{16 + 16}}{8}$$

$$= \frac{-4 \pm \sqrt{16 \cdot 2}}{8} = \frac{-4 \pm 4\sqrt{2}}{8} = \frac{4(-1 \pm \sqrt{2})}{8} = \frac{-1 \pm \sqrt{2}}{2}$$

$$x \approx \frac{-1 \pm 1.414}{2} = \begin{cases} \dfrac{0.414}{2} \approx 0.21 \\ \dfrac{-2.414}{2} \approx -1.21 \end{cases}$$

One solution is $\dfrac{-1 + \sqrt{2}}{2}$ (≈ 0.21), and the other is $\dfrac{-1 - \sqrt{2}}{2}$ (≈ -1.21).

20. One solution is $\dfrac{1 + \sqrt{3}}{3} \approx 0.91$, and the other is $\dfrac{1 - \sqrt{3}}{3} \approx -0.24$.

10

21. Let w = width (in ft)

Then $w + 2$ = length (in ft)

Area = $(w + 2)w$

$w^2 + 2w = 2$

$1w^2 + 2w - 2 = 0;\ a = 1,\ b = 2,\ c = -2$

$w = \dfrac{-(2) \pm \sqrt{(2)^2 - 4(1)(-2)}}{2(1)} = \dfrac{-2 \pm \sqrt{4 + 8}}{2}$

$= \dfrac{-2 \pm \sqrt{12}}{2} = \dfrac{-2 \pm 2\sqrt{3}}{2} = \dfrac{\overset{1}{2}(-1 \pm \sqrt{3})}{\underset{1}{2}} = -1 \pm \sqrt{3}$

$w = \begin{cases} -1 + \sqrt{3} \approx -1 + 1.732 \approx 0.73 \\ -1 - \sqrt{3} \approx -1 - 1.732 \approx -2.73 \end{cases}$

Reject $w = -1 - \sqrt{3}$; a length cannot be negative.

If $w = -1 + \sqrt{3},\ w + 2 = (-1 + \sqrt{3}) + 2 = 1 + \sqrt{3} \approx 2.73$.

The width is $(-1 + \sqrt{3})$ ft ≈ 0.73 ft, and the length is

$(1 + \sqrt{3})$ ft ≈ 2.73 ft.

22. Width = $(-2 + \sqrt{10})$ cm ≈ 1.16 cm;

length = $(2 + \sqrt{10})$ cm ≈ 5.16 cm

23. Let x = length of a side of square

$4x = x^2 - 6$

$x^2 - 4x - 6 = 0;\ a = 1,\ b = -4,\ c = -6$

$x = \dfrac{-(-4) \pm \sqrt{(-4)^2 - 4(1)(-6)}}{2(1)}$

$= \dfrac{4 \pm \sqrt{16 + 24}}{2} = \dfrac{4 \pm \sqrt{40}}{2}$

$= \dfrac{4 \pm 2\sqrt{10}}{2} = \dfrac{2(2 \pm \sqrt{10})}{2} = 2 \pm \sqrt{10}$

Area = x^2

Perimeter = $4x$

$x = \begin{cases} 2 + \sqrt{10} \approx 2 + 3.162 \approx 5.16 \\ 2 - \sqrt{10} \approx 2 - 3.162 \approx -1.16 \end{cases}$

Reject $2 - \sqrt{10}$; a length cannot be negative.

The length of a side is $2 + \sqrt{10}\ (\approx 5.16)$.

24. The length of a side is $2 + \sqrt{6} \approx 4.45$.

Exercises 10.6   (page 520)

1. $11^2 \overset{?}{=} 6^2 + 8^2$

$121 \overset{?}{=} 36 + 64$

$121 \neq 100$

The triangle is *not* a right triangle.

2. The triangle is *not* a right triangle.

3. $15^2 \overset{?}{=} 9^2 + 12^2$

$225 \overset{?}{=} 81 + 144$

$225 = 225$

The triangle *is* a right triangle.

4. The triangle *is* a right triangle.

5. $5^2 \overset{?}{=} (\sqrt{7})^2 + (\sqrt{18})^2$

$25 \overset{?}{=} 7 + 18$

$25 = 25$

The triangle *is* a right triangle.

6. The triangle *is* a right triangle.

7. $x^2 = 16^2 + 12^2$ **8.** 15   **9.** $x^2 + 2^2 = 6^2$

$x^2 = 256 + 144$ $x^2 + 4 = 36$

$x^2 = 400$ $x^2 = 32$

$x = \sqrt{400} = 20$ $x = \sqrt{32} = 4\sqrt{2}$

$x \approx 5.66$

10. $7\sqrt{3} \approx 12.12$ **11.** $x^2 = 2^2 + (\sqrt{5})^2$ **12.** 4

$x^2 = 4 + 5$

$x^2 = 9$

$x = \sqrt{9} = 3$

13. $x^2 + 3^2 = 5^2$ **14.** 5

$x^2 + 9 = 25$

$x^2 = 16$

$x = \sqrt{16} = 4$

15. $x^2 + 3^2 = (3\sqrt{3})^2$ **16.** $2\sqrt{2} \approx 2.83$

$x^2 + 9 = 27$

$x^2 = 18$

$x = \sqrt{18} = 3\sqrt{2} \approx 4.24$

17. $x^2 = 6^2 + 10^2$ **18.** $4\sqrt{13} \approx 14.42$

$x^2 = 36 + 100$

$x^2 = 136$

$x = \sqrt{136} = 2\sqrt{34} \approx 11.66$

19. $(x + 1)^2 + (\sqrt{20})^2 = (x + 3)^2$ **20.** 7

$x^2 + 2x + 1 + 20 = x^2 + 6x + 9$

$12 = 4x$

$x = 3$

21. Let w = width (in cm)

$(24)^2 + w^2 = (25)^2$

$576 + w^2 = 625$

$w^2 = 49$

$w = \sqrt{49}$

$w = 7$

The width is 7 cm.

22. 9 in.

23. Let x = length of one leg (in m)

Then $2x - 4$ = length of other leg (in m)

$10^2 = (2x - 4)^2 + x^2$

$100 = 4x^2 - 16x + 16 + x^2$

$0 = 5x^2 - 16x - 84$

$0 = (x - 6)(5x + 14)$

$\begin{array}{l|l} x - 6 = 0 & 5x + 14 = 0 \\ \quad x = 6 & \quad x = -\frac{14}{5} \quad \text{Reject} \\ 2x - 4 = 12 - 4 = 8 & \end{array}$

One leg is 6 m long, and the other is 8 m long.

24. Width = 9 yd; length = 12 yd

25. Let x = length of diagonal (in in.)

$x^2 = (\sqrt{6})^2 + (\sqrt{6})^2$

$x^2 = 6 + 6$

$x^2 = 12$

$x = \sqrt{12}$

$x = 2\sqrt{3}$

The diagonal is $2\sqrt{3}$ in. long.

26. $2\sqrt{7}$ m

27. Let x = length of side (in cm)

$(4\sqrt{2})^2 = x^2 + x^2$

$32 = 2x^2$

$16 = x^2$

$\sqrt{16} = x$

$x = 4$

A side is 4 cm long.

28. 5 cm

29. Let x = length of the wire (in ft)

$x^2 = 50^2 + 30^2$

$x^2 = 2,500 + 900$

$x^2 = 3,400$

$x = \sqrt{3,400} = 10\sqrt{34} \approx 58.31$

The wire needs to be 58.31 ft long.

(That's the *shortest* the wire could

be, without allowing any extra for

making connections, etc.)

30. 20 in.

10

31. In $\frac{1}{2}$ hr, Anne will have bicycled

$\left(8\dfrac{\text{mi}}{\text{hr}}\right)\left(\dfrac{1}{2}\text{ hr}\right)$, or 4 mi,

and Leticia will have bicycled

$\left(6\dfrac{\text{mi}}{\text{hr}}\right)\left(\dfrac{1}{2}\text{ hr}\right)$, or 3 mi.

Let x = distance between Anne and
       Leticia (in mi)

$x^2 = 4^2 + 3^2$
$x^2 = 16 + 9$
$x^2 = 25$
$x = \sqrt{25} = 5$

Anne and Leticia will be 5 mi apart after $\frac{1}{2}$ hr.

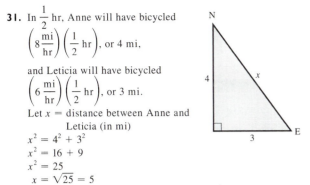

32. The planes will be 400 mi apart.

33. Let x = length of diagonal (in cm)
$(x)^2 = (41.6)^2 + (41.6)^2$
$x^2 = 1{,}730.56 + 1{,}730.56$
$x^2 = 3{,}461.12$
$x = \sqrt{3{,}461.12} \approx 58.83$
The diagonal is about 58.83 cm long.

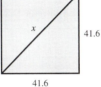

41.6

41.6

34. ≈ 2.66 m

Sections 10.1–10.6 Review Exercises (page 524)

(The checks will not be shown.)

1. $x^2 + x - 6 = 0$
$(x + 3)(x - 2) = 0$
$x + 3 = 0 \quad | \quad x - 2 = 0$
$x = -3 \quad | x = 2$
The solutions are -3 and 2.

2. $0, 49$

3. $x^2 - 2x - 4 = 0$; $a = 1, b = -2, c = -4$
$x = \dfrac{-(-2) \pm \sqrt{(-2)^2 - 4(1)(-4)}}{2(1)}$
$ = \dfrac{2 \pm \sqrt{4 + 16}}{2} = \dfrac{2 \pm \sqrt{20}}{2}$
$ = \dfrac{2 \pm 2\sqrt{5}}{2} = \dfrac{2(1 \pm \sqrt{5})}{2} = 1 \pm \sqrt{5}$
The solutions are $1 + \sqrt{5}$ and $1 - \sqrt{5}$.

4. $2 \pm \sqrt{3}$

5. $x^2 - 5x = 0$
$x(x - 5) = 0$
$x = 0 \quad | \quad x - 5 = 0$
$x = 5$
The solutions are 0 and 5.

6. $\pm\frac{5}{6}$

7. LCD $= 3(x - 2)$
$\dfrac{3(x - 2)}{1} \cdot \dfrac{(x + 2)}{3} = \dfrac{3(x - 2)}{1} \cdot \dfrac{1}{(x - 2)} + \dfrac{3(x - 2)}{1} \cdot \dfrac{2}{3}$
$(x - 2)(x + 2) = 3 + 2(x - 2)$
$x^2 - 4 = 3 + 2x - 4$
$x^2 - 2x - 3 = 0$
$(x - 3)(x + 1) = 0$
$x - 3 = 0 \quad | \quad x + 1 = 0$
$x = 3 \quad | x = -1$
The solutions are 3 and -1.

8. Not real

9. $5x + 10 = x^2 + 5x$
$x^2 = 10$
$x = \pm\sqrt{10}$
The solutions are $\sqrt{10}$ and $-\sqrt{10}$.

10. $\pm 2\sqrt{3}$

11. LCD $= x(x + 1)$
$\dfrac{x(x + 1)}{1} \cdot \dfrac{2}{x} + \dfrac{x(x + 1)}{1} \cdot \dfrac{x}{(x + 1)} = \dfrac{x(x + 1)}{1} \cdot \dfrac{5}{1}$
$2(x + 1) + x^2 = 5x(x + 1)$
$2x + 2 + x^2 = 5x^2 + 5x$
$4x^2 + 3x - 2 = 0$; $a = 4, b = 3, c = -2$
$x = \dfrac{-(3) \pm \sqrt{(3)^2 - 4(4)(-2)}}{2(4)} = \dfrac{-3 \pm \sqrt{9 + 32}}{8} = \dfrac{-3 \pm \sqrt{41}}{8}$
The solutions are $\dfrac{-3 + \sqrt{41}}{8}$ and $\dfrac{-3 - \sqrt{41}}{8}$.

12. $\dfrac{-1 \pm \sqrt{73}}{6}$

13. $3x^2 + 2x + 1 = 0$; $a = 3, b = 2, c = 1$
$x = \dfrac{-(2) \pm \sqrt{(2)^2 - 4(3)(1)}}{2(3)}$
$ = \dfrac{-2 \pm \sqrt{4 - 12}}{6} = \dfrac{-2 \pm \sqrt{-8}}{6}$
Not real

14. $\pm\dfrac{3\sqrt{5}}{5}$

15. $15^2 = 12^2 + x^2$
$225 = 144 + x^2$
$81 = x^2$
$\sqrt{81} = x$
$\phantom{\sqrt{8}}x = 9$

16. 12 in.

17. Let x = length of diagonal (in m)
$x^2 = (\sqrt{10})^2 + (\sqrt{10})^2$
$x^2 = 10 + 10$
$x^2 = 20$
$x = \sqrt{20}$
$ = 2\sqrt{5}$
The diagonal is $2\sqrt{5}$ m long.

18. 80 gal per min

19. Let x = altitude (in in.)
Then $x + 3$ = base (in in.)
Area $= \frac{1}{2}(x + 3)x$
$\frac{1}{2}(x + 3)x = 35$
$2\left(\frac{1}{2}\right)(x + 3)x = 2(35)$
$x^2 + 3x = 70$
$x^2 + 3x - 70 = 0$
$(x - 7)(x + 10) = 0$
$x - 7 = 0 \quad | \quad x + 10 = 0$
$x = 7 \quad | x = -10 \quad$ Reject
$x + 3 = 10 \quad |$ A length cannot be negative.
The altitude is 7 in., and the base is 10 in.

20. 8 ft

21. Let x = length of diagonal (in in.)
$x^2 = 8^2 + 4^2$
$x^2 = 64 + 16$
$x^2 = 80$
$x = \sqrt{80}$
$ = \sqrt{16 \cdot 5} = 4\sqrt{5}$
$x \approx 8.94$
The diagonal is $4\sqrt{5}$ in. (≈ 8.94 in.) long.

22. The width is $(-3 + \sqrt{15})$ ft ≈ 0.87 ft, and the length is $(3 + \sqrt{15})$ ft ≈ 6.87 ft.

Chapter 10 Diagnostic Test (page 527)

Following each problem number is the number (in parentheses) of the textbook section where that kind of problem is discussed.

1. (10.1) **a.** $5 + 3x^2 = 7x$
$3x^2 - 7x + 5 = 0$; $a = 3, b = -7, c = 5$

b.
$$3(x + 5) = 3(4 - x^2)$$
$$3x + 15 = 12 - 3x^2$$
$$3x^2 + 3x + 3 = 0; a = 3, b = 3, c = 3$$

2. (10.5)
$$x^2 = 8x + 15$$
$$x^2 - 8x - 15 = 0; a = 1, b = -8, c = -15$$
$$x = \frac{-(-8) \pm \sqrt{(-8)^2 - 4(1)(-15)}}{2(1)}$$
$$= \frac{8 \pm \sqrt{64 + 60}}{2} = \frac{8 \pm \sqrt{124}}{2}$$
$$= \frac{8 \pm 2\sqrt{31}}{2} = \frac{2(4 \pm \sqrt{31})}{2} = 4 \pm \sqrt{31}$$

(10.4) If completing the square is used:
$$x^2 - 8x = 15$$
$$x^2 - 8x + 16 = 15 + 16$$
$$(x - 4)^2 = 31$$
$$x - 4 = \pm\sqrt{31}$$
$$x = 4 \pm \sqrt{31} \quad \text{(same solutions as above)}$$
The solutions are $4 + \sqrt{31}$ and $4 - \sqrt{31}$.

3. (10.3)
$$5x^2 = 17x$$
$$5x^2 - 17x = 0$$
$$x(5x - 17) = 0$$
$$x = 0 \quad \bigg| \quad 5x - 17 = 0$$
$$5x = 17$$
$$x = \tfrac{17}{5}$$
The solutions are 0 and $\frac{17}{5}$.

4. (10.5) $x^2 + 2x + 6 = 0; a = 1, b = 2, c = 6$
$$x = \frac{-(2) \pm \sqrt{(2)^2 - 4(1)(6)}}{2(1)} = \frac{-2 \pm \sqrt{4 - 24}}{2}$$
$$= \frac{-2 \pm \sqrt{-20}}{2}$$
Not real

(10.4) If completing the square is used:
$$x^2 + 2x = -6$$
$$x^2 + 2x + 1 = -6 + 1$$
$$(x + 1)^2 = -5$$
$$x + 1 = \pm\sqrt{-5}$$
$$x = -1 \pm \sqrt{-5}$$
Not real

5. (10.3) $6x^2 = 54$
$$x^2 = \tfrac{54}{6} = 9$$
$$x = \pm 3$$
The solutions are 3 and -3.

6. (10.5) $3(x + 1) = x^2$
$$3x + 3 = x^2$$
$$0 = x^2 - 3x - 3; a = 1, b = -3, c = -3$$
$$x = \frac{-(-3) \pm \sqrt{(-3)^2 - 4(1)(-3)}}{2(1)}$$
$$= \frac{3 \pm \sqrt{9 + 12}}{2} = \frac{3 \pm \sqrt{21}}{2}$$
The solutions are $\dfrac{3 + \sqrt{21}}{2}$ and $\dfrac{3 - \sqrt{21}}{2}$.

(10.4) If completing the square is used:
$$x^2 - 3x = 3$$
$$x^2 - 3x + \tfrac{9}{4} = \tfrac{12}{4} + \tfrac{9}{4}$$
$$\left(x - \tfrac{3}{2}\right)^2 = \tfrac{21}{4}$$
$$x - \tfrac{3}{2} = \pm\sqrt{\tfrac{21}{4}}$$
$$x = \tfrac{3}{2} \pm \tfrac{\sqrt{21}}{2} \quad \text{(same solutions as above)}$$

7. (10.3) $x(x + 3) = x$
$$x^2 + 3x = x$$
$$x^2 + 2x = 0$$
$$x(x + 2) = 0$$
$$x = 0 \quad \bigg| \quad x + 2 = 0$$
$$x = -2$$
The solutions are 0 and -2.

8. (10.6) $(\sqrt{11})^2 + x^2 = (4)^2$
$$11 + x^2 = 16$$
$$x^2 = 5$$
$$x = \sqrt{5}$$

9. (10.6) Let x = length of diagonal (in m)
$$x^2 = (7)^2 + (7)^2$$
$$x^2 = 49 + 49$$
$$x^2 = 98$$
$$x = \sqrt{98} = 7\sqrt{2}$$
The diagonal is $7\sqrt{2}$ m long.

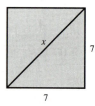

10. (10.6) Let x = width of rectangle (in ft)
Then $x + 6$ = length of rectangle (in ft)
$$x(x + 6) = 78$$
$$x^2 + 6x - 78 = 0; a = 1, b = 6, c = -78$$
$$x = \frac{-(6) \pm \sqrt{(6)^2 - 4(1)(-78)}}{2(1)} = \frac{-6 \pm \sqrt{36 + 312}}{2}$$
$$= \frac{-6 \pm \sqrt{348}}{2} = \frac{-6 \pm 2\sqrt{87}}{2} = \frac{2(-3 \pm \sqrt{87})}{2} = -3 \pm \sqrt{87}$$
We reject $-3 - \sqrt{87}$ because a length cannot be negative.
The width is $(-3 + \sqrt{87})$ ft, and the length is
$[(-3 + \sqrt{87}) + 6]$ ft $= (3 + \sqrt{87})$ ft.

Chapters 1–10 Cumulative Review Exercises   (page 527)

1. No   **2.** No   **3.** 0   **4.** Yes   **5.** Not defined

6. $9 - 2\sqrt{3}$ **7.**
$$
\begin{array}{r}
4x^2 + 3x - 2 \\
5x - 3 \\
\hline
-12x^2 - 9x + 6 \\
20x^3 + 15x^2 - 10x \\
\hline
20x^3 + 3x^2 - 19x + 6
\end{array}
$$
8. $26 - 2\sqrt{105}$

9. $\dfrac{x(\cancel{x + 3})}{(2x + 5)(\cancel{x + 1})} \cdot \dfrac{(\cancel{x + 1})(\cancel{x - 3})}{(\cancel{x + 3})(\cancel{x - 3})} = \dfrac{x}{2x + 5}$

10. $x - 11 - \dfrac{38}{x - 3}$

11. $\dfrac{(x + 4)(\cancel{x - 2})}{1} \cdot \dfrac{x(\cancel{x + 2})}{(\cancel{x + 2})(\cancel{x - 2})} = x(x + 4)$ **12.** $\dfrac{6x}{x^2 - 9}$

13. $\dfrac{x^2 + 3x - 2}{(x + 2)(x + 3)} - \dfrac{x - 1}{(2x + 1)(x + 3)}$
$$= \frac{(x^2 + 3x - 2)(2x + 1)}{(x + 2)(x + 3)(2x + 1)} - \frac{(x - 1)(x + 2)}{(2x + 1)(x + 3)(x + 2)}$$
$$= \frac{2x^3 + 7x^2 - x - 2}{(x + 2)(x + 3)(2x + 1)} - \frac{x^2 + x - 2}{(2x + 1)(x + 3)(x + 2)}$$
$$= \frac{(2x^3 + 7x^2 - x - 2) - (x^2 + x - 2)}{(x + 2)(x + 3)(2x + 1)}$$
$$= \frac{2x^3 + 7x^2 - x - 2 - x^2 - x + 2}{(x + 2)(x + 3)(2x + 1)}$$
$$= \frac{2x^3 + 6x^2 - 2x}{(x + 2)(x + 3)(2x + 1)}$$

14. $\dfrac{27\sqrt{2}}{2}$

10

(The checks will usually not be shown.)

15. $7 - 1(2x + 3) = 3x - 2 - 1(4x - 3)$　　**16.** $-\frac{2}{3}$
$7 - 2x - 3 = 3x - 2 - 4x + 3$
$4 - 2x = -x + 1$
$2x - 1 + 4 - 2x = 2x - 1 - x + 1$
$3 = x$
$x = 3$
The solution is 3.

17. $(\sqrt{5x + 3})^2 = (10x)^2$
$5x + 3 = 100x^2$
$0 = 100x^2 - 5x - 3$
$0 = (5x - 1)(20x + 3)$
$\begin{array}{c|c} 5x - 1 = 0 & 20x + 3 = 0 \\ 5x = 1 & 20x = -3 \\ x = \frac{1}{5} & x = -\frac{3}{20} \end{array}$

Check for $x = \frac{1}{5}$:　　*Check for* $x = -\frac{3}{20}$:
$\sqrt{5\left(\frac{1}{5}\right) + 3} = 10\left(\frac{1}{5}\right)$　　$\sqrt{5\left(-\frac{3}{20}\right) + 3} = 10\left(-\frac{3}{20}\right)$
$\sqrt{1 + 3} \stackrel{?}{=} 2$　　$\sqrt{-\frac{3}{4} + 3} \stackrel{?}{=} -\frac{3}{2}$
$\sqrt{4} = 2$　*True*　　$\sqrt{\frac{9}{4}} = -\frac{3}{2}$　*False*
The only solution is $\frac{1}{5}$.

18. $-\frac{3}{7}$　　**19.**　　$12x - 3 \leq 7x - 13$
$3 - 7x + 12x - 3 \leq 3 - 7x + 7x - 13$
$5x \leq -10$
$x \leq -2$

20. $-7, 2$

21. $4x^2 - 12x + 3 = 0$; $a = 4$, $b = -12$, $c = 3$
$x = \dfrac{-(-12) \pm \sqrt{(-12)^2 - 4(4)(3)}}{2(4)}$
$= \dfrac{12 \pm \sqrt{144 - 48}}{8}$
$= \dfrac{12 \pm \sqrt{96}}{8} = \dfrac{12 \pm 4\sqrt{6}}{8} = \dfrac{4(3 \pm \sqrt{6})}{8} = \dfrac{3 \pm \sqrt{6}}{2}$
The solutions are $\dfrac{3 + \sqrt{6}}{2}$ and $\dfrac{3 - \sqrt{6}}{2}$.

22. $2, -1$　　**23.** $7x^2 = 1$
$x^2 = \frac{1}{7}$
$x = \pm\sqrt{\dfrac{1}{7}} = \pm\sqrt{\dfrac{1 \cdot 7}{7 \cdot 7}} = \pm\dfrac{\sqrt{7}}{7}$
The solutions are $\dfrac{\sqrt{7}}{7}$ and $-\dfrac{\sqrt{7}}{7}$.

24. $6 + 2\sqrt{2}$　　**25.** $3(x^2 - 4x - 5) = 3(x - 5)(x + 1)$

26. $5(x + 3)(x - 3)$　　**27.** Not factorable

28. $(7x + 1)(x - 2)$

29. Let $x =$ number of inches corresponding to 12 ft
Let $y =$ number of inches corresponding to 18 ft
$\dfrac{\frac{1}{4} \text{ in.}}{1 \text{ ft}} = \dfrac{x \text{ in.}}{12 \text{ ft}}$　　$\dfrac{\frac{1}{4} \text{ in.}}{1 \text{ ft}} = \dfrac{y \text{ in.}}{18 \text{ ft}}$
$x = \frac{1}{4}(12) = 3$　　$y = \frac{1}{4}(18) = \frac{9}{2}$
The drawing will be 3 in. by $\frac{9}{2}$ in.

30. 300 mph

31. Let　　　$x =$ amount invested at 10.5% interest
Then $27,300 - x =$ amount invested at 7% interest
$0.105x + 0.07(27,300 - x) = 2,324$
$105x + 70(27,300 - x) = 2,324,000$
$105x + 1,911,000 - 70x = 2,324,000$
$35x + 1,911,000 = 2,324,000$
$35x + 1,911,000 - 1,911,000 = 2,324,000 - 1,911,000$
$35x = 413,000$
$x = 11,800$　Dollars invested at 10.5% interest
$27,300 - x = 15,500$　Dollars invested at 7% interest

32. $5.50 per saw for less expensive saws

Exercises 11.1　(page 534)

1. a. (3, 1): Start at the origin, move right 3 units, then up 1 unit.
b. (−4, −2): Start at the origin, move left 4 units, then down 2 units.
c. (0, 3): Start at the origin, move up 3 units.
d. (5, −4): Start at the origin, move right 5 units, then down 4 units.
e. (4, 0): Start at the origin, move right 4 units.
f. (−2, 4): Start at the origin, move left 2 units, then up 4 units.

2.　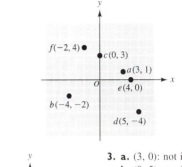

3. a. (3, 0): not in any quadrant
b. (0, 5): not in any quadrant
c. (−5, 2): II
d. (−4, −3): III

4. a. (4, 3): I
b. (−6, 0): not in any quadrant
c. (3, −4): IV
d. (0, −3): not in any quadrant

5. a. 5 because A is 5 units to the right of the y-axis.
b. −3 because C is 3 units to the left of the y-axis.
c. 3 because E is 3 units to the right of the y-axis.
d. 1 because F is 1 unit to the right of the y-axis.

6. a. 5　**b.** −2　**c.** 0　**d.** −5

7. a. 1 because F is 1 unit to the right of the vertical axis.
b. 2 because C is 2 units above the horizontal axis.

8. a. 5　**b.** −4

9.　**10.**

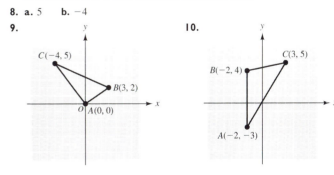

11. Yes; 0　　**12.** Yes; 0

Exercises 11.2　(page 539)

1. a. (0, 0) is not a solution, because $3(0) + 2(0) = 0 \neq 6$.
b. (0, 3) is a solution, because $3(0) + 2(3) = 0 + 6 = 6$.
c. (3, 0) is not a solution, because $3(3) + 2(0) = 9 \neq 6$.
d. (2, 0) is a solution, because $3(2) + 2(0) = 6 + 0 = 6$.

2. a. No **b.** No **c.** Yes **d.** Yes

3. a. $(0, 0)$ is a solution, because $3(0) + 4(0) = 0$.
 b. $(0, 3)$ is not a solution, because $3(0) + 4(3) = 0 + 12 \neq 0$.
 c. $(4, -3)$ is a solution, because $3(4) + 4(-3) = 12 - 12 = 0$.
 d. $(-4, 3)$ is a solution, because
 $3(-4) + 4(3) = -12 + 12 = 0$.

4. a. Yes **b.** No **c.** No **d.** Yes

5. a. $4(0) - 3y = 12$ **b.** $4x - 3(0) = 12$
 $ -3y = 12$ $ 4x = 12$
 $ y = -4$ $ x = 3$
 $ (0, -4)$ $ (3, 0)$
 c. $4(3) - 3y = 12$ **d.** $4x - 3(-4) = 12$
 $ 12 - 3y = 12$ $ 4x + 12 = 12$
 $ -3y = 0$ $ 4x = 0$
 $ y = 0$ $ x = 0$
 $ (3, 0)$ $ (0, -4)$

6. a. $(0, 3)$ **b.** $(1, 0)$ **c.** $(3, -6)$ **d.** $(2, -3)$

7. a. $2(0) - 5y = 0$ **b.** $2x - 5(0) = 0$
 $ -5y = 0$ $ 2x = 0$
 $ y = 0$ $ x = 0$
 $ (0, 0)$ $ (0, 0)$
 c. $2(2) - 5y = 0$ **d.** $2x - 5(2) = 0$
 $ 4 - 5y = 0$ $ 2x - 10 = 0$
 $ -5y = -4$ $ 2x = 10$
 $ y = \frac{4}{5}$ $ x = 5$
 $\left(2, \frac{4}{5}\right)$ $ (5, 2)$

8. a. $(0, 0)$ **b.** $(0, 0)$ **c.** $(3, -9)$ **d.** $(1, -3)$

9. a. $x - 5 = 0$ **b.** $(5, 5)$ **c.** $(5, 2)$ **d.** $(5, -2)$
 $ x = 5$
 $ (5, 0)$

10. a. $(0, -3)$ **b.** $(3, -3)$ **c.** $(-3, -3)$ **d.** $(5, -3)$

In Exercises 11–18, your answers may vary from these.

11. $(0, -5), (5, 0), (8, 3)$ **12.** $(0, 2), (-2, 0), (5, 7)$

13. $(0, 2), (6, 0), (3, 1)$ **14.** $(0, 1), (4, 0), (-4, 2)$

15. $(0, -3), \left(\frac{3}{5}, 0\right), (1, 2)$ **16.** $(0, 4), \left(-\frac{4}{3}, 0\right), (1, 7)$

17. $(0, 3), (2, 0), (-2, 6)$ **18.** $(0, -4), (6, 0), (12, 4)$

Exercises 11.3 (page 551)

1. $x + y = 3$
 $ y = 3 - x$
 If $x = 0, y = 3 - 0 = 3$.
 If $x = 1, y = 3 - 1 = 2$.
 If $x = 3, y = 3 - 3 = 0$.

x	y
0	3
1	2
3	0

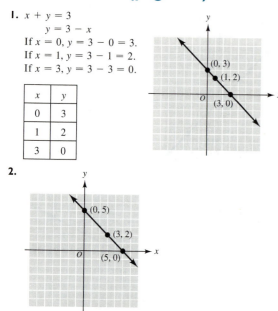

2.

3. $2x - 3y = 6$
 Intercepts:
 If $y = 0, x = 3$.
 If $x = 0, y = -2$.
 This gives the points $(3, 0)$
 and $(0, -2)$.
 Checkpoint: If $x = 2$,
 $y = -\frac{2}{3}$.

4.

5. x can be any number,
 but y is always 8.

6.

7. y can be any number,
 but x is always -2.

8.

9. If $x + 5 = 0, x = -5$;
 x is always -5.

10.

11. If $y + 5 = 0, y = -5$;
 y must be -5.

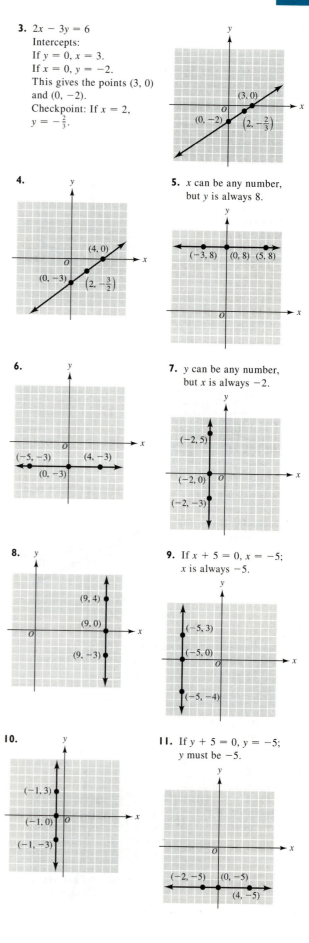

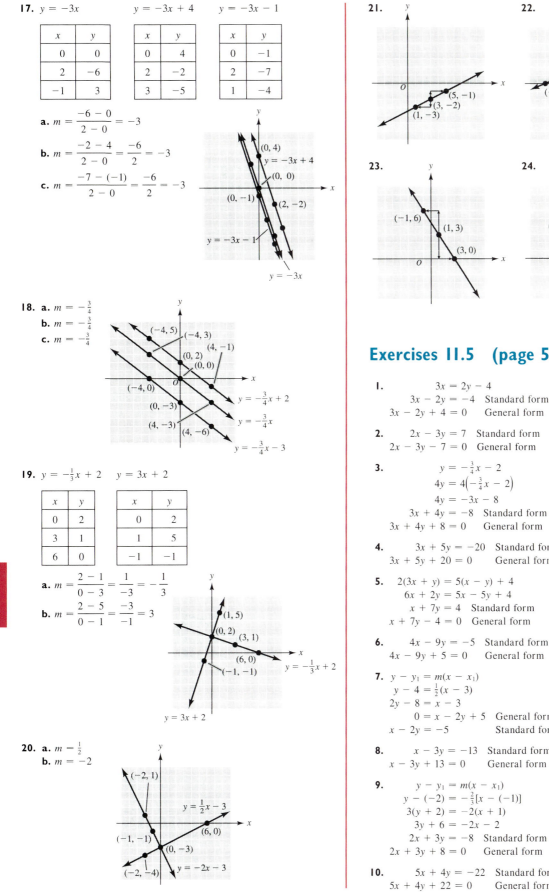

17. $y = -3x$ $y = -3x + 4$ $y = -3x - 1$

x	y
0	0
2	-6
-1	3

x	y
0	4
2	-2
3	-5

x	y
0	-1
2	-7
1	-4

a. $m = \dfrac{-6 - 0}{2 - 0} = -3$

b. $m = \dfrac{-2 - 4}{2 - 0} = \dfrac{-6}{2} = -3$

c. $m = \dfrac{-7 - (-1)}{2 - 0} = \dfrac{-6}{2} = -3$

18. a. $m = -\frac{3}{4}$
b. $m = -\frac{3}{4}$
c. $m = -\frac{3}{4}$

19. $y = -\frac{1}{3}x + 2$ $y = 3x + 2$

x	y
0	2
3	1
6	0

x	y
0	2
1	5
-1	-1

a. $m = \dfrac{2 - 1}{0 - 3} = \dfrac{1}{-3} = -\dfrac{1}{3}$

b. $m = \dfrac{2 - 5}{0 - 1} = \dfrac{-3}{-1} = 3$

20. a. $m = \frac{1}{2}$
b. $m = -2$

21.

22.

23.

24.

Exercises 11.5   (page 569)

1.
$$3x = 2y - 4$$
$3x - 2y = -4$ Standard form
$3x - 2y + 4 = 0$ General form

2. $2x - 3y = 7$ Standard form
$2x - 3y - 7 = 0$ General form

3.
$$y = -\tfrac{3}{4}x - 2$$
$$4y = 4\left(-\tfrac{3}{4}x - 2\right)$$
$$4y = -3x - 8$$
$3x + 4y = -8$ Standard form
$3x + 4y + 8 = 0$ General form

4. $3x + 5y = -20$ Standard form
$3x + 5y + 20 = 0$ General form

5.
$$2(3x + y) = 5(x - y) + 4$$
$$6x + 2y = 5x - 5y + 4$$
$x + 7y = 4$ Standard form
$x + 7y - 4 = 0$ General form

6. $4x - 9y = -5$ Standard form
$4x - 9y + 5 = 0$ General form

7.
$$y - y_1 = m(x - x_1)$$
$$y - 4 = \tfrac{1}{2}(x - 3)$$
$$2y - 8 = x - 3$$
$0 = x - 2y + 5$ General form
$x - 2y = -5$ Standard form

8. $x - 3y = -13$ Standard form
$x - 3y + 13 = 0$ General form

9.
$$y - y_1 = m(x - x_1)$$
$$y - (-2) = -\tfrac{2}{3}[x - (-1)]$$
$$3(y + 2) = -2(x + 1)$$
$$3y + 6 = -2x - 2$$
$2x + 3y = -8$ Standard form
$2x + 3y + 8 = 0$ General form

10. $5x + 4y = -22$ Standard form
$5x + 4y + 22 = 0$ General form

11. $y = mx + b$
$y = \frac{3}{4}x - 3$
$4y = 3x - 12$
$12 = 3x - 4y$ Standard form
$0 = 3x - 4y - 12$ General form

12. $2x - 7y = 14$ Standard form
$2x - 7y - 14 = 0$ General form

13. $y = mx + b$
$y = -\frac{2}{5}x + \frac{1}{2}$ LCD is 10
$10y = -4x + 5$
$4x + 10y = 5$ Standard form
$4x + 10y - 5 = 0$ General form

14. $20x + 12y = 9$ Standard form
$20x + 12y - 9 = 0$ General form

15. a. $4y = -3x - 12$, so $y = -\frac{3}{4}x - 3$.
b. $-\frac{3}{4}$ **c.** -3
d. $y - (-4) = -\frac{3}{4}(x - 2)$, or $y + 4 = -\frac{3}{4}(x - 2)$
e. $y - 5 = \frac{4}{3}[x - (-3)]$, or $y - 5 = \frac{4}{3}(x + 3)$

16. a. $y = -\frac{2}{5}x - 2$ **b.** $-\frac{2}{5}$ **c.** -2
d. $y + 4 = -\frac{2}{5}(x - 2)$ **e.** $y - 5 = \frac{5}{2}(x + 3)$

17. a. $3y = -2x + 9$, so $y = -\frac{2}{3}x + 3$.
b. $-\frac{2}{3}$ **c.** 3
d. $y - (-4) = -\frac{2}{3}(x - 2)$, or $y + 4 = -\frac{2}{3}(x - 2)$
e. $y - 5 = \frac{3}{2}[x - (-3)]$, or $y - 5 = \frac{3}{2}(x + 3)$

18. a. $y = -\frac{2}{5}x + 3$ **b.** $-\frac{2}{5}$ **c.** 3
d. $y + 4 = -\frac{2}{5}(x - 2)$ **e.** $y - 5 = \frac{5}{2}(x + 3)$

19. a. $3y = 5x - 30$, so $y = \frac{5}{3}x - 10$.
b. $\frac{5}{3}$ **c.** -10
d. $y - (-4) = \frac{5}{3}(x - 2)$, or $y + 4 = \frac{5}{3}(x - 2)$
e. $y - 5 = -\frac{3}{5}[x - (-3)]$, or $y - 5 = -\frac{3}{5}(x + 3)$

20. a. $y = \frac{4}{7}x - 4$ **b.** $\frac{4}{7}$ **c.** -4
d. $y + 4 = \frac{4}{7}(x - 2)$ **e.** $y - 5 = -\frac{7}{4}(x + 3)$

21. a. $2y = -x + 6$, so $y = -\frac{1}{2}x + 3$.
b. $-\frac{1}{2}$ **c.** 3
d. $y - (-4) = -\frac{1}{2}(x - 2)$, or $y + 4 = -\frac{1}{2}(x - 2)$
e. $y - 5 = \frac{2}{1}[x - (-3)]$, or $y - 5 = 2(x + 3)$

22. a. $y = -\frac{1}{5}x + 1$ **b.** $-\frac{1}{5}$ **c.** 1
d. $y + 4 = -\frac{1}{5}(x - 2)$ **e.** $y - 5 = 5(x + 3)$

23.
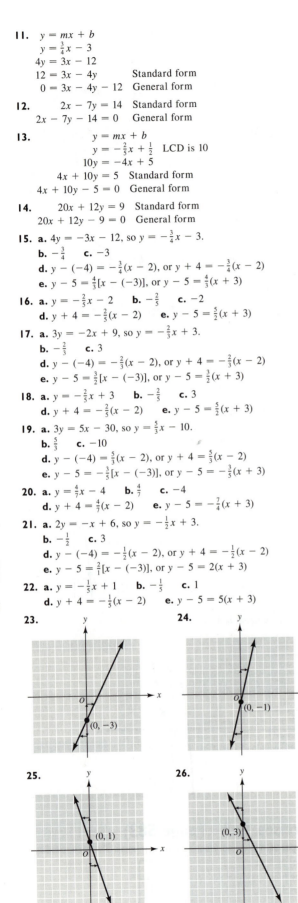

24.

25.

26.

27.
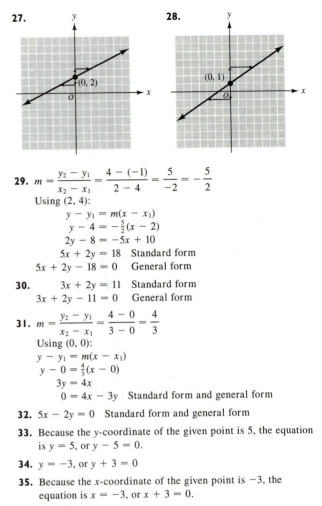

28.
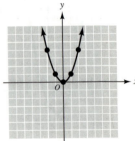

29. $m = \dfrac{y_2 - y_1}{x_2 - x_1} = \dfrac{4 - (-1)}{2 - 4} = \dfrac{5}{-2} = -\dfrac{5}{2}$
Using (2, 4):
$y - y_1 = m(x - x_1)$
$y - 4 = -\frac{5}{2}(x - 2)$
$2y - 8 = -5x + 10$
$5x + 2y = 18$ Standard form
$5x + 2y - 18 = 0$ General form

30. $3x + 2y = 11$ Standard form
$3x + 2y - 11 = 0$ General form

31. $m = \dfrac{y_2 - y_1}{x_2 - x_1} = \dfrac{4 - 0}{3 - 0} = \dfrac{4}{3}$
Using (0, 0):
$y - y_1 = m(x - x_1)$
$y - 0 = \frac{4}{3}(x - 0)$
$3y = 4x$
$0 = 4x - 3y$ Standard form and general form

32. $5x - 2y = 0$ Standard form and general form

33. Because the y-coordinate of the given point is 5, the equation is $y = 5$, or $y - 5 = 0$.

34. $y = -3$, or $y + 3 = 0$

35. Because the x-coordinate of the given point is -3, the equation is $x = -3$, or $x + 3 = 0$.

36. $x = -2$, or $x + 2 = 0$

Exercises 11.6   (page 576)

1. The x- and y-intercepts are (0, 0).
$y = x^2$
$y = (-2)^2 = 4$
$y = (-1)^2 = 1$
$y = 0^2 = 0$
$y = 1^2 = 1$
$y = 2^2 = 4$

x	y
-2	4
-1	1
0	0
1	1
2	4

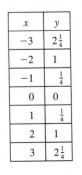

2. The x- and y-intercepts are (0, 0).

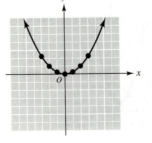

x	y
-3	$2\frac{1}{4}$
-2	1
-1	$\frac{1}{4}$
0	0
1	$\frac{1}{4}$
2	1
3	$2\frac{1}{4}$

21. $y = x^3 - 3x$
$y = (-3)^3 - (3)(-3)$
  $= -27 + 9 = -18$
$y = (-2)^3 - 3(-2)$
  $= -8 + 6 = -2$
$y = (-1)^3 - 3(-1)$
  $= -1 + 3 = 2$
$y = 0^3 - 3(0) = 0$
$y = 1^3 - 3(1) = -2$
$y = 2^3 - 3(2) = 2$
$y = 3^3 - 3(3) = 18$

x	y
−3	−18
−2	−2
−1	2
0	0
1	−2
2	2
3	18

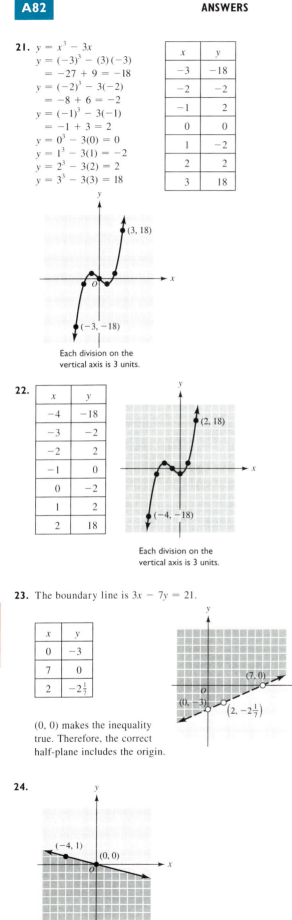

Each division on the
vertical axis is 3 units.

22.

x	y
−4	−18
−3	−2
−2	2
−1	0
0	−2
1	2
2	18

Each division on the
vertical axis is 3 units.

23. The boundary line is $3x - 7y = 21$.

x	y
0	−3
7	0
2	$-2\frac{1}{7}$

(0, 0) makes the inequality
true. Therefore, the correct
half-plane includes the origin.

24.

25. The boundary line is $y = 3$,
which is a horizontal line.
(0, 0) makes the inequality
false. Therefore, the correct
half-plane does not include
the origin.

26.

27. LCD = 12
$$(12)\left(\frac{y}{3}\right) - (12)\left(\frac{x}{4}\right) \le 1(12)$$
$$4y - 3x \le 12$$
The boundary line is
$4y - 3x = 12$.

x	y
0	3
−4	0
−2	$1\frac{1}{2}$

The correct half-plane includes the origin.

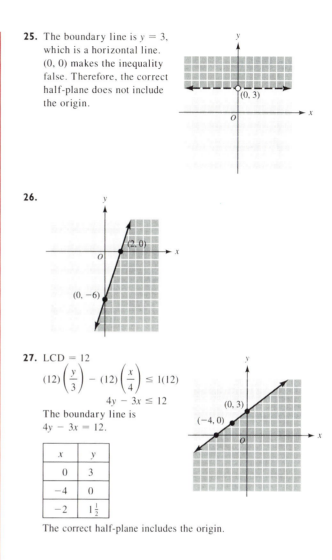

Chapter 11 Diagnostic Test
(page 591)

Following each problem number is the number (in parentheses) of
the textbook section where that kind of problem is discussed.

1. (11.1) **a.**

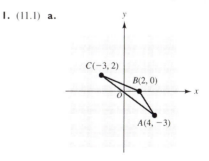

b. $D = (4, 5)$; $E = (-4, 1)$; $F = (-2, -6)$; $G = (5, -4)$

2. (11.2) **a.** (0, 0) is not a solution, because $5(0) - 3(0) = 0 \ne 15$.
b. (3, 0) is a solution, because $5(3) - 3(0) = 15$.
c. (0, 3) is not a solution, because
  $5(0) - 3(3) = -9 \ne 15$.
d. (0, −5) is a solution, because $5(0) - 3(-5) = 15$.
e. (6, 5) is a solution, because $5(6) - 3(5) = 15$.

3. (11.3) **a.** Since $y = -2$ is a horizontal line, the y-coordinate of every point on the line is -2.

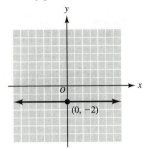

b. $y = -3x + 3$
If $x = 0$, $y = 3$.
If $y = 0$, $x = 1$.
Checkpoint:
If $x = 2$, $y = -3$.

4. (11.3) **a.** $y = 2x - 4$
If $x = 0$, $y = -4$.
If $y = 0$, $x = 2$.
Checkpoint:
If $x = 4$, $y = 4$.

b. $4x + 9y = 18$
If $x = 0$, $y = 2$.
If $y = 0$, $x = 4\frac{1}{2}$.
Checkpoint:
If $x = 2$, $y = 1\frac{1}{9}$.

5. (11.4) **a.** $m = \dfrac{y_2 - y_1}{x_2 - x_1} = \dfrac{1 - 2}{-3 - 5} = \dfrac{-1}{-8} = \dfrac{1}{8}$

b. $7y = -6x + 3$

$y = \dfrac{-6x + 3}{7}$

$y = -\dfrac{6}{7}x + \dfrac{3}{7}$

The slope is $-\frac{6}{7}$, and the y-intercept is $\frac{3}{7}$.

6. (11.5) **a.** $x = -3$ (the x-coordinate of every point on the line is -3), or $x + 3 = 0$

b. $y = 4$ (the y-coordinate of every point on the line is 4), or $y - 4 = 0$

7. (11.5) Using the slope-intercept form, we have

$$y = mx + b$$
$$y = 5x + 3$$
$$-5x + y - 3 = 0$$
$$5x - y + 3 = 0, \text{ or } 5x - y = -3$$

8. (11.4 and 11.5) **a.** $m = \dfrac{y_2 - y_1}{x_2 - x_1} = \dfrac{-4 - 2}{1 - (-3)} = \dfrac{-6}{4} = -\dfrac{3}{2}$

b. Using the point-slope form, we have

$$y - y_1 = m(x - x_1)$$
$$y - 2 = -\tfrac{3}{2}(x - [-3]) \quad \text{LCD is 2}$$

$$2(y - 2) = \overset{1}{\cancel{2}}\left(-\dfrac{3}{\underset{1}{\cancel{2}}}\right)(x + 3)$$

$$2y - 4 = -3(x + 3)$$
$$2y - 4 = -3x - 9$$
$$3x + 2y = -5,$$
$$\text{or } 3x + 2y + 5 = 0$$

9. (11.6)
$y = x^2 - x - 6$
$y = (-3)^2 - (-3) - 6 = 6$
$y = (-2)^2 - (-2) - 6 = 0$
$y = (-1)^2 - (-1) - 6 = -4$
$y = (0)^2 - (0) - 6 = -6$
$y = (1)^2 - (1) - 6 = -6$
$y = (2)^2 - (2) - 6 = -4$
$y = (3)^2 - (3) - 6 = 0$
$y = (4)^2 - (4) - 6 = 6$

x	y
-3	6
-2	0
-1	-4
0	-6
1	-6
2	-4
3	0
4	6

10. (11.7) The boundary line is $3x - 2y = 6$.

x	y
0	-3
2	0
4	3

$(0, 0)$ makes the inequality true. Therefore, the half-plane includes the origin.

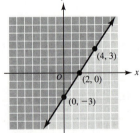

Chapters 1–11 Cumulative Review Exercises (page 591)

1. $A = \dfrac{h}{2}(B + b)$

$A = \dfrac{5}{2}(11 + 7)$

$A = \dfrac{5}{\underset{1}{\cancel{2}}}(\overset{9}{\cancel{18}}) = 45$

2. $\dfrac{9b^2}{49a^6}$

3. $3z - 5\overline{)6z^2 - z - 7}$ $\quad 2z + 3 \text{ R } 8$
$\underline{6z^2 - 10z}$
$9z - 7$
$\underline{9z - 15}$
8

4. $\dfrac{2x^2 + 1}{(x - 2)(x + 1)}$

5. $\dfrac{b+1}{a(a+b)} \cdot \dfrac{a(a+b)(a-b)}{(b+1)(b-2)} = \dfrac{a-b}{b-2}$ **6.** $\dfrac{xy-2}{y}$

7. LCD $= (x-6)(x-1)(2x+3)$

$$\dfrac{3(2x+3)}{(x-6)(x-1)(2x+3)} - \dfrac{4(x-6)}{(2x+3)(x-1)(x-6)}$$

$$= \dfrac{6x+9-(4x-24)}{(x-6)(x-1)(2x+3)}$$

$$= \dfrac{6x+9-4x+24}{(x-6)(x-1)(2x+3)}$$

$$= \dfrac{2x+33}{(x-6)(x-1)(2x+3)}$$

8. The slope is $-\frac{2}{3}$, and the y-intercept is $-\frac{5}{3}$.

9. LCD $= 18$

$$18\left(\dfrac{2x}{9}\right) - 18\left(\dfrac{11}{3}\right) = \dfrac{5x}{6}(18)$$

$$4x - 66 = 15x$$

$$-4x + 4x - 66 = -4x + 15x$$

$$-66 = 11x$$

$$\dfrac{-66}{11} = \dfrac{11x}{11}$$

$$x = -6$$

The solution is -6.

10. $k = \dfrac{-2h^2}{2hC-5}$ or $k = \dfrac{2h^2}{5-2hC}$

11. $y = 3x - 2$
If $x = 0$, $y = -2$.
If $y = 0$, $x = \frac{2}{3}$.
Checkpoint: If $x = 2$, $y = 4$.

12.

x	y
-1	5
0	0
1	-3
2	-4
3	-3
4	0
5	5

13. $m = \dfrac{y_2 - y_1}{x_2 - x_1} = \dfrac{-4-1}{2-7} = \dfrac{-5}{-5} = 1$

$$y - y_1 = m(x - x_1)$$

$$y - 1 = 1(x - 7)$$

$$y - 1 = x - 7$$

$$-x + y + 6 = 0$$

$$x - y - 6 = 0 \quad \text{General form}$$

$$x - y = 6 \quad \text{Standard form}$$

14. First answer: One number is $-\frac{1}{2}$ and the other is $3\frac{1}{2}$. Second answer: One number is $\frac{1}{2}$ and the other is $-3\frac{1}{2}$.

15. Let $x =$ number of lb of Delicious apples
Then $50 - x =$ number of lb of Jonathan apples

$$98x + 85(50 - x) = 4{,}458$$

$$98x + 4{,}250 - 85x = 4{,}458$$

$$13x = 208$$

$$x = 16$$

$$50 - x = 34$$

The mixture contains 16 lb of Delicious apples and 34 lb of Jonathan apples.

16. Altitude is 8 cm; base is 11 cm.

17. Let $x =$ number
Then $\dfrac{1}{x} =$ reciprocal of number

$$x + \dfrac{1}{x} = \dfrac{53}{14} \quad \text{LCD is } 14x$$

$$(14x)x + (14x)\left(\dfrac{1}{x}\right) = \left(\dfrac{53}{14}\right)(14x)$$

$$14x^2 + 14 = 53x$$

$$14x^2 - 53x + 14 = 0$$

$$(2x - 7)(7x - 2) = 0$$

$$\begin{array}{c|c} 2x - 7 = 0 & 7x - 2 = 0 \\ 2x = 7 & 7x = 2 \\ x = \frac{7}{2} & x = \frac{2}{7} \\ \frac{1}{x} = \frac{2}{7} & \frac{1}{x} = \frac{7}{2} \end{array}$$

First answer: The number is $\frac{7}{2}$ and its reciprocal is $\frac{2}{7}$.
Second answer: The number is $\frac{2}{7}$ and its reciprocal is $\frac{7}{2}$.

18. First answer: The integers are -2 and -1. Second answer: The integers are 3 and 4.

19. Let $x =$ number of cc of water
There will be $(10 + x)$ cc in the mixture.

There is 0% of disinfectant in water.

$$10(0.17) + x(0) = (10 + x)(0.002)$$

$$10(0.17) = (10 + x)(0.002)$$

$$10(170) = (10 + x)(2)$$

$$1{,}700 = 20 + 2x$$

$$1{,}680 = 2x$$

$$840 = x$$

$$x = 840$$

840 cc of water must be added.

20. 55 words per minute

21. Let $x =$ cost per lens for cheaper lenses
Then $x + 23 =$ cost per lens for better lenses
Total cost of cheap lenses $=$ Total cost of better lenses

$$18x = 15(x + 23)$$

$$18x = 15x + 345$$

$$3x = 345$$

$$x = 115$$

$$x + 23 = 138$$

The cheaper lenses cost \$115 each, and the more expensive ones cost \$138 each.

Exercises 12.1   (page 595)

1. a. $(0, 8)$ is not a solution, because in the second equation,
$2(0) - (8) = 0 - 8 = -8 \neq 2$.
  b. $(1, 0)$ is not a solution, because in the first equation,
$3(1) + (0) = 3 + 0 = 3 \neq 8$.
  c. $(2, 2)$ is a solution, because $3(2) + (2) = 6 + 2 = 8$ and
$2(2) - (2) = 4 - 2 = 2$.

2. a. Not a solution   **b.** A solution   **c.** Not a solution

3. a. $(3, 2)$ is not a solution, because in the first equation,
$2(3) + 3(2) = 6 + 6 = 12 \neq 6$.

b. (0, 2) is a solution, because $2(0) + 3(2) = 0 + 6 = 6$ and $4(0) + 6(2) = 0 + 12 = 12$.

c. (3, 0) is a solution, because $2(3) + 3(0) = 6 + 0 = 6$ and $4(3) + 6(0) = 12 + 0 = 12$.

4. a. A solution   **b.** A solution   **c.** A solution

5. a. (0, 4) is not a solution, because in the first equation, $(0) + (4) = 0 + 4 = 4 \neq 2$.

b. (0, 2) is not a solution, because in the second equation, $(0) + (2) = 0 + 2 = 2 \neq 4$.

c. (4, 0) is not a solution, because in the first equation, $(4) + (0) = 4 + 0 = 4 \neq 2$.

6. a. Not a solution   **b.** Not a solution   **c.** Not a solution

Exercises 12.2   (page 600)

1. (1) $\begin{cases} 2x + y = 6 \\ 2x - y = -2 \end{cases}$ (2)

(1) Intercepts (0, 6), (3, 0)
(2) Intercepts (0, 2), (−1, 0)
Solution: (1, 4)

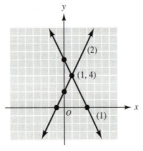

2.

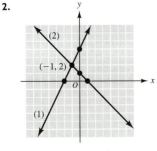

Solution: (−1, 2)

3. (1) $\begin{cases} x - 2y = -6 \\ 4x + 3y = 20 \end{cases}$ (2)

(1) Intercepts (−6, 0), (0, 3)
(2) Intercepts (5, 0), $\left(0, 6\frac{2}{3}\right)$
Solution: (2, 4)

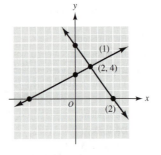

4.

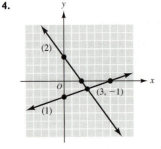

Solution: (3, −1)

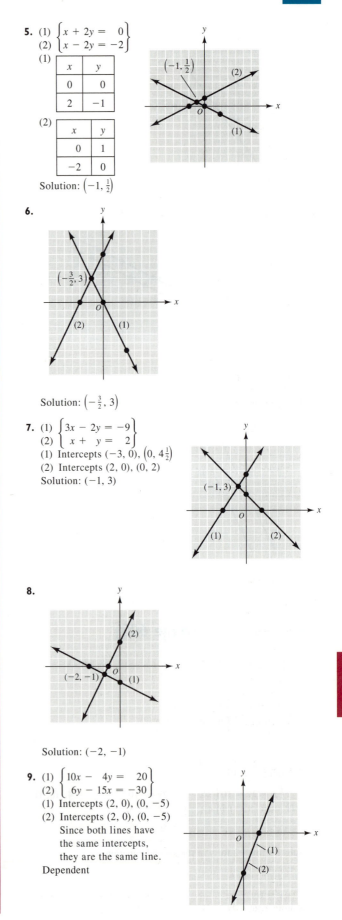

5. (1) $\begin{cases} x + 2y = 0 \\ x - 2y = -2 \end{cases}$ (2)

(1)

x	y
0	0
2	−1

(2)

x	y
0	1
−2	0

Solution: $\left(-1, \frac{1}{2}\right)$

6.

Solution: $\left(-\frac{3}{2}, 3\right)$

7. (1) $\begin{cases} 3x - 2y = -9 \\ x + y = 2 \end{cases}$ (2)

(1) Intercepts (−3, 0), $\left(0, 4\frac{1}{2}\right)$
(2) Intercepts (2, 0), (0, 2)
Solution: (−1, 3)

8.

Solution: (−2, −1)

9. (1) $\begin{cases} 10x - 4y = 20 \\ 6y - 15x = -30 \end{cases}$ (2)

(1) Intercepts (2, 0), (0, −5)
(2) Intercepts (2, 0), (0, −5)
Since both lines have the same intercepts, they are the same line.
Dependent

12

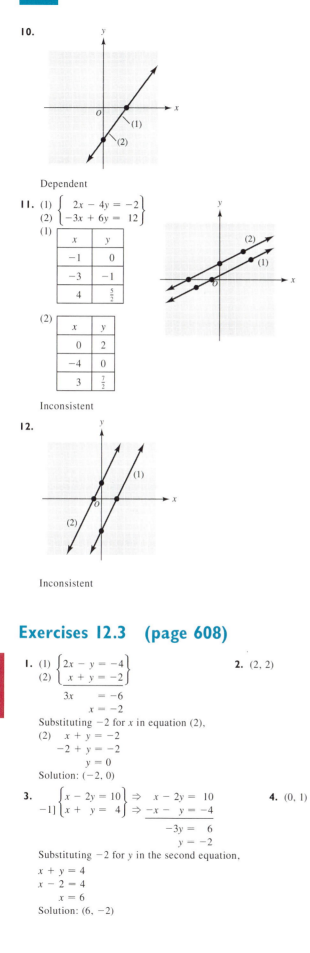

10.

Dependent

11. (1) $\begin{cases} 2x - 4y = -2 \\ -3x + 6y = 12 \end{cases}$
(2)

(1)

x	y
-1	0
-3	-1
4	$\frac{5}{2}$

(2)

x	y
0	2
-4	0
3	$\frac{7}{2}$

Inconsistent

12.

Inconsistent

Exercises 12.3   (page 608)

12

1. (1) $\begin{cases} 2x - y = -4 \\ x + y = -2 \end{cases}$
(2)

$\quad 3x \quad = -6$
$\quad\quad x = -2$

Substituting -2 for x in equation (2),
(2) $\quad x + y = -2$
$\quad -2 + y = -2$
$\quad\quad\quad y = 0$

Solution: $(-2, 0)$

2. $(2, 2)$

3. $\begin{array}{r} \\ -1] \end{array} \begin{cases} x - 2y = 10 \\ x + y = 4 \end{cases} \Rightarrow \begin{array}{r} x - 2y = 10 \\ -x - y = -4 \end{array}$

$\quad\quad\quad\quad\quad -3y = 6$
$\quad\quad\quad\quad\quad\quad y = -2$

Substituting -2 for y in the second equation,
$\quad x + y = 4$
$\quad x - 2 = 4$
$\quad\quad x = 6$

Solution: $(6, -2)$

4. $(0, 1)$

5. (1) $\begin{cases} x - 3y = 6 \\ 4x + 3y = 9 \end{cases}$
(2)

$\quad 5x \quad\quad = 15$
$\quad\quad x = 3$

Substituting 3 for x in equation (1),
(1) $x - 3y = 6$
$\quad 3 - 3y = 6$
$\quad\quad -3y = 3$
$\quad\quad\quad y = -1$

Solution: $(3, -1)$

6. $(1, 0)$

7. $\begin{array}{r} 2] \\ 1] \end{array} \begin{cases} 1x + 1y = 2 \\ 3x - 2y = -9 \end{cases} \Rightarrow \begin{array}{r} 2x + 2y = 4 \\ 3x - 2y = -9 \end{array}$

$\quad\quad\quad\quad\quad 5x \quad\quad = -5$
$\quad\quad\quad\quad\quad\quad x = -1$

Substituting -1 for x in the first equation,
$\quad x + y = 2$
$\quad -1 + y = 2$
$\quad\quad\quad y = 3$

Solution: $(-1, 3)$

8. $(-2, -1)$

9. $\begin{array}{r} 2] \\ 1] \end{array} \begin{cases} x + 2y = 0 \\ -2x + y = 0 \end{cases} \Rightarrow \begin{array}{r} 2x + 4y = 0 \\ -2x + y = 0 \end{array}$

$\quad\quad\quad\quad\quad\quad 5y = 0$
$\quad\quad\quad\quad\quad\quad y = 0$

Substituting 0 for y in the first equation,
$\quad x + 2y = 0$
$\quad x + 2(0) = 0$
$\quad\quad x = 0$

Solution: $(0, 0)$

10. Dependent

11. $\begin{array}{r} 3] \\ -4] \end{array} \begin{cases} 4x + 3y = 2 \\ 3x + 5y = -4 \end{cases} \Rightarrow \begin{array}{r} 12x + 9y = 6 \\ -12x - 20y = 16 \end{array}$

$\quad\quad\quad\quad\quad\quad -11y = 22$
$\quad\quad\quad\quad\quad\quad\quad y = -2$

Substituting -2 for y in the first equation,
$\quad 4x + 3y = 2$
$\quad 4x + 3(-2) = 2$
$\quad 4x - 6 = 2$
$\quad\quad 4x = 8$
$\quad\quad\quad x = 2$

Solution: $(2, -2)$

12. $(3, -2)$

13. $\begin{array}{r} 9] \quad 3] \\ -6] -2] \end{array} \begin{cases} 6x - 10y = 6 \\ 9x - 15y = -4 \end{cases} \Rightarrow \begin{array}{r} 18x - 30y = 18 \\ -18x + 30y = 8 \end{array}$

$\quad\quad\quad\quad\quad\quad 0 = 26 \quad \textit{False}$

Inconsistent

14. Inconsistent

15. $\begin{array}{r} 2] \\ 1] \end{array} \begin{cases} 3x - 5y = -2 \\ -6x + 10y = 4 \end{cases} \Rightarrow \begin{array}{r} 6x - 10y = -4 \\ -6x + 10y = 4 \end{array}$

$\quad\quad\quad\quad\quad\quad 0 = 0 \quad \textit{True}$

Dependent

16. Dependent

17. $\dfrac{x}{y} = \dfrac{5}{8} \Rightarrow 8x = 5y \Rightarrow 8x - 5y = 0$

$\dfrac{x + 3}{y + 4} = \dfrac{9}{14} \Rightarrow 14(x + 3) = 9(y + 4) \Rightarrow$

$\quad\quad\quad\quad 14x + 42 = 9y + 36 \Rightarrow 14x - 9y = -6$

$\begin{array}{r} 7] \\ -4] \end{array} \begin{cases} 8x - 5y = 0 \\ 14x - 9y = -6 \end{cases} \Rightarrow \begin{array}{r} 56x - 35y = 0 \\ -56x + 36y = 24 \end{array}$

$\quad\quad\quad\quad\quad\quad y = 24$

$8x = 5y$
$8x = 5(24)$
$\quad x = \dfrac{5(24)}{8} = 15$

Solution: $(15, 24)$

18. (14, 21)

19. $\dfrac{x+3}{y+7} = \dfrac{4}{3} \Rightarrow 3(x+3) = 4(y+7) \Rightarrow$

$$3x + 9 = 4y + 28 \Rightarrow 3x - 4y = 19$$

$\dfrac{x-1}{5-y} = \dfrac{2}{3} \Rightarrow 3(x-1) = 2(5-y) \Rightarrow$

$$3x - 3 = 10 - 2y \Rightarrow 3x + 2y = 13$$

$$\begin{aligned} &\begin{cases} 3x - 4y = 19 \\ 2 \begin{cases} 3x + 2y = 13 \end{cases} \end{cases} \begin{aligned} &\Rightarrow 3x - 4y = 19 \\ &\Rightarrow \underline{6x + 4y = 26} \\ &\;\; 9x = 45 \\ & x = 5 \end{aligned} \end{aligned}$$

$$\begin{aligned} 3x + 2y &= 13 \quad \text{(Equation 2)} \\ 3(5) + 2y &= 13 \\ 15 + 2y &= 13 \\ 2y &= -2 \\ y &= \frac{-2}{2} = -1 \end{aligned}$$

Solution: $(5, -1)$

20. $(-2, 3)$

Exercises 12.4 (page 614)

1. (1) $\begin{cases} 2x - 3y = 1 \\ x = y + 2 \end{cases}$ **2.** $(2, 7)$

Substituting $y + 2$ for x in equation (1):

$$\begin{aligned} (1) \quad 2x - 3y &= 1 \\ 2(y+2) - 3y &= 1 \\ 2y + 4 - 3y &= 1 \\ -y &= -3 \\ y &= 3 \end{aligned}$$

Substituting 3 for y in equation (2):

$$\begin{aligned} (2) \quad x &= y + 2 \\ x &= 3 + 2 = 5 \end{aligned}$$

Solution: $(5, 3)$

3. (1) $\begin{cases} 3x + 4y = 2 \\ y = x - 3 \end{cases}$ **4.** $(1, 3)$

Substituting $x - 3$ for y in equation (1):

$$\begin{aligned} (1) \quad 3x + 4y &= 2 \\ 3x + 4(x-3) &= 2 \\ 3x + 4x - 12 &= 2 \\ 7x &= 14 \\ x &= 2 \end{aligned}$$

Substituting 2 for x in equation (2):

$$y = x - 3 = 2 - 3 = -1$$

Solution: $(2, -1)$

5. (1) $\begin{cases} y = 2 - 4x \\ 7x + 3y = 1 \end{cases}$ **6.** $(3, -2)$

Substituting $2 - 4x$ for y in equation (2):

$$\begin{aligned} (2) \quad 7x + 3y &= 1 \\ 7x + 3(2 - 4x) &= 1 \\ 7x + 6 - 12x &= 1 \\ -5x &= -5 \\ x &= 1 \end{aligned}$$

Substituting 1 for x in equation (1):

$$y = 2 - 4(1) = -2$$

Solution: $(1, -2)$

7. (1) $\begin{cases} 4x - y = 3 \\ 8x - 2y = 6 \end{cases} \Rightarrow y = 4x - 3$

Substituting $4x - 3$ for y in equation (2):

$$\begin{aligned} (2) \quad 8x - 2y &= 6 \\ 8x - 2(4x - 3) &= 6 \\ 8x - 8x + 6 &= 6 \\ 6 &= 6 \quad \text{True} \end{aligned}$$

Dependent

8. Dependent

9. (1) $\begin{cases} x + 3 = 0 \\ 3x - 2y = 6 \end{cases} \Rightarrow x = -3$ **10.** $\left(-\frac{3}{5}, 4\right)$

Substituting -3 for x in equation (2):

$$\begin{aligned} (2) \quad 3x - 2y &= 6 \\ 3(-3) - 2y &= 6 \\ -9 - 2y &= 6 \\ -2y &= 15 \\ y &= \frac{15}{-2} = -7\tfrac{1}{2} \end{aligned}$$

Solution: $\left(-3, -7\tfrac{1}{2}\right)$

11. (1) $\begin{cases} x = 3y - 4 \\ 5y - 2x = 2 \end{cases}$ **12.** $(7, 13)$

Substituting $3y - 4$ for x in equation (2):

$$\begin{aligned} (2) \quad 5y - 2x &= 2 \\ 5y - 2(3y - 4) &= 2 \\ 5y - 6y + 8 &= 2 \\ -y &= -6 \\ y &= 6 \end{aligned}$$

Substituting 6 for y in equation (1):

$$x = 3y - 4 = 3(6) - 4 = 18 - 4 = 14$$

Solution: $(14, 6)$

13. (1) $\begin{cases} 8x + 4y = 7 \\ x = 2 - 2y \end{cases}$ **14.** $\left(\frac{3}{4}, \frac{1}{4}\right)$

Substituting $2 - 2y$ for x in equation (1):

$$\begin{aligned} (1) \quad 8x + 4y &= 7 \\ 8(2 - 2y) + 4y &= 7 \\ 16 - 16y + 4y &= 7 \\ 16 - 12y &= 7 \\ -12y &= -9 \\ y &= \frac{-9}{-12} = \frac{3}{4} \end{aligned}$$

Substituting $\frac{3}{4}$ for y in equation (2):

$$x = 2 - 2y = 2 - 2\left(\tfrac{3}{4}\right) = 2 - \tfrac{3}{2} = \tfrac{1}{2}$$

Solution: $\left(\tfrac{1}{2}, \tfrac{3}{4}\right)$

15. (1) $\begin{cases} 3x - 2y = 8 \\ 2y - 3x = 4 \end{cases} \Rightarrow 2y = 3x + 4$

$$y = \frac{3x + 4}{2}$$

Substituting $\dfrac{3x + 4}{2}$ for y in equation (1),

$$\begin{aligned} (1) \quad 3x - 2y &= 8 \\ 3x - \overset{1}{\cancel{2}}\left(\frac{3x + 4}{\underset{1}{\cancel{2}}}\right) &= 8 \\ 3x - 3x - 4 &= 8 \\ -4 &= 8 \quad \text{False} \end{aligned}$$

Inconsistent

16. Inconsistent

17. (1) $\begin{cases} 5x - 4y = 2 \\ y = 1 + 2x \end{cases}$ **18.** $(4, 1)$

Substituting $1 + 2x$ for y in equation (1),

$$\begin{aligned} (1) \quad 5x - 4y &= 2 \\ 5x - 4(1 + 2x) &= 2 \\ 5x - 4 - 8x &= 2 \\ -4 - 3x &= 2 \\ -3x &= 6 \\ x &= \frac{6}{-3} = -2 \end{aligned}$$

Substituting -2 for x in equation (2),

$$y = 1 + 2x = 1 + 2(-2) = 1 - 4 = -3$$

Solution: $(-2, -3)$

12

19. (1) $\begin{cases} 4x + 4y = 3 \\ 6x + 12y = -6 \end{cases} \Rightarrow 4x = 3 - 4y$
$$x = \frac{3 - 4y}{4}$$

20. $\left(2\frac{1}{2}, -2\frac{1}{3}\right)$

Substituting $\dfrac{3 - 4y}{4}$ for x in equation (2):

(2) $\qquad\qquad 6x + 12y = -6$

$$\overset{3}{\cancel{6}}\left(\frac{3 - 4y}{\underset{2}{\cancel{4}}}\right) + 12y = -6 \quad \text{LCD is } 2$$

$$\frac{\overset{1}{\cancel{2}}}{1} \cdot 3\left(\frac{3 - 4y}{\underset{1}{\cancel{2}}}\right) + 2(12y) = 2(-6)$$

$$9 - 12y + 24y = -12$$
$$12y = -21$$
$$y = -\frac{21}{12} = -\frac{7}{4} = -1\frac{3}{4}$$

Substituting $-\dfrac{7}{4}$ for y in $x = \dfrac{3 - 4y}{4}$:

$$x = \frac{3 - 4\left(-\frac{7}{4}\right)}{4} = \frac{3 + 7}{4} = \frac{10}{4} = 2\frac{1}{2}$$

Solution: $\left(2\frac{1}{2}, -1\frac{3}{4}\right)$

Exercises 12.5   (page 618)

(The checks will not be shown.)

1. Let $x =$ one number
Let $y =$ other number
(1) $\begin{cases} x + y = 80 \\ x - y = 12 \end{cases}$
$\qquad 2x \quad = 92$
$\qquad\quad x = 46$

Substituting 46 for x in equation (1):
$46 + y = 80$
$\qquad y = 34$
One number is 46, and the other is 34.

2. 47 and 43

3. Let $x =$ measure of first angle
Let $y =$ measure of second angle
(1) $\begin{cases} x + y = 90 \\ x - y = 60 \end{cases}$
$\qquad 2x \quad = 150$
$\qquad\quad x = 75$

Substituting 75 for x in equation (1):
$75 + y = 90$
$\qquad y = 15$
The angles measure 75° and 15°.

4. 106° and 74°

5. Let $x =$ amount invested at 4% interest
Let $y =$ amount invested at 3% interest
(1) $\begin{cases} x + y = 8{,}000 \\ 0.04x + 0.03y = 255 \end{cases}$
$-3]\quad x + y = 8{,}000 \Rightarrow -3x - 3y = -24{,}000$
$100]\ 0.04x + 0.03y = 255 \Rightarrow \underline{4x + 3y = 25{,}500}$
$\qquad\qquad\qquad\qquad\qquad x = 1{,}500$

Substituting 1,500 for x in equation (1):
$1{,}500 + y = 8{,}000$
$\qquad\qquad y = 6{,}500$
Dixie invested $1,500 at 4% interest and $6,500 at 3% interest.

6. Marty invested $1,200 at 5% interest and $4,800 at 4% interest.

7. Let $x =$ smaller number
Let $y =$ larger number
(1) $\begin{cases} 2x + 3y = 34 \\ 5x - 2y = 9 \end{cases}$
2] $2x + 3y = 34 \Rightarrow 4x + 6y = 68$
3] $5x - 2y = 9 \Rightarrow \underline{15x - 6y = 27}$
$\qquad\qquad\qquad\qquad 19x \quad = 95$
$\qquad\qquad\qquad\qquad\quad x = 5$

Substituting 5 for x in equation (1):
(1) $\quad 2x + 3y = 34$
$\qquad 2(5) + 3y = 34$
$\qquad 10 + 3y = 34$
$\qquad\qquad 3y = 24$
$\qquad\qquad\quad y = 8$
The smaller number is 5, and the larger one is 8.

8. 7 and 4

9. Let $x =$ number of lb of apple chunks
Let $y =$ number of lb of granola
(1) $\begin{cases} x + y = 6 \\ 4.24x + 2.10y = 16.88 \end{cases}$
$-210]\quad x + y = 6 \Rightarrow -210x - 210y = -1{,}260$
$100]\ 4.24x + 2.10y = 16.88 \Rightarrow \underline{424x + 210y = 1{,}688}$
$\qquad\qquad\qquad\qquad\qquad\qquad 214x = 428$
$\qquad\qquad\qquad\qquad\qquad\qquad\quad x = 2$

Substituting 2 for x in equation (1):
$2 + y = 6 \Rightarrow y = 4$
Jason bought 2 lb of apple chunks and 4 lb of granola.

10. 43 lb Grade A and 57 lb Grade B

11. Let $x =$ number of 18¢ stamps
Let $y =$ number of 22¢ stamps
(1) $\begin{cases} x + y = 22 \\ 0.18x + 0.22y = 4.48 \end{cases}$
$-18]\quad x + y = 22 \Rightarrow -18x - 18y = -396$
$100]\ 0.18x + 0.22y = 4.48 \Rightarrow \underline{18x + 22y = 448}$
$\qquad\qquad\qquad\qquad\qquad\qquad 4y = 52$
$\qquad\qquad\qquad\qquad\qquad\qquad y = 13$

Substituting 13 for y in equation (1):
$x + 13 = 22 \Rightarrow x = 9$
Don bought 13 22¢ stamps and 9 18¢ stamps.

12. 44 22¢ stamps and 6 45¢ stamps

13. Let $\qquad w =$ width (in ft)
Let $\qquad l =$ length (in ft)
Then $2w + 2l =$ perimeter
(1) $\begin{cases} 2w + 2l = 19 \\ l = w + 1.5 \end{cases}$
Substituting $w + 1.5$ for l in equation (1):
(1) $\qquad 2w + 2l = 19$
$\qquad 2w + 2(w + 1.5) = 19$
$\qquad 2w + 2w + 3 = 19$
$\qquad\qquad\qquad 4w = 16$
$\qquad\qquad\qquad\quad w = 4$

Substituting 4 for w in equation (2):
$l = w + 1.5$
$\ = (4) + 1.5 = 5.5$
The width is 4 ft, and the length is 5.5 ft, or 5 ft 6 in.

14. Width $= 5$ ft; length $= 7$ ft 6 in.

15. Let $n =$ numerator
Let $d =$ denominator
(1) $\dfrac{n}{d} = \dfrac{2}{3}$ $\qquad$ (2) $\dfrac{n + 4}{d - 2} = \dfrac{6}{7}$
$\qquad 2d = 3n$ $\qquad\qquad 6(d - 2) = 7(n + 4)$
$\qquad d = \dfrac{3n}{2}$ $\qquad\qquad 6d - 12 = 7n + 28$
$\qquad\qquad\qquad$ (3) $\qquad 6d = 7n + 40$

Substituting $\dfrac{3n}{2}$ for d in equation (3):

$$6\left(\frac{3n}{2}\right) = 7n + 40$$
$$9n = 7n + 40$$
$$2n = 40$$
$$n = 20$$

Substituting 20 for n in $d = \frac{3n}{2}$:

$$d = \frac{3(20)}{2} = 30$$

Therefore, the original fraction is $\frac{20}{30}$.

16. $\frac{36}{48}$

17. Let $x =$ speed of boat in still water (in mph)
Let $y =$ speed of current (in mph)
Note: We multiply both sides of (1) by $\frac{1}{7}$ and both sides of
(2) by $\frac{1}{9}$.

(1) $\begin{cases} 7(x+y) = 252 \\ (2) \; 9(x-y) = 252 \end{cases} \Rightarrow \begin{array}{c} x+y = 36 \\ x - y = 28 \end{array}$ Add

$$\begin{array}{c} 2x = 64 \\ x = 32 \end{array}$$

Substituting 32 for x in $x + y = 36$:
$$32 + y = 36 \Rightarrow y = 4$$
The speed of the boat in still water is 32 mph, and the speed
of the current is 4 mph.

18. The speed of the plane in still air is 120 mph. The speed of
the wind is 30 mph.

19. Let $t =$ cost of tie (in cents)
Let $p =$ cost of pin (in cents)

(1) $\begin{cases} t + p = 110 \\ (2) \; t = 100 + p \end{cases}$

Substituting $100 + p$ for t in equation (1):
$$(1) \qquad t + p = 110$$
$$100 + p + p = 110$$
$$2p = 10$$
$$p = 5$$
Substituting 5 for p in equation (2):
$$t = 100 + p = 100 + 5 = 105$$
The pin costs 5¢, and the tie costs $1.05.

20. 7 on upper branch; 5 on lower branch

Sections 12.1–12.5 Review Exercises (page 621)

1. (1) $\begin{cases} x + y = 6 \\ (2) \; x - y = 4 \end{cases}$
(1) Intercepts (6, 0), (0, 6)
(2) Intercepts (4, 0), (0, −4)
Solution: (5, 1)

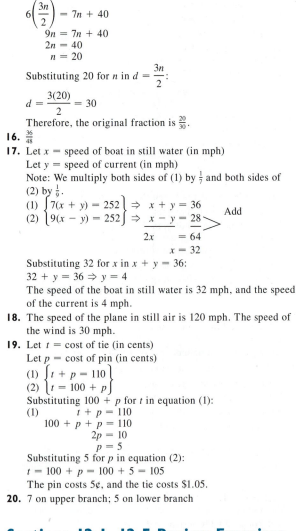

2.

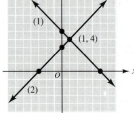

Solution: (1, 4)

3. $3x - y = 6 \qquad 6x - 2y = 12$

x	y
0	−6
2	0
4	6

x	y
0	−6
2	0
4	6

Dependent

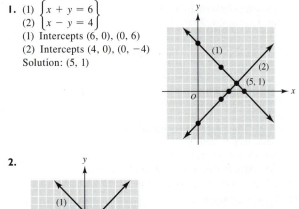

4. (3, 2)

5. $\begin{array}{r} 3] \\ -4] \end{array} \begin{cases} 4x - 8y = 4 \\ 3x - 6y = 3 \end{cases} \Rightarrow \begin{array}{r} 12x - 24y = 12 \\ -12x + 24y = -12 \end{array}$

$$\qquad\qquad\qquad\qquad 0 = 0 \; \textit{True}$$

Dependent

6. Inconsistent

7. (1) $\begin{cases} x = y + 2 \\ (2) \; 4x - 5y = 3 \end{cases}$

Substituting $y + 2$ for x in equation (2):
$$(2) \qquad 4x - 5y = 3$$
$$4(y + 2) - 5y = 3$$
$$4y + 8 - 5y = 3$$
$$-y = -5 \Rightarrow y = 5$$
Substituting 5 for y in equation (1):
$$x = 5 + 2 = 7$$
Solution: (7, 5)

8. (4, 9)

9. $\begin{cases} x + y = 4 \\ 2x - y = 2 \end{cases} \Rightarrow y = 4 - x$

Substituting $4 - x$ for y in equation (2):
$$(2) \qquad 2x - y = 2$$
$$2x - (4 - x) = 2$$
$$2x - 4 + x = 2$$
$$3x = 6$$
$$x = 2$$
Substituting 2 for x in $y = 4 - x$:
$$y = 4 - 2 = 2$$
Solution: (2, 2)

10. (−3, −2)

11. Let $x =$ larger number
Let $y =$ smaller number

(1) $\begin{cases} x + y = 84 \\ (2) \; x - y = 22 \end{cases}$ Add

$$\begin{array}{c} 2x = 106 \\ x = 53 \end{array}$$

Substituting 53 for x in equation (1):
$$(1) \quad x + y = 84$$
$$53 + y = 84$$
$$y = 31$$
The larger number is 53, and the smaller is 31.

12. $\frac{24}{30}$

13. Let $x =$ number of boxes of paper for laser printer
Let $y =$ number of boxes of paper for copy machine

(1) $-425]$ $\quad x + \quad y = 20 \quad \Rightarrow \quad -425x - 425y = -8,500$
(2) $100] \; 7.50x + 4.25y = 107.75 \Rightarrow \quad \underline{750x + 425y = 10,775}$

$$\qquad\qquad\qquad\qquad\qquad 325x = 2,275$$
$$\qquad\qquad\qquad\qquad\qquad\qquad x = 7$$

Substituting 7 for x in equation (1):
$$7 + y = 20 \Rightarrow y = 13$$
The manager purchased 7 boxes of paper for the laser
printer and 13 boxes of paper for the copy machine.

14. The speed of the boat in still water is 27 mph. The speed of
the current is 6 mph.

Chapter 12 Diagnostic Test (page 625)

Following each problem number is the number (in parentheses) of the textbook section where that kind of problem is discussed.

The checks will not be shown for Problems 1–7.

1. (12.2) (1) $\begin{cases} 2x + 3y = 6 \\ x + 3y = 0 \end{cases}$

$2x + 3y = 6$

x	y
0	2
3	0
-3	4

$x + 3y = 0$

x	y
0	0
-3	1
3	-1

Solution: $(6, -2)$

2. (12.3) (1) $\begin{cases} x + 3y = 5 \\ -x + 4y = 2 \end{cases}$ Add

$7y = 7$
$y = 1$

Substituting 1 for y in equation (1):
$x + 3(1) = 5 \Rightarrow x = 2$
Solution: $(2, 1)$

3. (12.3) $\begin{matrix} 2] \\ 3] \end{matrix} \begin{cases} 4x + 3y = 5 \\ 5x - 2y = 12 \end{cases} \Rightarrow \begin{matrix} 8x + 6y = 10 \\ 15x - 6y = 36 \end{matrix}$

$23x = 46$
$x = 2$

Substituting 2 for x in the first equation:
$4(2) + 3y = 5 \Rightarrow y = -1$
Solution: $(2, -1)$

4. (12.4) (1) $\begin{cases} 2x - 5y = -20 \\ x = y - 7 \end{cases}$
$2(y - 7) - 5y = -20$
$2y - 14 - 5y = -20$
$-3y - 14 = -20$
$-3y = -6$
$y = 2$
Substituting 2 for y in equation (2):
$x = (2) - 7 = -5$
Solution: $(-5, 2)$

5. (12.3) $\begin{matrix} 5] \\ 4] \end{matrix} \begin{cases} 3x + 4y = -1 \\ 2x - 5y = -16 \end{cases} \Rightarrow \begin{matrix} 15x + 20y = -5 \\ 8x - 20y = -64 \end{matrix}$

$23x = -69$
$x = -3$

Substituting -3 for x in the first equation:
$3(-3) + 4y = -1 \Rightarrow y = 2$
Solution: $(-3, 2)$

6. (12.3) $\begin{matrix} -2] \\ 1] \end{matrix} \begin{cases} 5x + y = 10 \\ 10x + 2y = 9 \end{cases} \Rightarrow \begin{matrix} -10x - 2y = -20 \\ 10x + 2y = 9 \end{matrix}$

$0 = -11$ *False*

Inconsistent (no solution)

7. (12.3) $\begin{matrix} 10]5] \\ 6]3] \end{matrix} \begin{cases} 6x - 9y = 3 \\ -10x + 15y = -5 \end{cases} \Rightarrow \begin{matrix} 30x - 45y = 15 \\ -30x + 45y = -15 \end{matrix}$

$0 = 0$ *True*

Dependent (many solutions)

8. (12.5) Let x = one number
Let y = other number
(1) $\begin{cases} x + y = 13 \\ x - y = 37 \end{cases}$ Add

$2x = 50$
$x = 25$

Substituting 25 for x in equation (1):
$25 + y = 13 \Rightarrow y = -12$
Check: $25 + (-12) = 13$
$25 - (-12) = 25 + 12 = 37$
The numbers are 25 and -12.

9. (12.5) Let x = speed of plane in still air (in mph)
Let y = speed of wind (in mph)
Then $x + y$ = speed with wind
and $x - y$ = speed against wind
Note: We will have small numbers to work with if we multiply both sides of (1) by $\frac{1}{11}$ and both sides of (2) by $\frac{1}{7.5}$.
(1) $\begin{cases} 11(x - y) = 1{,}650 \\ 7.5(x + y) = 1{,}650 \end{cases}$

$\frac{1}{11}]$ $11(x - y) = 1{,}650 \Rightarrow x - y = 150$ (3)
$\frac{1}{7.5}]$ $7.5(x + y) = 1{,}650 \Rightarrow x + y = 220$ (4)

$2x = 370$
$x = 185$

Substituting 185 for x in equation (4):
$185 + y = 220 \Rightarrow y = 35$
Check: The speed of the plane against the wind is

185 mph $- 35$ mph $= 150$ mph, and $150\frac{mi}{hr}(11 \text{ hr}) = 1{,}650$ mi.

The speed of the plane with the wind is 185 mph $+ 35$ mph

$= 220$ mph, and $220\frac{mi}{hr}(7.5 \text{ hr}) = 1{,}650$ mi.

The speed of the plane in still air is 185 mph, and the speed of the wind is 35 mph.

10. (12.5) Let x = number of cm in length
Let y = number of cm in width
(1) $\begin{cases} x = y + 3 \\ 2x + 2y = 98 \end{cases}$
Substituting $y + 3$ for x in equation (2):
$2(y + 3) + 2y = 98$
$2y + 6 + 2y = 98$
$4y + 6 = 98$
$4y = 92$
$y = 23$
Substituting 23 for y in (1):
$x = 23 + 3 = 26$
Check: The length of the rectangle is 3 cm more than its width. Its perimeter is $2(23 \text{ cm}) + 2(26 \text{ cm}) =$
46 cm $+ 52$ cm $= 98$ cm.
The length is 26 cm, and the width is 23 cm.

Chapters 1–12 Cumulative Review Exercises (page 625)

1. a. $2(3x^2 - 13xy - 10y^2) = 2(3x + 2y)(x - 5y)$
b. $5(x^2 - 4) = 5(x + 2)(x - 2)$

2. $11z^2 - 5z - 13$

3. $(8t)^2 - 2(8t)(5) + (5)^2 = 64t^2 - 80t + 25$

4. $-5x + 14$

5. $(x)(x^2 + 3x - 1) + (-4)(x^2 + 3x - 1)$
$= x^3 + 3x^2 - x - 4x^2 - 12x + 4 = x^3 - x^2 - 13x + 4$

6. $3x^2 - 6x + 1$ R 3, or $3x^2 - 6x + 1 + \dfrac{3}{x - 4}$

7. $\dfrac{5m}{m^2 - 9} \div \dfrac{10m^3 - 50m^2}{m^2 - 2m - 15} = \dfrac{5m}{(m + 3)(m - 3)} \cdot \dfrac{(m - 5)(m + 3)}{10m^2(m - 5)}$
$= \dfrac{1}{2m(m - 3)}$

8. $\dfrac{3x^2 + 16x + 5}{(x - 2)(2x + 3)}$

9. $\dfrac{5x}{(2x + 1)(x - 3)} - \dfrac{3}{(2x + 1)(x + 1)}$
$= \dfrac{5x(x + 1)}{(2x + 1)(x - 3)(x + 1)} - \dfrac{3(x - 3)}{(2x + 1)(x + 1)(x - 3)}$
$= \dfrac{5x^2 + 5x - 1[3x - 9]}{(2x + 1)(x - 3)(x + 1)} = \dfrac{5x^2 + 5x - 3x + 9}{(2x + 1)(x - 3)(x + 1)}$
$= \dfrac{5x^2 + 2x + 9}{(2x + 1)(x - 3)(x + 1)}$

10. $\dfrac{2s + 1}{s}$

11. $(-7)(4)(x^2x^4)(y^{-5}y^2) = -28x^6y^{-3} = -28x^6 \cdot \dfrac{1}{y^3} = -\dfrac{28x^6}{y^3}$

12. $\dfrac{b^5}{a}$

13. $\sqrt{25 \cdot 5} + 4\sqrt{4 \cdot 5} - \sqrt{9 \cdot 5} = 5\sqrt{5} + 4(2\sqrt{5}) - 3\sqrt{5}$
$= 5\sqrt{5} + 8\sqrt{5} - 3\sqrt{5} = 10\sqrt{5}$

14. $8 - 2\sqrt{7}$

15. $\dfrac{12(\sqrt{7} + 2)}{(\sqrt{7} - 2)(\sqrt{7} + 2)} = \dfrac{12(\sqrt{7} + 2)}{7 - 4} = \dfrac{12(\sqrt{7} + 2)}{3}$
$= 4(\sqrt{7} + 2) = 4\sqrt{7} + 8$

16. $2\frac{1}{2}$ **17.** $6x^2 - 13x + 5 = 0$ **18.** -5
$(2x - 1)(3x - 5) = 0$

$\begin{array}{c|c} 2x - 1 = 0 & 3x - 5 = 0 \\ 2x = 1 & 3x = 5 \\ x = \frac{1}{2} & x = \frac{5}{3} \end{array}$

19. $(\sqrt{11x - 18})^2 = (x)^2$
$11x - 18 = x^2$
$0 = x^2 - 11x + 18$
$0 = (x - 9)(x - 2)$

$\begin{array}{c|c} x - 9 = 0 & x - 2 = 0 \\ x = 9 & x = 2 \end{array}$

$\begin{array}{c|c} \textit{Check for } x = 9: & \textit{Check for } x = 2: \\ \sqrt{11(9) - 18} \overset{?}{=} 9 & \sqrt{11(2) - 18} \overset{?}{=} 2 \\ \sqrt{99 - 18} \overset{?}{=} 9 & \sqrt{22 - 18} \overset{?}{=} 2 \\ \sqrt{81} \overset{?}{=} 9 & \sqrt{4} \overset{?}{=} 2 \\ 9 = 9 & 2 = 2 \end{array}$

The solutions are 9 and 2.

20. $(1, 2)$ **21.** $2x - 6 \le 4$
$2x - 6 + 6 \le 4 + 6$
$2x \le 10$
$\dfrac{2x}{2} \le \dfrac{10}{2}$
$x \le 5$

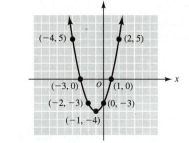

22. $x = \dfrac{3y - 1}{4 - y}$

23. $x^2 + x - 8 = 0$; $a = 1, b = 1, c = -8$
$x = \dfrac{-1 \pm \sqrt{(1)^2 - 4(1)(-8)}}{2(1)} = \dfrac{-1 \pm \sqrt{1 + 32}}{2} = \dfrac{-1 \pm \sqrt{33}}{2}$

24. $3x + 7y - 5 = 0$ (general form); $3x + 7y = 5$ (standard form)

25. $3x - 4y > 12$
The boundary line is $3x - 4y = 12$.

x	y
0	-3
4	0

The correct half-plane does not include the origin because
$3(0) - 4(0) \not> 12$
$0 \not> 12$

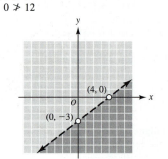

26. $y = x^2 + 2x - 3$

27. Let $\quad x$ = number of lb of stew meat
Then $5 - x$ = number of lb of back ribs
$2.10x + 1.20(5 - x) = 7.35$
$210x + 120(5 - x) = 735$
$210x + 600 - 120x = 735$
$90x = 135$
$x = \frac{135}{90} = \frac{3}{2}$, or $1\frac{1}{2}$
$5 - x = 5 - \frac{3}{2} = \frac{10}{2} - \frac{3}{2} = \frac{7}{2}$, or $3\frac{1}{2}$
Therefore, there is $1\frac{1}{2}$ lb of stew meat and $3\frac{1}{2}$ lb of back ribs.

28. First answer: The integers are -3 and -2. Second answer: The integers are 4 and 5.

29. Let $\quad x$ = Elaine's speed (in mph)
$\qquad\qquad$ (Elaine's time was 8 hr.)
Then $x + 6$ = Heather's speed (in mph)
$\qquad\qquad$ (Heather's time was 7 hr.)
$x \dfrac{\text{mi}}{\text{hr}}(8 \text{ hr})$ = Elaine's distance (in mi)
$(x + 6)\left(\dfrac{\text{mi}}{\text{hr}}\right)(7 \text{ hr})$ = Heather's distance (in mi)
$8x = 7(x + 6)$
$8x = 7x + 42$
$-7x + 8x = -7x + 7x + 42$
a. $\qquad\qquad x = 42 \quad$ Elaine's speed
b. $\qquad x + 6 = 48 \quad$ Heather's speed
c. $\qquad 8x = 8 \text{ hr}\left(42\dfrac{\text{mi}}{\text{hr}}\right) = 336 \text{ mi}$

Elaine's speed is 42 mph, Heather's speed is 48 mph, and the distance is 336 mi.

30. The width is 12 m, and the length is 17 m.

12

Exercises A.1   (page 630)

1. Yes, because it is a collection of objects or things.   2. Yes

3. Yes, because they have exactly the same elements.

4. $\{0, 1, 2\}$ 5. $\{\ \}$, since there are no digits greater than 9.

6. $\{0, 1, 2\}$ 7. $\{10, 11, 12, \ldots\}$ 8. $\{5\}$

9. $\{\ \}$, since there are no whole numbers greater than 4 and at the same time less than 5.

10. $\{0, 1, 2, 3\}$ 11. $2, a, 3$ 12. $0, 1, 2, 3, 4, 5, 6, 7, 8, 9$

13. a. $n(\{1, 1, 3, 5, 5, 5\}) = n(\{1, 3, 5\}) = 3$. This set has three elements: 1, 3, and 5.
    b. $n(\{0\}) = 1$. This set has one element: 0.
    c. $n(\{a, b, g, x\}) = 4$
    d. $n(\{0, 1, 2, 3, 4, 5, 6, 7\}) = 8$
    e. $n(\varnothing) = 0$. The empty set has no elements.

14. $\varnothing$ has no elements; $\{0\}$ has one element—namely, 0. Since the sets do not have exactly the same elements, they are not equal.

15. a. The set of digits is finite, because when we count its elements, the counting comes to an end. In this case the counting ends at 10.
    b. The set of whole numbers is infinite, because when we attempt to count its elements, the counting never comes to an end.
    c. Finite; the counting ends at 7.
    d. Finite; if we started counting the books in the ELAC library, we could eventually finish counting them.

16. a. True   b. True   c. False   d. False

17. a. True   b. False   c. False   d. True

18. a. False   b. True   c. True   d. False

Exercises A.2   (page 632)

1. a. $\{3, 5\}$ is a subset of M because each of its elements, 3 and 5, is an element of M.
   b. $\{0, 1, 7\}$ is not a subset of M because elements 0 and 7 are not elements of M.
   c. $\varnothing$ is a subset of M because $\varnothing$ is a subset of every set.
   d. $\{2, 4, 1, 3, 5\}$ is a subset of M because each element of C is an element of M.

2. a. Yes   b. Yes   c. Yes   d. No

3. $\{R, G, Y\}$ All the subsets with three elements
   $\{R, G\}, \{R, Y\}, \{G, Y\}$ All the subsets with two elements
   $\{R\}, \{G\}, \{Y\}$ All the subsets with one element
   $\{\ \}$ All the subsets with no elements

4. $\{\square, \triangle\}, \{\square\}, \{\triangle\}, \{\ \}$

Exercises A.3   (page 633)

1. a. $\{1, 5, 7\} \cup \{2, 4\} = \{1, 2, 4, 5, 7\}$
      $\{1, 5, 7\} \cap \{2, 4\} = \{\ \}$
   b. $\{a, b\} \cup \{a, x, y, z\} = \{a, b, x, y, z\}$
      $\{a, b\} \cap \{a, x, y, z\} = \{a\}$
   c. $\{\ \} \cup \{k, 2\} = \{k, 2\}$
      $\{\ \} \cap \{k, 2\} = \{\ \}$
   d. $\{river, boat\} \cup \{boat, streams, down\}$
      $= \{river, boat, streams, down\}$
      $\{river, boat\} \cap \{boat, streams, down\} = \{boat\}$

2. a. $\{1, 2, 3, 4, 5, 6, 7\}$ b. $\{1, 2, 3, 4, 5, 6, 7\}$ c. $\{\ \}$
   d. $\{6\}$

3. C and D are disjoint because they have no member in common; that is, $C \cap D = \varnothing$.

4. a. $P = \{c, d, k\}$ b. $Q = \{f, j, k\}$
   c. $P \cup Q = \{c, d, f, j, k\}$ d. $P \cap Q = \{k\}$
   e. $U = \{a, c, d, e, f, g, h, j, k\}$

5. a. $X \cap Y = \{5, 11\}$ because these are the only elements in both X and Y.
   b. $Y \cap X = \{5, 11\}$
   c. Yes, because they have exactly the same elements.

6. a. $K \cap L = \{4, b\}$ b. $n(K \cap L) = 2$
   c. $L \cup M = \{b, m, n, t, 4, 6, 7\}$ d. $n(L \cup M) = 7$

7.

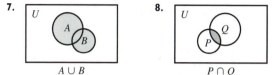

$A \cup B$ $P \cap Q$

9. $R \cap S$ because the shaded area is in both R and S.

10. $Y \cup Z$

Exercises B.1   (page 642)

1. $\dfrac{\overset{2}{\cancel{6}}}{\underset{3}{\cancel{9}}} = \dfrac{2}{3}$ 2. $\dfrac{3}{5}$ 3. $\dfrac{49}{24}$ (already in lowest terms)

4. $\dfrac{7}{8}$ 5. $\dfrac{5}{2} = 2\overset{2\,R\,1\;=\;2\frac{1}{2}}{)5}$ 6. $1\dfrac{3}{8}$

7. $\dfrac{47}{25} = 25\overset{1\,R\,22\;=\;1\frac{22}{25}}{)47}$ 8. $2\dfrac{4}{11}$
$\qquad\qquad\qquad\quad \underline{25}$
$\qquad\qquad\qquad\quad 22$

9. $3\dfrac{7}{8} = \dfrac{3 \cdot 8 + 7}{8} = \dfrac{24 + 7}{8} = \dfrac{31}{8}$

10. $\dfrac{25}{9}$ 11. $2\dfrac{5}{6} = \dfrac{2 \cdot 6 + 5}{6} = \dfrac{12 + 5}{6} = \dfrac{17}{6}$ 12. $\dfrac{43}{5}$

13. LCD $= 12$; $\dfrac{2}{3} = \dfrac{8}{12}$ 14. $\dfrac{1}{2}$
$\qquad\qquad\qquad \underline{+\dfrac{1}{4} = \dfrac{3}{12}}$
$\qquad\qquad\qquad\qquad\ \ \dfrac{11}{12}$

15. LCD $= 10$; $\dfrac{7}{10} = \dfrac{7}{10}$ 16. $\dfrac{7}{20}$ 17. $\dfrac{\overset{1}{\cancel{3}}}{\underset{4}{\cancel{16}}} \cdot \dfrac{\overset{5}{\cancel{20}}}{\underset{3}{\cancel{9}}} = \dfrac{5}{12}$
$\qquad\qquad\quad \underline{-\dfrac{2}{5} = -\dfrac{4}{10}}$
$\qquad\qquad\qquad\qquad\ \ \dfrac{3}{10}$

18. $\dfrac{9}{28}$ 19. $\dfrac{4}{3} \div \dfrac{8}{9} = \dfrac{4}{\underset{1}{\cancel{3}}} \cdot \dfrac{\overset{3}{\cancel{9}}}{\underset{2}{\cancel{8}}} = \dfrac{3}{2}$ 20. 4

21. LCD $= 15$; $2\dfrac{2}{3} = 2\dfrac{10}{15}$ 22. $7\dfrac{3}{4}$
$\qquad\qquad\qquad \underline{+1\dfrac{3}{5} = 1\dfrac{9}{15}}$
$\qquad\qquad\qquad\qquad 3\dfrac{19}{15} = 4\dfrac{4}{15}$

23. LCD $= 10$; $5\dfrac{4}{5} = 5\dfrac{8}{10}$ 24. $2\dfrac{1}{6}$
$\qquad\qquad\qquad \underline{-3\dfrac{7}{10} = -3\dfrac{7}{10}}$
$\qquad\qquad\qquad\qquad\ \ 2\dfrac{1}{10}$

25. $4\dfrac{1}{5} \cdot 2\dfrac{1}{7} = \dfrac{\overset{3}{\cancel{21}}}{\underset{1}{\cancel{5}}} \cdot \dfrac{\overset{3}{\cancel{15}}}{\underset{1}{\cancel{7}}} = 9$ 26. $10\dfrac{1}{2}$

27. $2\dfrac{2}{5} \div 1\dfrac{1}{15} = \dfrac{12}{5} \div \dfrac{16}{15} = \dfrac{\overset{3}{\cancel{12}}}{\underset{1}{\cancel{5}}} \cdot \dfrac{\overset{3}{\cancel{15}}}{\underset{4}{\cancel{16}}} = \dfrac{9}{4} = 2\dfrac{1}{4}$ 28. $\dfrac{2}{3}$

29. $\dfrac{\frac{5}{8}}{\frac{5}{6}} = \dfrac{5}{8} \div \dfrac{5}{6} = \dfrac{\overset{1}{\cancel{5}}}{\underset{4}{\cancel{8}}} \cdot \dfrac{\overset{3}{\cancel{6}}}{\underset{1}{\cancel{5}}} = \dfrac{3}{4}$ 30. $\dfrac{1}{2}$

31. 34.5
1.74
18.
0.016
————
54.256

32. 94.745

33. 356.40
34.67
———
321.73

34. 336.56

35. $100 \times 7.45 = 745$

36. 3,540

37. $\dfrac{46.8}{100} = 0.468$

38. 0.0895

39. 94.78 (2 decimal places)
70.9 (1 decimal place)
———
8 5 3 0 2
6 6 3 4 6
————
6719.902 ≈ 6,719.9

40. 2,798.5

41. $7.25_\wedge \overline{)6.00_\wedge 700}$ ≈ 0.83
0.8 2 8
5 8 0 0
——
20 70
14 50
——
6 200
5 800
——
400

42. 1.71

43. $5\frac{3}{4} = \frac{23}{4} = 4\overline{)23.000}$ 5.75

44. 4.6

45. $0.65 = \dfrac{65}{100} = \dfrac{13}{20}$

46. $\frac{1}{4}$

47. $5.9 = 5 + .9 = 5 + \frac{9}{10} = 5\frac{9}{10}$

48. $4\frac{3}{10}$

49. $4.70_\wedge = 470\%$

50. 0.1%

51. $0.25_\wedge 8 = 25.8\%$

52. 1,200%

53. $_\wedge 03.5\% = 0.035$

54. 0.0002

55. $1_\wedge 57\% = 1.57$

56. 0.178

57. $\frac{1}{8} = 0.12_\wedge 5 = 12.5\%$

58. 60%

59. $\frac{7}{12} \approx 0.58_\wedge 33 = 58.33\%$

60. 14.29%

61. $35\% = \frac{35}{100} = \frac{7}{20}$

62. $\frac{4}{5}$

63. $12\% = \frac{12}{100} = \frac{3}{25}$

64. $\frac{9}{50}$

Exercises C.1   (page 648)

1. The set is a function, because no two ordered pairs have the same first element but different second elements. The domain is {4, 2, 3, 0}.

2. The set is a function. The domain is {6, −2, 4, 0}.

3. The set is not a function, because the pairs (2, −5) and (2, 0) have the same first element but different second elements.

4. The set is not a function.

5. The set is a function, because no two ordered pairs have the same first element but different second elements. The domain is {−8, 3, 6, 9}.

6. The set is a function. The domain is {−3, −1, 0, 3}.

7. The domain is the set of all real numbers except 12. (12 is an excluded value.)

8. The domain is the set of all real numbers except 2.

9. The domain is the set of all real numbers.

10. The domain is the set of all real numbers.

11. The radicand must be ≥ 0.
$x - 5 \geq 0$
$x \geq 5$
The domain is the set of all real numbers greater than or equal to 5.

12. The domain is the set of all real numbers greater than or equal to 10.

13. There is a restriction: $x \geq 0$. Therefore, the domain is the set of all real numbers greater than or equal to 0.

14. The domain is the set of all real numbers greater than or equal to −3.

15. a. No   **b.** Yes   **c.** No

16. a. No   **b.** No   **c.** Yes

Exercises C.2   (page 650)

1. $f(x) = (x + 2)^2$
a. $f(0) = ([0] + 2)^2 = (2)^2 = 4$
b. $f(2) = ([2] + 2)^2 = (4)^2 = 16$
c. $f(-3) = ([-3] + 2)^2 = (-1)^2 = 1$
d. $f(1) = ([1] + 2)^2 = (3)^2 = 9$

2. a. 9   **b.** 25   **c.** 1   **d.** 49

3. $g(x) = x^2 + 4x + 4$
a. $g(0) = (0)^2 + 4(0) + 4 = 0 + 0 + 4 = 4$
b. $g(2) = (2)^2 + 4(2) + 4 = 4 + 8 + 4 = 16$
c. $g(-3) = (-3)^2 + 4(-3) + 4 = 9 - 12 + 4 = 1$
d. $g(1) = (1)^2 + 4(1) + 4 = 1 + 4 + 4 = 9$

4. a. 9   **b.** 25   **c.** 1   **d.** 49

5. $F(x) = x^2 + 4$
a. $F(0) = (0)^2 + 4 = 0 + 4 = 4$
b. $F(2) = (2)^2 + 4 = 4 + 4 = 8$
c. $F(-3) = (-3)^2 + 4 = 9 + 4 = 13$
d. $F(1) = (1)^2 + 4 = 1 + 4 = 5$

6. a. 9   **b.** 13   **c.** 25   **d.** 25

7. $f(x) = 3x^2 + x - 1$
a. $f(0) = 3(0)^2 + (0) - 1 = 0 + 0 - 1 = -1$
b. $f(-2) = 3(-2)^2 + (-2) - 1 = 12 - 2 - 1 = 9$
c. $f(5) = 3(5)^2 + (5) - 1 = 75 + 5 - 1 = 79$
d. $f(1) = 3(1)^2 + (1) - 1 = 3 + 1 - 1 = 3$

8. a. −5   **b.** 4   **c.** 39   **d.** 0

9. $f(x) = x^3 - 1$
a. $f(0) = (0)^3 - 1 = 0 - 1 = -1$
b. $f(-1) = (-1)^3 - 1 = -1 - 1 = -2$
c. $f(2) = (2)^3 - 1 = 8 - 1 = 7$
d. $f(1) = (1)^3 - 1 = 1 - 1 = 0$

10. a. 1   **b.** −7   **c.** 2   **d.** 0

11. $g(t) = \dfrac{4t^2 + 5t - 1}{t + 2}$

a. $g(0) = \dfrac{4(0)^2 + 5(0) - 1}{(0) + 2} = \dfrac{0 + 0 - 1}{2} = -\dfrac{1}{2}$

b. $g(-1) = \dfrac{4(-1)^2 + 5(-1) - 1}{(-1) + 2} = \dfrac{4 - 5 - 1}{1} = -2$

c. $g(2) = \dfrac{4(2)^2 + 5(2) - 1}{(2) + 2} = \dfrac{16 + 10 - 1}{4} = \dfrac{25}{4}$ or $6\dfrac{1}{4}$

d. $g(4) = \dfrac{4(4)^2 + 5(4) - 1}{(4) + 2} = \dfrac{64 + 20 - 1}{6} = \dfrac{83}{6}$ or $13\dfrac{5}{6}$

12. a. $\frac{1}{3}$ **b.** −1   **c.** −1   **d.** 1

INDEX

A94